Handbook of
Optical and
Laser Scanning

OPTICAL ENGINEERING

Founding Editor
Brian J. Thompson
University of Rochester
Rochester, New York

1. Electron and Ion Microscopy and Microanalysis: Principles and Applications, *Lawrence E. Murr*
2. Acousto-Optic Signal Processing: Theory and Implementation, *edited by Norman J. Berg and John N. Lee*
3. Electro-Optic and Acousto-Optic Scanning and Deflection, *Milton Gottlieb, Clive L. M. Ireland, and John Martin Ley*
4. Single-Mode Fiber Optics: Principles and Applications, *Luc B. Jeunhomme*
5. Pulse Code Formats for Fiber Optical Data Communication: Basic Principles and Applications, *David J. Morris*
6. Optical Materials: An Introduction to Selection and Application, *Solomon Musikant*
7. Infrared Methods for Gaseous Measurements: Theory and Practice, *edited by Joda Wormhoudt*
8. Laser Beam Scanning: Opto-Mechanical Devices, Systems, and Data Storage Optics, *edited by Gerald F. Marshall*
9. Opto-Mechanical Systems Design, *Paul R. Yoder, Jr.*
10. Optical Fiber Splices and Connectors: Theory and Methods, *Calvin M. Miller with Stephen C. Mettler and Ian A. White*
11. Laser Spectroscopy and Its Applications, *edited by Leon J. Radziemski, Richard W. Solarz, and Jeffrey A. Paisner*
12. Infrared Optoelectronics: Devices and Applications, *William Nunley and J. Scott Bechtel*
13. Integrated Optical Circuits and Components: Design and Applications, *edited by Lynn D. Hutcheson*
14. Handbook of Molecular Lasers, *edited by Peter K. Cheo*
15. Handbook of Optical Fibers and Cables, *Hiroshi Murata*
16. Acousto-Optics, *Adrian Korpel*
17. Procedures in Applied Optics, *John Strong*
18. Handbook of Solid-State Lasers, *edited by Peter K. Cheo*
19. Optical Computing: Digital and Symbolic, *edited by Raymond Arrathoon*
20. Laser Applications in Physical Chemistry, *edited by D. K. Evans*
21. Laser-Induced Plasmas and Applications, *edited by Leon J. Radziemski and David A. Cremers*
22. Infrared Technology Fundamentals, *Irving J. Spiro and Monroe Schlessinger*
23. Single-Mode Fiber Optics: Principles and Applications, Second Edition, Revised and Expanded, *Luc B. Jeunhomme*
24. Image Analysis Applications, *edited by Rangachar Kasturi and Mohan M. Trivedi*
25. Photoconductivity: Art, Science, and Technology, *N. V. Joshi*
26. Principles of Optical Circuit Engineering, *Mark A. Mentzer*
27. Lens Design, *Milton Laikin*
28. Optical Components, Systems, and Measurement Techniques, *Rajpal S. Sirohi and M. P. Kothiyal*

29. Electron and Ion Microscopy and Microanalysis: Principles and Applications, Second Edition, Revised and Expanded, *Lawrence E. Murr*
30. Handbook of Infrared Optical Materials, *edited by Paul Klocek*
31. Optical Scanning, *edited by Gerald F. Marshall*
32. Polymers for Lightwave and Integrated Optics: Technology and Applications, *edited by Lawrence A. Homak*
33. Electro-Optical Displays, *edited by Mohammad A. Karim*
34. Mathematical Morphology in Image Processing, *edited by Edward R. Dougherty*
35. Opto-Mechanical Systems Design: Second Edition, Revised and Expanded, *Paul R. Yoder, Jr.*
36. Polarized Light: Fundamentals and Applications, *Edward Collett*
37. Rare Earth Doped Fiber Lasers and Amplifiers, *edited by Michel J. F. Digonnet*
38. Speckle Metrology, *edited by Rajpal S. Sirohi*
39. Organic Photoreceptors for Imaging Systems, *Paul M. Borsenberger and David S. Weiss*
40. Photonic Switching and Interconnects, *edited by Abdellatif Marrakchi*
41. Design and Fabrication of Acousto-Optic Devices, *edited by Akis P. Goutzoulis and Dennis R. Pape*
42. Digital Image Processing Methods, *edited by Edward R. Dougherty*
43. Visual Science and Engineering: Models and Applications, *edited by D. H. Kelly*
44. Handbook of Lens Design, *Daniel Malacara and Zacarias Malacara*
45. Photonic Devices and Systems, *edited by Robert G. Hunsberger*
46. Infrared Technology Fundamentals: Second Edition, Revised and Expanded, *edited by Monroe Schlessinger*
47. Spatial Light Modulator Technology: Materials, Devices, and Applications, *edited by Uzi Efron*
48. Lens Design: Second Edition, Revised and Expanded, *Milton Laikin*
49. Thin Films for Optical Systems, *edited by Francoise R. Flory*
50. Tunable Laser Applications, *edited by F. J. Duarte*
51. Acousto-Optic Signal Processing: Theory and Implementation, Second Edition, *edited by Norman J. Berg and John M. Pellegrino*
52. Handbook of Nonlinear Optics, *Richard L. Sutherland*
53. Handbook of Optical Fibers and Cables: Second Edition, *Hiroshi Murata*
54. Optical Storage and Retrieval: Memory, Neural Networks, and Fractals, *edited by Francis T. S. Yu and Suganda Jutamulia*
55. Devices for Optoelectronics, *Wallace B. Leigh*
56. Practical Design and Production of Optical Thin Films, *Ronald R. Willey*
57. Acousto-Optics: Second Edition, *Adrian Korpel*
58. Diffraction Gratings and Applications, *Erwin G. Loewen and Evgeny Popov*
59. Organic Photoreceptors for Xerography, *Paul M. Borsenberger and David S. Weiss*
60. Characterization Techniques and Tabulations for Organic Nonlinear Optical Materials, *edited by Mark G. Kuzyk and Carl W. Dirk*
61. Interferogram Analysis for Optical Testing, *Daniel Malacara, Manuel Servin, and Zacarias Malacara*
62. Computational Modeling of Vision: The Role of Combination, *William R. Uttal, Ramakrishna Kakarala, Spiram Dayanand, Thomas Shepherd, Jagadeesh Kalki, Charles F. Lunskis, Jr., and Ning Liu*

63. Microoptics Technology: Fabrication and Applications of Lens Arrays and Devices, *Nicholas Borrelli*
64. Visual Information Representation, Communication, and Image Processing, *edited by Chang Wen Chen and Ya-Qin Zhang*
65. Optical Methods of Measurement, *Rajpal S. Sirohi and F. S. Chau*
66. Integrated Optical Circuits and Components: Design and Applications, *edited by Edmond J. Murphy*
67. Adaptive Optics Engineering Handbook, *edited by Robert K. Tyson*
68. Entropy and Information Optics, *Francis T S. Yu*
69. Computational Methods for Electromagnetic and Optical Systems, *John M. Jarem and Partha P. Banerjee*
70. Laser Beam Shaping, *Fred M. Dickey and Scott C. Holswade*
71. Rare Earth Doped Fiber Lasers and Amplifiers: Second Edition, Revised and Expanded, *edited by Michel J. F. Digonnet*
72. Lens Design: Third Edition, Revised and Expanded, *Milton Laikin*
73. Handbook of Optical Engineering, *edited by Daniel Malacara and Brian J. Thompson*
74. Handbook of Imaging Materials: Second Edition, Revised and Expanded, *edited by Arthur S. Diamond and David S. Weiss*
75. Handbook of Image Quality: Characterization and Prediction, Brian W. Keelan
76. Fiber Optic Sensors, *edited by Francis T S. Yu and Shizhuo Yin*
77. Optical Switching/Networking and Computing for Multimedia Systems, *edited by Mohsen Guizani and Abdella Baftou*
78. Image Recognition and Classification: Algorithms, Systems, and Applications, *edited by Bahram Javidi*
79. Practical Design and Production of Optical Thin Films: Second Edition, Revised and Expanded, *Ronald R. Willey*
80. Ultrafast Lasers: Technology and Applications, *edited by Martin E. Fermann, Almantas Galvanauskas, and Gregg Sucha*
81. Light Propagation in Periodic Media: Differential Theory and Design, *Michel Neviere and Evgeny Popov*
82. Handbook of Nonlinear Optics: Second Edition, Revised and Expanded, *Richard L. Sutherland*
83. Polarized Light: Second Edition, Revised and Expanded, *Dennis Goldstein*
84. Optical Remote Sensing: Science and Technology, *Walter Egan*
85. Handbook of Optical Design: Second Edition, *Daniel Malacara and Zacarias Malacara*
86. Nonlinear Optics: Theory, Numerical Modeling, and Applications, *Partha P. Banerjee*
87. Semiconductor and Metal Nanocrystals: Synthesis and Electronic and Optical Properties, *edited by Victor I. Klimov*
88. High-Performance Backbone Network Technology, *edited by Naoaki Yamanaka*
89. Semiconductor Laser Fundamentals, *Toshiaki Suhara*
90. Handbook of Optical and Laser Scanning, *edited by Gerald F. Marshall*

Additional Volumes in Preparation

Handbook of Optical and Laser Scanning

edited by
Gerald F. Marshall
Consultant in Optics
Niles, Michigan, U.S.A.

MARCEL DEKKER, INC. NEW YORK · BASEL

Although great care has been taken to provide accurate and current information, neither the author(s) nor the publisher, nor anyone else associated with this publication, shall be liable for any loss, damage, or liability directly or indirectly caused or alleged to be caused by this book. The material contained herein is not intended to provide specific advice or recommendations for any specific situation.

Trademark notice: Product or corporate names may be trademarks or registered trademarks and are used only for identification and explanation without intent to infringe.

Library of Congress Cataloging-in-Publication Data
A catalog record for this book is available from the Library of Congress.

ISBN: 0-8247-5569-3

This book is printed on acid-free paper.

Headquarters
Marcel Dekker, Inc., 270 Madison Avenue, New York, NY 10016, U.S.A.
tel: 212-696-9000; fax: 212-685-4540

Distribution and Customer Service
Marcel Dekker, Inc., Cimarron Road, Monticello, New York 12701, U.S.A.
tel: 800-228-1160; fax: 845-796-1772

Eastern Hemisphere Distribution
Marcel Dekker AG, Hutgasse 4, Postfach 812, CH-4001 Basel, Switzerland
tel: 41-61-260-6300; fax: 41-61-260-6333

World Wide Web
http://www.dekker.com

The publisher offers discounts on this book when ordered in bulk quantities. For more information, write to Special Sales/Professional Marketing at the headquarters address above.

Copyright © 2004 by Marcel Dekker, Inc. All Rights Reserved.

Neither this book nor any part may be reproduced or transmitted in any form or by any means, electronic or mechanical, including photocopying, microfilming, and recording, or by any information storage and retrieval system, without permission in writing from the publisher.

Current printing (last digit):
10 9 8 7 6 5 4 3 2 1

PRINTED IN THE UNITED STATES OF AMERICA

*With gratitude to my wife, Irene,
colleagues, and friends.*

*To the memory of my parents,
Ethelena and Albert,
brothers, Donald and Edward, and
sisters, Andrée and Kathleen.*

Preface

Optical and laser beam scanning is the controlled deflection of a light beam, visible or invisible. The aim of *Handbook of Optical and Laser Scanning* is to provide application-oriented engineers, managerial technologists, scientists, and students with a guideline and a reference to the fundamentals of input and output optical scanning technology and engineering. This text has its origin in two previous books, *Laser Beam Scanning* (1985) and *Optical Scanning* (1991). Since their publication, many advances have occurred, which has made it necessary to update and include the changes of the past decade. This book brings together the knowledge and experience of 27 international specialists from England, Japan, and the United States.

Optical and laser scanning technology is a comprehensive subject that encompasses not only the mechanics of controlling the deflection of a light beam, but also all aspects that affect the imaging fidelity of the output data that may be recorded on paper or film, displayed on a monitor, or projected onto a screen. A scanning system may be an input scanner, an output scanner, or one that combines both of these functional attributes. A system's imaging fidelity begins with, and depends on, the accurate reading and storage of the input information—the processing of the stored information—and ends with the presentation of the output data. Optical scanning intimately involves a number of disciplines: optics, material science, magnetics, acoustics, mechanics, electronics, and image analysis, with a host of considerations.

The continuous and rapid changes in technological developments preclude the publication of a definitive book on optical and laser scanning. The contributors have accomplished their tasks painstakingly well, and each could have written a volume on his own particular subject. This book can be used as an introduction to the field and as an invaluable reference for persons involved in any aspect of optical and laser beam scanning.

To assist the international scientific and engineering readership, measured quantities are expressed in dual units wherever possible and appropriate; the secondary units are in parentheses. The metric system takes precedence over other systems of units, except where it does not make good sense. A serious effort has been made for a measure of uniformity throughout the book with respect to terminology, nomenclature, and symbology. However, with the variety of individual styles from 27 contributing authors who are scattered across the Northern Hemisphere, I have placed greater importance on the unique contributions of the authors than on form.

The chapters are arranged in a logical order beginning with the laser light source and ending with a glossary. Chapters 1 through 3 cover three basic scanning systems topics: gaussian laser beam characterization, optical systems for laser scanners, and scanned image quality. Chapters 4 through 7 cover aspects of monogonal (single mirror-facet) and polygonal scanning system design, including bearings. Chapters 8 and 9 discuss aspects of galvanometric and resonant scanning systems, including flexure pivots. Chapters 10 through 14 cover holographic, optical disk, acousto-optical, electro-optical scanning systems, and thermal printhead technology. A useful glossary of scanner terminology follows Chapter 14.

Gerald F. Marshall

Acknowledgments

My appreciation goes to all of the contributors for their support and patience, without which this book would not have been possible. I especially thank those with whom I have worked closely in preparing this manuscript: Thomas Johnston, Stephen Sagan, Donald Lehmbeck, Emery Erdelyi, Chris Gerrard, Jean "Coco" Montagu, David Brown, Timothy Good, Tetsuo Saimi, Reeder Ward, Timothy Deis, Seung Ho Baek, Daniel Hass, and Alan Ludwiszewski.

I would like to acknowledge my former supervisors, Stanford Ovshinsky and Peter Klose at Energy Conversion Devices Inc., Rochester Hills, Michigan. The knowledge, insight, and support of these two talented individuals helped to bring this book to fruition.

Apart from an innate interest in mathematics and physics, my love of learning was stimulated by the encouragement and perseverance of two dedicated grammar school teachers, whom I'll forever remember with gratitude: The Reverend J. C. Harris, Salesian of Don Bosco, and Sister Virgilius, Religious of the Sacred Heart of Mary. I also thank my university physics educators, H. T. Flint, H. S. Barlow, and W. F. "Bill" Williams, who significantly contributed to my enjoyment and success throughout my career as a physicist.

In closing, I thank my colleague and friend Leo Beiser—a specialist in the field of laser beam scanning—for his guidance and suggestions for my chapter, as well as while working with me in organizing and cochairing many scanning conferences.

Contents

Preface v
Acknowledgments vii
Contributors xi

1. Characterization of Laser Beams: The M^2 Model 1
   Thomas F. Johnston, Jr. and Michael W. Sasnett

2. Optical Systems for Laser Scanners 71
   Stephen F. Sagan

3. Image Quality for Scanning 139
   Donald R. Lehmbeck and John C. Urbach

4. Polygonal Scanners: Components, Performance, and Design 265
   Glenn Stutz

5. Motors and Controllers (Drivers) for High-Performance Polygonal Scanners 299
   Emery Erdelyi and Gerald A. Rynkowski

6. Bearings for Rotary Scanners 345
   Chris Gerrard

7. Preobjective Polygonal Scanning 385
   Gerald F. Marshall

8. Galvanometric and Resonant Scanners ... 417
   Jean Montagu

9. Flexures Pivots for Oscillatory Scanners ... 477
   David C. Brown

10. Holographic Barcode Scanners: Applications, Performance, and Design ... 509
    Leroy D. Dickson and Timothy A. Good

11. Optical Disk Scanning Technology ... 551
    Tetsuo Saimi

12. Acousto-Optic Scanners and Modulators ... 599
    Reeder N. Ward, Mark T. Montgomery, and Milton Gottlieb

13. Electro-Optical Scanners ... 665
    Timothy K. Deis, Daniel D. Stancil, and Carl E. Conti

14. Multichannel Laser Thermal Printhead Technology ... 711
    Seung Ho Baek, Daniel D. Haas, David B. Kay, David Kessler, and Kurt M. Sanger

Glossary ... 769
Alan Ludwiszewski

Index ... 799

Contributors

Seung Ho Baek, Ph.D.   Eastman Kodak Company, Rochester, New York, U.S.A.

David C. Brown, Ph.D.   GSI Lumonics, Inc., Billerica, Massachusetts, U.S.A.

Carl E. Conti   Consultant, Hammondsport, New York, U.S.A.

Timothy K. Deis, B.S., M.S.   Consultant, Pittsburgh, Pennsylvania, U.S.A.

LeRoy D. Dickson, Ph.D.   Wasatch Photonics, Inc., Logan, Utah, U.S.A.

Timothy A. Good, M.S.   Metrologic Instruments, Inc., Blackwood, New Jersey, U.S.A.

Emery Erdelyi, B.S.E.E.   Axsys Technologies, Inc., San Diego, California, U.S.A.

Chris Gerrard, B.Sc.   Westwind Air Bearings Ltd., Poole, Dorset, United Kingdom

Milton Gottlieb, B.S., M.S., Ph.D.   Consultant, Carnegie Mellon University, Pittsburgh, Pennsylvania, U.S.A.

Daniel D. Haas, Ph.D.   Eastman Kodak Company, Rochester, New York, U.S.A.

Thomas F. Johnston, Jr., Ph.D.   Optical Physics Solutions, Grass Valley, California, U.S.A.

David B. Kay, Ph.D.   Eastman Kodak Company, Rochester, New York, U.S.A.

David Kessler, Ph.D.   Eastman Kodak Company, Rochester, New York, U.S.A.

Donald R. Lehmbeck, B.S., M.S.   Xerox Corporation, Webster, New York, U.S.A.

Alan Ludwiszewski, B.S.   Ion Optics, Inc., Waltham, Massachusetts, U.S.A.

Gerald F. Marshall, B.Sc., F.Inst.P.   Consultant in Optics, Niles, Michigan, U.S.A.

Jean Montagu, M.S.   Clinical MicroArrays, Inc., Natick, Massachusetts, U.S.A.

Mark T. Montgomery, B.S., M.S.   Direct2Data Technologies, Melbourne, Florida, U.S.A.

Gerald A. Rynkowski   Axsys Technologies, Inc., Rochester Hills, Michigan, U.S.A.

Stephen F. Sagan, M.S.   Agfa Corporation, Wilmington, Massachusetts, U.S.A.

Tetsuo Saimi, M.D.   Matsushita Electric Industrial Co., Ltd., Kadoma, Osaka, Japan

Kurt M. Sanger, B.S., M.S.   Eastman Kodak Company, Rochester, New York, U.S.A.

Michael W. Sasnett, M.S.E.E.   Optical System Engineering, Los Altos, California, U.S.A.

Daniel D. Stancil, Ph.D.   Carnegie Mellon University, Pittsburgh, Pennsylvania, U.S.A.

Glenn Stutz, B.S., M.S., M.B.A.   Lincoln Laser Company, Phoenix, Arizona, U.S.A.

John C. Urbach, B.S., M.S., Ph.D.[†]   Consultant, Portola Valley, California, U.S.A.

Reeder N. Ward, B.S., M.S.   Noah Industries, Inc., Melbourne, Florida, U.S.A.

[†]Deceased

1

Characterization of Laser Beams: The M^2 Model

THOMAS F. JOHNSTON, Jr.

Optical Physics Solutions, Grass Valley, California, U.S.A.

MICHAEL W. SASNETT

Optical System Engineering, Los Altos, California, U.S.A.

1 INTRODUCTION

The M^2 model, in the characterization of laser beams, is currently the preferred way of quantitatively describing a laser beam including its propagation through free space and lenses; specifically as ratios of its parameters with respect to the simplest theoretical gaussian laser beam. In addition the present chapter describes the measuring techniques for reliably determining – in each of the two orthogonal propagation planes – the key interrelated spatial parameters of a laser beam; namely, the beam waist diameter $2W_0$, the Rayleigh range z_R, the beam divergence Θ, and waist location z_0.

2 HISTORICAL DEVELOPMENT OF LASER BEAM CHARACTERIZATION

In 1966, six years after the first laser was demonstrated, a classic review paper[1] by Kogelnik and Li of Bell Telephone Laboratories was published that served as the standard reference on the description of laser beams for many years. Here the $1/e^2$-diameter definition for the width of the fundamental-mode, gaussian beam was used.[2] The more complex transverse irradiance patterns, or transverse modes, of laser beams were identified with sets of eigenfunction solutions to the wave equation, including diffraction, giving the electric fields of the beam modes. These solutions came in two forms: those with rectangular symmetry were described mathematically by Hermite–gaussian functions; those with cylindrical symmetry were described by Laguerre–gaussian functions. Eigenfunctions form basis-sets in which arbitrary field distributions can be expanded, and in principle, any beam could be decomposed into a weighted sum of the electric fields of these modes. Mathematically, for this expansion to be unique, the phases of the electric fields must be known; this is difficult at optical frequencies. Irradiance measurements alone, where the phase information is lost in squaring the E-fields, could not allow determination of the expansion coefficients. This "in principle but not in practice" description of light beams was all that was available and seemed to be all that was needed for several succeeding years.

Beam diameters were measured by scanning an aperture across the beam to detect the transmitted power profile. Apertures used were pinholes, slits, or knife-edges, the beam diameter being defined for the latter as twice the distance between the 16% and 84% transmission points, as this width agreed with the $1/e^2$-diameter for a fundamental mode beam. Commercial laser beams were specified as being pure fundamental mode, the lowest order or zero–zero transverse electromagnetic wave eigenfunction, "TEM_{00}."

In 1971 a short note was published by Marshall[3] introducing the M^2 factor, where M was the multiplier giving the factor by which the diameter of a beam, consisting of a mixture of higher-order modes, was larger than the diameter of the fundamental mode of the same laser resonator. Marshall's interest lay with industrial lasers and he was pointing out that laser beams with large M^2 values did not cut or weld well because of their larger focused spot sizes, or small depth of fields if focused more tightly with large f-number optics. Higher M^2 beams were thus of lower beam quality. No discussion was given of how to measure M^2 and the concept languished thereafter for several years.

From the late 1970s and into the 1980s, Bastiaans,[4] Siegman,[5,6] and others developed theories of bundles of light rays at narrow angles to an axis based on the Fourier transform relationship between the irradiance and the spatial frequency (or ray-angle) distributions to account for the propagation of the bundle. Such a bundle of rays is a beam. The beam diameter was defined as the standard deviation of the irradiance distribution (now called the second-moment diameter, when multiplied by four), and the square of this diameter was shown to increase as the square of the propagation distance – an expansion law for the diameter of hyperbolic form. These theories could be tested by measuring just the beam's irradiance profile along the propagation path.

In about 1987, one of us designed a telescope to locate a beam waist for an industrial CO_2 laser at a particular place in the external optical system. The design was based on measurements showing where the input beam waist was located and on blind faith that the laser datasheet claim for a TEM_{00} beam was correct. This telescope provided nothing like the expected result. Out of despair and disorientation came the energy to make more beam measurements and from these measurements came the realization that the factor that limited the maximum distance between the telescope and the beam waist it produced was exactly the same factor by which actual focus-spot diameter at the work surface exceeded the calculated TEM_{00} spot diameter. That factor was M^2 and when used in modified Kogelnik and Li equations, design of optical systems for multimode beams became possible.[7] This ignited some interest in knowing more about laser beams than had previously been considered sufficient. Laser datasheets that claimed "TEM_{00}" were no longer adequate.

In the 1980s, commercial profilers[8] reporting a beam's $1/e^2$ diameter became ubiquitous. By the end of the 1980s, experience with commercial profilers and these theories converged with the development[6] of the theoretical M^2 model and a commercial instrument[9] to measure the beam quality based on it, which first became available in 1990. The time to determine a beam's M^2 value dropped from half a day to half a minute.

With high accuracy M^2 measurements more readily available in the early 1990s, the reporting of a beam's M^2 value became commonplace, and commercial lasers with good beams were now specified[10] as having $M^2 < 1.1$. The International Organization for Standards (ISO) began committee meetings to define standards for the spatial characterization of laser beams, ultimately deciding on the beam quality M^2 value based on the second-moment diameter as the standard.[11] This diameter definition has the best theoretical support, in the form of the Fourier transform theories of the 1980s, but suffers from being sensitive to noise on the profile signal, which often makes the measured diameters unreliable.[12,13] That led to the development in 1993 of rules[14] to convert diameters measured with the more forgiving methods into second-moment diameters for a large class of beams.

The M^2 model as commercially implemented does not cover beams that twist as they propagate in space, that is, those with general astigmatism.[15,16] The earlier Fourier transform theories and their more recent extensions do, however, and allow for ten constants[17] needed to fully characterize a beam (adding to the six used in the M^2 model). Recently, in 2001, the first natural beam[18] (as opposed to a test beam artificially constructed) was measured by Nemes *et al.* that required all ten constants for its complete description.

For the present, several recommendations can be listed for characterizing a beam:

1. Use the six-constant M^2 model for the measurement plan.
2. Use a beam diameter measurement method that gives reliable results, and convert that M^2 value into the ISO standard units at the end.
3. Watch for developments, particularly new software, that could make the direct measurement of second-moment diameters acceptably reliable.
4. Watch for practical developments to make the ten-constant measurements easier.

3 ORGANIZATION OF THIS CHAPTER

Section 1 provides an historical introduction to the field. This outlines how the field developed to its present state and anticipates where it is going.

The technical discussion begins in Sec. 4 by explaining the M^2 model. This mathematical model built around the quantity M^2 (variously called the beam quality, times-diffraction-limit number, or the beam propagation factor) describes the real, multi-mode beams that all lasers produce and how their properties change when propagating in free space.

This discussion is continued in Sec. 5 covering the transformation of such a beam through a lens. Section 6 explains the different methods used to define and measure beam diameters, and how measurements made with one method can be converted into the values measured with one of the other methods. This includes the standard diameter definition recommended by ISO, the second-moment diameter, and the experimental difficulties encountered with this method.

The technical development continues in Sec. 7 where the logic and precautions needed in measuring the beam quality M^2 are presented. Thoroughly discussed is the "four-cuts" method (a cut is a measurement of a beam diameter), the simplest way to obtain an accurate M^2 value.

Section 8 discusses the common and possible types of beam asymmetry that may be encountered in three dimensions when the propagation constants for the two orthogonal (and usually independent) propagation planes are combined. The concept of the "equivalent cylindrical beam" is introduced to complete the technical development of the M^2 model. Propagation plots for beams with combinations of asymmetries are illustrated. A short discussion follows of "twisted beams," those with general astigmatism, which are not covered in the M^2 model, but require a beam matrix of ten moments of second order. This second-order beam matrix theory is a part of the underpinnings of the ISO's choice of the noise-sensitive second-moment diameter as the "standard."

Section 9 applies the M^2 model to an analysis of a stereolithography laser-scanning system. Using results of earlier sections, by working backward from assumed perturbations or defects in the scanned beam at the work surface, the deviations in beam constants at the laser head that would produce them are found. An overview of the M^2 model, in Sec. 10, concludes the text.

A glossary follows, explaining the technical terms used in the field, with the references ending the chapter.

4 THE M^2 MODEL FOR MIXED MODE BEAMS

In laser beam-scanning applications, the main concern is having knowledge of the beam spot-size – the transverse dimensions of the beam – at any point along the beam path. The mixed mode ($M^2 > 1$) propagation equations are derived as extensions of those for the fundamental mode, so pure modes and particularly the fundamental mode are the starting point.

4.1 Pure Transverse Modes: The Hermite–Gaussian and Laguerre–Gaussian Functions

Lasers emit beams in a variety of characteristic patterns or transverse modes that can occur as a pure single mode or, more often, as a mixture of several superposed pure modes. The transverse irradiance distribution or beam profile of a pure mode is the square of the electric field amplitude where this amplitude is described mathematically by Hermite–gaussian functions if it has rectangular symmetry, or by a Laguerre–gaussian function if it has circular symmetry.[1,2,5,19] These functions when plotted reproduce the familiar spot patterns – the appearance of a beam on an inserted card – first photographed in Ref. 20 and shown in Refs. 1 and 19. Computed spot patterns are displayed in Fig. 1. The computations were done in Mathematica for the first six cylindrically symmetric modes, in order of increasing diffraction loss for a circular limiting aperture. These modes are the solutions to the wave equation for a bundle of rays propagating at small angles (paraxial rays) to the z-axis, under the influence of diffraction and are of the general forms[1,2,7]

$$U_{mn}(x, y, z) = H_m(x/w)H_n(y/w)u(x, y, z) \tag{1a}$$

or

$$U_{pl}(r, \varphi, z) = L_{pl}(r/w, \varphi)u(r, z) \tag{1b}$$

In Eq. (1a), $H_m(x/w)H_n(y/w)$ represents a pair of Hermite polynomials, one a function of x/w, the other of y/w, where x, y are orthogonal transverse coordinates and w is the radial scale parameter. In Eq. (1b), $L_{pl}(r/w, \varphi)$ represents a generalized Laguerre polynomial, a function of the r, φ transverse radial and angular coordinates. These polynomials have

The M² Model

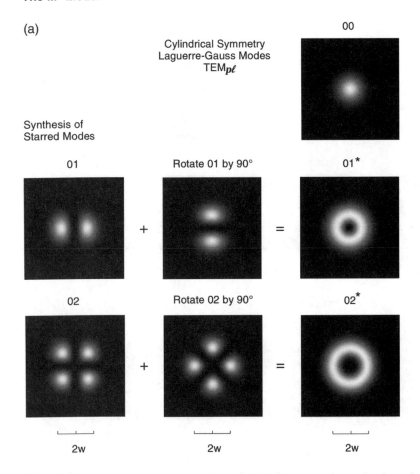

Figure 1 Computed spot patterns for cylindrically-symmetric modes in order of increasing diffraction loss for a circular limiting aperture. The subscript numbers *pl* above each image indicate the mode order. Starred modes are constructed as shown, as the sum of a pattern with a copy of itself rotated by 90°: (a) first three modes. (*Continued*)

no dependence on the propagation distance z other than through the dependence $w(z)$ in x/w, y/w, or r/w. The $w(z)$ dependence describes the beam convergence or divergence. The other function u is the gaussian

$$u = (2/\pi)^{1/2} \exp[-(x^2 + y^2)/w^2] = (2/\pi)^{1/2} \exp[-r^2/w^2] \tag{2}$$

Because the radial gaussian function splits into a product of two gaussians, one a function of x, the other of y, the Hermite–gaussian function splits into the product of two functions, one in x/w only and the other in y/w only, each of which is independently a solution to the wave equation. This has the consequence that beams can have independent propagation parameters in the two orthogonal planes (x, z) and (y, z).

These functions of the transverse space coordinates consist of a damping gaussian factor, limiting the beam diameter, times a modulating polynomial that pushes light

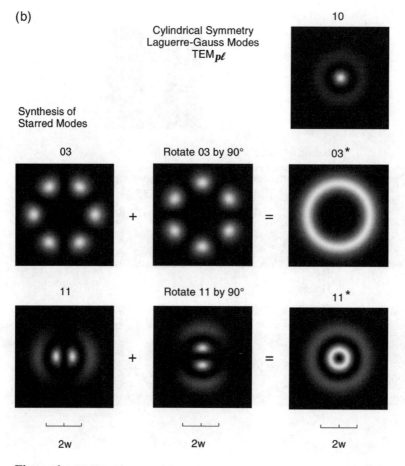

Figure 1 (b) Next three modes.

energy out radially as polynomial orders increase. The order numbers m, n of the Hermite polynomials, or p, l of the Laguerre polynomial of the pure mode also determine the number of nodes in the spot pattern, for which the modes are named. They are designated as transverse electromagnetic modes, or $TEM_{m,n}$ for a mode with m nodes in the horizontal direction and n nodes in the vertical direction, or $TEM_{p,l}$ for a mode with p nodes in a radial direction – not counting the null at the center if there is one – and l nodes going angularly around half of a circumference. Figures 2(a–f) show the theoretical beam irradiance profiles for the six pure modes from Fig. 1. Because these are the six lowest loss modes,[21,22] they are commonly found in real laser beams. The modes as shown all originate in the same resonator – they all have the same radial scale parameter $w(z)$. The addition of an asterisk to the mode designation—a "starred mode"—signifies a composite of two degenerate (same frequency) Hermite–gaussian modes, or as here, Laguerre–gaussian modes in space and phase quadrature to form a mode of radial symmetry. This is explained in Ref. 20, discussed in Ref. 5, p. 689, and shown in Fig. 1 for a mode pattern with an azimuthal variation ($l \neq 0$) as the addition of the mode with a copy of itself after a 90° rotation, to produce a smooth ring-shaped pattern.

The M² Model

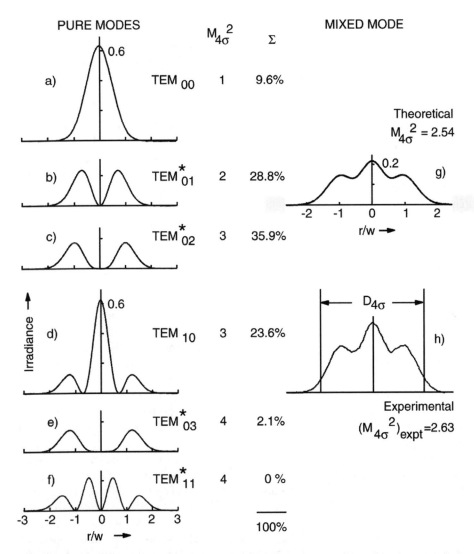

Figure 2 Pinhole profiles, on the left, of the same six low-order, radial, pure modes (a) to (f) as the first figure, summed on the right, to produce the theoretical mixed mode (g) using the mode fractions Σ. The beam quality $M_{4\sigma}^2$ for the mixed mode is the sum of the products of the $M_{4\sigma}^2$ for each pure mode, times the mode fraction Σ for that mode. The mode profile (g) matches the experimental pinhole profile (h).

The simplest mode is the TEM_{00} mode, also called the lowest order mode or fundamental mode of Fig. 1 and Fig. 2(a), and consists of a single spot with a gaussian profile (here L_{pl} is unity). The next higher order mode has a single node [Fig. 1 and Fig. 2(b)] and is appropriately called the "donut" mode, symbol TEM_{01}^*. The next two "starred" modes spots look like a donuts with larger holes, the spot pattern of the TEM_{10} mode looks like a target with a bright center, and the TEM_{11}^* mode spot looks like a target with dark center, (Fig. 1 and Fig. 2). All higher order modes have a larger beam diameter than the fundamental mode. The six pure modes of Fig. 2 are shown with the vertical scale normalized such that when integrated over the transverse plane, each contains unit power.

The physical reason that Hermite–gaussian and Laguerre–gaussian functions describe the transverse modes of laser beams is straightforward. Laser beams are generated in resonators by the constructive interference of waves multiply reflected back and forth along the beam axis. For this interference to be a maximum, permitting a large stored energy to saturate the available gain, the returned wave after a round trip of the resonator, should match the transverse profile of the initial wave. These functions are the eigenfunctions of the Fresnel–Kirchhoff integral equation used to calculate the propagation of a paraxial rays with diffraction included.[5,19] In other words, these are precisely the beam irradiance profiles that in propagating and diffracting maintain a self-similar profile, allowing after a round trip, maximum constructive interference and gain dominance.

4.2 Mixed Modes: The Incoherent Superposition of Pure Modes

While a laser may operate in a close approximation to a pure higher order mode, for example, by a scratch or dust mote on a mirror forcing a node and suppressing a lower order mode with an irradiance maximum at that location, actual lasers tend to operate with a mixture of several high-order modes oscillating simultaneously. The one exception is lasing in the pure fundamental mode in a resonator with a circular limiting aperture, where the aperture diameter is critically adjusted to exclude the next higher order (donut) mode. Each pure transverse mode has a unique frequency different from that for adjacent modes by tens or hundreds of MHz. This is usually beyond the response bandwidth of profile measuring instruments so any mode interference effects are invisible in such measurements.

Figure 2(g) shows a higher order mode synthesized by mixing the five lowest order modes of Figs 2(a–e) in a sum with the weightings shown in the column labeled Σ. These weights – also called mode fractions – were chosen by a fitting program to match the result to the experimental pinhole profile (see below) of Fig. 2(h). In the experiment[14] the number of transverse modes oscillating and their orders were known (by detecting the radio-frequency transverse mode beat notes in a fast photodiode). This information was used in the fitting procedure. The laser was a typical 1 m long argon ion laser operating at a wavelength of 514 nm, except that a larger than normal intracavity limiting aperture diameter was used to produce this mode mixture.

Because the polynomials of Eq. (1) have no explicit dependence on z, the profiles and widths of the modes in a mixture remain the same relative to each other and specifically to the fundamental mode as the beam propagates. This means that however the diameter $2W$ of a mixed mode beam is defined (several alternatives are discussed in Sec. 6 below), if this diameter is M times larger than the fundamental mode diameter at one propagation distance, it will remain so at any distance:

$$W(z) = Mw(z) \tag{3}$$

This equation introduces the convention that upper case letters are used for the attributes of high order and mixed modes and lower case letters used for the underlying fundamental mode.

The M² Model

4.3 Properties of the Fundamental Mode Related to the Beam Diameter

The attributes of the simplest beam, a fundamental mode with a round spot (a cylindrically symmetric or stigmatic beam) are reviewed in Figs 3 and 4. The beam profile varies as the transverse irradiance distribution and is given by the function of gaussian form[1,2] [Fig. 3(a)]:

$$I(r/w) = I_0 \exp[-2(r/w)^2] \qquad (4)$$

The symbol I denotes a detector signal proportional to irradiance (and by using I instead of E, the recommended symbol for irradiance, avoids confusion with the electric field of the beam). The peak irradiance is I_0, and the radial scale parameter w introduced in Eq. (1) can now be identified as the distance transverse to the beam axis at which the irradiance value falls to $1/e^2$ (13.5%) of the peak irradiance. This $1/e^2$-diameter definition, introduced[1] in the early 1960s, has been universally used since, with one exception. The exception is in the field of biology where the fundamental mode diameter is defined as the radial distance to drop to $1/e$ (36.8%) of the central peak value, making biological beams a diameter $2w' = \sqrt{2}w$ instead of $2w$. Many different beam diameter definitions have been used subsequently for higher order modes (these are discussed below), but they all share one common property: when applied to the fundamental mode, they reduce to the traditional $1/e^2$-diameter.

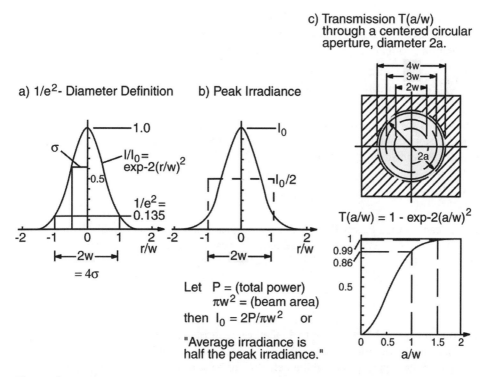

Figure 3 Properties of the fundamental mode relating to beam diameter.

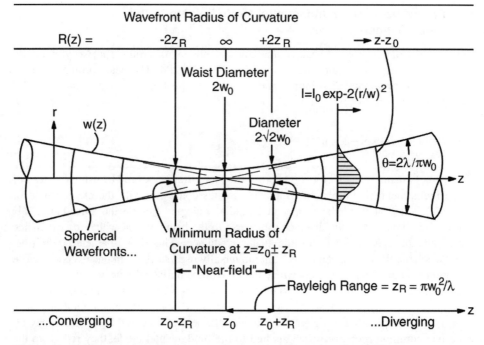

Figure 4 Propagation properties of the pure gaussian, fundamental-mode beam. The wavefront curvatures are exaggerated to show their variation with propagation distance.

Tables of the gaussian function are usually[23] listed under the heading of the normal distribution, normal curve of error, or Gauss distribution and are of the form

$$I(x) = [1/\sigma(2\pi)^{1/2}] \exp(-x^2/2\sigma^2) \tag{5}$$

where σ is the standard deviation of the gaussian distribution. Comparing Eq. (4) and Eq. (5) shows that the $1/e^2$-diameter is related to the standard deviation σ of the irradiance profile, as defined in Eq. (5), as

$$2w = 4\sigma \tag{6}$$

For a beam of total power P, the value of the peak irradiance I_0 is found[5] by integrating Eq. (4) over the transverse plane (yielding I_0 times an area of $\pi w^2/2$) and equating this to P. The result

$$I_0 = 2P/\pi w^2 \tag{7}$$

is easily remembered by noting that "the average irradiance is half the peak irradiance." This is a handy, often-used simplification allowing the actual beam profile to be replaced by a round flat-topped profile of diameter $2w$ for back-of-the-envelope conceptualizations [see Fig. 3(b)].

If the gaussian beam is centered on a circular aperture of diameter $2a$ the transmitted fraction $T(a/w)$ of the total beam power is given by a similar integration[5] over the cross-sectional area as [see Fig. 3(c)]

$$T(a/w) = 1 - \exp[-2(a/w)^2] \tag{8}$$

This gives a transmission fraction of 86.5% for an aperture of diameter $2w$, and 98.9% for one of diameter $3w$. In practice, a minimum diameter for an optic or other aperture to pass the beam and leave it unaffected is $4.6w$ to $5w$ to reduce the sharp edge diffraction ripples overlaid on the beam profile to an amplitude of $<1\%$.[5] It is interesting to note that for a low power, visible, fundamental mode beam, the spot appears to be a diameter of about $4w$ to the human eye viewing the spot on a card.

The transmission of a fundamental mode beam past a vertical knife-edge is also readily computed [see Fig. 8(c) in Sec. 6.1]. The knife edge transmission function is $T(x/w) = 0$ for $x' \leq x$, $T = 1$ for $x' > x$, where x is the horizontal distance of the knife edge from the beam axis and x' is the horizontal distance integration variable. In Eq. (4), the substitution $r^2 = x^2 + y^2$ is made, the integration over y yields multiplication by a constant, and the final integration over x' is expressed in terms of the error function as

$$T(x/w) = (1/2)[1 \pm \text{erf}(\sqrt{2}x/w)], \quad + \text{ if } x < 0, \quad - \text{ if } x > 0 \tag{9}$$

The error function[23] of probability theory in Eq. (9) is defined as

$$\text{erf}(t) = (2/\sqrt{\pi}) \int_0^t \exp(-u^2)\,du \tag{10}$$

and is tabulated in many mathematical tables. The $1/e^2$-diameter of a fundamental mode beam is measured with a translating knife-edge by noting the difference in translation distances of the edge $(x_1 - x_2)$ that yield transmissions of 84.1 and 15.9%. By Eq. (9) this separation equals w, and the beam diameter is twice this difference [see Fig. 8(c), later].

4.4 Propagation Properties of the Fundamental Mode Beam

The general properties expected for the propagation of a gaussian beam can be outlined from simple physical principles. As predicted by solving the wave equation with diffraction, a bundle of focused paraxial rays converges to a *finite* minimum diameter $2w_0$, called the waist diameter. The full angular spread, θ, of the converging and, on the other side, diverging beam is proportional to the beam's wavelength λ divided by the minimum diameter, $\theta \propto \lambda/2w_0$.[10] A scale length z_R for spread of the beam, is the propagation distance for the beam diameter to grow an amount comparable to the waist diameter, or $z_R\theta \sim w_0$, giving $z_R \propto w_0^2/\lambda$. Because the rays of the bundle propagate perpendicularly to the wavefronts (surfaces of constant phase), at the minimum's location the rays are parallel by symmetry and the wavefront there is planar. At large distances $z - z_0$ from the waist diameter location at z_0 – the propagation axis is z – the wavefronts become Huygen's wavelets diverging from z_0 with wavefront radii of curvature $R(z)$, and eventually become plane waves. Since the wavefronts are plane at the minimum diameter at the waist and at large distances on either side, but converge and diverge through the waist, there must be points of maximum wavefront curvature (minimum radius of curvature) to either side of z_0.

The actual beam propagation equations describing the change in beam radius $w(z)$ and radius of curvature $R(z)$ with z, are derived[1,2,5] as solutions to the wave equation in the complex plane and show all of these features. They are (Fig. 4)

$$w(z) = w_0\sqrt{[1 + (z - z_0)^2/z_R^2]} \qquad (11)$$
$$R(z) = (z - z_0)[1 + z_R^2/(z - z_0)^2] \qquad (12)$$
$$z_R = \pi w_0^2/\lambda \qquad (13)$$
$$\theta = 2\lambda/\pi w_0 = 2w_0/z_R \qquad (14)$$

and

$$\psi(z) = -\tan^{-1}(z/z_R) \qquad (15)$$

In these equations, the minimum beam diameter $2w_0$ (the waist diameter) is located at z_0 along the propagation axis z. A plot of $w(z)$ vs. z, beam radius vs. propagation distance [Eq. (11)] is termed the axial profile or propagation plot and is a hyperbola. The scale length for beam expansion, z_R, is termed the Rayleigh range [Eq. (13)] and has the expected dependence on λ and w_0. The radius of curvature $R(z)$ of the beam wavefront, as given by Eq. (12), has the expected behavior. At large distances from the waist – the region termed the "far-field" – and where $|z - z_0| \gg z_R$, the radius of curvature first becomes $R \to (z - z_0)$ and then becomes plane when $|R| \to \infty$ as $|z - z_0| \to <\infty$, and also is plane at $(z - z_0) = 0$. By differentiating Eq. (12) and equating the result to zero the points of minimum absolute value of the radius of curvature are found to occur at $z - z_0 = \pm z_R$ and have the values $R_{min} = \pm 2z_R$. The full divergence angle θ develops in the far-field, the beam envelope is asymptotic to two straight lines crossing the axis at the waist location (Fig. 4). Finally, $\psi(z)$ is the phase shift[5,24] of the laser beam relative to that of an ideal plane wave. It is a consequence of the beam going through a focus (the waist), the gaussian beam version of the Gouy phase shift.[24]

By Eq. (11), the diameter $2w(z)$ of the beam increases by a factor $\sqrt{2}$ (and for a round beam the cross-sectional area doubles) for a propagation distance $\pm z_R$ away from the waist (Fig. 4). This condition is often used to define the Rayleigh range z_R,[5,25] but another significant condition is that at these two propagation distances the wavefront radius of curvature goes through its extreme values ($|R| = R_{min}$). The Rayleigh range can be defined as half the distance between these curvature extremes. The region within a Rayleigh range of the waist is defined as the "near-field" region. Within this region, wavefronts flatten as the waist is approached and outside they flatten as they recede from the waist. A positive lens placed in a diverging beam and moved back towards the source waist will encounter ever steeper wavefront curvatures so long as the lens remains out of the near-field. On the lens output side, the transformed waist moves away from the lens, moving qualitatively as a geometrical optics image would. When the lens enters the near-field region still approaching the source waist, ever flatter wavefronts are encountered and then the transformed waist *also* approaches the lens. The laser system designer who misunderstands this unusual property of beams will have unpleasant surprises. Many laser systems have undergone emergency redesign when prototype testing revealed this counterintuitive focusing behavior! In many ways, Rayleigh range is the single most important quantity in characterizing a beam [note that this is a factor in all of Eqs (11–15)]. It will be shown in the next section that measurement of a beam's Rayleigh range is the basis for measuring the beam quality M^2 of a mixed mode beam.

As the lowest order solution to the wave equation, the fundamental mode with a gaussian irradiance profile of a given $1/e^2$-diameter $2w_0$ is the beam of lowest divergence, at the limit set by diffraction,[10] of any paraxial bundle with that minimum diameter. Confining a bundle to a smaller diameter proportionally increases, by diffraction, the divergence angle of the bundle, and the product $2w_0\theta$ is an invariant for any mode. The smallest possible value, $4\lambda/\pi$, is achieved only by the fundamental mode. This is just the Uncertainty Principle for photons – laterally confining a photon in the bundle increases the spread of its transverse momentum and correspondingly the divergence angle of the bundle. This limit cannot be achieved by real-world lasers, but sometimes it is closely approached. Helium–neon lasers, especially the low-cost versions with internal mirrors (no Brewster windows), are wonderful sources of beams within 1 or 2% of this limit. Aside from the wavelength, which must be known to specify any beam, the ideal, round, (stigmatic) fundamental mode beam is specified by only two constants: the waist diameter $2w_0$ and its location z_0 (or equivalents such as z_R and z_0). This will no longer be true when mixed modes are considered.

As noted at the beginning of this section the propagation constants for the (x, z) and (y, z) planes are independent and can be different. In each plane, the rays obey equations exactly of the same form[6] as Eqs. (11–15) with subscripts added indicating the x or y plane. For beams with pure (but different) gaussian profiles in each plane, two more constants are introduced for a total of four required to specify the beam. If $z_{0x} \neq z_{0y}$ (different waist locations in the two principal propagation planes) the beam exhibits simple astigmatism; if $2w_{0x} \neq 2w_{0y}$ (different waist diameters) the beam has asymmetric waists. (See Fig. 15 in Sec. 8 for illustrations of these and other beam asymmetries.)

4.5 Propagation Properties of the Mixed Mode Beam: The Embedded Gaussian and The M^2 Model

In Sec. 4.2 a mixed mode was defined as the power-weighted superposition of several higher order modes originating in the same resonator, each with the same underlying gaussian waist radius w_0 determining the radial scale length $w(z)$ in their mode functions [Eqs. (1), (2)]. This underlying fundamental mode, with w_0 fixed[5] by the radii of curvature and spacing of the resonator mirrors, is called the embedded gaussian for that resonator regardless of whether or not the mixed mode actually has some fundamental mode content. To treat the mixed mode case, use is made[7] of the fact that its diameter is everywhere (for all z) proportional to the embedded gaussian diameter. From Eq. (3) the substitution $w(z) = W(z)/M$ in Eqs. (11)–(15) yields the mixed mode propagation equations

$$W(z) = W_0 \sqrt{[1 + (z - z_0)^2/z_R^2]} \tag{16}$$

$$R(z) = (z - z_0)[1 + z_R^2/(z - z_0)^2] \tag{17}$$

$$Z_R = \pi W_0^2/M^2\lambda = z_R \tag{18}$$

and

$$\Theta = 2M^2\lambda/\pi W_0 = 2W_0/z_R = M\theta \qquad (19)$$

The mixed mode, a sum of transverse modes with different optical frequencies, no longer has a simple expression for the Gouy phase shift analogous to Eq. (15). The convention followed here is that upper case quantities refer to the mixed mode and lower case quantities refer to the embedded gaussian.

Many of the properties of the fundamental mode beam carry over to the mixed mode one (Fig. 5). Since $W_0 = Mw_0$, substitution in the middle part of Eq. (19) gives the last part, the mixed mode divergence is M times that of the embedded gaussian. Similarly, the beam propagation profile $W(z)$ also has the form of a hyperbola (one M times larger) with asymptotes crossing at the waist location. The Rayleigh ranges are the same for both mixed and embedded gaussian modes as substituting $W_0 = Mw_0$ in the middle of Eq. (18) shows, so the radii of curvature and the limits of the near-field region are the same for both. The mixed mode beam diameter still expands by a factor of $\sqrt{2}$ in a propagation distance of z_R away from the waist location z_0, the starting diameter W_0 is just M times larger.

In considering propagation in the independent (x, z) and (y, z) planes, there are now two new constants needed to specify the beam, M_x^2 and M_y^2, for a total of six required constants. In making up the mixed mode, the Hermite–gaussian functions summed in the two planes need not be the same or have the same distribution of weights, making

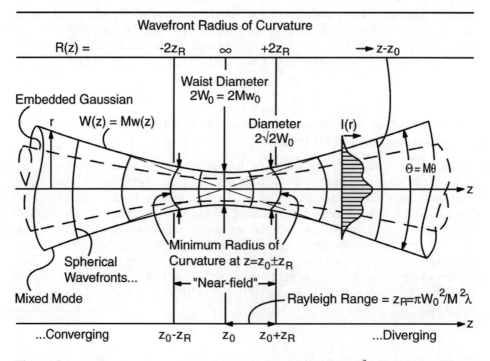

Figure 5 Propagation properties of the mixed mode beam drawn for $M^2 = 2.63$. The embedded gaussian is the fundamental mode beam originating in the same resonator. The wavefront curvatures are exaggerated to show their variation with propagation distance.

The M² Model

$M_x^2 \neq M_y^2$ a possibility. In this case the beam is said to have divergence asymmetry since $\Theta \propto M^2$ by the first part of Eq. (19).

It might be asked, why are these Eqs (16)–(19) termed the "M^2 model" (and not the "M model")? There are two reasons. The first is that the embedded gaussian is buried in the mixed mode profile, and cannot be measured independently, making it difficult to directly determine M. The mixed mode diameter still grows by $\sqrt{2}$ in a propagation distance z_R from the waist location, so z_R can be found from several diameter measurements fitted to a hyperbolic form. The waist diameter $2W_0$ can also be measured, thus giving directly, by Eq. (18),

$$M^2 = \pi W_0^2 / \lambda z_R \qquad (20)$$

This is how M^2 is in fact measured, the practical aspects of which will be discussed in Sec. 7. [As an aside, note that Eq. (20) shows that M^2 scales as the square of the beam diameter; this is used later in the discussion of conversions between different diameter definitions, Sec. 6.4.]

The second reason is the more important one: M^2 is an invariant of the beam, and is conserved[26] as the beam propagates through ordinary non-aberrating optical elements. Like the fundamental mode beam whose waist diameter–divergence product was conserved, the same product for the mixed mode beam is

$$(2W_0)\Theta = (2W_0) 2M^2 \lambda / \pi W_0 = M^2 (4\lambda / \pi) \qquad (21)$$

This is larger by the factor M^2 than the invariant product for a fundamental mode.

Equation (21) can be rearranged to read

$$M^2 = \Theta / (2\lambda / \pi W_0) = \Theta / \theta_n \qquad (22)$$

Here $\theta_n = 2\lambda / \pi W_0$ is recognized as the divergence of a fundamental mode beam with a waist diameter $2W_0$, the same as the mixed mode beam. This is called the normalizing gaussian; it has an M times larger scale constant $W_0 = Mw_0$ in its exponential term than the embedded gaussian and it would *not* be generated in the resonator of the mixed mode beam. It does represent the diffraction-limited minimum divergence for a ray bundle constricted to the diameter $2W_0$. Thus by Eq. (22) the invariant factor M^2 can be seen to be the "times-diffraction-limit" number referred to in the literature.[5] This also identifies M^2 as the inverse beam quality number, the highest quality beam being an idealized diffraction limited one with $M^2 = 1$, while all real beams are at least slightly imperfect and have $M^2 > 1$.

The value of the M² model is twofold. Once the six constants of the beam are accurately determined (by fitting propagation plot data for each of the two independent propagation planes) they can be applied by the system designer to predict accurately the behavior of the beam throughout the optical system before it is built. The spot diameters, aperture transmissions, focus locations and depths of field, and so on, can all be found for the vast majority of existing commercial lasers. The second value is that there are commercial instruments available that efficiently measure and document a beam's constants in the M² model. This permits quality control inspection of the lasers at final test, or whenever there is a system problem and the laser is the suspected cause. Defective optics can introduce aberrations in the beam wavefronts. If inside the laser they increase M^2 by forcing larger amounts of high divergence, high-order modes in the mixed mode sum. If outside the resonator, they also adversely affect M^2. Measurement of the beam quality during system assembly, after each optic is added to detect a downstream increase in M^2, can aid in quality control of the overall optical system.

Beams excluded from the model as described are those whose orthogonal axes rotate or twist about the propagation axis (called beams with general astigmatism[15,16,27]) such as might come from lasers with nonplanar ring or out-of-plane folded resonators. The symmetry of the beam is determined by the symmetry of the resonator. Fortunately, few commercial lasers produce beams having these characteristics. An overview of the full range of symmetry possibilities for laser beams is discussed in Sec. 8.3.

The fact that M^2 is not unique, that is, that a given value of M^2 can be arrived at by a variety of different higher order modes or mode weights in the mixed mode, is sometimes stated to be a deficiency of the M^2 model. This is also its strength. It is a simple predictive model that does not require measurement and analysis to determine the mode content in a beam. In the evolution of beam models, the original discussion[1,2] pointed out that as eigenfunctions of the wave equation, the full (infinite) set of Hermite–gaussian or Laguerre–gaussian functions [Eq. (1)] describing the electric field of the beam modes form an orthonormal set. As such they could model an arbitrary paraxial light bundle with a weighted sum. This is true only if the phases of the E-fields are kept in the sum, and measuring the phase of an optical wave generally is a difficult matter. Summing the irradiances (the square of the E-fields) breaks the orthonormality condition and for years it was not obvious that a simple model relying only on irradiance measurements was possible. Then in the 1980s, methods based on Fourier transforms of irradiance and ray angular distributions of light bundles were introduced[4,6] that showed that as far as predictions of beam diameters in an optical system were concerned, irradiance profile measurements would (usually) suffice. The M^2 model was born, and commercial instruments[10] for its application soon followed. Later we realized that modes "turn on" in a characteristic sequence as diffraction losses are reduced in the generating resonator. This makes a given M^2 correspond to a unique mode mix in many common cases after all (see Sec. 6.4).

5 TRANSFORMATION BY A LENS OF FUNDAMENTAL AND MIXED MODE BEAMS

Knowledge of how a beam is transformed by a lens is not only useful in general, but in particular, a lens is used to gain an accessible region around the waist for the measurements of diameters that give M^2 (see Sec. 7). This transformation is discussed next.

In geometrical optics a point source at a distance s_1 from a thin lens produces a spherical wave whose radius of curvature is R_1 at the lens (and whose curvature is $1/R_1$), where $R_1 = s_1$. In traversing the lens, this curvature is reduced by the power $1/f$ of the lens (f is the effective focal length of the lens) to produce an exiting spherical wave of curvature $1/R_2$ according to the thin lens formula

$$1/R_2 = 1/R_1 - 1/f \tag{23}$$

The M^2 Model

An image of the source point forms at the distance R_2 from the lens from convergence of this spherical wave. Note that the conventions used in Eq. (23) are the same as in Eq. (17), namely, the beam always travels from left to right, converging wavefronts with center of curvature to the right have negative radii, and diverging wavefronts with centers to the left have positive radii. [The usual convention in geometrical optics[28] is that converging wavefronts leaving the lens are assigned positive radii, which would put a minus sign on the term $1/R_2$ of Eq. (23).]

The quantities used in the beam-lens transform are defined in Fig. 6. Following Kogelnik[1] the beam parameters on the input side of the lens are designated with a subscript 1 (for "1-space") and on the output side with a subscript 2 (for "2-space"). The principal plane description[28] of a real (thick) lens is used, in which the thick lens is replaced by a thin one acting at the lens principal planes H1, H2. Rays between H1 and H2 are drawn parallel to the axis by convention, and waist locations z_{01} and z_{02} are measured from H1 and H2, respectively (with distances to the right as positive for z_{02} and distances to the left as positive for z_{01}).

A lens inserted in a beam makes the same change in wavefront curvature as it did in geometrical optics [Eq. (23)], but the wavefront R_2 converges to a waist of finite diameter $2W_{02}$ at a distance z_{02} given by Eq. (17). For each of the two independent propagation planes, there are three constants required to specify the transformed beam, and three constraints needed to fix them. The lens should be aberration-free (typically, used at $f/20$ or smaller aperture) and, if so, the beam quality does not change in going through it, giving the first condition $M_2^2 = M_1^2$. The second constraint is that the wavefront curvatures match, between the input curvature modified by the lens [Eq. (23)], and the transformed beam at the same location as specified by the transformed beam constants through Eq. (17). A beam actually has two points with the same magnitude and sign of the curvature, one inside the near-field region of that sign and one outside, which differ in beam diameters. The ambiguity as to which point is matched is removed by the third constraint, that the beam diameter is unchanged in traversing the (thin) lens.

When the equations these three constraints define are solved for the transformed waist diameter and location, the result[1,29–31] is given (in modern notation; that is, the notation used in a commercial M^2 measuring instrument[9]) in terms of the transformation

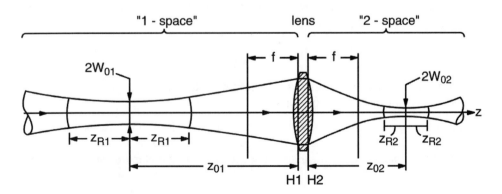

Figure 6 Definitions of quantities used in the beam-lens transform.

constant Γ as follows:

$$\Gamma = f^2/[(z_{01} - f)^2 + z_{R1}^2] \tag{24}$$

$$M_1^2 = M_2^2 = M^2 \tag{25}$$

$$W_{02} = \sqrt{\Gamma} W_{01} \tag{26}$$

$$z_{R2} = \Gamma z_{R1} \tag{27}$$

$$z_{02} = f + \Gamma(z_{01} - f) \tag{28}$$

A set of these equations applies to each of the two principal propagation planes (x, z) and (y, z).

The transform equations [Eqs. (24)–(28)] are not as simple as in geometrical optics because of the complexity of the way the beam wavefront curvatures change with propagation distance [Eq. (17)]. Like the image and object distances in geometrical optics, the transformed beam waist location depends on the input waist location, but also depends, as does the wavefront curvature, on the Rayleigh range of the input beam. The most peculiar behavior as the waist to lens distance varies is when the input focal plane of the lens moves within the near-field of the incident beam, $|z_{01} - f| < z_{R1}$. Then the slope of the z_{02} vs. z_{01} curve turns from negative to positive (in geometrical optics the slope of the object to image distance curve is always negative). This sign change can be demonstrated by substituting Eq. (24) into Eq. (28) and differentiating the result with respect to z_{01}. As the lens continues to move closer to the input waist, the transformed waist location also moves closer to the lens, exactly opposite to what happens in geometrical optics. In the beam-lens transform, the input and transformed waists are *not* images of each other (in the geometrical optics sense). Despite the intransigence of beam waists, the object–image relationship of beam diameters at conjugate planes on each side of the lens does apply just as in geometrical optics. A good modern discussion of the beam-lens transform is given in O'Shea[32] (where his parameter $\alpha^2 = \Gamma$ here).

A pictorial description of the beam-lens transform is given by a figure in Ref. 30, redrawn here as Fig. 7. Variables normalized to the lens focal length f are used to show how the transformed waist location z_{02}/f varies with the input waist location z_{01}/f. The input Rayleigh range z_{R1}/f (also normalized) is used as a parameter and several curves are plotted for different values. The anomalous slope regions of the plot are evident. The geometrical optics thin lens result, Eq. (23), is recovered when the input Rayleigh range becomes negligible, $z_{R1}/f = 0$ (the condition for a point source), and the slopes of both wings of the curve are always negative.

5.1 Application of the Beam-Lens Transform to the Measurement of Divergence

An initial application of the beam-lens transform equations is to show that the divergence of the input beam Θ_1 in 1-space of Fig. 6 can be determined by measuring the beam diameter $2W_f$ at precisely one focal length behind the lens exit plane H2 in 2-space from the equation

$$\Theta_1 = 2W_f/f \tag{29}$$

The M^2 Model

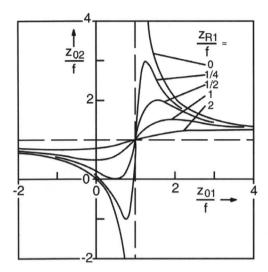

Figure 7 Parametric plots of the transformed waist location as a function of the input waist location for the beam-lens transform, with f as the lens focal length and the Rayleigh range z_{R1} of the input beam as parameter.

This result is independent of where the lens is placed in the input beam. This follows by finding in 2-space the diameter $2W_f$ at $z_2 = f$ [from Eq. (16)] and substituting Eqs (19), (24), and (28)

$$2W_f = 2W_{02}\{[1 + (f - z_{02})^2/z_{R2}^2]\}^{1/2} = 2W_{02}(f/z_{R2})(1/\Gamma^{1/2})$$
$$= 2W_{01}(f/z_{R2})(1/\Gamma) = 2W_{01}(f/z_{R1}) = \Theta_1 f$$

which is the same as Eq. (29). In Fig. 3(b) of Ref. 25 there is an illustration showing how the transform equations operate to keep the output beam diameter one focal length from the lens fixed at the value $\Theta_1 f$ despite variations in the input waist location, z_{01}. The measurement method implied by Eq. (24) is the simplest way to get a good value for the beam divergence Θ_1. Care should be taken to use a long enough focal length that the beam diameter is large enough for the precision of the measurement technique.

5.2 Applications of the Beam-Lens Transform: The Limit of Tight Focusing

When the aperture of a short focal length lens is filled on the input side, the smallest possible diameter output waist is reached and this is called the limit of tight focusing. This limit is characterized by (a) the beam diameter at the lens being given by $2W_{\text{lens}} = \Theta_2 f$, (b) the output waist being near the focal plane $z_{02} = f$, and (c) there being a short depth of field at the focus, $z_{R2}/f \ll 1$. Applying Eq. (29) in the reverse direction gives the 2-space divergence as the ratio of the beam diameter $2W_{1f}$ at f to the left of the lens, to the focal length, $\Theta_2 f = 2W_{1f}$. By condition (a) this means $2W_{1f} = 2W_{\text{lens}}$ or that there is little change in the input beam diameter over a propagation distance f. That makes the first condition characterizing the tight focusing case equivalent to $z_{R1}/f \gg 1$. Then from Eq. (19),

$$2W_{lens} = 2\lambda M^2 f/\pi W_{02}$$

or

$$2W_{02} = 2\lambda M^2(f/\pi W_{lens}) = 2\lambda M^2(f/\#) \tag{30}$$

for the tight focusing limit. Here Siegman's definition[5] is used that a lens of diameter D_{lens} is filled for a fundamental mode beam of diameter πW_{lens} (this degree of aperture filling gives <1% clipping of the beam), thus $f/\pi W_{lens} = f/D_{lens} = (f/\#)$. The depth of field of the focus is $z_{R2} = \pi W_{02}^2/\lambda = 4\pi M^4(f/\#)^2$. This generalizes a familiar result[5] for a fundamental mode beam to the $M^2 \neq 1$ case.

Marshall's point[3] (from 1971) is made by Eq. (30), that a higher order mode beam focuses to a larger spot by a factor of M^2, with less depth of field, and therefore cuts and welds less well than a fundamental mode beam.

5.3 The Inverse Transform Constant

The transform equations work equally well going from 2-space to 1-space, with one transformation constant the inverse of the other,

$$\Gamma_{21} = 1/\Gamma_{12} \tag{31}$$

This obviously is true by symmetry but the algebraic proof is left to the reader.

6 BEAM DIAMETER DEFINITIONS FOR FUNDAMENTAL AND MIXED MODE BEAMS

It has been said that the problem of measuring the cross-sectional diameter of a laser beam is like trying to measure the diameter of a cotton ball with a pair of calipers. The difficulty is not in the precision of the measuring instrument, but in deciding what is an acceptable definition of the edges.

Unlike the fundamental mode beam where the $1/e^2$-diameter definition is universally understood and applied, for mixed modes a number of different diameter definitions[7] have been employed. The different definitions have in common that they all reduce to the $1/e^2$-diameter when applied to an $M^2 = 1$ fundamental mode beam, but when applied to a mixed mode with higher order mode content, they in general give different numerical values. As M^2 always depends on a product of two measured diameters, its numerical value changes also as the square of that for diameters. It is all the same beam, but different methods provide results in different currencies; one has to specify what currency is in use and know the exchange rate.

Since the recommendation[11] by the ISO committee on beam widths to standardize on the second-moment definition for the beam diameter, there has been increasing agreement among laser users to do so. This definition, discussed below, has the best analytical and theoretical support but is difficult experimentally to measure reproducibly because of sensitivity to small amounts of noise in the data. The older methods therefore persist and the best strategy[25] at present is to use the more forgiving methods for the multiple diameter measurements needed to determine M^2. Then at one propagation distance, carry out a careful diameter measurement by the second-moment definition to provide a conversion factor. This conversion factor can then be applied to obtain standardized diameters at any distance z in the beam. This strategy will evolve in the future as instrument makers respond to the ISO committee's choice and devise algorithms and direct methods for ready and accurate computations of second-moment diameters.

6.1 Determining Beam Diameters From Irradiance Profiles

Beam diameters are determined from irradiance profiles, the record of the power transmitted through a mask as a function of the mask's translation coordinate transverse to the beam. A sufficiently large linear power detector is inserted in the beam, with a uniformly sensitive area to capture the total power of the beam. Detection sensitivity should be adequate to measure $\approx 1\%$ of the total power, and response speed should allow faithful reproduction of the time-varying transmitted power. The mask is mounted on a translation stage, placed in front of the detector, and moved or scanned perpendicularly to the beam axis to record a profile. An instrument that performs these functions is called a beam profiler. In a useful version based on a charge-coupled-device (CCD) camera devices, the masking is done on electronic pixel data under software control.

The beam propagation direction defines the z-axis. The scan direction is usually along one of the principal diameters of the beam spot and commercial profilers are mounted to provide rotation about the beam axis to facilitate alignment of the scan in these directions. The principal diameters for an elliptical spot are the major and minor axes of the ellipse (or the rectangular axes for a Hermite–gaussian mode). The principal propagation planes (x, z) and (y, z) are defined as those containing the principal spot diameters. The beam orientation is arbitrary and in general may require rotation of coordinates to tie it to the laboratory reference frame. It is assumed this rotation is known, and without loss of generality to give simple descriptive terminology in this discussion, here the z-axis is taken to be horizontal, the principal propagation planes as the horizontal and vertical planes in the laboratory, with the scan along the x-axis. If the mask requires centering in the beam (e.g., a pinhole) to find the principal diameter, it is mounted on a y-axis stage as well and x-scans at different y-heights taken to determine the widest one at the beam center. Alternatively, a mirror directs the beam onto the profiler and the spot is put at different heights to find the beam center by tipping the mirror about a horizontal rotation axis. If the beam is repetitively pulsed and detected with an energy meter, the stage is moved in increments between pulses. If a CCD camera is the detector, a scan line is the readout of sequential pixels and no external mask is required in front of the camera. A CCD camera generally requires a variable attenuator[33] inserted before the camera to set the peak irradiance level just below the saturation level of the camera for optimum resolution of the irradiance value on the ordinate axis of the profile.

The results of this process are irradiance profiles such as shown in Fig. 8 for two pure modes, the fundamental mode in the first row and the donut mode in the second, where three scans are calculated for each, one for a pinhole (first column), a slit (second column), and a knife-edge (third column) as masks. The traditional definitions used to extract diameters from these profiles are the same for the pinhole and slit. This is to normalize the scan to the highest peak as 100%, then to come down on the scan to an ordinate level at $1/e^2$ (13.5%) and measure the diameter—or clip-width—as the scan width between these

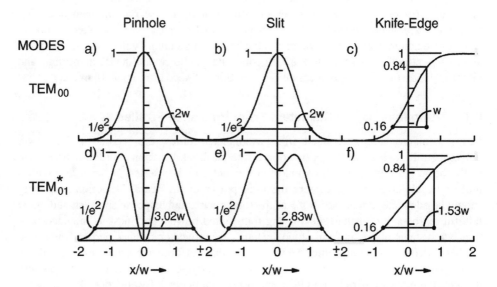

Figure 8 Theoretical beam profiles (irradiance vs. translation distance) from a scanning pinhole (a) and (d), slit (b) and (e), and knife-edge (c) and (f) cutting the fundamental and donut modes, illustrating that different methods give different diameters for higher order mode beams. The knife-edge diameter is defined as *twice* the translation distance between the 15.9% and 84.1% cut points.

crossing points (called clip-levels or clip-points and shown as dots in Fig. 8). The symbols D_{pin} and D_{slit} are used for these two diameters. For the knife-edge diameter (symbol D_{ke}) the definition is to take the scan width between the 15.9 and 84.1% clip-points and double it, as this rule produces the $1/e^2$-diameter when applied to the fundamental mode.

As shown in Fig. 8, the diameter results for the donut mode (TEM_{01}^*) are all larger than the $2w$ diameter of the fundamental mode, as expected. However, the answers for the three different methods for the donut mode – and, in general, for all higher order modes – are all different! The ratio of the donut mode to fundamental mode diameter is 1.51, 1.42, and 1.53 by the pinhole, slit, and knife-edge methods, respectively. The reason, obviously, is that traces of different shapes are produced by the different methods. The pinhole cuts the donut right across the hole and records a null at the center; the slit extends across the whole spot and records a transmission dip in crossing the hole but never reaches zero due to the contribution of the light above and below the hole. Even higher transmission results with the knife-edge and here the donut profile differs from the fundamental one only in being less steeply sloped (the spot is wider) and having slight inflections of the slope around the hole at the 50% clip-point, the beam center.

There are two other common definitions. The first is the diameter of a circular aperture giving 86.5% transmission when centered on the beam. It is variously called the variable aperture diameter, the encircled power diameter, or the "power in the bucket" method, and designated by the symbol D_{86}. The last is the second-moment diameter, defined as four times the standard deviation of the radial irradiance distribution recorded by a pinhole scan, and designated by the symbol $D_{4\sigma}$. For the ratio of donut mode to

The M² Model

fundamental mode diameters, these definitions give 1.32 and 1.41, respectively, also different from the three other values above.

After discussion of some common considerations (Sec. 6.2), these five diameter definitions are evaluated below (Sec. 6.3), leading to the summary given in Table 1.

6.2 General Considerations in Obtaining Useable Beam Profiles

Five questions are important in evaluating what beam diameter method is best for a given application.

1. *How important is it to resolve the full range of irradiance variations?* Only a pinhole scan (or its near equivalent, a CCD camera snapshot read out pixel by pixel) shows the full range, but this is not of significance in some applications, for example, where the total dose of light delivered is integrated in an absorber.

2. *How important is it to use a method that is insensitive to the alignment of the beam into the profiler?* If the test technician cannot be relied on to carefully center the beam on the profiler, the slit or knife-edge methods still give reliable results, but not the other methods. With a CCD camera there is a trade-off between alignment sensitivity and accuracy. For best accuracy, a magnifying lens – of known magnification – can be placed in front of the camera to fill the maximum number of pixels, but then the camera becomes somewhat alignment sensitive.

3. *With what accuracy and repeatability is the diameter determined?* The amount of light transmitted by the mask determines the signal-to-noise ratio of the profile and ultimately answers the question. The methods based on a pinhole scan (D_{pin}, $D_{4\sigma}$, and CCD cameras) suffer from low light levels in this regard. On the other hand, a laser beam is generated in a resonator subject to microphonic perturbations, making the beam jitter in position and the profile distort typically by about 1% of the beam diameter, so that a greater instrument measurement accuracy is usually not significant.

4. *Is the convolution error associated with the method significant?* The convolution error is the contribution to the measured diameter due to the finite dimensions of the scan aperture, either the diameter H of a pinhole or width S of a slit. A 10-micron focused spot cannot be accurately measured with a pinhole of 50-micron diameter. The distortion of a pinhole profile of a fundamental mode is shown in Fig. 9(a) as a function of the ratio of hole diameter to the mode width $H/2w$. The peak amplitude drops and a slight broadening occurs as $H/2w$ increases. The central 100% peak amplitude point is "washed out" or averaged to a lower value in the profile by the sampling of lower amplitude regions nearby as the finite diameter pinhole scans across the center as Fig. 9(b) indicates. The reduction in peak amplitude of the convoluted profile is like lowering the clip-level below 13.5% on the original profile: the measured diameter becomes larger. Very similar profile distortions occur with a slit scan as a function of $S/2w$; here S is the slit width. The ratio of the measured width including this convolution error to the correct width is plotted in Fig. 9(c) for the pinhole (H) and slit (S). This gives the rule of thumb for pinhole scans: to keep the error in the measured diameter to 1% or less, keep the pinhole diameter H to one-sixth or less of $2w$, that is, $H < w/3$. The corresponding rule[34] for slits is the measured diameter is in error by <1% if the width S is 1/8 or less of $2w$. For modes like TEM_{10} of Fig. 2(d) with a feature (the central peak) narrower than that of the fundamental mode the H or S should be no bigger than the same fractions of that feature's width. (Note, McCally[34] uses the biologist's definition of $1/e$ clip-points for the fundamental mode diameter, a factor $1/\sqrt{2}$ smaller than our $1/e^2$-diameter; his results require conversion.)

Table 1 Properties of Mixed Mode Diameter Definitions

Diameter symbol	Scan aperture and name	Diameter definition	Conversion constant $c_{i\sigma}$ to $D_{4\sigma}$	Alignment sensitive?	Resolution of $I(r)$ peaks	Convolution error?	Signal-to-noise ratio	Comments
D_{pin}	Pinhole (dia. H)	Separation of clip-points $1/e^2$ down from highest peak	0.805	Yes	High	Yes, if $H/2w > 1/6$	Low	Shows best details of irradiance peaks
D_{slit}	Slit (width S)	Separation of clip-points $1/e^2$ down from highest peak	0.950	No	Medium	Yes if $S/2w > 1/8$	Medium	Directly measured diameter is close to $D_{4\sigma}$
D_{ke}	Knife-edge	Twice the separation of 15.9%, 84.1% clip-points	0.813	No	Low	None	High	Most robust diameter experimentally (vs. noise and spot structure, see text)
D_{86}	Variable aperture ("power in the bucket")	Diameter of centered circular aperture passing 86.5% of total power	1.136	Yes	Low	None	High	Works well only on round beams. Computed readily on CCD cameras. Used on kW lasers
$D_{4\sigma}$ or $D_2\sqrt{(2\sigma)}$	Second-moment (linear or radial)	Four times the standard deviation of irradiance distribution from a pinhole scan	1	Yes	High	Yes, as for pinhole scan	Low	*ISO standard diameter.* Susceptible to error from noise on wings of the profile. Supported best by theory. Computed readily on CCD cameras
N/A	CCD camera	Various custom algorithms	N/A	No	Medium	Yes	Low	Computes all of above diameter definitions with appropriate software

The M² Model

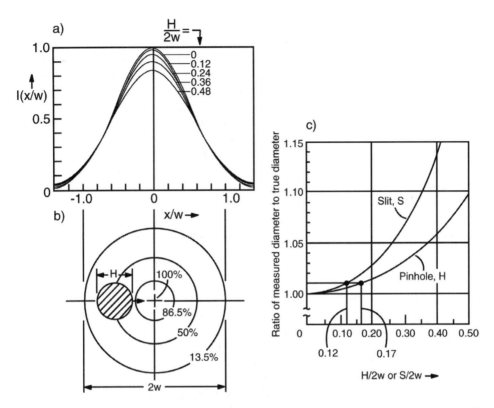

Figure 9 Convolution of the theoretical fundamental mode profile in a scan with a pinhole or slit of finite dimensions (H, diameter of the pinhole; S, width of the slit; $2w$, the $1/e^2$-diameter of the mode). (a) Distortion of the shape and width of the pinhole profile as $H/2w$ increases. (b) Plan view of the pinhole scan showing "washout" of the 100% amplitude point. For the pinhole shown, $H/2w = 0.24$, corresponding to the third curve down from the top in (a). (c) Convolution error, or ratio of the measured diameter $2w_{\mathrm{meas}}$ to the true diameter $2w$, as a function of $H/2w$ for the pinhole and $S/2w$ for the slit.

Distortion of the profile can be a more subtle effect and can give misleading results. When measuring a predominantly TEM_{01}^* focused beam through the waist region, for example, a pinhole profiler will at first show the expected trace, with a dip in the middle like Fig. 8(d) or (e). This will change to one with a central peak as in Fig. 8(a) at the propagation distance along the beam where the pinhole is no longer small compared to the beam diameter. The donut hole can fall through the pinhole!

Convolution errors are a concern normally only when working with focused beams, as when measuring divergence by the method of Sec. 5.1. Generally, however, it is desirable to go to the far-field, reached by working in 2-space at the focal plane behind an inserted lens, to obtain a true (undistorted) profile. The beam coming out of the laser has a "diffractive overlay," low-amplitude high-divergence light diffracted from the mode limiting internal aperture, overlaid on the main beam. The resulting interference can significantly distort the profile, even at <1% amplitude of the diffracted light. It is the E-fields that interfere; for an irradiance $I = E^2$ overlaid by a $0.01\,E^2$ distorting component, the E-fields add and subtract as $E \pm 0.1\,E$ at the interference peaks and valleys.

The resulting fringe contrast ratio, $I_{\text{peak}}/I_{\text{valley}} = [(1.1)/(0.9)]^2 = 1.49$ is a significant distortion to the profile even though the power in the diffractive overlay is insignificant. Moving the profiler some distance away from the output end of the laser spreads the diffractive overlay rapidly compared to the beam expansion, but often several meters additional distance is required. This leaves the use of a lens to reach the far-field as the answer, and convolution distortion then must be dealt with.

Aligning a small diameter (e.g., 10 microns) pinhole to a small (e.g., 100 microns) focused spot is another problem. The search time can be very long if done manually, so having a fast update rate – 10 scans a second is good – provided by commercial instruments can be a major aid. Some instruments[9] have electronic alignment systems to facilitate finding the overlap of small pinhole and small beam.

Knife-edges have no convolution error to the extent that they are straight (razor blades are straight[8] to <2 microns deviation over 1000 microns length). The circular aperture of the encircled power method is usually a precision drilled hole and has no convolution error so long as it is accurately round and made in a material much thinner than the hole diameter (to avoid occultation error).

5. *Are the diameter measurements along the propagation path free of discontinuities and abrupt changes?* Consider making many diameter measurements along the propagation axis, and fitting the data to a hyperbola to find the beam's Rayleigh range and beam quality. Discontinuities in the data will make a poor fit and final result. Such discontinuities can arise[35] with the $1/e^2$-clip level diameter definitions with mixed modes with low peaks on the edges, as in Fig. 2(g), only lower. As the mode mixture changes to bring the outer peaks near the clip level, the measured diameter can jump from the separation of the outer peaks of the profile to the width of the central peak as amplitude noise perturbs the profile. Similarly, for a mixed mode with rectangular symmetry, as azimuth is continuously changed from the major principal plane direction towards the minor one, the relative amplitude of the outermost peaks of the profile can drop.[35] The clip-point then can jump discontinuously with perturbing noise when the height is near the clip-level. Only D_{pin} and D_{slit} are subject to this difficulty.

This last question can be rephrased to ask, is the diameter definition readable by a machine? A human observer will notice an outer peak of height near the clip-level causing the profiler readout to fluctuate, and correct the situation by adjusting the mode mixture, the azimuth, or the clip-level. A machine will take the bad data in, and produce unreliable results. When a lot of diameter data needs to be gathered, as in measuring a propagation plot to determine M^2, automated machine data acquisition is desirable. In this regard, the knife-edge diameter is best, as it always produces an unambiguous monotonic trace for all higher order and mixed modes.

6.2.1 How Commercial Profilers Work

Commercial profilers[8] typically use the $1/e^2$-diameter definition with pinhole and slit masks, and occasionally will report an incorrect diameter due to the "not entirely machine-readable" defect of these definitions. These profilers use a rotating drum to carry a slit or pinhole mask smoothly and rapidly (at a 10 Hz repetition rate) over a detector inserted into the drum. On the first pass through the laser spot, the electronics remembers the 100% signal level, and on the second pass when the 13.5% clip-level is crossed as the signal rises, a counter is started. This counts the angular increments of drum motion from an angular encoder, which when multiplied by the known drum radius, gives the mask translation in spatial increments of 0.2 microns. When the clip-level is passed as the signal falls, the counter is stopped and the value of the beam diameter – total counts times spatial increment – is reported. Actually, what is reported on the digital readout is an average selected by the user of the last two to twenty measurements, to slow the report rate down to what can be read. If a pure donut mode is scanned with the pinhole version of this instrument [the profile of Fig. 8(d)], the counter starts at the clip-level dot on the left ($x/w = -1.51$) but stops as the falling clip-level is met at the left edge of the donut hole ($x/w = -0.16$). The scan continues and the counter turns on again as the clip-level is passed with the rising signal at the right edge of the donut hole ($x/w = +0.16$), because the drum has not completed a revolution to reset the counter for a new measurement. Finally, the counter turns off again at the rightmost clip-level dot ($x/w = +1.51$), and the diameter reported is the actual diameter minus the width of the hole at the clip-level height, an error of about -11%. This difficulty usually goes unnoticed because the dips in mixed mode profiles do not often go as low as 13.5%.

Commercial profilers, because of their speed and accuracy, provide an advantage for frequent beam diameter measurements over the traditional practice of manually driving a translation stage with a razor blade mounted on it. Focused beams in particular need the high instrument accuracy to resolve the small focused spot and provide the real-time update rate to acquire signal. With a signal linearity range of 10^4 and a spatial resolution (including slit convolution error) of 0.3 microns over a 9 mm scan range (30,000 spatial resolution elements) one of these small, simple profilers brings an impressive potential of 3×10^8 information bits to the problem of measuring a beam diameter. Compare this to a typical CCD camera of 9 mm scan line length, 20 micron pixel spacing (450 spatial resolution elements), and $10^2 \to 10^3$ linearity range, for a total of $5 \times 10^4 \to 10^5$ information bits. It is understandable why in measuring beam quality M^2, profiler-based instruments surpass camera-based ones in speed and accuracy. The camera, of course, has its own advantages of giving a two-dimensional map of the laser spot's irradiance peaks and is able to measure beams from pulsed lasers.

6.3 Comparing the Five Common Methods for Defining and Measuring Beam Diameters

The discussion below and Table 1 summarize the properties of the five diameter definitions.

6.3.1 D_{pin}, Separation of $1/e^2$-Clip-Points of a Pinhole Profile

The pinhole scan reveals the structure of the irradiance variations across the beam spot with the greatest accuracy and detail, but does so working with a low light signal level and it is subject to convolution error with focused spots. To minimize convolution error, several pinholes of diameters H (10 micron and 50 micron pinholes are common) are used to keep $H < w/3$ where w here is the fundamental mode radius or smallest feature radius for a higher order mode beam. The pinhole method requires accurate centering of the beam on the scan line of the pinhole and this makes it less adaptable to a machine measurement. This diameter definition also can give ambiguous results if the profile contains secondary peaks of a height close to the clip-level. The pinhole profile provides the basic data from which the second-moment diameter is calculated. Be sure the rule for the profile to be free of convolution error is met first!

6.3.2 D_{slit}, Separation of $1/e^2$-Clip-Points of a Slit Profile

The slit scan does not require centering of the beam spot and works at a medium light signal level, but does not reveal as much detail of the irradiance variations [compare Fig. 8(d) and (e)]. This method is subject to convolution error with focused spots; S, the slit width, should satisfy $S/2w < 1/8$ with $2w$ as the smallest feature size of the profile. It too can give ambiguous results on profiles with secondary peaks near the clip-level. Without performing a conversion explained below, this diameter definition produces a result closest to the standard second-moment diameter of the three other methods.

6.3.3 D_{ke}, Twice the Separation of the 15.9% and 84.1% Clip-Points of a Knife-Edge Scan

The knife-edge does not require centering of the beam spot and works at a high light signal level, but reveals almost no detail of the irradiance variations [compare Fig. 8(d) and (f)]; only the slight inflection points in the slope of the knife-edge profile show that there are any irradiance peaks at all. All modes give a simple slanted S-shaped profile. There generally is no convolution error with this method, and there are no diameter ambiguities when secondary peaks are present. Experimentally, it is the most robust diameter measurement and is least affected by beam pointing jitter and power fluctuations, making this method fully machine readable. This diameter is the basic one measured in the most common commercial instrument[9] designed to automatically measure propagation plots and all six beam parameters.

6.3.4 D_{86}, Diameter of a Centered Circular Aperture Passing 86.5% of the Total Beam Power

Unlike the other diameter measurements, the variable aperture diameter passes light in both the x- and y-transverse planes simultaneously and cannot be used to separately measure the two principal diameters; it works best with round beams. It must also be centered in the beam for accurate results. While an iris or variable aperture can be used, more frequently sets of precision fixed apertures are used instead. A metal plate drill gauge, with some of the plate milled away on the back side of the gauge to reduce its thickness to less than the smallest aperture size to eliminate occultation error, is a convenient tool. The two diameters bracketing the 86.5% transmission point are first found, and the final result computed by interpolation. Alternatively, if there is a long propagation length available, an aperture with a transmission near 86.5% may be moved along the beam to locate the distance where that diameter produces precisely this transmission. This diameter definition is used mainly for two reasons. For high-power lasers – for instance CO_2 lasers in the kilowatt range – little diagnostic analytical instrumentation is available that can absorb this power. A water-cooled copper aperture, however, can still be safely inserted in front of a power meter to give some quantification of the beam diameter. The second reason is that this diameter is readily computed from the output of a CCD camera and is available on camera instrumentation, with the computation locating the beam centroid, making physical centering of the camera unnecessary.

6.3.5 $D_{4\sigma}$, Four Times the Standard Deviation of the Pinhole Irradiance Profile

This diameter is computed from a pinhole irradiance profile, which for accuracy should be free of convolution error and diffractive overlay. For a beam with a rectangular

The M² Model

cross-sectional symmetry described by a weighted sum of Hermite–gaussian modes the calculation proceeds by finding the rectangular moments of the profile treated as a distribution function. The zeroth-moment gives the total power P of the beam, the first-moment the centroid, and the second-moment leads to the variance σ^2 of the distribution:

$$\text{Zeroth-moment or total power} \quad P = \int_{-\infty}^{\infty} \int_{-\infty}^{\infty} I(x, y) \, dx \, dy \tag{32}$$

$$\text{First-moment or centroid} \quad \langle x \rangle = (1/P) \int_{-\infty}^{\infty} \int_{-\infty}^{\infty} x I(x, y) \, dx \, dy \tag{33}$$

$$\text{Second-moment} \quad \langle x^2 \rangle = (1/P) \int_{-\infty}^{\infty} \int_{-\infty}^{\infty} x^2 I(x, y) \, dx \, dy \tag{34}$$

$$\text{Variance of the distribution} \quad \sigma_x^2 = \langle x^2 \rangle - \langle x \rangle^2 \tag{35}$$

$$\text{Linear second-moment diameter} \quad D_{4\sigma x} = 4\sigma_x \tag{36}$$

This last equation comes from the requirement that the second-moment diameter reduce to the $1/e^2$-diameter when applied to a fundamental mode beam, as explained in arriving at Eq. (6). A precisely similar set of equations holds for the moments in the vertical plane (y, z) to define a vertical principal plane centroid and diameter [Eqs. (33)–(36) with x and y interchanged]:

$$\text{Linear second-moment diameter} \quad D_{4\sigma y} = 4\sigma_y \tag{37}$$

A similar set of moment equations defines a radial second-moment diameter, applicable to beams with cylindrical symmetry described by a weighted sum of Laguerre–gaussian functions. Here the pinhole x-scan profile is split in half at the centroid point, $\langle x \rangle$ and the half-profile is taken as the radial variation of the cylindrically symmetric beam. In the transverse radial coordinate plane (r, θ), the origin is the center of the beam spot defined by the centroid $(\langle x \rangle, \langle y \rangle)$ and given by the rectangular first moments [Eq. (33)].

$$\text{Zeroth-moment or total power} \quad P = \int_0^{2\pi} \int_0^{\infty} I(r, \theta) r \, dr \, d\theta \tag{38}$$

$$\text{Radial second-moment} \quad \langle r^2 \rangle = (1/P) \int_0^{2\pi} \int_0^{\infty} r^3 I(r, \theta) \, dr \, d\theta \tag{39}$$

$$\text{Variance of the distribution} \quad \sigma_r^2 \equiv \langle r^2 \rangle \tag{40}$$

$$\text{Radial second-moment diameter} \quad D_{2\sqrt{2}\sigma r} = 2\sqrt{2}\sigma_r \tag{41}$$

This last equation derives from the requirement that the linear and radial variances are related[6] by:

$$\sigma_x^2 + \sigma_y^2 = \sigma_r^2 \tag{42}$$

Then for a cylindrically symmetric mode $\sigma_x = \sigma_y$, yielding $2\sigma_x^2 = \sigma_r^2$ or $\sigma_x = (1/\sqrt{2})\sigma_r$. Since for a fundamental mode beam $2w = 4\sigma_x$, from the radial mode description of that beam, there results[6] $2w = 4(1/\sqrt{2})\sigma_r = 2\sqrt{2}\sigma_r$, which is Eq. (41). By mixing modes, combinations of Hermite–gaussian modes can be made to have the same irradiance profiles as Laguerre–gaussian modes, and vice versa. Therefore, for compactness the symbols $D_{4\sigma}$ or $M_{4\sigma}^2$ will be used for either linear or radial second-moment quantities unless there is a need to specifically distinguish a quantity as a radial moment.

6.3.6 Sensitivity of $D_{4\sigma}$ to the Signal-to-Noise Ratio of the Pinhole Profile

The experimental difficulties in evaluating these integrals with noise on the profile signal comes from the weighting by a high power of the transverse coordinate in the second-moment calculation, by the square in the linear case [Eq. (34)], and by the cube in the radial case [Eq. (39)]. Take as a typical example a measurement of a fundamental mode spot with a CCD camera, using 256 counts to digitize the irradiance values, and 128 counts used to digitize half the integration range of the transverse coordinate. In the linear case, one noise count (0.4% noise) at the edge of the range – at the 128th transverse count – is weighted by the factor $1 \times (128)^2 = 16,384$ in the integration, vs. 256×1 counts for the central peak. The contribution of this single noise count is 64 times that of the pixel at the central peak in the integration. In the radial case, the one noise count at the limiting transverse pixel makes a contribution $(128)^3/256 = 8192$ times that of the pixel at central peak. A good discussion of the high sensitivity of the second moment diameter to noise on the wings of the profile is given in Ref. 12. There the second-moment and knife-edge methods are compared for five simulated modes, and the knife-edge found to be considerably more forgiving and in line with common expectations.

To manage this sensitivity to noise, a background trace is recorded (a blank profiler scan, with the beam blocked) and later subtracted from the signal trace to reduce noise on the wings. Additionally, limits on the transverse coordinates, over which the integration is performed, are adjusted out to three to four beam diameters, and the constancy of the computed second moment is observed. This is to find the integration limits that are just wide enough to yield stable second-moment values. When the width setting is judged correct, the measurement should be repeated to check reproducibility.

In one commercial instrument[9] two additional checks are built in to the second-moment calculation. The first check is used for the radial second moment and consists of comparing the second moment calculated from the right half-profile, to that from the left half-profile. If the beam is indeed cylindrically symmetric and the contribution from noise on the profile is negligible, the ratio of these two calculations is unity. The second check is an option to use in the calculation called "noise clip ON/OFF." In the wings of the profile where the signal is near zero, noise counts vary the trace above and below the average (no signal) level, and the low-noise pixels acquire a negative sign when the background is subtracted. This is desirable; these negative noise pixels help cancel positive ones, but it is straightforward for the processor in the instrument to clip these pixels to a zero value with the "noise clip" option turned ON. The size of the resulting change in the calculated second-moment diameter provides a test of how large the contribution is from noise in the wings.

It is also recommended, when measuring a second-moment diameter, to vary the sources of noise on the laser beam. Check that the resonator alignment is peaked, the sources of microphonics impinging on the laser are minimized, the laser is warmed up and bolted down to the stable table, and so on, and watch for variations in the second-moment diameter. A more complete analysis[9] of the effect of noise on diameter measurements showed that the standard deviation over the mean of ten repeated second-moment diameter measurements was 5 to 10 times larger than that for knife-edge measurements of the same

The M² Model

beam at (low) signal-to-noise levels from 50 down to 10. With these precautions required in interpreting $D_{4\sigma}$ results, it is fair to say that the second moment as currently implemented is not a "machine readable" diameter definition.

6.3.7 Reasons for $D_{4\sigma}$ Being the ISO Choice of Standard Diameter

Since there is considerable experimental difficulty in measuring second-moment diameters, why is this definition the one recommended[11] as the standard by the International Organization for Standards? The primary answer is that this definition is the one best supported by theory. The general theories of the propagation of ray bundles[4,6,19] are based on the Fourier transform relationship[6] between the irradiance distribution and angular spatial-frequency distribution. These show two essential requirements are met if the beam width is defined by the second-moment diameter [Eq. (36)]. The beam width is rigorously defined[6] for all realizable beams [excluding only those with discontinuous edges[6], for which the integration Eq. (34) may not converge] and the square of this width (the variance) increases as a quadratic function of the free space propagation distance away from the waist. That is, $D_{4\sigma}(z)$ increases with z according to the hyperbolic form [Eq. (16)]. All other diameter definitions gain legitimacy in propagation theory by being shown to be proportional to the second-moment diameter.

A third feature of the second-moment diameter is that the beam quality M^2 values calculated using it turn out to be integers for either the pure, rectangular-symmetry Hermite–gaussian modes, or the pure, cylindrical-symmetry Laguerre–gaussian modes. Thus, not only for the fundamental mode is $M_{4\sigma}^2 = 1$, which happens by definition, but for the next higher order mode, the donut mode, $M_{4\sigma}^2 = 2$, and so on counting up by unity each time the mode order increases. In general[6] the formulas are

Hermite–gaussian modes $\quad M_{4\sigma}^2 = (m + n + 1)$ (43)

Laguerre–gaussian modes $\quad M_{2\sqrt{2}\sigma}^2 = (2p + l + 1)$ (44)

where m, n are the order numbers of the Hermite polynomials, and p, l the order numbers for the generalized Laguerre polynomials associated with the modes as before [Eq. (1)]. For the six modes shown in Fig. 2, of increasing order from (a) to (f), the values are $M_{4\sigma}^2 = 1, 2, 3, 3, 4, 4$, respectively. The integers $(m + n + 1)$ or $(2p + l + 1)$ are termed the mode order numbers, and they also determine the mode's optical oscillating frequency. Modes with the same frequency are termed degenerate. As the mode order number increases, the degree of degeneracy increases, there being three degenerate pure modes each for $(2p + l + 1) = M^2 = 5$ or 6, four for $M^2 = 7$ or 8, five for $M^2 = 9$ or 10, and so on. The diameters of the pure modes in second-moment units are just the square root of the mode order numbers times the fundamental mode diameter [by Eq. (3)]:

Pure Hermite–gaussian modes $\quad D_{4\sigma}/2w = \sqrt{(m + n + 1)}$ (45)

Pure Laguerre–gaussian modes $\quad D_{2\sqrt{2}\sigma}/2w = \sqrt{(2p + l + 1)}$ (46)

Another consequence of the pure modes having integer values of beam quality is that for mixed modes, the $M_{4\sigma}^2$ value is simply the power-weighted sum of the integer $M_{4\sigma}^2$ values of the component modes. Finding integers like this in a physical theory is strong indication that the quantities have been defined and measured "the way nature intended."

Another reason for the ISO committee's choice of $D_{4\sigma}$ as the diameter standard is that the committee members were aware that conversion formulas were available to permit diameters measured according to the other definitions to be put in standard form. These formulas are discussed in the next section.

The last line of Table 1 refers to CCD camera properties. A CCD camera together with frame-grabber electronics and appropriate software can be a universal instrument capable of providing diameter measurements according to any or all of the definitions. Affordable cameras do not provide a very large dynamic range for irradiance levels (useful range $\sim 100:1$) compared to that for a silicon detector ($\sim 10^4$) but good variable attenuators are readily available[33] to allow camera operation just below saturation to make the most of the range that exists. Spatial resolution of 5–10 micron per pixel may not be adequate for direct measurement of focused beams, but flexibility, ease of use, and quick access to colorful two-dimensional irradiance maps make it an attractive choice for beams of ~ 0.5 mm and up. Imaging optics can be used if necessary to measure smaller beams. As this technology continues to improve, it could become superior to all the older methods of measuring beam diameters.

6.3.8 Diameter Definitions: Final Note

It is important to emphasize that the M^2 model can be applied using any reasonable definition of beam diameter as long as the definition is used consistently both in making measurements and interpreting calculated values. Results will then be meaningful and reliable.

In fact, there can be cases where it is important to use a "nonstandard" diameter definition. For example, there is a trend toward steeper sides and flattened tops as M^2 increases. The effect becomes pronounced for M^2 values above 10 and at 50 or more, profiles can be aptly described[5] as a "top hat" shape. The diameter of such a beam becomes unambiguous and it makes sense to abandon the standard definitions ($D_{4\sigma}$, D_{86}, and so on) and just measure the particular diameter. The good news is that for such beams, pinhole scans would show the diameter at half-maximum irradiance to be insignificantly different from that at the $1/e^2$ level. The aperture size that passes 86.5% of the total power will not provide as meaningful a result in this situation as the aperture that transmits 95% of the power. The latter would likely be little different in size from the one that passes 98%. Curve fitting to a series of D_{95} measurements will yield a set of valid parameters describing the beam but this defines a new "currency" and one must stay consistent and not mix these diameters with those arrived at by a different method or definition.

6.4 Conversions Between Diameter Definitions

For a diameter conversion algorithm to be widely applied, it must be normalized, with the natural normalization being the diameter of the fundamental mode generated in the same resonator as the measured beam, the embedded gaussian. Using Eq. (3), this essentially changes the problem of converting diameters into one of converting M^2 values.

The conversion rules that are now part of the ISO beam widths document[11] were first derived empirically and later found to have theoretical support. They apply to cylindrically-symmetric modes generated in a resonator with a circular limiting aperture and an approximately uniform gain medium. In this case, if $M^2_{2\sqrt{2}\sigma}$ is known, then the mixture and relative amplitudes of the modes oscillating can also be reasonably estimated.

6.4.1 Is M^2 Unique?

Determining the fractions of the pure modes in a mixture for a cylindrically-symmetric beam from the beam quality alone seems unlikely at first, because the beam quality M^2 is not unique in the mathematical sense. Consider the case of a beam with $M^2 = 1.1$ in second-moment units. An experienced laser engineer might guess the likely composition is 90% fundamental mode ($M^2 = 1$) and 10% donut mode ($M^2 = 2$) to give $M^2 = (0.9) + 2(0.1) = 1.1$ for the mixed mode, and she/he would be right. For a beam of $M^2 = 5$, however, the problem is much harder. The number of possible modes above threshold makes for an infinite range of possible mix fractions within the $M^2 = 5$ constraint.

Our empirical results showed, however, that for a class of lasers with round beams described above, M^2 was unique at least up to values of $M_{4\sigma}^2 = 3.2$.[14] In these resonators, diffraction losses and spatial mode competition in saturating the gain determine the mixed mode composition. As the circular limiting aperture is opened – as the Fresnel number of the resonator is increased – some modes grow and others decrease in a predictable and reproducible way, such that for each M^2 there is a unique known mode mixture. Furthermore, this knowledge has allowed us to establish mathematical rules for inter-conversion of beam diameters between the various measurement definitions.

6.4.2 Empirical Basis for the Conversion Rules

We acquired the empirical data[14] by using an argon ion laser set up to give beams with a large range of M^2 values as a function of the diameter of the circular mode-limiting aperture. By varying this aperture diameter and the gain by adjusting the laser tube's current, values of $M_{2\sqrt{2}\sigma}^2$ from 1 to 2.5 were covered with the green line at 514 nm; the upper limit was increased to 3.2 by changing to the higher gain of the 488 nm blue line. As the blue line was generated in the same resonator, the blue beam diameters here could be scaled by multiplying the square root of the ratio of the wavelengths, a factor of 1.027, for comparison to the green line diameters. The beam from this laser was split to feed an array of monitoring equipment. A radio-frequency photodiode and rf spectrum analyzer indicated how many modes and what mode orders were oscillating. Profiles were recorded with a commercial slit and pinhole profiler[8] and a commercial beam propagation analyzer[9] to obtain knife-edge diameters, M_{ke}^2, and radial second-moment diameters. A CCD camera and software computed the variable aperture diameter. In front of the camera, a lens provided a known (1.47 times) magnification to fill an adequate number of pixels, and a variable attenuator set the light level.

As the laser's internal aperture was opened and the beam diameter enlarged, the mode spot alternated from one with a peak in the center to one with a dip at the center in over one-and-a-half cycles as shown in the profiles of Fig. 10(b). Seven aperture settings were chosen spanning the range of M^2 values, two giving the highest central peaks (A and E), two at the deepest dips (C and F) and three transitional ones (AP, named the "perturbed A-mode," B, and D). The full set of diagnostic data at these settings was recorded. Knowing the number of modes oscillating and the mode orders at each setting from the rf spectrum, trial mode mixtures were assumed. The resulting theoretical profiles were adjusted[14] to match the experimental pinhole profiles. An example is Fig. 2, where the theoretical mixed mode profile, (g), is matched to experimental profile, (h), which is the same as Mode E in Fig. 10(b).

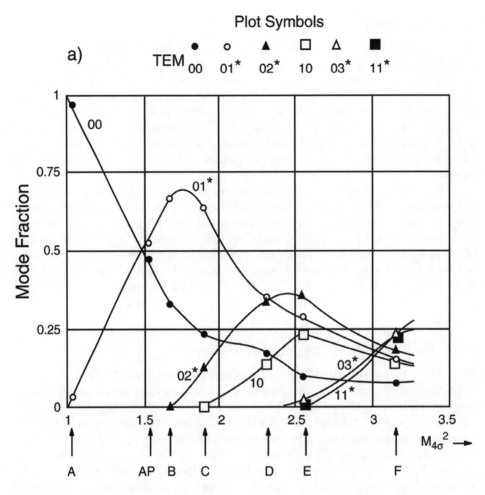

Figure 10 Observed mode fractions for a beam from a resonator with a limiting circular aperture. As the aperture diameter increases, $M_{4\sigma}^2$ follows, with the mode fractions changing in a characteristic fashion as higher order modes come above threshold. (a) The mode fractions as a function of $M_{4\sigma}^2$. (*Continued*)

Once the TEM$_{0n}^*$ modes were included[14] in the mode mix, good matches of profiles were found. These modes are like the donut mode, for which $n = 1$, but with increasingly larger holes in the center as their order $(n + 1)$ increases. Because they have $p = 0$ they are "all null" (nearly zero in amplitude) in the middle. They make the most of the r^3 weighting factor in the second-moment integral to reach a given second-moment diameter $Mw = \sqrt{(2p + l + 1)}w$ at the smallest radius, resulting in the lowest tails[14] to their profiles of all modes of the same order number. They thus have the lowest diffraction loss for a limiting circular aperture and always oscillate first among pure modes of the same order with an increasing Fresnel number. It was noted in Ref. 20 that in this aperture-opening process there was a gradual extinction of a mode of lower order soon after a mode of next higher order reached threshold. This is clearly a gain competition effect won by the higher order mode. A possible physical reason of general applicability discussed in Ref. 20 was that the

The M² Model

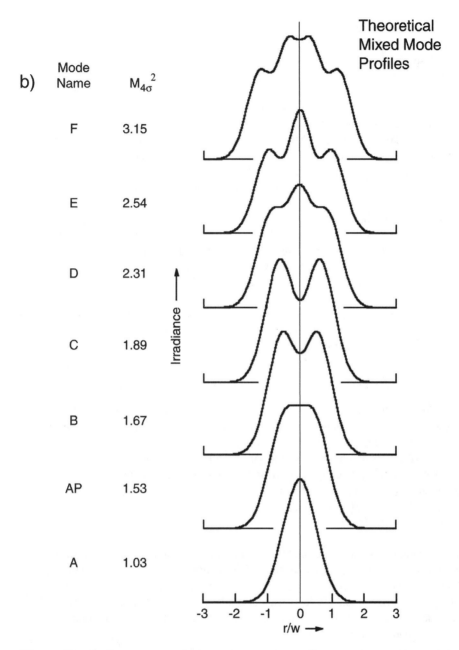

Figure 10 (b) The computed pinhole profiles and their $M_{4\sigma}^2$ values for the characteristic set of mixed modes A to F measured to determine the mode fractions.

larger spatial extent of the higher order mode provided access to a region of gain not addressed by the competing lower order mode.

The final mode fractions for the seven mixed modes were determined using a Mathematica function called SimpleFit made available by Wolfram Research. These fractions are plotted in Fig. 10(a) as a function of the resultant beam quality $M_{4\sigma}^2$ for the mixed modes. The modes turn on in the order of increasing diffraction loss as shown by McCumber[21] and then gradually extinguish, as predicted. At each value of $M_{4\sigma}^2$ for this argon ion laser there is a characteristic set of oscillating modes, mode fractions, and mode profiles [Fig. 10(b)]. Here for every M^2 value there is a unique mixture of modes. From all the data gathered, simple conversion rules given below in the next section between diameter definitions were derived. Over the range measured of $M_{4\sigma}^2 = 1$ to 3.2, the error to convert knife-edge, slit, and variable aperture diameters to second-moment diameters was $\pm 2\%$ (one standard deviation). This is a $\pm 4\%$ error in converting M^2. The error was $\pm 4\%$ for conversion of pinhole diameters to second-moment diameters.

We then tested the rules on other lasers[14] within this M^2 range and found that knife-edge diameter measurements converted to second-moment diameters agreed with directly measured second-moment diameters within $\pm 2\%$. The conversion error is defined as the fraction in excess of unity of the $D_{4\sigma}$ diameter obtained by the conversion rule, over that obtained directly from the variance of the irradiance profile, expressed in percent. The knife-edge diameter conversion subsequently was tested on three other gas lasers at $M_{4\sigma}^2 = 4.2, 7.5,$ and 7.7 and found to remain valid to $\pm 2\%$. However, a test[25] on a pulsed Ho:YAG laser at $M_{4\sigma}^2 = 13.8$ gave a conversion error of -9%; this is thought to be due to the strong transient thermal lensing in this medium affecting the spatial gain saturation. This consistency in the face of an extrapolation by a factor of two indicates that these conversion rules are fairly robust, valid to the stated accuracy, and that the mixed modes on which they are based exist in this large class of lasers. Apparently, for many lasers, M^2 is unique.

6.4.3 Rules for Converting Diameters Between Different Definitions

The empirical results showed there was a linear relationship between $M_i = \sqrt{M_i^2}$ and the square root of the second-moment beam quality $M_{4\sigma} = \sqrt{M_{4\sigma}^2}$, where M_i is the square root of the beam quality obtained by method "i" where i can signify any of the other definitions. Since all the diameter definitions give the same result for the fundamental mode beam in which the beam quality is unity, the linear relationship can be expressed with a single proportionality constant $c_{i\sigma}$ in the form

$$M_{4\sigma} - 1 = c_{i\sigma}(M_i - 1) \tag{47}$$

for the conversion from the method "i" to second-moment quantities. This form ensures that the linear plot of $M_{4\sigma}$ vs. M_i passes through the origin with no offset term and that only the slope constant c is required to define the relationship.

In the same resonator, the fundamental mode diameter is given by the ratio of the mixed mode diameter to M. This is true independently of what diameter definition is used, and thus a second relationship is

$$D_i/M_i = 2w = D_{4\sigma}/M_{4\sigma} \tag{48}$$

Here D_i is the diameter obtained by method "i". Substituting Eq. (48) into Eq. (47) yields

$$D_{4\sigma} = (D_i/M_i)[c_{i\sigma}(M_i - 1) + 1] \tag{49}$$

The values of the conversion constants $c_{i\sigma}$ are listed in Table 1 to convert from the diameter definitions summarized there to the second-moment diameter, $D_{4\sigma}$.

The M² Model

Since each of the other diameter methods is linearly related to the second-moment diameter, they all are linearly related. The conversion constants between the other methods can be obtained from those for the second-moment conversions. Let one of the other methods be denoted by subscript "j". From Eq. (47) there results

$$(M_{4\sigma} - 1) = c_{i\sigma}(M_i - 1) = c_{j\sigma}(M_j - 1)$$

therefore

$$(M_i - 1) = (c_{j\sigma}/c_{i\sigma})(M_j - 1)$$

By definition of a conversion constant for method $i \to j$,

$$(M_i - 1) = c_{ji}(M_j - 1)$$

Hence

$$c_{ji} = (c_{j\sigma}/c_{i\sigma}) \tag{50}$$

This gives the conversion constants between any two methods in Table 1, by taking the ratios of their constants for conversion to the second-moment values. Note that Eq. (50) also implies that $c_{ji} = 1/c_{ij}$, which is also useful.

The values for the $c_{i\sigma}$ constants in Table 1 are an improvement over our earlier results[14] that were incorporated in the ISO beam-test document.[11] More experimental data later became available, but also it was realized once the mode fractions were determined experimentally that the conversion constants could then be calculated from theory alone. From the mixed mode set A to F defined by the mode fractions of Fig. 10(a), each of the theoretical diameters D_i for the different methods were calculated. By Eq. (3), these were converted to M_i values. Then plots of $M_{4\sigma} - 1$ vs. $M_i - 1$ were least-squares curve-fit to determine by Eq. (47) the values of $c_{i\sigma}$ listed in Table 1. The fit for the slope $c_{i\sigma}$ was for one parameter only with the intercept forced to be zero. This gives an internally consistent set of $c_{i\sigma}$ so that Eq. (50) is valid.

7 PRACTICAL ASPECTS OF BEAM QUALITY M² MEASUREMENT: THE FOUR-CUTS METHOD

The four-cuts method means measuring the beam diameter at four judicious positions, the minimum number – as explained below – to permit an accurate determination of M^2. To execute this method well, several subtleties should first be understood.

The simplest way to measure M would be to take the ratio of the mixed mode beam diameter to that of the embedded gaussian, from Eq. (3), $M = W/w$, except that the embedded gaussian is inaccessible by being enclosed inside the mixed mode. However, both beams have the same Rayleigh range. By measuring z_R and the waist diameter $2W_0$ for the accessible mixed mode, the beam quality is determined through Eq. (20):

$$M^2 = \pi W_0^2/\lambda z_R \tag{20}$$

The general approach is to measure beam diameters $2W_i$ at multiple locations z_i along the propagation path and least-squares curve fit this data to a hyperbolic form to determine z_R and $2W_0$. But even by taking this computer-intensive approach, unreliable values will sometimes result unless a number of subtle pitfalls[25] (often ignored) are avoided on the way to good ($\pm 5\%$) M^2 values. The pitfalls are highlighted in *italics* as they are encountered below.

Well-designed commercial instruments[9] avoid these pitfalls, and a button push yields a good answer. For the engineer performing the measurement on his or her own, and who can start by roughly estimating the beam's waist diameter and location (using burn paper, a card inserted in the beam, or a profiler slid along the propagation axis) a minimum effort, logical, quick method exists that circumvents the subtle difficulties. This is the method[25] of "four-cuts," the subject of this section.

The first pitfall is avoided by realizing that in the M^2 model the beam divergence is no longer determined by the inverse of the waist diameter alone (as it is for a fundamental mode) but has the additional proportionality factor M^2:

$$\Theta = 2M^2\lambda/\pi W_0 \tag{19}$$

The implications of this additional degree of freedom are that the beam waist must be measured directly, not inferred from a divergence measurement. Consider the propagation plots shown in Fig. 11(a). Several beams are plotted, all with the same values of M^2/W_0 and therefore the same divergence, but with different M^2 [accomplished by having the Rayleigh range proportional to W_0, see the second form of Eq. (19) in Sec. 4]. From measurements all far from the waist it would be impossible to distinguish between these curves to fix M^2. On the other hand, in Fig. 11(b) are propagation plots for several beams with the same waist diameters but different M^2 and therefore divergences. Here $\Theta \propto M^2$ and by Eq. (18), $z_R \propto 1/M^2$. Measurements all near the waist could not distinguish these curves to fix the divergences. Both near- and far-field diameter measurements are needed to measure M^2.

Any of the diameter measurement methods can be used to define an M^2 value, and *the next pitfall is avoided by staying in one currency*, and do not mix, for instance, the knife-edge divergence measurement with the laser manufacturer's quoted $D_{4\sigma}$ (second moment) waist value. Consistently use the most reliable diameter-measurement method you have available, and in the end convert the your results to values in the standard $D_{4\sigma}$ currency.

7.1 The Logic of the Four-Cuts Method

The four-cuts method starts with the error estimate by your best method for measuring diameters, and uses that to set the tolerances on all other measurements. Let diameters be determined to a fractional error g,

$$g = (2W_{\text{meas}}/2W) - 1 \tag{51}$$

where $2W_{\text{meas}}$ is the measured diameter, and $2W$ the correct diameter. It is assumed g is small, usually 1–2%. This will yield a fractional precision h for the beam quality of $h = 3\text{–}5\%$ since M^2 varies as the product of two diameters, with a small error added for

The M² Model

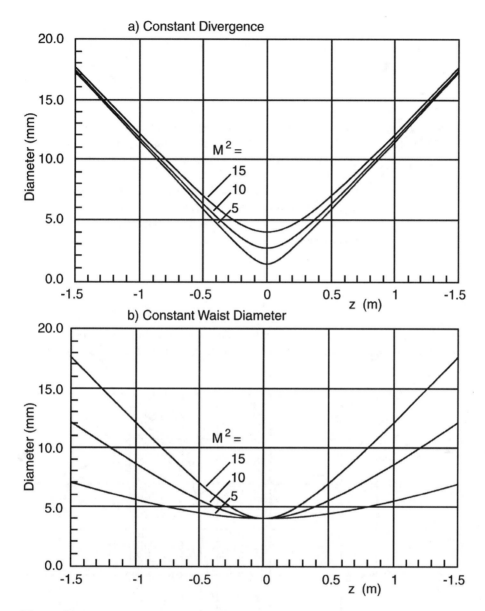

Figure 11 Beams of constant divergence (a) and constant waist diameter (b) to illustrate the consequence of $M^2 \neq 1$. The beam must be sampled in both near- and far-fields to distinguish these possibilities. The curves are drawn with values appropriate for a beam of $\lambda = 2.1$ microns (redrawn from Ref. 25).

a required lens transform (discussed below). The term "cut" is used for a diameter measurement, after the common use of a knife-edge scan cutting across the beam to fix a diameter. Let us define the normalized or fractional propagation distance from the waist as

$$\eta(z) = (z - z_0)/z_R \tag{52}$$

Let the fractional error in locating the waist be η_0. For this miss in cut placement in measuring the waist diameter $2W_0$ to cause an error of less than g, Eq. (16) gives

$$\sqrt{(1+\eta_0^2)} < g + 1 \quad \text{or} \quad \eta_0 < \sqrt{(2g)} \tag{53}$$

for $g \ll 1$. If $g = 0.01$, then $h_0 < \sqrt{(0.02)} \cong 1/7$. *The tolerable error in locating z_0 is $1/7$ of a Rayleigh range* for a 1% precision in diameter measurements.

To locate the waist to this precision, beam cuts must be taken far enough away from the waist to detect the growth in beam diameter with distance. At the waist location the diameter change with propagation is nil; to precisely locate a waist requires observations far from it where the diameter variation can be reliably detected. On both sides of the waist, cuts must be made at a sizeable fraction of the Rayleigh range.

To find the optimum cut distances, look at the fractional change Q in beam diameter vs. normalized propagation distance

$$Q \equiv (1/W)\, dW/d\eta = \eta/(1+\eta^2) \tag{54}$$

Figure 12 is a plot of this function, Eq. (54), in which it is easy to see that the maximum fractional change of Q occurs at $\eta = \pm 1$. By making cuts within -2 to -0.5 and $+0.5$ to $+2.0$ Rayleigh ranges from the waist corresponding to η equal to these same numerical values, 80% of the maximum fractional change is available. This will significantly

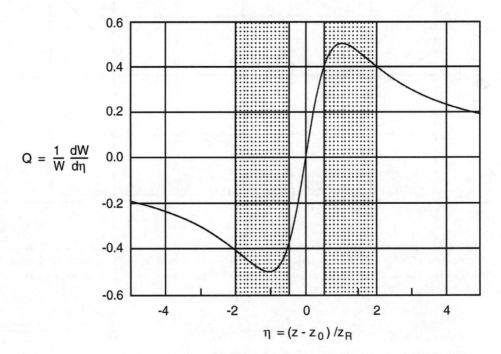

Figure 12 The fractional change Q in beam diameter as a function of the normalized propagation distance from the waist. Cuts made to locate the waist in the shaded regions benefit from a fractional change of 80% or more of the maximum change. This requires a minimum of one Rayleigh range of access to the beam around the waist location (redrawn from Ref. 25).

The M^2 Model

enhance the reliability of the position determination over that made using diameters from less than $0.5 z_R$ away from the waist. *An accessible span of at least a Rayleigh range centered on the waist is needed for diameter measurements.*

Note that Fig. 12 highlights the physical significance of the propagation locations one Rayleigh range to either side of the waist. The wavefront curvature is largest in absolute magnitude there, resulting in the fractional change in diameter, Q, with propagation coordinate z, reaching extremes of ± 0.5 there as well.

7.1.1 Requirement of an Auxiliary Lens to Make an Accessible Waist

Most lasers have their beam waists located internally where they are inaccessible. Therefore, an accessible auxiliary waist related to the inaccessible one is achieved by inserting a lens or concave mirror into the beam, and making the M^2 measurement on the new beam. Then the constants found are transformed back through the lens to determine the constants for the original beam. *This requirement to insert a lens, and then transform through the lens back to the original beam constants, is an often-ignored pitfall in making accurate beam measurements.*

The temptation is to use what is available, and just measure the beam on the output side of the output coupler. Usually this means the data is all on the diverging side of the waist. The problem is that nothing in this data constrains the waist location very well. In the curve fit, small errors in the measured diameters will send the waist location skittering back and forth to the detriment of the extrapolation to find the waist diameter. Inserting a lens and making a beam that is accessible on both sides of its waist is a significantly more reliable procedure.

There are three constants (z_{02}, $2W_{02}$, M^2) needed to fix the 2-space beam shown in Fig. 6 for one of the principal propagation planes, so, in principle, only three cuts should suffice, but then one of them would have to be at $|\eta_0| < 1/7$. The location of this narrow range $z_{02} \pm z_{R2}/7$ is at this point unknown. Therefore four cuts are used, the first an estimated Rayleigh range z_{R2} to one side of the estimated waist location z_{02}, the second and third at about 0.9 and 1.1 times this estimated Rayleigh range to the other side (see Fig. 13). These cut locations and the diameters determined there are labeled by their cut numbers $i = 1, 2, 3$. Between z_2 and z_3 there is a diameter that matches $2W_1$ and the location z_{match} of this is determined by interpolation:

$$z_{match} = z_2 + (z_3 - z_2)(W_1 - W_2)/(W_3 - W_2) \tag{55}$$
$$z_4 = z_{02} = (z_1 + z_{match})/2 \tag{56}$$

The waist is located exactly halfway between z_1 and z_{match}, and the fourth cut is made there at z_4 to directly measure the waist diameter $2W_{02} = 2W_4$ of the 2-space beam and complete the minimum data to determine M^2.

7.1.2 Accuracy of the Location Found for the Waist

If the locating cuts (1, 2, and 3 of Fig. 13) are within the ranges specified from $|Q| > 0.4$ and the diameters are measured to the fractional error g, then the error in the normalized waist location η_0 is no worse than $g/Q = 2.5\%$. This is much less (since g is small) than the tolerance $\sqrt{(2g)} = 14.1\% = 1/7$ determined from inequality Eq. (53). The measured waist diameter is then correct to the fractional error g.

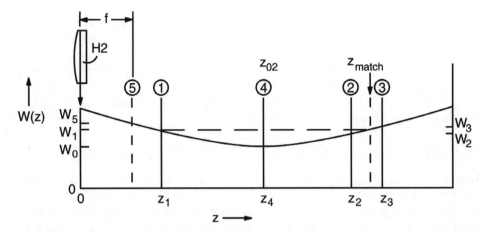

Figure 13 The four-cuts method. Shown is the beam propagation plot in 2-space, behind the inserted auxiliary lens; the circled numbers indicate the order of the cuts made to locate the waist. The propagation distance z_{match} gives the waist location as halfway between these equal diameters (redrawn from Ref. 25).

The fractional error in measurement of diameters, g, when divided by the fractional change in diameter with normalized propagation, Q, gives the fractional error in normalized waist location, $\eta_0 = g/Q$. The plot of Fig. 12 is thus actually a quantitative version of the statement "to precisely locate a null requires observations far from the null" when locating the waist. Diameter measurements inside the range $z_{02} \pm z_R/2$ quickly lose any ability to contribute precision in locating the waist as here Q drops to zero.

There is much value in locating the waist as accurately as the diameter-measurement tolerance will allow in that it reduces the number of unknown constants to be determined by curve fitting from three to two. The number of terms in the curve fit drops by a factor of four, and the remaining terms are made more accurate. Some of these terms depend on the distance from the waist to the ith-cut location, $z_i - z_{02}$, either squared or raised to the fourth power. It is often useful to take a fifth cut at $z_5 = f$ as shown by the vertical dashed line in Fig. 13. This cross-checks the input beam divergence by Eq. (29) and balances the number of points on either side of the auxiliary waist at z_{02} to improve the curve fit.

7.2 Graphical Analysis of the Data

The data, which consist of a table of four or five cut locations and their beam diameters for each of the two independent principal propagation planes is next plotted. A sample plot for the $\lambda = 2.1$ micron Ho:YAG laser beam analyzed in Ref. 25 is shown in Fig. 14. There it was found that with as few data points as required in the four-cuts method, and with the initial waist location and Rayleigh range estimates close to the final values (within $\sim 10\%$), a simple and quick graphical analysis is as accurate as a curve fit. Generally, with more points as in commercial instrumentation, a weighted least-squares curve fit of the data to a hyperbolic form is required,[25] discussed in Sec. 7.3. The curve fit also generates a sum of residuals for a statistical measure of the goodness of fit.

In the graphical analysis after the points are plotted, smooth curves of approximately hyperbolic form are laid in symmetrically about the known waist locations for each

The M² Model

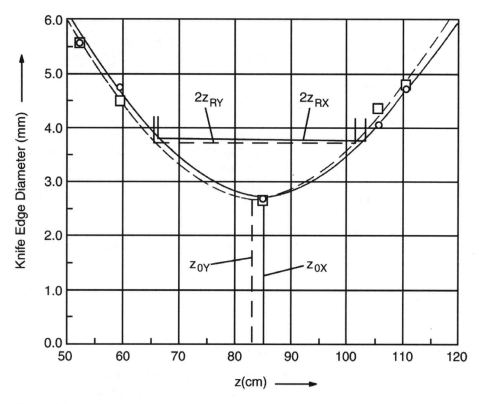

Figure 14 An example of graphical analysis of propagation data for the auxiliary beam in 2-space. The chords give the Rayleigh ranges for the x- and y-planes. They are drawn at ordinates on the plot $\sqrt{2}$ larger than the waist diameters located at z_{0x} and z_{0y} (redrawn from Ref. 25).

principal propagation plane, here in Fig. 14 with a French curve. Next, horizontal chords are marked off at heights $\sqrt{2}$ times the waist diameters $2W_4$ to intersect the smooth curves. The distance between these intersection points on each curve are twice the Rayleigh ranges $2z_{Rx}$, $2z_{Ry}$ respectively, and these lengths are measured off the plot for use in Eq. (20) with $2W_{0x} = 2W_{4x}$ (and $2W_{0y} = 2W_{4y}$) to determine M_x^2 (and M_y^2) for the auxiliary 2-space beam. For the data of Fig. 14 the results were $z_{Rx} = 17.6$ cm and $z_{Ry} = 17.8$ cm, resulting in knife-edge beam qualities $M_x^2 = 15.4$ and $M_y^2 = 14.9$.

These results are termed the initial graphical solution and can be improved to give the corrected graphical solution by using the fact that a better estimate of the waist diameter is available than just the closest measured point. By the propagation law, Eq. (16), if the miss distance of the closest point (cut 4) is η_0 then the best estimate of the corrected waist diameter is

$$2W_{02} = 2W_4/\sqrt{(1+\eta_0^2)} \tag{57}$$

The corrected solution uses the Rayleigh range and waist values from the initial graphical solution in Eq. (57) to obtain a corrected waist diameter, and plots a chord at a height of $\sqrt{2}$ times this diameter to determine a corrected length $2z_R$ and M^2 from Eq. (20). In the example of Fig. 14, the chords shown are the corrected chords; only the y-axis data changed slightly to $z_{Ry} = 17.3$ cm and $M_y^2 = 15.2$. After curve fitting the same data, the fractional rms error (goodness of fit) for the five diameter points were the same at $<1.9\%$.

This good accuracy is a consequence of the four-cuts strategy. The waist diameter is directly measured and if the initial estimate for the Rayleigh range is close, the other cuts give data points near the intersection points of the chords fixing the $2z_R$ values on the plot. The graphical analysis then amounts to an analog interpolation to find the best positions for the intersection points.

There are two last steps. The first is to transform the 2-space data back to 1-space to get the constants for the original beam, using Eqs. (24)–(28). This adds a small fractional error to the end result due to the uncertainties in z_{02} and z_{R2}, which contribute a slight uncertainty to the transformation constant Γ of Eq. (24) (in the example of Ref. 25, a 2% error in Γ, 1% additional error in transformed diameters).

The second step is to convert these knife-edge measurements of Fig. 14 to standard second-moment units as done in Table 3 of Ref. 25. The beam of Ref. 25 is the one that did not work well with the conversion rules of Sec. 6. Instead the conversion of $M_{ke}^2 = 15.4$ to $M_{4\sigma}^2 = 13.8$ was done by comparing measurements at cut 5, the focal plane of the auxiliary lens, of the knife-edge diameter to the second-moment diameter calculated from a pinhole scan. This gave the ratio $D_{ke}/D_{4\sigma} = 1.055$ or a factor of $1/(1.055)^2 = 0.897$ for the M^2 conversion.

7.3 Discussion of Curve-Fit Analysis of the Data

A complete numerical example of a full weighted least-squares curve fit to analyze the four-cuts data, or a larger data set, is given in Ref. 25 and need not be repeated. There are some subtle pitfalls to avoid in using curve fits on beam propagation data and these are briefly discussed.

A least-squares curve fit is the only general way to account for all the data properly. A common mistake is *to use the wrong function for the curve fit*, which necessitates a discussion of what is the correct one. The fit should be to a hyperbolic form, Eq. (16), but that is not all. It also should be a weighted curve fit, with the weight of the ith squared residual in the least-squares sum being the inverse square power of the measured diameter $2W_i$.

There are three reasons for this choice of weighting. The first is that in general in a weighted curve fit, the weights[36] should be the inverse squares of the uncertainties in the original measurements. For many lasers, the fractional error in the measured diameter is observed to increase with the diameter; this is probably due to the longer time it takes to scan a larger diameter. The spectrum of both amplitude noise and pointing jitter on a beam tend to increase towards lower frequencies and longer measurement times give this noise a greater influence.

The second reason arises from an empirical study[25] of different weightings one of us did during the development of a commercial M^2 measuring instrument.[9] Amplitude noise was impressed on the beam of a fundamental mode ion laser with a known $M_{4\sigma}^2 = 1.03$, by rapid manual dithering of the tube current while the instrument's 30 second data gathering run[9] was under way. (Note, the ModeMaster[9] gathers 260 knife-edge cuts in each principal propagation plane during the 30 second focus pass to record the propagation plot of the auxiliary beam.) The same data was then fitted to a hyperbola five times, with five different weighting factors. The weights were the measured diameter raised to the nth power, $(2W_i)^n$, with $n = -1, -0.5, 0, +0.5$, or $+1$. The weight with $n = 0$ is unity or equal weight for all data points. Data runs were repeated many times with increasing noise amplitude, and the resulting M^2 values for all five weighting schemes were compared each time. The unity or negative power weightings gave stable M^2 values within 3% of the correct value up to 5% peak-to-peak amplitude noise. The positive power weightings $n = +0.5$ gave 4–5% and $n = +1$ gave 12–19% errors in M^2, respectively, at this noise level. With larger noise amplitudes, the positive power weightings gave errors that grew rapidly and nonlinearly.

A common curve-fitting technique is to use a polynomial fit for the square of the beam diameter vs. propagation distance. This may be convenient but it could give an unsatisfactory result. This technique takes advantage of the wide availability of polynomial curve fit software, and the fact that the square of Eq. (16) gives a quadratic for $W(z)^2$ as a function of z. However, look at what this does. Let $2W_i$ be the measured ith diameter, and $2W'_i$ be the exact diameter with the small deviation between them $2\delta_i = 2W_i - 2W'_i$. In the W^2 polynomial curve fit, the ith term is

$$(W_i)^2 = (W'_i + \delta_i)^2 = (W'_i)^2 + 2W'_i\delta_i$$

making the residual

$$(W_i)^2 - (W'_i)^2 = 2W'_i\delta_i$$

The residual from the exact polynomial curve is weighted in the fit by $2W'_i$, a positive power $(+1)$ of W'_i, and so will give unstable results if there is more than a few percent amplitude noise on the beam. At the time of completion of the 1995 ISO document[11] on beam test procedures, this difficulty with a polynomial curve fit was unrecognized, and a polynomial fit is (incorrectly) recommended there.

The third reason for an inverse-power weighting is that mathematically the least fractional error results for a ratio quantity like $M^2 = \Theta/\theta_n$ in Eq. (22) if the fractional errors from the denominator and numerator roughly balance. The residuals from the more numerous cuts far from the waist – the points giving the measurement of divergence Θ, or numerator – would swamp with equal weighting the fewer (or single) cut at the waist – the point(s) giving the divergence of the normalizing gaussian, or denominator. An inverse square weighting approximately halves the influence of the three or four far points, compared to the unity weighting at the waist in the four-cuts method, giving the desired rough balance.

7.4 Commercial Instruments and Software Packages

There are three main commercial instruments for measuring beam quality and a host of less well developed others. The first is the original[9,35] system designed as a beam propagation analyzer and believed at this time to be the most fully developed, the ModeMaster™ from Coherent, Inc., Portland, OR. The basic diameter measurements are achieved with two orthogonal knife-edge cuts. Both principal propagation planes are measured nearly simultaneously on a drum spinning at 10 Hz behind auxiliary lens. As with any spinning drum profiler, measurements are restricted to continuous-wave laser beams or very high repetition rate, $\sim$100 kHz, pulsed lasers. The lens moves to carry the auxiliary beam through the plane of the knife-edges to assemble 260 cuts in each principal propagation plane making up a pair of propagation plots for the auxiliary beam in a 30 second "focus" pass. A curve-fit, with an inverse-diameter weighting, to a hyperbola is done and the fitted parameters are transformed through the lens by the on-board processor to present a data report for the original beam (see Fig. 17 in Sec. 9). The drum also carries two pinholes, each of different diameter, giving pinhole profiles that are processed to give direct second-moment diameters. The instrument also measures beam-pointing stability. Electronic alignment aides are included.

Another profiler-based instrument is the ModeScan™ from Photon, Inc (San Jose, CA). This modular package automates the on-beam-axis motion of their slit profiler behind a fixed auxiliary lens. These 10 Hz rotating drum profilers are commonly available and this system can be acquired to upgrade an existing profiler to become a beam propagation analyzer. The system addresses CW laser beams. The software is designed to "train" the instrument to repeat an operation, once it is first set up, for optical system quality control use.

A CCD camera-based instrument is the M2-200 Beam Propagation Analyzer from Spiricon, Inc. (Logan, UT), the only one that operates with pulsed laser beams. This uses a fixed lens and moves an optical delay line to essentially scan the detection surface through the auxiliary beam. The PC computer attached to the system calculates the second-moment diameters[37] directly from the CCD profiles after automatically subtracting background noise. A curve fit to a hyperbola is done,[37] and the results transformed through the lens to present the constants of the original beam. The CCD camera gives the system versatility in computations but brings the limitations already discussed for cameras of limited dynamic range, fewer pixels in a profile, and overall poorer accuracy compared to the analog instruments.

8 TYPES OF BEAM ASYMMETRY

In the previous sections, the means for the spatial characterization of laser beams were established. This section looks at commonly found beam shapes and others that are possible. The three common types of beam asymmetry are depicted in Fig. 15. These are the pure forms but mixtures of all three are common in real beams.

8.1 Common Types of Beam Asymmetry

The first is *simple astigmatism* [Fig. 15(a)], where the waist locations for the two orthogonal principal propagation planes do not coincide, $z_{0x} \neq z_{0y}$, but $W_{0x} = W_{0y}$, and $M_x^2 = M_y^2$. Because here the waist diameters and beam qualities are the same for the principal propagation planes, so are the divergences, $\Theta_x \propto M_x^2/W_{0x} = M_y^2/W_{0y} \propto \Theta_y$ [see Eq. (19)]. This makes the beams round in the converging and diverging far-fields. At the two waist planes the beam cross-sections are elliptical, one oriented in the vertical and the other the horizontal plane, with the minor diameters equal. Midway between the waists, the beam becomes round like the "circle of least confusion" point in the treatment of astigmatism[28] in geometrical optics. The simple astigmatic beam is characterized by three round cross-sections, at the distant ends and midpoint, with orthogonally oriented elliptical cross-sections in between.

The window frame inserts of Fig. 15 show the wavefront curvatures, which are spherical in the far-field, cylindrical at the waist planes, with one cylindrical axis horizontal, the other vertical, and saddle-shaped at the midpoint between the waists. The wavefront curvatures determine the nature of the focus when a lens is inserted.

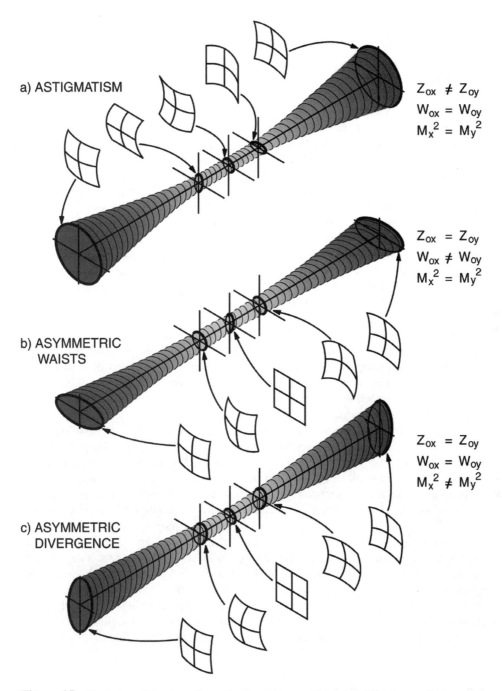

Figure 15 Depiction of the three-dimensional appearance of the three basic types of asymmetry for a mixed mode beam: (a) astigmatism, (b) asymmetric waist diameters, and (c) asymmetric divergence. The window inserts show the wavefront curvatures along the beam path (redrawn from Ref. 25).

Simple astigmatic beams can be generated in resonators with three spherical mirrors, with one used off-axis to give an internal focus,[38] unless there is astigmatic compensation built in as with a Brewster plate of the correct thickness[38] added to the focusing arm. Many diode lasers are astigmatic but with the other two types of asymmetry as well because the channeling effects in the plane parallel to the junction differ from those in the plane perpendicular to it, giving two different effective source points for the parallel and perpendicular wavefronts. Beams formed using angle-matched second harmonic generation can be astigmatic due to walk-off in the phase-matching plane of the beam in the birefringent doubling crystal. The diode lasers in laser pointers frequently have a large astigmatism, as large as the Rayleigh range for the high divergence axis.

The next type of beam asymmetry is *asymmetric waists* [Fig. 15 (b)], where the waist diameters are unequal. Because of the different waist diameters but with equal beam qualities, in the far-fields where divergence dominates, the cross sections are elliptical with the long axes of the ellipses (shown as horizontal) perpendicular to the long axis of the waist ellipse (here vertical). In between there are round cross-sections at planes symmetrically placed around the waist location – the same geometry as in Fig. 15(a), with the ellipses and circles interchanged. The wavefronts are plane at the waist, and ellipsoidal everywhere else, with curvatures at the round cross-sections in the ratio of the square of the waist diameters.

Lasers having an out-of-round gain medium are likely to produce beams with asymmetric waists. A solid-state laser pumped from the end by an elliptical beam from a diode laser is an example. Mode selection is by the combined effects of gain aperturing and absorption in the unpumped regions. The resonant beam shape will mimic the geometry of the pumped region. Beam walk-off from angle-matched nonlinear processes can also produce asymmetric waists.

The third type is *asymmetric divergence* [Fig. 15(c)], where the beam qualities differ in the principal propagation planes to give proportionally different divergence angles, $\Theta_x \propto M_x^2 \neq M_y^2 \propto \Theta_y$, but $W_{0x} = W_{0y}$, and $z_{0x} = z_{0y}$. The simplest description of this beam is that it has a mode of higher order in one principal propagation plane than in the other. In the far-field, cross-sections are elliptical as in case (b), but the beam is round only at the waist plane. The wavefronts are plane at the waist and ellipsoidal everywhere else and the Rayleigh ranges are different in the two principal propagation planes.

A CW dye laser using a high-viscosity dye jet provides an example of pure asymmetric divergence.[39] The pump-beam spot was round, but the heat it deposited in the dye stream was cooled differentially by the flow. In the flow direction the temperature gradient was smoothed by the forced convection but in the other direction a more severe thermal gradient existed, causing an aberration that resulted in $M_{4\sigma y}^2 = 1.51$ for that plane compared to $M_{4\sigma x}^2 = 1.06$ for the plane parallel to the flow with negligible aberration. Because of the round pump beam, waist asymmetry was only $2W_{0y}/2W_{0x} = 1.06$.

8.2 The Equivalent Cylindrical Beam Concept

Beams with combinations of these asymmetries can be depicted with superposed (x, z) plane and (y, z) plane propagation plots as shown in Fig. 16(a). More generally there is a propagation plot $W(\alpha, z)$ for each azimuth angle α around the propagation axis z. The angle α is measured from the x-axis and $W(\alpha, z)$ lies in the plane containing α and the z-axis. The three-dimensional beam envelope shown in Fig. 15(a), (b), or (c) is called

The M² Model

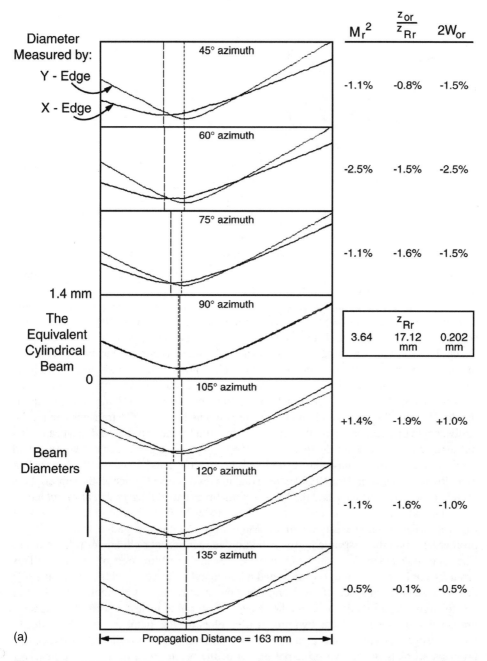

Figure 16 (a) Experimental propagation plots with beam diameters measured by orthogonal knife-edges for a beam with both astigmatism and waist asymmetry. The percentage variation of the constants of the equivalent cylindrical beam, computed from the plots for each instrument azimuth, are listed in the right-hand columns. The small variations demonstrate the constants are independent of the azimuth of the two orthogonal cutting planes intersecting the beam caustic surface. The constants of the equivalent cylindrical beam, in the box, correspond to the cuts at an instrument azimuth of 90°. (*Continued.*)

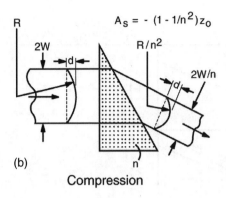

Figure 16 (b) Diagram showing how a half-Brewster prism introduces both astigmatism A_s and waist asymmetry $W_{0y}/W_{0x} = n$ to the beam.

the beam caustic surface, and is swept out by $W(\alpha, z)$ as the azimuth angle α rotates through a full circle from 0 to 2π.

For beams with combinations of moderate asymmetries, it is convenient to define an *equivalent cylindrical* beam. This is a beam with cylindrical symmetry – with a round spot for all z – and the real beam asymmetries are treated as deviations from this round beam. The constants defining this equivalent cylindrical beam are the best average of the beam constants for the two independent principal propagation planes. Many problems can be treated with just this simpler equivalent beam. In particular, it has been predicted theoretically (A.E. Siegman, personal communication, 1990) and demonstrated experimentally[40] that the centered circular aperture computed to give 86% transmission for the equivalent cylindrical beam has this same transmission for the out-of-round real beam. The minimum aperture sizes for the real beam after propagation in free space can be computed using just the three equivalent cylindrical beam constants. Because the equivalent cylindrical beam is round for all z like the radial modes discussed in Sec. 4, the subscript r is used to denote its constants, and the beam is sometimes called the *equivalent radial mode*.

The equivalent cylindrical beam is best understood by considering the plots of Fig. 16(a). These were measured with the ModeMaster beam propagation analyzer.[9] The profiler built into this instrument uses two knife-edge masks at right angles to each other. They are mounted on a rotating drum at 45° to the scan direction of the drum. This arrangement is equivalent to a vertical and a horizontal knife-edge, each scanned at $1/\sqrt{2}$ times the actual scan speed of the drum, when the analyzer head's azimuth angle is set to 45° to align one knife-edge with the horizontal. Each run to measure beam diameters vs. propagation distance produces two propagation plots for the diameters at right angles to the two edges. Normally the analyzer head's azimuth angle is adjusted to record the propagation plots in the two principal planes of the beam. For Fig. 16(a), the analyzer head's azimuth angle was incremented in 15° steps through 90° and new sets of propagation plots recorded for each increment, generating the seven plots shown.

The asymmetric beam of Fig. 16(a) was formed by inserting a Brewster-angle half-prism[41] in the cylindrically symmetric beam Mode E of Fig. 10(b) and Fig. 2(g) and (h). The prism was oriented as in Fig. 16(b) to produce a compression of the beam diameter in one dimension in the (x, z) plane. The prism thus introduces astigmatism and waist asymmetry to the beam. From Fig. 16(b) the incoming wavefront of radius of curvature R

The M² Model

has a sagitta of the arc, $d = W^2/2R$, which remains unchanged upon the transit of the prism while the beam diameter is compressed. For the Brewster angle prism, it can be shown[41] that the exiting beam diameter is smaller by the factor $1/n$, where n is the index of refraction of the prism material, here silica with $n = 1.46$. The radius of curvature exiting the prism is thus R/n^2. The M^2 of the beam is unchanged in traversing the prism. From these three conditions both the reduced waist diameter in the x-direction and the astigmatic distance introduced in the exiting beam can be determined (using Eqs. 9.10 and 9.11 from Ref. 7 and a little algebra) to be $2W_0/n$ and $A_s = (z_{0y} - z_{0x}) = -(1 - 1/n^2)z_0$, where z_0 is the propagation distance from the input waist location to the prism.

The propagation plots of Fig. 16(a) are for the directly measured internal beam, behind the lens of the beam propagation analyzer. These were used because the beam diameter and propagation distance scales of the internal plots remain the same as the instrument azimuth is varied and this facilitates comparison of the plots. Notice in the top 45° instrument-azimuth plot, because the internal plots are shown, the axis with the n-times larger divergence and $1/n$-times smaller waist is the y-axis, interchanged with the compressed x-axis of the external beam by the beam-lens transform equations of Sec. 5.

As the instrument azimuth angle moves around from the initial 45° value (which measures the principal propagation planes for this beam) to 90°, the plots from the two orthogonal edges coalesce into a single "average" curve, then separate with continuing azimuth increments. The plots at 135° are identical to the 45° plots with the x-edge and y-edge curves interchanged. The dashed and dotted vertical lines on each plot locate the waists for the x-edge and y-edge curves, respectively. The beam constants for the symmetric, 90° azimuth plots are those for the equivalent cylindrical beam.

To visualize this process of cutting the beam caustic surface with two orthogonal planes, then rotating the azimuth of the cutting planes, look at Fig. 15(c). The initially vertical (y-edge) plane is cutting the caustic in its highest divergence plane, and moves towards a lower divergence $W(\alpha, z)$ plot as the azimuth is incremented. The initially horizontal (x-edge) plane is cutting the caustic in its lowest divergence plane, and moves toward a higher divergence $W(\alpha, z)$ plot as the azimuth is incremented. When the cutting planes reach 45° azimuth to the principal planes of the beam, the orthogonal propagation plots match as they would for a round beam with no asymmetries.

Siegman[42] gives the following expressions for the beam constants of the equivalent cylindrical beam in terms of the six constants of the real beam:

$$z_{0r} = \left(\frac{M_x^4 W_{0y}^2}{M_x^4 W_{0y}^2 + M_y^4 W_{0x}^2}\right)z_{0x} + \left(\frac{M_y^4 W_{0x}^2}{M_x^4 W_{0y}^2 + M_y^4 W_{0x}^2}\right)z_{0y} \tag{58}$$

$$W_{0r}^2 = W_{0x}^2 + W_{0y}^2 + \left(\frac{1}{\pi^2}\right)\left(\frac{M_x^4 M_y^4}{M_x^4 W_{0y}^2 + M_y^4 W_{0x}^2}\right)\lambda^2(z_{0x} - z_{0y})^2 \tag{59}$$

and

$$M_r^2 = \frac{W_{0r}^2}{4}\left(\frac{M_x^4}{W_{0x}^2} + \frac{M_y^4}{W_{0y}^2}\right) \tag{60}$$

The columns of numbers in Fig. 16(a) demonstrate that the beam constants of the equivalent cylindrical beam are the same when computed from plots for any azimuth, a necessary condition for the equivalent cylindrical beam concept to be useful. The equivalent cylindrical beam quality, waist location, and waist diameter were computed for each azimuth increment from the plots shown, and normalized to the constants measured for the 90° azimuth shown in the box. The percentage errors for these measurements are given in the three columns; the magnitudes of all errors are no larger than 2.5% and are within the instrument measurement tolerances.

From Eq. (58) and Eq. (59) for an astigmatic beam the equivalent cylindrical waist lies between the two astigmatic waists, and the square of the cylindrical waist diameter exceeds the sum of the squares of the two astigmatic waist diameters. For a beam with no astigmatism ($z_{0x} = z_{0y}$) the equivalent cylindrical beam constants become

$$W_{0r}^2 = W_{0x}^2 + W_{0y}^2 \tag{61}$$

$$M_r^4 = \left(\frac{W_{0x}^2 + W_{0y}^2}{4W_{0x}^2}\right)M_x^4 + \left(\frac{W_{0x}^2 + W_{0y}^2}{4W_{0y}^2}\right)M_y^4 \tag{62}$$

For a beam with no astigmatism and no waist asymmetry the equivalent cylindrical beam quality is

$$M_r^4 = (M_x^4 + M_y^4)/2 \tag{63}$$

A beam of this type with different values of M_x^2 and M_y^2 will have a round spot at the waist plane, but not in the far-field as illustrated in Fig. 15(c).

8.3 Other Beam Asymmetries: Twisted Beams, General Astigmatism

The shape of a beam caustic surface is determined by the straight-line paths of rays where they emerge at the margin of the particular beam. Such shapes are all examples of ruled surfaces and those depicted in Fig. 15 are hyperboloids. In principle, any paraxial ensemble of light rays (i.e., a beam) will be enclosed by a ruled surface. Another example is a taut ribbon, and these surfaces can be twisted. Imagine the shapes of Fig. 15 as taut, flexible membranes. Start with a shape similar to Fig. 15(b) except with *all* of the cross-sections being horizontally elongated ellipses (a beam with *both* asymmetric waists and divergence). Mentally rotate the far-field ellipses to vertical, the distant one by +90° and the foreground one by −90° azimuth, keeping the waist ellipse horizontal. In propagating from $z = -\infty$ to $+\infty$ the elliptical cross-sections of this beam twist through 180° of azimuth. Such a twisted beam can physically be realized and is said to have *general astigmatism*.[15,16] Here all spots can be ellipses,[15] a waist location is defined by a cross-section having a uniform phasefront,[16] and the Rayleigh range is defined as the distance of propagation away from the waist that increments[16] the Gouy phase by $\pi/4$.

Such beams are produced by nonorthogonal[5] optical systems, e.g., two astigmatic elements in cascade with azimuth angles that differ by something other than 0° or 90°.

Rays in the (x, z) and the (y, z) planes are coupled and cannot be analyzed independently. The general theory for spatial characterization of such beams uses ray matrices weighted by the Wigner density function[4,17] averaged over a four-dimensional geometrical optics "phase space." Rays are described by 4×1 column vectors; each vector gives the position x, y and slope u $(=\theta_x)$, v $(=\theta_y)$ of the ray at the location z along the propagation axis. There are 16 possible second-order moments of these variables; they propagate in free space with a quadratic expansion law.[6,26] The square of the second-order moment diameter, $D_{4\sigma}^2$, is such a second-order moment and this is the theoretical support this diameter definition enjoys. The beam matrix **P**, the 4×4 array of these 16 second-order moments, then fully characterizes the beam with general astigmatism.

The 16 possible second moments can be listed as

$\langle x^2 \rangle$; $\langle xy \rangle$; $\langle xu \rangle$; $\langle xv \rangle$; $\langle y^2 \rangle$; '$\langle yx \rangle$'; $\langle yu \rangle$; $\langle yv \rangle$;

$\langle u^2 \rangle$; '$\langle ux \rangle$'; '$\langle uy \rangle$'; $\langle uv \rangle$; $\langle v^2 \rangle$; '$\langle vx \rangle$'; '$\langle vy \rangle$'; '$\langle vu \rangle$'.

However, by symmetry $\langle xy \rangle = \langle yx \rangle$, and so on, so that only ten of these are independent, and in the list those in single quotes are redundant. The moments containing only spatial variables $\langle x^2 \rangle$, $\langle xy \rangle$, $\langle y^2 \rangle$ can be evaluated as the variances of irradiance pinhole profiles in the proper direction; the $\langle xy \rangle$ profile is at 45° to the x- or y-axes. The moments containing the angular variables cannot be evaluated directly, but are found by inserting optics (usually a cylindrical lens) and measuring downstream irradiance moments at appropriate propagation distances.

From these moments, beam constants are calculated. The first six are the familiar set $2W_{0x}$, $2W_{0y}$, z_{0x}, z_{0y}, M_{0x}^2, M_{0y}^2. The other four address the rate of twist of the phasefront and spot pattern with propagation distance, the generalized radii of curvature of the wavefronts, and the orbital angular momentum[43] carried per photon by the beam. Beam classes are defined[44] by values of invariants calculated from the 10 second-order moments. A simple association of the resultant shapes for the beam envelopes with each class is *not* immediately available from these definitions.

Twisted phase beams have been generated by inserting an appropriate computer-designed diffractive optical element into an ordinary beam.[45,46] The first report of a beam from an "ordinary" laser (not one deliberately perturbed to produce a twisted phase) that required all ten matrix elements for its characterization, was recent.[18] Nonorthogonal beams can be expected to arise in nonorthogonal resonators,[5] such as in twisted ring resonators. Until instruments are developed[17] to measure all elements of the beam matrix **P** and then used to characterize beams from many lasers, the now hidden fraction of laser beams with general astigmatism will not be clear. The techniques discussed in previous sections are the methods that can be used together with various auxiliary optics to measure these second-order moments.

9 APPLICATIONS OF THE M² MODEL TO LASER BEAM SCANNERS

This section applies previous concepts and results to determine appropriate specifications for a laser used in an industrial scanning system, by working backward from the beam properties needed at the work surface. This example shows how parts of the M^2 model interact in a system design and how they can be used individually to solve simpler problems.

9.1 A Stereolithography Scanner

The example analyzes an actual stereolithography scanning system, shown in Fig. 17. A multimode ultraviolet beam (of 325 nm wavelength) writes on a liquid photopolymer surface under computer control, selectively hardening tiny volume elements of plastic to build up a three-dimensional part. After a 1/4 mm thick slice of the part is completed, a jack supporting the growing part inside the vat of liquid lowers the part and brings it back to 1/4 mm below the surface for the next slice to be written. Parts of great complexity can be formed overnight directly from their CAD file specifications. This process is called stereolithography and it has spawned the "rapid-prototyping" industry.

Beam characteristics for the laser are shown in the data report of Fig. 17. Many of the scanning system design elements used in this analysis are available in the literature.[47–50] The beam from the laser is expanded in an adjustable telescope that also focuses the spot on the liquid surface at the optimum beam spot size[47] for the solidification process, $2W_{02} = 0.25$ mm $\pm$ 10% at the vat, measured with a slit profiler.

Notice first that the system geometry defines a maximum M^2 for the laser beam in this application. The rapidly moving y-scan mirror benefits from a low moment of inertia and has a small diameter A. This is the minimum diameter needed to just pass the expanded beam incident on the mirror, making the beam diameter at the mirror $2W_A$ smaller than A only by some safety factor γ or $2W_A = A/\gamma$. From this mirror, the beam is focused on the liquid surface below at the throw distance T shown in Fig. 17. The maximum convergence angle of the beam focused on the vat surface is therefore $\Theta_2|_{max} = A/\gamma T$ (a larger angle would overfill the y-scan mirror). The focused beam waist diameter, given above, is $2W_{02}$; a diffraction limited beam of that waist diameter – a normalizing gaussian – has a divergence angle of $\theta_n = 2\lambda/\pi W_{02}$. This defines a maximum M^2 for this application by Eq. (22) of $M^2|_{max} = \Theta_2|_{max}/\theta_n = \pi W_{02} A / 2\lambda \gamma T$.

This may be evaluated in two different ways. From scaling a photograph of the system,[49] an estimate for T can be made as between 0.6 and 0.7 m, or a reasonable value is $T = 0.65$ m. The y-scan mirror diameter A is likely that of a small standard substrate, such as $A \sim 7.75$ mm, and a likely safety factor is about $\gamma \sim 1.5$, yielding $2W_A \sim 5.2$ mm and $M^2_{slit} \sim 4.8$. This rough estimate is refined below. This beam quality is given in slit "units" because this is the currency for the focal diameter at the vat; the assumption being made that the value of $2W_A$ used is also in slit units for this estimate.

Alternatively, the beam diameter A/γ can be determined working back from the vat to the y-scan mirror since it is known that the laser of Fig. 17 is designed for this application and that the measured data (given in the figure) are within the nominal beam specifications. Those measurements are in knife-edge currency[9] (see Sec. 9.4). Once the knife-edge waist diameter at the vat is found, so is $\theta_n = 2\lambda/\pi W_{02}$ and $2W_A = T\Theta = TM^2\theta_n$, all in knife-edge units. A diameter conversion is thus required to bring the diameter at the waist into knife-edge units for a consistent currency.

9.2 Conversion to a Consistent Knife-Edge Currency

By Eq. (48), for any diameter definition, i, the ratio D_i/M_i equals the embedded gaussian diameter $2w$ and therefore the conversion from slit to knife-edge diameters at the vat is just $D_{ke} = D_{slit}(M_{ke}/M_{slit})$. The square root of the beam quality M_{ke} is known from the report, $M_{ke} = \sqrt{(5.24)} = 2.289$. Here the R or "round beam" column value was used, the equivalent cylindrical beam constants as discussed later in Sec. 9.4. To determine M_{slit}, use is made of the expression just above Eq. (50) relating any M_i to any M_j for different

The M² Model

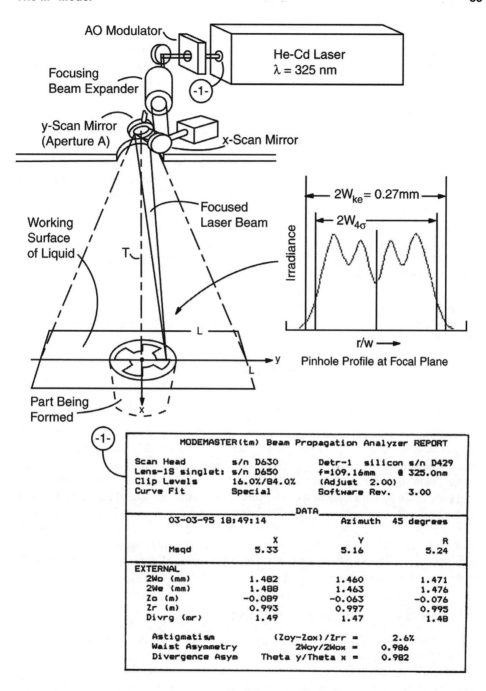

Figure 17 A stereolithography scanning system based on a helium–cadmium ultraviolet multimode laser. The pinhole focal plane profile (upper inset) shows the irradiance profile at the surface of the liquid photopolymer. The printout from the commercial beam propagation analyzer (lower inset) applies to the beam at the laser output, location (-1-). Laser data courtesy of Melles Griot, Inc.

diameter definitions i and j. This conversion formula requires knowledge of the M^2 of the starting method j; M^2 is known here from knife-edge measurements, so $j =$ knife-edge. The desired ending method is a slit measurement, $i =$ slit. Then Eq. (50) gives the required conversion constant, in terms of the conversion constants to second-moment diameters from Table 1, as:

$$c_{\text{ke} \to \text{slit}} = c_{ji} = c_{j\sigma}/c_{i\sigma} = (c_{\text{ke} \to \sigma})/(c_{\text{slit} \to \sigma}) = (0.813)/(0.950) = 0.856$$

This gives $(M_{\text{slit}} - 1) = 0.856(M_{\text{ke}} - 1) = 1.103$, thus $M_{\text{slit}} = 2.103$ and $M_{\text{slit}}^2 = 4.423$. Then Eq. (48) yields the focal diameter at the vat in knife-edge measurements, $2W_{02\text{ke}} = 0.272$ mm, a knife-edge to slit diameter ratio of 1.088 for this beam. The "normalizing gaussian" divergence angle above is then evaluated as $\theta_n = 1.521$ mr, the maximum convergence angle is larger than θ_n by $M^2 = 5.24$, making the beam diameter at the y-scan mirror $2W_A = TM^2\theta_n = 5.180$ mm, all in knife-edge units.

For comparison, using the knife-edge to second-moment conversion constant from Table 1 and Eq. (47) gives the second-moment beam quality and beam diameter at the vat of $M_{4\sigma}^2 = 4.19$ and $2W_{02}|_{4\sigma} = 0.243$ mm. The irradiance profile in Fig. 17 shows the relative size of the second-moment diameter to the knife-edge diameter. It is evident that the former would require a larger safety factor γ than the latter if used in estimating a safe minimum mirror aperture.

For the remainder of this section, diameters are all from knife-edge measurements and for simplicity the subscripts indicating this are suppressed.

9.3 Why Use a Multimode Laser?

What is the advantage of a multimode laser in this application? First, the critical optic, the scan mirror of diameter A, required for the larger multimode beam diameter is of reasonable size, so it is possible to use one here. The significant advantage is seen from the product data sheet for this laser (Melles Griot Model 74 Helium–Cadmium laser): with single isotope cadmium used in the laser (the X models on the data sheet) the multimode power is 55 mW, the fundamental mode power is 13 mW, *a ratio of 4.2 times*. With natural isotopic mix cadmium, the numbers are 40 mW and 8 mW, *a ratio of 5 times*. Thus the laser's output power is roughly proportional to its M^2, making the multimode laser considerably smaller and less expensive than a fundamental mode laser would be at the power level required for this application.

9.4 How to Read the Laser Test Report

Notice that the beam quality number used above was from the "R" column (for round mode) of the report shown in Fig. 17. These are the beam constants for the equivalent cylindrical beam discussed in Sec. 8.2, the best theoretical average[6,40,42] of the X and Y column constants for the two principal propagation planes on the report. Since there is less than 4% difference between M_x^2 and M_y^2, it is appropriate to use the average values in the R column and treat the beam as round for this exercise. The fact that the report is all in knife-edge units is signified by the "clip-levels" line reading 16%/84% (adjust: times 2.00) as explained in Ref. 9. The EXTERNAL label means these constants are for the original beam external to the instrument, after the lens transform has been done from the constants measured for the INTERNAL auxiliary beam inside. Next, listed for the two principal

The M² Model

propagation planes in the X and Y columns, and the equivalent cylindrical beam in the R column, are: the external beam waist diameter $2W_0$; the beam diameter $2W_e$ at the instrument's reference surface (its entrance plane for the beam); the waist location z_0 with respect to the reference surface with negative values being back towards the laser; the Rayleigh range z_R; and the beam divergence. Lastly, the significant beam asymmetry ratios are listed, with the astigmatism normalized to the equivalent cylindrical beam's Rayleigh range.

The whole report is readily converted into a different currency with a diameter a factor of τ larger, if desired, by multiplying the M^2 values by τ^2, the diameters and the divergence by τ, and leaving the z_0, z_R, values and the asymmetry ratios unchanged.

9.5 Replacing the Focusing Beam Expander with an Equivalent Lens

The beam expander of Fig. 17, when left at a fixed focus setting, can be replaced with an equivalent thin lens placed at the y-scan mirror location, with the laser moved back a distance z_{01} from the lens, as shown in Fig. 18(b). The propagation over z_{01} expands the beam to match the spot size at this mirror. To find z_{01}, Eq. (16) is used to get the propagation distance for the required beam expansion of $\rho = 2W_A/2W_{01} = (5.180 \text{ mm})/(1.471 \text{ mm}) = 3.521$. Here $2W_{01}$ is the laser's waist diameter in 1-space, on the input side of the equivalent lens, from the report of Fig. 17. From Eq. (16), $\rho = \sqrt{[1 + (z_{01}/z_{R1})^2]}$, yielding $z_{01}/z_{R1} = \sqrt{[\rho^2 - 1]}$, and with the 1-space Rayleigh range $z_{R1} = 0.995$ m taken from the report, there results $z_{01} = 3.538$ m also shown in Fig. 18(b).

The equivalent lens focal length $f_{\text{equiv}} \equiv f$ is next properly chosen to bring this beam to a focus at the vat. Since the waists on either side of the equivalent lens are known, by Eq. (26), the required transformation constant Γ is also known. This then gives by Eq. (24) a quadratic equation solvable for f

$$\Gamma = (2W_{02}/2W_{01})^2 = f^2/[(z_{01} - f)^2 + z_{R1}^2]^2 \tag{64}$$

yielding

$$f = z_{01}[\Gamma/(\Gamma - 1)]\{1 \pm \sqrt{[1 - [(\Gamma - 1)/\Gamma](1 + z_{01}^2/z_{R1}^2)]}\} \tag{65}$$

Inserting $2W_{02} = 0.272$ mm and $2W_{01} = 1.471$ mm in Eq. (64) gives $\Gamma = 0.03422$, and this in Eq. (65) produces $f = f_{\text{equiv}} = 0.5511$ m. In what follows, a precise value of z_{02} which corresponds to T in Fig. 17 is needed and the only value at hand is the previous estimate of $T = 0.65$ m for the y-scan mirror to vat distance. A precise value is needed consistent with the quantities used in the lens transform z_{01} and f_{equiv} above. This is given by Eq. (28), now that Γ, z_{01}, and f_{equiv} are known, as $z_{02} = 0.6472$ m. This also shows that the original estimate of T was reasonable. The nominal values for the quantities involved in the equivalent lens transform are shown in Fig. 18(b). The effect of perturbing the nominal values is studied below.

9.6 Depth of Field and Spot Size Variation at the Scanned Surface

With the equivalent lens transform defined, questions relating the input beam to the scanned beam can be answered. First, what is the amount of defocus over the scanned field? From Eq. (27), the Rayleigth range in 2-space at the vat is $z_{R2} = \Gamma z_{R1} = 3.404$ cm.

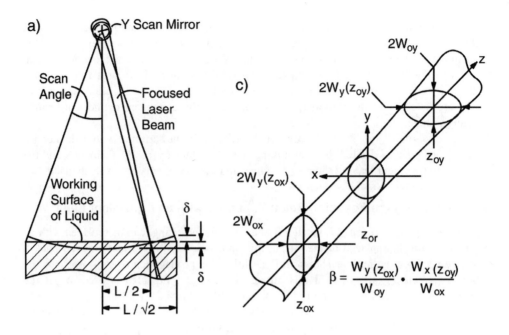
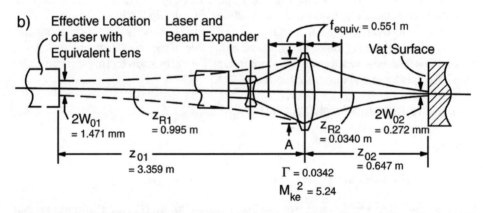

Figure 18 Analysis of the stereolithography system. (a) Optimum focus to minimize spot size change over the working surface. For clarity the scan angle shown is larger than the actual scan angle of $\pm 15°$. (b) Replacement of the focusing beam expander with an equivalent thin lens of focal length f_{equiv}. Parameters for the equivalent lens transform of the unperturbed beam are shown. (c) Definition of an out-of-roundness parameter, $\beta \equiv$ (quadratic ratio of astigmatic diameters) for the focal region of an astigmatic beam. The quantities shown are all for the vat-side focal region or 2-space.

The longest radial scan distance is to the corner of the square vat of side length $L = 250$ mm (Fig. 17), a distance of $\sqrt{2}L/2 = L/\sqrt{2}$. The variation in distance from the y-scan mirror to the corner of the square vat's working surface is $\sqrt{[T^2 + (L/\sqrt{2})^2]} - T = 2.371$ cm, or 0.696 times the vat side Rayleigh range. By Eq. (16), from the center of the vat to the far corner, the spot size of the beam will grow by a factor of $\sqrt{[1 + (0.696)^2]} = 1.219$. The simplest way to reduce this range is to focus the

The M^2 Model

beam at the middle of a side edge of the vat, at a radial distance of $L/2$ [see Fig. 17 and Fig. 18(a)]. This splits the defocus amount 2δ equally among the corners and the middle, to 11% maximum change in spot diameter on the liquid surface over the scanned field. Equivalently, the liquid level could be raised by δ. However, for simplicity the focal distance will be left at T here for the remainder of this analysis.

9.7 Laser Specifications to Limit Spot Out-of-Roundness on the Scanned Surface

Next, the inverse lens transform, from the vat side, back to the laser side, is used to transfer scanning beam specifications into laser beam specifications. From Eq. (31), for the transform equations going from 2-space to 1-space, use the inverse transformation constant $\Gamma_{21} = 1/\Gamma_{12}$.

Since the transformation constant depends on both the input waist location and the Rayleigh range, in general, beams without astigmatism but with some other asymmetry when transformed become astigmatic as the results below show. The plan, starting from the nominal, round, equivalent cylindrical beam of Fig. 17 transformed to a round beam at the vat, is to perturb the beam at the vat to have a $\sim 10\%$ out-of-round spot. This beam is then transformed back to the laser to see which 1-space variables change and by how much, to account for the perturbation on the scanned beam side of the lens. The 10% out-of-roundness of the scanned spot is deemed acceptable because that amount of growth in spot diameter over the field was found acceptable above.

The perturbations are made as equal changes of opposite sign in the two independent propagation planes. For example, 10% out-of-roundness due to waist asymmetry is accomplished with a +5% change in W_{02y} and a -5% change in W_{02x}. The resulting changes in the 1-space beam constants are not completely symmetrical, and this illustrates the nonlinearity of the beam-lens transform. The effect of perturbing a constant in only one principal propagation plane is given directly in the following tables by the percentage changes, the column in parentheses, for 1-space shown for that plane. Because the propagation planes are independent, so are the percentage changes in each plane.

9.7.1 Case A: 10% Waist Asymmetry

Assume $2W_{02x}$ is reduced 5% (to 0.259 mm), and $2W_{02y}$ is raised 5% (to 0.286) to give a *waist asymmetry* different from one by 10%. To calculate the effect on the input beam, first the new Rayleigh ranges for the beam at the vat are found as $z_{R2x} = 3.088$ cm (reduced 10%) and $z_{R2y} = 3.753$ cm (increased 10%). For each of these a new $1/\Gamma$ for the inverse transform is computed from Eq. (24), followed by the remaining constants through Eqs. (26)–(28). The results for the 1-space beam constants, and their percentage change shown in parentheses are summarized in Table 2. The initial value of $1/\Gamma$ is 29.2259. In the table A_s/z_{Rr} stands for the normalized astigmatism $A_s/z_{Rr} = (z_{01y} - z_{02x})/z_{R1r}$, where z_{R1r} is the Rayleigh range in 1-space of the equivalent cylindrical beam.

The +10% pure waist asymmetry at the vat (i.e., accompanied by no astigmatism or divergence asymmetry) for the most part transforms through the lens to a corresponding +8% waist asymmetry at the laser. The same is true for the divergence asymmetry. The different waist diameters at the vat generate different Rayleigh ranges there and in the lens transform produce a -12% normalized astigmatism at the laser. Specify the laser to have less than these asymmetries to keep the scanned beam out-of-roundness below 10%.

Table 2 Laser Constants Corresponding to *10% Waist Asymmetry* at the Scanned Surface

Quantity	x	y	Ratios (y/x)	Ratio was:
$1/\Gamma$	29.816 (+2.0%)	28.540 (−2.4%)	0.957 (−4.3%)	1
$2W_{01}$ (mm)	1.415 (−3.8%)	1.526 (+3.8%)	1.079 (+7.9%)	1
z_{01} (m)	3.416 (+1.7%)	3.294 (−2.0%)	$A_s/z_{Rr} = -12.3\%$	0
z_{R1} (m)	0.921 (−7.5%)	1.071 (+7.7%)	1.163 (+16.4%)	1
Θ_1 (mr)	1.537 (+3.8%)	1.425 (−3.7%)	0.927 (−7.3%)	1

9.7.2 Case B: 10% Divergence Asymmetry

Here it is assumed M_x^2 is reduced 5% and M_y^2 is increased by 5% to give a +10% change in the 2-space *divergence asymmetry* without changing the waist asymmetry $W_{02y}/W_{02x} = 1$. By Eq. (18) or (19) the Rayleigh ranges on the vat side of the lens change inversely with their M^2 to make them $z_{R2x} = 3.088$ cm, $z_{R2y} = 3.753$ cm. Applying Eqs. (24)–(28) to each principal plane produces Table 3, the results for the 1-space beam constants and their percentage change.

The divergence asymmetry of the beam at the vat carries through the lens to the laser, and implies some astigmatism is necessary at the laser (but half as much as Case A) to get pure divergence asymmetry at the vat.

9.7.3 Case C: 12% Out-of-Roundness Across the Scanned Surface Due to Astigmatism

A little discussion is required to define an out-of-roundness parameter for the focal region of an astigmatic beam in general before applying the concept to the focus at the vat. It has already been shown (Sec. 9.6) that the path length to the liquid surface changes over the scanned field by 2.37 cm. This path change causes a spot size variation of 21.9% if the spot is focused at the center of the vat, and 11% if focused, to reduce the variation, at the middle of a vat edge. On top of this, there is an out-of-round change in the spot if the beam is astigmatic. The fastest change of shape with z of the elliptical spots in a beam with pure astigmatism [Fig. 16(a)], takes place between the two astigmatic waists in the focal region, where the beam in the stereolithography system is working. Suppose the astigmatic distance $z_{02y} - z_{02x}$ is matched to the path length change of 2.37 cm but the edge focus is used to split this distance [see Fig. 18(a)]. This makes the largest path between an astigmatic waist and the liquid working surface anywhere in the field be 1.19 cm.

Table 3 Laser Constants Corresponding to *10% Divergence Asymmetry* at the Scanned Surface

Quantity	x	y	Ratios (y/x)	Ratio was:
$1/\Gamma$	28.896 (−1.1%)	29.532 (+1.0%)	1.022 (+2.2%)	1
$2W_{01}$ (mm)	1.463 (−0.6%)	1.478 (+0.5%)	1.011 (+1.1%)	1
z_{01} (m)	3.328 (−0.9%)	3.389 (+0.9%)	$A_s/z_{Rr} = +6.1\%$	0
z_{R1} (m)	1.033 (+3.8%)	0.958 (−3.8%)	0.927 (−7.3%)	1
Θ_1 (mr)	1.416 (−4.3%)	1.544 (+4.3%)	1.091 (+9.1%)	1

The M^2 Model

Then from Eq. (16) and Fig. 18(c)

$$W_{2x}(z_{02r})/W_{02x} = \sqrt{[1 + (1.19/3.40)^2]} = 1.059$$

where z_{02r} is the equivalent cylindrical beam waist location halfway between the x and y waist locations. The spot at the vat only goes to 5.9% out-of-round, but the orientation of the out-of-round ellipse is along the y-axis in the corners where the liquid is below z_{02r} and along the x-axis in the middle of the field where the liquid is above z_{02r}. This can have an unpleasant effect on the part, because the texture of the x- and y-formed surfaces varies. Therefore, an adequate out-of-round parameter for the focal region of an astigmatic beam can be defined [Fig. 18(c)] as $\beta \equiv$ quadratic ratio of astigmatic diameters, where the product of the x-direction and y-direction out-of-round diameter ratios is

$$\beta = [W_{2y}(z_{02x})/W_{02y}][W_{2x}(z_{02y})/W_{02x}] \tag{66}$$

The ratios are evaluated at the two astigmatic waist locations as indicated in Eq. (66). From the above, $\beta = (1.059)^2 = 1.12$ for an astigmatic distance equal to the scanned depth of field; this is taken here as "12% out-of-roundness due to astigmatism" for the final example.

The calculations proceed in this example with $z_{02x} = (0.6472 - 0.0119)$ m $= 0.6353$ m and $z_{02y} = (0.6472 + 0.0119)$ m $= 0.6591$ m, with the other 2-space beam parameters left at their unperturbed values of Fig. 18(b). Table 4 gives the results for the 1-space beam.

This type of asymmetry, transformed back to the laser side of the equivalent lens, is devastating to the 1-space beam constants. More correctly, it would take devastating input beam characteristics to produce this large a "quadratic-ratio-of-astigmatic-diameters" parameter. There are large percentage changes in 1-space waist asymmetry, astigmatism, and divergence asymmetry. Actual lasers with asymmetries this large would be rejected by the laser manufacturer, and the scanner manufacturer would not have to deal with them. Lasers with sufficient beam asymmetry to give $\beta = 1.12$ at the scanned surface would not make it into the field.

In conclusion, the strictest specifications found for the laser to meet upon incoming testing were from Case A, yielding 10% waist asymmetry at the vat surface. To stay below this out-of-roundness at the vat, the analysis gave bounds at the laser of less than 12% normalized astigmatism and less than 8% waist and divergence asymmetry. These values were easily met by the laser tested and reported in Fig. 17. In an actual situation of setting

Table 4 Laser Constants Corresponding to a *12% Out-of-Roundness Across the Scanned Surface Due to Astigmatism* ($\beta = 1.121$)

Quantity	x	y	Ratios (y/x)	Ratio was:
$1/\Gamma$	36.794 (+25.9%)	23.708 (−18.9%)	0.644 (−35.6%)	1
$2W_{01}$ (mm)	1.651 (+12.2%)	1.345 (−9.9%)	0.803 (−19.7%)	1
z_{01} (m)	3.651 (+8.7%)	3.110 (−7.4%)	$A_s/z_{Rr} = -52.4\%$	0
z_{R1} (m)	1.253 (+25.9%)	0.807 (−18.9%)	0.644 (−35.6%)	1
Θ_1 (mr)	1.318 (−11.0%)	1.641 (+10.9%)	1.216 (+24.6%)	1

laser specifications, several more examples should be run, including cases starting on the laser side and calculating the asymmetries that result in the scanning beam. Readers journeying this far into this applications section should now have sufficient analytical tools provided by the M^2 model to complete those calculations themselves.

10 CONCLUSION: OVERVIEW OF THE M^2 MODEL

The M^2 model is the simple concept that real, mixed mode beams can be described by generalizing the equations describing the fundamental mode beam. The mixed mode beam is of larger diameter – for all propagation distances z – by the factor M than the fundamental mode beam from the same resonator, the embedded gaussian beam. Thus the generalization takes the form of replacing the $1/e^2$-radius w of the embedded gaussian beam by W/M, where W is the radius of the mixed mode beam. This replacement generalizes both the beam propagation and beam-lens transform equations.

The mixed mode, with waist diameter $2W_0$, being larger than the embedded gaussian by the factor M for all z, diverges at an M times larger rate. All diffraction limited beams have a gaussian irradiance profile, and one of waist diameter $2W_0$, being M times larger than the embedded gaussian diverges at a rate $1/M$ as fast as the embedded gaussian. Hence the mixed mode divergence is M^2 times larger than the diffraction limit. This identifies M^2 as a beam invariant unchanging in free space propagation or transmission through non-aberrating lenses, and as a measure of the mixed mode beam quality. An M^2 of unity is the highest quality, a diffraction-limited beam, and beams with larger values have increasing degrees of higher order mode content or wavefront aberration.

To apply this analytical description of a mixed mode beam, its M^2 must first be measured, and here the simplicity of the ideas becomes more complex. The measurement requires finding the scale length for expansion of the beam diameter with propagation, z_R, the Rayleigh range. Several diameters at well chosen z locations on both sides of the waist are determined and this data is fit to the correct hyperbolic form. The fit gives three beam constants – the beam quality, the waist diameter, and the waist location – for each independent and orthogonal principal propagation plane.

The first additional complexity is that different definitions give different numerical values for the diameters for the mixed mode and the higher order modes it contains. Beam diameters are still measured from beam irradiance profiles, but different profiling masks (pinholes, slits, knife-edges, or centered-circular-apertures) all give different shaped profiles for higher order modes and hence different diameters. Care is required to keep track of which measurement method is in use and to not mix results from different methods. A standard diameter definition – the second-moment diameter, four times the standard deviation of a pinhole irradiance distribution of the beam – has been recommended by the International Organization for Standardization. However, this diameter is computation-intensive and difficult to measure reliably because of sensitivity to noise on the wings of the profile. Therefore, conversion rules have been developed applicable to cylindrically-symmetric mixed mode beams permitting measurements done in one diameter method to be converted to those from another. The basis of these rules is our observation that higher order modes turn on and off in a characteristic sequence as the diameter of the circular limiting aperture in the resonator is opened. This associates with the increasing second-moment M^2 a unique set of mode fractions, allowing accurate rules to be derived.

The second additional complexity is that the diameter measurements and curve-fits done to determine M^2 may give unreliable answers unless several pitfalls in the process are avoided. Chief among these is that the mixed mode waist must be accurately located and its diameter physically measured and not inferred or assumed. Since the waists of most lasers are buried inside the resonator, this requires the use of an auxiliary lens to form an auxiliary beam with an accessible waist. The constants determined for this auxiliary beam then are transformed back to those for the original beam by means of the beam-lens transform equations. Commercial instruments that do this automatically are available.

Beams with pure forms of the classic asymmetries have been illustrated, those with only astigmatism, or waist asymmetry, or divergence asymmetry. Beams with combinations of asymmetries may be represented by pairs of propagation plots, one for each principal propagation plane. Beam asymmetries can also be interpreted as deviations from a theoretical "best weighted average" round beam, the equivalent cylindrical beam. There are also beams not directly covered in the M^2 model whose principal propagation planes twist in space like a twisted ribbon – beams with "general astigmatism."

Lastly, the M^2 model was demonstrated by analyzing an actual laser beam scanning system use in stereolithography. Asymmetries causing out-of-round spots on the scanned surface were analytically projected back through the delivery system to determine the size of the corresponding asymmetries at the laser source.

There are many applications of the M^2 model. It allows quantification of mode specifications and provides a basis to test to these specifications. Its use permits design of multimode lasers and their beam delivery systems. The beam transformations occurring in nonlinear optics can be better analyzed. The divergence of a high M^2 laser beam can be matched into the acceptance angle of a high numerical aperture fiber, to take advantage of the lower cost per unit of output power of a multimode laser. These are just a few of many applications, all with the back-up of commercial instrumentation to make the beam measurement process easy and efficient. This chapter has provided the analytical tools to make these applications realities.

ACKNOWLEDGMENTS

The authors would like to thank Prof. Emeritus A. E. Siegman, Stanford University, for many years of enlightening interactions on this subject. Thanks also, for helpful discussions, to Gerald Marshall, the editor of this book, who always has another intriguing question, and to G. Nemes of Astigmat, who taught us about beams with general astigmatism. Lastly, David Bacher and John O'Shaughnessy of Melles Griot, Inc., and especially Gerald Marshall contributed very helpful and constructive reviews of this manuscript.

GLOSSARY

Astigmatism, general The property of beams having elliptical cross-sections for all z, with the principal axes of the ellipses rotating with propagation along the beam axis (nonorthogonal beams; "twisted" beams).

Astigmatism, normalized The difference in waist locations for the two independent principal propagation planes divided by the Rayleigh range of the equivalent cylindrical beam, $A_s/z_{Rr} = (z_{0y} - z_{0x})/z_{Rr}$, usually expressed in percent.

Astigmatism, simple   Having different waist locations in the two principal propagation planes, $z_{0x} \neq z_{0y}$.

Asymmetric divergence   Having different divergence angles $\Theta_x \neq \Theta_y$ in the two principal propagation planes.

Asymmetric waists   Having different waist diameters in the two principal propagation planes, $2W_{0x} \neq 2W_{0y}$.

Beam caustic surface   The envelope of the beam swept out by rotating the curve of the beam radius $W(z)$ vs. propagation distance about the propagation axis z. When a plane containing the z-axis and at an angle α to the x-axis cuts the caustic surface, the intersection gives the propagation plot for azimuth α. See the discussion of Fig. 16(a).

Beam, equivalent cylindrical   A cylindrically-symmetric beam constructed mathematically in the M^2 model from the beam constants measured in the two principal propagation planes of an asymmetric beam; see the explanation of Fig. 16(a). The propagation plot for the equivalent cylindrical beam is obtained from the beams of Fig. 15 by slicing them along the z-axis and at a 45° inclination to the x- or y-axes. The best cylindrically-symmetric average beam for a beam with asymmetry. The subscript r is used to denote the constants for this beam, for round or radial symmetry.

Beam, gaussian   A uniphase beam with spherical wavefronts whose transverse irradiance profiles everywhere have the form of a gaussian function. Such an idealized beam would be diffraction limited, $M^2 = 1$, a condition that can only be approached by real beams.

Beam, idealized   The abstract mathematical description of a beam (which can have $M^2 = 1$).

Beam propagation analyzer   An instrument that measures beam diameters as a function of propagation distance, displays the $2W(z)$ vs. z propagation plot, and curve-fits this data to a hyperbola to determine beam quality M^2, waist location z_0, and waist diameter $2W_0$.

Beam propagation constant: M^2   So called because replacing the fundamental mode radius $w(z)$ in its propagation equation by $w(z) = W(z)/M$ predicts the propagation of the mixed mode, of radius $W(z)$.

Beam quality   The quantity M^2, so called because a real beam has M^2 times the divergence of a diffraction limited beam of the same waist diameter, see "normalizing gaussian."

Beam, real   An actual beam; all have slight imperfections and thus $M^2 > 1$.

Clip-width   Distance between the points at a specified fraction of the height of the highest peak on an irradiance profile, such as $1/e^2 = 13.5\%$.

Conversions, beam diameter   Empirical rules derived for beams of cylindrical symmetry, to convert diameters measured by one method to those measured by another, such as slit diameters to knife-edge diameters.

Convolution error   Contribution to the measured diameter from the finite size of the scanning aperture. Important consideration for pinhole and slit measurements.

Cut   Beam diameter measurement, from the cutting action of a profiler's scanning aperture.

Diameter, $1/e^2$ Beam diameter defined by the aperture translation distance between clip-points on an irradiance profile at a height of $13.5\% = 1/e^2$ relative to the highest peak at 100%.

The M^2 Model

Diffractive overlay Interference from high angle rays overlapping the beam, diffracted from the limiting aperture in the resonator. This can distort profiles taken within a Rayleigh range of the laser output coupler.

Eigenfunctions A set of functions f_n associated with a linear operator $\mathbf{Q}$ satisfying $\mathbf{Q}f_n = c_n f_n$, where the c_n are scalar constants (the eigenvalues). Because of this self-replicating property these functions occur in many physical problems; for example, the laser mode functions also describe the harmonic oscillator and hydrogen wave functions in quantum mechanics.

Embedded gaussian The fundamental mode of the resonator that generates a mixed mode beam. The mixed mode beam diameter is, for all z, M times larger than the embedded gaussian beam diameter.

Far-field The propagation region(s) of a beam many Rayleigh ranges away from the waist locations. In the far-field, the transverse extent of the beam grows at a constant rate with increasing distance from the waist.

Four-cuts method The simplest method for determining M^2, requiring only four well-chosen diameter measurements straddling the waist location and at the waist location.

Fresnel number The square of the radius of the limiting aperture in a resonator, divided by the mirror separation and the wavelength. As the aperture is opened and this number increases, modes of higher order oscillate and join the mix of modes.

Gaussian A mathematical function of the form $\exp(-x^2)$; see also "beam, gaussian."

Hermite–gaussian function An eigenfunction of the wave equation including diffraction, that describes beams of rectangular symmetry. Has the form of a gaussian function times a pair of Hermite polynomials of orders (m, n).

Invariant, beam A quantity that is unchanged by propagation in free space or transmission through ordinary, non-aberrating, optical elements (lenses, Brewster windows, etc.).

Irradiance The power per unit cross-sectional area of the beam.

Laguerre–gaussian function An eigenfunction of the wave equation including diffraction, that describes beams of cylindrical symmetry; of the form of a gaussian function times a generalized Laguerre polynomial of order (p, l).

M^2 The product of waist diameter times divergence angle, normalized so this irreducible minimum is always unity, regardless of wavelength. A beam invariant. Also the "times diffraction limit" number, the beam quality, and the beam propagation factor.

Mode The characteristic frequencies and transverse irradiance patterns of beams formed in laser oscillators, described by Hermite–gaussian and Laguerre–gaussian functions, denoted by the symbols $\text{TEM}_{m,n}$, $\text{TEM}_{p,l}$ with m, n or p, l the order numbers of the function's polynomials.

Mode, degenerate Modes with the same optical frequency, and therefore, order numbers.

Mode, donut A starred mode, TEM_{01}^*, with the second-lowest diffraction loss through a circular limiting aperture, and an irradiance profile with a hole (null) in the center (see Fig. 1).

Mode, fundamental The TEM_{00} mode, with a gaussian irradiance distribution, a single-spot peaked profile, and with $M^2 = 1$ in the limit of perfection. The lowest-order mode. The smallest diameter beam from a given resonator. Mode with the lowest diffraction loss through a circular limiting aperture.

Mode, higher order Any mode of order number greater than that of the fundamental mode.

Mode, longitudinal   A mode of frequency $q(c/2L)$, where c is the speed of light and q is a large integer equal to the number of beam wavelengths that fit in the round trip path $2L$ of the resonator. The $(q + 1)$th longitudinal mode has a frequency $(c/2L)$ higher than the qth; each longitudinal mode is associated with a given transverse mode.

Mode, lowest order   The fundamental mode, of order number one.

Mode, mixed   An incoherent superposition of pure modes, all from the same resonator, with a diameter $2W$ that is M times larger for all z than $2w$, the fundamental mode diameter from the set. Also called a real beam as only idealized beams have $M^2 = 1$ (indicating zero higher order mode content).

Mode order number   For Hermite–gaussian modes $(m + n + 1)$; for Laguerre–gaussian modes, $(2p + l + 1)$; the order numbers determine the mode frequencies and phase shifts, and give the mode's beam quality $M^2_{4\sigma}$ measured in second-moment units.

Mode or spot pattern   The two-dimensional pattern of the irradiance distribution as would be viewed on a flat surface inserted normally in the beam.

Mode, pure   Any transverse mode that is *not* a mixture of modes of different orders.

Mode, starred   A circularly symmetric mode that is a composite of two degenerate modes combined in space and phase quadrature, that is, superposed with a copy of itself after a 90° rotation (see Fig. 1).

Mode, transverse   A mode, designated by the symbols TEM$_{m,n}$, TEM$_{p,l}$, whose transverse irradiance distribution is described by the Hermite–gaussian or Laguerre–gaussian functions of m, n or p, l order numbers.

Near-field   The beam propagation region(s) within a Rayleigh range from the waist location.

Noise-clip option   A test of the sensitivity to noise of the second-moment diameter computed from a pinhole profile, consisting of discarding any profile data with negative values after subtraction of the background, and looking for a change in the computed diameter.

Normalizing gaussian   A diffraction limited, idealized gaussian beam of the same waist diameter as a mixed mode real beam, whose divergence is used as the denominator in a ratio with the real beam's divergence to compute the real beam's M^2.

Paraxial   Close to the beam axis. Referring to a bundle of rays propagating at angles small enough with respect to the axis that the angle and its tangent are essentially equal.

Power-in-the-bucket   Alternate term for D_{86}, the variable-aperture beam diameter definition.

Principal diameters (of an elliptical spot)   The major and minor axes of the ellipse.

Principal propagation planes, independent   The two perpendicular planes containing the major and minor axes of an elliptical beam spot (x- and y-axes) and the propagation axis (z). In the M^2 model the three propagation constants in each of these two planes are independent.

Profile   The record of transmitted power vs. translation distance of an aperture or mask scanned across the beam.

Profile, knife-edge   A profile taken with a knife-edge mask, giving a tilted S-shaped curve.

Profile, pinhole   A profile taken with a pinhole aperture. Capable of showing all the irradiance highs and lows but requiring careful centering of the beam on the scanned track of the pinhole. Signal-to-noise ratio and convolution error are inversely dependent on the pinhole diameter, making the hole diameter an important consideration.

The M² Model

Profiler An instrument for measuring beam diameters, which scans a mask (pinhole, slit, or knife-edge) through the beam. Displays the profile, and (usually) reports the beam diameter on a digital readout as the scan distance, or clip-width, between pre-set clip-points on the profile.

Profile, slit A profile taken with a slit aperture, showing something of the irradiance highs and lows and *not* requiring centering of the beam to the scanned track. Signal-to-noise ratio and convolution error are counter-dependent on the slit width, making it an important consideration.

Propagation constants The set of parameters: waist diameter $2W_0$, waist location z_0, and beam quality M^2, in each of the two principal propagation planes that define how the transverse extent of a beam changes as it propagates.

Propagation plot The plot of beam diameter vs. propagation distance, $2W(z)$ vs. z. For the beams covered in the M^2 model, the form of this plot is a hyperbola.

Rayleigh range The propagation distance z_R from the waist location to where the wavefront reaches maximum curvature. Also the distance from the waist where the beam diameter has increased by $\sqrt{2}$. The scale length for beam expansion with propagation, $z_R = \pi W_0^2 / M^2 \lambda$.

Resonator The aligned set of mirrors providing light feedback in a closed path through the gain medium in a laser. Since the wavefront curvatures and surface curvatures must match at the mirrors, the resonator determines the mode properties of the beam.

Scan Movement of a mask or aperture transversely across a beam while recording the transmitted power; see "cut."

Second-moment diameter $D_{4\sigma}$, equal to four times the standard deviation, σ, of the transverse irradiance distribution obtained from a pinhole profile.

Second moment, linear The integral over the transverse plane of the square of the linear coordinate times the irradiance distribution, for example, $\langle x^2 \rangle$, used in calculating the variance of the distribution $\sigma^2 = \langle x^2 \rangle - \langle x \rangle^2$.

Second moment, radial The integral over the transverse plane of the square of the radial coordinate times the irradiance distribution measured outwardly from the centroid of the spot, for example, $\langle r^2 \rangle$, used in calculating the variance of the distribution $\sigma_r^2 = \langle r^2 \rangle$. In the integration r^3 weights the distribution since the area element is $dA = r\, dr\, d\theta$.

Spot The two-dimensional irradiance distribution or cross-section of a beam as seen on a flat surface normal to the beam axis.

Stigmatic Describes a beam that maintains a round cross-section as it propagates, or, more formally, a beam that maintains a rotationally symmetric irradiance distribution in free space. The opposite of astigmatic where cross-sections are elliptical at some locations z.

TDL, times diffraction limit number The number of times larger the divergence of a real beam is than that of a diffraction limited beam (called the normalizing gaussian) of the same waist diameter; $TDL = \Theta/\theta_n = M^2$. Also the factor by which a real-beam waist diameter is larger than that for a gaussian beam ($M^2 = 1$) converging at the same numerical aperture (NA).

TEM$_{mn}$ (For Transverse ElectroMagnetic wave). A symbol used to designate a transverse mode of rectangular symmetry described by a Hermite–gaussian function with polynomial orders m, n.

TEM$_{pl}$ (For Transverse ElectroMagnetic wave). A symbol used to designate a transverse mode of cylindrical symmetry described by a Laguerre–gaussian function with polynomial orders p, l.

Variable-aperture diameter D_{86} (or D_{xx}) the diameter of a centered circular aperture passing 86.5% (or xx%) of the total power in the beam.

Waist, beam   Minimum diameter for a beam. The wavefront is planar at the waist.

Waist diameters $2W_{0x}, 2W_{0y}$, the minimum diameters in each principal propagation plane.

Waist locations z_{0x}, z_{0y}, the points along the propagation axis where the minimum (waist) diameter(s) of the beam in each of the independent principal propagation planes are located.

Wave equation   Propagation of paraxial rays including the effect of diffraction are described by either the Fresnel–Kirchhoff diffraction integral equation of Boyd and Gordon[2] or the simple scalar wave equation used by Kogelnik and Li [1] and both have the Hermite–gaussian and Laguerre–gaussian functions as eigenfunction solutions.

REFERENCES

1. Kogelnik, H.; Li, T. Laser beams and resonators. Appl. Opt. 1966, *5*, 1550–1567.
2. Boyd, G.D.; Gordon, J.P. Confocal multimode resonator for millimeter through optical wavelength masers. Bell System Technical J. 1961, *40*, 489–508.
3. Marshall, L. Applications à la mode. Laser Focus 1971, *7*(4), 26–29.
4. Bastiaans, M.J. Wigner distribution function and its application to first-order optics. J. Opt. Soc. Am. 1979, *69*, 1710–1716.
5. Siegman, A.E. *Lasers*; University Science Books: Sausalito, CA, 1986; ISBN 0-935702-11-3.
6. Siegman, A.E. New developments in laser resonators. Proc. SPIE 1990, *1224*, 2–14.
7. Sasnett, M.W. Propagation of multimode laser beams – the M^2 factor. In *The Physics and Technology of Laser Resonators*; Hall, D.R., Jackson, P.E., Eds.; Adam Hilger: New York, 1989; Chapter 9, ISBN 0-85274-117-0.
8. Johnston, T.F., Jr.; Fleischer, J.M. Calibration standard for laser beam profilers: method for absolute accuracy measurement with a Fresnel diffraction test pattern. Appl. Opt. 1996, *35*, 1719–1734.
9. The Coherent, Inc., ModeMaster™. The manual for this instrument containing much useful information on this subject is available upon request from Coherent Instruments Division. 7470 S. W. Bridgeport Road, Portland, OR 97224.
10. Johnston, T.F., Jr. M^2 concept characterizes beam quality. Laser Focus 1990, *26*(5), 173–183.
11. Test methods for laser beam parameters: beam widths, divergence angle, and beam propagation factor, ISO/TC 172/SC9/WG1, ISO/DIS 11146, 1995, available from Deutsches Institut fur Normung, Pforzheim, Germany.
12. Lawrence, G.N. Proposed international standard for laser-beam quality falls short. Laser Focus World 1994, *30*(7), 109–114.
13. Sasnett, M. et al. Toward an ISO beam geometry standard. Laser Focus World 1994, *30*(9), 53.
14. Johnston, T.F., Jr.; Sasnett, M.W.; Austin, L.W. Measurement of "standard" beam diameters. In *Laser Beam Characterization*; Mejias, P.M., Weber, H., Martinez-Herrero, R., Gonzales-Urena, A., Eds.; SEDO: Madrid, 1993; 111–121.
15. Arnaud, J.A.; Kogelnik, H. Gaussian light beams with general astigmatism. Appl. Opt. 1969, *8*, 1687–1693.
16. Mansuripur, M. Gaussian beam optics. Optics and Photonics News 2001, *12*(1), 44–47.
17. Nemes, G.; Siegman, A.E. Measurement of all ten second-order moments of an astigmatic beam by the use of rotating simple astigmatic (anamorphic) optics. J. Opt. Soc. Am. 1994, *11*, 2257–2264.
18. Serna, J.; Encinas-Sanz, F.; Nemes, G. Complete spatial characterization of a pulsed doughnut-type beam by use of spherical optics and a cylinder lens. J. Opt. Soc. Am. 2001, *18*, 1726–1733.

19. Silfvast, W.T. *Laser Fundamentals*; Cambridge University Press: New York, 1996; Chapter 10, ISBN 0-521-55617-1.
20. Rigrod, W.W. Isolation of axi-symmetric optical resonator modes. Appl. Phys. Lett. 1963, 2, 51–53.
21. McCumber, D.E. Eigenmodes of a symmetric cylindrical confocal laser resonator and their perturbation by output-coupling apertures. Bell System Technical J. 1965, 44, 333–363.
22. Koechner, W. *Solid-State Laser Engineering*, 5th Ed.; Springer-Verlag: NY, 1999; Fig. 5.10.
23. Wolfram, S. *The Mathematica Book*, 3rd Ed.; Cambridge University Press: Cambridge, UK, 1996; ISBN 0-521-58889-8, 745, 763 pp.
24. Feng, S.; Winful, H.G. Physical origin of the Gouy phase shift. Opt. Lett. 2001, 26, 485–489.
25. Johnston, T.F., Jr. Beam propagation (M^2) measurement made as easy as it gets: the four-cuts method. Appl. Opt. 1998, 37, 4840–4850.
26. Belanger, P.A. Beam propagation and the ABCD ray matrices. Opt. Lett. 1991, 16, 196–198.
27. Serna, J.; Nemes, G. Decoupling of coherent Gaussian beams with general astigmatism. Opt. Lett. 1993, 18, 1774–1776.
28. Hecht, E. *Optics*, 2nd Ed.; Addison-Wesley Publishing Co.: Menlo Park, CA, 1987; ISBN 0-201-11609-X.
29. Kogelnik, H. Imaging of optical modes – resonators with internal lenses. Bell System Technical J. 1965, 44, 455–494.
30. Self, S.A. Focusing of spherical gaussian beams. Appl. Opt. 1983, 22, 658–661.
31. Herman, R.M.; Wiggins, T.A. Focusing and magnification in gaussian beams. Appl. Opt. 1986, 25, 2473–2474.
32. O'Shea, D.C. *Elements of Modern Optical Design*; John Wiley & Sons: New York, 1985; ISBN 0-471-07796-8, 235–237.
33. Wright, D.L.; Fleischer, J.M. Measuring Laser Beam Parameters Using Non-Distorting Attenuation and Multiple Simultaneous Samples. US Patent No. 5,329,350, 1994.
34. McCally, R.L. Measurement of Gaussian beam parameters. Appl. Opt. 1984, 23, 2227.
35. Sasnett, M.W.; Johnston, T.F., Jr. Apparatus for Measuring the Mode Quality of a Laser Beam. US Patent No. 5,100,231, March 31, 1992.
36. Taylor, J.R. *An Introduction to Error Analysis*; University Science: Mill Valley, CA, 1982; ISBN 0-935702-10-5.
37. Green, L. Automated measurement tool enhances beam consistency. Laser Focus World 2001, 37(3), 165–166.
38. Kogelnik, H.; Ippen, E.P.; Dienes, A.; Shank, C.V. Astigmatically compensated cavities for CW dye lasers. IEEE J. Quant. Electron. 1972, 3, 373–379.
39. Johnston, T.F., Jr.; Sasnett, M.W. The effect of pump laser mode quality on the mode quality of the CW dye laser. SPIE Proceedings 1992, 1834, Optcon Conference, Boston, 1992, Paper #29.
40. Johnston, T.F., Jr.; Sasnett, M.W. Modeling multimode CW laser beams with the beam quality meter. OPTCON, Boston, MA, 5 November 1990, Paper OSM 2.4.
41. Firester, A.H.; Gayeski, T.E.; Heller, M.E. Efficient generation of laser beams with an elliptic cross section. Appl. Opt. 1972, 11, 1648–1649.
42. Siegman, A.E. Laser beam propagation and beam quality formulas using spatial-frequency and intensity-moment analysis, distributed to the ISO Committee on test methods for laser beam parameters, August 1990, 32 p.
43. Simpson, N.B.; Dholakia, K.; Allen, L.; Padgett, M.J. Mechanical equivalence of spin and orbital angular momentum of light: and optical spanner. Opt. Lett. 1997, 22, 52–54.
44. Nemes, G.; Serna, J. Laser beam characterization with use of second order moments: an overview. In *DPSS Lasers: Applications and Issues, OSA TOPS*; 1998; Dowley, M.W. Ed.; 17, 200–207.
45. Piestun, R. Multidimensional synthesis of light fields. Optics and Photonics News 2001, 12(11), 28–32.

46. Kivsharand, Y.S.; Ostrovskaya, E.A. Optical vortices. Optics and Photonics News 2002, *13*(4), 24–28.
47. Partanen, J.P.; Jacobs, P.F. Lasers for stereolithography. In *OSA TOPS on Lasers and Optics for Manufacturing*, Tam, A.C., Ed. Vol. 9, pp. 9–13. Optical Society of America: Washington, DC, 1997.
48. Partanen, J. Lasers for solid imaging. Optics and Photonics News 2002, *13*(5), 44–48.
49. Ibbs, K.; Iverson, N.J. Rapid prototyping: new lasers make better parts, faster. Photonics Spectra 1997, *31*(6), 4 p.
50. SLA 250/30 Product Data Sheet from 3D Systems, 26081 Avenue Hall: Valencia, CA 91355.

2

Optical Systems for Laser Scanners

STEPHEN F. SAGAN

Agfa Corporation, Wilmington, Massachusetts, U.S.A.

1 INTRODUCTION

This chapter builds on the original work of Robert E. Hopkins and David Stephenson on optical systems for laser scanners[1] to provide yet another perspective. The goal of this chapter is to provide the background knowledge that will help develop an insight and intuition for optical designs in general and scanning systems in particular. Combined with a familiarity with optical design tools, these insights will help lead to optical designs with higher performance and fewer components. Design issues and considerations for holographic scanning systems are discussed in detail.

The interaction between optical requirements and constraints imposed on optical systems for laser scanners is discussed. The optical components that many applications of laser scanning depend on to direct and focus the laser beam are discussed, including lenses, mirrors, and prisms.

The optical invariant, first-order issues, and third-order lens design theory as they relate to scanning systems are presented as a foundation to the layout and design of the optical systems. Representative optical systems, with their characteristics, are listed along with drawings showing the lenses and ray trajectories. Some of the optical systems used for scanning that require special methods for testing and quality control are discussed.

2 LASER SCANNER CONFIGURATIONS

Optical system configurations for laser scanners can vary in complexity from a simple collimated laser source and scanner to one including beam conditioning optical components, modulators, cylinders, anamorphic optical relays, laser beam expanders, multiple scanners, and anamorphic optical components for projecting the scanned beam.

The scanned laser beam can be converging, diverging, or collimated. Figure 1 illustrates the three basic scanning configurations: *objective*, *post-objective*, and *pre-objective* scanning.[2]

2.1 Objective Scanning

The objective scanning configuration, where the objective, laser source, image plane, or a combination of these is moved, is the least common method of optical scanning. Objective scanning is accomplished by rotating about a remote axis as illustrated in Fig. 1 (or translating in a linear fashion) a focusing objective across the collimated beam. The moving objective can be a reflective mirror, refractive lens, or diffractive element (such as a holographic disc). The fundamentals of holographic scanning will be described beginning in Sec. 11.

2.2 Post-objective Scanning

The post-objective scanning configuration requires one of the simplest optical systems because it works on-axis. The rotation axis of the scanner can be orthogonal to the optical axis as with a galvanometer (as illustrated in Fig. 1) or coaxial, as in the case of a monogon scanner.

For many low-resolution applications (barcode scanners, for instance), simple lenses are sufficient to expand or begin focusing the beam prior to being scanned. As system

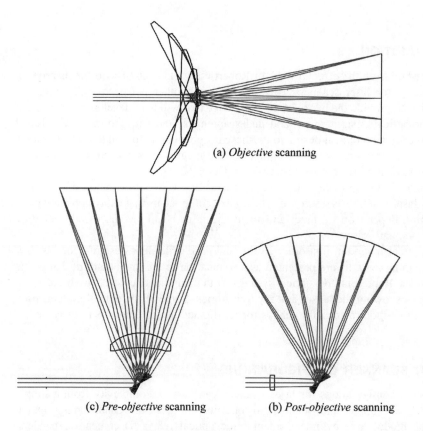

Figure 1 Three basic scanning configurations.

resolution requirements increase, larger numerical apertures and better optical correction will require additional lens elements and element complexity (such as doublets for spherical and color correction). The disadvantage of post-objective scanning is that the focal plane is curved, requiring an internal drum surface.

From an optical viewpoint, internal drum scanning offers high resolution over large formats with relatively simple optics. The laser beam, lens elements, and monogon scanner can be mounted coaxially into a carriage, with the scanner rotated about the optical axis of the incident laser beam. In such a system, the scanned spot would trace a complete circle on the inside of a cylinder. Translating either the carriage (scanning optical subsystem) or the drum will generate a complete two-dimensional raster. This type of system is ideal for inspecting the inside surface of a tube or writing documents inserted on the inside of a drum.

2.3 Pre-objective Scanning

In a pre-objective scanning configuration the beam is first scanned into an angular field and then usually imaged onto a flat surface. The entrance pupil of the scan lens is located at or near the scanning element. The clearance from the scanner to the scan lens is dependent on the entrance pupil diameter, the input beam geometry, and the angle of the scanned field. The complexity of the scan lens is dependent on the optical correction required over a finite scanned field, that is, spot size, scan linearity, astigmatism, and depth of focus.

Pre-objective scanners are the most commonly seen systems; these systems often require multi-element flat-field lenses. The special conditions described in the next few sections must be considered during the design of these lenses.

3 OPTICAL DESIGN AND OPTIMIZATION: OVERVIEW

Computers and software packages available to the optical designer for the layout, design/optimization, and analysis of optical systems (including developments in global optimization and synthesis algorithms) can be very powerful tools. Despite these advances, the most important tools available to the optical designer are simply a calculator, pen and paper, and a keen understanding of the first- and third-order fundamentals. These fundamentals provide key tools for back-of-the-envelope assessment of the issues and limitations in the preliminary phases of an optical design.

A successful design begins with an appropriate starting point including: (a) a list of the system specifications to scope the design problem, assess its feasibility, and guide the design process (see example list given in Table 1); (b) a first-order layout of the system configuration – the position of optical component groups, the aperture stop, and intermediate pupils and images; and (c) the selection of candidate design forms for the design of the optical component groups. Parameters that are entirely dependent on other specifications (in other words redundant) can be listed as reference parameters to provide further clarity.

The important fundamentals in the design of an optical system are:

1. First-order parameters, particularly the optical invariant;
2. First-order diffraction theory;
3. Third-order aberrations;

and then the rest.

Understanding the fundamentals can often mean the difference between achieving a simple "relaxed" design (with fewer optical components and a reduced sensitivity to

Table 1 Example List of Optical Specification for a Scanning System

Parameter	Specification or goal
1. Image format (line length)	216 mm
2. Wavelength	770–795 nm
3. Nominal $1/e^2$ spot size	26 μm diameter, ±10% (~1000 DPI)
4. Spot size variation	<4% (reference)
5. RMS wavefront error	<1/30 wave
6. Scan linearity ($F-\Theta$ distortion)	<1% (<0.2% over ±25°)
7. Scanned field angle	±30°
8. Effective focal length	206 mm (reference)
9. F-number	$F/26$ (reference)
10. Depth of focus	>1 mm (reference)
11. Overall length	335 mm
12. Scanner clearance	25 mm
13. Image clearance	270 mm
14. Optical throughput	>50% (including source truncation)

DPI, dots per inch.

fabrication and alignment errors), and a complicated "stressed" design, which meets the nominal performance goals but is difficult to assemble to meet as-built specifications.

A relaxed design will have low net third-order aberrations with reduced and distributed individual surface contributions to minimize induced higher order aberrations that can affect the performance of the as-built system. Lens elements in a relaxed design will generally bend with the marginal or chief ray, based on the intermediate speed and field of view demands of the system. Figure 2 shows a microscope objective where most air/glass surfaces are bent to minimize marginal ray angles of incidence and therefore minimize individual surface aperture dependent aberrations and a wide field of view fisheye objective where most surfaces are bent to minimize chief ray angles of incidence, minimizing individual surface field dependent aberrations. The ability to recognize which design forms work better over the field and which work better over the aperture will help in developing relaxed design forms.

Spherical surfaces are naturally easier to fabricate and test. However, aspheric surfaces (as a design variable) can be used to gain insight into what is holding back a design, or help find a new design form. Their moderate use can save weight and space, or they can often be replaced later in the design process with additional spherical elements. Aspheric surfaces can also be over used, with surfaces competing for correction during the optimization process, leading to overly complex and tolerance-sensitive design solutions.

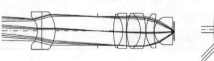   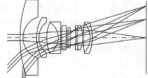

Microscope Objective, Plan-Achromat   Fisheye Lens, U.S. Pat. 4,412,726

Figure 2 Example lens designs configured for aperture (microscope objective at left) and primarily field (wide-field fisheye objective at right).

Design variables, such as surface curvatures, thickness between surfaces, and glass types, and optimization constraints appropriate to the design should be used. Too many variables and/or constraints, particularly conflicting ones, will limit the optimization convergence and performance of the design. Glass type and element thickness are often weak design variables. When glass variables are important, parameters such as cost, availability, production schedule, weight, and transmission, in addition to baseline performance must be considered in their final selection. Changing the glass map boundaries during the optimization process (allowing a wider range early in the design) can lend insight into possible alternative solutions. Vendor glass maps and catalogs are useful reference tools during the selection of glass types.

Anamorphic optical systems using combinations of cylindrical, toroidal, and anamorphic surfaces (with different radii in X and Y directions) can add more degrees of freedom and lens complexity, but are substantially more difficult to fabricate and test.

4 OPTICAL INVARIANTS

The optical invariant is defined at any arbitrary plane in a medium with refractive index n as a function of the paraxial marginal ray height and angle (y_m and nu_m) and paraxial chief ray height and angle (y_c and nu_c), as illustrated in Fig. 3 and given by the relationship

$$I = (y_m nu_c - y_c nu_m) \tag{1}$$

The optical invariant, as the name implies, is a constant throughout the optical system, provided it is not modified by discontinuities in the optical system such as diffusers, gratings, or other discontinuities such as vignetting apertures. The optical invariant is typically calculated at the object, aperture stop, or final image of the system, conveniently defined by the product of the object (or image) height times marginal ray angle or pupil height times chief ray angle. At the aperture stop or a pupil plane the chief ray height y_c is equal to zero, and the optical invariant reduces to

$$I = y_m nu_c \tag{2}$$

where the chief ray angle term (nu_c) is the paraxial half field or scan angle. At the object or an image plane the marginal ray height y_m is equal to zero, and the optical invariant reduces to

$$I = -y_c nu_m \tag{3}$$

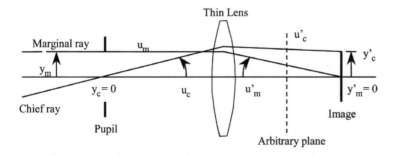

Figure 3 Paraxial Marginal and Chief rays for a simple lens.

where the marginal ray angle term (nu_m) is the paraxial equivalent of the sine of the cone half angle in air of the light focused on the image plane, known as the *numerical aperture* (*NA*). These reduced invariant equations are very useful when dealing with the optical properties at intermediate images or pupil conjugates within the system.

The *f*-number of a lens, defined as the lens focal length F divided by the design entrance aperture diameter (D_L), is also used to describe the image cone angle, with the relationship between numerical aperture and *f*-number ($F/\#$) for infinite conjugates given by

$$F/\# = F/D_L = \frac{1}{2NA} \tag{4a}$$

This relationship is clear for a collimated object, but at finite conjugates the lens *f*-number no longer describes the operating *f*-number, which is simply defined by the relative aperture

$$F/\# = \frac{1}{2NA} \tag{4b}$$

Most scan lenses operate in collimated space and it is convenient to use the $F/\#$ to describe the image-side cone angle. It is this relative aperture definition that will be used throughout this section.

4.1 The Diffraction Limit

Most scanning systems are required to perform at or very near the diffraction limit. The fundamental limit of performance for an imaging system of focal length F, illuminated by a uniform plane-wave of wavelength λ and truncated by an aperture of diameter D is defined by the Airy disk first ring diameter

$$d = \frac{2.44\lambda}{D/f} = 2.44\lambda(F/\#) \tag{5}$$

This diffraction limit is an optical invariant that determines the resolution in both the spatial and angular domain, and can be thought of as a *spot-invariant* (or spot-divergence product) that can be rewritten as

$$d(2NA) = 2.44\lambda \tag{6}$$

The fundamental diffraction limit for an ideal Gaussian beam with no truncation is defined by the *waist-invariant* (or waist-divergence product)

$$w_0 \theta_{1/2} = \lambda/\pi \tag{7}$$

where w_0 is the radius of the beam waist and $\theta_{1/2}$ is the half divergence angle in the far-field (where z is much greater than $\pi w_0 \lambda$) at the $1/e^2$ level for an ideal Gaussian beam of wavelength λ. Defined in terms of the $1/e^2$ waist diameter and full divergence angle, the

Optical Systems for Laser Scanners

waist-invariant becomes

$$d_0\theta = 4\lambda/\pi = 1.27\lambda \tag{8}$$

4.2 Real Gaussian Beams

Laser scanning systems typically use a near-Gaussian input beam. The degree to which the beam is Gaussian (TEM00) depends on the type of laser and the quality of the beam. Siegman[3] has shown that real laser beams (irregular or multimode) can be described analytically by simply knowing the near-field beamwidth radius W_0 and far-field half divergence angle $\Theta_{1/2}$, defined for each as the standard deviations measured in two orthogonal planes coincident with the axis of propagation. The product of these parameters defines the real beam waist-invariant and is proportional to the Gaussian beam diffraction limit given by

$$W_0\Theta_{1/2} = (\lambda/\pi)M^2 \tag{9}$$

where the factor M^2 defines the "times diffraction limit". When comparing beams of equal waist or divergence, the real beam divergence or waist, respectively, will be greater than the diffraction limit by a factor M^2, as illustrated in Fig. 4. The waist-invariant of a real beam will always be greater than the Gaussian diffraction limit.

Engineers developing scanning systems often use the concept of spot diameter. The specifications will call for a spot diameter measured at a specified intensity level, typically the $1/e^2$ and the 50% intensity levels. The maximum allowable growth of spot size across the length of the scan line is also included in the specification. Measurements by commercially available instruments that measure spot profile with a scanning slit will differ from the calculated point-spread function of the point image because the spot profile is determined by integrating the irradiance as the slit passes over the point image. This line-spread function measurement of the Airy disc does not have zeros in the irradiance distribution and is a more appropriate measure of integrated exposure when the spot is constantly moving during the exposure.

4.3 Truncation Ratio

Laser scanning systems typically use a near-Gaussian input beam with some truncation. *Truncation* means that a hard aperture restricts the diameter extent of the Gaussian beam,

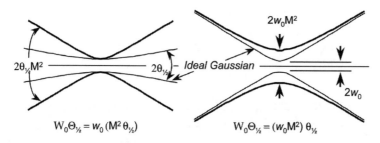

Figure 4 Relationship between ideal and real Gaussian beams.

usually located in the input collimator. The truncation ratio (W) is the ratio of the diameter of the Gaussian beam D_B (usually defined at the $1/e^2$ irradiance level), to the diameter of the truncating aperture D_L, defined as

$$W = D_B/D_L \tag{10}$$

Figure 5 shows how the image of a diffraction-limited beam is affected by different truncation ratios.

It is important to remember that a scan lens does not have a fixed aperture stop and that the lens diameters are actually much larger than the design aperture to pass the oblique ray bundles of the scanned beam. The pre-scan collimated beam, often called the feed beam, usually determines the aperture. The diameter of the beam should be no larger than the diameter of the largest beam for which the lens can provide the required image quality, which is usually diffraction limited. This is called the *design aperture* and is the value to use for D_L, when calculating truncation ratio W, but does not refer to the actual physical diameter of the scan lens.

Extending the definition of the diffraction limit to include the effect of the truncation ratio leads to the definition

$$d_x = \frac{k_x \lambda}{2NA} = k_x \lambda (F/\#) \tag{11}$$

The value of k_x depends on the truncation ratio W and the level of irradiance in the image spot used to measure the diameter of the image. Figure 6 shows how the value of k_x and consequently the diameter of the image of a point source is affected by different amounts

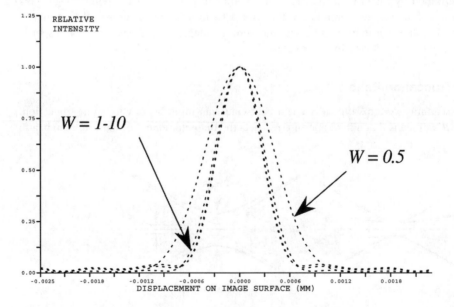

Figure 5 Point spread for a perfect wavefront and various truncations.

Optical Systems for Laser Scanners

of the truncation ratio W. The figure also shows two criteria for the image diameter d, one for the $1/e^2$ irradiance level and another for the 50% irradiance level. Equations for these two cases may be found in Ref. 4.

$$k_{FWHM} = 1.021 + 0.7125/(W - 0.2161)^{2.179} - 0.6445/(W - 0.2161)^{2.221} \quad (12)$$

$$k_{1/e^2} = 1.6449 + 0.6460/(W - 0.2816)^{1.821} - 0.5320/(W - 0.2816)^{1.891} \quad (13)$$

A truncation ratio of 1 generally provides a reasonable trade-off between spot diameter and conservation of total energy (86.5%). With $W = 1$, the following equations can be used to estimate a spot diameter:

$$d_{1/e^2} = 1.83\lambda(F/\#) \quad (14)$$

and

$$d_{50} = 1.13\lambda(F/\#) \quad (15)$$

There are several points to consider when deciding what truncation ratio to use. It would appear that the Gaussian beam spot with a 1.83λ diameter dependence is smaller than the uniform beam Airy disc with a 2.44λ diameter dependence. However, the Gaussian beam diameter formula refers to the 13.5% irradiance level in the image, while the Airy disc formula relates to the diameter of the first zero in irradiance. Figure 5 illustrates this with the irradiance distributions of the near uniform illumination of the $W = 10$ curve approaching that of an Airy disc pattern that is narrower than the truncated

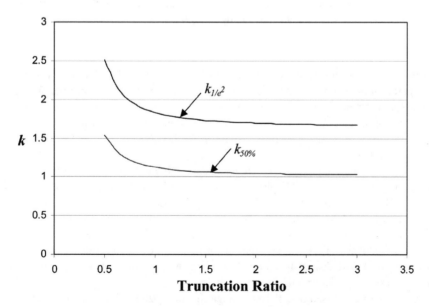

Figure 6 Effect of truncation ratio on relative spot diameter.

Gaussian illumination beam of the $W = 1$. On the other hand, the Airy disc image has more energy out in the wings of the image than does the Gaussian beam.

It is clear that the heavier truncation ratios ($W \gg 1$) yield smaller spot sizes, but they also suffer the flare from the diffraction rings formed by the truncated beam. For this reason, many designers believe that lower values of W (in the 0.5–1.0 range) are a better compromise, providing more light with less danger of image flare.

5 PERFORMANCE ISSUES

This section describes the terminology and unique image requirements of laser scan-lens design that are not typical factors in the design for most photographic objectives.

5.1 Image Irradiance

There are subtle differences to be considered in the calculation of image irradiance produced by a scan lens, compared to that of a normal camera lens. In galvanometer and polygon laser beam scanners, while the design aperture stop of scan lenses should be located on or near the deflecting mirror surface, these turning mirrors do not alter the circular diameter of the incoming beam as the deflection changes. This is different from a camera lens, which has a fixed aperture stop perpendicular to the lens optical axis. The oblique beam in a camera lens is foreshortened by the cosine of the angle of obliquity on the aperture stop. Designing the scan lens with a slightly larger entrance pupil (by the inverse cosine of the field angle) will provide a good first-order solution.

Most lens design programs do not automatically take this aperture effect into account, so at some point in the design process it will be necessary to use the proper tilts in the design program to maintain the beam diameter at each field angle to be optimized. This can be done in the multiconfiguration (or zoomed) setup available in most of the commercial design programs. The design program then optimizes several versions of the design simultaneously. Section 8.4 discusses in greater detail how multiconfiguration design procedures can be used in scan-lens design.

5.2 Image Quality

Addressability is an important term widely used in laser scanning. It refers to the least resolvable separation between two independent addressable points on a scan line. When the concept of spot diameter is used to describe optical performance, it is difficult to know how close the two spots can be to recognize them as separate points. Electrical engineers tend to think in terms of Fourier analysis, suggesting the concept of the modulation transfer function (MTF).[5]

The MTF specification can offer advantages in describing the optical performance of laser scanners. Figure 7 shows MTF plots for diffraction-limited images formed with truncation ratios W of 10 (near Airy disc), 2, 1, and 0.5. It is clear that the lower values of truncation have higher MTF for the low frequencies. The best value for W is close to 1. At this truncation ratio the MTF is highest, up to 43% of the design aperture theoretical cutoff frequency. This suggests that the principle of design to follow is to use a value of W close to 1 and design to as small an $F/\#$ as possible, consistent with the performance and cost considerations.

The foregoing rule is based on a perfect image. In attempting to increase the MTF at frequencies below 43% of the cutoff frequency, problems with aberration eventually occur

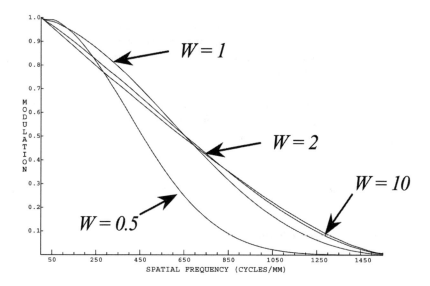

Figure 7 Modulation transfer function (MTF) curves for a perfect image with truncation ratio W.

in the large design apertures required. Fortunately, the small values of W mean that the intensity of the rays near the edge of the aperture is reduced, so the acceptable tolerance on the wavefront aberrations can be relaxed. It is not as easy to give a rule-of-thumb tolerance on the wavefront errors because it depends on the type of aberration. The higher order aberrations near the edges of the pupil will have less effect than will lower order aberrations, such as out-of-focus or astigmatism errors.

Specifying the performance of the optical system of a laser scanner and the appropriate measure, be it point-spread or line-spread function based spot size or MTF, can be a point of confusion. In terms of writing an image with independent image points, the point-spread function is a convenient concept. The lens point-spread function would be evaluated at several points along the scan, and the written spot sizes determined by the exposure profile defined by the point spread; that is, as the exposure level increases or decreases, the observable spot size determined by the irradiance level in the point-spread function also increases or decreases. The effect of exposure on spot size depends on the type of image being written – analog gray scale or digital half tone – and the response of the medium being written on. Often the choice of intensity level used to define the design spot size is $1/e^2$ because it draws more attention to energy pushed into side lobes of the point spread that can deteriorate the performance of a real lens with aberrations. Sometimes even side lobes above 5% are of consequence to the written image performance. The specification of full-width half maximum (FWHM) most often relates to the final written product.

A real lens with aberrations will spread the energy of an imaged spot beyond the Airy disc. This redistribution of energy reduces the irradiance in the center of the point-spread function. A measure of that redistribution is the Strehl ratio, the ratio of the spot peak intensity relative to the diffraction limit. The Strehl ratio is a convenient measure of the lens image quality during the design process (along with RMS wavefront error) and it is useful in calibrating normalized point-spread function calculations when evaluating a lens over several field angles.

The problem with the above concept is that the laser beam is constantly moving as it writes, smearing the imaged spot along the scan line. As the spot moves, the beam irradiance level is also being modulated to write the information required. This as-used writing process would point to the MTF as an appropriate measure of performance. However, the use of MTF assumes that the recording medium records irradiance level linearly over the complete range of exposures. This may or may not be true. It is important for the lens designer to discuss these differences with the system designer to ensure that all parties understand the issues and trade-offs.

It usually pays to be conservative and overdesign by at least 10% on initial ventures into laser scanning system development. The time to be most critical of a new design is in the first prototype and in the testing of the first complete system.

5.3 Resolution and Number of Pixels

The total number of pixels along a scan line is a measure of the optical achievement, given by

$$n = \frac{L}{d} \\ = \frac{2\theta F}{k\lambda F/D_L} \\ = \frac{2\theta D_L}{k\lambda} \quad (16)$$

where n = number of pixels, L = length of scan, d = spot diameter, D_L = diameter of lens design aperture, θ = scan half angle (radians), and F = scan lens focal length. The criterion for spot diameter will largely depend on the media sensitivity and its response to the $1/e^2$ or the 50% irradiance level.

5.4 Depth of Focus Considerations

Another important consideration in laser scanning systems is the *depth of focus* (DOF). The classical DOF for a perfectly spherical wavefront is given by

$$\text{DOF} = \pm 2\lambda (F/\#)^2 \quad (17)$$

This widely used criterion is based on a one-quarter wave departure from a perfect spherical wavefront. A similar criterion defined for a Gaussian beam as the optimum balance between beam size and depth of focus is given by the Raleigh range

$$Z_R = \pi w_0^2 / \lambda \quad (18)$$

where Z_R is the distance along the beam axis on either side of the beam waist at which the wavefront has a minimum radius of curvature of

$$R_{\min} = 2Z_R \quad (19a)$$

Optical Systems for Laser Scanners

and the transverse $1/e^2$ beam radius is

$$w_R = \sqrt{2}w_0 \tag{19b}$$

Each of these generalized criteria [Eqs (17) and (18)] serve their particular purpose, but many system specifications state that the spot size diameter must be constant within 10% (or even less) across the entire scan line. Additionally, manufacturers of scanning systems often impose a lower limit to the tolerable depth of focus. There is no simple formula to relate depth of focus to this requirement, but spot size or MTF calculations made for several focal plane positions can provide the pertinent data. Figure 8 shows the MTF curves of an $F/5$ parabola under the following conditions.

- A. The perfect image with uniform 632.8 nm wavelength irradiance across the entire design aperture ($W = 1000$).
- B. The perfect image with $W = 0.85$.
- C. The same image as A, but with a focal shift of 0.063 mm. This corresponds to a wavefront error of half of a wave at the maximum design aperture.
- D. The same image and truncation as B, but with a focal shift of 0.063 mm.
- E. The image from a parabola with aspheric deformation added to introduce a half-wave of fourth-order wavefront error at the edge of the design aperture; $W = 1000$, no focus shift.
- F. The same as E with $W = 0.85$, no focus shift.

The truncation value of $W = 0.85$ was used for this example instead of 1.0 in order to help reduce the influence of aberrated rays near the edge of the design aperture. These curves show that the depth of focus is slightly improved by truncating at this value. They

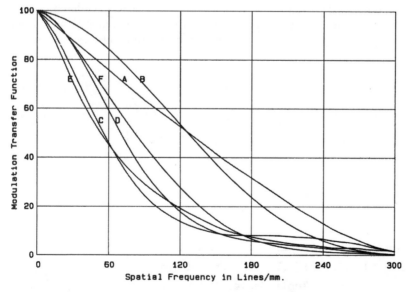

Figure 8 Effect of focus shift, spherical aberration, and truncation ratio on modulation transfer function (MTF) (from Ref. 1).

also show that one half-wave of spherical aberration does not have as serious an effect on the depth of focus as does an equivalent amount of focus error. Therefore it is most important to reduce the Petzval curvature and astigmatism in a scan lens, because these aberrations cause focal shift errors.

5.5 The F–Θ Condition

In order to maintain uniform exposure on the material being scanned, the constant power image spot must move at a constant velocity. As the scanner rotates through an angle $\theta/2$, the reflected beam is deflected through an angle θ, where the angle θ is measured from the optical axis of the scan lens. Because polygon scanners rotate at a constant velocity, the reflected beam will rotate at a constant angular velocity. The scanning spot will move along the scan line at a constant velocity if the displacement of the spot is linearly proportional to the angle θ. The displacement H of the spot from the optical axis should follow the equation

$$H = F\theta \tag{20}$$

where the constant F is the approximate focal length of the scan lens. Figure 9 is the distortion in an $F-\theta$ lens relative to that of a normal lens corrected for linear distortion (F-tan θ) plotted over scan angle. The curve's departure from the straight line represents the distortion required of an $F-\theta$ lens for a constant scan velocity. As the field angle increases, a classical distortion-free lens image points too far out on the scan line, causing the spot to move too fast near the end of the scan line. Fortunately, typical scan lenses begin with negative (barrel) third-order distortion – the image height curve laying below the F-tan θ curve. The distortion can be designed to match the $F-\theta$ image height at the edge of the field or balanced over the image. For a given distortion profile, the plus and minus departures from the ideal $F-\theta$ height over the scanned image can be balanced for minimum plus and minus departures by scaling the value of F used in defining the data rate. Scaling the data rate effectively scales the pixel spatial frequency written at the image plane. The focal length that minimizes the departures from linearity is called the *calibrated focal length*.

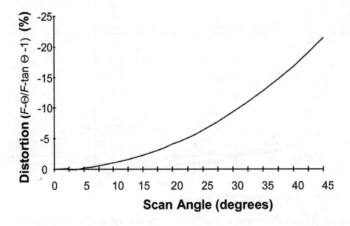

Figure 9 Error between F–tan θ and $F-\theta$ distortion correction.

When the field angle is as large as $\pi/6$ radians (30°), the residual departures from linearity may still be too large for many applications. Balancing negative third-order distortion against positive fifth-order distortion can further reduce the departures. Lens designers will recognize this technique as similar to the method of reducing zonal spherical aberration by using strongly collective and dispersive surfaces, properly spaced. In this case the zonal spherical aberration of the chief ray must be reduced. Figure 10 shows an example of this correction.

This high-order correction should not be carried too far, since the velocity of the spot begins to change rapidly near the end of the scan and may result in unacceptable changes in exposure or pixel placement. It may also begin to distort the spot profile, turning a circular spot into an elliptical one. This *local distortion* results in a change in the resolution or spatial frequencies near the end of the scan line. A standard observer can resolve frequencies of 10 line pairs/mm (254 lines/in.), but is even more sensitive to variations of frequency in a repetitive pattern. Variations of frequency as small as 10% may be detected by critical viewing. The linearity specification is often expressed as a *percent error* (the spot position error divided by the required image height). For example, the specification often reads that the $F-\theta$ error must be less than 0.1%. This means that the deviations must be smaller and smaller near the center of the scan line. It is not reasonable to specify such a small error for points near the center of the scan. The proper specification should state rate of change of the scan velocity and the allowable deviation from the ideal of Fig. 9. More detail on this subject may be found in Ref. 4.

6 FIRST- AND THIRD-ORDER CONSIDERATIONS

The optical system in a scanner should have a well-considered first-order layout. This means that the focal lengths and positioning of the lenses should be determined before any aberration correction is attempted. Most of the optical systems to be discussed in this

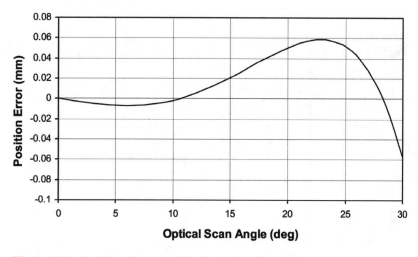

Figure 10 $F-\theta$ linearity error minimized with calibrated focal length, third-order and fifth-order distortion.

section will first be described as groups of thin lenses. The convention used for thin lenses is described in most elementary books on optics.[6,7]

The graphical method shown in Fig. 11 is useful for a discussion of determining individual and total system focal lengths. The diagram shows an axial ray that is parallel to the optical axis. This represents a collimated beam entering the lens. The negative lens "a" refracts the axial beam upward to the positive lens "b." The positive lens "b" then refracts the ray to the axis at the focal plane at F_{2ab}, which is the writing plane for the laser beam.

The second focal point of the negative lens is at F_{2a}. This point is located by extending the refracted axial ray backward from the negative lens until it meets the optical axis. The second focal point of the positive lens (F_{2b}) may be determined by drawing a construction line through the center of the positive lens parallel to the axial ray as it passes between the positive and negative lenses. Because the two lines are parallel, they must come to focus in the focal plane of the positive lens. The focal lengths of the two lenses are now determined. The front and back focal lengths of each lens are equal because the lenses are in air. The focal points F_{1a}, F_{2a} and F_{1b}, F_{2b} are now located. The diagram also shows the construction for finding the second principal point P_2. The distance P_2 to F_{2ab} is the focal length F of the negative–positive lens combination.

The chief ray is next traced through the two-element system. This is done using the concept that two rays that are parallel on one side of a lens must diverge or converge to the second focal plane of the lens. The chief ray enters the lens system after it passes through the entrance pupil (or aperture stop) of the system. For scan lenses, the entrance pupil is usually located at the scanning element. Note that the entrance pupil is located in front of the lens, which is in contrast to a photographic lens where the entrance pupil is usually virtual (located on the image side of the front lens) and the aperture stop is usually located between the lens elements. This is the primary reason why a photographic lens should not be used as a scan lens. It is also one of the reasons why scan lenses are limited in the field angles they can cover.

The completed diagram labels the lens focal lengths. The system focal length is 80.79, $F_a = -55.42$, and $F_b = 48.63$. The *Petzval curvature* is given by the sum of the

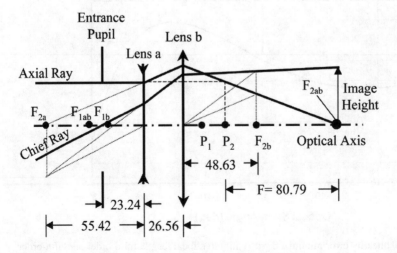

Figure 11 Graphical solutions to a system of thin lenses.

Optical Systems for Laser Scanners

power of each lens element divided by its index of refraction as

$$P = \sum_i \Phi_i/n_i = \Sigma_i 1/(F_i n_i) \tag{21}$$

The Petzval radius $(1/P)$ is 3.3 times the focal length and it is curved towards the lens. This is not flat enough for an $F/20$ system when the lens has to cover a long scan line. Equation (22) described in Sec. 6.4 provides a formula for estimating the required Petzval radius for a given system. When the Petzval radius is too short, the field has to be flattened by introducing positive astigmatism, which will cause an elliptically shaped writing spot. The Petzval radius is a fundamental consideration in laser scan lenses and becomes a major factor that must be reckoned with in systems requiring small spot sizes. Small spots require a large numerical aperture or a small $F/\#$. Observations to be made from this layout include:

1. The distance from the entrance pupil to the lens is 23.24 or 29% of the focal length of the scan lens.
2. If the entrance pupil is moved out toward F_{1ab}, the chief ray will emerge parallel to the optical axis and the system will be telecentric. This condition has several advantages, but lens "b" must be larger than the scan length and the large amount of refraction in lens "b" will introduce negative distortion, making it difficult to also meet the $F-\theta$ condition.
3. Reducing the power of the negative lens or decreasing the spacing between the lenses will allow for a longer distance between the lens and the entrance pupil, but this will introduce more inward-curving Petzval curvature.

This brief discussion illustrates some of the considerations involved in establishing an initial layout of lenses for a scanner. One must decide, on the basis of the required spot diameter and the length of scan, what the Petzval radius has to be in order to achieve a uniform spot size across the scan length. When field flattening is required, it is necessary to introduce more negative power in the system. The most effective way to do this is to insert a negative lens at the first, second, or both focal points of a positive-focal-length scan lens. In these positions they do not detract from the focal length of the positive lens, so the Petzval curvature can be made to be near zero when the negative lens has approximately the same power as the positive lens. The negative lens at the second focal point, however, must have a diameter equal to the scan length, and it will introduce positive distortion if it is displaced from the focal plane. This distortion will make it difficult to meet the $F-\theta$ condition. A negative lens located at the first focal point of the lens is impractical, since there would be no distance between the lens and the position of the scanning element.

The next best thing to do is to place a single negative lens between the positive lens and the image plane. When the negative lens and the positive lens have equal but opposite focal lengths and the spaces between the lenses are half the focal length of the original single lens, then the focal lengths of two lenses are $+0.707F$ and $-0.707F$. The system with the positive lens in front is a telephoto lens, and the one with a negative lens in front is an inverted telephoto. The telephoto lens has a long working distance from the first focal point to the lens, while the inverted telephoto has a long distance from the rear lens to the image plane. The question now is, "which is the better form to use for a scan lens?"

It is well known that a telephoto lens has positive distortion, while the inverted telephoto lens has negative distortion. Scan lenses that have to be designed to follow the

$F-\theta$ condition must have negative distortion. This suggests that the preferred solution is with the negative lens first, even though it makes a much longer system from the last lens to the focal plane and the entrance pupil distance is considerably shorter. Most of the scan lenses in use are a derivative of this form of inverted telephoto lens, employing a negative element on the scanner side of the lens.

Often the clearance required for the scanning element causes aberration correction problems. A telecentric design provides more clearance. Strict telecentricity may introduce too much negative distortion because the positive lens has to bend the chief ray through a large angle. When there is a tight tolerance on the $F-\theta$ condition it is better to move the scanner (aperture stop) closer to the first lens. Experience has shown that it is difficult to achieve an overall length of the system (from the scan element to the image plane) of less than 1.6 times the lens focal length. The characteristics of several scan lenses are described in Ref. 4; few have a smaller ratio. In cases where the distance from scanning element to the first lens surface has to be longer than the focal length, it is advantageous to use the telephoto configuration. However, it will be difficult to make the lens meet the $F-\theta$ condition. Systems like this have been used for galvanometer scanning. It is particularly useful for XY scanning systems where more space is needed between the aperture stop and the lens.

The lenses used in the above example are extreme lenses to illustrate the two cases. In most designs the Petzval radius is not set to infinity. A Petzval radius of 10 to 50 times the focal length is usually all that is needed. The two lenses are also usually made of different glass types in order to follow the Petzval rule: to increase the Petzval radius, the negative lenses should work at low aperture and have a low index of refraction and the positive lenses should work at high aperture and have a high index of refraction. It has been pointed out[8] that if the incoming beam is slightly diverging, instead of collimated, it increases the radius of the Petzval surface. The diverging beam in effect adds positive field curvature. The idea has occasionally been used in systems, but the focus of the collimator lens has to be set at the correct divergence – not as convenient to set as strictly collimated.

Some lenses that are required to image small spot diameters (2–4 μm) use negative lenses on both sides of the positive lens to correct the Petzval curvature. Examples are shown in Sec. 12.9.

6.1 Correction of First-Order Chromatic Aberrations

The correction of axial and lateral chromatic aberrations illustrated in Figs. 12(a) and (b), respectively, is usually a challenge with scan lenses because the aperture stop is remote from the scan lens. Some system specifications call for simultaneous scanning of two or more wavelengths. These lenses have to be color-corrected at multiple wavelengths for no change in focus or focal length – that is, designed to be achromatic. Axial (or longitudinal) color, a marginal ray aberration, is a variation of focus with wavelength and is directly proportional to the relative aperture and is independent of field. Lateral (or transverse) color, a chief ray aberration, is a variation of lens magnification or scale with wavelength and is directly proportional to the field.

The simplest way to correct axial and lateral color is to make each element into an achromatic cemented doublet. To make a positive lens achromatic it is necessary to have a positive and negative lens with glass of different dispersions. The positive lens should have low-dispersion glass and the negative lens should have high-dispersion glass. The

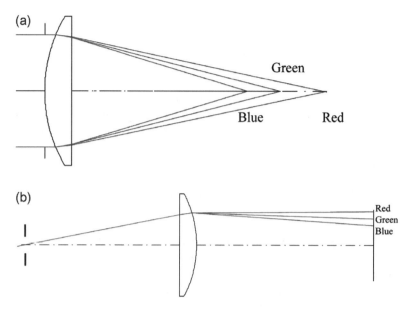

Figure 12 (a) Axial color aberration (focus change with wavelength); (b) Lateral color aberration (magnification change with wavelength) with a remote pupil.

negative focal length lens reduces the positive power, so the positive lens power must be approximately double what it would be if not achromatized.

This procedure halves the radii so the thickness must be increased in order to maintain the lens diameter. In scan lenses, the lens diameters are determined by the height of the chief ray, so the lenses are much larger in diameter than indicated by the axial beam. As thickness is increased to reach the diameters needed, the angles of incidence on the cemented surfaces increase, resulting in higher order chromatic aberrations. When the angles of incidence in an achromatic doublet become too large, the doublet has to be split up and made into two achromatic doublets. It is safe to say that asking for simultaneous chromatic correction can more than double the number of lens elements.

Materials used for the lenses, mirrors, and mounting can be affected by environmental parameters such as temperature and pressure. For broadband systems or systems where wavelength can vary over time and/or temperature, the chromatic variation in the third-order aberrations are often the most challenging aberrations to correct. While achromats corrected for primary color use glasses with dissimilar chromatic dispersion, achromats also corrected for secondary color in addition use glasses with similar partial dispersion (i.e., glasses with similar rate of change in dispersion with wavelength). Where glasses with similar dispersion are impractical or not available, an additional element to form a triplet is used to synthesize the glass relationships needed to correct the higher order chromatic aberrations.

Some specifications ask for good correction for a small band of wavelengths where small differences due to color can be corrected by refocus or by moving the elements. These systems do not need full color correction, and they can be designed to meet other more demanding requirements. The highest performance scan lenses are usually used with strictly monochromatic laser beams.

6.2 Properties of Third-Order Aberrations

The ultimate performance of any unconstrained optical design is almost always limited by a specific aberration that is an intrinsic characteristic of the design form. Familiarity with the aberrations and lens forms is still an important ingredient in a successful design optimization. Understanding of the aberrations helps designers to recognize lenses that are incapable of further optimization, and gives guidance in what direction to push a lens that has strayed from the optimal configuration. Table 2 summarizes the dependence of third- and selected fifth-order aberrations on aperture and field.

An understanding of the source of aberrations and their elimination comes from third-order theory. A detailed description of the theory is beyond the scope of this chapter, but can be found in Refs. 6,10–12. The following discussion will touch on these aberrations with the intent to provide a familiarity and some rules of thumb as guidelines for the design of scan lenses.

Third-order theory describes the lowest-order monochromatic aberrations in an optical system. Any real system will usually have some balance of third-order and higher order aberrations, but the basic third-order surface-by-surface contributions are important to understand. These aberrations are illustrated in Fig. 13 and briefly described below.

6.2.1 Spherical Aberration

This aberration is a result of a lens with different focal lengths for different zones of the aperture, a consequence of greater deviations of the sine of the angle and paraxial angle. It is an aperture-dependent aberration (varying with the cube of the aperture diameter) that causes a rotationally symmetrical blurred image of a point object on the optical axis. In rotationally symmetric optical systems it is the only aberration that occurs on the optical axis, but, if present, it will also appear at every object point in the field – in addition to other field aberrations.

6.2.2 Coma

This aberration is the first asymmetrical aberration that appears for points close to the optical axis. It is a result of different magnifications for different zones of the aperture. Coma gets its name from the shape of the image of a point source – the image blur is in the form of a comet. The coma aberration blur varies linearly with the field angle and with the square of the aperture diameter.

Table 2 Aperture ($F/\#$) and Field (θ) Dependence of Third- and Fifth-Order Transverse Aberrations

Transverse aberration	Third-order	Fifth-order
Spherical	$(F/\#)^{-3} \theta^0$	$(F/\#)^{-5} \theta^0$
Coma	$(F/\#)^{-2} \theta^1$	$(F/\#)^{-4} \theta^1$
Astigmatism	$(F/\#)^{-1} \theta^2$	$(F/\#)^{-1} \theta^4$
Field curvature	$(F/\#)^{-1} \theta^2$	$(F/\#)^{-1} \theta^4$
Distortion	$(F/\#)^{0} \theta^3$	$(F/\#)^{0} \theta^5$

Source: Ref. 9.

Optical Systems for Laser Scanners

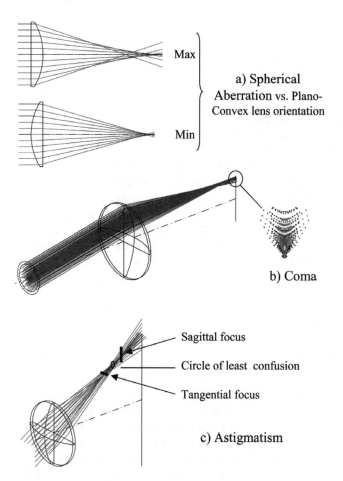

Figure 13 Aberrations: (a) spherical, (b) coma, and (c) astigmatism.

6.2.3 Astigmatism

When this aberration is present, the meridional fan of rays (the rays shown in a cross-sectional view of the lens) focuses at the *tangential focus* as a line perpendicular to the meridional plane. The sagittal rays (rays in a plane perpendicular to the meridional plane) come to a different line focus perpendicular to the tangential line image. This focus position is called the *sagittal focus*. Midway between the two focal positions, the image is a circular blur with a diameter proportional to the numerical aperture of the lens and the distance between the focal lines. The third-order theory shows that the tangential focus position is three times as far from the Petzval surface as the sagittal focus. This is what makes the Petzval field curvature so important. If there is Petzval curvature, the image plane cannot be flat without some astigmatism. The astigmatism and the Petzval field sags both increase proportional to the square of the field. They increase faster than coma and become the most troublesome aberrations as the field (length of scan) is increased.

6.2.4 Distortion

Distortion is a measure of the displacement of the real chief ray from its corresponding paraxial reference point (image height $Y = F \text{-} \tan \theta$) and is independent of f-number. Distortion does not result in a blurred image and does not cause a reduction in any measure of image quality (such as MTF). In an aberration-free design, the center of the energy concentration is on the chief ray. The third-order displacement of the chief ray from the paraxial image height varies with the cube of the image height. The percent distortion varies as the square of the image height.

Earlier it was noted that the distortion has to be negative in order to meet the $F\text{-}\theta$ condition. Third-order distortion refers to the displacement of the chief ray. If the image has any order of coma, it is not rotationally symmetric. The position of the chief ray may not represent the best concentration of energy in the image; there may be a displacement. Here the specification for linearity of scan becomes difficult. If there is a lack of symmetry in the image, then how does one define the error? If MTF is used as a criterion, this error is a phase shift in the tangential MTF. If an encircled energy criterion is used, then what level of energy should be used? When a design curve of the departure from the $F\text{-}\theta$ condition is provided, it usually refers to the distortion of the chief ray. The designer must therefore attempt to reduce the coma to a level that is consistent with the specification of the $F\text{-}\theta$ condition, or use an appropriate centroid criterion.

6.3 Third-Order Rules of Thumb

Collective surfaces[7] almost always introduce negative spherical aberration. A collective surface bends a ray above the optical axis in a clockwise direction as shown in Fig. 14. There is a region where a collective surface introduces positive spherical aberration. This occurs when the axial ray is converging to a position between the center of curvature of the surface and its aplanatic point. When a converging ray is directed at the aplanatic point the angles of the incident and refracted rays, with respect to the optical axis, satisfy the sine condition $U/U' = \sin U / \sin U'$, and no spherical or coma aberrations are introduced. Unfortunately this condition is usually not accessible in a scan lens. Surfaces with positive spherical aberration are important because they are the only sources of positive astigmatism.

Dispersive surfaces always introduce positive spherical aberration. A dispersive surface bends rays above the axis in a counterclockwise direction. In order to correct spherical aberration it is necessary to have dispersive surfaces that can cancel out the under correction from the collective surfaces.

Coma can be either positive or negative, depending on the angle of incidence of the chief ray. This makes it appear that the coma should be relatively easy to correct, but

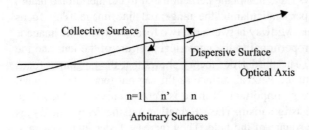

Figure 14 Simplest example of collective and dispersive surfaces.

in the case of scan lenses it is difficult to correct the coma to zero. The primary reason is that the aperture stop of the lens is located in front of the lens. This makes it more difficult to find surfaces that balance the positive and negative coma contributions.

As the field increases, astigmatism dominates the correction problem. The astigmatism introduced by a surface always has the same sign as the spherical aberration. When the lenses are all on one side of the aperture stop, this makes it difficult to control astigmatism and coma. A lens with a positive focal length usually has an inward-curving field so the astigmatism has to be positive. This is the reason that a designer must have surfaces that introduce positive spherical aberration. Because distortion is an aberration of the chief ray, surfaces that are collective to the chief ray will add negative distortion and dispersive surfaces will add positive distortion.

6.4 Importance of the Petzval Radius

Even though the Petzval curvature is a first-order aberration, it is closely related to the third-order because of the 3:1 relation with the tangential and sagittal astigmatism. It is not possible to eliminate Petzval curvature by merely setting up the lens powers so that the Petzval sum is zero. By doing this, the lens curves become so strong that higher order aberrations are introduced, causing further correction problems. For this reason it is important to set up the initial design configuration with a reasonable Petzval field radius, and the designer should continually note the ratio of the Petzval radius to the focal length.

An estimate of the desired Petzval radius for a flatbed scan lens can be derived based on third-order astigmatism and depth of focus. Eliminating third-order astigmatism, the tangential and sagittal fields will coincide with the Petzval surface. The maximum departure of this Petzval surface from a flat image plane defined over a total scan line length L is given by the sag[11,13]

$$\delta z = -L^2/8 * [\text{Petzval curvature}]$$

Setting δz equal to the total depth of focus from Eq. (17) yields the relationship

$$4\lambda(F/\#)^2 = -L^2/8 * [\text{Petzval curvature}]$$

The Petzval radius relative to the lens focal length F is then given by

$$[\text{Petzval radius}]/F = \frac{-L^2}{[32\lambda(F/\#)^2 F]} \qquad (22)$$

Section 9 describing some typical scan lenses lists this ratio in Table 6 as a guideline for each application. The ratio is only an approximation. Lenses operating at large field angles or small $F/\#$ values will have high-order aberrations not accounted for in the equation. Depending on the type of correction, the final designed ratio may be higher or lower than given by the above equation. Furthermore, negative lenses working at low aperture and positive lenses working at large aperture reduce the Petzval field curvature. The negative lenses should have a low index of refraction, and the positive lenses should have a high index of refraction, atypical for an achromatized optical system.

In most monochromatic scan lenses, the negative lens will have a lower index of refraction than the positive lenses. The positive lenses will usually have an index of refraction above 1.7, while the negative lenses will usually have values around 1.5. Lens design programs can vary the index of refraction during the optimization. Occasionally in a lens with three or more elements, the optimized design violates this rule and one of the positive lenses turns out to have a lower index of refraction than the others. This may mean the design has more than enough Petzval correction, so the index of one of the elements is reduced in order to correct other aberrations, or it may mean that one of the positive lenses is no longer necessary. To remove such an element during the optimization process, distribute its net power (by adding or subtracting curvature to one or both neighboring surfaces) and optimize for a few iterations using curvatures and a few constraints to reinitialize the design before proceeding with the full optimization.

7 SPECIAL DESIGN REQUIREMENTS

This section discusses specific optical design requirements for different types of laser scanners.

7.1 Galvanometer Scanners

Galvanometer scanners are used extensively in laser scanning. Their principal disadvantage is that they are limited in writing velocity. Their many advantages from an optical perspective are:

- The scanning mirror can rotate about an axis in the plane of the mirror. The mirror can then be located at the entrance pupil of the lens system and its position does not move as the mirror rotates.
- The $F-\theta$ condition is often not required, for the shaft angular velocity of the mirror can be controlled electronically to provide uniform spot velocity.
- The galvanometer systems are suitable for X and Y scanning.

Galvanometer scanners provide the easiest way to design an XY scanning system. The two mirrors, however, have to be separated from each other, and this means the optical system has to work with two separated entrance pupils, with considerable distance between them. This in effect requires that the lens system be aberration corrected for a much larger aperture than the laser beam diameter. A system demanding both a large aperture and large field angle will have different degrees of distortion correction for the two directions of scan. In principle the distortions can be corrected electronically, but this adds considerable complexity to the equipment.

An alternate approach is to use a telescope (afocal) relay system with one scanner placed at the entrance pupil and the other placed at the exit pupil of the telescope relay. The telescope adds complexity and field curvature to the design, but also an intermediate image that can be useful in dealing with the field curvature. Systems requiring a large number of image points should avoid extra relay lenses that add Petzval field curvature, the aberration that often limits optical performance in scanning systems.

For precision scanning, any wobble the galvanometer mirror may have can be corrected with cylindrical optics, as described in the next section on polygon scanning.

7.2 Polygon Scanning

Some precision scanning system require extreme uniformity of scanning velocity, sometimes as low as 0.1%, with the addressability of a few microns. These requirements of high-speed scanning velocities force systems into high-speed rotating elements that scan at high uniform velocity. Polygon and holographic scanners are most commonly used in these applications.

Special design requirements that must be considered in the design of lens systems for polygons that effect optical quality are: scan line bow, beam displacement, and cross-scan errors.

7.2.1 Bow

The incoming and exiting beams must be located in a single plane that is perpendicular to the polygon rotation axis. Error in achieving this condition will displace the spot in the cross-scan direction by an amount that varies with the field angle. This results in a curved scan line, which is said to have *bow*. The spot displacement as a function of field angle is given by the equation

$$E = F \sin \alpha (1/\cos\theta - 1) \tag{23}$$

where F is the focal length of the lens, θ is the field angle, and α is the angle between the incoming beam and the plane that is perpendicular to the rotation axis.

The optical axis of the focusing lens should be coincident with the center of the input laser beam, hereafter referred to as the *feed beam*. Any error will introduce bow. The bow introduced by the input beam not being in the plane perpendicular to the rotation can be compensated for, to some extent, by tilting the lens axis. Some system designers have suggested using an array of laser diodes to simultaneously print multiple rasters. Only one of the diodes can be exactly on the central axis, so all other diode beams will enter and exit the scanner out of the plane normal to the rotation axis, so bow will be introduced. The amount will increase for diodes farther away from the central beam. There is no simple remedy to this problem.

7.2.2 Beam Displacement

A second peculiarity of the polygon is that the facet rotation occurs around the polygon center rather than the facet face. This causes a facet displacement and a displacement of the collimated beam as the polygon rotates, as illustrated in Fig. 15. This displacement of the incoming beam means that the lens must be well corrected over a larger aperture than the laser beam diameter. A comprehensive treatment of the center-of-scan locus for a rotating prismatic polygonal scanner is given in Ref. 14.

7.2.3 Cross-Scan Errors

Polygons usually have pyramidal errors in the facets as well as some axis wobble. These errors cause cross-scan errors in the scan line. These errors must be corrected to a fraction of a line width, typically on the order of a $1/4$ to a $1/10$ of a line width. When designing a 2400 DPI (dots per inch) high-resolution system, the tolerable error can be less than 1 μm. A system with no cross-scan error correction using a 700 mm focal length lens would require pyramid errors no greater than 1.4 μradians.

Correction methods for cross-scan errors due to polygon pyramidal errors can include deviation of the feed beam, the use of cylindrical and anamorphic lenses to focus on the polygon, an anamorphic collimated beam at the polygon, or use of a retro-reflecting prism to auto correct.

Deviating the feed beam to anticipate the cross-scan errors at the polygon that are predictable and measurable can be accomplished by tilting a mirror, moving lens, or

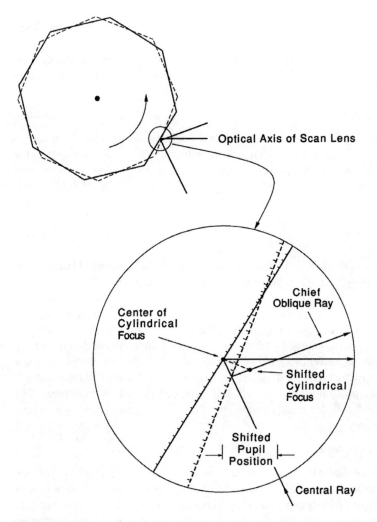

Figure 15 Facet rotation around the polygon center causes a translation of the facet, resulting in a beam displacement (from Ref. 1).

steering with an acousto-optic deflector (AOD). This method cannot correct for the random errors caused by polygon bearing wobble and is therefore limited in its application.

The diagram in Fig. 16 shows how the use of cylindrical lenses can reduce the effects of wobble in the facet of a polygon. The top figure illustrates the in-scan plane, showing the length of the scan line. The lower section shows the cross-scan plane, where the laser beam is focused on the facet of the mirror by a cylindrical lens in the collimated beam. It then diverges as it enters the focusing lens. The scan lens with rotational symmetry cannot focus the cross-scan beam to the image plane without the addition of a cylindrical lens. The focal length and position of the cylindrical lens depends on the distance from the polygon facet to the all-spherical scan lens and the numerical aperture of the cylindrical lens that focuses a line image on the facet.

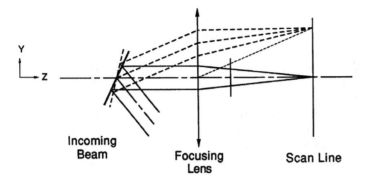
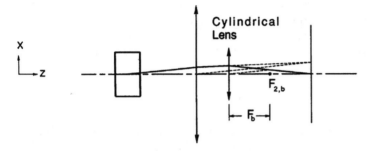

Figure 16 The use of a cylindrical lens to focus a line on the facet can reduce the cross-scan error caused by facet wobble (from Ref. 1).

In order to form a round image in the scan plane, the beam in the cross-scan plane must focus with the same numerical aperture as in the in-scan plane. The ratio of the cross-scan NA at the facet to the cross-scan NA at the scanned image defines the cross-scan magnification of the lens. The selection of cylinder powers before the scan, between the scanner and scan lens, and/or between the scan lens and image plane will affect the sensitivity to wobble errors at the edges of the depth of field. Generally a cross-scan magnification of near 1 : 1 optimizes the correction in the presence of this polygon facet displacement from the line focus.

When the facet rotates to direct the light to the edge of the scan, the distance to the spherical lens increases. In the cross-scan plane the optical system is focusing the beam from a finite object distance. When the facet rotates to direct the light to the edge of the scan, the object distance increases so there is a conjugate change. As the scan spot moves from the center of scan, the object distance in the cross-scan plane also increases. The image conjugate distance is therefore shortened. The consequence of this is that astigmatism is introduced in the final image with the sagittal focal surface made inward curving. To compensate for this, the all-spherical focusing lens must be able to introduce enough positive astigmatism to eliminate the total astigmatism.

It has been shown[15] that placing a toroidal lens between the facet and the all-spherical lens may reduce this induced astigmatism. The toroidal surface in-scan radius of curvature should be located near the facet. In the cross-scan plane the curve should be adjusted to collimate the light. However, this solution bears with it the cost of special tooling, and imposes severe procurement, testing, and alignment challenges.

An anamorphic collimated beam at the polygon combined with a scan lens having a short cross-scan focal length and long in-scan focal length, as illustrated in Fig. 17, can be used to reduce the effects of facet wobble. The reduction in sensitivity relative to no correction is simply the ratio of cross-scan to in-scan focal lengths, accomplished by adding cross-scan cylindrical lenses to modify the all-spherical in-scan lens to an inverse telephoto cross-scan configuration. The feed beam is likewise compressed in the cross-scan plane to provide the necessary round beam converging on the image plane. The diagrams show the two focal lengths of the scan lens as F_{yz} and F_{xz}. In its simplest terms, the feed beam is compressed with an inverted cylinder beam expander and a comparable cylinder beam expander is placed before, distributed across, or after the scan lens. This system does introduce some conjugate shift astigmatism, but it eliminates the bow error, because collimated light is incident on the facet.

Placing the negative cylinder close to the all-spherical focusing lens and placing the positive cylinder as close to the image plane as is practical reduce the cylindrical lens powers. The position of the positive cylindrical lens, however, must consider such things as bubbles or defects on the surfaces of the lens. The beam size is extremely small when the lens is placed close to the focal plane and the entire beam can be blocked with a dust particle.

Systems using a retro-reflective prism (with 90° roof edge) that reflects the scanning beam back onto the facet face before it passes to the scan lens have been built to correct for facet wobble. The optical error introduced by the pyramidal or wobble error in the polygon is canceled on the second pass. Unfortunately the facet face has to be more than twice the aperture required to reflect the beam in a single reflection system to keep the retro-reflective beam on the facet. Consequently this configuration has low scan efficiency and limited uses.

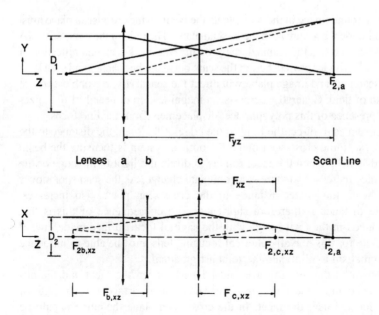

Figure 17 An anamorphic beam incident on the facet will also reduce cross-scan error (from Ref. 1).

7.2.4 Summary

Axis wobble and pyramidal error cause serious problems by introducing cross-scan errors. There are ways to reduce the cross-scan errors, but many other challenges are introduced. The use of cylinders results in procurement and alignment issues. The conjugate shift is difficult to visualize because the entire line image on the facet is not in focus and the analysis can be complex. The only way to determine accurately the combination of all the effects – pyramidal error in the facets, translation of the facet during its rotation, bow tie effect, and the conjugate shift – is to raytrace the system and simulate the precise locations of the facet as it turns through the scanning positions. This can be done using the *multi configuration modes* available in most optical design programs. The multi-configuration technique of design is discussed in more detail in the following sections.

7.3 Polygon Scan Efficiency

Figure 18 shows one facet of a polygon with feed beam for scanning. The parameters and relationships that determine the limits of scan efficiency and minimum size for a polygon scanner (assuming no facet tracking) are D = beam diameter, β = nominal feed beam offset angle, ε = facet edge roll zone, $\alpha = 2\pi/$(no. of facets N) = angular extent of facet, and $\delta \sim [D/\cos(\beta) + \varepsilon]/r$ = angular extent of beam plus roll zone, where the scan efficiency limit for a given polygon is

$$\eta_s = 1 - (\delta + \varepsilon/r)/\alpha \tag{24}$$

and the minimum polygon circumscribed radius is

$$r > \frac{[D/\cos(\beta) + \varepsilon]}{[\alpha^*(1 - \eta_s)]} \tag{25}$$

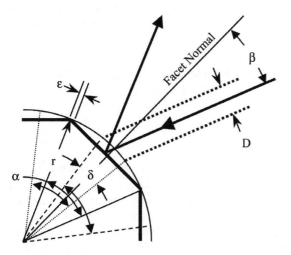

Figure 18 Diagram of a polygon facet in its central position and parameters describing the maximum beam diameter that can be reflected with no vignetting through a scan angle at peak efficiency.

Given the circumscribed radius of the polygon, feed beam angle, and facet scan angle, and assuming the edge roll zone (unusable part of the facet clear aperture) is negligible, the maximum beam diameter that can be supported without vignetting is given by the equation

$$D < r\cos(\beta)[\alpha^*(1 - \eta_s)] \tag{26}$$

The circumscribed cord defined by polygon facet less the cords defined by the feed beam footprint on the circumscribed circumference and roll zone will limit the useful polygon rotation.

A sample polygon and scan lens design might have the following specifications: 2000 DPI with a 12.7 μm $1/e^2$ spot diameter; wavelength of 0.6328 μm; scan length L of 18 in. (457.2 mm); an eight-sided polygon with facet angle α of 0.7854 radians, scan efficiency η_s of 60%, and feed beam angle β of 30°.

The derived system parameters are: $F/\#$ of the lens for this spot diameter is $F/11$ [given by Eq. (14)]; scan lens focal length F is 485.1 mm [given by $L/(2\alpha^*\eta_s)$]; and beam diameter D is 44.1 mm. The required circumscribed radius of the polygon is greater than 162 mm, with a facet face width of 2.8 times the diameter of the incoming beam.

To scan a given number of image points on a single scan line, there is a trade-off between the scan angle θ and the diameter of the feed beam [Eq. (16)]. To achieve compactness with a smaller polygon, a smaller feed beam would be required along with a greater scan angle. The search for system compactness drives the field angle to larger and larger values. Scan angles above 20° increase the difficulty in correcting the $F/\#$. The conflict can be somewhat resolved by using a smaller angle of incidence β, but then there may be interference between the incident beam and the lens mount. A compromise between these variables requires close cooperation between the optical and mechanical engineering effort.

Figure 19 illustrates the relationship between the number of polygon facets, polygon diameter, and scan efficiency. The previous example is illustrated in the eight facet plots. Polygons typically work at around 50% scan efficiency because of the feed beam diameter required by the resolution coupled with the rotating polygon scanners size limitations and cost considerations. A comprehensive treatment of the relationship between the incident beam, the scan axis, and the rotation axis of a prismatic polygonal scanner is given in Ref. 16.

Note: Care should be taken in the specification of the *scan angle*. Without a clear definition it can be interpreted as the mechanical scan angle, the optical scan angle, or even the optical half scan angle.

7.4 Internal Drum Systems

As stated before, internal drum scanning systems are least demanding on the optical system, because the lenses do not have to cover a wide field. Most of the burden is shifted to the accurate mechanical alignment of the turning mirror. The concept of the internal drum scanner can be applied to a flat-bed scanner by using a flat-field lens. The system then becomes the equivalent of a pyramidal scanner with only one facet. All of these systems have common alignment requirements.

The nominal position of the turning mirror is usually set at 45° (0.785 radians) with respect to the axis of rotation. It does not have to be exactly 45° as long as the collimated feed optical beam enters parallel to the axis of the rotation. The latter condition is needed

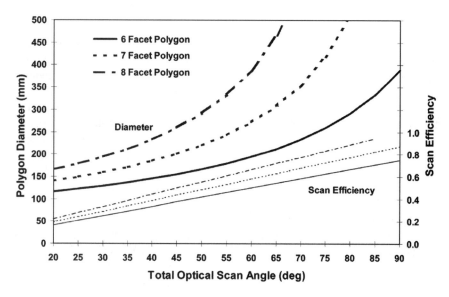

Figure 19 Polygon diameter and scan efficiency vs. total optical scan angle ($D = 44.1$ mm, $\varepsilon = 0$ mm, and $\beta = 30°$).

to eliminate bow in the scan line. In a perfectly aligned system the ray that passes through the nodal points of the lens must meet the deflecting mirror on the axis of rotation. When the lens is placed in front of the turning mirror, the second nodal point of the lens must be on the rotation axis of the mirror. When the lens is placed after the turning mirror, its first nodal point must be on the ray that intersects the mirror on the rotation axis. There is some advantage in placing the lens between the mirror and the recording plane. In this position the lens has a shorter focal length, and the bow resulting from any error in the nodal point placement of the lens is reduced.

7.5 Holographic Scanning Systems

From an optical designer's perspective, holographic scanning systems have an advantage, as the need for wobble correction can be reduced significantly without resorting to cylindrical components. Conversely, they usually require some bow correction, and if used with laser diodes (which exhibit wavelength shifts), they require significant color correction. Line bow correction can be achieved by using a prism (or grating) component after the holographic scanning element and/or adding complex holograms, reflective and refractive optical components to the lens system. The prism component introduces bow to balance out the bow in the same way that a spectrographic prism adds curvature to the spectral lines. The lens can be tilted and decentered as an alternative method for reducing the bow. Holographic scanning systems are discussed in greater detail in Sec. 11.

8 LENS DESIGN MODELS

Regardless of the ultimate complexity of a scanning system, a simple model is often the best starting point for the design of a scan lens, and sometimes all that is needed. The exception is when adapting or tweaking a previous complex design model for minor changes in wavelength, scan length, or resolution. In a simple model, the actual method by which the beam deflection is introduced is not included in the lens design; the beam deflection method is assumed to introduce only angular motion; and it neglects any beam displacement that may occur due to the deflection method. The lens is modeled and optimized to perform at several field angles, in much the same way a standard photographic objective is optimized. The main differences are in the external placement of the aperture stop where the chief ray for each field passes through its center as if scanning and the optimization with distortion constraints for $F-\theta$ linearity. A detailed example demonstrating the simple model is developed in Sec. 8.1 with additional examples provided in Sec. 9.1 through 9.7.

In practice, the reason that all parallel bundles appear to pivot about the center of this external aperture stop surface is that a fixed beam is incident on a beam deflector rotating in proximity to the stop. If the mechanical rotation axis of this deflector intersects the plane of the mirror facet and the optical axis of the scan lens, then the simple model is accurate. This is the case for galvanometer-based systems, where the mechanical rotation axis is close enough to the plane of the mirror facet that the deviation from the simple model is negligible.

At some point in the design process it must be decided how rigorously the geometry of the moving deflector needs to be modeled. At the very least, the design should be analyzed for the actual angular motion with the rotation of the pupil and the displacement of the beam to determine as-built performance of a design and validate the effectiveness of the simple model. This can sometimes be accomplished without modeling the actual scanner, as described below.

Multifaceted holographic deflectors do not suffer from beam displacement, even though their mechanical rotation axis is some distance from the active region of the facets. Where dispersion from the hologon is not an issue, a simple model can suffice for the design of a lens for a hologon-based scan system, although complex truncation and multifacet illumination effects are ignored.

8.1 Anatomy of a Simple Scan Lens Design

The following is a description of a scan lens design, which begins with the design specifications outlined in Table 3 and describes the evolution of a design to meet these specifications.[17] The first five numbered specifications (scan line length, wavelength, resolution, image quality, and scan linearity) are the very minimum required to begin the optical design. The laser source for this exercise is a diode operating near Gaussian TEM00 with a wavelength that can drift with temperature and power, and emit over a bandwidth of 25 nm. Parameters listed with no numbers are provided as reference or potential additional specifications.

Beginning with the resolution requirement of 300 DPI based on $1/e^2$ and a nominal wavelength of 780 nm, the ideal Gaussian waist size (pixel size) is

$2w_0 = 84.7 \,\mu\text{m}$

with a numerical aperture defined by

$NA = \theta_{1/2} = \lambda/(\pi w_0)$
$= 0.006$

Optical Systems for Laser Scanners

Table 3 Specifications for a Scan Lens Example

Parameter	Specification or goal
1. Image format (line length)	216 mm (8.5 in.)
2. Wavelength	780 + 15, −10 nm
3. Resolution ($1/e^2$ based)	300 DPI (600 DPI goal)
4. Wavefront error	<1/20 wave RMS
5. Spot size and variation over scan	TBD ±20% (±10%, goal)
6. Scan linearity ($F-\Theta$ distortion)	<1% (<0.1%, <0.03%)
7. Depth of field	No specific requirement
8. Telecentricity	No scan length vs. DOF control
9. Optical scan angle	TBD (±15 to 45°)
10. F-number	Defined by resolution
11. Effective focal length	... by resolution and scan angle
12. Overall length	<500 mm
13. Scanner clearance	>25 mm
14. Image clearance	>10 mm
15. Scanner requirements	TBD (type, size)
16. Packaging	TBD
17. Operating/storage temperature	TBD

TBD, to be determined (at a later date); DOF, depth of focus.

Setting the optical scan angle at ±15° (0.26 radians) for the initial (back-of-the-envelope) calculations, the required focal length for a 216 mm scan line is determined by

$$F\theta = 216\,\text{mm}$$
$$F = 216/(2 \times 0.26) = 415\,\text{mm}$$

Experience has shown that the focal length arrived at in this first pass would likely result in a system that is too long (when considering scanner to lens clearance, the thickness of real lenses, and the image distance). To shorten the length of the optical system the optical scan angle is increased to ±20° (0.35 radians). The new required focal length then becomes

$$F = 309\,\text{mm}$$

and the required design aperture diameter is

$$\text{EPD} = F(2NA) = 3.7\,\text{mm}$$

The simple two-lens configuration shown in Fig. 20(a) comprises a concave-plano flint (Schott® F2) element and plano-convex crown (Schott SK16) element (both by Schott Glass Technologies Inc., Duryea, PA) with powers appropriate for axial color correction selected as a starting point. Scaled for focal length and focused, the composite RMS wavefront error (weighted average over the field) is 0.038 waves. While at first glance this wavefront error appears to meet the requirements, further examination of the ray aberration and field performance plots illustrated in Figs 20(b) and (c) indicate substantial astigmatism limiting performance at the edge of the field and distortion that is far from $F-\theta$.

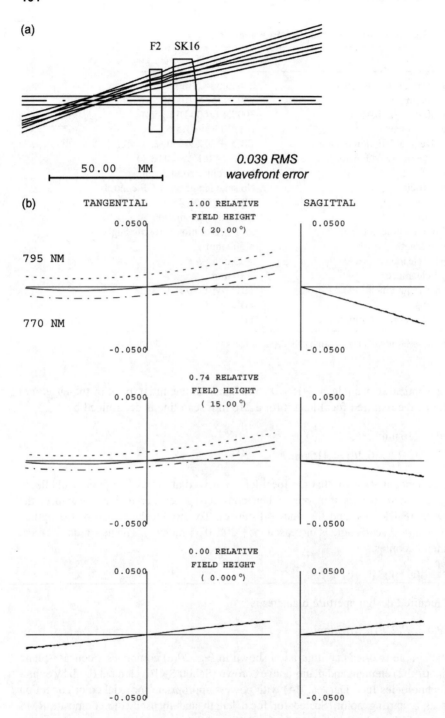

Figure 20 (a) Two-element starting scan lens design; (b) ray aberration plots for starting design; and (c) field performance plots for starting design.

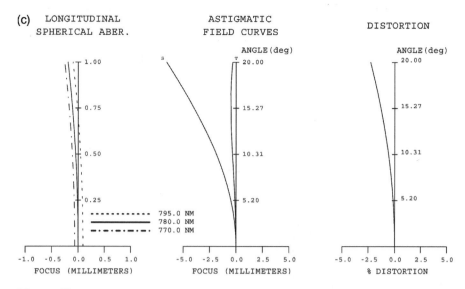

Figure 20 *Continued.*

The horizontal plot axis in Fig. 20(b) is relative aperture and the vertical plot axis is the transverse ray error at the image plane. The slope of the shallow curve is a measure of focus shift and a change in slope over the relative field and between the sagittal and tangential curves is a measure of the field curvature and astigmatism. These ray aberration plots clearly show astigmatism (indicated by the slope difference between sagittal and tangential curves at 15 and 20° field angles) and lateral color (indicated by the displacement of the tangential curves, a change in image height, for the extreme wavelengths at the 15 and 20° field angles).

The field performance plots in Fig. 20(c) also indicate a very small amount of axial color and spherical aberration (displaced longitudinal curves for each wavelength of the left plot), substantial astigmatism (indicated by the departure of the sagittal field curve from the nearly flat tangential curve of the center plot), and the distortion that is closer to F–tan Θ than F–Θ (seen in the distortion curve on the right).

Optimizing this starting design with surface curvatures as variables for best spot performance while maintaining scan length with a constraint (but no distortion controls) yields the bi-convex bi-concave configuration illustrated in Fig. 21. The RMS wavefront error improved after the first round of computer optimization iterations by a less than desirable balance of the astigmatism with higher order aberrations (center field plots). The air interface between elements was deleted, leaving the three surfaces of a doublet for a second round of optimization (to test the design for a simpler solution), resulting in performance slightly better than the starting design for RMS wavefront error and astigmatism.

Adding glass variables such as index of refraction and dispersion and adding F–θ constraints, weighted rather than absolute for a more stable convergence, results in no significant change in performance, as illustrated in Fig. 22. Splitting the power of the positive crown element between the doublet and an additional plano-convex lens provides more design variables for the next round of optimization iterations. The result is much better correction of both the astigmatism and F–θ distortion with an RMS wavefront

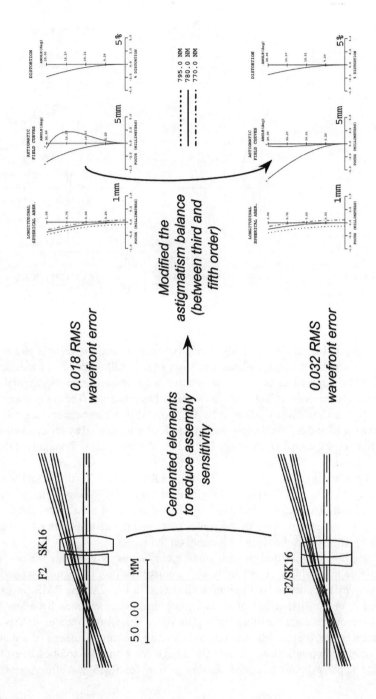

Figure 21 First and second design iterations with no distortion controls.

Optical Systems for Laser Scanners

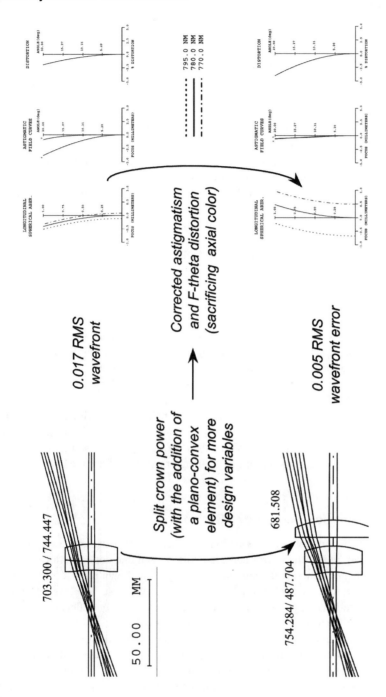

Figure 22 More design iterations (adding weighted $F-\theta$ distortion controls).

error of 0.005 waves, but at the sacrifice of some axial color correction (a shift in focus with wavelength). This trade-off is acceptable if most of the wavelength variation is from diode to diode, where focus can be used to accommodate the different laser wavelengths, and the wavelength variations from changes in junction temperature due to power and ambient temperature are controlled to a few degrees. Selecting real glasses for the final design iterations results in the design and performance as illustrated in Fig. 23, with no noticeable change in performance.

Raising the performance requirement target of this example design lens to the resolution goal of 600 DPI (double the original specification) requires twice the numerical aperture and hence twice the design aperture diameter. Design iterations for the higher resolution $F-\theta$ lens begins by adding back glass variables to deal with the pupil aberrations introduced by the larger aperture, illustrated in Fig. 24. The performance after these first iterations deteriorated to a wavefront error of 0.037 waves RMS, predominantly from spherical aberration (the tangential S-shaped curve) with a bit of coma at the edge of the field (indicated by the asymmetry in the full field tangential curve).

The performance target is raised yet again with an increase in the scan angle specification to $\pm 30°$ from $\pm 20°$. The design for this exercise starts with the previous lens, scaled by the ratio of scan angles (20/30) from pupil to image including the design diameter. The performance after several more iterations and the selection of real glass types, illustrated in Fig. 25, is improved to a 0.028 RMS wavefront error, by improving spherical aberration, axial color and astigmatism (with the help of the 2/3 scaling of the design aperture), and more linear distortion resulting in better $F-\theta$ correction. The final design prescription for the scan lens example is listed in Table 4 with performance specifications listed in Table 5. The calibrated $F-\theta$ scan distortion is plotted in Fig. 26. It shows a linearity better than 0.1% over scan and better than 1% locally. The selection of LAFN23 in this latest design is not ideal for its availability or glass properties. Later iterations to finalize this preliminary design should explore glass types that optimize the availability, cost, and transmission along with image quality performance.

8.2 Multiconfiguration Using Tilted Surfaces

There is often no substitute for the introduction of a mirror surface in the lens design model to simulate the scan, where the mirror surface is tilted from one configuration (or "zoom position") to another to generate the angular scanning of the beam prior to the scan lens. Modeled in this way, there is no object field angle in the usual sense. The beam prior to the rotating mirror is stationary. When a beam of circular cross-section reflects off the plane mirror, the reflected beam will have the same cross-section.

That is not the case in a simple model involving object field angles and a fixed external circular stop, where bundles from off-axis field angles will be foreshortened in the scan direction to an ellipse by the cosine of the field angle. As the field angle approaches 30° this effect becomes increasingly significant. At some stage in the design process the software can be tricked into enlarging the bundle in the scan direction to compensate for this foreshortening, but not all analyses may run using this work around. In particular, diffraction calculations typically rely on tracing a grid of rays that are limited by defined apertures in the optical system, which would defeat the intended effect of simple tricks.

A constant beam cross-section at all scan angles can also be maintained by simply bending the optical axis at the entrance pupil in a multiconfiguration. This is often a

Optical Systems for Laser Scanners

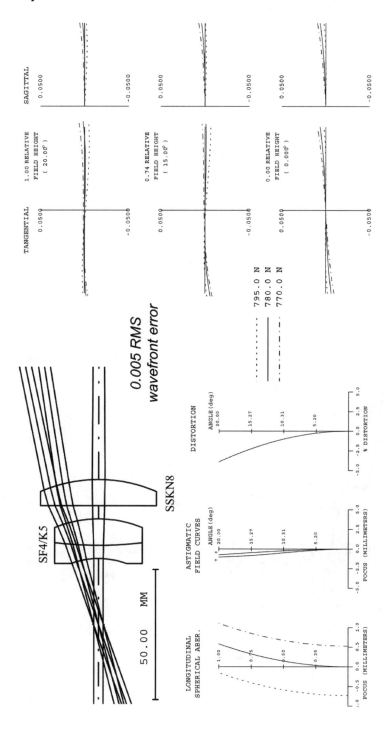

Figure 23 Final design iteration (with real glass types).

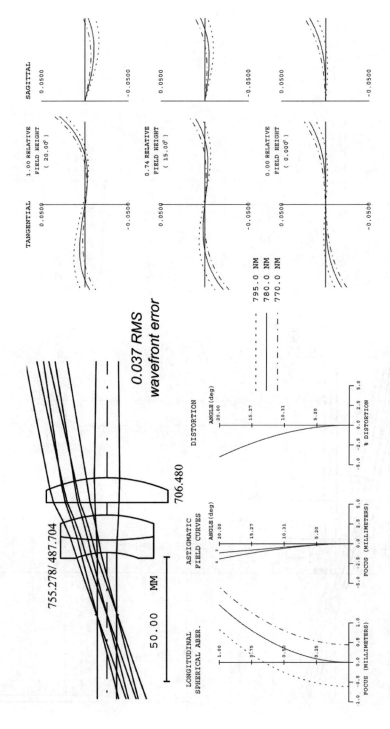

Figure 24 New design iterations (with previous $F-\theta$ lens and twice the design aperture diameter).

Optical Systems for Laser Scanners

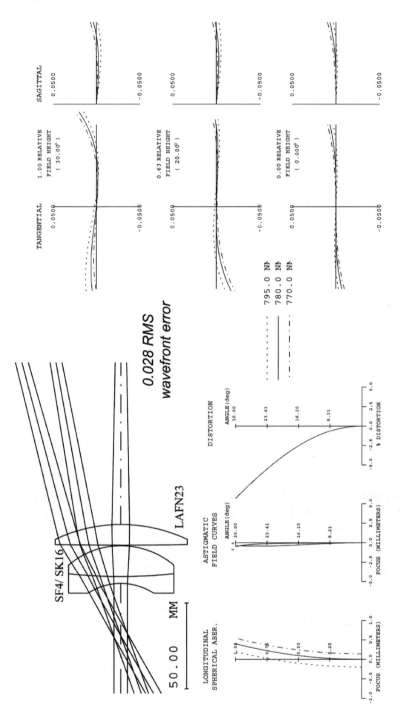

Figure 25 Final design iteration (with selected glasses).

Table 4 Final Design Prescription for the Scan Lens Example

Surface	Radius	Thickness	Glass type
1		30.3512	
2	−36.5662	4.0000	SF4
3	310.0920	18.0000	SK16
4	−49.8537	0.3276	
5	2541.4236	12.0000	LAFN23
6	−94.6674	270.3002	
Image		0.0462	

Table 5 Final Design Specifications for the Scan Lens Example

Parameter	Specification or goal
1. Image format (line length)	216 mm (8.5 in.)
2. Wavelength	770–795 nm
3. Resolution ($1/e^2$ based)	26 μm (~1000 DPI)
4. Wavefront error	<1/30 wave RMS
5. Scan linearity ($F-\theta$ distortion)	<1% (<0.2% over ±25°)
6. Scanned field angle	±30°
7. Effective focal length	206 mm (calibrated)
8. F-number	f/26
9. Overall length	335 mm
10. Scanner clearance	25 mm
11. Image clearance	270 mm

good compromise in modeling complexity that does not require the use of reflective surfaces, but maintains the integrity of the scanned beam optical properties.

Holographic deflectors may not faithfully emulate a tilted mirror. In such systems the beam cross-section does change as a function of angle, where a circular input beam results in an elliptical output whose orientation changes with scan angle.

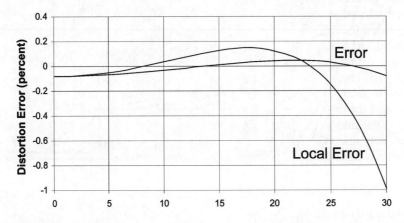

Figure 26 Final $F-\theta$ scan linearity.

Optical Systems for Laser Scanners

8.3 Multiconfiguration Reflective Polygon Model

A polygon in a lens design model is simply a mirror at the end of an arm. All rotation is performed about the end of the arm opposite the mirror. The length of the arm defined by the inscribed radius of the polygon, the location of the arm's rotation axis relative to the scan lens, and the amount of rotation about the pivot can all be (with care) optimized during the design. The facet shape is defined by an appropriate clear aperture specification on the mirror surface.

Because the mechanical rotation axis rarely intersects the optical axis of the scan lens and the mirror facet is far from the rotation axis, it is best just to apply a rigorous model that automatically accounts for the complicated mirror surface tilts, displacements, and aperture effects of the polygon scan complex geometry.

Specific pupil shifts and aperture effects could be computed and specified for each configuration, but this does not exploit the full potential of a multi-configuration optical design program. When the model is defined in a general fashion, the actual constructional parameters of the polygon or parameters governing its interface with the feed beam and scan lens may be optimized simultaneously as the scan lens is being designed, particularly in the final stages of design. This often leads to better system solutions than simply combining devices in some preconceived way.

If the location of the entrance pupil is defined as where the chief ray intersects the optical axis of the scan lens, then the axial position of the pupil shifts with polygon rotation angle (and scan angle). This axial pupil shift requires the lens to be well corrected over an aperture larger than just the feed beam diameter. The effect is greater for systems with polygons having fewer facets, larger optical scan angles, and/or larger feed beam offset angles. High-aperture (large NA) scan lenses are especially susceptible to this effect.

Determining where the real entrance pupil is for each configuration, or how much larger of a beam diameter the lens should really be designed to accommodate, is difficult. It is especially difficult if the polygon is to be designed at the same time as the scan lens! When this pupil shift is included in the lens design model by rigorously modeling the polygon geometry, the effect on lens performance can be accurately assessed. More importantly, the lens being designed may be desensitized to expected pupil shifts and, if the polygon is being designed, the pupil shifts may be minimized. As maximum scan angle is approached, the facet size may be insufficient to reflect the entire beam. Asymmetrical truncation or vignetting occurs, which can modify the shape of spot at the image plane. Accurate aperture modeling is especially important for accurate diffraction-based spot profile calculations. In a rigorous polygon model, by putting aperture specifications on the surface that represents the reflective facet surface, all vignetting by the facet as a function of polygon rotation will be automatically accounted for.

By having a rigorous polygon model implemented, it is also possible to further evaluate what actually happens to the section of the beam that misses the facet. Classifying the rays that miss the facet as vignetted is really an oversimplification. In reality these rays will likely reflect off the tip and adjacent facet. This stray light beam may enter the lens and find its way to the image surface. Stray light problems can ruin a system. Double-pass systems are especially susceptible to this design flaw. The multiconfiguration setup can be used to evaluate stray light problems and suggest baffle designs.

8.4 Example Single-Pass Polygon Setup

The key to exploiting the multiconfiguration design method is to include the rotation axis when modeling the polygon. A brief overview of such a model begins with the definition of the expanded laser beam, followed by a first fold mirror, the defined polygon with its rotation axis, and finally the scan lens.

The following CODE V* sequence file (*CODE V is a trademark of Optical Research Associates, Pasadena, CA) for the lens illustrated in Fig. 27, shows one way to set up a polygon and lens in a commercial lens design program. Standard catalog components were chosen: the six-sided polygon is Lincoln Laser's PO06-16-037, and the two-element sectioned scan lens is Melles Griot's LLS-090. The combination will create 300 DPI output using a laser diode source.

8.4.1 Multiconfiguration Code V Lens Prescription

```
RDM; LEN
TITLE "LINCOLN PO-6-16-37 MELLES GRIOT LLS-090 90mm F/50 31.5-deg P-468"
EPD 1.8145
PUX 1.0    ;PUY 1.0    ;PUI 0.135335
DIM M
WL  780
YAN 0.0
S0    0.0   0.1e20                               ! Surface 0        "A"
S     0.0   50.8                                 ! Surface 1
   STO
S     0.0   -25.4  REFL                          ! Surface 2
   XDE 0.0;  YDE 0.0;  ZDE 0.0;  BEN
   ADE -31.5; BDE 0.0;  CDE 0.0;
S     0.0   0.0                                  ! Surface 3        "B"
   XDE 0.0;  YDE 0.0;  ZDE 0.0;
   ADE 63.0;  BDE 0.0;  CDE 0.0;
```

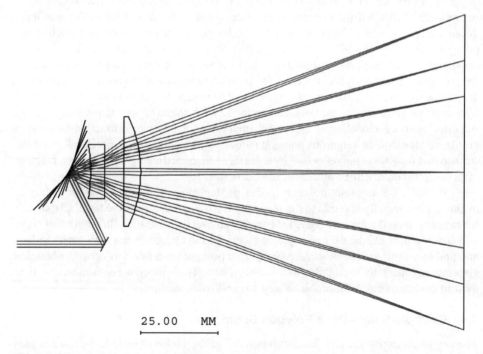

Figure 27 A multiconfiguration, single-pass polygon system.

Optical Systems for Laser Scanners

```
S     0.0   -16.892507                          ! Surface 4      "C"
S     0.0    0.0                                ! Surface 5
  XDE 0.0;  YDE 10.351742;  ZDE 0.0;
  ADE -31.5;  BDE 0.0;  CDE 0.0;
S     0.0   19.812                              ! Surface 6      "D"
  XDE 0.0;  YDE 0.0;  ZDE 0.0;
  ADE -15.5;  BDE 0.0;  CDE 0.0;
S     0.0   -19.812   REFL                      ! Surface 7
S     0.0    0.0                                ! Surface 8      "E"
  XDE 0.0;  YDE 0.0;  ZDE 0.0;  REV
  ADE -15.5;  BDE 0.0;  CDE 0.0;
S     0.0   16.892507                           ! Surface 9      "F"
  XDE 0.0;  YDE 10.351742;  ZDE 0.0;  REV
  ADE -31.5;  BDE 0.0;  CDE 0.0;
S     0.0    7.0                                ! Surface 10     "G"
S   -49.606  4.5   SK16_SCHOTT                  ! Surface 11
S     0.0    6.35                               ! Surface 12
S     0.0    5.35  SFL6_SCHOTT                  ! Surface 13
S   -38.633  104.340988                         ! Surface 14
  PIM
SI    0.0   -0.633188
ZOOM 7                                          !                "H"
ZOOM ADE S6 -15.5 -11 -7.5 0 7.5 11 15.5
ZOOM ADE S8 -15.5 -11 -7.5 0 7.5 11 15.5
GO
CA                                              !                "I"
CIR S2  2.5        ;CIR S2  EDG  2.5
REX S7  4.7625     ;REY S7  11.43
REX S7  EDG  4.7625 ;REY S7  EDG  11.43
REX S11  5.0       ;REY S11  5.1
REX S12  5.0       ;REY S12  7.6
REX S13  5.0       ;REY S13  14.9
REX S14  5.0       ;REY S14  15.8
GO
```

8.4.2 Lens Prescription Model

Start with a collimated, expanded laser beam of the required diameter and fold its path with a $-31.5°$ mirror tilt to obtain the desired feed angle of $63°$ with respect to the planned optical axis of the scan lens. The aperture stop should be defined prior to the polygon (preferably on surface 1). Any truncation of the Gaussian input beam should be done at the aperture stop. Do not flag the polygon surface as the aperture stop, because some software will automatically ray-aim each bundle to pass through the center of the stop surface.

Define a reference point that will be on the optical axis of the scan lens. It is convenient to have its location where the facet would intersect the axis when the polygon is rotated for on-axis evaluation. The surface should be tilted $63°$ so that any subsequent thickness would be along the optical axis.

Go to the polygon rotation center and tilt so that any subsequent thickness would be radial from the polygon center toward the facet surface. To get to the polygon center from the reference point, use a combination of surface axial and transverse decenters (traveling in right angles for simplicity). It is convenient to choose a tilt that will cause the polygon to be rotated into position for on-axis evaluation. Here, we must translate -16.9 mm away from the lens along the optical axis and then decenter up along Y with $YDE = 10.4$ mm.

Tilting about X in the YZ-plane with ADE = $-31.50°$ ($63°/2$) points the surface normal to the facet.

Before going to the polygon facet, any additional tilt about X (ADE) is specified. This is a multiconfiguration parameter: each configuration will have a different value specified for this additional tilt. Here, ADE 15.50 is specified to cause the polygon to rotate into position for maximum scan on the negative side. This is really the polygon shaft rotation angle; for nonpyramidal polygons the reflection angle (scan angle) changes at twice the rate of the shaft angle. Once the image surface is defined, the system will be defined to have seven configurations, and a different ADE value for this surface will be specified for each (step "H"). Translate to the facet using the polygon inscribed radius for thickness (19.8 mm). Now reflect. This reflection occurs at the first real surface that the beam encounters since the fold mirror. All other surfaces have been "dummy" surfaces where no reflection or refraction takes place. It is on this reflective surface that aperture restrictions may be defined to describe the shape of the facet. Use a thickness specification on this surface to go back to the polygon center following the reflection (-19.8 mm), to maintain the integrity of the first-order optical path lengths.

Some commercial software programs (such as CODE V) have a "return" surface that one could now use to get back to the reference point defined in step B, prior to defining the scan lens. Here, a more conservative approach is taken that can be used with any software package. Undo the additional polygon shaft rotation that was done in step D. The REV flag in CODE V internally negates the angle.

Continuing to move back to the reference point defined in step B, undo the polygon shaft tilt that sets it for on-axis evaluation (ADE = $31.5°$), decenter down along Y, back to the scan lens axis (YDE = -10.4 mm), and translate toward the lens along its axis to the reference. This places the mechanical axis back to the same location that it was in step B, before the reflection off the polygon. Here, the REV flag changes the signs of the tilt and decenter specifications and performs the tilt before the decenter.

Define the scan lens. The reference surface defined in step B, and returned to in steps E and F, is approximately the location of the entrance pupil. The thickness at the image surface is the focus shift from the paraxial image plane.

Having now specified a valid single-configuration system to the software, the system is redefined to have seven configurations (ZOOM 7) and the parameters that change from one configuration to another are listed. Owing to the way that the polygon was modeled, rotating it to a different position is simply a matter of changing the parameter that represents the shaft rotation angle. ADE is on surfaces 6 and 8.

Since these are catalog components, the clear apertures are available and are specified here. The rectangular aperture specifications for the polygon facet are given for surface 7.

8.5 Dual-Axis Scanning

When more than one galvanometer is used to generate a two-dimensional scan at the image plane it is usually necessary to use a multiconfiguration setup. Each galvanometer defines an apparent pupil and net effect of their physical separation creates a very astigmatic pupil, where X-scan and the Y-scan do not originate at the same location on the optical axis. In modeling these scanners, the optical axis or reflective surfaces representing the galvanometer mirrors may be tilted. The latter approach is usually worth the effort in order to visualize the problem and avoid mechanical interferences.

Optical Systems for Laser Scanners

9 SELECTED LASER SCAN LENS DESIGNS

Scan lenses with laser beams have been considered almost from the time of the development of the first laser. Laboratory models for printing data transmitted from satellites were underway in the late 1960s. Commercial applications began coming out in the early 1970s, and laser printers became popular in the early 1980s. The range of applications is steadily increasing. With the development of new laser sources (including violet and UV) and high-precision manufacturing processes for complex surfaces (including plastics), the field of lens design offers new challenges and oppor-tunities with room for new design concepts.

The lenses presented in this section were selected to show what appears to be the trend in development. The spot diameters are getting smaller, the scan lengths longer, and the speed of scanning higher. Some of the lenses near the bottom of the list are beginning to exhaust our present design and manufacturing capabilities. The newest requirements are reaching practical limits on the size of the optics and the cost of the fabrication and mounting of the optics. It appears that the future designs will have to incorporate mirrors and lenses with large diameters, and new methods for manufacturing segmented elements will be needed.

The lenses shown in this section start with some modest designs for the early scanners and progress to some of the latest designs. Two of the designs were obtained from patents. This does not mean that they are fully engineered designs. The rest of the designs are similar paper designs. This means that the designer has the problem "boxed in" — where all the aberrations are in tolerance and under control. The next, however, is only the beginning of the engineering task of preparing the lenses for manufacture. This phase is a lengthy process of making sure that all the clear apertures will pass the rays, and that the lenses are not too thick or too thin. The glass types selected have to be checked with availability and cost, and the experience of the shop working with the glass. The design has to be reviewed to consider how the lenses are to be mounted. Some of the lenses may require precision bevels on the glass or a redesign may avoid this costly step. This section also includes a few comments about the designs with regard to practicality. Table 6 contains a summary of attributes for the selected lenses.

In the following lens descriptions all the spot diameters refer to the diameter of the spot at the $1/e^2$ irradiance level. The number of spots on the line are calculated assuming contiguous spots packed adjacent to each other at the $1/e^2$ irradiance level.

9.1 A 300 DPI Office Printer Lens (λ = 633 nm)

Figure 28 presents a patent (U.S. Patent 4,179,183, Tateoka, Minoura; December 18, 1979) assigned to Canon Kabushiki Kaisha. The patent contains a lengthy description of the design concepts used in developing a whole series of lenses. Fifteen designs are offered with the design data along with plots of spherical aberration, field curves, and the linearity of scan. The design shown is example 6 of the patent. The design data were set up and evaluated, and the results agree well with the patent. The focal length is given as 300 mm. The aberration curves appear to be given for the paraxial focal plane, and the linearity is shown to be within 0.6% over the scan. However, if one selects a calibrated focal length of 301.8 mm (11.8 in.) and shifts the focus 2 mm (0. 079 in.) in back of the paraxial focus, it appears that the lens is well corrected to within $\lambda/4$ OPD (optical path difference) and linear to within 0.2%. A similar lens may have been used in early Canon laser printer engines. This lens has an exceedingly wide angle for a scan lens. It has a great advantage in

Table 6 Summary of Attributes for the Selected Scan Lenses

Section	F	F/#	L	ROAL	RFWD	d	L/d	RBcr	RPR	REPR	NS/I	NO.el
9.1	300	60	328	1.4	0.13	70	4,700	0.66	−11	−5	370	2
9.2	100	24	118	1.4	0.06	28	4,300	0.34	−12	−12	920	3
9.3	400	20	310	1.6	0.17	23	13,000	0.49	−26	−30	1,100	3
9.4	748	17	470	1.4	0.06	20	23,000	0.29	−15	−50	1,200	3
9.5	55	5	29	2.1	0.44	5.8	5,000	0.84	−32	−24	4,300	3
9.6	52	2	20	4.2	0.39	4	5,000	1.0	−56	−16	6,350	14
9.7	125	24	70	2.3	0.8	20	3,500	0.5	−4.5	−5	1,200	5

F, focal length of the lens (mm); $F/\#$, F-number (ratio F/D); L, total length of scan line (mm); ROAL, overall length from the entrance pupil to the image plane relative to the focal length (mm/mm); RFWD, front working distance relative to the focal length F; d, diameter of the image of a point at the $1/e^2$ irradiance level (μm); L/d, number of spots of a scan line; RBcr, paraxial chief ray bending relative to the input half angle; RPR, ratio of the Petzval radius to the lens focal length; REPR, estimated Petzval radius relative to the lens focal length from Eq. (22); NS/I, number of spots per inch; and NO.el, number of lens elements.

the design of a compact scanner. This printer meets the needs of 300 DPI, which is quite satisfactory for high-quality typewriter printing of its time. The secret of the good performance of this lens is the airspace between the positive and negative lenses. There are strong refractions on the two inner surfaces of the lens, which means the airspace has to be held accurately and the lenses must be well centered.

9.2 Wide-Angle Scan Lens ($\lambda = 633$ nm)

The lens in Fig. 29 has a 32° half-field angle. It has a careful balance of third-, fifth-, and seventh-order distortion, so that at the calibrated focal length it is corrected to be $F-\theta$ to

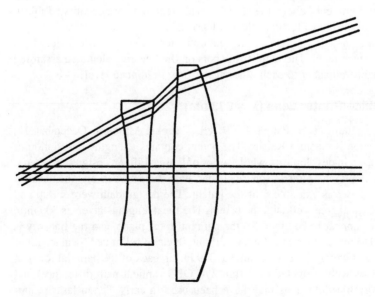

Figure 28 Lens 1: U.S. Patent 4,179,183 Tateoka, Minoura; $F = 300$ mm, $F/60$, $L = 328$ mm.

Optical Systems for Laser Scanners

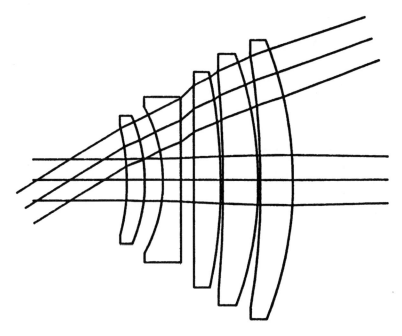

Figure 29 Lens 2: U.S. Patent 4,269,478 Maedo, Yuko; $F = 100$ mm, $F/24$, $L = 118$ mm.

within 0.2%. To do this the lens uses strong refraction on the fourth lens surface and refraction on the fifth surface to achieve the balance of the distortion curve. The airspace between these two surfaces controls the balance between the third- and fifth-order distortion. This would also be a relatively expensive lens to manufacture. The design may be found in U.S. Patent 4,269,478; it was designed by Haru Maeda and Yuko Kobayashi and assigned to Olympus Optical Co., Japan.

9.3 Semi-Wide Angle Scan Lens, ($\lambda = 633$ nm)

The lens in Fig. 30 shows how lowering the $F/\#$ to 20 and increasing the scan length increases the sizes of the lenses. This lens is a Melles Griot product designed by David Stephenson. It is capable of writing 1096 DPI and is linear to better than 25 µm. The large front element is 128 mm in diameter. It will transmit 2.8 times as many information points as the first lens. This lens requires modest manufacturing techniques, but, as shown in the diagram, the negative lens may be in contact with the adjacent positive lens. Either the airspace should be increased, or careful mounting has to be considered.

9.4 Moderate Field Angle Lens with Long Scan Line ($\lambda = 633$ nm)

The lens in Fig. 31 was designed by Robert E. Hopkins for a holographic scanner. It has a half scan angle of 18°, covers a 20 in. scan length, and can write a total of 23,100 image points. It has a short working distance between the holographic scan element and the first surface of the lens. This makes it more difficult to force the $F-\theta$ condition to remain within 0.1%. The working distance was kept short in order to keep the lens diameters as small as

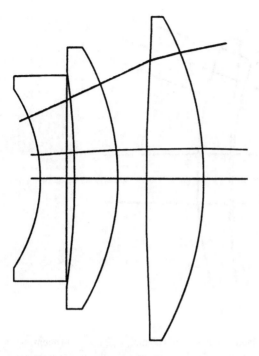

Figure 30 Lens 3: Melles Griot, Designer D. Stephenson; $F = 400$ mm, $F/20$, $L = 300$ mm.

possible. The largest lens diameter is 110 mm. Lens performance could not be improved without making the elements considerably larger. Adding more lense elements does not help. The lens performs well in the design phase, but, since it is relatively fast for a scan lens, the small depth of focus makes the lens sensitive to manufacture and mounting. The only way to make the lens easier to build is to improve the Petzval field curvature, and this requires more separation between the positive and negative lenses. The result is that the lens becomes longer and larger in diameter. Another way is to increase the distance between the scanner and the lens. This will also increase the lens size. We believe that this lens is close to the boundary of what can be done with a purely refracting lens for scanning a large number of image points. To extend the requirements will require larger lenses. It may be possible to combine small lenses with a large mirror close to the focal plane, but costs would have to be carefully considered. Photographic-quality shops can build this lens with attention to mounting.

9.5 Scan Lens for Light-Emitting Diode ($\lambda = 800$ nm)

In Fig. 32 is another lens designed by Robert E. Hopkins to perform over the range of wavelengths from 770 to 830 nm. It was designed to meet a telecentricity tolerance of $\pm 2°$. The lens could not accommodate the full wavelength range without a slight focal shift. If the focal shift is provided for, the diodes may vary their wavelength from diode to diode over this wavelength region, and the lens will perform satisfactorily. This lens was not fully engineered for manufacture. It would be necessary to consider carefully how the negative–positive glass-to-glass contact combination would be mounted.

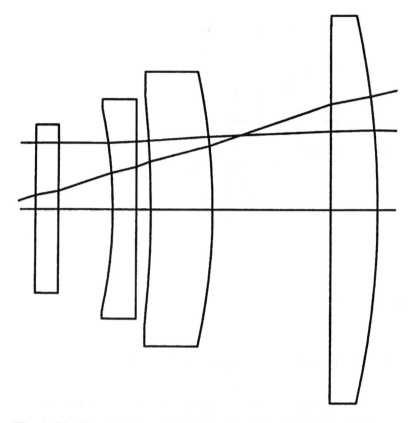

Figure 31 Lens 4: Designer R. Hopkins; $F = 748$ mm, $F/17$, $L = 470$ mm.

9.6 High-Precision Scan Lens Corrected for Two Wavelengths ($\lambda = 1064$ and 950 nm)

The Melles Griot lens in Fig. 33 was designed for a galvanometer XY scanner system. It is capable of positioning a 4 μm spot anywhere within a 20 mm diameter circle. The spot is addressed with 1064 nm energy with simultaneous viewing of the object at 910–990 nm. The complexity of the design is primarily due to the need for the two-wavelength operation, especially the broad band around 946 nm. The thick cemented lenses require glass types with different dispersions. This lens requires precision fabrication and assembly to realize the full design potential.

9.7 High-Resolution Telecentric Scan Lens ($\lambda = 408$ nm)

The lens in Fig. 34 was designed for a violet laser diode to image 1200 DPI over a small telecentric field. The design comprises three spherical elements and two cross cylinders to provide optical cross-scan correction and a precise telecentric field well corrected for $F-\Theta$ distortion. This lens form resembles Lens 5 in Fig. 32, before the addition of the cross-scan correction. The careful selection of optical glasses for the violet and UV are particularly important for transmission and dispersion. Many of the new eco-friendly glasses can have significant absorption around 400 nm and below. For example, the internal transmittance

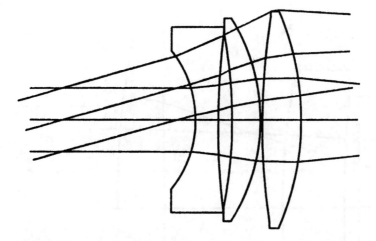

Figure 32 Lens 5: Designer R. Hopkins; $F = 55$ mm, $F/5$, $L = 29$ mm.

at 400 nm for a 10 mm thick Schott SF4 element is 0.954, while the new glass, N-SF4 internal transmittance is 0.79.

10 SCAN LENS MANUFACTURING, QUALITY CONTROL, AND FINAL TESTING

The first two designs shown in the previous section require tolerances that are similar to quality photographic lenses. Compared to the lens diameters, the beam sizes are small, and so surface quality is generally not difficult to meet unless scan linearity and distortion are critical performance parameters. Designs 2 and 3 do not require the highest quality precision lenses, but will require an acceptable level of assembly precision. Because the scan line uses only one cross-sectional sweep across the lenses, the yield of acceptable lenses can be improved by rotating the lenses in their cell, avoiding defects in the individual elements by finding the best line of scan. However, this requires that each lens be appropriately mounted into the scanning system, highlighting the need for good communications between the assemblers and the lens builders.

Lenses 5, 6, and 7 require lens fabricators capable of a precision build to achieve the expected performance of design. Surfaces need better than quarter-wave surface quality,

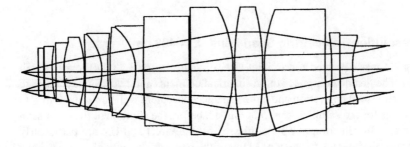

Figure 33 Lens 6: Designer D. Stephenson; $F = 50$ mm, $F/2$, $L = 20$ mm.

Optical Systems for Laser Scanners 123

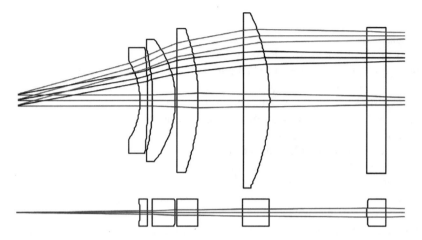

Figure 34 Lens 7: Designer S. Sagan; $F = 125$ mm, $F/24$, $L = 70$ mm.

and must be precision-centered and mounted. Precision equipment will be required to maintain quality control through the many steps to fabricate and mount such lenses.

Precision lenses need special equipment for testing the performance of the finished lens assembly, using appropriate null tests and/or scanning of the image with a detector to measure the spot diameter in the focal plane. Attention must also be paid to the straightness of the detector measurement plane and motion relative to the lens axis. Typically the collimated beam should be directed into the lens exactly the way it will be introduced into the final scanning system. If the image beam intersects the image plane at an angle, the entire beam should pass through the scanner and to the detector. If the detector needs to be rotated relative to the image plane, a correction of the as-measured spot size to an as-used spot size has to be made. If the image is relayed with a microscope objective, the numerical aperture should be large enough to collect all the image cone angles.

Because scan lenses are usually designed to form diffraction-limited images, it is recommended that the lenses be tested using a laser beam with uniform intensity across the design aperture of the lens. The image of the point source should be diffraction-limited and its expected dimensions predictable. The image can be viewed visually or measured for profile and diameters via scanning spot or slit. Departures from a spherical wavefront as small as a $1/10$ of a wavelength are easily detected. It is also possible to detect the effects of excessive scattered light.

11 HOLOGRAPHIC LASER SCANNING SYSTEMS

Holographic scanning systems were first developed in the late 1960s, in part through government-sponsored research for image scanning (reading) of high-resolution aerial photographs. Its application to image scanning and printing for high-resolution business graphics followed in the 1970s with efforts dominated by IBM and Xerox.[18] Subsequent developments in both the holographic process (design and fabrication) and laser technology (from commercializing of the HeNe laser to low-cost diode lasers) have helped broaden its application into commercial and industrial systems.

Applications include: low-resolution point-of-sale barcode scanners; precision noncontact dimensional measurement, inspection and control of the production of high-tech optical fibers, medical extrusions and electrical cables; medium resolution (300–600 DPI) desk-top printers; and high-resolution (1200 DPI and up) direct-to-press marking engines.

Advantages of holographic scanners over traditional polygon mirror scanners include lower mass, less windage, and reduced sensitivity to scan disc errors such as jitter and wobble. Advancements in replication methods to fabricate scanning discs with surface holograms have also helped lower cost.

Disadvantages of holographic scanners include limited operating spectral bandwidth, deviation from classical optical design methods, and the introduction of cross-scan errors in simple design configurations. These issues must be managed during the design process through the configuration of the optical design and its ability to balance the image aberrations.

This section deals with the optical design of holographic scanning systems, beginning with the basics of a rotating holographic scanner and then developing its use with other components into more complex systems.[19]

11.1 Scanning with a Plane Linear Grating

A simple scanning system in terms of both the complexity of the holographic optical element (HOE) and its configuration is illustrated in Fig. 35.[20] This system comprises a collimated laser beam incident on a hologram (nonplane grating the product of the interference of beams in a two-point construction). The performance characteristics of such a rudimentary scanner (line straightness, length, scan linearity, etc.) will depend on the angle of incidence at the hologram and the deviation by the hologram. These performance characteristics can be best understood through modeling of a simple plane linear diffraction grating (PLDG) rotated to generate the scan, the PLDG being equivalent to a hologram constructed from two collimated beams.

Deviation of the beam by the grating (for a grating perpendicular to a plane containing the incident beam and scan disc rotation axis) is the sum of the input incidence angle θ_i and output exit angle θ_o, as illustrated in Fig. 36. These angles are derived from the grating Eq. (21) for a given grating or hologram fringe spacing d, the wavelength of the light λ_0, and the diffraction order m,

$$\sin \theta_i + \sin \theta_o = m\lambda_0/d \tag{27}$$

As with other types of scanning systems, errors such as line bow, scan disc wobble, eccentricity, axis longitudinal vibration, and disc tilt and wedge can affect scanned image position.

11.2 Line Bow and Scan Linearity

The typical purpose of a laser scanning system is to generate a straight line of points by moving a focused beam across a focal plane at a linear rate relative to the scanner rotation. A method for deriving a nearly straight line scan from a PLDG in a disclike configuration was developed independently by C. J. Kramer in the United States and M. V. Antipin and N. G. Kiselev in the former Soviet Union.[18]

The scanning configurations they developed operate at the Bragg condition where the nominal input angle θ_i and output angle θ_o are nearly 45°. The Bragg condition is

Optical Systems for Laser Scanners

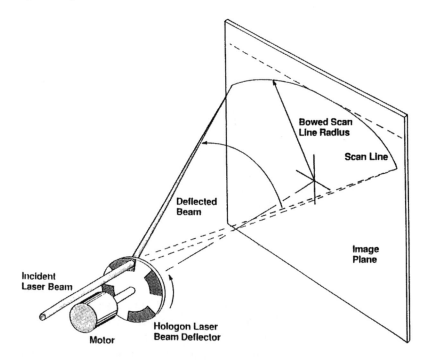

Figure 35 Cindrich-type holographic scanning (from Ref. 20).

where the input beam and diffracted output beam are at equal angles relative to the diffracting surface. Operating near this Bragg condition minimizes the effects of scan disc wobble and operating near 45° minimizes line bow (cross-scan departure from straightness). For the first diffraction order $m = 1$, the grating or hologram fringe spacing d would be given by the reduced equation

$$d = \lambda_0/\sqrt{2} \tag{28}$$

The actual optimum angle will depend on the scan length and the degree of line bow correction desired. The dependence of the scan line bow and scan linearity (in-scan position error relative to the scan angle) on the Bragg angle is illustrated in Figs 37(a) and

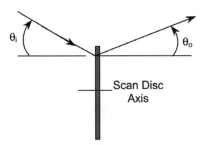

Figure 36 Scan disc input and output angles.

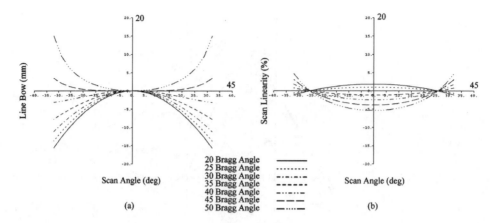

Figure 37 Effect of Bragg angle on line bow (a) and scan linearity (b).

(b) for a design wavelength of 786 nm. In a 45° monochromatic corrected configuration, the chromatic variation of the line bow and scan length for a ± 1 nm change in wavelength results in the departures illustrated in Figs 38(a) and (b), respectively. These errors (line bow, scan linearity, and their chromatic variation) are significant compared to the resolution of the printer system and must be balanced in the designs to allow the use of cost effective laser diodes with diffractive optical components.

11.3 Effect of Scan Disc Wobble

In general, holographic optical elements (HOEs) are best designed to operate in the Bragg regime to minimize the effects of scan disc wobble. Wobble is the random tilt of the scan disc rotation axis due to bearing errors. The error in the diffraction angle ε resulting from a wobble angle δ indicated in Fig. 39 is given by the modified grating equation

$$\varepsilon = \arcsin[m\lambda_0/d - \sin(\theta_i + \delta)] + \delta - \theta_o \tag{29}$$

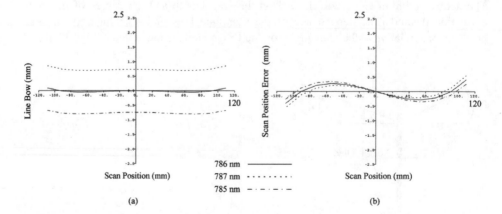

Figure 38 Chromatic variation of line bow (a) and scan length (b).

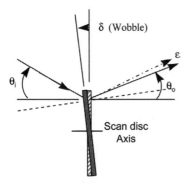

Figure 39 Scan disc wobble and beam deviation.

This angular error produces a cross-scan displacement in the scanned beam, affecting the position of a measurement being made or the position of a point being written. That displacement error is a product of the angle error and the focal length of the projection optics. In systems with traditional reflective scanners (such as a polygon), this angular error is twice the wobble tilt error unless it is optically compensated using anamorphic optical methods, which greatly increases complexity and cost. Scan jitter in the scanner rotation will likewise generate twice the in-scan position errors. The precision of the scan degrades as the motor bearings wear and mirror wobble increases.

In a holographic scanning system, the cross-scan image error can be minimized to hundredths or even thousandths of the disc wobble error and the in-scan image error is approximately equal to the disc rotation (jitter) error. In a classical polygon scanning system the image errors are twice the wobble and jitter errors. The effect on output angle error of a tilt in the scan disc rotation axis is shown in the curves of Fig. 40 derived from the modified

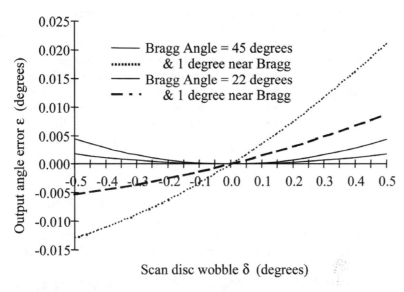

Figure 40 Effect of input angle and disc wobble on output scan angle.

grating Eq. (29). Two Bragg angle and two near-Bragg angle design configurations operating near the minimum line bow condition of 45° and arbitrary angle of 22° are plotted.

12 NONCONTACT DIMENSIONAL MEASUREMENT SYSTEM USING HOLOGRAPHIC SCANNING

Opto-electronic systems for noncontact dimensional measurements have been available to industry since the early 1970s. Major applications include the production and inspection of linear products like wire, cable, hose, tubing, optical fiber, and metal, plastic and rubber extruded shapes.

Linear measurement instruments have used two basic technologies to perform the measurement. The first systems used a linearly scanned laser beam and a collection system with a single photodetector. As the telecentric beam moves across the measuring zone, the object to be measured blocks the laser beam from the collection system for a period of time. By knowing the speed of the scan and the time the beam is blocked, the dimension of the object can be calculated by a microprocessor and then displayed. Later, as video array technology developed, systems using a collimated light source and a linear charge coupled device (CCD) or photodiode array were introduced. These systems use an incandescent lamp or an LED with a collimating lens to produce a highly collimated beam of light that shines across the measuring zone. The object to be measured creates a shadow that is cast on a linear array of light-sensing elements. The count of dark elements is scaled and then displayed.

In most scanning systems, a laser beam is scanned by reflecting the beam off a motor-driven polygon mirror. The precision of the scan degrades as the motor bearings wear, increasing mirror wobble and scan jitter. Otherwise, the overall accuracy of a laser scanning system can be quite good, because resolution is based on a measurement of time, which can be made very precisely.

The electronic interface to a scanning dimensional measurement system is well understood and has been used in various applications for years. Assuming that the scanning beam travels through the measuring zone at a constant, nonvarying speed, improvements in system performance are limited to: (1) maximizing the reference clock speed, (2) further dividing the reference clock speed by delay lines, capacitive charging, or other electronic techniques, and (3) minimizing the beam on/off detection error.

Video array systems that use incandescent lamps drive the lamp very hard, which produces substantial heat and reduces the lamp life. Systems that use a solid-state source, such as an LED, are very efficient and require much less power, but also emit less photon energy. The quality of the shadow image depends on the entire optical system. Relatively large apertures are required to collect a sufficient number of photons to achieve the needed signal-to-noise ratio. The physical dimensions and element size in the video array limit the resolution and achievable accuracy of the system. System reliability, however, is typically high, and the mean time between failures can be long if a solid-state light source is used. An LED/video array system is all solid state and has no moving parts to wear out.

The application of holographic scanning to noncontact dimensional measuring systems provides the opportunity to take advantage of the positive points in scanning systems and the high reliability of the all-solid-state video array systems, while avoiding some of the problems found in each.

A polygon mirror is manufactured one facet at a time, but a holographic disc can be replicated, like a compact disc or CD ROM. A holographic disc can be produced with 20 to

Optical Systems for Laser Scanners

30 facets at a fraction of the cost of producing a comparably high-precision polygon mirror with as many facets. Furthermore, other holographic components that might be used in the system, such as a prescan hologram, can also be replicated.

12.1 Speed, Accuracy, and Reliability Issues

As line speeds have increased and tolerances have narrowed, the need to provide not only diameter measurements but also flaw detection has grown dramatically. To provide fast response in process control and surface flaw detection, a measurement system must make many scans per second. The number of scans per second that can be made by a measurement system is determined by the number of facets, the speed of the motor, and the data rate capability of the analog to digital (A/D) converter. With the high-speed electronics that are available today, the motor speed and thus motor lifetime and cost are the limiting factors when designing high-speed measurement devices. Typically, polygon mirror scanners have one-third the number of facets (or fewer) than a holographic disc, so the motor must run three times faster (or more) to produce the same number of scans per second as the holographic scanner. Consequently, the trade-off between speed and lifetime/cost of the motor for holographic disc scanners is considerably more attractive than polygon mirror scanners.

Traditional laser-based systems scan from 200–600 times/s/axis and provide limited single-scan information. The holographic disc in the Holix® Gage by Target Systems Inc., Salt Lake City, UT, has 22 segments as compared to 2–8 sides on a typical polygon mirror, with single-scan-based flaw detection possible at 2833 scans per second per axis.[22]

The ability to scan faster means more diameter measurements can be made over a given length of the test object in a given time interval. With this resolution increase, the system is more likely to detect surface flaws, as illustrated in Fig. 41. In these holographic scanning systems there is no need to average groups of scans to compensate for the surface irregularities in a manufactured polygon mirror. This is a major advancement in the ability to detect small surface flaws that can be missed by traditional gages.

In reflective scanners, any tilt error in the mirror can cause twice the error in the output beam. In transmission holographic scanners, the beam is diffracted rather than reflected, and the beam error on output is much smaller than the tilt of the disc axis.

As the number of facets increase on a polygon mirror, another problem can develop. Because of the finite size of the laser beam and the required scan angle, there is a minimum size for each facet on the mirror. As the size of each facet is increased, the distance from the center of rotation to the facet surface increases (the polygon mirror gets larger in

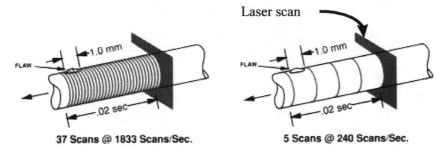

Figure 41 Detection of surface flaws with multiple scans.

diameter). As the facet surface moves farther from the center of rotation, the virtual location of the scan center point shifts during the scan (pupil shift). This means the scan center point is not maintained at the focal point of the scan lens (which is designed to generate a telecentric scan) for the entire duration of the scan. The telecentricity errors from this shift in the pupil reduce the measurement accuracy or the depth of the measurement zone. Holographic disc scanners have no pupil shift, and therefore, are not limited by these errors.

12.2 Optical System Configuration

The optical system illustrated in Fig. 42 offers scan and laser spot size performance in the measurement zone that, together with a proprietary processing algorithm, can provide repeatable measurements to within one micro-inch. The optical design comprises only the basic components required to scan a line: a laser diode, a collimating lens, a prescan holographic optical element (HOE), a scanning HOE, a parabolic mirror (the scan lens in this system), a collecting lens, and fold mirrors to provide the desired packaging.

The laser diode is mounted in a metal block that is thermally isolated from the instrument frame. A thermoelectric cooler (TEC) or heater and temperature controller are used to maintain the operating temperature of the laser diode over a narrow range of $\sim 0.5°$ C. Temperature control is required to prevent "mode hops" in the laser diode as the temperature changes. The wavelength shift of a mode hop can be 0.5–1 nm with temperature change, as illustrated in Fig. 43. The temperature control setpoint is selected to center the laser diode between mode hops and prevent the diode from changing modes and abruptly shifting the emission wavelength. This is a critical feature of the system because diffractive elements are very sensitive to changes in wavelength. Although corrected for a wavelength drift in the ± 0.1 nm class, like many other HOE-based optical systems, it cannot accommodate mode hops.

The diverging beam emitted from the laser diode is apertured and quasi-collimated by a typical laser diode collimator. The ratio of the collimator focal length to the parabolic mirror focal length sets the magnification for the projected laser spot width in the measurement zone. The beam is then diffracted by a stationary prescan HOE and diffracted again by a rotating scan disc HOE. The first-order diffracted beam exits the scan

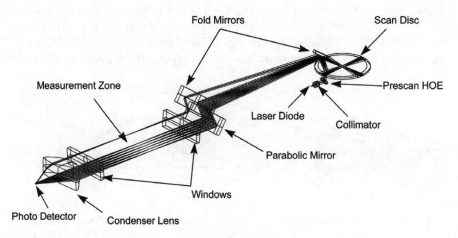

Figure 42 Single axis HOE optical system configuration.

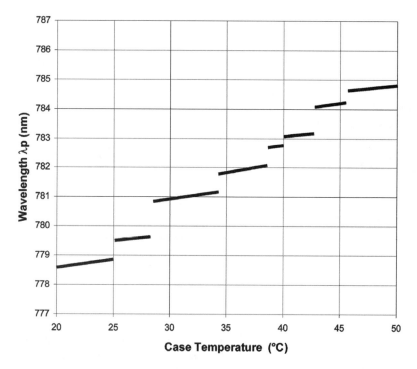

Figure 43 Laser diode characteristic wavelength vs. temperature dependence (SHARP 5 mW, 780 nm, 11 × 29° divergence).

HOE at an angle about 22° off normal. As the disc rotates, the orientation of the diffractive structure in the scan HOE changes, causing the beam to scan from side to side. The ratio between disc rotation angle and scan angle is approximately 1 : 1. The zero-order beam is not diffracted by the HOE and continues straight to a photodiode that is used to monitor disc orientation.

The use of replication to manufacture production holographic discs shown in Fig. 44 provides an economical way to reproduce discs with other features in addition to a large number of facets. Disc sectors are two-point construction or multiterm $(x, y, \ldots, x^n y^m)$ phase HOEs. All of the HOE component or surfaces can be produced by photo resist on glass substrates, embossing (replication) using polymers, or injection/compression molding in plastics.

The discs are designed with a gap between two adjacent facets that is much narrower than a holographic facet, and that has no grating on it. When the gap rotates over the laser beam, the beam is not diffracted and continues straight to a photodiode. This signal is used to identify the orientation of the disc. The data processor can use this synchronization pulse to associate a unique set of calibration coefficients with each facet. This allows variations in optical parameters from one facet to another to be calibrated out.

The diffracted first-order beam is folded by two mirrors, reflected by the scan mirror, and then continues out through a window to the measurement zone. The parabolic mirror serves two purposes. First, it provides a nearly telecentric scan. Secondly, it focuses the beam so that the minimum in-scan waist is in the center of the measurement zone. After the measurement zone, the beam passes through a second window and is collected by a

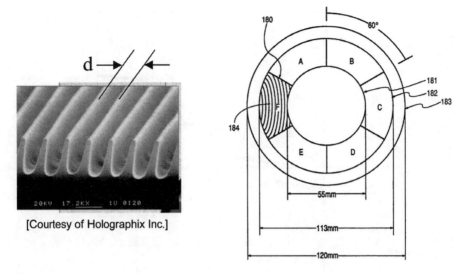

Figure 44 Replicated holographic scan disc (courtesy of Holographix, Inc.).

condenser lens that focuses it onto a single photodetector. The collecting lens is an off-the-shelf condenser lens with an aspheric surface to minimize the spherical aberration generated by the nearly $F/1$ operating condition. The ratio of the condenser to scan lens focal length determines the size of the beam (an image of the beam at the scan disc) on the photodetector.

The inherent chromatic variations of scan length and scan bow in a holographic scanner are corrected to acceptable levels by the Holographix patented configuration of Fig. 45.[23] The HOE scan disc introduces both line bow and chromatic aberrations (in- and cross-scan). The prescan HOE is used to introduce additional cross-scan chromatic error, which, when coupled with the bow correction provided by the tilted curved mirror (a rotationally symmetric parabola for the noncontact dimensional measurement system), produces a chromatically corrected in-scan beam. The cross-scan corrector hologram in the noncontact dimensional measurement system can be eliminated at the expense of cross-scan chromatic correction, illustrated in Fig. 46. However, the magnitude of the cross-scan error is less than 100 microns, and does not affect the diameter measurement.

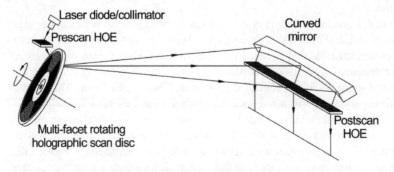

Figure 45 Holographix, Inc., U.S. patent 5,182,659.

Optical Systems for Laser Scanners

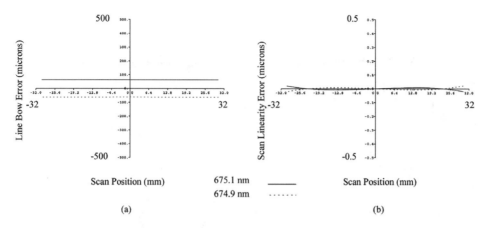

Figure 46 Chromatic variations of line bow (a) and scan position (b).

The foregoing description is of a single measurement axis instrument. For many applications a single instrument with two measurement axes is used. This configuration includes two lasers and collimating systems that use a single scan disc at separate locations (at 90°) on the radius. The resulting two scanning beams continue and pass through separate mirrors and optics to produce two orthogonal telecentric scans. Separate collection lenses and photodetectors complete the optical paths. This system, with two measuring axes, is used to measure the nonroundness of round objects, two dimensions of nonround objects, and to view the object's surface from more directions for more complete surface flaw detection. The discs are designed with facets of a specific size, so systems that use a single disc to produce multiple scan axes will be compatible with the orientation of the optical paths.

12.3 Optical Performance

The single measurement axis instrument can be optimized to provide a relatively large measurement zone while maintaining telecentricity and a consistent beam size in the scan direction. The profile and consistency of the in-scan beam width are shown in Fig. 47 for a 50 mm gage system over a ±20 mm measurement zone. A key to providing this performance is orientation of the laser diode astigmatism with the slow axis parallel to the scan, also optimizing scan efficiency. The $1/e^2$ in-scan beam width over the measurement zone is approximately 210 microns, with centroid stability better than one micron and a monochromatic telecentricity better than 0.04 mrad (as-designed). The profile of the elliptical spot is Gaussian with small diffraction side lobes that are the result of truncation by the collimating lens aperture. The cross-scan spot size is typically much smaller, but is not controlled except by the first-order configuration. The design is optimized for an operating wavelength of 675 ± 0.1 nm anywhere within a setpoint of 670–680 nm. The subtle variations in the spot width are static for each system. Also static is the scan nonlinearity (distortion). These static variations are calibrated over the inspection zone to provide measurement repeatability of 30 micro-inches for the 50 mm and 3 micro-inch for the 7 mm Holix gages.

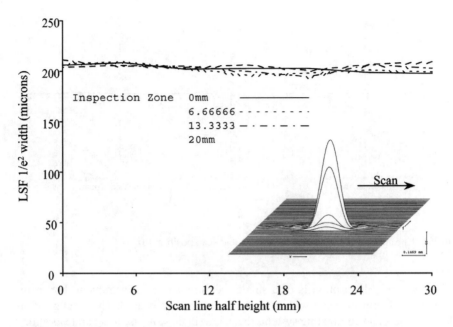

Figure 47 Line spread function based in-scan spot width.

13 HOLOGRAPHIC LASER PRINTING SYSTEMS

The configuration of a lower performance scanning system based on the Holographix patent is shown in Fig. 48. This design is for a 300 DPI system comprising a laser diode source, a collimating lens, a prescan HOE, a holographic scan disc, a tilted concave cylindrical mirror, a postscan HOE, and assorted fold mirrors for packaging. This configuration provides a method for balancing errors of line bow, scan linearity, and their significant chromatic variations to the resolution requirements of the printer system, allowing the use of cost-effective laser diodes coupled with diffractive optical components.

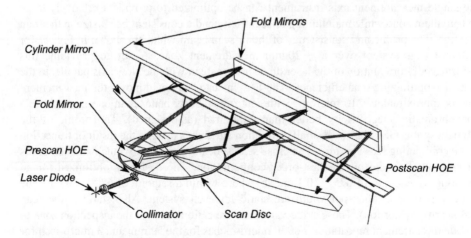

Figure 48 Holographix laser printer optical system.

Optical Systems for Laser Scanners

This configuration uses a scan disc holographic optical element (HOE) at other than the minimum line bow condition (to introduce a prescribed amount of line bow) that both focuses and scans the beam. A prescan HOE is used to introduce additional chromatic cross-scan error that, coupled with the bow correction provided by a tilted curved mirror, produces a corrected in-scan beam. The postscan HOE completes the correction by balancing the cross-scan compounded errors and focusing the beam. The unique configuration yields a system with significantly and acceptably reduced sensitivity to changes in laser wavelength (due to mode hops during scan, wavelength drift over temperature, and variation from diode to diode).

Typical system performance specifications (beam size, line bow, scan linearity, and change in line bow and scan linearity over wavelength and scan) are listed in Table 7. As the system's performance requirements increase, tighter control of the system aberrations (particularly field curvature and astigmatism) is required. Additional design variables such as higher order terms on the scan disc and postscan HOEs provide the degrees of freedom necessary to achieve the line bow, linearity, beam quality, and chromatic variations specifications. These laser scanning system designs are nearly telecentric in image space to avoid position errors over the in-use depth of focus.

Since the original 300 DPI systems were developed, these designs have been continually refined and improved. The latest 600 DPI designs actually have a larger depth of focus, smaller package, reduced material cost, increased ease of assembly and alignment, and better chromatic correction. These benefits have been obtained using an optimization process that increases the complexity of the recording parameters of the HOEs while decreasing the complexity of the rest of the system.

Over the years, Holographix has developed proprietary alignment and recording techniques to fabricate complex HOEs for both transmission and reflection. As complexity has increased, so has the cost of recording of the "master" HOE. However, with the development of HOE mastering and replication techniques in which several thousand replicas can be made from one master, the increased cost of the masters becomes insignificant. Development of holographic scanning systems has included refinement of production/replication processes to achieve higher diffraction efficiencies at lower cost. Diffraction efficiencies of replicated HOEs averaging 80% provide for high system throughput.

The optical design and tolerancing of systems with holographic optical elements is nontrivial. Optimization of the optical system requires controlling (and sometimes limiting) the HOE degrees of freedom to minimize a feedback interaction between HOEs. The tolerancing of the optical designs for as-built performance requires the tolerancing of each HOE construction setup and then introducing the as-built HOEs into the tolerancing of the optical design. This multiple configuration/layer tolerancing can be modeled in Code V to predict the as-built performance of these complicated systems.

14 CLOSING COMMENTS

The significant trends in refinement of laser scanning systems are ever-increasing scan lengths and larger number of spots per inch. The design examples show present-day practical boundaries for scan lenses. These boundaries continue to slowly expand with new concepts and ever-increasing precision, using more combinations of refractive lenses,

Table 7 Typical Performance Specifications for Holographic Laser Printer

Parameter	Specification
1. System configuration	• Commercial laser diode • Collimator • Prescan HOE • 65 mm scan disc with HOE • Cylindrical mirror • Post-scan HOE
2. Laser diode	
(i) Center wavelength	670, 780, 786 (nm)
(ii) Wavelength drift	± 1 nm
(iii) FWHM divergence	11H × 29V degrees
(iv) Astigmatism	7 microns
3. Wavelength accommodation	± 10 nm

4. Beam diameter at focal plane (nominal, best focus)	300 DPI: ($1/e^2$)
	  In-scan: 80 ± 10 microns
	  Cross-scan: 100 ± 20 microns
	600 DPI: ($1/e^2$)
	  In-scan: 50 ± 10 microns
	  Cross-scan: 50 ± 10 microns
	1200 DPI: (FWHM)
	  In-scan: 20 ± 5 microns
	  Cross-scan: 25 ± 5 microns

5. Scan line	
(i) Total length	300/600 DPI: 216 mm 1200 DPI: 230 mm
(ii) Linearity (w.r.t. scan disc rotation)	300/600 DPI: $\pm 1\%$ 1200 DPI: $\pm 0.03\%$
(iii) Chromatic variation of line length (over ± 1 nm)	300 DPI: <20 microns 600 DPI: <5 microns 1200 DPI: <5 microns
(iv) Bowing (microns)	300/600 DPI: <300 microns 1200 DPI: <25 microns
(v) Chromatic variation of line bow (over ± 1 nm)	300 DPI: <20 microns 600 DPI: <10 microns 1200 DPI: <5 microns
(vi) Telecentricity	<4°

diffractive elements, and mirrors with anamorphic power – both nearer the scanner and closer to the image plane. Methods for economically manufacturing these elements continue to be developed, allowing for larger diameter refracting lenses and reduced cost lens segments and mirrors.

If not limited by the optical invariant, optical systems are generally limited by the precision of lens fabrication, assembly, alignment, and testing. Availability of fabrication houses with special equipment for unconventional surface types such as cylinders and toroids, in combination with spherical and aspheric surfaces are limited. As usual, the market to pay for the special equipment and tooling has to be large enough to support the investment.

ACKNOWLEDGMENTS

I would like to thank my good friends and colleagues James Harder, Eric Ford, David Rowe, and Torsten Platz for their review and inputs in preparing this chapter. Special thanks to my friend and mentor Gerald Marshall, whose support over many years and conferences has played a significant role in my association with the scanning community. And last but not least, my deepest love and appreciation to my wife Maria for her encouragement and patience.

REFERENCES

1. Hopkins, R.E.; Stephenson, D. Optical systems for laser scanners. In *Optical Scanning*; Marshall, G.F., Ed.; Marcel Dekker: New York, 1991; 27–81.
2. Beiser, L. *Unified Optical Scanning Technology*; John Wiley & Sons: New York, 2003.
3. Siegman, A.E. *Lasers*; University Science Books: Mill Valley, California, 1986.
4. Melles Griot. *Laser Scan Lens Guide*; Melles Griot: Rochester, NY, 1987.
5. Wetherell, W.B. The calculation of image quality. In *Applied Optics and Optical Engineering*; Academic Press: New York, 1980; Vol. 7.
6. Hopkins, R.E.; Hanau, R. *MIL-HDBK-141*; Defense Supply Agency: Washington, DC, 1962.
7. Kingslake, R. *Optical System Design*; Academic Press: New York, 1983.
8. Hopkins, R.E.; Buzawa, M.J. Optics for Laser Scanning, SPIE 1976, *15*(2), 123.
9. Thompson, K.P. *Methods for Optical Design and Analysis – Seminar Notes*; Optical Research Associates: California, 1993.
10. Kingslake, R. *Lens Design Fundamentals*; Academic Press: New York, 1978.
11. Smith, W.J. *Modern Optical Engineering*; McGraw-Hill: New York, 1966.
12. Welford, W.T. *Aberrations of Symmetrical Optical Systems*; Academic Press: London, 1974.
13. Levi, L. *Applied Optics, A Guide to Optical System Design/Volume 1*; John Wiley & Sons: New York, 1968; 419 pp.
14. Marshall, G.F. Center-of-scan locus of an oscillating or rotating mirror. In *Recording Systems: High-Resolution Cameras and Recording Devices and Laser Scanning and Recording Systems*, Proc. SPIE Vol. 1987; Beiser, L., Lenz, R.K., Eds.; 1987; 221–232.
15. Fleischer; Latta; Rabedeau. IBM Jrnl. of Res. and Dev. 1977, *21*(5), 479.
16. Marshall, G.F. Geometrical determination of the positional relationship between the incident beam, the scan-axis, and the rotation axis of a prismatic polygonal scanner. In *Optical Scanning 2002*, SPIE Proc. Vol. 4773; Sagan, S., Marshall, G., Beiser, L., Eds.; 2002; 38–51.
17. Sagan, S.F. *Optical Design for Scanning Systems*; SPIE Short Course SC33, February 1997.
18. Beiser, L. *Holographic Scanning*; John Wiley & Sons: New York, 1988.
19. Sagan, S.F.; Rowe, D.M. Holographic laser imaging systems. SPIE Proceedings 1995, *2383*, 398.
20. Kramer, C.J. Holographic deflector for graphic arts system. In *Optical Scanning*; Marshall, G.F., Ed.; Marcel Dekker: New York, 1991; 240 pp.
21. O'Shea, D.C. *Elements of Modern Optical Design*; John Wiley & Sons: New York, 1985; 277 pp.
22. Sagan, S.F.; Rosso, R.S.; Rowe, D.M. Non-contact dimensional measurement system using holographic scanning. SPIE Proc. 1997, *3131*, 224–231.
23. Clay, B.R.; Rowe, D.M. Holographic Recording and Scanning System and Method. U.S. Patent 5,182,659, January 26, 1993.

3

Image Quality for Scanning

DONALD R. LEHMBECK

Xerox Corporation, Webster, New York, U.S.A.

JOHN C. URBACH[†]

Consultant, Portola Valley, California, U.S.A.

1 INTRODUCTION

1.1 Imaging Science for Scanned Imaging Systems

This chapter presents some of the basic concepts of image quality and their application to scanned imaging systems. For readers familiar with the previous edition we have added discussions about image quality factors associated with binary representation of scanned images, color imaging and color management, losses from image compression, and digital cameras. In keeping with the handbook theme of this revised edition, we have also added a section on psychometric testing methods, pointers to the developing industry standards in image quality, as well as more reference data and charts. New references and other technical details have been added throughout.

    The emphasis in this chapter will be on the input scanner. Output scanners and diverse systems topics will be dealt with mainly by inference, since many input scanner considerations and metrics are directly applicable to the rest of a complete electronic scanned imaging system. The chapter is organized as 10 major sections moving from the basic concepts and phenomena of image scanning and color, through practical aspects of image quality, to performance of input scanners that produce multilevel (gray) signals and then the special but common case of binary scanned images. This is followed by sections on very specific topics: various summary measures of imaging performance and specialized image processing. To assist the reader, psychophysical measurement methods used to evaluate image quality and some reference data and charts have been added.

[†]Deceased.

1.1.1 Scope

We, like so many others, follow in the path pioneered over a half century ago by the classic 1934 paper of Mertz and Gray.[1] Without going into the full mathematical detail of that paper and many of its successors, we attempt to bring to bear some of the modern approaches that have been developed both in image quality assessment and in scanned image characterization. Many diverse technologies used in scanned imaging systems are addressed throughout the rest of this book. We cannot address the explicit effects of any of these on quality because they provide an enormous array of choices and trade-offs. Building on a more general foundation of imaging science, we shall attempt to provide a framework in which to sort out the many image quality issues that depend on these choices.

It is our intent not to show that one scanner or technique is better than another, but to describe the methods by which each scanning system can be evaluated to compare to other systems and to assess the technologies used in them. This chapter therefore deals primarily with such matters as the sharpness or graininess of an image and not with such hardware issues as the surface finish of an aluminum mirror, uniformity of a drive motor, or the efficiency of charge transfer in a charge-coupled device (CCD) imager.

Scanning is considered here in the general context of electronic imaging. An electronic imaging system often consists of an input scanner that converts an optical image into an electrical signal (often represented in digital form). This is followed by electronic hardware and software for processing or manipulation of the signal and for its storage and/or transmission to an output scanner or display. The latter converts the final version of the signal back into an optical (visible) image, typically for transient (soft) or permanent (hard copy) display to a human observer.

1.1.2 The Literature

Considerable research, development, and engineering have occurred over the last decade since our earlier chapter and only a very small portion is referenced in the following pages. A few general references of note are provided as Refs. 2–14 and elementary tutorials in Refs. 15 and 16. Other more specific work of importance, but which is not discussed later, includes: image processing appropriate to scanner image quality,[17–19] digital halftoning,[20,21] color imaging,[22–25] and various forms of image quality assessment.[26–32]

While the focus here is on imaging modules and imaging systems, scanners may, of course, be used for purposes other than imaging, such as digital data recording. We believe that the imaging science principles used here are sufficiently general to enable the reader with a different application of a scanning system to infer appropriate knowledge and techniques for these other applications.

1.1.3 Types of Scanners

All input scanners convert one- or (usually) two-dimensional image irradiance patterns into time-varying electrical signals. Image integrating and sampling systems, such as those found in many forms of electronic cameras and electronic copying devices, have sensors such as a CCD array. The signals produced by these scanners can be in one of two general forms, either (a) binary output (a string of on and off pulses), or (b) gray-scale output (a series of electrical signals whose magnitude varies continuously).

The term *digital* here refers to a system in which each picture element (pixel) must occupy a discrete spatial location; an analog system is one in which a signal level varies continuously with time, without distinguishable boundaries between individual picture elements. A two-dimensional analog system is usually only analog in the more rapid direction of

Image Quality for Scanning

scanning and is discrete or "digital" in the slower direction, which is made up of individual raster lines. Television typically works in this fashion. In one form of solid-state scanner, the array of sensors is actually two-dimensional with no moving parts. Each individual detector is read out in a time sequence, progressing one raster line at a time within the two-dimensional matrix of sensors.

In other systems a solid-state device, arranged as a single row of photosites or sensors, is used to detect information one raster line at a time. In these systems either the original image is moved past the stationary sensor array, or the sensor array is scanned across the image to obtain information in the slow scan direction.

Many new devices in digital photography employ totally digital solid-state scanners using two-dimensional sampling arrays. In our judgment, they are the most commonly encountered forms of input scanners today. The reader should be able to infer many things about the other forms of scanners from these examples.

1.2 The Context for Scanned Image Quality Evaluation

Building blocks for developing a basic understanding of image quality in scanning systems are shown in Fig. 1. The major elements of a generalized scanning system are on the left, with the evaluation and analysis components on the right. This chapter will deal with all of these elements and it is therefore necessary to see how they all interact.

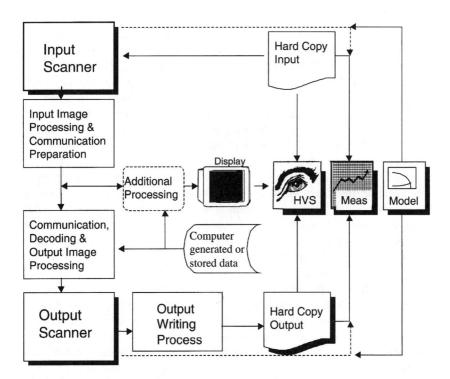

Figure 1 The elements of scanned imaging systems as they interact with the major methods of evaluating image quality. "HVS" refers to the human visual system. "Meas" refers to methods to measure both hard copy and electronic images and "Models" refers to predicting the imaging systems performance, not evaluating the images *per se*.

The general configuration of scanning systems often requires two separate scanning elements. One is an input scanner to capture, as an electronic digital image, an input analog optical signal from an original scene (object), shown here as a hard copy input, such as a photograph. The second scanning element is an output scanner that converts a digital signal, either from the input scanner or from computer-generated or stored image data, into analog optical signals. These signals are rendered suitable for writing or recording on some radiation-sensitive medium to create a visible image, shown here as hard copy output. The properties of this visible image are the immediate focus of image quality analysis. It may be photographic, electrophotographic, or something created by a variety of unconventional imaging processes. The output scanner and recording process may also be replaced by a direct marking device, such as a thermal, electrographic, or ink jet printer, which contains no optical scanning technology and therefore lies outside the scope of this volume. Nonetheless, its final image is also subject to the same quality considerations that we treat here.

It is to be noted that the quality of the output image is affected by several intermediate steps of image processing. Some of these are associated with correcting for the input scanner or the input original, while others are associated with the output scanner and output writing process. These are mentioned briefly throughout, with the digital halftoning process, described in Sec. 2.2.3, cited as a major example of a correction for the output writing. Losses or improvements associated with some forms of data communication, and compression are very important in a practical sense, especially for color. These are briefly reviewed in Sec. 7.1. Additional processing to meet user preferences or to enable some particular application of the image must also be considered a part of the image quality evaluation. A few examples are given throughout. A comprehensive treatment of image processing is beyond the scope of this chapter but several references are given at the end of this chapter to help the reader learn more about this critical area of scanned imaging.

The assessment of quality in the output image may take the form of evaluation by the human visual system (HVS) and the use of psychometric scaling (see Sec. 8) or by measurement with instruments as described in parts of Sec. 3–5. One can also evaluate measured characteristics of the scanners and integrated systems or model them to try to predict, on average, the quality of images produced by these system elements. (Both of these hardware characterizations are also described in parts of Sec. 3–5.) The description of overall image quality (Sec. 6) tends to focus on the models of systems and their elements, not the images themselves. For some purposes, for example, judging the quality of a copier, the comparison between the input and output images is the most important way of looking at image quality, whether it be by visual or measurement means. For other applications it is only the output image that counts. In some cases, the most common visual comparison is between the partially processed image, as can only be seen on the display, and either the input original or the hard copy output. In most cases, the evaluation criteria depend on the intended use of the image. A display of the scanned image in a binary (black or white) imaging mode reveals some interesting effects that carry through the system and often surprise the unsuspecting observer. These are covered in Sec. 5. Physical and visual measurements evaluate output and input images, hence the arrows in Fig. 1 flow from hardcopy toward these evaluation blocks. Models, however, are used mostly to synthesize imaging systems and components and may be used to predict or simulate performance and output. Hence the "model" arrows flow toward the system components.

The nonscanner components for electronic image processing and the analog writing process play a major role in determining quality and hence will be unavoidably included in any realistic HVS or measurement evaluation of the quality of a scanned image or imaging

Image Quality for Scanning

system. Models of systems and components, on the other hand, often ignore the effects of these components and the reader is cautioned to be aware of this distinction when designing, analyzing, or selecting systems from the literature.

A model has been described by P. Engeldrum[11,33–35] called the Image Quality Circle, which ties all of these evaluations together and expands them into a logical framework to evaluate any imaging system. This is shown in Fig. 2 as the circular path connecting the oval and box shapes, along with the three major assessment categories from Fig. 1, namely the HVS, Measurements, and Models. In his model, the HVS category above is expanded to show a type of model he calls "visual algorithms," which predict human perceived attributes of images from physical image parameters. Examples of perceptions would include such visual subjective sensations as darkness, sharpness, or graininess (i.e., "nesses"). These are connected to physical measurements of densities, edge profiles, or halftone noise, respectively, made on the images used to evoke these subjective responses. In Engeldrum's analysis, the rest of what we call the HVS and brain combination includes "image quality models," which predict customer preferences based on relationships among the perceived

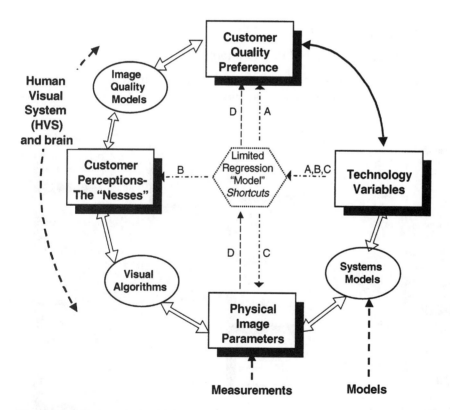

Figure 2 An overall framework for image quality assessment, composed of the elements connected by the outline arrows, known as the "Image Quality Circle" (from Refs. 11 and 35) and the inner "spokes" which illustrate four commonly used, but limited, regression model shortcuts as paths A, B, C, and D. The latter were not proposed by Engeldrum as part of the Image Quality Circle model, but added here to illustrate how selected examples given in Sec. 6 fit the framework. The connection to HVS, measurement and model elements of Fig. 1 are indicated by the labels and heavy dashed lines that surround the figure.

attributes. This purely subjective dimension of individuals is often not included in the "brain" functions normally associated with HVS, therefore it is mentioned explicitly here. The methodologies to enable these types of analysis generally fall into the realm of psychometrics (quantifying human psychological or subjective reactions). They will be reviewed in Sec. 8.

Many authors (Sec. 6) have attempted to short-circuit this framework, following the dashed "spokes" we have added to the circle in Fig. 2. These create regression models using psychometrics that directly connect physical parameters (path D) or technology variables with overall image quality models (path A) or preferences (path C). These have been partially successful, but, having left out some of the steps around the circle, they are very limited, often applying only to the circumstances used in their particular experiment. When these circumstances apply, however, such abbreviated methods are valuable. Following all the steps around the circle leads to a more complete understanding and more general models that can be adapted to a variety of situations where preferences and circumstances may be very different. The reader needs to be aware of this and judge the extent of any particular model's applicability to the problem at hand.

2 BASIC CONCEPTS AND EFFECTS

2.1 Fundamental Principles of Digital Imaging

The basic electronic imaging system performs a series of image transformations sketched in Fig. 3. An object such as a photograph or a page with lines and text on it is converted

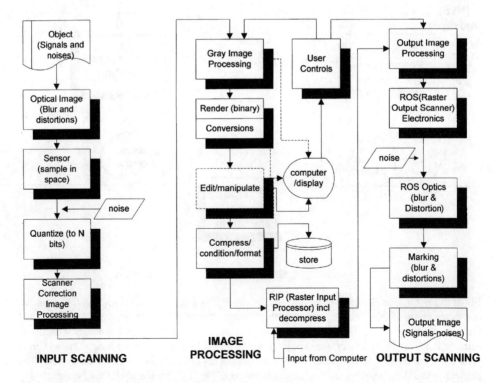

Figure 3 Steps in typical scanning electronic reprographic system showing basic imaging effects.

Image Quality for Scanning

from its analog nature to a digital form by a *raster input scanner* (RIS). It becomes "digital" in distance where microscopic regions of the image are each captured separately as discrete pixels; that is, it is *sampled*! It is then quantized, in other words, digitized in level, and is subsequently processed with various strictly digital techniques. This digital image is transformed into information that can be displayed or transmitted, edited or merged with other information by the *electronic and software subsystem* (ESS). Subsequently a *raster output scanner* (ROS) converts the digital image into an analog form; that is, it is *reconstructed*, typically through modulating light falling on some type of photosensitive material. The latter, working through analog chemical or physical processes, converts the analog optical image into a reflectance pattern on paper, or into some other display as the final output image.

What follows assumes optical output conversion, but direct-marking processes, involving no optics (e.g., ink jet, thermal transfer, etc.) can be treated similarly. Therefore, while one often thinks of electronic imaging or scanned imaging as a digital process, we are really concerned in this chapter with the imaging equivalent of analog to digital (A/D) and digital to analog (D/A) processes. The digital processes occur between as image processing. In fact that is where we become familiar with the scanned imaging characteristics because that is one place where we can take a look at a representation of the image, that is, in a computer.

2.1.1 Structure of Digital Images

Before considering all the system and subsystem effects, let us turn our attention to the microscopic structure of this process, paying particular attention to the A/D and sampling domain of the input scanner. Sampled electronic images were first studied in a comprehensive way by Mertz and Gray.[1]

To understand how sampling works, let us examine Fig. 4. It illustrates four different aspects of the input scanning image transformations. Part (a) shows the microscopic reflectance profile representative of an input object: there is a sharp edge on the left, a "fuzzy" edge (ramp), and a narrow line. Part (b) shows the optical image, which is a blurred version of the input object. Note that the relative heights of the two pulses are now different and the edges are sloping that were previously straight. Part (c) represents the blurred image with a series of discrete signals, each being centered at the position of the arrows. This process is referred to as *sampling*.

Each sample in part (c) has some particular height or gray value associated with it (scale at right). When these individual samples can be read as a direct voltage or current, that is they can have any level whatsoever, then the system is analog. When an element in the sensor output circuit creates a finite number of gray levels such as 10, 128, or even 1000, then the signal is said to be *quantized*. (When a finite number of levels is employed and is very large, the quantized signal resembles the analog case.) Being both sampled and quantized in a form that can be manipulated by a digital processor makes the image *digital*. Each of these individual samples of the image is a *picture element*, often referred to as a *pixel* or pel. A sampled and *multilevel* (>2) quantized image is often referred to as a *gray-scale image* (a term also used in a different context to describe a continuous tone analog image). When the quantization is limited to *two levels*, it is termed a *binary image*. Image processing algorithms that manipulate these different kinds of images can be "bit constrained" to the number of levels appropriate to the image bit depth (another expression for the number of levels), that is, integer arithmetic. This is effectively equivalent to many

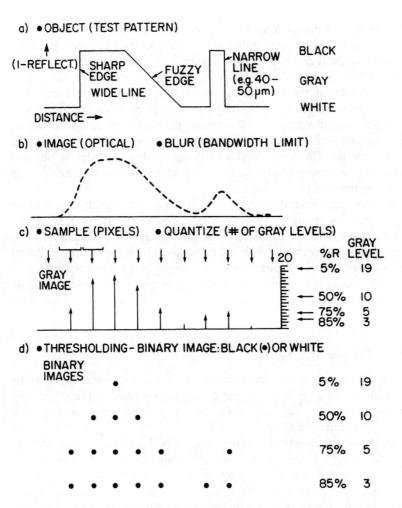

Figure 4 Formation of binary images, illustrating how a single, blurred electronic image of a small continuous tone test object could yield many different binary images depending on the threshold selected.

digital image processing circuits. Alternatively, algorithms may be floating point arithmetic, the results of which are quite different from the bit constrained operations.

A common and simple form of image processing is the conversion from a gray to a binary image as represented in part (d) of Fig. 4. In this process a threshold is set at some particular gray level, and any pixel at or above that level is converted to white or black. Any pixel whose gray value is below that level is converted to the other signal, that is, black or white, respectively. Four threshold levels are shown in part (c) by arrows on the gray-level scale at the right. Results are depicted in part (d) as four rows, each being a raster from the different binary images, one for each of the four thresholds. In part (d), each black pixel is represented by a dot, and each white pixel is represented by the lack of a dot. (It is common to depict pixels as series of contiguous squares in a lattice representing the space of the image. They are better thought of as points in time and space that can have any number of dimensions, attributes, and properties.)

Image Quality for Scanning

Each row of dot patterns shows one line of a sampled binary image. These patterns are associated with the location of the sampling arrows, shown in part (c), the shape of the blur, and the location of the features of the original document. Notice at the 85% threshold, the narrow line is now represented by two pixels (i.e., it has grown), but the wider and darker pulse has not changed in its representation. It is still five pixels wide. Notice that the narrow pulse grew in an asymmetric fashion and that the wider pulse, which was asymmetric to begin with, grew in a symmetric fashion. These are quite characteristic of the problems encountered in digitizing an analog document into a finite number of pixels and gray levels. It can be seen that creating a thresholded binary image is a highly nonlinear process. The unique imaging characteristics resulting from thresholding are discussed in detail in Sec. 5.

Figure 5 represents the same type of process using a real image. The plot is the gray profile of the cross-section of a small letter "I" for a single scan line. The width of the letter is denoted at various gray levels, indicated here by the label "threshold" to indicate where one could select the potential black to white transition level. The reader can see that the width of the binary image can vary anywhere from 1 to 7 pixels, depending on the selection of threshold.

Figure 6(a) returns to the same information shown in Fig. 4, except that here we have doubled the frequency with which we sampled the original blurred optical image. There are now twice as many pixels, and their variation in height is more gradual. In this particular instance, increased resolution is responsible for the binary case detecting the narrow pulse at a lower level (closer to 0% threshold). This illustration shows the general results that one would expect from increasing the spatial density at which one samples the

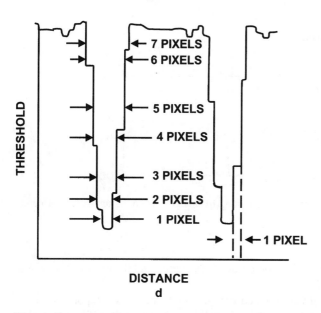

Figure 5 An actual scanned example of a gray scan line across the center of a letter "I". A different representation of the effect shown in step (c) in Fig. 4. Here the sample points are displayed as contiguous pixels. The width of one pixel is indicated. The image is from a 400 dpi scan of approximately a six-point Roman font.

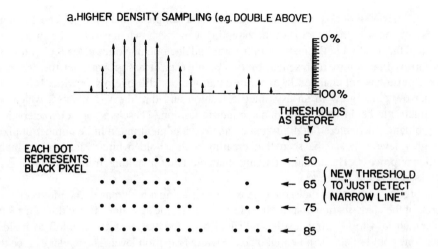

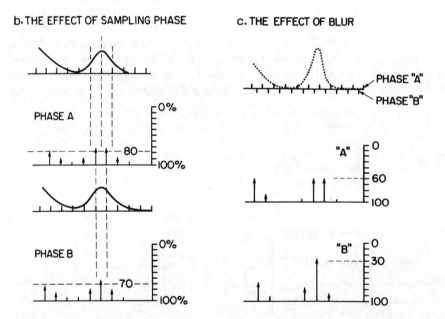

Figure 6 The effects of (a) doubling the resolution, (b) changing sampling phase, (c) sharpening the optical image.

image; that is, one sees somewhat finer detail in both the gray and the binary images with higher sampling frequency.

This is, however, not always the case when examining every portion of the microstructure. Let us look more closely at the narrower of the two pulses [Fig. 6(b)]. Here we see the sampling occurring at two locations, shifted slightly with respect to each other. These are said to be at different sampling phases. In phase A the pulse has been sampled in such a way that the separate pixels near the peak are identical to each other in their

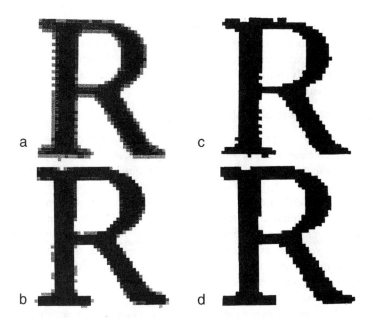

Figure 7 Digital images of a 10-point letter "R" scanned at 400 dpi showing quantization and sharpening effects. Parts (a) and (c) were made with normal sharpness for typical optical systems and parts (b) and (d) show electronic enhancement of the sharpness (see Fig. 29). Parts (a) and (b) are made with 2 bits/pixel, that is, four levels including white, black and two levels of gray. Parts (c) and (d) are 1 bit/pixel images, that is, binary with only black and white where the threshold was set between the two levels of gray used in (a) and (b). Note the thickening of some strokes in the shaper image and the increased raggedness of the edges in the binary images. Some parts of the sharp binary images are also less ragged.

intensity, and in phase B one of the pixels is shown centered on the peak. When looking at the threshold required to detect the information in phase A and phase B, different results are obtained for a binary representation of these images. Phase B would show the detection of the pulse at a lower threshold (closer to ideal) and phase A, when it detects the pulse, would show it as wider, namely as two pixels in width.

Consider an effect of this type in the case of an input document scanner, such as that used for facsimile or electronic copying. While the sampling array in many input scanners is constant with respect to the document platen, the location of the document on the platen is random. Also the locations of the details of any particular document within the format of the sheet of paper are random. Thus the phase of sampling with respect to detail is random and the type of effects illustrated in Fig. 6 would occur randomly over a page. There is no possibility that a document covered with some form of uniform detail can look absolutely uniform in a sampled image. If the imaging system produces binary results, it will consistently exhibit errors on the order of one pixel and occasionally two pixels of edge position and line width. The same is true of a typically quantized gray image, except now the errors are primarily in magnitude and may, at higher sampling densities, be less objectionable. In fact, an analog gray imaging process, sampling at a sufficiently high frequency, would render an image with no visible error (see the next subsection). Continuing with the same basic illustration, let us consider the effect of blur. In Fig. 6(c) we have sketched a less blurred image in the region of the narrower pulse and now show two sampling phases A and B, as before, separated by half a pixel width. Two things should be noted. First, with higher sharpness (i.e., less blur), the threshold at which detection occurs is higher. Secondly, the effect of sampling phase is much larger with the sharper image. Highly magnified images in Fig. 7 illustrate some of these effects.

2.1.2 The Sampling Theorem and Spatial Relationships

By means of these illustrations we have shown the effects of sampling frequency, sampling phase, and blur at an elementary level. We now turn our attention to the more formal description of these effects in what is known as the sampling theorem. For these purposes we assume that the reader has some understanding of the concepts of Fourier analysis or at least the frequency-domain way of describing time or space, such as in the frequency analysis of audio equipment. In this approach, distance in millimeters is transformed to frequency in cycles per millimeter (cycles/mm). A pattern of bars spaced 1 mm apart would result in 1 cycle/mm as the fundamental frequency of the pattern. If the bars were represented by a square wave, the Fourier series showing the pattern's various harmonics would constitute the frequency-domain equivalent.

Figure 8 has been constructed from such a point of view. In Fig. 8(a) we see a single-raster profile of an analog input document (i.e., an object) represented by the function $f(x)$. This is a signal extending in principle to $\pm\infty$ and contains, upon analysis, many different frequencies. It could be thought of as a very long microreflectance profile across an original document. Its spectral components, that is, the relative amplitudes of sine waves that fit this distribution of intensities, are plotted as $F(\mu)$ in Fig. 8(b). Note that there is a maximum frequency in this plot of amplitude vs. frequency, at w. It is equal to the reciprocal of λ (the wavelength of the finest detail) shown in Fig. 8(a). This is the highest frequency that was measured in the input document. The frequency w is known as the bandwidth limit of the input document. Therefore the input document is said to be band-limited. This limit is often imposed by the width of a scanning aperture that is performing the sampling in a real system.

We now wish to take this analog signal and convert it into a sampled image. We multiply it by $s(x)$, a series of narrow impulses separated by Δx as shown in Fig. 8(c). The product of $s(x)$ and $f(x)$ is the sampled image, and that is shown in Fig. 8(e). To examine this process in frequency space, we need to find the frequency composition of the series of impulses that we used for sampling. The resulting spectrum is shown in Fig. 8(d). It is, itself, a series of impulses whose frequency locations are spaced at $1/\Delta x$ apart. For the optical scientist this may be thought of as a spectrum, with each impulse representing a different order; thus the spike at $1/\Delta x$ represents the first-order spectrum, and the spike at zero represents the zero-order spectrum. Because we multiplied in distance space in order to come up with this sampled image, in frequency space, according to the convolution theorem, we must convolve the spectrum of the input document with the spectrum of the sampling function to arrive at the spectrum of the sampled image. The result of this convolution is shown in Fig. 8(f).

Now we can see the relationship between the spectral content of the input document and the spacing of the sampling required in order to record that document. Because the spectrum of the document was convolved with the sampling spectrum, the negative side of the input document spectrum $F(\mu)$ folds back from the first-order over the positive side of the zero-order document spectrum. Where these two cross is exactly halfway between the zero- and first-order peaks. It is a frequency $(1/2\Delta x)$ known as the Nyquist frequency. If

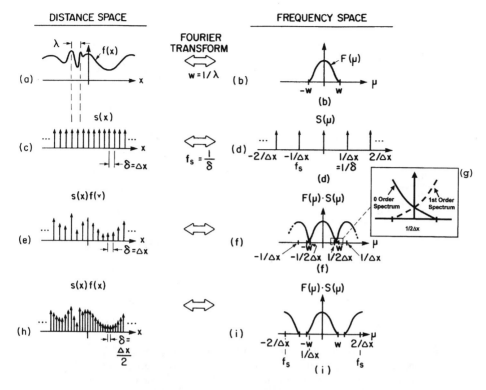

Figure 8 The Fourier transformation of images and the effects of sampling frequency. The origin and prevention of aliasing: (a) original object; (b) spectrum of object; (c) sampling function; (d) spectrum of sampling function; (e) sampled object; (f) spectrum of sampled object; (g) detail of sampled object spectrum; (h) object sampled at double frequency; and (i) spectrum of object sampled at double frequency. (From Ref. 88.)

we look at the region in Fig. 8(g) between zero and the Nyquist frequency, the region reserved for the zero-order information, we see that there is "contamination" from the negative side of the first order down to the frequency $[(1/\Delta x) - w]$, where w is the band limit of the signal. Any frequency above that point contains information from both the zero and the first order and is therefore corrupted or mixed, often referred to as *aliased*.

Should one desire to avoid the problem of aliasing, one must sample at a finer sampling interval, as shown in Fig. 8(h). Here the spacing is one half that of the earlier sketches, and therefore the sampling frequency is twice as high. This also doubles the Nyquist frequency. This merely separates the spectra by spreading them out by a factor of 2. Since there is no overlap of zero and first orders in this example, one can recover the original signal quite easily by simply filtering out the higher frequencies representing the orders other than zero. This is illustrated in Fig. 9, where a rectangular function of width $\pm w$ and amplitude 1 is multiplied by the sampled image spectra, resulting in recovery of the original signal spectra. When inversely Fourier transformed, this would give the original signal back [compare Figs. 9(e) and 8(a)].

We can now restate Shannon's[36] formal sampling theorem, [sometimes referred to as the Whittaker – Shannon Sampling Theoerm (R. Loce, personal communication,

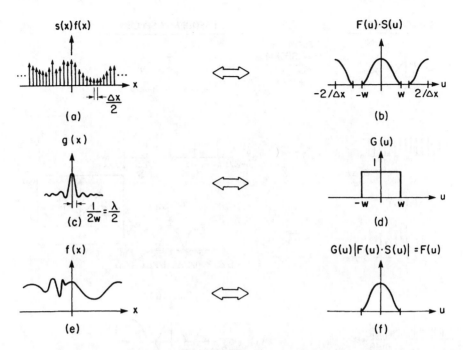

Figure 9 Recovery of original object from properly sampled imaging process: (a) object sampled at double frequency [from Fig. 8(h)]; (b) spectrum of "a" [from Fig. 8(i)]; (c) spread function for rectangular frequency filter function; (d) rectangular frequency function; (e) recovered object function; (f) recovered object spectrum. (From Ref. 88.)

2001)] in terms that apply to sampled imaging: if a function $f(x)$ representing either an original object or the optical/aerial image being digitized contains no frequencies higher than w cycles/mm (this means that the signal is band-limited at w), it is completely determined by giving its values at a series of points $<1/2w$ mm apart. It is formally required that there be no quantization or other noise and that this series be infinitely long; otherwise windowing effects at the boundaries of smaller images may cause some additional problems (e.g., digital perturbations from the presence of sharp edges at the ends of the image). In practice, it needs to be long enough to render such windowing effects negligible.

It is clear from this that any process such as imaging by a lens between the document and the actual sampling, say by a CCD sensor, can band-limit the information and ensure accurate effects of sampling with respect to aliasing. However, if the process of band-limiting the signal in order to prevent aliasing causes the document to lose information that was important visually, then the system is producing restrictions that would be interpreted as excessive blur in the optical image. Another way to improve on this situation is, of course, to increase the sampling frequency, that is, decrease the distance between samples.

We have shown in Fig. 9 that the process of recovering the original spectrum is accomplished by a filter having a rectangular shape in frequency space [Fig. 9(d)]. This filter is known as the reconstruction filter and represents an idealized reconstruction process. The rectangular function has a $(\sin x)/x$ inverse transform in distance space [Fig. 9(c)], whose zero crossings are at $\pm N\Delta x$ from the origin where $N = 1, 2, \ldots$

Image Quality for Scanning

Rectangular and other filters with flat modulation transfer functions (MTFs) are difficult to realize in incoherent systems. This comes about because of the need for negative light in the sidelobes (in distance space). A reconstruction filter need not be precisely rectangular in order to work. It should be relatively flat and at a value near 1.0 over the bandwidth of the signal being reconstructed (also difficult and often impossible to achieve). It must not transmit any energy from the two first-order spectra. If the sampling resolution is very high and the bandwidth of the signal is relatively low, then the freedom to design the edge of this reconstruction filter is relatively great and therefore this edge does not need to be as square. From a practical point of view the filter is often the MTF of the output scanner, typically a laser beam scanner, and is not usually a rectangular function but more of a gaussian shape. A nonrectangular filter, such as that provided by a gaussian laser beam scanner, alters the shape of the spectrum that it is trying to recover. Because the spectrum is multiplied by the reconstructing MTF, this causes some additional attenuation in the high frequencies, and a trade-off is normally required in practical designs.

2.1.3 Gray Level Quantization: Some Limiting Effects

Now that we have seen how the spatial or distance dimension of an input image may be digitized into discrete pixels, we explore image quantization into a finite number of discrete gray levels. From a practical standpoint this quantization is accomplished by an A/D converter, which quantizes the signal into a number of gray levels, usually some power of 2. A popular quantization is 256 levels, that is, 8 bits, which lends itself to many computer applications and standard digital hardware. There may be good reasons for other quantizations, higher or lower, to optimize a design or a system.

From an engineering perspective, one needs to understand the limits on the useful number of quantization levels. This should be based upon noise in the input as seen by the system or upon the ultimate output goal of how many distinguishable gray levels can be seen by the human eye. Both approaches have been explored in the literature and involve complex calculations and experimental measurements.

Use of the *HVS response* with various halftoning methods represents an *outbound limit* approach to defining practical quantization limits for scanned imaging. The "visual limit" results shown in Fig. 10[37] plot the number of visually distinguishable gray levels against the spatial frequency at which they can be seen. This curve was derived from a very conservative estimate of the visual system frequency response and may be thought of as an upper limit on the number of gray levels required by the eye. Plotted on the same curve are performance characteristics for 20 pixels/mm (500 pixels/in.) digital imaging systems that produce 3 bits/pixel and 1 bit/pixel (binary) images. These were obtained by use of a generalized algorithm to create halftone patterns (see Sec. 2.2.3 and Ref. 38) at different spatial frequencies. The binary limit curve, added here to Roetling's, graph, shows the number of effective gray levels for each frequency whose period is two halftone cells wide. The 3 bit limit assumes each halftone cell contributes 2^3 gray values, including black and white.

Roetling[37] integrated the visual response curve to find an average of 2.8 bits/pixel as a good upper bound for the eye itself. Note that his general halftoning approach, using 3 bits/pixel and 20 pixels/mm (500 pixels/in.) also approximates the visual limit in the important mid-frequency region. Specialized halftoning techniques[6,38] may produce different and often more gray levels per pixel at the lower frequencies.

Another approach to setting quantization limits is to examine the noise in the input, assuming in so doing that the quantization is input bound and not output bound by the

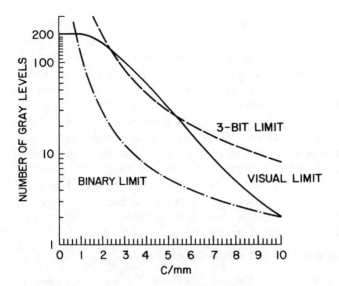

Figure 10 Example of outbound quantization limits, using visually distinguishable number of gray levels vs. spatial frequency, with corresponding 1 (binary) (from Ref. 37) and 3 bit/pixel limits.

visual process as in the foregoing approach. A range of photographic input was selected as examples of a practical lower limit (best) on input noise. The basic principle for describing the useful number M of gray levels in a photograph involves quantizing its density scale into steps whose size is based on the noise (granularity) of that photographic image[39] when scanned by the digital imaging process. In simplified terms this can be described as

$$M = \frac{L}{2k\sigma_a} \quad (1)$$

where L = the density range of the image, σ_a = measured standard deviation of density using aperture area = A, and k = the number of standard deviations in each distinguishable level.

The question being addressed by this type of quantization is how reliably one wants to be able to determine the specific tone in a given part of the input picture from a reading of a single pixel. For some purposes, where the scanned image is used to extract radiometric information from a picture,[39] the reliability must be high, for other cases such as simply copying a scene for artistic purposes it can be much lower. To precisely control a digital halftone process (see later) it must be fairly high.

Photographic noise is approximately random uncorrelated noise. To a first order, photographic noise (granularity) is the standard deviation of the density fluctuations. It is directly proportional to the square root of the effective detection area,[40,41] a of a measuring instrument or scanner-sensor, that is, Selwyn's law:

$$\sigma_a = S(2a)^{1/2} \quad (2)$$

where S is a proportionality constant defined as the Selwyn granularity. It is also proportional to the square root of the mean density, that is, Siedentopf's relationship,[40,42] in

Image Quality for Scanning

an ideal film system. In practical cases, as is done here, the density relationship must be empirically determined. Figure 11 shows the number of distinguishable gray levels reported in the literature by various authors for various classes of films obtained by directly measuring granularity as a function of density. They are reported at apertures that are approximately equivalent in size to the smallest detail the film could resolve, that is, the diameter of the film spread function. For a real world example, assume that 35 mm film images are enlarged perfectly by a high-quality 3.3 × enlarger. The conversion to the number of distinguishable gray levels per pixel is based on assuming Selwyn's law, a reliability of 99.7% ($\pm 3\sigma_a$ or $k = 6$) and that any nonlinear relationship between granularity and density scales as the aperture size changes. The actual scanner aperture is reduced by 3.3 × in its two dimensions to resemble directly scanning the film.

Four specific films were selected, each representative of a different class, three of which are black and white films: (a) an extremely fine-grained microfilm, (b) a fine-grained amateur film, (c) a high-speed amateur film.[43] A special purpose color film was also included.[45] Despite now being obsolete, these films still represent a reasonable cross-section of photographic materials. A 3.3 × enlargement was selected as typical of consumer practice, roughly giving a 3.5″ × 5″ print from a 35 mm negative. The reciprocal of this magnification is used to scale the scanner aperture back to film

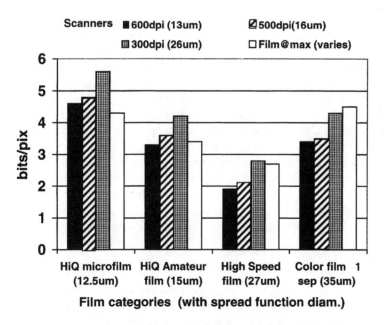

Figure 11 Example of inbound quantization limits, using the number of distinguishable gray levels, in bits/pixel, for input consisting of 3.3 × enlargements (3 × 5 in. prints of 35 mm film) from four example films (from Refs. 43 and 45) scanned by four generic types of systems indicated by their scanning resolutions. Color film is for a single separation, others are black and white films. The limiting blur in μm for the first three scanners is given in the parentheses after the scan frequency. It is the sensor aperture width scaled to the film size. The fourth scanner has variable resolution set by a scaled aperture width adjusted to equal the width of each film blur function (spread function), shown in parentheses with the film type. Assumes a 99.7% confidence on distinguishability using Eq. (1) (i.e., with $k = 6$).

dimensions. Two popular scanner resolutions of 600 and 300 dpi were selected. The corresponding sensor "aperture" widths in μm, scaled to the film, are noted in parentheses in the key at the top of each figure. The width is the inverse of the sampling period. A third scanner aperture, equivalent to that in the Roetling visual calculations, was used for one case, that is, a 20 samples/mm (500 samples/in.) scanning system with an aperture of 50 × 50 μm (2 × 2 mils). The fourth situation, called "Film @ max" describes the number of levels resulting from scanning the film with an aperture that matches the blur (spread function) for the film, given in the film category label in parentheses at the bottom of each figure. These approximate calculations are an oversimplification of the photographic and enlarging processes, ignoring significant nonlinearities and blurring effects, but they provide a rough first-order analysis.

Examination of the charts suggests that a practical range of inbound quantization limits (IQLs) for pictorial images is approximately anywhere from 2 to 4 bits/pixel (microfilm is not made for pictorials). For typical high-quality reproduction, then, an input bound limit is a little over 3 bits/pixel at 600 dpi using the three standard deviation criterion. This compares with the rate of 2.8 bits/pixel found by Roetling for a visual outbound quantization limit (OQL). Recent work by Vaysman and Fairchild,[44] limited to an upper frequency of 300 dpi by their printer selection, also found, through psychophysical studies, that 3 bits/pixel/color was a useful system optimum for reproducing color pictures.

One may ask, then, why are there so many input scanners operating at 8, 10, or even 12 bits/pixel? There are at least two answers. First, these scanner specifications are often greatly exaggerated for commercial reasons, reflecting only the performance of an internal analog to digital converter, not the true detection-noise-limited capability of the scanner. Secondly, when the specifications are technically valid, there can be system level justification for such performance. The reason for this is, in part, that the input scanners are only the first element in the system. Many printers are nominally limited to 1 bit/pixel (ink or no ink). Carrying extra information, up front, which essentially preserves the details of the input noise, often enables downstream image processing of the input to better compensate for the printer limitations and to optimize the information capacity of the system. As we shall see later, this involves spatial resolution trade-offs, analyzing structures to segment them, and gathering statistics. In addition, several other factors, which we did not explore here, but which are covered later (see sections on system response, halftone response, gray scanner tone reproduction and noise, and summary measures of image quality), can lead to higher or lower effective quantization levels at different places in the system:

- Lowering pixel level reliability; for example, reducing reliability from 99.7%, where one may use the distinguishable levels to control tone reproduction very accurately or to perform radiometric measurements, needing $\pm 3\sigma$, where $k = 6$, to a case of $\pm 1\sigma$, where $k = 2$ for 68% reliability [Eq. (1)]. The latter might be the case when tone reproduction control is not a concern and subjective quality is more important. This creates three times as many levels. Multiplication is represented as addition in log units. To convert to bits, we take $\log_2 3 = 1.6$, that is, add 1.6 bits to the IQL (see also "noise" later Sec. 2.2.4 and Sec. 4.3).
- Considering a different "effective sampling," commonly at halftone resolution; for example, one looks at many scanned pictures reproduced as halftones. Here the user ignores the screen pattern, adjusting viewing or other factors to create a

Image Quality for Scanning

lower effective sampling, justifying the idea that the sampling may be comparable to that created by the halftone itself. In one such case, a 100 dpi halftone cell has an "effective aperture" 36 times the area of the 600 dpi actual sampling aperture. This decreases noise by the multiplicative factor of the square root of the ratios of aperture area, according to Selwyn's Law in Eq. (2). This increases the number of levels by $6\times$, again converting to $\log_2 6 = 2.6$, that is, adding 2.6 bits to the IQL (see also "halftones" in Sec. 2.2.3).
- System blurring, which enlarges the effective aperture (realistic increase of up to $2\times$ over the ideal sampling apertures, that is, adding up to 1 bit to the IQL) (see blur and MTF later in Sec. 2.2.1 and Sec. 4.2).

To this point it can be seen that various practical considerations might considerably change the effective inbound limit for photographic originals. If all the effects add, this could increase the limits shown in Fig. 11 by adding as much as 5 bits. Continuing on, there are many other factors affecting quantization:

- Trade-off between gray quantization and spatial resolution can add or subtract 1 to 3 bits (see discussion on information capacity later in Sec. 6.7).
- Many selective operations to reduce levels are often carried out off line, after capture of the maximum possible number of levels.
- System noise is combined with input noise for realistic limits (see Sec. 4.3).
- Recording enough gray differences at some resolution to capture nonlinear eye response in shadows (approximately 10–11 bits total in linear space). See ΔE^* and color models later in Sec. 2.2.5. Note, for example, that to change from L^* of 9 to 8, that is, one just noticeable difference (1.0 JND, see p. 236), requires one-quarter of a gray level in an 8-bit system or 1 gray level in a 10-bit system.

When writing the output of a laser scanner to film, the distinguishable density level analysis given above, used in reverse, can also provide an outbound or system quantization limit. Here the laser spot size is the effective aperture area in Eqs. (1) and (2), provided it is larger than the film spread function. Being aware of the inbound limits, the system options and the outbound limits as an endpoint give a framework for robust engineering of image quality. Information capacity approaches extend these concepts (see Sec. 6.7).

2.2 Basic System Effects

2.2.1 Blur

Blur, that is, the spreading of the microscopic image structure, is a significant factor in determining the information in an image and therefore its quality. In the input scanner, blur is caused by the optical system, the size and properties of the light-sensing element, other electronic elements, and by mechanical and timing factors involved in motion. This blur determines whether the system is aliased. Roughly speaking, if the image of a point (the profile of which is called the point spread function) spreads over twice the sampling interval, the system is unaliased. The spreading also determines the contrast of fine details in the gray video image prior to processing. The cascading of these elements can be described conveniently by a series of spatial frequency responses [see later under modulation transfer functions (MTFs) for a detailed discussion] or other metrics that relate generally to the sharpness of optical images. It can be compensated for, in certain aspects, by subsequent electronic or computer image processing.

Blur in an output scanner is caused by the size of the writing spot, for example, the laser beam waist at focus, by modulation techniques and by the spreading of the image in any marking process such as xerography or photographic film. It is also affected by motion of the beam relative to the data rate and by the rate of motion of the light-sensitive receptor material. Output scanner blur more directly affects the appearance of sharpness in the final hard copy image that is presented to the human visual system than does blur in the input scanner. Overall enhancement of the electronic input scanned image can, however, draw visual attention to details of the output image unaffected by blur limitations of either scanner.

Blur for the total system, from input scanner through various types of image processing to output scanner and then to marks on paper, is not easily cascaded, because the intervening processing of the image information is extremely nonlinear. This nonlinearity may give rise to such effects as a blurred input image looking very sharp on the edges of a binary output print because of the small spot size and low blur of the marking process. In such a case, however, the edges of square corners look rounded and fine detail such as serifs in text or textures in photographs may be lost. Conversely, a sharp input scan printed by a system with a large blurring spot would appear to have fuzzy edges, but the edge noise due to sampling would have been blurred together and would be less visible than in the first case. Moiré, from aliased images of periodic subjects caused by low blur relative to the sample spacing, however, would still be present in spite of output blur. (Note, superposition of periodic patterns such as a halftoned document (see Sec. 2.2.3) and the sampling grid of a scanner results in new and often striking periodic patterns in the image commonly called Moiré patterns (see Bryngdahl[46]). Once aliased, no amount of subsequent processing can remove this periodic aliasing effect from an image.) The popular technologies called "anti aliasing" deal with a different effect of undersampling, namely that binary line images exhibit strong visible staircase or jaggie effects on slanted lines when the output blur and sampling are insufficient for the visual system. These techniques nonlinearly "find" the stair steps and locally add gray pixels to reduce the visibility of the jaggie (see Sec. 7.2 and Fig. 44).[149] Aliasing is also known as spurious response.[12]

It is apparent, then, that blur can have both positive and negative impacts on the overall image quality and requires a careful trade-off analysis when designing scanners.

2.2.2 System Response

There are four ways in which electronic imaging systems display tonal information to the eye or transmit tonal information through the system:

1. By producing a signal of varying strength at each pixel, using either amplitude or pulse-width modulation.
2. By turning each pixel on or off (a two-level or binary system; see Sec. 5).
3. By use of a halftoning approach, which is a special case of binary imaging. Here, the threshold for the white–black decision is varied in some structured way over very small regions of the image, simulating continuous response. Many, often elaborate, methods exist for varying the structure; some involve multiple pixel interactions (such as error diffusion; see the end of Sec. 2.2.3) and others use subpixels (such as high addressability, extensions of the techniques mentioned in Sec. 7.2).
4. By hybrid halftoning combining the halftone concept in (3) with the variable gray pixels from (1) (e.g., see Refs 37 and 38).

Image Quality for Scanning

From a hardware point of view, the systems are either designed to carry gray information on a pixel-by-pixel basis or to carry binary (two-level) information on a pixel-by-pixel basis. Because a two-level imaging system is not very satisfactory in many applications, some context is added to the information flow in order to obtain pseudo-gray using the halftoning approach.

Macroscopic tone reproduction is the fundamental characteristic used to describe all imaging systems' responses, whether they are analog or digital. For an input scanner it is characterized by a plot of an appropriate, macroscopic output response, as a function of some representation of the input light level. The output may characteristically be volts or digital gray levels for a digital input scanner and intensity or perhaps darkness or density of the final marks-on-paper image for an output scanner. The correct choice of units depends upon the application for which the system response is being described. There are often debates as to whether such response curves should be in units of density or optical intensity, brightness, visual lightness or darkness, gray level, and so on. For purposes of illustration, see Fig. 12.

Here we have chosen to use the conventional photographic characterization of output density plotted against input density using normalized densities. Curve A shows the case of a binary imaging system in which the output is white or zero density up to an input density of 0.6, at which point it becomes black or 2.0 output density. Curve B shows what happens when a system responds linearly in a continuous fashion to input density. As the input is equal to the output here, this system would be linear in reflectance, irradiance, or even Munsell value (visual lightness units).

Curve C shows a classic abridged gray system attempting to write linearly but with only eight levels of gray. This response becomes a series of small steps, but because of the choice of density units, which are logarithmic, the sizes of the steps are very different. Had we plotted output reflectance as a function of input reflectance, the sizes of the steps would have been equal. However, the visual system that usually looks at these tones operates in a more or less logarithmic or power fashion, hence the density plot is more representative of the visual effect for this image. Had we chosen to quantize in 256 gray levels, each step

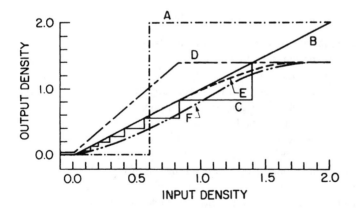

Figure 12 Some representative input/output density relationships for (A) binary imaging response; (B) linear imaging response; (C) stepwise linear response; (D) saturation – limited linear response; (E) linear response with gradual roll-off to saturation; (F) idealized response curve for best overall acceptability.

shown would have been broken down into 32 smaller substeps, thereby approximating very closely the continuous curve for B.

When designing the system tone reproduction, there are many choices available for the proper shape of this curve. The binary curve, as in A, is ideal for the case of reproducing high-contrast information because it allows the minimum and maximum input densities considerable variation without any change to the overall system response.

For reproducing continuous tone pictures, there are many different shapes for the relationship between input and output, two of which are shown in Fig. 12. If, for example, the input document is relatively low contrast, ranging from 0 to 0.8 density, and the output process is capable of creating higher densities such as 1.4, then the curve represented by D would provide a satisfactory solution for many applications. However, it would create an increase in contrast represented by the increase in the slope of the curve relative to B, where B gives one-for-one tone reproductions at all densities. Curve D is clipped at an input density greater than 0.8. This means that any densities greater than that could not be distinguished and would all print at an output density of 1.4.

In many conventional imaging situations the input density range exceeds that of the output density. The system designer is confronted with the problem of dealing with this mismatch of dynamic ranges. One approach is to make the system respond linearly to density up to the output limit; for example, following curve B up to an output density of 1.4 and then following curve D. This generally produces unsatisfactory results in the shadow regions for the reasons given earlier for curve D. One general rule is to follow the linear response curve in the highlight region and then to roll off gradually to the maximum density in the shadow regions starting perhaps at a 0.8 output density point for the nonlinear portion of the curve as shown by curve E. Curve F represents an idealized case approximating a very precisely specified version arrived at by Jorgenson.[47] He found the "S"-shaped curve resembling F to be a psychologically preferred curve among a large number of the curves he tried for lithographic applications. Note that it is lighter in the highlights and has a midtone region where the slope parallels that of the linear response. It then rolls off much as the previous case toward the maximum output density at a point where the input density reaches its upper limit.

2.2.3 Halftone System Response and Detail Rendition

One of the advantages of digital imaging systems is the ability to completely control the shape of these curves to allow the individual user to find the optimum relationship for a particular photograph in a particular application. This can be achieved through the mechanism of digital halftoning as described below. Historically important studies of tone reproduction, largely for photographic and graphic arts applications, include those of Jones and Nelson,[48] Jones,[49] Bartleson and Breneman,[50] and two excellent review articles, covering many others, by Nelson.[51,52] Many recent advances in the technology of digital halftoning have been collected by Eschbach.[6]

The halftoning process can be understood by examination of Fig. 13. In the top of this illustration two types of functions are plotted against distance x, which has been marked off into increments 1 pixel in width. The first functions are three uniform reflectance levels, R1, R2, and R3. The second function $T(x)$ is a plot of threshold vs. distance, which looks like a series of up and down staircases, that produces the halftone pattern. Any pixels whose reflectance is equal to or above the threshold is turned on, and any that is below the threshold for that pixel is turned off.

Image Quality for Scanning 161

ANALYSIS OF THE HALFTONE PROCESS

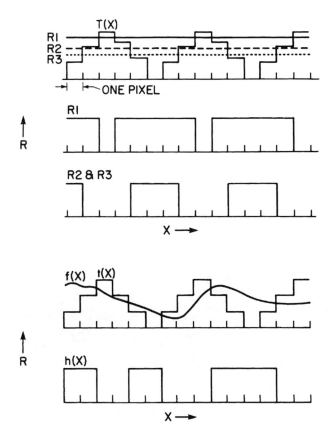

Figure 13 Illustration of halftoning process. Each graph is a plot of reflectance R vs. distance X. $T(x)$ is the profile of one raster of the halftone threshold pattern, where image values above the pattern are turned on (creates black in system shown) by the halftone thresholding process. R1, R2, and R3 represent three uniform images of different average reflectances shown at the top as uniform input and in the middle of the chart as profiles of halftone dots after halftone thresholding. $f(x)$ represents an image of varying input reflectance and $t(x)$ is a different threshold pattern. $h(x)$ is the resulting halftone dot profile, with dots represented, here, as blocks of different width illustrating image variation.

Also sketched in Fig. 13 are the results for the thresholding process for R1 on the second line and then for R2 and R3 on the third line. The last two are indistinguishable for this particular set of thresholding curves. It can be seen from this that the reflectance information is changed into width information and thus that the method of halftoning is a mechanism for creating dot growth or spatial pulse width modulation over an area of several pixels. Typically, such threshold patterns (i.e., screens) are laid out two-dimensionally. An example is shown in Fig. 14.

This thresholding scheme emulates the printer's 45° screen angle, which is considered to be favorable from a visual standpoint because the 45° screen is less visible (oblique effect[3]) than the same 90° screen. Other screen angles may also be conveniently

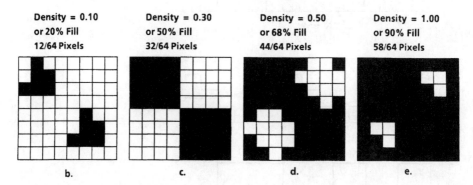

Figure 14 Example of two-dimensional quantized halftone pattern, with illustrations of resulting halftone dots at various density levels.

generated by a single string of thresholds and a shift factor that varies from raster to raster.[53,54] The numbers in each cell in the matrix represent the threshold required in a 32-gray-level system to turn the system on or off. The sequence of thresholds is referred to as the dot growth pattern. At the bottom, four thresholded halftone dots (Parts b–e) are shown for illustration. There are a total of 64 pixels in the array but only 32 unique levels. This screen can be represented by 32 values in a 4 × 8 pixel array plus a shift factor of 4 pixels for the lower set of 32, which enables the 45° screen appearance as illustrated. It may also be represented by 64 values in a single 8 × 8 pixel array, but this would be a 90° screen. It is also possible to alternate the thresholding sequence between the two 4 × 8 arrays, where the growth pattern in each array is most commonly in a spiral pattern, resulting in two unique sets of 32 thresholds for an equivalent of 64 different levels and preserving the screen frequency as shown. This screen is called a "double dot." The concept is sometimes extended to four unique dot growth patterns and hence is named a "quad dot." Certain percent area coverage dot patterns in these complex multicentered dot structures generate very visible and often objectionable patterns.

What is perhaps less obvious is that high spatial frequency information can be recorded by this type of halftoning process. This can be seen in Fig. 15. Part a is the threshold matrix, identical to that used in Fig. 14. It is also sometimes called the halftone

Image Quality for Scanning

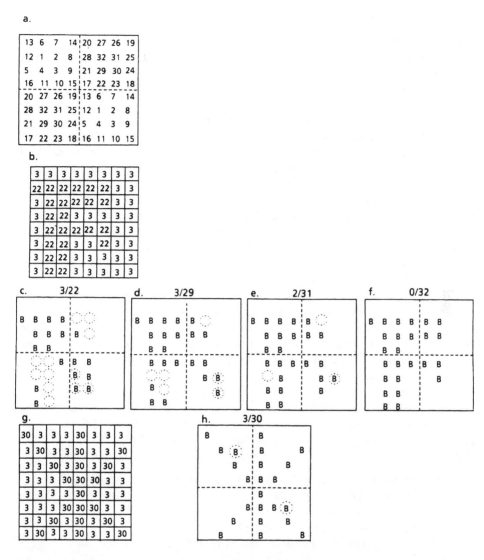

Figure 15 Illustration of the preservation of image detail finer than the scale of the halftone pattern ("partial dotting"): (a) the threshold matrix for screen used in (c)–(f) and (h); (b) pattern of gray pixels in the image leading to the binary screened images in (c)–(f) showing the explicit combination of gray values, 3 and 22, used for (c). The alternate gray image values for (d)–(f) are shown above each figure; (g) gray values in image leading to (h). If gray value in image (b) or (g) is larger than or equal to the screen element in (a), turn pixel black. An empty dotted circle indicates an error of not writing black. B indicates a black pixel, blank is white in (c)–(f) and (h). A circled B indicates an error of writing black when white is desired.

screen or dot growth pattern. Part b of the figure is the pixel-by-pixel gray level map representing the original image as it has been captured by an electronic imaging system. The gray levels of 3 and 22 represent background and foreground levels or light and dark pixels. They are the only ones available in this image, and they are laid out somewhat in the shape of the letter "F." This is a medium contrast image. In part c we see the results of the halftoning process using the threshold matrix of part a applied to the image in part b. Pixels marked "B" show where the image value was higher than the halftone screen value. In parts d, e, and f we see the resultant halftone images as the contrast of the electronic image increases by specific gray level assignments of the foreground and background pixels. These might be caused by increases in sharpness or the gain and offset of the electronic characteristic curve or a different contrast original object being scanned. The contrast progresses from 3 and 29 to 2 and 31, and eventually to the maximum of 0 and 32. Dotted circles indicate errors in correctly reproducing background or foreground pixels as black or white. It is seen that as the contrast of the input increases the errors made by the screen in detecting all of the necessary components go down. One and 2 pixel lines and serifs are readily reproduced by this process. This phenomenon (ability to resolve structures finer than the halftone screen array or cell size) has been described as "partial dotting" by Roetling[55] and others. The illustrations in parts g and h of the same figure show three narrow 1 pixel wide lines (whose contrast is defined by 30 for the line and 3 for the background field) at 0 and 45° angles. The latter shows two phase shifts with respect to the halftone threshold matrix. It can be seen, here, that angle and shift information can also be recorded through the process of screening of high-contrast detail.

The halftone matrix described represented 32 specific thresholds in a specific layout. There are many alternatives to the size and shape of the matrix, the levels chosen, the spatial sequence in which the thresholds occur, and arrangements of multiple, uniquely different matrices in a grouping called a super cell. The careful selection of these factors gives good control over the shape of the apparent tone reproduction curve, granularity, textures, and sharpness in an image, as is seen below.

There are also many other methods for converting binary images into pseudo-gray images using digital halftoning methods of a more complex form.[56,57] These include alternative dot structures, that is, different patterns of sequences in alternating repeat patterns, random halftoning, and techniques known as error diffusion. In his book *Digital Halftoning*, Ulichney[58] describes five general categories of halftoning techniques:

1. Dithering with white noise (including mezzotint).
2. Clustered dot ordered dither.
3. Dispersed dot ordered dither (including "Bayer's dither").
4. Ordered dither on asymmetric grids.
5. Dithering with blue noise (actually error diffusion).

He states that "spatial dithering is another name often given to the concept of digital halftoning. It is perfectly equivalent, and refers to any algorithmic process that creates the illusion of continuous tone images from the judicious arrangement of binary picture elements." The process described in Figs 14 and 15 falls into the category of a clustered dot ordered dither method (category 2) as a classical rectangular grid on a 45° base.

There is no universally best technique among these. Each has its own strengths and weaknesses in different applications. The reader is cautioned that there are many important aspects of the general halftoning process that could not be covered here. (See Ref. 38 for a summary of digital halftoning technology and many references, and Ref. 59 for many practical aspects of conventional halftoning for color reproduction.) For example, the densities described in Fig. 14 only apply to the case of perfect reproduction of the illustrated pixel maps on non-light-scattering material using perfect, totally black inks. In reality, each pattern of pixels must be individually calibrated for any given marking process. The spatial distribution interacts with various noise and blurring characteristics of

Image Quality for Scanning

output systems to render the mathematics of counting pixels to determine precise density relationships highly erroneous under most conditions. This is even true for the use of halftoning in conventional lithographic processes, due to the scattering of light in white paper and the optical interaction of ink and paper. These affect the way the input scanner "sees" a lithographic halftone original. Some of these relationships have been addressed in the literature, both in a correction factor sense[59,60] and in a spatial frequency sense.[61-63] All of these methods involve various ways of calculating the effect that lateral light scattering through the paper has on the light reemerging from the paper between the dots.

The effects of blur from the writing and marking processes involved in generating the halftone, many of which may be asymmetric, require individual density calibrations for each of the dot patterns and each of the dithering methods that can be used to generate these halftone patterns. The control afforded through the digital halftoning process by the careful selection of these patterns and methods enables the creation of any desired shape for the tone reproduction curve for a given picture, marking process, or application.

2.2.4 Noise

Noise can take on many forms in an electronic imaging system. First there is the noise inherent in the digital process. This is generally referred to as either sampling noise associated with the location of the pixels or quantization noise associated with the number of discrete levels. Examples of both have been considered in the earlier discussion. Next there is electronic noise associated with the electronic components from the sensor to the amplification and correction circuits. As we move through the system, the digital components are generally thought to be error-free and therefore there is usually no such noise associated with them.

Next, in a typical electronic system, we find the raster output scanner (ROS) itself, often a laser beam scanner. If the system is writing a binary file, then the noise associated with this subsystem is generally connected with pointing of the beam at the imaging material and is described as jitter, pixel placement error, or raster distortion of some form (see the next subsection). Under certain circumstances, exposure variation produces noise, even in a binary process. For systems with gray information, there is also the possibility that the signals driving the modulation of exposure may be in error, so that the ROS can also generate noise similar to that of granularity in photographs or streaks if the error occurs repeatedly in one orientation. Finally we come to the marking process, which converts the laser exposure from the ROS into a visible signal. Marking process noise, which generally occurs as a result of the discrete and random nature of the marking particles, generates granularity.

An electronic imaging system may enhance or attenuate the noise generated earlier in the process. Systems that tend to enhance detail with various types of filters or adaptive schemes are also likely to enhance noise. There are, however, processes (see Sec. 7.2) that search through the digital image identifying errors and substitute an error-free pattern for the one that shows a mistake.[64,149] These are sometimes referred to as noise removal filters.

Noise may be characterized in many different ways, but in general it is some form of statistical distribution of the errors that occur when an error-free input signal is sent into the system. In the case of imaging systems, an error-free signal is one that is absolutely uniform, given a noise-free, uniform input. Examples would include a sheet of white microscopically uniform paper on the platen of an input scanner, or a uniform series of laser-on pulses to a laser beam scanner, or a uniform raster pattern out of a perfect laser beam scanner writing onto the light-sensitive material in a particular marking device. A typical way to measure noise for these systems would be to evaluate the standard deviation of the output signal in whatever units characterize it. A slightly more complete analysis would break this down into a spatial frequency or time–frequency distribution of fluctuations. For example, in a photographic film a uniform exposure would be used to generate images whose granularity was measured as the root-mean-square fluctuation of density. For a laser beam scanner it would be the root-mean-square fluctuation in radiance at the pixel level for all raster lines.

In general, certain factors that affect the signal aspect of an imaging system positively, affect the noise characteristics of that imaging system negatively. For example, in scanning photographic film, the larger the sampled area, as in the case of the microdensitometer aperture, the lower the granularity [Eq. (1)]. At the same time, the image information is more blurred, therefore producing a lower contrast and smaller signal level. In general the signal level increases with aperture area and the noise level (as measured by the standard deviation of that signal level) decreases linearly with the square root of the aperture area or the linear dimension of a square aperture. It is therefore very important when designing a scanning system to understand whether the image information is being noise limited by some fundamentals associated with the input document or test object or by some other component in the overall system itself. An attempt to improve bandwidth, or otherwise refine the signal, by enhancing some parts of the system may, in general, do nothing to improve the overall image information, if it is noise in the input that is limiting and that is being equally "enhanced." Also, if the noise in the output writing material is limiting, then improvements upstream in the system may reach a point of diminishing returns.

In designing an overall electronic imaging system it should be kept in mind that noises add throughout the system, generally in the sense of an RSS (root of the sum of the squares) calculation. The signal attenuating and amplifying aspects, on the other hand, tend to multiply throughout the system. If the output of one subsystem becomes the input of another subsystem, the noise in the former is treated as if it were a signal in the latter. This means that noise in the individual elements must be appropriately mapped from one system to the other, taking into account various amplifications and nonlinearities. In a complex system this may not be easy; however, keeping an accurate accounting of noise can be a great advantage in diagnosing the final overall image quality. We expand on the quantitative characterization of these various forms of signal and noise in the subsequent parts of this chapter.

2.2.5 Color Imaging

Color imaging in general and especially digital color imaging have received considerable attention in the literature in recent years.[4,5,13,22,23] An elementary treatment is given below covering a few major points important to scanning and image quality. See Ref. 5, 22 or 23 for a recent broad overview and literature survey of digtal color imaging and Ref. 65 for a classic review of more traditional color reproduction systems and colorimetry.

Fundamentals

There are two basic methods of creating images, including digital images, in color, called additive and subtractive methods.

In an *additive color* system one creates the appropriate color image pixels by combining red (R), green (G), or blue (B) micro-sized lights, that is, pixels of varying intensities. Roughly equal amounts of each produce the sensation of "white" light on viewing. This applies to many self-luminous displays such as a CRT/TV or liquid crystal displays. The pixels must be small enough that the eye blurs them together. The eye detects these signals using sensors called "cones" in the retina. They have sensitivies to long, medium, and short wavelength regions of the spectrum, referred to as rho, gamma, and beta cones, respectively, as illustrated in Fig. 16. In turn these are associated with the HVS sensations of red, green, and blue.

In the second method of color imaging, called *subtractive color*, light is removed from otherwise white light by filters that subtract the above components one at a time. Red is removed by a cyan (C) filter, green by a magenta (M) filter, and blue by a yellow (Y) filter. For an imaging system, these filters are created by an imagewise distribution of transparent colorants created pixel by pixel in varying amounts. They are laid down color layer by color layer. The "white" light may come from a projector as in the case of transparencies or from white room light reflected by a white sheet of paper with the imagewise distribution of transparent colorants bonded to it. Here the subtraction occurs once on the way to the paper and then a second time after reflection on the way to the eye. Color photographic reflection prints and color offset halftone printing both use this method.

A digital color imaging system, designed to capture the colors of an original object, breaks down light reflected (or transmitted) from the object into its R, G, and B components by a variety of possible methods. It uses separate red, green, and blue image capture systems and channels of image processing, which are eventually combined to form a full color image.

The visual response involves far more than just the absorption of light. It involves the human neurological system and many special processes in the brain. The complexity of

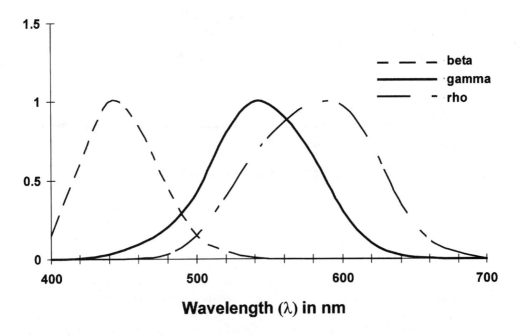

Figure 16 Approximate sensitivity of the eye, normalized for equal area under the curves.

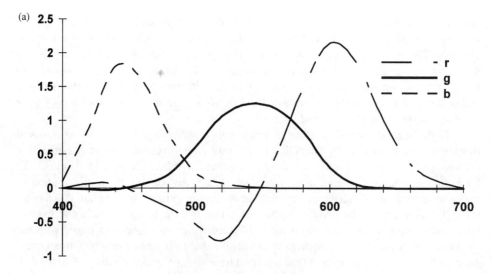

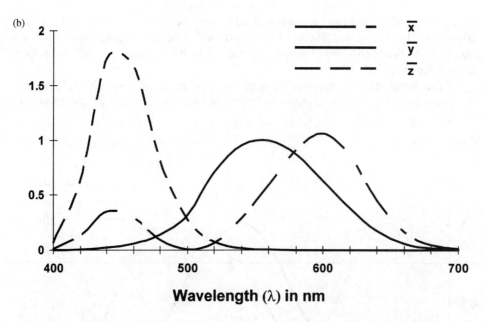

Figure 17 Color matching functions: (a) example of a directly measured result (from Ref. 13); (b) a transformed result chosen as the CIE Standard Observer for 2° field of view.

this can be appreciated by observing the results of simple color matching experiments, in which an observer adjusts the intensities of three color primaries until their mixture appears to match a test color. Such experiments, using monochromatic test colors, lead to the development of a set of color matching functions for specific sets of colored light sources and specific observer conditions. Certain monochromatic colors require the subtraction of colored light (addition of the light to the color being matched) in order to create a match. Color matching experiments are described extensively in the literature[3,9,13,22] and provide the foundation to the science of colorimetry.

Two such sets of color matching functions are shown in Fig. 17(a) and (b). The first set reports experimental results using narrow band monochromatic primaries. Note the large negative lobe on the third curve of "a," showing the region where "negative light" is needed, that is, where the light must be added to the color under test to produce a match. The second set has become a universally accepted representation defining the CIEs (Commission Internationale de l'Eclairage) 1931 2° Standard Colorimetric Observer. It is a linear transformation of standardized color matching data, carefully averaged over many observers and is representative of 92% of the human population having normal color vision. This set of functions provides the standardization for much of the science of the measurement of color, in other words, important colorimetry standards.

This overly simplistic description goes beyond the scope of this chapter to explain. Ideally the information recorded by a color scanner should be equivalent to that seen by an observer. In reality, the transparent colorant materials used to create images are not perfect. Significant failures stem from the nonideal shapes of the spectral sensitivities of the capturing device and the nonideal shapes of the spectral reflectance or transmittance of the colorants. Practical limitations in fabricating systems and noise also restrict the accuracy of color recording for most scanners. Ideal spectral shapes of sensitivities and filters would allow the system designer to better approximate the HVS color response. For example, an input original composed of conventional subtractive primaries such as real magenta (green absorbing) ink, not only absorbs green light, but also absorbs some blue light. Different magentas have different proportions of this unwanted absorption. Similar unwanted absorptions exist in most cyan and, to a lesser extent, in most yellow colorants. These unwanted characteristics limit the ability of complete input and output systems to reproduce the full range of natural colors accurately. Significant work has been carried out recently to define quality measures for evaluating the color quality of color recording instruments and scanning devices.[25]

Colorimetry and Chromaticity Diagrams

This leads to two large problem areas in color image quality needing quantification, namely (a) that the color gamuts of real imaging systems are limited, and (b) that colors which appear to match under one set of conditions appear different by some amount under another set of circumstances. This is conveniently described by a color analysis tool from the discipline of colorimetry (the science of color measurement) called a *chromaticity diagram*, shown in Fig. 18. It describes color in a quantitative way. It can be seen, in this illustration, that the monitor display is capable of showing different colors from a particular color printer. It is also possible, with this diagram, to show the color of an original. Note that a color gamut is the range of colors that can be produced by the device of interest as specified in some three- or more dimensional color space. It is important to note that a two-dimensional representation, like that shown here, while very helpful, is only a part of the whole three-dimensional color space (Fig. 19). Variations derived from the chromaticity diagram, and the equations that define it, however, provide a basis for much of the literature that describes color image quality today. It is designed to facilitate description of small color differences, for example, between an original and a reproduced color or two different reproductions of the same color.

The reader must be warned, however, that the actual perception of colors involves many psychophysical and psychological factors beyond those depicted in this diagram.[3] It is, however, a useful starting point. It describes any color in an image or a source and is

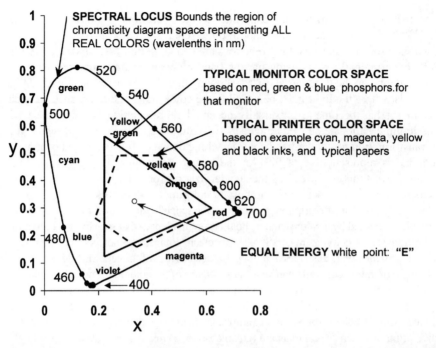

Figure 18 The x, y chromaticity diagram. Variations derived from it, and the equations that define it, provide a basis for much of the literature that describes color image quality today. It is designed to facilitate description of small color differences such as an original and a reproduced color or two different reproductions of the same color. Examples of the differences between possible colors at a given lightness formed in two different media, a printer and monitor, are shown (from Ref. 16, which cites data from X-Rite Inc). A more precise chromaticity diagram is shown in Fig. 47.

often the starting point in many of the thousands of publications on color imaging. There are also many different transformations of basic chromaticity diagram, a few primary examples of which we will describe here.

For the purposes of this chapter the basic equations used to derive the chromaticity diagram and to transform it provide an introduction to color image quality measurement. The outer, horseshoe-shaped curve, known as the "spectral locus," represents the most saturated colors possible, those formed by monochromatic sources at different wavelengths. All other possible colors lie inside this locus. Whites or neutrals by definition are the least saturated colors, and lie nearer the center of the horseshoe-shaped area. The colors of selected broad-spectrum light sources are shown later in Fig. 48 using a different form of chromaticity diagram (u', v' coordinates) described below. Saturation (a perceptual attribute, that is, "ness") of any color patch (transparent or reflection) can be estimated on this chart by a physical measure called excitation purity. It can be seen as the relative distance from the given illumination of the patch to the horseshoe limit curve along a vector. The dominant wavelength (approximate correlate with perceptual attribute of hue) is given by the intersection of that vector with the spectral locus. The lightness of the color is a third dimension, not shown, but is on an axis perpendicular to the plane of the diagram (coming out of the page). Use of dominant wavelength and purity to describe colors in the x, y version of the chromaticity diagram is shown in Fig. 47 in Sec. 9. Different light sources may be used

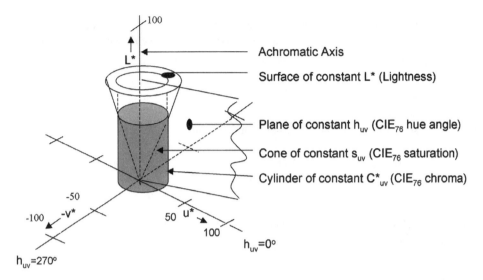

Figure 19 $L^*u^*v^*$ color space showing 1976 CIE colorimetric quantities of lightness, hue angle, saturation, and chroma, organized in a fashion similar to the Munsell color space (from Ref. 66).

but standard source "C" (See Fig. 49) was chosen here. These correlates are only approximate because lines of constant hue are slightly curved in these spaces.

To understand the chromaticity coordinates, x and y, return to Fig. 17(b). From these curves for $\bar{x}, \bar{y}, \bar{z}$, the spectral power of the light source $S(\lambda)$, and the spectral reflectance (or transmittance) of the object $R(\lambda)$, one can calculate

$$X = k \sum_{\lambda=380}^{780} S(\lambda)R(\lambda)\bar{x}(\lambda) \tag{3a}$$

$$Y = k \sum_{\lambda=380}^{780} S(\lambda)R(\lambda)\bar{y}(\lambda) \tag{3b}$$

$$Z = k \sum_{\lambda=380}^{780} S(\lambda)R(\lambda)\bar{z}(\lambda) \tag{3c}$$

where k is normally selected to make $Y = 100$ when the object is a perfect white, that is, an ideal, nonfluorescent isotropic diffuser with a reflectance equal to unity throughout the visible spectrum. The spectral profile of several standard sources is given later in Fig. 49.

These results are used to calculate the *chromaticity coordinates* in the above diagram as follows:

$$x = \frac{X}{(X+Y+Z)} \tag{4a}$$

$$y = \frac{Y}{(X+Y+Z)} \tag{4b}$$

$$z = \frac{Z}{(X+Y+Z)} \tag{4c}$$

One of the most popular transformations is the CIE $L^*a^*b^*$ version (called CIELAB for short) which is one of most widely accepted attempts to make distances in color space more uniform in a visual sensation sense. Here

$$L^* = 116(Y/Y_n)^{1/3} - 16 \tag{5}$$

which represents the achromatic lightness variable, and

$$a^* = 500[(X/X_n)^{1/3} - (Y/Y_n)^{1/3}] \tag{6}$$
$$b^* = 500[(Y/Y_n)^{1/3} - (Z/Z_n)^{1/3}] \tag{7}$$

represent the chromatic information, where X_n Y_n Z_n are the X, Y, Z tristimulus value of the *reference white*. Color differences are given as

$$\Delta E^*_{ab} = [(\Delta L^*)^2 + (\Delta a^*)^2 + (\Delta b^*)^2]^{1/2} \tag{8}$$

In practical terms, results where $\Delta E^*_{ab} = 1$ represent approximately one just noticeable visual difference (see Sec. 8). However, the residual nonlinearity of the CIELAB chromaticity diagram, the remarkable adaptability of the human eye to many other visual factors, and the effect of experience require situation-specific experiments. Only such experiments can determine rigorous tolerance limits and specifications. Color appearance models that account for many such dependencies and nonlinearities have been developed.[3,66] Attempts to standardize the methodology have been developed by CIE TC1-34 as CIECAM97s and proposed CIECAM02. (See Appendix A of Ref. 3).

Many other important sets of transformations have evolved over the last several decades and are described in the color literature.[9] They especially vary in the way the chromatic information is represented.[66,67] Three additional systems are described here. In the first two, called the $L^*u'v'$ and the $L^*u^*v^*$ systems

$$u' = 4x/(-2x + 12y + 3) \quad \text{and} \quad u^* = 13L^*(u' - u'_n) \tag{9a}$$
$$v' = 9/(-2x + 12y + 3) \quad \text{and} \quad v^* = 13L^*(v' - v'_n) \tag{9b}$$

These have historical significance, are easy to compute, and are often preferred by certain expert groups. The reader will observe that the color transformations carrying an * are all normalized to the reference white (values with subscript n), which is an important acknowledgement that perceived colors are actually highly dependent on the color and lightness of the surround. All of the color spaces defined by these equations are only approximately uniform, each having its own unique attributes and hence different advocates.

Another very important color description tool is the Munsell system in which painted paper chips of different colors have been arranged in a three-dimensional cylindrical coordinate system. The vertical axis represents *value* (akin to lightness) the radius represents *chroma*, and the angular position around the perimeter is called *hue*. These have been carefully standardized and are very popular as color references. To create a similar cylindrical coordinates description for colorimetry, the equations above are

Image Quality for Scanning

rewritten in polar form, resulting in a chroma and hue angle as follows:

$$C^*_{uv} = (u^{*2} + v^{*2})^{1/2} \quad \text{or} \quad C^*_{ab} = (a^{*2} + b^{*2})^{1/2} \tag{10a}$$

$$h_{uv} = \arctan(u^*/v^*) \quad \text{or} \quad h_{ab} = \arctan(b^*/a^*) \tag{10b}$$

In $L^*u^*v^*$ space (called CIELUV) one can also obtain a simple correlate of saturation called CIE 1976 u, v saturation, s_{uv}, where

$$s_{uv} = 13[(u' - u'_n)^2 + (v' - v'_n)^2]^{1/2} \tag{10c}$$

See Fig. 48, later, for the chromaticity diagram in $L^*u'v'$ space.

3 PRACTICAL CONSIDERATIONS

Several overall systems design issues are of some practical concern, including the choice of scan frequency as well as motion errors and other nonuniformities. They will be addressed here in fairly general terms.

3.1 Scan Frequency Effects

As digital imaging evolved in the previous decade, it had generally been thought that the spatial frequency, in raster lines or pixels per inch, which is used either to create the output print or to capture the input document, is a major determinant of image quality. Today there is a huge range of scan frequencies emanating from a huge range of products and applications from low-end digital cameras and fax machines, through office scanners and copiers, to high-end graphic arts scanners, all used with a plethora of software and hardware image processing systems that enlarge and reduce and interpolate the originally captured pixel spacings to something else. Then, other systems with yet additional processing and imaging affects are employed to render the image prior to the human reacting to the quality. It is only at this point in the process, where all the signal and noise effects roll up that the underlying principles from other parts of this chapter can be used to quantify overall image quality. Needless to say, scan frequency or pixel density is only one of these effects, and to assert it is *the* dominant effect is questionable in all but the most restrictive of circumstances. Yet it is an important factor and many type A shortcut experiments have attempted to address the connection between the technology variable of pixel density and various dimensions of overall image quality.

If a scanned imaging system is designed so that the input scanning is not aliased and the output reconstruction faithfully prints all of the information presented to it, then the scan frequency tends to determine the blur, which largely controls the overall image quality in the system. This is frequently not the case, and, as a result, scan frequency is not a unique determinant of image quality. In general, however, real systems have a spread function or blur that is roughly equivalent to the sample spacing, meaning they are somewhat aliased and that blur correlates with spacing. However, it is possible to have a large spot and much smaller spaces (i.e., unaliased), or vice versa (very aliased). The careful optimization of the other factors at a given scan frequency may have a great deal more influence on the information capacity of any electronic imaging system and therefore on the image-quality performance than does scan frequency itself. To a certain extent, gray information can be readily exchanged for scan frequency. We shall subsequently explore this further when dealing with the subject of information content of an imaging system.

In the spirit of taking a snapshot of this huge and complex subject, Fig. 20 summarizes three types of practical findings, two about major applications of scanned or digital images, namely digital photography and graphic arts–digital reprographics, and one simplification of human perception. The curves in the lower graph (solid dots) show results of two customer acceptability experiments with digital photography, varying camera resolution and printing on 8 bpp contone printers (A1 from Ref. 86, A2 from Ref. 92). Experiments on digital reprographics are shown by the curves with the open symbols, which suggest acceptable enlargement factors for input documents scanned at various resolutions and printed at various output screen resolutions. Finally we can put this in perspective by noting, as triangles along the frequency axis, the resolution limitations of the HVS at normal and close inspection viewing distances using modest 6% and very sensitive 1% contrast detection thresholds. Returning to Fig. 2, both applications are type A methods, while the HVS limits were inferred from visual algorithms.

A fairly general practice is to design aliased systems in order to achieve the least blur for a given scan frequency. Therefore, another major effect of scan frequency concerns the interaction between periodic structures in the input and the scanning frequency of the

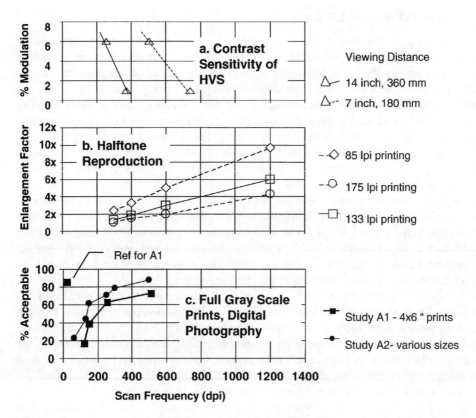

Figure 20 Summary of practical findings about sampling frequency in graphic arts for halftones (from Ref. 15) in digital photography (from Refs. 86 and 192) and related HVS contrast sensitivity reference values (from Ref. 3).

Image Quality for Scanning

system that is recording the input information. These two interfere, producing beat patterns at sum and difference frequencies leading to the general subject of moiré phenomena. Hence, small changes in scan frequency can have a large effect on moiré.

Input and output scanning frequencies also affect magnification. A 300 pixels/in. (11.8 pixels/mm) electronic image printed at 400 pixels/in. (15.7 pixels/mm) is only three-quarters as large as the original, while one printed at 200 pixels/in (7.87 pixels/mm) appears to be enlarged 1.5×. One-dimensional errors in scan frequency cause anamorphic magnification errors.

One of the major considerations in selecting output scan frequency is the number of gray levels required from a given range of halftone screens. Recall the discussion of Fig. 14. Dot matrices from 4×4 to 12×12 are shown in Table 1 at a range of frequencies from 200 to 1200 raster lines per inch (7.87–47.2 raster lines/mm). For example, a 10×10 matrix of thresholds can be used to generate a 50 gray level, 45° angle screen (two shifted 5×10 submatrices) whose screen frequency is shown in the ninth column in Table 1. Also indicated in the table is the approximate useful range for the visual system.

Table 1 Relationship among halftone matrix size (given in pixels), maximum possible number of gray levels in the halftone, and output scan frequency (in pixels/inch). Entries are given in halftone dots/inch measured along the primary angle (row 2) of the halftone pattern. Dot types are given as (see quadrants of Fig. 14): (A) the conventional 45° halftone where quadrants Q1 = Q4, Q2 = Q3; (B) conventional 90° halftone where Q1 = Q2 = Q3 = Q4. Expansions of the number of halftone gray levels show three new types: (C) = Type A except Q3 and Q4 thresholds are set at halfway between those in Q1 and Q2 (45° double dot), (D) where Q1 = Q4, but Q2 and Q3 thresholds are set halfway between those in Q1 (90° double dot); (E) where Q1 through Q4 thresholds are each set to generate intermediate levels among each other (90° quad dot). The number of gray values includes one level for white.

Matrix in pixels	4 x 4	3 x 3	4 x 4	6 x 6	5 x 5	8 x 8	6 x 6	10 x 10	12 x 12
Angle	45°	90°	90°	45°	90°	45°	90°	45°	45°
No of gray levels - Type*: Conventional	9 - A	10 - B	17 - B	19 - A	26 - B	33 - A	37 - B	51 - A	73 - A
Expanded** 2x	17 - C	19 - D	33 - D	37 - C	51 - D	65 - C	73 - D	101 - C	145 - C
Expanded** 4x	33	41 - E	65 - E	73	101 - E	129	145 - E	201	289
Scan freq. in pixels/in ↓									
1200	426	400	300	282	240	212	200	170	141
1000	352	333	250	236	200	176	167	142	118
800	284	267	200	188	160	142	133	114	94
600	212	200	150	141	120	106	100	85	71
500	176	166	125	118	100	88	84	71	59
400	142	133	100	94	80	71	67	57	47
300	106	100	75	71	60	53	50	43	36
200	71	67	50	47	40	35	33	28	24

Practical upper limit (~200 LPI)
Typical halftone (~100 LPI)
Halftone dot, practical lower limit (~65 LPI)

* Type refers to specific halftone structures A-E (see caption) where appropriate
** Number of gray levels is increased by 2x or 4x over conventional by gray pixels or multi-centered dots

The range starts at a lower limit of 65 dots/in. (2.56 dots/mm) halftone screen, formerly found in newspapers. This results in noticeably coarse halftones and has recently moved into the range of 85 dots/in. (3.35 dots/mm) to 110 dots/in. (4.33 dots/mm) in modern newspapers. The upper bound represents a materials limit of around 175 dots/in. (6.89 dots/mm), which is a practical limit for many lithographic processes. This table assumes that the pixels are binary in nature. If a partially gray or high addressability output imaging system is employed then the number of levels in the table must be multiplied by the number of gray levels or subpixels per pixel appropriate to the technology.

3.2 Placement Errors or Motion Defects

Since the basic mode of operation for most scanning systems is to move or scan rapidly in one direction and slowly in the other, there is always the possibility of an error in motion or other effect that results in locating pixels in places other than those intended. Figure 21 shows several examples of periodic raster separation errors, including both a sinusoidal and a sawtooth distribution of the error. These are illustrated at 300 raster lines/in. (11.8 lines/mm) with ± 10 through ± 40 μm (± 0.4 through ± 1.6 mils) of spacing error, which refers to the local raster line spacing and not to the error in absolute placement accuracy. Error frequencies of 0.33 cycles/mm (8.4 cycles/in.) and 0.1 cycles/mm (2.5 cycles/in.) are illustrated.

For input scanners, which convert an analog signal to a digital one, the error takes the form of a change in the sampling of the analog document. Since sampling makes many

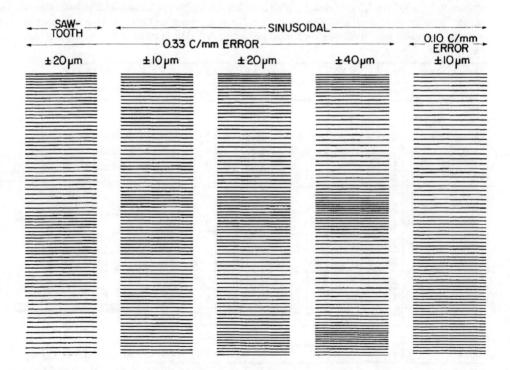

Figure 21 Enlarged examples of rasters with specified image motion variation at 300 dpi.

Image Quality for Scanning

mistakes, the sampling errors due to motion nonuniformity are most visible in situations where the intrinsic sampling error is made to appear repeatable or uniform, and the motion error, therefore, appears as an irregular change to an otherwise uniform pattern. Long angled lines that are parallel to each other provide such a condition because each line has a regular periodic phase error associated with it, and a motion error would appear as a change to this regular pattern. Halftones that produce moiré are another example, except that the moiré pattern is itself usually objectionable so that a change in it is not often significant.

In patterns with random phase errors such as text, the detection of motion errors is more difficult. Effects that are large enough to cause a two-pixel error would be perceived very easily; however, effects that produce less than one-pixel error on average would tend to increase the phase errors and noise in the image generally and would therefore be perceived on a statistical basis. Many identical patterns repeated throughout a document would provide the opportunity to see the smaller errors as being correlated along the length of the given raster line that has been erroneously displaced, and would therefore increase the probability of seeing the small errors.

Motion errors in an output scanner that writes on some form of image-recording material can produce several kinds of defects. In Table 2, several attributes of the different types of raster distortion observables are shown. The first row in the matrix describes the general kind of error, that is, whether it is predominantly a pixel placement error or predominantly a developable exposure effect or some combination of the two. The second row is a brief word description or name of the effect that appears on the print. The third row describes the spatial frequency region in cycles/mm in which this type of error tends to occur. The next row indicates whether the effect is best described and modeled as one-dimensional or two-dimensional. Finally, a graphical representation of an image with the

Table 2 The Effects of Motion Irregularities, Defects, or Errors on the Appearance of Scanned Images

Error Category	Pixel Placement Error		Combination of Both		Developable Exposure Effects		
Effect on print	Spacing Nonuniformity	Character Distortion/ Fast Scan Jitter	Halftone Nonuniformity	Line Darkness Nonuniformity	Structured Background	Ragged/ Structured Edges	Unsharp Line-text-edge Images
Typical frequencies of motion error (c/mm)	< 0.5	0.5 – 2	0.1 – 6	0.005 – 2	0.005 – 8	1 – 8	4 – 20+
Dimensions of effect	1D	2D	2D (1D)	1D (2D)	1D	2D	2D
Scanner type	input/output	input/output	input/output	input/output	output	output	output
Example of defect	(image)	Y	(image)	nnı iıiı	(image)	(image)	K
Example with little or no defect	(image)	Y	(image)	nn iiii		(image)	K
Comment	Shows local reduction, magnification is also possible		Illustrates 2 levels, bottom case has very small defect		Shows "write white" system	Shows strong low & high freq. – approx 3 x 3 mm sample	Examples: higher magnification than at left – eye blurs structures shown

specific defect is shown in the top row, while the same image appears in the bottom row without the defect.

The first of the columns on the left is meant to show that if the frequency of the error is low enough then the effect is to change the local magnification. A pattern or some form of texture that should appear to have uniform spacings would appear to have nonuniform spacings and possibly the magnification of one part of the image would be different from that of another. The second column is the same type of effect except the frequency is much higher, being around 1 cycle/mm (25 cycle/in.). This effect can then change the shape of a character, particularly one with angled lines in it, as demonstrated by the letter Y.

Moving to the three righthand columns, which are labeled as developable exposure effects, we have three distinctly different frequency bands. The nature and severity of these effects depends in part on whether we are using a "write white" or "write black" recording system and on the contrast or gradient of the recording material. The first of these effects is labeled as structured background. When the separation between raster lines increases and decreases, the exposure in the region between the raster lines where the gaussian profile writing beams overlap increases or decreases with the change. This gives an overall increase or decrease in exposure, with an extra large increase or decrease in the overlap region. Since many documents that are being created with a laser beam scanner have relatively uniform areas, this change in exposure in local areas gives rise to nonuniformities in the appearance in the output image.

In laser printers, for example, the text is generally presented against a uniform white background. In a positive "write white" electrophotographic process, such as is used in many large xerographic printers, this background is ideally composed of a distribution of uniformly spaced raster lines that expose the photoreceptor so that it discharges to a level where it is no longer developable. As the spacing between the raster lines increases, the exposure between them decreases to a point where it no longer adequately discharges the photoreceptor, thereby enabling some weak development fields to attract toner and produce faint lines on a page of output copy. For this reason among others, some laser printers use a reversal or negative "write black" form of electrophotography in which black (no light) output results in a white image. Therefore, white background does not show any variation due to exposure defects, but solid dark patches often do.

The allowable amplitude for these exposure variations can be derived from minimum visually perceivable modulation values and the gradient of the image recording process.[68,69] In the spatial frequency region near 0.5 cycles/mm (13 cycles/in.), where the eye has its peak response at normal viewing distance, an exposure modulation of 0.004–0.001 $\Delta E/E$ has been shown to be a reasonable goal for a color photographic system with tonal reproduction density gradients of 1–4.[70]

If the frequency of the perturbation is of the order 1–8 cycles/mm (25–200 cycles/in.), and especially if the edges of the characters are slightly blurred, it is possible for the nonuniform raster pattern to change the exposure in the partially exposed blurred region around the characters. As a result, nonuniform development appears on the edge and the raggedness increases as shown by the jagged appearance of the wavy lines in column 7. The effects are noticeable because of the excursions produced by the changes in exposure from the separated raster lines at the edges of even a single isolated character. The effect is all the more noticeable in this case because the darkened raster lines growing from each side of the white space finally merge in a few places. The illustration here, of course, is a highly magnified version of just a few dozen raster lines and the image contained within them.

In the last column we see small high-frequency perturbations on the edge, which would make the edge appear less sharp. Notice that structured background is largely a one-dimensional problem, just dealing with the separation of the raster lines, while character distortion, ragged or structured edges, and unsharp images are two-dimensional effects showing up dramatically on angled lines and fine detail. In many cases the latter require two dimensions to describe the size of the effect and its visual appearance.

Visually apparent darkness for lines in alphanumeric character printing can be approximately described as the product of the maximum density of the lines in the character times their widths. It is a well-known fact in many high-contrast imaging situations that exposure changes lead to line width changes. If the separation between two raster lines is increased, the average exposure in that region decreases and the overall density in a write white system increases. Thus, two main effects operate to change the line darkness. First, the raster information carrying the description of the width of the line separates, writing an actually wider pattern. Secondly, the exposure level decreases, causing a further growth in the line width and to some extent causing greater development, that is, more density. The inverse is true in regions where the raster lines become closer together. Exposure increases and linewidth decreases.

If these effects occur between different strokes within a character or between nearby characters, the overall effect is a change in the local darkness of text. The eye is generally very sensitive to differences of line darkness within a few characters of each other and even within several inches of each other. This means that the spatial frequency range over which this combination of stretching and exposure effect can create visual differences is very large, hence the range of 0.005–2 cycles/mm (0.127–50 cycles/in.). Frequencies listed in the figure cover a wide range of effects, also including some variation of viewing distance. They are not intended as hard boundaries but rather to indicate approximate ranges.

Halftone nonuniformity follows from the same general description given for line darkness nonuniformity except that we are now dealing with dots. The basic effect, however, must occur in such a way as to affect the overall appearance of darkness of the small region of an otherwise uniform image. A halftone works on the principle of changing a certain fractional area coverage of the halftone cell. If the spatial frequency range of this nonuniformity is sufficiently low, then the cell size changes at the same rate that the width of the dark dot within the cell changes. Therefore the overall effect is to have no change in the percent area coverage and only a very small change in the spacing between the dots. Hence, the region of a few tenths to several cycles/mm (several to tens of cycles/in.) is the domain for this artifact. It appears as stripes in the halftone image.

The allowable levels for the effects of pixel placement errors on spacing nonuniformity and character distortion depend to a large extent upon the application. In addition to application sensitivity, the effects that are developable or partially developable are highly dependent upon the shape of the profile of the writing spot and upon amplification or attenuation in the marking system that is responding to the effects. Marking systems also tend to blur out the effects and add noise, masking them to a certain extent.

3.3 Other Nonuniformities

There are several other important sources of nonuniformity in a raster scanning system. First, there is a pixel-to-pixel or raster-to-raster line nonuniformity of either response in the case of an input scanner or output exposure in the case of an output scanner. These generally appear as streaks in an image when the recording or display medium is sensitive to exposure variations. These, for example, would be light or dark streaks in a printed halftone or darker and lighter streaks in a gray recorded image from an input scanner looking at a uniform area of an input document. A common example of this problem in a rotating polygon output scanner is the effect of facet-to-facet reflectivity variations in the polygonal mirror itself. The exposure tolerances described for motion errors above also apply here.

Another form of nonuniformity is sometimes referred to as *jitter* and occurs when the raster synchronization from one raster line to another tends to fail. In these cases a line drawn parallel to the slow scan direction appears to oscillate or jump in the direction of the fast scan. These effects, if large, are extremely objectionable. They will manifest themselves as raggedness effects or as unusual structural effects in the image, depending upon the document, the application, and the magnitude and spatial frequency of the effect.

3.3.1 Perception of Periodic Nonuniformities in Color Separation Images

Research on the visibility of periodic variations in the lightness of 30% halftone tints of cyan, magenta, yellow, and black color image separations printed on paper substrates has been translated into a series of guidelines for a specification for a high-quality color print engine.[71] (Fig. 22). They were chosen to be slightly above the onset of visibility. Specifically they are set at $\{[1/3] \times [(2 \times$ "visible but subtle threshold") + ("obvious threshold")]\}$ and adjusted for a wider range of viewing distances and angles than during the experiments, which were at 38–45 cm. These guidelines are given in terms of colorimetric lightness units on the output prints. Visibility specifications must ultimately be translated into engineering parameters. We have selected the traditional CIE-$L^*a^*b^*$ metrics version for illustration. These also tend to shows the smallest, most demanding

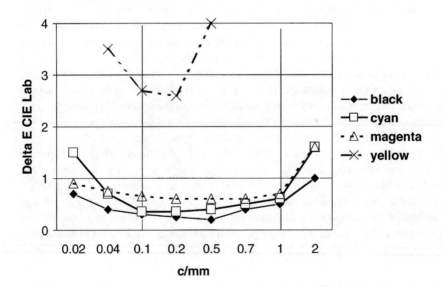

Figure 22 Guidelines for specification of periodic nonuniformities in black, cyan, magenta, and yellow color separations as indicated, in terms of ΔE derived from CIE $L^*a^*b^*$, plotted against the effective spatial frequency of the periodic disturbance (from Ref. 71).

Image Quality for Scanning

ΔEs. Guidelines developed in terms of ΔE for other color difference metrics (CMC 2:1 and CIE-94) have also been developed,[71] and show different visual magnitudes, by as much as a factor of 2.

To translate these in to a guidelines for the approximate optical scanner exposure variation, the ΔE values in this graph must be divided by the slope of the system response curve, in terms of $\Delta E/\Delta$ exposure, for the color separation of interest. Exposure, H, is the general variable of interest since it is the integrated effect of intensity and time variations, both of which can result from the scanner errors discussed in the pages above. The system response would be approximated by the cascaded (multiplied) slopes of the responses of all the intermediate imaging systems between the scanner and the resulting imaging media assuming the small signal theory approximation to linearity of the cascaded systems. If a system is linear in terms of input exposure variations vs. output reflectance, and the black separation is of interest, then

$$\Delta E = \Delta L^* \tag{11}$$

Given $Y/Y_n = R$ (normalized reflectance factor) in Eq. (5) earlier, then

$$L^* = 116 R^{1/3} - 16 \tag{12}$$

and solving the equation for *small* differences

$$\Delta L^* = (38.7 R^{-2/3}) \Delta R \tag{13}$$

Some simplifying assumptions can now be used to approximate the magnitude of the worst case visual guidelines for the optical scanner exposure variations. Using the Murray Davies equation[59] and assuming a solid area density of Goodman's toner of 1.3 gives a reflectance of $R \cong 0.71$ for the 30% area coverage used in her experiments (ignoring light scattering in the substrate).

Setting $\Delta R = \Delta H_r$ (for a system with gain = 1.0 where H_r = relative exposure, i.e., in normalized units) under the linearity assumption yields

$$\Delta E = \Delta L^* \cong 48.4 \Delta H_r \tag{14}$$

As an example note that for frequencies near 0.5 cycles/mm, $\Delta E \cong 0.2$, implying $\Delta H_r \cong 0.004$.

Specific relationships for exposure and reflectance, and for any of the other colorimetric units described in this research should be developed for each real system. The linear gain = 1.0 assumption shown here should not be taken for granted. The reader is also reminded that these results are for purely sinusoidal errors of a single frequency and a single color and that actual nonuniformities occur in many complex spatial and color forms.

4 CHARACTERIZATION OF INPUT SCANNERS THAT GENERATE MULTILEVEL GRAY SIGNALS (INCLUDING DIGITAL CAMERAS)

In this section we will discuss the elementary theory of performance measurements and various algorithms or metrics to characterize them, the scanner factors that govern each, some practical considerations in the measurements, and visual effects where possible. Generally speaking, this is the subject of analyzing and evaluating systems that acquire sampled images. Originally this was explored as analog sampled images in television, most notably by Schade[72,73] in military and in early display technology.[74] As computer and digital electronics technology grew, this evolved into the general subject of evaluating digital sampled image acquisition systems, which include various cameras and input scanners. Modern scanners and cameras are different only in that a scanner moves the imaging element to create sampling in one direction while the camera imaging element is static, electronically sampling a two-dimensional array sensor in both dimensions. This topic can be divided into two areas. The first concerns scanners and cameras that generate output signals with a large number of levels (e.g., 256), where general imaging science using linear analysis applies.[12] The second deals with those systems that generate binary output, where the signal is either on or off (i.e., is extremely nonlinear) and more specialized methods apply.[75] These are discussed in Sec. 5.

In recent years, the advent of digital cameras and the plethora of office, home, and professional scanners has promoted wide interest in the subject of characterizing devices and systems that produce digital images. Also, several commercially available image analysis packages have been developed for general image analysis, many using scanners or digital cameras, often attached to microscopes or other optical image magnification systems. Components of these packages and the associated technical literature specifically address scanner analysis or calibration.[76-78] A variety of standards activities have evolved in this area.[79-82] Additional related information is suggested by the literature on evaluating microdensitometers.[83,84] These systems are a special form of scanners in which the sensor has a single aperture of variable shape. Much of this work relates to transmitted light scanners but reflection systems have also been studied.[85] Methods for evaluating digital cameras and commercially available scanners for specific applications have been described by many authors.[82,86,87]

4.1 Tone Reproduction and Large Area Systems Response

Unlike many other imaging systems, where logarithmic response (e.g., optical density) is commonly used, the tonal rendition characteristics of input scanners are most often described by the relationship between the output signal (gray) level and the input reflectance or brightness. This is because most electronic imaging systems respond linearly to intensity and therefore to reflectance. Three such relationships are shown in Fig. 23. In general these curves can be described by two parameters, the offset, O, against the output gray level axis and the gain of the system Γ, which is defined in the equation in Fig. 23. Here g is the output gray level, and R is the relative reflectance factor. If there is any offset, then the system is not truly linear despite the fact that the relationship between reflectance and gray level may follow a straight-line relationship. This line must go through the origin to make the system linear.

Often the maximum reflectance of a document will be far less than the 1.0 (100%) shown here. Furthermore, the lowest signal may be significantly higher than 1 or 2% and may frequently reach as much as 10% reflectance. In order to have the maximum number of gray levels available for each image, some scanners offer an option of performing a histogram analysis of the reflectances of the input document on a pixel-by-pixel or less frequently sampled basis. The distribution is then examined to find its upper and lower limits. Some appropriate safety factor is provided, and new offset and gain factors are

Image Quality for Scanning

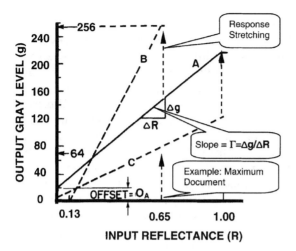

Figure 23 Typical types of scanner input responses, illustrating the definitions of "gain" (i.e., slope), "offset", and "response stretching".

computed. These are applied to stretch out the response to cover as many of the total (256 here) output levels as possible with the information contained between the maximum and minimum reflectances of the document.

Other scanners may have a full gray-scale capability from 4 to 12 bits (16–4096 levels). In the figure, curve C is linear, that is, no offset and a straight-line response up to a reflectance of 1.0 (100%), in this case yielding 128 gray levels. Curve A would represent a more typical general purpose gray response for a scanner while curve B represents a curve adjusted to handle a specific input document whose minimum reflectance was 0.13 and whose maximum reflectance was 0.65. Observe that neither of these curves is linear. This becomes very important for the subsequent forms of analysis in which the nonlinear response must be linearized before the other measurement methods can be applied properly. This is accomplished by converting the output units back to input units via the response function.

It is also possible to arrange the electronics in the video processing circuit so that equal steps in gray are not equal steps in reflectance, but rather are equal steps in some units that are more significant, either visually or in terms of materials properties. A logarithmic A/D converter is sometimes used to create a signal proportional to the logarithm of the reflectance or to the logarithm of the reciprocal reflectance (which is the same as "density"). Some scanners for graphic arts applications function in this manner. These systems are highly nonlinear, but may work well with a limited number of gray levels.

Many input scanners operate with a built-in calibration system that functions on a pixel-by-pixel basis. In such a system, for example, a particular sensor element that has greater responsivity than others may be attenuated or amplified by adjusting either the gain or the offset of the system or both. This would ensure that all photosites (individual sensor elements) respond equally to some particular calibrated input, often, as is common with most light measuring devices such as photometers and densitometers, using both a light and dark reflectance reference (e.g., a white and black strip of paint).

It is possible in many systems for the sensor to be significantly lower or higher in responsivity in one place than another. As an example, a maximum responsivity sensor may perform as shown in curve A while a less sensitive photosite may have the response shown in curve C. If curve C was captured with the same A/D converter at the same settings (as is often the case in high-speed integrated circuits), the maximum signal range it contains has only 120 gray levels. A digital multiplier can operate upon this to effectively double each gray level, thereby increasing the magnitude of the scale to 220 or 240, depending upon how it handles the offset. Note that if some of the elements of a one-dimensional sensor responded as curve C, others as A, with the rest in between, then this system would exhibit a kind of one-dimensional granularity or nonuniformity, whose pattern depends upon the frequency of occurrence of each sensor type. This introduces a quantization error varying spatially in 1-pixel-wide strips, and ranging, for this example, from strips with only 120 steps to others with 240 steps, yet covering the same distribution of output tones.

An ideal method for measuring tone reproduction is to scan an original whose reflectance varies smoothly and continuously from near 0 to near 100%, or at least to the lightest "white" that one expects the system to encounter. The reflectance is evaluated as a function of position, and the gray value from the scanner is measured at every position where it changes. Then the output of the system can be paired with the input reflectance at every location and a map drawn to relate each gray response value to its associated input reflectance. A curve like Fig. 23 can then be drawn for each photosite and for various statistical distributions across many photosites.

Most scanners operate with sufficiently small detector sites or sensor areas that they respond to input granularity. Thus, a single pixel or single photosite measurement will not suffice to get a solid area response to a so-called uniform input. Some degree of averaging across pixels is required, depending upon the granularity and noise levels of the input test document and the electronic system.

The use of a conventional step tablet or a collection of gray patches, where there are several discrete density levels, provides an approximation to this analysis but does not allow the study of every one of the discrete output gray levels. For a typical step tablet with approximately 20 steps of 0.15 reflection density, half of the gray values are measured by only two steps, 0.15 and 0.3 density (or 50% reflectance). Thus a smoothly varying density wedge is more appropriate for the technical evaluation of an electronic input scanner.

However, suitable wedges are difficult to fabricate repeatably and the use of uniform patches is common in many operations. See, for example, Ref. 87 and the ISO standard IT8 target in Fig. 51. Wedges are essential, however, to accurately evaluate binary scanning (see Sec. 5.2).

Because of the usual straight line input to output relationship as shown in Fig. 23, one may describe scanner response fairly accurately as

$$g = \Gamma R + O_A \tag{15}$$

using the slope Γ and the offset O_A as indicated in the figure. Note that this includes the A/D response. For systems that scan transparent documents, the reflectance axis is readily changed to a transmittance axis.

Setting the maximum point equal to 100% input reflectance is often a waste of gray levels since there are no documents whose real reflectance is 100%. A value somewhere between 70 and 90% would be more representative of the upper end of the range of real

Image Quality for Scanning 185

documents. Some systems adjusting automatically to the input target are therefore difficult to evaluate. They are highly nonlinear in a way that is difficult to compensate. See Gonzalez and Wintz[88] for an early discussion of automatic threshold or gray scale adjustment and Hubel[82] for more recent comments on this subject as it relates to color image quality in digital cameras. Most amateur and some professional digital cameras fall into this automatic domain.[82] A system that finds this point automatically is optimized for each input differently and is therefore difficult to evaluate in a general sense.

An offset in the positive direction can be caused either by an electronic shift or by stray optical energy in the system. If the electronic offset has been set equal to zero with all light blocked from the sensor, then any offset measured from an image can be attributed to optical energy. Typical values for flare light, the stray light coming through the lens, would range from just under 1% to 5% or more of full scale.[85] While offset from uniform stray light can be adjusted out electronically, signals from flare light are document-dependent, showing up as errors in a dark region only when it is surrounded by a large field of white on the document. Therefore, correction for this measured effect in the particular case of an analytical measurement with a gray wedge or a step tablet surrounded by a white field may produce a negative offset for black regions of the document that are surrounded by grays or dark colors. If, however, the source of stray light is from the illumination system, the optical cavity, or some other means that does not involve the document, then electronic correction is more appropriate. Methods for measuring the document-dependent contribution of flare have been suggested in the literature.[85,87,89] Some involve procedures that vary the surround field from black to white while measuring targets of different widths;[85] others use white surround with different density patches.[87]

A major point of confusion can occur in the testing of input scanners and many other optical systems that operate with a relatively confined space for the illumination system, document platen, and recording lens. This can be thought of as a type of integrating cavity effect. In this situation, the document itself becomes an integral part of the illumination system, redirecting light back into the lamp, reflectors, and other pieces of that system. The document's contribution to the energy in the illumination depends on its relative reflectance and on optical geometry effects relating to lamp placement, document scattering properties, and lens size and location. In effect the document acts like a position-dependent and nonlinear amplifier affecting the overall response of the system. One is likely to get different results if the size of the step tablet or gray wedge used to measure it changes or if the surround of the step tablet or gray wedge changes between two different measurements. It is best, therefore, to make a variety of measurements to find the range of responses for a given system. These effects can be anywhere from a few percent to perhaps as much as 20%, and the extent of the interacting distances on the document can be anywhere from a few millimeters to a few centimeters (fraction of an inch to somewhat over one inch). Relatively little has been published on this effect because it is so design specific, but it is a recognized practical matter for measurement and performance of input scanners. An electronic correction method exists.[90,91]

4.2 MTF and Related Blur Metrics

We will now return to the subject of blur. Generally speaking, the factors that affect blur for any type of scanner include (Table 3): the blur from optical design of the system, motion of the scanning element during one reading, electronic effects associated with the rise time of

Table 3 Factors Affecting Input Scanner Blur and Pointers to Useful MTF Curves That Describe Selected Cases

Solid-state scanners
- Lens aberrations (see Fig. 56 for useful equation to fit) as functions of wavelength (see Fig. 55 if diffraction-limited, e.g., some microscope optics), field position, orientation, focus distance
- Sensor: Aperture dimensions (see Fig. 53), charge transfer efficiency (CCD), charge diffusion, leaks in aperture mask
- Motion of sensor during reading (Fig. 53)
- Electronics rise time (measured frequency response)

Flying spot laser beam scanner
- Spot shape and size at document (Gaussian case see Fig. 54)
- Lens aberrations (as above)
- Polygon aperture or equivalent
- Motion during reading (Fig. 53)
- Sensor or detector circuit rise time (measured frequency response)

the circuit, the effective scanning aperture (sensor photo site) size, and various electro-optical effects in the detection or reading out of the signal. The circuits that handle both the analog and the digital signals, including the A/D converter, may have some restrictive rise times and other frequency response effects that produce a one-dimensional blur.

To explore the analysis of these effects, refer back to Fig. 4. A primary concept begins with a practical definition of an ideally narrow line object and the image of it. Imagine that the narrow line object profile shown at the top right of Fig. 4(a) were steadily reduced in width until the only further change seen in the resulting image Fig. 4(b) is that the height of the image peak changes but not the width of its spreading. This is a practical definition of an ideally narrow line source. Under these conditions we would say that the peak of the image on the right of Fig. 4(b) was a profile of the *line spread function* for the imaging system. [It is also seen at higher sampling resolution in Fig. 6(b).]

To be completely rigorous about this definition of the line spread function, we would actually use a narrow white line rather than a black line. If the input represented a very fine point in two-dimensional space we would refer to its full two-dimensional image as a *point spread function*. This spreading is a direct representation of the blur in any point in the image and can be convolved with the matrix of all the pixels in the sampled image to create a representation of the blurred image. The line spread function is a one-dimensional form of the spreading and is usually more practical from a measurement perspective. In the case illustrated, the line spread function after quantization would be shown in Fig. 4(c) as the corresponding distribution of gray pixels.

There are several observations to be made about this illustration, which underscore some of the practical problems encountered in typical measurements. First, the quantized image in part (c) is highly asymmetric while the profile of the line shown in parts (a) and (b) appears to be more symmetric. This results from sampling phase and requires that a measurement of the line spread function must be made, adjusting sampling phase in some manner [Fig. 6(b)]. This is especially important in the practical situation of evaluating a fixed sampling frequency scanner. Secondly, note the limited amount of information in any one phase. It can be seen that the smooth curve representing the narrow object in Fig. 4(b) is only represented by three points in the sampled and quantized image.

The averaging of several phases would improve on this measurement, increasing both the intensity resolution and the spatial resolution of the measurement. One of the easiest ways to do this is to use a long narrow line and tip it slightly relative to the sampling grid so that different portions along its length represent different sampling phases. One can then collect a number of uniformly spaced sampling phases, each being on a different scan line, while being sure to cover an integer number of complete cycles of sampling phase. One cycle is equivalent to a shift of one complete pixel. The results are then combined in an interleaved fashion, and a better estimate of the line spread function is obtained. (This is tantamount to increasing the sampling resolution, taking advantage of the one-dimensional nature of the test pattern.) This is done by plotting the recorded intensity for each pixel located at its properly shifted absolute position relative to the location of the line. To visualize this consider the two-phase sampling shown in Figs 6(b) and (c). There the resulting pixels from phase A could be interleaved with those from phase B to create a composite of twice the spatial resolution. Additional phases would further increase effective resolution.

In the absence of nonlinearities and nonuniformities, the individual line spread functions associated with each of the effects in Table 3 can be mathematically convolved with each other to come up with an overall system line spread function.

4.2.1 MTF Approaches

For engineering analysis, use of convolutions and measurements of spread functions are often found to be difficult and cumbersome. The use of an *optical transfer function* (OTF) is considered to have many practical advantages from both the testing and theoretical points of view. The optical transfer function is the Fourier transform of the line spread function. This function consists of a modulus to describe normalized signal contrast attenuation (or amplification), and a phase to describe shift effects in location, both given as a function of spatial frequency. The signal is characterized as the modulation of the sinusoidal component at the indicated frequency. Therefore the contrast altering function is described as a *modulation transfer function* (MTF). The value of optical transfer function analysis is that all of the components in a linear system can be described by their optical transfer functions, and these are multiplied together to obtain the overall system response. The method and theory of this type of analysis has been covered in many journal articles and reference books.[12,40,92,93]

Certain basic effects can be described in analytic form as MTFs and a few of these are indicated in Table 3 and illustrated in Sec. 9 in Figs 53–56, plotted in logarithmic form to facilitate graphical manipulation. Several photographic MTF curves are plotted in Fig. 57 to provide a reference both as a range of input signals for film scanners or a range of output filters that transform optical signals to permanently readable form. One may also consider using these with an enlargement factor for understanding input of photographic prints to a desktop or graphic arts scanner. (For example, a spatial frequency of 10 cycles/mm on an 8× enlarged print is derived from the 80 cycles/mm pattern on the film. Therefore the film MTF at 80 cycles/mm is an upper limit input signal for an 8 × 10 in. enlargement of a 35 mm film.) Other output MTFs would involve display devices such as monitors, projection systems, analog response ink systems, and xerographic systems. Obtaining the transform of the line spread function has many of the practical problems associated with measuring the line spread function itself plus the uncertainty of obtaining an accurate digital Fourier transform using a highly quantized input.

There are several commonly used methods for measuring the optical transfer function. These include:

1. Measuring images of narrow lines using appropriate compensation for finite widths.
2. Directly measuring images of sinusoidal distributions of radiation.[94,95]
3. Harmonic analysis of square wave patterns[87,94,95]
4. Taking the derivative of the edge profile in the image of a very sharp input edge. This generates the line spread function, and then the Fourier transform is taken, taking care to normalize the results properly.[12,81] (Table 10, ISO TC42, WG18)
5. Spectral analysis of random input (e.g., noise) targets with nearly flat spatial frequency spectrum.

It should also be mentioned at this point that for most characterizations of imaging systems the modulus, that is, the MTF, is more significant than the phase. The phase transfer function, however, may be important in some cases and can be tracked either by careful analysis of the relative location of target and image in a frequency-by-frequency method or by direct computation from the line spread function.

In general, these methods involve the use of input targets that are not perfect. They must have spatial frequency content that is very high. The frequency composition of the input target is characterized in terms of the modulus of the Fourier transform, $M_{in}(f)$, of its spatial radiance profile. The frequency decomposition of the output image is similarly characterized, yielding $M_{out}(f)$. Dividing the output modulation by the input modulation yields the modulation transfer function as

$$MTF(f) = M_{out}(f)/M_{in}(f) \tag{16}$$

The success of this depends upon the ability to characterize both the input and the output accurately.

A straightforward method to perform this input and output analysis involves imaging a target of periodic intensity variations and measuring the modulation on a frequency-by-frequency basis. If the target is a set of pure sine waves of reflectance or transmittance, that is, each has no measurable harmonic content, and the input scanner is linear, then the frequency-by-frequency analysis is straightforward. Modulation of a sinusoidal distribution is defined as the difference between the maximum and minimum divided by their sum. The modulation is obtained directly, measuring the maximum and minimum output gray values g', and the corresponding input reflectance (or transmittance or intensity) values, R, of Eq. (17) for each frequency pattern. Expanding the numerator and denominator for Eq. (16) and the case of sinusoidal patterns and linear systems yields

$$MTF(f) = \frac{-[g'_{max}(f) - g'_{min}(f)]/[g'_{max}(f) + g'_{min}(f)]}{[R_{max}(f) - R_{min}(f)]/[R_{max}(f) + R_{min}(f)]} \tag{17}$$

where the prime is used to denote gray response that has been corrected for any nonlinearity as described below.

Figures 24 and 25 show an example of this process. In Fig. 24 we see the layout of a representative periodic square-wave test target (aR) and a sinusoidal test target (aL) which exhibits features of well-known patterns[96] available today in a variety of forms (from

Image Quality for Scanning

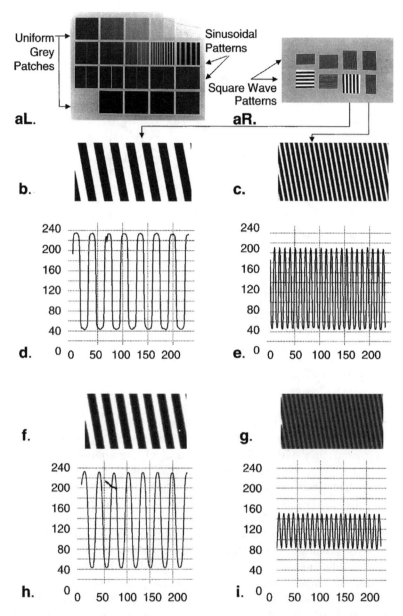

Figure 24 Example of images and profiles used in MTF analysis. Part (aL) shows a full pattern of gray patches and sinusoidal reflectance distributions at various frequencies. Part (aR) shows the frequency components of a square-wave test chart. Note that the bars are slanted slightly to facilitate measuring at different sampling phases. The figures on the left, (b), (d), (f), and (h), come from a low-frequency square-wave pattern as indicated by the arrow. The figures on the right, (c), (e), (g), and (i) are from a higher frequency square wave. Enlargements of the test patterns in (aR) are shown in parts (b) and (c). Slightly blurred images after scanning (as might be seen on a display of the scanner output) are shown in parts (f) and (g). Profiles of each of these images are displayed beneath them in parts (d), (e), (h), and (i), respectively. Because these are square-wave test patterns, special analysis of these patterns is required to compensate for effects of harmonics as described in the text. The reader should ignore small moiré effects caused by the reproduction process used to print this illustration.

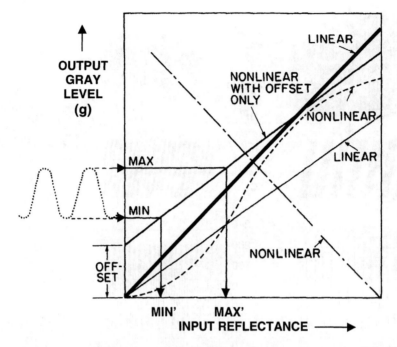

Figure 25 Examples of linear and nonlinear large-area response curves with an illustration of output modulation correction for offset using effective gray response at max' and min'.

Sine Patterns, Penfield, NY[97]). The periodic distributions of intensity (reflectance) are located in different blocks in the center of the pattern. Uniform reflectance patterns of various levels are placed in the top and bottom rows of the sinusoid to enable characterizing the tone response. A similar arrangement of uniform blocks is used with the square waves but not shown here. This enables correcting for its nonlinearities should there be any. Parts (b) and (c) show enlargements of parts of the square wave pattern selecting a lower and a higher frequency. Parts (f) and (g) are enlargements of a gray image display of the electronically captured image of the same parts of the test target. Parts (d), (e), (h), and (i) show profiles of the patterns immediately above them.

To calculate a modulation transfer function, the modulation of each pattern is measured. For sinusoidal input patterns, one can use Eq. (17) directly, finding the average maximum and minimum for many scan lines for each separate frequency. These modulation ratios, plotted on a frequency-by-frequency basis, describe the MTF. For square wave input, the input and output signals must be Fourier transformed into their spatial frequency representations and only the amplitudes of the fundamental frequencies used in Eq. (17). Schade[94] offers a method to compute the MTF by measuring the modulations of images of each square wave directly (i.e., the square wave response) and then unfolding for assumed perfect input square waves without taking the transforms.

From a practical standpoint it is important to tip the periodic patterns slightly as seen in parts (d) and (e) to cover the phase distributions as described above under the spread function discussion. A new higher resolution image can be calculated by interleaving data points from the individual scan lines, each of which is phase-shifted with respect to the sine or square wave.

Image Quality for Scanning

Figure 25 shows several examples of linear and nonlinear response curves. It describes correcting the output of an MTF analysis [i.e., using g' in Eq. (17)] for the case of nonlinearity with offset. Here the maximum and minimum values for the sine waves are unfolded through the response curve to arrive at minimum and maximum input reflectances, that is, the linear variables. For a full profile analysis, as needed for a Fourier transform method, and to obtain corrected modulation, each output gray level must be modified by such an operation.

If the response curve for the system was one of those indicated as linear in Fig. 25, then no correction is required. It is important to remember that while the scanner system response may obey a straight-line relationship between output gray level and the reflectance, transmittance, or intensity of the input pattern, it may be offset due to either optical or electronic biases (e.g., flare light, electronic offset, etc.). This also represents a nonlinearity and must be compensated.

As the frequency of interest begins to approach the sampling frequency in an aliased input scanning system, the presence of sampling moiré becomes a problem. This produces interference effects between the sampling frequency and the frequency of the test pattern. If the pattern is a square wave, this may be from the higher harmonics (e.g., $3\times$, $5\times$ the fundamental). When modulation is computed from sampled image data using maxima and minima in Eq. (17), errors may arise. There are no harmonics for the sinusoidal type of patterns, a distinct advantage of this approach.

See Fig. 26 for an example of these phase effects on a representative MTF curve. It shows errors for test sine waves whose period is a submultiple of the sampling interval. Consider the case where the sinusoidal test pattern frequency is exactly one-half the sampling frequency, that is, the Nyquist frequency. In this case, when the sampling grid lines up exactly with the successive peaks and valleys of the sine wave, we get a strong signal indicating the maximum modulation of the sine wave (point A). When the sampling grid lines up at the midpoint between each peak and valley of the sinusoidal image (phase shifted $90°$ relative to the first position), each data point will be the same, and no modulation whatever results (point B). There is no right or wrong answer to the question of which phase represents the true sine wave response, but the analog or highest value is often considered as the true modulation transfer function. Each phase may be considered as having its own sine wave response. Reporting the maximum and minimum frequency response or reporting some statistical average are both legitimate approaches, depending upon the intended use of the measurement. It is common practice to represent the average or maximum and the error range for the reported value.

The analog modulation transfer function, on the other hand, is only given as the maximum curve, representing the optical function before sampling. Therefore, the description of upper and lower phase boundaries for sine wave response shows the range of errors in the measurement of the modulation transfer function which one might get for a single measurement. This strongly suggests the need to use several phases to reduce error if the analog MTF is to be measured. Mathematically, phase errors may be thought of as a form of microscopic nonstationarity complicating the meaning of MTF for sampled images at a single phase. The use of information from several phases reduces this complication by enabling one to approximate the correct analog MTF that obeys the principle of stationarity.

In the case of a highly quantized system, meaning one having a relatively small number of gray levels, quantization effects becomes an important consideration in the design and testing of the input scanner. The graph in Fig. 27 shows the limitation that quantization step size, E_q, imposes on the measurement of the modulation transfer

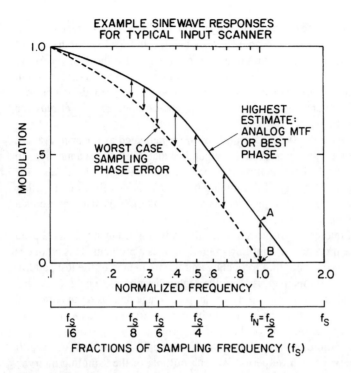

Figure 26 Example of the possible range of measured sine wave response values of an input scanner, showing the uncertainty resulting from possible phase variations in sampling.

function using sine waves. The number of gray levels used in an MTF calculation can be maximized by increasing the contrast of the sinusoidal signal that is on the input test pattern. It can also be increased by repeated measurements in which some analog shifts in signal level are introduced to cause the quantization levels to appear in steps between the previous discrete digital levels and therefore at different points on the sinusoidal distributions. The latter could be accomplished by changing the light level or electronic gain.

It is also important to note that because the actual modulation transfer function can vary over the field of view, a given measurement may only apply to a small local region over which the MTF is constant. (This is sometimes called an isoplanatic patch or stationary region within the image.) To further improve the accuracy of this approach, one can numerically fit sinusoidal distributions to the data points collected from a measurement, using the amplitude of the resultant sine wave to determine the average modulation. Taking the Fourier transform of the data in the video profile may be thought of as performing this fit automatically. The properly normalized amplitude of the Fourier transform at the spatial frequency of interest would, in fact, be the average modulation of the sine wave that fits the video data best.

The approach involving the application of square-wave test patterns (as opposed to sinusoidal ones, which have intrinsic simplicity as an advantage) has been shown by Newell and Triplett[95] to have significant practical advantages. They also show square-wave analysis has excellent accuracy when all-important details are carefully considered, especially the sampling nature of the analysis and the noise and phase effects.

Image Quality for Scanning

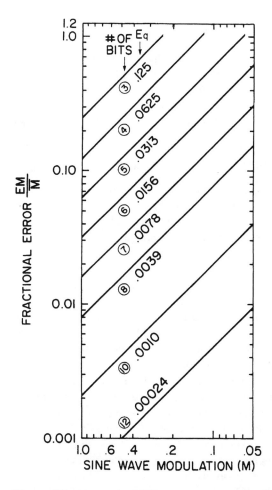

Figure 27 Errors in MTF measurements, showing the effects of modulation at various quantization errors. EM is in zero to peak units. $\bar{M}$ is average modulation. The numbers in the circles on each line indicate the system quantization in bits. E_q is the size of quantization step, where a full-scale signal = 1.0.

Square-wave test patterns are commonly found in resolving power test targets and are much easier to fabricate than sine waves, because the pattern exists as two states, foreground (e.g., black bars) and background (e.g., white bars). Two levels of gray bars may also be used depending on desired contrast. Fourier transform analysis and paying attention to the higher harmonics have been particularly effective.[87,95] It has been shown that the general Discrete Fourier Transform (DFT) algorithms where the length of the input can be altered is much better suited to MTF analysis than use of the Fast Fourier Transform (FFT) where the required power of two sampling points are a limitation.

One successful practice[95] included tipping the bar patterns to create a one pixel phase shift over 8 scan lines and averaging over approximately 30 scan lines to reduce noise. The DFT was used and tuned to the precise frequency of the given bar pattern by changing the number of cycles of the square wave being sampled and using approximately

1000 data points. An improvement in MTF accuracy of several percent was demonstrated using the DCT over the more common FFT.

It is generally advisable to measure the target's actual harmonic content rather than to assume that it will display the theoretical harmonics of a perfect mathematical square wave. Likewise, other patterns of known spectral content can be calibrated and used. Edge analysis techniques are also popular.[80,81,98] Using similar care such as a 5° slanted edge, a standardized algorithm, and specification on the edge quality, these achieve good accuracy too.

4.2.2 The Human Visual System's Spatial Frequency Response

As a matter of practical interest, several spatial frequency response measurements of the human visual system are shown in Fig. 28. These provide a reference to compare to system MTFs. The work of several authors is included.[99–105] The curves shown have all been normalized to 100% at their respective peaks to provide a clearer comparison. Except for the various normalizing factors, the ordinates are analogous to a modulation transfer factor of the type described by Eq. (17). However, MTFs are applicable only to linear systems, which the human eye is not. The visual system is in fact thought to be composed of many independent, frequency-selective channels,[106,108] which, under certain circumstances, combine to give an overall response as shown in these curves. It will be noted that the response of the visual system has a peak (i.e., modulation amplification relative to lower frequencies) in the neighborhood of 6 cycles/degree (0.34 cycles/milliradian) or 1 cycle/mm (25 cycles/in.) at a standard viewing distance of 340 mm (13.4 inches). The variations among these curves reflect the experimental difficulties inherent in the measurement task and may also illustrate the fact that a nonlinear system such as human vision cannot be characterized by a unique MTF.[107] For this reason such curves are called contrast sensitivity functions (CSFs) and not MTFs. For readers desiring a single curve, the luminance CSF reported by Fairchild[3] is given in Sec. 9, Fig. 58. While similar in shape to many curves in Fig. 28, it displays a greater range of responses and also shows the red–green and blue–yellow chromatic contrast sensitivity functions.

4.2.3 Electronic Enhancement of MTFs: Sharpness Improvement

These visual frequency response curves suggest that the performance of an imaging system could be improved if its frequency response could be increased at certain frequencies. It is not possible with most passive imaging systems to create amplification at selected frequencies. The use of electronic enhancement, however, can impart such an amplified response to the output of an electronic scanner. Amplification here is meant to imply a high-frequency response that is greater than the very low-frequency response or greater than unity (which is the most common response at the lowest frequencies). This can be done by convolving the digital image with a finite-impulse response (FIR) electronic filter that has negative sidelobes on opposite sides of a strong central peak. The details of FIR filter design are beyond the scope of this chapter, but the effects of two typical FIR filters on the system MTF are shown in Fig. 29.

4.3 Noise Metrics

Noise in an input scanner, whether the scanner is binary or multilevel gray, comes in many forms (see Sec. 2.2). A brief outline of these can be found in Table 4. Various specialized methods are required in order to discriminate and optimize the measurement of each.

Image Quality for Scanning

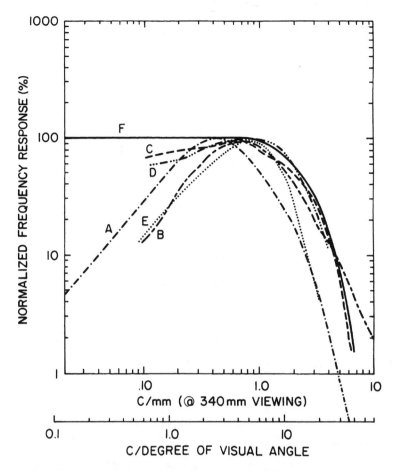

Figure 28 Measured spatial frequency response of the human visual system, showing the effects of experimental conditions on the range of possible results. Findings are presented according to (A) Campbell [99], (B) Patterson [105] (Glenn *et al.* [103]), (C) Watanabe *et al.* [102], (D) Hufnagel (after Bryngdahl) [107], (E) Gorog *et al.* [100], (F) Dooley and Shaw [104]. All measurements are normalized for 100% at peak and for 340 mm viewing distance. Note the universal visual angle scale at the bottom. See Fig. 58 (from Fairchild [3] for a seventh and more recent curve showing a larger response range and the two chromatic channels.

In this table we see that there are both fixed and time-varying types of noise. They may occur in either the fast or the slow scan direction and may either be additive noise sources or multiplicative noise sources. They may be either totally random or they may be structured. In terms of the spatial frequency content, the noise may be flat (white), that is, constant at all frequencies to a limit, or it may contain dominant frequencies, in which case the noise is said to be colored. These noise sources may be either random or deterministic; in the latter case, there may be some structure imparted to the noise.

The sources of the noise can be in many different components of the overall system, depending upon the design of the scanner. Instances of these may include the sensor of the radiation, or the electronics, which amplify and alter the electrical signal, including, for example, the A/D converter. Other noise sources may be motion errors, photon noise in

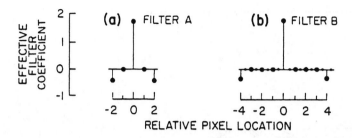

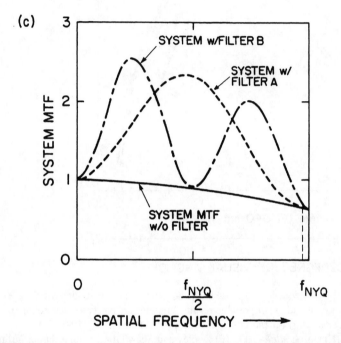

Figure 29 Two examples of enhancement of a scanning system MTF, using electronic finite-impulse response (FIR) filters in conjunction with an input scanner: (a) the one-dimensional line spread function for "filter A"; (b) line spread function for "filter B", which is the same shown as in (a) but 4 pixels wider; (c) shows the effect of these filters on the MTF of an aliased high-quality scanner.

low-light-level scanners, or noise from the illuminating lamp or laser. Sometimes the optical system, as in the case of a laser beam input scanner, may have instabilities that add noise. In many scanners there is a compensation mechanism to attempt to correct for fixed noise. This typically utilizes a uniformly reflecting or transmitting strip of one or more different densities parallel to the fast scan direction and located close to the input position for the document. It is scanned, its reflectance(s) is memorized by the system, and it is then used to correct or calibrate either the amplifier gain or its offset or both.

Such a calibration system is, of course, subject to many forms of instabilities, quantization errors, and other kinds of noise. Since most scanners deal with a digital signal in one form or another, quantization noise must also be considered.

Image Quality for Scanning

Table 4 Types and Sources of Noise in Input Scanners

Category	Type
Distribution	Fixed with platen, time varying
Type of operation	Multiplicative, additive
Spatial frequency	Flat (white), colored
Statistical distribution	Random, structured, image-dependent
Orientation	Fast scan, slow scan, none (two-dimensional)
Sources	Sensor, electronic system, motion error
	Calibration error, photons, lamp, laser
	Controls, optics

In order to characterize noise in a gray output scanner, one needs to record the signal from a uniform input target. The most challenging task is finding a uniform target with noise so low that the output signal does not contain a large component due to the input or document noise. In many of these scanners, the system is acting much like a microdensitometer, which reacts to such input noise as the paper fibers or granularity in photographic, lithographic, or other apparently uniform samples.

The basic measurement of noise involves understanding the distribution of the signal variation. This involves collecting several thousand pixels of data and examining the histogram of their variation or the spatial frequency content of that variation. Under the simplifying assumptions that we are dealing with noise sources that are linear, random, additive, and flat (white), a typical noise measurement procedure would be to evaluate the following expression:

$$\sigma_s^2 = \sigma_t^2 - \sigma_o^2 - \sigma_m^2 - \sigma_q^2 \tag{18}$$

where σ_s = the standard deviation of the noise for the scanner system (s); σ_t = the total (t) standard deviation recorded during the analysis; σ_o = the standard deviation of the noise in the input object (o) measured with an aperture that is identical to the pixel size; σ_m = the standard deviation of the noise due to measurement (m) error; and σ_q = the standard deviation of the noise associated with the quantization (q) error for those systems that digitize the signal. This equation assumes that all of the noise sources are independent. Removing quantization noise is an issue of whether one wants to characterize the scanner with or without the quantization effects, since they may in fact be an important characteristic of a given scanner design. The fundamental quantization error[109] is

$$\sigma_q^2 = \frac{2^{-2b}}{12} \tag{19}$$

where b = the number of bits to which the signal has been quantized.

The second and third terms in Eq. (18) give the performance of the analog portion of the measurement. They would include the properties of the sensor amplification circuit and the A/D converter as well as any other component of the system that leads to the noise noted in the table above. The term σ_q^2 characterizes the digital nature of the scanner and, of course, would be omitted for an analog scanning system.

Equation (18) is useful when the noise in the system is relatively flat with respect to spatial frequency or when the shape of the spatial frequency properties of all of the subsystems is similar. If, however, one or more of the subsystems involved in the scanner is contributing noise that is highly colored, that is, has a strong signature with respect to spatial frequency, then the analysis needs to be extended into frequency space. This approach uses Wiener or power spectral analysis.[40,110] Systems with filters of the type shown in Fig. 29 would exhibit colored (spatial frequency dependent) noise. A detailed development of Wiener spectra is beyond the scope of this chapter. However, it is important here to realize its basic form. It is a particular normalization of the spatial frequency distribution of the *square* of the signal fluctuations. The signal is often in optical density (D), but may be in volts, current, reflectance, and so on. The normalization involves the area of the detection aperture responsible for recording the fluctuations. Hence units of the Wiener spectrum are often $[\mu m\, D]^2$ and can be $[\mu m\, R]^2$. (See Ref. 110; the latter units are more appropriate for scanners because they respond linearly to reflectance, or, more generally, to irradiance.)

5 EVALUATING BINARY, THRESHOLDED, SCANNED IMAGING SYSTEMS

5.1 Importance of Evaluating Binary Scanning

Many output scanners accept only binary signals, that is, on or off signals for each pixel or subpixel. This translates to only black or white pixels on rendering. Therefore, there are many input scanners or image processing systems that generate binary images. A binary thresholded image may be generated directly by the scanner, or reduced to this state through image processing just prior to delivery to an output scanner. It may also be the degenerate state of inappropriate gray or dithered image processing in which signals are overamplified in a variety of ways, to look like thresholded images. Irrespective of how they are generated, binary thresholded renderings remain an important class of images today and often produce image characteristics that are surprising to the uninitiated. Understanding and quantifying this type of imaging becomes an important part of the evaluation of the overall input scanner to output printer system.

As noted earlier, there are two types of binary digital images, either thresholded or dithered signals. To a first order, dithered systems (halftoned or error diffused) can be evaluated in a way similar to that used to evaluate full gray systems, with one simplifying assumption. The underlying concept is that, within the effective dither region over which a halftone dot is clustered or the error is diffused, the viewer does not notice the pattern, so image detail is roughly invisible. As a result, these systems are primarily evaluated using instruments and methods whose resolution is equal to or larger than the effective dither region and hence are confined largely to tone rendition and some forms of image noise. It is possible to perform a one-dimensional analysis of dithered edges, for example, where the length of a narrow evaluation slit covers a great many effective dither regions. Extensive discussions of these measurement approaches are beyond the scope of this chapter as they involve the evaluation of the image processing algorithms and marking technologies more than the fundamentals of the scanner image quality. Some of the basic underlying principles are discussed in Sec. 2.2.3 under halftone system response and detail rendition. Limited discussion follows here under Sec. 5.4.3.

To understand and evaluate binary images, a few new concepts are explored and appropriate analytic methods developed.[111]

5.1.1 Angled Lines and Line Arrays

To adequately describe performance over a range of sampling phases, especially when fine structures that may be rendered in a binary form are being evaluated, it is important that the structures must be measured at a large number of sampling phases. In other words the evaluation is repeated several times with respect to the input pattern at positions predetermined to create images at different sampling phases. These may be produced by shifting the components by n pixels + various fractions of a pixel. For measurements that are essentially one-dimensional in nature, tilting lines or rectilinear patterns by a few degrees on the test pattern generates a continuum of phases along the edge or other feature of the designated structure. This was discussed earlier for MTF measurements. Without tilting serious errors may be encountered when, for example, a fine line may be imaged in one test as two pixels wide, and on another random test and therefore at another sampling phase, it may be imaged as one pixel wide.

5.2 General Principles of Threshold Imaging Tone Reproduction and Use of Gray Wedges

Since the output is not gray video (e.g., gray implies precise to 1 part in 256 as typical in the previous section while binary is precise to 1 part in 2), the testing process must be modified to compensate. For system response, that is, tone reproduction, testing can best be accomplished by having smooth calibrated structures that allow finding the on–off binary transitions to a small fraction, say 1 part in 200, of an input characteristic like reflectance.

The use of a calibrated gray wedge is often required. This device resembles a photographic step tablet except that it varies smoothly from a very low density to a very high density without steps, that is, it is a wedge of uniformly changing lightness as compared to perhaps 15 discrete logarithmically spaced levels in a typical step tablet. Ideally it would vary linearly in reflectance or transmittance as a function of distance, but the physical means for creating wedges often make them somewhat logarithmic. Therefore one may require extra length to insure adequate resolution of gray in the compressed end. Accurate measurement of transmittance or reflectance vs. distance from some reference mark on the wedge is used to calibrate the pattern, as in the top of Fig. 30. Distances in the image then correspond to linear signals.

The distance at which the wedge turns from black to white (or is 50% black and 50% white pixels for a noisy image) is measured for a given *gray threshold* and converted to a *reflectance (or transmittance) threshold*, the fundamental linear tone reproduction value for all testing.

Most testing of binary systems must use threshold as an independent variable. As will be seen below, many imaging performance characteristics are extremely sensitive to the threshold level. It is important, therefore, to specify this in both terms of reflectance, determined as indicated above, and in terms of the input gray level. This is essential in order to fully understand latitudes, repeatability, and uniformity. If the irradiance of the document or responsivity of the sensor in an input scanner varies in time or in space, then knowing the reflectance threshold at the place or time a specific attribute is observed becomes the key to sorting out the exact cause and effect relationships.

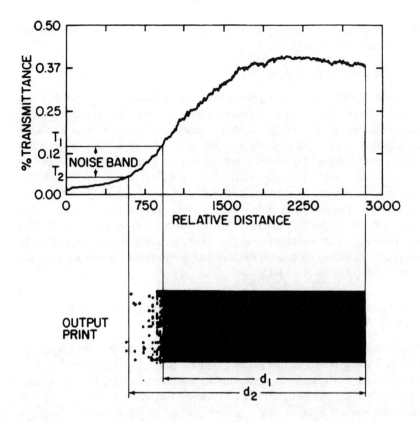

Figure 30 Transmittance profile of a gray wedge and a corresponding output print (binary image), both as function of distance in arbitrary units. Smallest dots at left are individual pixels.

5.2.1 Underlying Characteristic Curve

If one is trying to determine the underlying characteristic curve of the scanner, a series of specified reflectances can be determined along the wedge. The gray level of the threshold setting that creates the black to white transition at each specified reflectance is then plotted. This describes the underlying characteristic curve of the binary system.

Because of noise in the typical system, including noise on the input document, the location along the length of the wedge where the image changes from white to black will not be a sharp straight line. Rather it will be a region of noise as shown in Fig. 30. Typically the middle of this transition region is identified as the reflectance at which the threshold is set. For a highly nonlinear distance vs. reflectance characteristic, a small offset should be considered.

5.2.2 Dithered (Halftone or Error Diffused) Tonal Response

For binary imaging systems that are to represent multiple tonal values, one must deal with the measurement of a dithered response. Here the input document is either already halftoned and the system is trying to reproduce the dots (a matter of detail rendition, not tone reproduction), or it is a fine screened or continuous tone document (e.g., photograph or original painting). For these cases the binary system converts it to a halftone using

Image Quality for Scanning

a repeating matrix of spatially varying thresholds. Measurement of this performance typically uses a densitometer or spectrophotometer with a large aperture to measure densities or spectral reflectances of an input step tablet. One then captures the digital image of the step tablet and computes the percent area coverage, that is, the fractional number of pixels per halftone cell (or per unit area for error diffused images) that are turned either black or white for each input reflectance (see Fig. 38 in Sec. 5.4.3 for a halftone example). Because of noise, it is required that many halftone cells (or a large area) be averaged in order to get a reasonable representation of each step. One must average an integer number of halftone cells in order to make an accurate calculation of this performance.

If the input document was already halftoned, various sources of sampling errors produce moiré, which make the output so nonuniform that it becomes a meaningless task to measure response.

5.3 Binary Imaging Metrics Relating to MTF and Blur

Given the on–off nature of a binary thresholded image, a linear approach such as MTF analysis does not work. To deal with this nonlinearity we can pose three specific types of questions about imaging performance: (a) *Detectability*: what is the smallest isolated detail that the system can detect? (b) *Discriminability of fine detail*: what is the finest, most complex small structure or fine texture that the system can handle (legibility, resolving power)? (c) *Fidelity of reproduction*: for the larger details and structures, how do the images compare to the original input such as some reasonable width line? To create a specific metric in each of these categories, one defines a specific test object or test pattern that relates to the imaging application. One then defines a set of rules or criteria by which to judge performance against that pattern.

Further, one describes the statistics needed to characterize a sampled imaging system in which the phase relationship between the input pattern and the sampling grid of the input scanner varies. For the world of document processing and office copying, the following metrics have been found reasonably satisfactory. The reader is encouraged to use these as examples for consideration as he extends the thought processes to his own patterns and applications.

5.3.1 Detectability Metrics

Fine line detectability for document scanning is measured using an array of fine lines spaced several millimeters apart on a white background and varying in specific increments of both width and density. Line width increases from 15 to 50% from line to line can be used, depending on the precision requirements for this type of test, with 30% change being a good general purpose increment. Densities above background varying between 0.2 and 1.5 in a few steps are reasonable. For some practical applications this can be expanded to include light lines on a black background or various colors and density lines on a background of various colors and densities. One also must have a strategy for defining a threshold, a method for varying the phase of the document with respect to the imaging system, and criteria for judging performance. Some examples are shown in Table 5.

Variations on the above approaches may also be useful; for example, requiring the threshold for methods (a) and (b) above to be fixed by another image quality criterion such as the threshold required for maximum resolving power or minimum line width growth (see below) or best reproduction of a particular low-frequency halftone. The examples

Table 5 Description of Binary Detectability Metrics

Case	Name	Threshold	Line character	Criteria for judging performance (across phases)
A	Line width detection	Fixed (e.g., 50% of white)	Array of line widths and contrasts	Narrowest width of a given contrast (e.g., >50:1 reflectance black) showing detection for the observed % of phases
B	Line detection probability	Fixed	Single line width and contrast	% probability of detection across all phases
C	Detection threshold for lines	Variable	Single line width and contrast	Threshold to achieve greater than a specified detection probability
D	Detection threshold range for lines	Variable	Single line width and contrast	Range of thresholds to achieve a specified line detection probability range <100%

used above all relate to lines but could equally well relate to some other symbol of significant practical importance. Small squares or discs (dots) of varying widths and density are frequently encountered in certain applications and are therefore appropriate for them.

5.3.2 Line Fidelity Metrics

Given that an image has detected an object, the next logical question to ask is how much does it look like the original object, that is, what is its fidelity? Since we are dealing with binary imaging systems, it is clear that the intensity variable has been ruled out as a factor in any image quality considerations (i.e., each pixel is either on or off). We are therefore dealing with the fidelity of the spatial dimensions of the object and are concerned about error in the widths of the image compared to widths of the object. There are a number of ways to utilize concepts similar to those described above under detectability. These are shown in Table 6.

To measure *line width fidelity*, one merely needs create an image of a line of some standard width at many phases and orientations, measuring the width of the image in pixels. As before, one measures the line at various phases and positions along its length, treating the variations along the length as additional errors due to phase.

An application of great interest is the fidelity of lines originally having widths typical for text or line drawings (say 250 to 500 μm wide) with these lines being oriented in several different directions. The results are tabulated separately for each of several orientations, giving the average. See Case A in Table 6.

Some measure of the distribution of widths is important, such as the range, the standard deviation, or the actual *line width probability distribution* in terms of percentage. See Case B in Table 6. For example, an image of the 500 μm line scanned at 300 dpi (11.8 line/mm) with 85 μm pixels might show 10% at 5 pixels, 40% at 6 pixels, and 50%

Table 6 Description of a Few Binary Line Width Fidelity Metrics

Case	Name	Threshold	Line character	Criteria for judging performance (across all phases)
A	Line width fidelity	Fixed	Single line width and contrast	Difference between average width of image of the line (in number of pixels converted to distance and then averaged) and the measured width of the input line
B	Line width probability distribution	Fixed	Single line width and contrast	Distribution of pixel widths across all phases
C	Width-threshold profile for lines	Variable	Single line width and contrast	A plot: Average line width in pixels vs. threshold
D	Fidelity threshold range	Variable	Single line width and contrast	Range of thresholds to achieve the correct *average* line width across a large number of sampling phases

at 7 pixels. One would report the line width fidelity computing the average width as 6.4 pixels, converting it to distance as 544 μm and then reducing by 500 μm to show a difference of 44 μm.

The fidelity metric can be defined as either a difference from the original or as a ratio. In the example an effective growth of 8.8% (544 divided by 500 = 1.088) would be found to the original line width. The present authors prefer to use differences.

The concept of fidelity can be extended to two dimensions by scanning a two-dimensional symbol whose shape and size are related to the input in a broad general way. One then performs a template matching operation between the two-dimensional array of pixels in the image and the original object or a perfect representation of it. This would produce (in the output) differences or ratios of widths in each of several orientations with respect to features of the two-dimensional object being monitored. For many applications, while the line width changes caused by a low MTF can be brought back by changes in the threshold, the rounding of corners, as in a square or triangular input object cannot be "fixed" by changes in threshold. Thus template matching at corners for fine details such as serifs becomes a valuable extension of the fidelity constructs. Detailed implementation for the many possible two-dimensional shapes is beyond the scope of this chapter. An important cautionary note for the use of such measures is that the inde-pendent x and y control over MTF and the square sampling grid produce an orientation sensitivity that can be pronounced at various angles for many two-dimensional patterns.

5.3.3 Resolving Power

Resolving power is a commonly used descriptor of image quality for nearly every kind of imaging system. Its application to binary electronic imaging systems is therefore appealing. However, because of its extreme sensitivity to threshold and test pattern design, it must be applied with great care to prevent misleading results. Its primary value is in understanding performance for fine structures. The metrics noted above apply to isolated detail, while resolving power tends to emphasize the ability to distinguish many closely spaced details. In general, it can often be considered as an attempt to measure the cutoff frequency, that is, the maximum frequency for the MTF of a system. Binary systems are so nonlinear that even an approximate frequency-by-frequency MTF analysis cannot be considered.

The basic concept of a resolving power measurement is to attempt (through somewhat subjective visual evaluations) to detect a pattern in the thresholded video, which, to some level of confidence, resembles the pattern presented in the test target. For example, one may establish a criterion of 75% confidence that the image represents five black bars and four white spaces at the appropriate spacing. Values of 50, 95, 100% or any other confidence could also be used. As in all the metrics above, each judgment must be measured over a wide range of sampling phases for the bar pattern, resulting in an appropriate average confidence over all phases. Tipping the bar target so the length of the bars intercepts an integer number of sampling phases is again convenient.

One must be aware of some unique situations in binary resolving power measurements. The image of the just-resolved pattern will appear as crooked and sometimes incomplete black bars of one or two pixels in width. This is due to sampling phase. The bars will be separated by occasional white pixels and joined by occasional black pixels representing the noise in the process. Because the resolving power target is a periodic structure being imaged by another periodic structure, there will be a sampling moiré superimposed on the basic frequency.

In a binary imaging system, resolving power is not only a function of the test pattern modulation, but also of the specific reflectances of the white and black spaces, which interact strongly with the threshold selected for the measurement.

The reflectance of the white space between the bars of the original document is strongly determined by the scattering of light in the substrate on which the pattern has been made. For substrates that scatter a lot of light a long distance, the black parts of the image block significant amounts of light in regions between the black parts, thus reducing the light emitted by the "white" substrate at that point. Hence the white appears gray. This can be explained by the MTF models of the light scattering in paper.[61-63]

Binary imaging systems have extremely powerful contrast enhancement properties under the right circumstances. Selecting exactly the right threshold, one between the light and dark part of a resolving power image, amplifies a one or two percent modulation of the optical or gray electronic image to an on–off pattern (i.e., 100% modulation) that can be easily resolved in the video bit map.

Because it is possible to detect these low contrast patterns, it is also common to detect the situation known as pseudo- or spurious resolution. Here the blurring due to the input scanner is in a particular form that causes the light bars of the pattern to turn dark, and the dark bars to turn light. The condition for this situation is easily understood if one envisions a scanning aperture that is 1.5 times the width of a dark and light bar pair in a resolution target. Clearly the scanning aperture cannot resolve the bars. However, when

Image Quality for Scanning

centered on the dark bars, the system integrates two light and one dark bars, giving a lighter average. When centered on the light bars it averages two dark bars for a darker response. This turns a five black bar pattern into a four black bar pattern. If the viewer is not counting bars he may mistake this for a resolved pattern.

As a result of all these cautions, it is clear that resolving power by itself should not have the universal appeal for binary digital imaging systems that it may have for many other imaging situations. Nonetheless, it is a convenient tool for understanding how a system interacts with structured patterns. As in the cases above, there are many strategies available for using this metric. These would include either:

- Varying the threshold and noting the pattern that is resolved (see Fig. 31), fixing the bar spacing of interest, and looking for the threshold at which it is detected;
- Fixing the threshold and varying the contrast of a fixed bar spacing, noting the contrast at which the target fails to be resolved.

It should be noted that there is a strong dependence on angular orientation. Unlike resolving power in a conventional optical system, a nonzero or non-90° orientation may in fact perform better because of the independent MTFs in the x- and y-directions and the rectangular sampling grid.

Resolving power test targets come in many forms and these forms make a significant difference in the results, as noted earlier. Some of these are illustrated in Fig. 32. Only the coarser patterns are imaged in this illustration and no attempt should be made by the reader

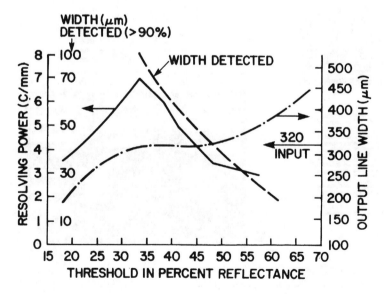

Figure 31 Plots of line width detectability, fidelity, and resolving power as functions of threshold setting in a binary imaging system. (from Ref. 111). Arrows on each curve indicate which axis represents the ordinate for that curve. "Output line width" in μm is for images of 320 μm "input line width" as noted by arrow at right. The "Width Detected" curve refers to the left inside axis and is given in μm of the input line width, which is detected at the designated threshold for >90% of the sampling phases.

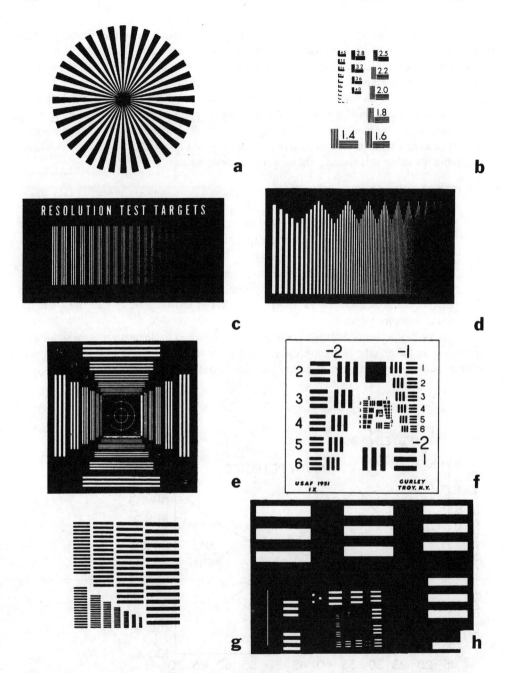

Figure 32 Images made with various bar pattern test targets in common use for measurement of resolving power and related metrics. See text for identification and description of each type.

to use these images for testing. They are illustrations only. Two general forms exist: those with discrete changes in the bar spacing and those with continuous variation in the bar spacing. In the former category there are several fairly commonly encountered types, designed for visual testing, namely the NBS lens testing type (e), NBS microcopy test chart type (b), the US Air Force type (f), the Cobb chart type (2 bars, not shown) and finally, the ANSI Resolving Power test patterns (h). This form also includes the extended square-wave types as represented by either Ronchi rulings (not shown) or ladder charts (not shown), which are simply larger arrays of the Ealing test pattern (g) which shows 15 bars of each square wave. Machine-readable forms are also useful where the modules are arranged in a pattern that can be scanned in a single straight line, as in (c). The differences between these can be seen in the aspect ratio of the bars in the various patterns, the number of bars per frequency, the layout of the pattern itself, whether it is a spiral or a rectangle displaying progression of spacing, and the actual numerical progression of different frequency patterns within the target. In many cases low contrast versions or reverse polarity (white and black parts are switched) are also available.

The second major class, the continuously varying frequency pattern, is exemplified by the Sayce chart (d) and the radial graded frequency chart (a). The Sayce patterns are particularly useful for automated readout, provided the appropriate phase information is obtained (coordinates of each black bar) to prevent the pseudo-resolution phenomena described above. The radial graded frequency bar charts are especially useful for visually evaluating performance in some display contexts, since they provide both continuously variable frequency and orientation information in great detail from a single compact pattern. Variations of each pattern exist in which the rate of change of spatial frequency with respect to distance differs and where various hints about the frequency at a given location are provided.

5.3.4 Line Imaging Interactions

A strategy for evaluating line and text imaging against all of the above metrics is to establish a fixed threshold that optimizes system performance for one of the major categories, such as line width fidelity, and then to report performance for the other variables, such as detectability and resolving power, at that threshold. One may also choose to plot detectability, fidelity, and resolving power as a function of threshold on a single plot in order to observe the relationships among the three and find an optimum threshold, trading off one against the other. This is illustrated in Fig. 31. Such a plot provides several useful perspectives relating to the effects of blur on a binary system. It is clear that the maximum fidelity occurs between 35 and 45% threshold while the maximum fine-line detectability keeps growing as the threshold drops below 30%. The resolving power has a distinct maximum at about 33%. Such a plot is different for each system and is governed by the shape of the underlying MTF curves and the various nonlinear interactions produced by image processing and the electronics.

Figures 33(a)–(d) show enlarged displays of binary electronic images illustrating practical examples of what would be evaluated for each of these metrics at various thresholds. In these displays a pixel is represented as a tiny square of dimensions equal to the pixel spacing. Parts (a) and (b) show the effect of threshold on detectability. Observe the narrower lines in each of Figs (a1) to (a4) where threshold is varied over a total range of 12%. Line width detection varies from 50% for the 50 μm line [to the left of the "5" in image (a4)] at threshold 45% reflectance to 90% for the 100 μm line [left of "10" in image (a1)] at threshold 33% reflectance. There is a 1° tip between the upper and lower array of lines to display the effects of different phases. The wider lines in parts (a) and (b) represent arrays useful for fidelity measurements, but require greater enlargement for the actual measurement. Part (c) shows greater enlargements that enable the counting of pixels to determine line width fidelity. (A display as shown later

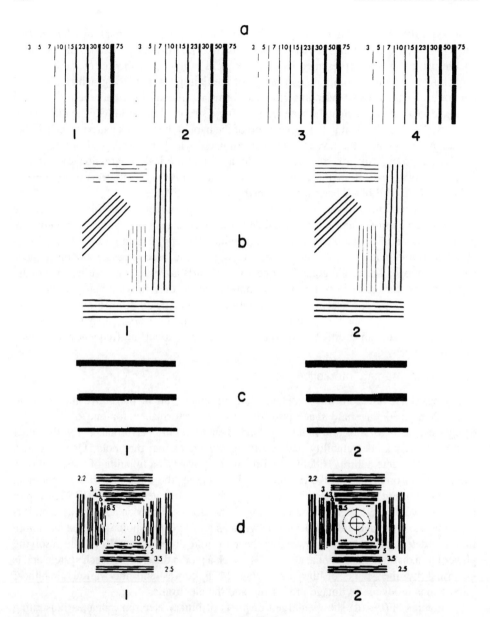

Figure 33 Enlarged displays of binary electronic images illustrating characteristics observed during evaluations for each metric in Fig. 31 as described in the text. Parts (a) and (b) illustrate fine line detectability, part (c) shows line width fidelity, and part (d) illustrates resolving power.

in Fig. 35 for another metric is an ideal magnification and form for counting pixels.) Part (d) of Fig. 33 shows examples of images of resolving power patterns. As an example of interpretation, note in this figure that a 75% criterion for probability-of-resolving-the-pattern would resolve 6 cycles/mm on the left figure and 3.5 cycles/mm on the right figure for the vertical bars. Threshold increases from left to right.

Image Quality for Scanning

5.4 Binary Metrics Relating to Noise Characteristics

Conventional approaches to measuring the amplitude of the noise fluctuations using various statistical measures of the distribution are not appropriate for binary systems. In these systems the noise shows up as pixels that are of the wrong polarity; that is, a black pixel that should have been white or a white pixel that should have been black. In general it is the distribution and location of these errors that needs to be characterized. The practical approach to this problem is to examine the noise in a context equivalent to the main applications of interest for the binary imaging system. The resulting metrics include:

1. The range of uncertainty associated with determining the threshold using a gray wedge as described above in Sec. 5.2, which has led to the *gray wedge metric for noise*.
2. The noise seen on edges of lines and characters, which has led to the *line edge range metric for noise*.
3. The characterization of noise in a halftone image, that is, *halftone granularity*.

These are all described below.

5.4.1 Gray Wedge Noise

Returning to Fig. 30 shows the transmittance profile of a gray wedge as a function of distance. The thresholded image of that gray wedge is shown below this profile with the x-axis lined up to correspond to the position in the profile plot. The right-hand edge of the gray wedge is a high contrast step whose location can be accurately determined in both the plot and the image. By measuring the distance d_1 from this edge to the location in which the image of the wedge begins to break up, and then the distance d_2 to the location at which the random noise becomes all white, one can determine a region (of length $d_2 - d_1$) along the wedge in which the thresholding operation becomes uncertain. This is the noise region.

These distances can be converted by means of the profile of the document into a noise band, that is, a range of linear variables (reflectances or transmittances or relative threshold values) represented by T_1 and T_2 in the figure.

To make a statistically satisfactory measure of noise, a probability distribution is used with the criteria for determining the positions d_1 and d_2. As illustrated, these are the point where the signal is 95% black and the point where the signal is 95% white. Under the assumption of normally distributed noise this would represent plus or minus approximately two standard deviation limits on the noise distribution. Similarly, the halfway point or the 50% black/white point may be used as an estimate of the mean signal for signal-to-noise calculations. This would also be the threshold point as described earlier. The figure shows roughly a 95% noise band for a highly noisy system.

To fully characterize a binary system with this metric, one plots the width of the noise band in effective transmittance as a function of the independent variable, threshold (converted to transmittance).

Using the linear variable (here transmittance) to designate both threshold and noise bandwidth is desirable since most scanners are fundamentally fairly linear.

The granularity (Wiener spectrum) of the gray wedge contributes significantly to the noise. It should be removed from the calculation by using an RSS (square root of the sum of squares) technique as described earlier for gray systems. One must keep track of the aperture area for this granularity measurement. It should be set equal to that of the imaging system under test in order to perform the RSS calculation correctly.

5.4.2 Line Edge Noise Range Metric

The noise associated with the edges of lines is an important type of noise to be directly evaluated for many practical reasons. It can be seen from the discussions above on blur and MTF that the image of every line has a microscopic gray region associated with it where the intensity falls off gradually from the white surround field into the black line. For the image of the edge of a line oriented at a very small angle to the sampling matrix, in one scan line at the edge of the line there is a distribution of gray varying along the edge of the line. It gradually increases from white to black, as the tilted line approaches the pixel whose area is completely covered by the line. This scan line appears white until it reaches a fractional coverage required by the threshold, and then changes to black. However, the gray signal at the edge of this line acts much like that associated with the wedge in the previous metric. As the edge approaches the transition point where the threshold causes a change in the binary signal, the probability for an error resulting from noise increases. Thus the binary signal along the length of this slightly tipped line acts much like the signal for the wedge in the previous example, oscillating between black and white. This provides the basis for a second metric, which we refer to as the line edge noise range metric.

In Fig. 34, a slightly tilted input line is shown relative to several scan lines. The binary video bit map for this line is shown in the lower part of the figure. Vertical lines mark the location at which the edge of the line makes a transition from the center of one raster line to the next. In the video bit map this transition is noisy and the two ranges in which this uncertainty of the black to white transition exists are indicated as N_1 and N_2. The centers of these noisy transition regions are marked by the transition lines at X_1 and X_2 and are separated by the distance ΔX. The metric can be applied to edges of lines, or edges of solids, or any straight edge and is therefore generically referred to as "edge noise range"

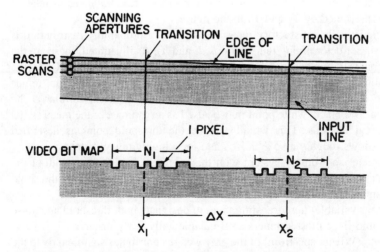

Figure 34 Scanning of a slightly tilted line, with the corresponding binary video bit map image, showing noise effects, which define edge noise range (ENR).

Image Quality for Scanning

or ENR and simply defined as

$$ENR = \frac{\sum_{i=1}^{n} N_i}{\sum_{i=1}^{n-1} \Delta X_i} \qquad (20)$$

The numerator and the denominator are averages over a large number of transitions along one or more constantly sloping straight lines. ΔX is the number of pixels per "step." The range N is determined by subtracting the pixel number of the first white pixel in the black region from that of the last black pixel in the white region along the length of the line in each transition region. A worked example is shown in Fig. 35 using an actual image from a graphic arts scanner.

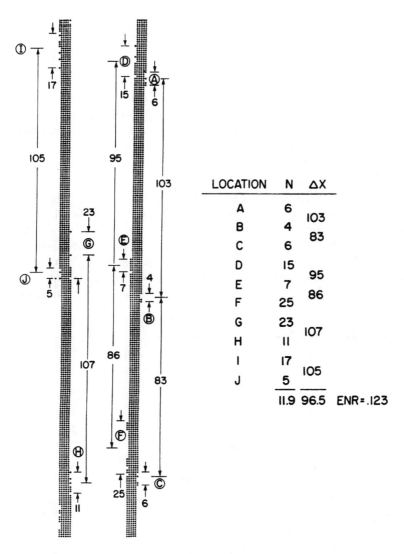

Figure 35 Experimental examples of edge noise range (ENR) for several scan line transitions.

Figures 36 and 37 show the relationships among ENR, the step ratio ΔX, and the percentage of RMS noise in the imaging system, assuming additive white Gaussian noise distributions (private communication, J.C. Dainty, 1984) These are not intuitive relationships. For example, it should be noted from Fig. 30 that an increase of a factor of 2 in RMS noise for a given angle line produces a line edge noise range increase of anywhere from $2\frac{1}{2}$ to $4\times$ depending upon the slope of the line and the exact noise level. It is noted that the noise is highly dependent on the angle of the line. Gradually sloping lines not only produce a larger absolute range but also a larger fractional range.

Here the MTF of the imaging system was considered to be perfect. The effect of blur, that is, decreases in MTF, is to increase the magnitude of ENR above those values shown. This is because the distribution of gray in the vicinity of the transition point is a function of the MTF of the system. Therefore a more blurred or lower MTF system produces a longer region of gray transition and hence a larger range on the noise for this metric.

It must also be noted that document noise will create extra fluctuations along the edge of the line and also increase the length of the range. Therefore for practical considerations very sharp low noise edges should be used on test targets for this metric. On the other hand, noise on the edges of input documents, which are considered typical for the application, can be included in the test target. This serves to include it in the measurement for a given application. This may be especially valuable for a thresholding system, which is nonlinear. Here, a simple RSS technique for compensating, or predicting, the effects of document noise may not work correctly.

5.4.3 Noise in Halftoned or Screened Digital Images

Scanning a typical photograph and applying a halftone screen of the type described earlier results in a bit map in which some of the arrangements of pixels in the halftone cells do not follow the prescribed growth pattern for the screen (see Fig. 38). The noise in the scanning system itself can produce changes in the effect of threshold at each one of the sites within the halftone cell. Some of these errors are introduced by the partial dotting mechanism, described earlier, when the granularity of the otherwise uniform input document, which was scanned to create the image, is of sufficient contrast to change the structure inside

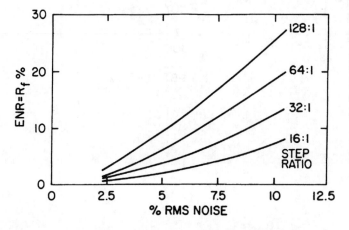

Figure 36 Relationship between LENR and RMS noise for various step ratios.

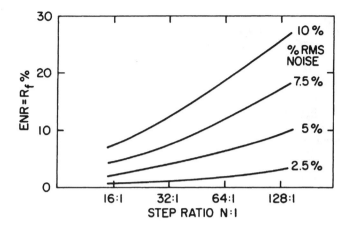

Figure 37 Relationship between LENR and step ratio for various values of RMS noise.

areas formed by individual halftone cell's threshold matrix. See Sec. 2.2.3 and Figs. 13–15 for a review of these mechanisms.

One way of evaluating this type of noise is to create images of a series of perfectly uniform patches of differing density and process them through the halftoning method of choice. One then measures the RMS fluctuation in the percent area coverage for the resulting halftone cells, one patch at a time. To the extent that the output system is insensitive to the orientation of the bit map inside the halftone cell, this fluctuation becomes a reasonable measure of the granularity of the digital halftone pattern. For the electronic image, it can be calculated with a simple computer program that searches out each halftone cell and calculates its area coverage, collecting the statistics over a large number of halftone cells.

There are many image analysis packages on the market that will find particles in a digital image and evaluate their statistical distribution. They are found in biology, medicine, or metallurgy software applications. In this case a "particle" is a halftone dot whose area corresponds to the number of pixels. A simple calibration of the software tools correlating the area it reports to the number of pixels the user counts for a few test halftone dots of different sizes is advised. In using such a tool, one must be careful not to include the statistics of the fractional halftone cells at the edge of a field of dots caused by the cropping of a larger field of dots with such a tool. Halftone dots are often rotated 45°, creating partial size dots on all edges. Scanning from dot to dot on the diagonal fixes this. Editing out the partial dots created by the edge of the measured area also eliminates the source of difficulty.

To the extent that the output marking process does not respond the same to equal area coverage of different orientations of pixels, this process is not applicable. The analysis should be made in the context of the intended output display or printing device such as a laser printer. Such output subsystems usually create their own noise in addition to that of the input scanner. Hence granularity of a halftone input pattern in many practical situations is a matter for systems analysis, which corrects for the characteristics of the subsystems in the imaging path. If the image has been printed, a microdensitometer can be set up with an aperture that exactly covers one halftone cell, then scanned along rows of halftone dots. The RMS fluctuations or the low-frequency components of the noise spectra

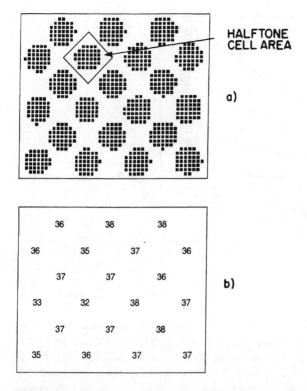

Figure 38 Noise (granularity) in a binary halftone image. part (a) Bitmap of the halftone rendering of a scanned image of a uniform area on an original. Part (b) Number of black pixels in each cell where the average number of black pixels per cell is 36.4 and the estimate of the standard deviation is 1.56 pixels.

can then be evaluated. This is sometimes referred to as *aperture filtered granularity* measurement.

From all of the above measurement methods for binary imaging systems, it can be seen that the nonlinear nature of these systems creates the need for many specialized metrics to describe performance in the context of the imaging application. These non-linearities make summary measures of image quality, like those that follow, difficult to apply directly. There are strong noise and MTF interactions as well as the obvious nonlinearity associated with the thresholded tone reproduction. The nature of the halftone response can provide a quasi-linear system at the macroscopic level, but at the microscopic level it produces its own significant distortions. It is at the microstructure level that one needs to understand the binary image in order better to translate the bit map characteristics required for a laser printer or other output device.

6 SUMMARY MEASURES OF IMAGING PERFORMANCE

Many attempts have been made to take the general information on image quality measurement described above and reduce it to a single measure of imaging performance. These often take the form of shortcut "D" in Fig. 2. While none of the resulting measures

Image Quality for Scanning

provides a single universal figure of merit for image quality, each brings additional insight to the design and analysis of particular imaging systems. Each has achieved some level of success in a limited range of applications. Perceived image quality, however, is a psychological reaction to a complex set of trade-offs and visual stimuli. There is a very subjective, application-oriented aspect to this reaction that does not readily lend itself to analytical description. Section 8 on psychometric measurement methods describes the general approach to assessing the human response in such studies. In some instances where the application is sufficiently well bounded by identifiable visual tasks, an overall measure of image quality can be found. This narrowly applicable metric is a result of psychological and preference research and usually involves extensive measurement of a limited class of imaging variables. We have never found such a figure of merit with broad applicability.

Instead, in an attempt to help the engineer control or design his systems, we shall describe a number of metrics. Each individual metric, in many cases, is suited to optimizing one or two subsystems and is valuable in its own right. For given applications, several of these can be combined to explore a trade-off space and perhaps construct a measure of overall image quality for that application. We shall describe some examples of how these have been constructed so that the reader has a good starting point for his own applications. Since genuine image quality attributes are really preference features, user or market research is usually required and will not be covered in this section. The building blocks are offered here to enable the reader to build his own regression equation or other means for connecting physical quantification and description of an image with its subjective value.

The metrics are described here in their general form, as they would apply to analog imaging systems such as cameras and film. They have not been particularized (except in a few cases) to the digital imaging conditions in order to simplify their treatment in this summary section. To the extent that the scanning systems in question are unaliased and have a large number of gray levels associated with them, the direct application of analog metrics is valid. In general, it should be remembered that digital imaging systems are not symmetric in slow and fast scan orientations in either noise or spatial frequency response (MTF). Therefore, what is given below in one-dimensional units must be applied in both dimensions for successful analysis of a digital input scanner. These concepts can be extended to an entire imaging system with little modification if the subsequent imaging modules, such as a laser beam scanner, provide gray output writing capability and generate no significant sampling or image conversion defects of their own (i.e., they are fairly linear). Since full gray scale input scanners are usually linear, most of these concepts can be applied to them, with the qualification that some display or analysis technique is required to convert the otherwise invisible electronic image to a visual or numerical form.

All of these measures involve the concept of the signal-to-noise ratio. Some deal directly with the terms described above, while others are oriented toward a particular application, and still others use more generalized constructs. Tailoring the metric to a specific application involves finding those signal and noise characteristics that are most relevant to the intended application.

Most of these summary measures can be described by curves like the illustrative ones shown in Fig. 39. Here we show some common measures, generically symbolized by F for both signal and noise, such as intensity, modulation or (modulation)^2 plotted as a function of spatial frequency f. A signal $S(f)$ is shown generally decreasing from its value at 0 spatial frequency to the frequency f_{max}. A limiting or noise function $N(f)$ is plotted on

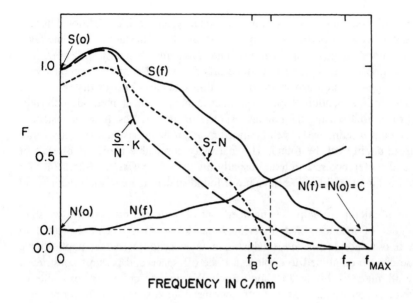

Figure 39 Signal, $S(f)$, noise, $N(f)$, and various measures of the relationships between them, plotted as functions of spatial frequency. Various critical frequencies are noted as points on the frequency axis.

the same graph starting at a point below the signal; it too varies in some fashion as spatial frequency increases. The various unifying constructs (metrics) involve very carefully considered approaches to the relationship between S and N, to their respective definitions, and to the frequency range over which the relationship is to be considered, along with the frequency weighting of that relationship

6.1 Basic Signal-to-Noise Ratio

The simplest of all signal-to-noise measures is the ratio of the mean signal level $S(0)$ to the standard deviation $N(0)$ of the fluctuations at that mean. If the system is linear and the noise is multiplicative, this is a useful single number metric. If the noise varies with signal level, then this ratio is plotted as a function of the mean signal level to get a clearer picture of performance. Hypothetical elementary examples of this are shown in Fig. 40, in which are plotted both the multiplicative type of noise at 5% of mean signal level (here represented as 100) and additive noise of 5% with respect to the mean signal level. It can be seen why such a distinction is important in evaluating a real system. It should be noted that in some cases, multiplicative or additive noise might vary as a function of signal level for some important design reason.

In comparing signals to noise, one must also be careful, to ensure that the detector area over which the fluctuations are collected is appropriate for the application to which the signal-to-noise calculation pertains. This could be the size of the input or output pixel, of the halftone cell, or of the projected human visual spread function. The data must also be collected in the orientation of interest. In general, for scanned imaging systems there will be a different signal-to-noise ratio in the fast scanning direction than in the slow scanning direction.

Image Quality for Scanning

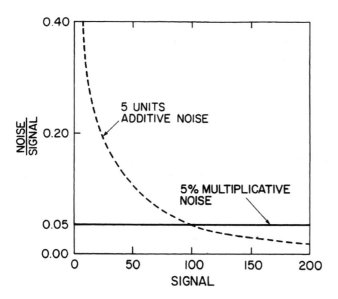

Figure 40 Relative noise (noise/signal) as a function of signal level for additive and multiplicative noise.

6.2 Detective Quantum Efficiency and Noise Equivalent Quanta

When low light levels or highly noise limited situations occur, it is desirable to apply the concepts of detective quantum efficiency (DQE) and noise equivalent quanta (NEQ). These fundamental measurements have been extensively discussed in the literature.[40]

The expression for DQE and its relationship to NEQ is

$$DQE = \frac{NEQ}{q} = \frac{A(\log_{10} e)^2}{W(0)} \frac{\gamma^2}{q} \qquad (21)$$

This form of the equation relates to a typical photographic application in which the signal is in density units. Density is the negative $\log_{10}$ of either transmission or reflectance, and γ is the slope of the relationship between the output density and the input signal (expressed as the $\log_{10}$ of exposure). W is the Wiener spectrum of the noise given here at 0 spatial frequency (where it is equal to $A\sigma_s^2$, see Eq. 18), and q is the number of exposure quanta collected by the detector whose area is A. For a more general case, Eq. (21) is rewritten as Eq. (22), where the units have been expressed in terms of a general output unit, O which could represent voltage, current, number of counts, gray level, and so on.

$$DQE = \frac{q(dO/dq)^2}{\overline{\Delta O^2}} \qquad (22)$$

where $\overline{\Delta O^2}$ is the mean-square output fluctuation [40, p. 155; 112]. The term $W(0)$ contains the factor $1/q^2$, which, through manipulation, leads to q in the numerator of Eq. (22). The term in parentheses is simply the slope of the characteristic curve of the detection system (see Fig. 23) indicating the change in the output divided by the change in the input.

It is noted that this is related to the gain, Γ, of the system (Fig. 23) and appears as a squared factor in the equation. If we set this gain to a constant r in arbitrary output units, and assume the distribution of fluctuations obeys normal statistics, then we can rewrite Eq. (22) as

$$DQE = r^2 q / \sigma_0^2 \qquad (23)$$

where σ_0 represents an estimate of the standard deviation of the distribution of the output fluctuations. It can now be seen that this expression differs in a significant way from the simple construct used above, namely the mean divided by the standard deviation. Here the average signal level q is divided by the square of the standard deviation (i.e., the variance) and contains a modifier that is related to the characteristic amplification factors associated with a particular detection system, namely r, which also enters as a square. It should also be pointed out that detective quantum efficiency is an absolute measure of performance, since q is an absolute number of exposure events, that is, number of photons or quanta. Returning to Eq. (21), we can now see that under these simplifying assumptions, the noise equivalent quanta can be represented by

$$NEQ = \frac{r^2 q^2}{\sigma_0^2} \qquad (24)$$

The concepts just described can be extended to the rest of the spatial frequency domain in the form

$$NEQ = \frac{A r^2 q^2 [MTF(f_x, f_y)]^2}{W(f_x, f_y)} \qquad (25)$$

Provided the response characteristics of the system are linear, Eq. (25) reduces to Eq. (26), since $r \times q$ is equal to O, the output in whatever units are required.

$$NEQ = \frac{A O^2 [MTF(f_x, f_y)]^2}{W(f_x, f_y)} \qquad (26)$$

For illustration purposes, Fig. 39 shows all of the above constructs. For DQE, the generic signal measure $S(0)$ can be thought of as the numerator of Eqs. (22) and (23) and $N(0)$ may be thought of as the denominator. Similarly, for noise equivalent quanta, $S(0)$ represents the numerator and $N(0)$ the denominator of Eq. (24). For the spatial-frequency-dependent version of noise equivalent quanta, $S(f)$ represents the numerator of Eq. (25) or Eq. (26) and $N(f)$ represents the denominator of the same equations. The large-dashed curve in Fig. 39, KS/N, represents the ratio of $S(f)$ to $N(f)$ normalized to $S(0)$. K is an arbitrary constant, which, for this illustration, is set equal to 0.1. It is seen that this function continues to the cutoff frequency f_{max}. It should also be observed that this relationship between the signal characteristic and the noise characteristic can vary with signal level, as shown in Fig. 40, and hence a full functional description requires a three-dimensional plot, making S the third axis of an expanded version of Fig. 39.

Image Quality for Scanning

6.3 Application-Specific Context

The above descriptions are frequently derived from the fundamental physical characteristics of various imaging systems, but the search for the summary measure of image quality usually includes an attempt to arrive at some application-oriented subjective evaluation, correlating subjective with objective descriptions. Applications that have been investigated extensively include two major categories: those involving detection and recognition of specific types of detail and those involved in presenting aesthetically pleasing renderings of a wide variety of subject matter. These have centered on a number of imaging constraints, which can usually be grouped into the categories of display technologies and hard-copy generation. Many studies of MTF have been applied to each.[74,87,113,118,119,122,123] All of these studies are of some interest here. Note that modern laser beam scanning tends to focus on the generation of hard copy where the raster density is hundreds of lines per inch and thousands of lines per image compared with the hundreds of lines per image for early CRT technology used in the classical studies of soft display quality.

6.4 Modulation Requirement Measures

One general approach characterizes $N(f)$ in Fig. 39 as a "demand function" of one of several different kinds. Such a function is defined as the amount of modulation or signal required for a given imaging and viewing situation and a given target type. In one class of applications, the curve $N(f)$ is called the threshold detectability curve and is obtained experimentally. Targets of a given format but varying in spatial frequency and modulation are imaged by the system under test. The images are evaluated visually under conditions and criteria required by the application. Results are stated as the input target modulation required (i.e., "demanded") for being "just resolved" or "just detected" at each frequency. It is assumed that the viewing conditions for the experiment are optimum and that the threshold for detection of any target in the image is a function of the target image modulation, the noise in the observer's visual system, and the noise in the imaging system preceding the observer. At low spatial frequencies this curve is limited mostly by the human visual system, while at higher frequencies imaging system noise as well as blur may determine the limit.

One such type of experiment involves measuring the object modulation required to resolve a three-bar resolving power target. For purposes of electronic imaging, it must be recalled that the output video of an input scanner cannot be viewed directly, and therefore any application of this method must be in the systems context, including some form of output writing or display. This would introduce additional noise restrictions. The output could be a CRT display of some type, such as a video monitor with gray-scale (analog) response. Another likely output would be a laser beam scanner writing on xerography or on silver halide film or paper. The details for measuring and using the demand function can be found in work by Scott[114] for the example of photographic film and in Biberman,[74] especially Chapter 3 for application to soft displays.

6.5 Area Under the MTF Cure (MTFA) and Square Root Integral (SQRI)

Modulation detectability, while useful for characterizing systems in task-oriented applications, is not always useful in predicting overall image quality performance for a broad range of imaging tasks and subject matters. It has been extended to a more general form through the concepts of the threshold quality factor[115] and area under the MTF curve (MTFA).[116,117] These concepts were originally developed for conventional photographic systems used in military photo-interpretation tasks.[115] They have been generalized to electro-optical systems applications for various forms of recognition and image-quality evaluation tasks, mostly involving soft displays.[74] The concept is quite simple in terms of Fig. 39. It is the integrated area between the curves $S(f)$ and $N(f)$ or, equivalently, the area under the curve labeled $S-N$. In two dimensions, this is

$$MTFA = \int_0^{f_{cx}} \int_0^{f_{cy}} [S(f_x, f_y) - N(f_x, f_y)] \, df_x \, df_y \qquad (27)$$

where S is the MTF of the system and N is the modulation detectability or demand function as defined above, and f_{cx} and f_{cy} are the two-dimensional "crossover" frequencies equivalent to f_c shown in Fig. 39.

This metric attempts to include the cumulative effects of various stages of the scanner, films, development, the observation process, the noise introduced into the perceived image by the imaging system, and the limitations imposed by psychological and physiological aspects of the observer by building all these effects into the demand function $N(f)$. Extensive psychophysical evaluation and correlation has confirmed the usefulness of this approach[117] for recognition of military reconnaissance targets, pictorial recognition in general, and for some alphanumeric recognition.

Related approaches using a visual MTF weighting have been successfully applied to a number of display evaluation tasks, showing good correlation with subjective quality.[118] Many studies examine differences in quality where noise factors are relatively constant. One of these is the square root integral (SQRI) model of Barten.[119,120] Here, the demand function is specified by a general contrast sensitivity of the human visual system and the comparison of the quality of two images of interest is specified in JND units (see Sec. 8 for a definition of JND, a just noticeable difference).

$$J = \frac{1}{\ln 2} \int_0^{W_{max}} \sqrt{\frac{M(w)d(w)}{M_t(w)w}} \qquad (28)$$

where $M(w)$ is the cascaded MTF of the image components, including that of the display and $M_t(w)$ is threshold modulation transfer function of the human visual system, both in units of angular spatial frequency w. Results are to be interpreted with the understanding that 1 JND is "practically insignificant." It is equal to a 75% correct response in a paired comparison experiment. Note that 3 JND is "significant," and 10 JND is "substantial."[119,121] The $M_t(w)$ term describes the HVS as the threshold contrast for detecting a grating of angular frequency w as follows:

$$1/M_t(w) = aw \exp(-bw)\sqrt{1 + c \exp(bw)} \qquad (29)$$

where

$$a = \frac{540(1 + 0.7/L)^{-0.2}}{1 + 12/[s_w(1 + w/3)^2]} \qquad (30a)$$

$$b = 0.3(1 + 100/L)^{0.15} \qquad (30b)$$

$$c = 0.6 \qquad (30c)$$

Image Quality for Scanning

L is the display luminance in cd/m² and s_w is the display size or width in degrees. These equations have been shown to have high correlation with perceived quality over a wide range of display experiments,[119] one of which is shown in Fig. 41 below. Here the resolution, size, and subject matter of projected slides were varied and the equation was fit to the data.

It is noted by Barten that noise should be taken into account in the modulation threshold function, which is done by using a root sum of squares method of a weighted noise modulation factor to $M_t(w)$.[121] Other authors have expanded on these concepts, extending them to include more fundamentals of visual mechanisms.[122–124]

6.6 Measures of Subjective Quality

Several authors have explored the broader connection between objective measures of image quality and overall aesthetic pictorial quality for a variety of subject matters encountered in amateur and professional photography.[125–131] The experiments to support these studies are difficult to perform, requiring extremely large numbers of observers to obtain good statistical measures of subjective quality. The task of assessing overall quality is less well defined than the task of recognizing a particular pattern correctly, as evaluated in most of the studies cited above. It would appear that no single measurement criterion has become universally accepted by individuals or organizations working in this area. Below we shall discuss a few of the key descriptors, but we do not attempt to list them all.

Many of the earlier studies tended to focus on the signal or MTF-related variable only. In one such series of studies,[125,126] S in Fig. 33 is defined as the modulation of reflectance on the output print (for square waves) divided by the modulation on the input document (approximately 0.6 for these experiments). The quality metric is defined as the spatial frequency at which this ratio falls to 0.5. This is indicated in the figure by

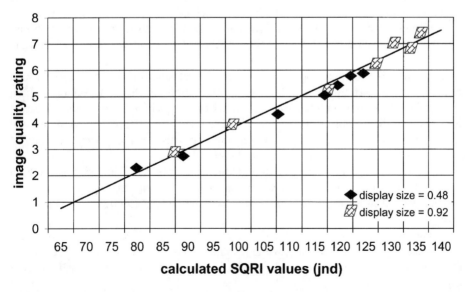

Figure 41 Linear regression between measured subjective quality and calculated SQRI values for projected slides of two different sizes, as indicated, illustrating the good fit. (From Ref. 119.)

the frequency f_b for the curve S as drawn. In these studies, a landscape without foreground was rated good if this characteristic or critical frequency was 4–5 cycles/mm (100–125 cycles/in.), but for a portrait 2–3 cycles/mm (50–75 cycles/in.) proved adequate. Viewing distance was not a controlled variable. By using modulation on the print and not simply MTF, the study has included the effects of tone reproduction as well as MTF. Granularity was also shown to have an effect, but was not explicitly taken into account in the determination of critical frequency.

Several studies have shown that the visual response curve discussed earlier can be connected with a measure of $S(f)$ to arrive at an overall quality factor. See, for example, system modulation transfer acutance (SMT acutance) by Crane[130] and an improvement by Gendron[131] known as cascaded area modulation transfer (CMT) acutance. One metric, known as the subjective quality factor (SQF),[127] defines an equivalent passband based on the visual MTF having a lower (initial) cutoff frequency at f_i and an upper (limiting) frequency of f_l. Here, f_i is chosen to be just below the peak of the visual MTF, and f_l is chosen to be four times f_i (two octaves above it). For prints that are to be viewed at normal viewing distance [i.e., about 340 mm (13.4 in.)], this range is usually chosen to be approximately 0.5–2.0 cycles/mm (13–50 cycles/in.).

The MTF of the system is integrated as follows.

$$SQF = \int_{f_x=0.5}^{2} \int_{f_y=0.5}^{2} S(f_x, f_y) \, d(\log_{10} f_x) \, d(\log_{10} f_y) \tag{31}$$

This function has been shown to have a high degree of correlation with pictorial image quality over a wide range of picture types and MTFs. It is possible that a demand function similar to that described in the MTF concepts above could be applied to further improve the performance. The SQF metric is applied to the final print as it is to be viewed and may be scaled to the imaging system, when reduction or enlargement is involved, by applying the appropriate scaling factor to the spatial frequency axis.

It should be noted that there is a significant difference between the upper band limit of this metric at 2 cycles/mm (50 cycles/in.) and the critical frequency described above in Biedermann's work for landscapes, which is in the 4–5 cycles/mm (100–125 cycles/in.) region. But there is good agreement for the portrait conclusions of the earlier work, which cites an upper critical frequency of 2–3 cycles/mm (50–75 cycles/in.). Authors of both metrics acknowledge the importance of granularity or noise without directly incorporating granularity into their algorithms. Granger[132] discusses some effects of granularity and digital structure in the context of the SQF model, but calls for more extensive study of these topics before incorporating them into the model.

It is clear that when the gray content and resolution of the digital system are high enough to be indistinguishable from an analog imaging system, then these techniques, which are general in nature, should be applicable. The quantization levels at which this equivalence occurs vary broadly. Usually 32 to 512 levels of gray suffice, depending on noise (higher noise requires fewer levels), while resolution values typically range from 100 to 1000 pixels/in. (4–40 pixels/mm), depending on noise, subject matter, and viewing distance.

Another fairly typical approach to quantifying overall subjective image quality involves measuring the important attributes of a set of images made under a range of technology variables of interest and then surveying a large number of observers, usually customers for the products using the technology. They are asked for their overall subjective reaction to each image. A statistical regression is then performed between the measured attributes and the average subjective score for each image. This is the "type D shortcut" illustrated in Fig. 2. An equation describing quality is derived using only the most important terms in the regression, that is, those that describe most of the variance. The "measures" may also be visual perceptions, that is, the "nesses," in which case the result is Engeldrum's "image quality models,"[11,34] but must include all the factors that could have any reasonable bearing on quality. Sometimes the technology variables themselves are used (type A shortcut, Fig. 2). This makes the resulting equation less general in its applicability but gives immediate answers to product questions.

Below is an example of an *image quality model* [133] selecting five visual perception attributes from a list of 10 general image quality attributes[134] to describe a series of 48 printed color images from lithography, electrophotography, inkjet, silver halide, and dye diffusion, under a wide variety of conditions. A linear regression against overall preferences of 61 observers yielded the following equation.

$$\text{Avg. Preference} = 8.8 \text{ Color Rendition} + 5.5 \text{ MicroUniformity} \\ + 4.4 \text{ Effective Resolution} + 3.5 \text{ MacroUniformity} \\ + 1.9 \text{ Gloss Uniformity} \qquad (32)$$

A plot showing the Preference vs. a fitted three-dimensional surface for the top two correlates is given in Fig. 42.

Regression equations between the physical image parameters and customer preference have been developed in many different imaging environments. An example

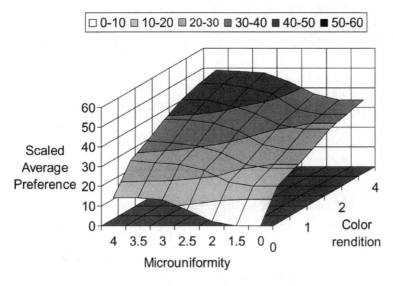

Figure 42 Illustration of the multivariate nature of a typical image quality model showing relationship between scaled image quality preference and two of several variables: color rendition and microuniformity (from Ref. 133).

exploring a wide range of images from color copiers and two specific input originals (a portrait and a map)[135] generated the following regression equation for image quality (IQ), which predicts ratings with a correlation coefficient of 0.9 17:

$$\text{Portrait IQ rating} = 0.393 \times 10^{-8} (C^* \text{ of red } 100\%)^{5.2}$$
$$+ 69.51 \times \exp(-0.125 \times \text{ graininess of cyan } 60\%)$$
$$- 0.000173 \times (H^\circ \text{ of } 1.0 \text{ Neutral Solid } - 305.0)^2$$
$$- 0.409 \times (C^* \text{ of blue } 10\% - 4.90)^2$$
$$+ 47.7 \times \exp(-0.0766 \times \text{ graininess of skin color })$$
$$- 0.0197 \times (C^* \text{ of blue } 40\% - 23.5)^2$$
$$- 0.0452 \times (C^* \text{ of Cyan } 70\% - 36.8)^2$$
$$- 15.22$$

F-values: 43.6, 21.9, 14.3, 13.5, 11.5, 7.5, 6.6

They used 799 kinds of image quality metrics, 35 observers, and a seven-point ordinal scale from very poor to very good. This was translated to a preference score. A separate but related equation for their map document showed an F-value of 64 for the graininess of 0.3 density neutral. Line widths and line densities were also found to be significant.

This work has been translated into a benchmarking activity on preference by applying these types of equations to a variety of products at various release times. They indicate trends for the quality of photography and electrophotography as shown in Fig. 43. Curve A shows the electrophotography trend line estimated with linear regression. Curve B projects the highest scoring electrophotography products using a line parallel to the trend line, A, and starting from the most recent point on the benchmark curve.

Numerous analyses of digital photography have been performed in recent years, some taking the form of customer research on preference using statistical regression against technology variables. These are examples of the type A shortcuts in Fig. 2. A study comparing perceived image quality and acceptability of photographic prints of images

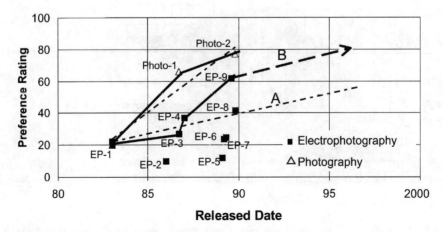

Figure 43 Example of using an image quality model to perform color image quality benchmarking for predicting industry trends (from Ref. 135).

Image Quality for Scanning

from different resolution digital capture devices[86] directly compared perceived quality by a range of individuals with varying experience in photography and computers. Photographic prints (4 × 6 inch) showing optimum tone and color rendering were used as output for viewing. Their results and others were given earlier in Fig. 20.

6.7 Information Content and Information Capacity

There are numerous articles in the imaging science literature that analyze imaging systems in terms of information capacities and describe their images as having various information contents.

Using basic statistics of noise and spread function concepts from Sec. 2.1.3 a simple description of image information is given by Eq. (33), below.[39,43,45] It defines image information, H, as

$$H = a^{-1} \log_2 \left[\frac{\text{probability of "density message" being correct}}{\text{probability of a specific density as input}} \right] \tag{33}$$

where a is the area of the smallest resolvable unit in the image (i.e., 2 × 2 pixels based on unaliased sampling from the sampling theorem) and the log factor is from the classic definition of information in any message,[136] here being messages about density (any other signal units can be used if done so consistently and they constitute a meaningful message in some context). To convert this into more useful terms let

$$H = a^{-1} \log_2 \left[\frac{p}{1/M} \right] \underset{p \to 1}{\cong} \left[\frac{\log_2 M}{a} \right] \tag{34}$$

where the numerator is set equal to p, the probability that a detected level within a set of levels is actually the correct one (i.e., the reliability), and M is the number of equally probable distinguishable levels (i.e., the quantization) from Eq. (1) in Sec. 2.1.3. Assuming a high reliability such that p approaches unity, the simplification on the right results. The standard deviation of density in Eq. (1) must be measured with a measuring tool whose aperture area is equal to a.

An approximation useful in comparing different photographic materials uses the standard deviation of density σ_a at a mean density of approximately 1 to 1.5 and Eq. (35) results:

$$H = a^{-1} \log_2 \left[\frac{L}{6\sigma_a} \right] \tag{35}$$

where k was set to ± 3 (=6), leading to $p = 0.997$ (~1); L is the density range of the imaging material.

Since the standard deviation of density is strongly dependent on the mean density level, it is more accurate and also common practice to measure the standard deviation at several average densities and segment the density scale into adjacent, empirically determined, unequal distinguishable density levels. These levels are separated by k standard deviations of density as measured for each specific level.[39,43,45] If the input scanner itself is very noisy, then the σ_a term must represent the combined effects of both input noise and scanner noise. This was covered in Sec. 4.3 (see Refs. 39, 43, and 45 for further information).

Another approach uses all of the spatial frequency based concepts developed above for the MTF and the Wiener spectrum and can incorporate the human visual system as well. It produces results in bits/area that are directly related to the task of moving electronic image data from an input scanner to an output scanner or other display. Much of the research in this area began on photographic processes, but has also been applied to electronic scanned imaging. Both are addressed here. The basic equation for the spatial frequency based information content of an image is given[137] by

$$H_i = \frac{1}{2} \int_{-\infty}^{\infty} \log_2 \left[1 + \frac{\Phi_S(f)}{\Phi_N(f)} \right] df \tag{36}$$

where H_i is the information content of the image, Φ_S is the Wiener spectrum of the signal, Φ_N is the Wiener spectrum of the system noise, and f is the spatial frequency, usually given in cycles per millimeter. This equation is in one-dimensional form for simplicity, in order to develop the basic concepts. For images, these concepts must, as usual, be extended to two dimensions. Unlike the work dealing with the photographic image, the assumption of uniform isotropic performance cannot be used to simplify the notation to radial units. For digital images the separation of the orthogonal x- and y-dimensions of the image must be preserved.

Alternative methods for calculating information capacity do not include explicit spatial frequency dependence but do explicitly handle probabilities.[39,43,45,138] They served as the basis for our discussion of quantization and Eq. (34), is rewritten as

$$H_i = N \log_2 (pM) \tag{37}$$

where N is the number of independent information storage cells per unit area. It may be set equal to the reciprocal of the smallest effective cell area of the image, e.g., a number of pixels or the spread function. Here, p is the reliability with which one can distinguish the separate messages within an information cell, and M is the number of messages per cell. M is determined by the number of statistically different gray levels that can be distinguished in the presence of system noise at the reliability p, using noise measurements made with the above cell area.

Generalizing Eqs. (34) and (1) to the "generic" units of Fig. 39, L is set equal to S_0 and σ_a is set equal to σ_s for the maximum signal and its standard deviation, respectively. We select a spread function for an *unaliased* system equal to 2 pixels by 2 pixels and translate this to frequency space using the reciprocal of the sampling frequencies f_{sx} and f_{sy} in the x- and y-directions. This gives a generalized, sampling-oriented version of Eqs. (34) and (37) as

$$H_i = \frac{f_{sx}f_{sy}}{4} \log_2 \left[\frac{S_0}{k\sigma_s} \right] \tag{38}$$

where S and σ_s are measured in the same units. k can be set to determine the reliability for a given application. Values from $2^{[45]}$ to $20^{[43]}$ have been proposed for k for different applications; 6 is suggested here, making $p = 0.997$. This assumes that σ_s is a constant

Image Quality for Scanning

(i.e., additive noise) at all signal levels. If not, then the specific functional dependence of σ_s on S must be accounted for in determining the quantity in the brackets, measuring the desired number of standard deviations of the signal at each signal level over the entire range.[39,45] While this approach predicts text quality and resolving power[138] and deals with the statistical nature of information, it does not (as noted above) permit the strong influence of spatial frequency to be handled explicitly.

Equation (36) may be expanded to illustrate the impact of the MTF on information content, giving

$$H_i = \frac{1}{2} \int_{-\infty}^{\infty} \log_2 \left[1 + \frac{K^2 \Phi_i(f) |MTF(f)|^2}{\Phi_N(f)} \right] df \qquad (39)$$

where $\Phi_i(f)$ is the Wiener spectrum of the input scene or document and $MTF(f)$ is the MTF of the imaging system (assumed linear) with all its components cascaded. At this point we need to begin making some assumptions in order to carry the argument further. The constant K in the equation is actually the gain of the imaging system. It converts the units of the input spectrum into the same units as the spectral content of the noise in the denominator. For example, a reflectance spectrum for a document may be converted into gray levels by a K factor of 256 when a reflectance of unity (white level) corresponds to the 256th level of the digitized (8-bit) signal from a particular scanner; the noise spectrum is in units of gray levels squared.

Various authors have gained further insight into the use of these general equations. Some of those investigating photographic applications have extended their analysis to allow for the effect of the visual system;[128] others attempted to apply some rigor to the terms in the equation that are appropriate for digital imaging.[139,140] Others have worked on image quality metrics for digitally derived images,[129] but some have tended to focus on the relationship to photointerpreter performance.[141]

Several of these authors have suggested that properly executed digital imagery does not appear to be greatly different from standard analog imagery in terms of subjective quality or interpretability. One almost always sees these images using some analog reconstruction process to which many analog metrics apply. It therefore seems reasonable to combine some of this work into a single equation for image information and to hypothesize that it has some direct connection with overall image quality when applied to a scanner whose output is viewed or printed by an approximately linear display system. It must also be assumed that the display system noise and MTF are not significant factors or can be incorporated into the MTF and noise spectra by a single cascading process. A generic form of such an equation is given below as Eq. (40) without the explicit functional dependencies on frequency in order to show and explain the principles that follow (expanding on the analysis in Ref. 128).

$$H_i = \frac{1}{2} \int_{-\infty}^{\infty} \int_{-\infty}^{\infty} \log_2 \left[1 + \frac{K^2 \Phi_i |MTF|^2 R_1^2}{\{1 + 12(f_x^2 + f_y^2)\}\{[\Phi_a + \Phi_n + \Phi_q]R_2^2 + \Phi_E\}} \right] df_x df_y \qquad (40)$$

Let us begin by examining the numerator. Several authors have attempted to multiply the modulation transfer function of the imaging system by a spatial frequency response function for the human visual system to arrive at an appropriate weighting for the signal part of Eq. (39). Kriss and his coworkers[128] observed that a substantial increase in the enhancement beyond the eye's peak response produced larger improvements in overall picture quality than did equivalent increases in enhancement at the peak of the eye's response. The pictures with large enhancement at the eye's peak response were "sharper," but were also judged to be too harsh. These results indicate that the human visual system does not act as a passive filter and that it may weigh the spatial frequencies beyond the peak in the eye's response function more than those at the peak.

Lacking a good model for the visual system's adaptation to higher frequencies as described above, Kriss et al. proposed the use of the reciprocal of the eye frequency response curve as a weighting function, $R_1(f)$, that could be applied to the numerator. The conventional eye response $R_2(f)$ should be applied in the denominator to account for the perception of the noise, since the eye is not assumed to enhance noise but merely to filter it. The noise term, Φ_N, in the earlier equation has been replaced by the expression in the square brackets and multiplied by $R_2^2(f)$. The reciprocal response, R_1, is set equal to 0.0 at 8 cycles/mm (200 cycles/in.) in order to limit this function.

Next let us examine the noise effects themselves. A major observation is that noise in the visual system, within one octave of the signal's frequency, tends to affect that signal. It can be shown that the sum in the first curly brackets in the denominator of Eq. (40) provides a weighting of noise frequencies appropriate to this one-octave frequency-selective model for the visual system.[128] Several authors[108,123,124,142] describe frequency-selective models of the visual system. The present construct for noise perception was first described by Stromeyer and Julesz.[143] A term for the Wiener spectrum of the noise in the visual process, Φ_E, has been added to the second factor in the denominator to account for yet one more source of noise. It is not multiplied by the frequency response of the eye, since it is generated after the frequency-dependent stage of the visual process.

The factor in square brackets in the denominator contains three terms unique to the digital imaging system.[140] These are Φ_a, the Wiener spectrum of the aliased information in the passband of interest; Φ_n, the Wiener spectrum of the noise in the electronic system, nominally considered to be fluctuations in the fast scan direction; and Φ_q the quantization noise determined by the number of bits used in the scanning process. We have thus combined in Eq. (40) important information from photographic image quality studies, including vision models and psychophysical evaluation, with scanning parameters pertinent to electronic imaging.

The study of information capacity, information content, and related measures as a perceptual correlate to image quality for digital images is an ongoing activity. By necessity it is focused on specific types of imaging applications and observer types. For example, an excellent database of images and related experiments on quality metrics was built for aerial photography as used by photointerpreters.[141]

Experiments correlating subjective quality scores with the logarithm of the basic information capacity, taking the log of H_i as defined in Eqs. (1) and (34) showed correlation of 0.87 and greater for subjective quality of pictorial images.[144] Specific MTF and quantization errors were studied. The results were normalized by the information content of the original.

By use of various new combinations of the same factors discussed above, it was possible to obtain even higher correlations. A digital quality factor was defined[144] as

$$DQF = \left[\frac{\int_0^{f_n} MTF_s(f) MTF_v(f) \, d(\log f)}{\int_0^{10} MTF_v(f) \, d(\log f)}\right] \times \log_2\left[\frac{L}{L/M + 2\sigma}\right]$$

where we retain the one-dimensional frequency description used for simplicity by the original authors and the subscripts "s" and "v" refer to the system under test and the visual process, respectively. L is the density range of the output imaging process, f_n is the Nyquist frequency, M is the number of quantization levels, and σ is the RMS granularity of the digital image using a 10×1000 μm microdensitometer slit. The first factor is related to the subjective quality factor (SQF) described in Eq. (32), and the second factor is related to the fundamental definition of image information capacity in Eq. (34). A correlation coefficient of 0.971 was obtained for these experiments, using student observers and pictures showing a portrait together with various test patterns. It must be noted, however, that information capacity or any of these information-related metrics cannot be accepted, without psychophysical verification, as a general measure of image quality when different imaging systems or circumstances are to be compared.[14,145] Since systems models are used to determine MTFs and information capacities and hence arrive at useful descriptions of technology variables, these are good examples of the type A shortcut regression models described in Fig. 2, but are restricted to the limitations of such regression shortcuts.

In conclusion, this brief overview of specific quality metrics should give the reader some perspectives on which ones may be best suited to his or her needs The variety of these metrics, and the considerable differences among them, are evidence of the inherent diversity of imaging applications and requirements. Given this diversity, together with the large and rapidly expanding range of imaging technologies, it is hardly surprising that no single universal measure of quality has been found.

7 SPECIALIZED IMAGE PROCESSING

Most scanned images either begin or end in a digital form that needs to be efficiently managed in the larger context of a computer system, often in a network with other devices. This brings other dimensions to scanned image quality, namely the need to control the size of the files and the quality of the scanned images beyond the devices themselves. Controlling the file size is the subject of image compression.[10,146–148] Compression is an image quality issue because several methods do so at the expense of image quality, with lossy compression being one example and reduced sampling vs. increased gray resolution, that is, resolution enhancement,[149] being another. Finally there is the color management challenge: finding a method to ensure that a color scanned image created by any of a number of scanners will look well when printed on any of a number of differently designed or maintained color printing devices.[5,13]

7.1 Lossy Compression

Image compression is a technology of finding efficient representation[146,147] for digital images to:

1. Reduce the size and cost of computer memory and disk drive space required for their storage;
2. Reduce the bandwidth and or time needed to send or receive images in a communication channel; and
3. Improve effective access time when reading from storage systems.

The need to improve storage is easily seen in the graphic arts business, where an

8.5 × 11 in., 600 × 600 in., 32 bit color image is approximately 10^9 (or a billion) bits/ image. Even the good quality portable amateur still cameras require 6+ megabytes (1 byte = 8 bits) per color image. Needless to say, transmitting such large files or accessing them takes tremendous amounts of time or bandwidth. Many standards groups are actively trying to create order out of the plethora of possible compression methods in order to reduce the number and types of tools needed to work in our highly interactive world of communications and networks.

There are generally two types of compression: lossless and lossy. Lossless takes advantage of better ways to encode highly redundant spatial or spectral information in the image, such as many contiguous white pixels in a text document. Results vary from compression ratios well over 100 : 1 on some text to 1.5 : 1 or less on many pictures. Group 3 and Group 4 facsimile ("fax") standards, established by the CCITT (Consultative Committee of the International Telephone and Telegraph, now ITU-T, the Telecommunication Standardization sector of the International Telecommunication Union) are perhaps the best known and apply only to binary images.[148] Other standards include JBIG,[10,150] which is especially important for black and white halftones where it achieves about 8 : 1 compression while best CCITT methods actually expands file size by almost 20% over the uncompressed version.[10,151]

All compression involves several different operations from transformation of the data to allow for efficient coding (e.g., discrete cosine transform) to the actual symbol-encoding step where many technologies have developed. The latter include Huffman,[152] LZ,[153–155] and LZW[156] encoding, which are often cited as important parts of complete compression schemes.

Lossy compression is important from an image quality perspective since it removes information contained in the original image and therefore potentially causes a reduction in image quality to gain a compression advantage. Sometimes lossy compressions are said to be "*visually* lossless" in that they only give up information about the original that they claim cannot be detected by the HVS. Simply invoking binary imaging, for example, is an excellent method of compression, which is visually lossless when scanning ordinary black text on a white substrate at high resolution. It reduces a gray image from 8 bits to one and preserves all the edge information if it is high enough in resolution while throwing away all the useless gray levels in between. It does not work well on a photograph, where the primary information is in the tones that are all lost! Most lossy compression methods are very complex, involving advanced signal processing and information theory[148] beyond the scope of this chapter.

The best-known lossy compression technique is called JPEG (after the Joint Photographic Experts Group formed under the joint direction of ISO-IEC/JTC1/SC2/ WG10 and CCITT SGVIII NIC in 1986).[10] It is aimed at still-frame, continuous tone, monochrome, and color images. In the case of JPEG, the underlying algorithm is a discrete cosine transform (DCT) of the image one 8 × 8 pixel cell at a time. It then makes use of the frequency-dependent quantization sensitivity of the eye (Fig. 10) to alter the quantization of the signal on a frequency-by-frequency basis within each cell.

Many lossy compression methods are adjustable depending on the users' needs, so that the amount of compression is proportional to the amount of loss. They can be adjusted to a visually lossless state or to some acceptable state of degradation for a given user or design intent. The JPEG technique is adjustable by programming a table of coefficients in frequency space, called a Q table, which specifies the quantization at each of several spatial frequency bands. It can also be adjusted using a scaling factor applied to the Q

Image Quality for Scanning

table. Psychometric experiments (see Sec. 8) should be employed to determine acceptable performance in making such changes, using the exact scanning and marking methods of interest. There are many other features of the JPEG approach that cannot be covered here. It has routinely been able to show an order of magnitude better compression over raw continuous tone pictures[10] with very little to no apparent visual loss of quality.

As noted earlier, compression is often aimed at improving communication of data and as such it is closely linked with file formats. In recent years a heavy focus on both the Internet and fax[151,157] has led to significant progress. JPEG and GIF have become widely used in the Internet,[157] where, in a greatly simplified view, it is seen that the former is lossy in spatial terms while the latter is lossy in color terms.

Color fax standards[151,158] have recently been developed in which the color, gray, and bitonal information is encoded into multiple layers for efficient transmission and compression. These are formally known as TIFF-FX formats and generally fall into the broad category of mixed raster content or MRC.[151]

A new standard, JPEG 2000[158a] is being developed, which, in addition to several other improvements, will also utilize wavelets as an underlying technology and includes several optional file formats called the JP family of file formats. One, the JPM file format[159] with extension .jpm is aimed at compression of compound images, those having multiple regions each with differing requirements for spatial resolution and tonality. It employs this multiple layer approach. Mixed Raster Content (MRC) formats allow the optimization of image quality, color quality via good color management, and best compression, all in one package. The base mode of MRC decomposes a mixed content image into three layers: a bitonal (binary) Mask layer, and color Foreground and Background layers. The wavelet approach in JPEG 2000 causes less objectionable artifacts than the DCT-based baseline JPEG.[159]

7.2 Nonlinear Enhancement and Restoration of Digital Images

The characteristics of a scanned image may be altered in nonlinear ways to enable its portability between output devices of different resolutions while maintaining image quality and consistency of appearance. This may also be done to improve quality by reducing sampling effects or otherwise enhancing image appearance when compared to a straightforward display or print of the bit map. These are the general goals of digital image enhancement and restoration, topics that have been covered extensively in the literature and pursued by many imaging and printer corporations. They have been summarized by Loce and Dougherty.[149] Many of the techniques fall in the domain of morphological image processing,[2,160] which treats images as collections of well-defined shapes and operates on them with other well-defined shapes. It is most often used with binary images where template matching, that is, finding an image shape that matches the filter shape and then changing some aspect of the image shape, is a good general example. Two particular examples illustrate some of the underlying concepts.

"Anti-aliasing" is a class of operations in which "jaggies" or staircases (i.e., sampling artifacts or "aliased" digital images of tilted lines) in binary images are reduced to a less objectionable visual form. In Fig. 44 the staircased image of a narrow line is analyzed by a filter programmed to find the jaggies (template matching) and then operated on, pixel by pixel, to replace certain all-white or all-black edge pixels with new pixels, each at an appropriate level of gray, in this case one of three levels. The gray pixels may be printed using conventional means of gray writing such as varying exposure on a

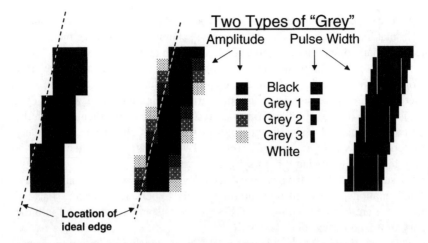

Figure 44 Anti-aliasing by the amplitude and pulse width methods. Pixels narrowed by pulse width changes are shown separated from the full pixels by a narrow white line only to illustrate where each is located.

continuous tone printing medium at the output stage. This may be thought of as amplitude modulation. A similar but often more satisfactory effect, producing sharper edges, can be achieved by using high addressability or pulse width modulation in conjunction with printing processes having an inherently sharp exposure threshold rather than continuous tone response.

Methods to evaluate the prints to determine the reduction of the appearance of the jaggies involve scanning along the edge of a line containing the effects of interest with a long microdensitometer slit whose length covers the space from the middle of the black line to the clear white surround. The resulting reflection profile is proportional to the excursions of the edge. It indicates the additional effects of the printing and measurement processes on decreasing or increasing the jaggies and can be analyzed for its visually significant components against an appropriate contrast sensitivity function.[161] Some of these components are random based on the marking process, others are periodic based on the angle of the line and the resulting frequency of the staircase effect.

Figure 45 shows an example of several practical effects of such enhancement and restoration on an italic letter "b." The upper figure is a representation of a conventional bit map of the original computer generated letter. Note the jaggies or staircase on the straight but tilted stroke at the left and a variety of undesirable effects throughout the character. Using the observation window employed by Hewlett Packard's RET (Resolution Enhancement Technology)[149,162,163] as shown on the left, roughly 200 pixel-based templates are compared to the surrounding pixels for each individual pixel in the original "b," a part of which is shown here. A decision is made regarding how large a mark, if any, should replace that pixel, based on a series of rules developed for a particular enhancement scheme, in this case the RET algorithm. The mark in this case is created by modifying the width of the pulse in the horizontal dimension as illustrated. The resulting map of full and width-modulated pixels is shown in the lower part of the letter "b." Note that some of the narrow pixels can be positioned left or right. This is called pulse width *position* modulation, PWPM.

When the individual pulses are blurred and developed by the marking process, they will tend to merge into the body of the letter, both physically and visually, to produce even

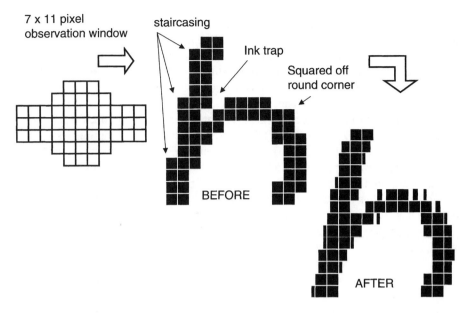

Figure 45 An example of resolution enhancement on a portion of an italic letter "b" (from Ref. 162).

smoother edges than shown as a bit map image here. There are many similar techniques patented prior to and following the above and sold by other companies such as Xerox,[164] IBM,[165] Destiny,[166] and DP-Tek,[167] now owned by Hewlett Packard, to name only a few, each with its own special features. They all create the effect on the HVS equivalent to that provided by a higher resolution print and many enhance the images in other ways as well, such as removing ink traps or sharpening the ends of tapered serifs.[149]

7.3 Color Management

Color measurement systems, as discussed earlier in Sec. 2.2.5, are the key to managing color reproduction in any situation. The advances in scanned color imaging systems that separated input and output scanning devices and inserted networks, electronic image archives, monitors and pre-printing (pre-press) software in between have made it desirable to automate the management of accurate or pleasing (not the same) color reproduction. This in turn has meant automating or at least standardizing and carefully controlling the objective measurement of the color performance of the many input, output, and image manipulation devices and a variety of methods for insuring consistency.[5,13]

The basic concept is to encode, transmit, store and manipulate images in a device-independent form, carrying along additional information to enable decoding the files at the step just before rendering to an output device, that is, just before making it device-specific. CIELAB (i.e., L*a*b*) based reference color space, above, is commonly used to relate characteristics of both of these types of devices to an objective standard. Standardized operating system software, operating with standardized tools and files accomplished this, but is beyond the scope of this chapter. See Refs 22 and 23 and the papers cited in them for more details and Ref. 16 for a practical guide to using the tools that are available at this time.

Today, a common ANSI standard target known as IT-8.7 (see Fig. 51) is manufactured by Kodak (shown), Fuji, and Agfa, each using their own photographic dyes. It is scanned by the scanner of interest into a file of red, green, and blue (RGB) pixels. It is also measured with a spectrophotometer to determine the CIE L*a*b* values for all 264 patches. Color management software then compares both results and constructs a *source profile* of the scanner color performance.

A well-known example of a color management system is the approach organized by the International Color Consortium (ICC). It has created a standard attempting to serve as a cross-platform device profile format to be used to characterize color devices. It enables the device-independent encoding and decoding primarily developed for the printing and pre-press industries, but allows for many solutions providers.

This profile, often called an "ICC profile" if it follows the Consortium's proforma, is a lookup table that is carried with all RGB files made with that scanner. It is useful for correction as long as nothing changes in the scanner performance or setup. Similarly, a destination profile is created, typically for a printer or a monitor. Here, known computer-generated patterns of color patches are displayed or printed, and measured with a spectrophotometer in L*a*b*. Again, a comparison between the known input and the output is performed by the color management software, which creates a lookup table as a *destination profile*.

The color management architecture incorporates two parts. The first part is the *profiles* as described above. They contain signal processing transforms plus other material and data concerning the transforms and the device. Profiles provide the information necessary to convert device color values to and from values expressed in a color space known as a *profile connection space* (L*a*b* in the ICC example). The second basic part is the *color management module* (CMM), which does the signal processing of the image data using the profiles.

Progress in color management and the ICC in particular have pulled together an important set of structures and guidelines.[5,13] These enable an open color management architecture that has made major improvements. Of course, gamut differences like those in Fig. 18, are not a problem that color management, *per se*, can ever solve. It is also important to note that drift in the device characteristics between profile calibrations cannot be removed. It is reported[172] that (averaging over a wide range of colors) rotogravure images in a long run show $\Delta E^*_{ab} = 3.0$ and for offset $\Delta E^*_{ab} = 5.5$, (i.e., the range for 90% of images) while they report for *input scanners* $\Delta E^*_{ab} = 0.4$. They also report that the use of color management and ICC profiles improved system results from $\Delta E^*_{ab} = 9$ down to 5, and suggest in general, with good processes, that this is inherently as good as one can achieve. Similarly, Chung and Kuo[173] found they could achieve an $\Delta E^*_{ab} = 6.5$ as the average for the best scenario in color matching experiments using ICC profiles for a graphic arts application. Control over specific colors or small color ranges can show much tighter tolerances than these. There is still a great deal of analysis and work that must be carried out to make color management more universal, easier, and more successful.[168–171]

8 PSYCHOMETRIC MEASUREMENT METHODS USED TO EVALUATE IMAGE QUALITY

8.1 Relationships Between Psychophysics, Customer Research, and Psychometric Scaling

As one attempts to develop a scanned imaging system, there are usually some image quality questions that cannot be answered by previous experience or by reference to the literature. Often this reduces to a question of determining quantitatively how "something" new *looks visually* for "some task". It is a problem because no one else has ever evaluated the "something" or never used it for "some task" or both. We give the reader at least some pointers to the basic visual scaling discipline and tools to attack his own specialized problems.

As the Image Quality Circle[11,33] and the full framework in Fig. 2 indicates, there are many places where one needs to quantify the human visual responses. Sometimes this is in the short cut
paths connecting technology variables (the "something") directly to customer quality preferences for "some tasks" through *customer research*. Sometimes, it is in creating a more thorough understanding by developing visual algorithms, which connect the physical image parameters, that is, attributes (other types of "something"), with the fundamental human perceptions of these attributes. The science of developing these latter connections is referred to as *psychophysics*. The underlying discipline for doing both engineering-oriented customer research and psychophysics is psychometric *psychometric scaling*. Hundreds of good technical papers, chapters, and whole books have been written on these subjects, but are often overlooked in imaging science and engineering for a variety of reasons. Many of the papers cited in this chapter draw on the rich resources of psychometric scaling disciplines in certain large corporations, government agencies, and universities to develop their algorithms. Engeldrum[11] has recently distilled many of the basic disciplines and compiled many of the classic references into a useful book and software toolkit for imaging systems development.

8.2 Psychometric Methods

There are many classes of psychometric evaluation methods, the selection of which depends on the nature of the imaging variable and the purpose of the evaluation. We can only describe them at a high level in this section. Figure 46 describes a framework for considering psychometric experiments, starting with two fundamental purposes, at the left, each of which breaks down into three basic approaches and six types of data.

The way in which the sample preparation is done, observer (called "respondent" in market research) quantity and selection methods, and the numbers of images shown can all be very different, depending on the purpose. In general the customer-user experiments require significantly more care in all areas, are restricted to user-like displays of relatively few images, and require several dozen to hundreds of respondents. They tend to focus on quantifying the "Customer Quality Preference" block in Fig. 2.

Visual sciences experiments on psychophysics and perception are useful for developing the image quality models and especially the visual algorithms of Fig. 2 and the comparisons between the HVS and measurements indicated in Fig. 1. Here smaller numbers of observers, from a few to several dozen, are often deemed adequate. These observers are often experts or technical personnel and can be told to overlook certain defects in samples and concentrate on the visual characteristic of interest. Such observers can be asked to try more fatiguing experiments. These are often broken into several visits to the laboratory, something not possible with customer research.

In general, experiments with good statistical design should be used, in which a targeted confidence level is established. It is common practice in many customer and general experiments to seek 95% confidence intervals (any basic statistics book[174] will provide equations and tables to enable this, provided the scaling method is properly classified as shown below). This requires estimating the size of the standard deviation

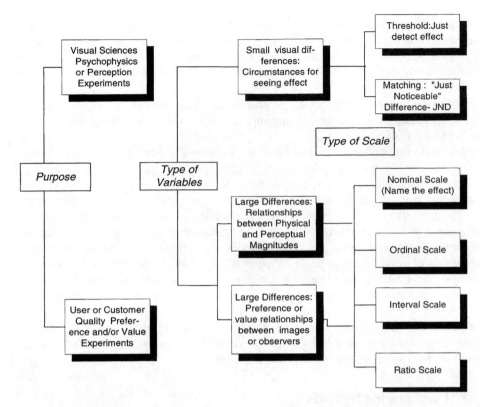

Figure 46 Psychometric experiments for diverse purposes, grouped in two classes here, can be further classified into three types of variables, which in turn lead to a few basic but significantly different types of scales.

between observers and using it along with the confidence interval equation to determine the number of independent observations that translate to the number of observers. The experimenters in visual sciences can use fewer observers than a customer researcher because the visual sciences use variables and trained observers that have much better agreement, that is, smaller standard deviations. Also, these experimenters may require less statistical confidence because they are often more willing to use other technical judgment factors such as models and inferences from other work.

For either purpose, the decisions regarding basic approach must be determined by looking at the types of variables and the types of scales to be built. If the goal is to determine when some small signal or defect (such as a faint streak) is just visually detected, *threshold* scales are developed. They show the probability of detection compared to the physical attribute(s) of the image samples or the observation variables. One may wish to compare readily visible signals, such as images of well-resolved lines, trying to distinguish when one is just visually darker than another. This involves determining the probability, in *matching* experiments, of what levels of a variable(s) cause two images to be seen as *just noticeably different (JND)*, that is, just do not match each other. Dvorak and Hamerly[175] and Hamerly[176] give examples of JND scaling for text and solid area image qualities.

Image Quality for Scanning

These experiments often explore fundamental mechanisms of vision and can draw on a relatively small number of observers in well-controlled experimental situations using electronic displays with side-by-side image comparisons. The temptation to substitute an easily controlled electronic display experiment for one in which the imaging media is identical to the actual images of interest (e.g., photographic transparencies viewed by projection, or xerographic prints viewed in office light) must be carefully weighed in each situation. Various forms of image noise, display factors affecting human vision (especially adaptation), as well as visual and psychological reference cues picked up from the surround, are often important enough to outweigh the ease of electronic display methods.

When the magnitude of visible variables is large and the goal is to compare quality attributes over a large range, as in the bottom two "variables" boxes in Fig. 46, then a decision about the mathematical nature of the desired scale and the general nature and difficulty of the experimental procedure becomes important. The four basic types of scales[177] shown here were developed by Stevens[177,178] and are shown in increasing order of "mathematical power" in Figure 46 and as row headings in Table 7.

There is an abundance of literature on the theory and application of scaling methods,[11] some of which are indicated in the table as column headings. Additional general references include Refs [175–190]. Below is a very brief summary of the methods listed in Table 7 to assist the reader in beginning to sort through these choices. Here we assume the samples are "images," but they could just as well be patches of colored chips, displays on a monitor, pages of text, or any other sensory stimulus.

8.3 Scaling Techniques

8.3.1 Identification

In this simple scaling method, observers group images by identifying names for some attributes and collecting images with those attributes. The resulting nominal scales are useful in organizing collections of images into manageable categories.

8.3.2 Rank Order

Observers arrange a set of images according to decreasing or increasing amount of the perceived attribute.[11,179,181,185] A median score for the group is frequently used to select the rank for each sample. Agreement between observers can be tested to understand the nature of the data by calculating the coefficient of concordance or the rank order coefficient.[188]

8.3.3 Category

Observers simply separate the images into various categories of the attribute of interest, often by sorting into labeled piles. This is useful for a large number of images, many of which are fairly close in attributes, so that there are some differences of opinion over observers or over time as to which category is selected. Interval scales can be obtained if the samples can be assumed to be normally distributed on the perceived attribute.[190]

8.3.4 Graphical Rating

The observers score the magnitude of the image attribute of interest by placing an indicator on a short line scale that has defined endpoints for that attribute. The mean of the positions on the scale for all observers is used to get a score for each image.

Table 7 Types of Psychometric Scales and Scaling Techniques Used to Create Them

Scale	Properties	Some popular scaling techniques									
		Identify by name	Rank order	Category	Graphical rating	Paired comparison	Partition scaling	Hybrid	Magnitude estimation	Ratio estimation	Semantic differential	Likert
Nominal	Names of categories/classes	♦										
Ordinal	Ordered along variables, determines "greater than" or "less than", gives arbitrary distances on variable scale		♦	♦		♦		♦			♦	♦
Interval	Ordinal scale + magnitude of differences quantified. $y = ax + b$ (equality of intervals, any linear transformation is OK, mean, standard deviation, coefficient of correlation are valid)			♦	♦	▽	♦	▽			▽	

(continued)

Table 7 Continued

Table 7 Continued

		Some popular scaling techniques								
Scale	Properties	Identify by name	Rank order	Category	Graphical rating	Paired comparison	Partition scaling	Hybrid	Magnitude estimation	Ratio estimation	Semantic differential	Likert
Ratio	Interval scale + "none" of attribute is assigned 0 response $y = ax$ (equality of ratios, interval operation and coefficient of variation are valid)					▽		▽	◆	◆		
Notes		$N =$ large		$N =$ large		$N < 9$		Many types See text	$N =$ large	Complex		Attitude surveys
References	Examples using the scaling method indicated		11, 179, 181, 185	11, 190	11	11, 182, 188, 191	3		11, 178, 180, 185	3, 186	181	181

◆, common method; ▽, less common and/or more difficult analysis process to obtain this type of scale.

8.3.5 Paired Comparison

All images are presented to all observers in all possible pairwise combinations, usually one pair at a time, sometimes with a reference. The observer selects one of the pair as having more of the attribute of interest. If there are N different images then there are $N(N-1)/2$ pairs The proportion of observers for which each particular image is selected over each other image is arrayed in a matrix. The average score for each image (i.e., any column in the matrix) is then computed to determine an ordinal scale.[11,182,188,191] If it is assumed that the perceived attributes are normally distributed, then, as with the category method, an interval scale can be determined. This is done using Thurstone's Law of Comparative Judgement[190] in which six types of conditions for standard deviations describing the datasets are used to construct tables of Z-Deviates,[11,188] from which interval scales are directly obtained.

8.3.6 Partition Scaling

The observer is given two samples, say S1 and S9, and asked to pick a third sample from the set, whose magnitude of the appearance variable under test is halfway between the two samples; call it S5 in this case. Next he finds a sample halfway between S1 and S5, call it S3, then he finds one between S5 and S9, calling it S7, and so on, until he has built a complete interval scale using as many samples and as fine a scale as desired.[3]

8.3.7 Magnitude Estimation

The observer is asked to directly score each sample for the magnitude of the attribute of interest.[11,178,180,185,186] Often, the observer is given a reference image at the beginning of his scoring process, called an anchor, whose attribute of interest is identified with a moderately high, easy to remember score, such as 100. His scores are based on the reference and he his coached in various ways to use values that reflect ratios. This process implies that a zero attribute gets a zero response and hence generates a ratio scale. However actual observations sometimes are more in line with an interval scale, and this needs to be checked after the test.

8.3.8 Ratio Estimation

This test may be done by selecting samples that bear specific ratios to a reference image. The experimenter does *not* assign a value to the reference. Alternatively, the observers may be shown two or more specific images at a time and asked to state the apparent ratios between them for the attribute of interest.[3]

8.3.9 Semantic Differential

Typically used for customer research.[181] The image attributes of interest are selected and a set of bipolar adjectives is developed for the attributes. For example, if the attribute class were tone reproduction, the adjectives could be such pairs as darker–lighter, high contrast–low contrast, good shadow detail–poor shadow detail. Each image in the experiment is then rated on a several point scale between each of the pairs. Each scale is treated as an interval scale and the respondents' scores for each image and each adjective pair are averaged. A profile is then displayed.

8.3.10 Likert Method

Typically used for customer research and attitude surveys. A series of statements about the image quality attributes of a set of images is provided (e.g., "The overall tones are perfect in this images," "the details in the dark parts of this image are very clear"). The respondents are then asked to rate each statement on the basis of the strength of their personal feelings about it: strongly agree $(+2)$; agree $(+1)$; indifferent (0); disagree (-1); strongly disagree (-2). Note signs on numbers reverse for negative statements. The statements used in the survey are often selected from a larger list of customer statements. A previous set of judges may be was used to determine those statements that produce the greatest agreement in terms of scores assigned to this set of images.

8.3.11 Hybrids

There are many approaches that combine the better features of these different methods to enable handling different experimental constraints and obtaining more accurate or more precise results. A few are noted here:

1. *Paired Comparison for Ratio*: Paired comparisons reduced to an interval scale that is fairly precise and transformed to accommodate a separate ratio technique (accurate but less precise) to set a zero. This gives a highly precise ratio scale.[188]
2. *Paired Comparison Plus Category*: The quality of each paired comparison is evaluated by the observer, using something like a Likert scale below. A seven-level scale from strongly prefer "left" (e.g., $+3$) to strongly prefer "right" (e.g., -3) is used.[191]
3. *Paired Comparison Plus Distance* using distance (e.g., linear scale on a piece of paper) to rate the magnitude of the difference between each pair, giving the same information as the graphical rating methods discussed earlier, but with the added precision of paired comparison.
4. *Likert and Special Categories*: A variety of nine-point symmetrical (about a center point) word scales can provide categories of preferences that are thought to be of equal intervals. One scale attributed to Bartleson[185] goes from: Least imaginable "...ness" → very little "...ness" → mild "...ness" → moderate "...ness" → average "...ness" → moderate high "...ness" → high "...ness" → very high "...ness" → highest imaginable "...ness." Another similar scale is 1 = Bad, 2 = Poor, 3 = Fair, 4 = Good, 5 = Excellent. Many other such scales are found in the literature.

8.4 Practical Experimental Matters Including Statistics

Each of these techniques has been used in many imaging studies, each with special mathematical and procedural variations well beyond the scope of this chapter. A short list of common procedural concerns is given in Table 8 (from literature[3,9,13,22] plus a few from the authors' experience).

The * items represent a dozen practical factors that must *always* be considered in designing nearly any major experiment on image quality or attributes of images.

The statistical significance of the results are often overlooked but cannot be stressed enough. For an interval scaling experiment that samples a continuous variable like darkness, standard deviations and means and subsequent confidence intervals on the

Table 8 Factors that Should be Considered in Designing Nearly Any Major Experiment on Image Quality or on Attributes of Images

Most important	Important
Complexity of observer task*	State of adaptation
Duration of observation sessions*	Background conditions
Illumination level*	Cognitive factors (many)
Image content*	Context
Instructions*	Control and history of eye movements
Not leading the observer in preference experiments*	Controls
Number of images*	Feedback (positive and negative effects)
Number of observers*	Illumination color
Observer experience*	Illumination geometry
Rewards*	Number of observation sessions
Sample mounting/presentation/ identification methods*	Observer acuity
Statistical significance of results*	Observer age
	Observer motivation
	Range effects
	Regression effects
	Repetition rate
	Screening for color vision deficiencies
	Surround conditions
	Unwanted learning during the experiment

*See text.

responses can be calculated in straightforward ways to determine if the appearances of two samples are statistically different or to determine the quality of a curve fit. (See any statistics book on the confidence interval for two means given an estimate of the standard deviations for each sample's score, or to determine the confidence for a regression.)

In detection experiments it is often desired to know if two scanned images, which gave two different percentages of observers who saw a defect or an attribute, are significantly different from each other (market researchers call such experiments sampling for attributes). This involves computing confidence intervals for proportions and therefore estimating standard errors for proportions, a procedure less commonly encountered in engineering. If $p =$ the fraction of observers detecting an attribute, $q =$ the fraction not detecting an attribute (note $p + q = 1.0$) and $n =$ number of observers, assuming n is a very small fraction of the population being sampled, then the standard error for proportions is

$$S_p = [(p \times q)/n]^{0.5} \tag{41}$$

and the confidence interval around p is

$$CI = Z \times S_p \tag{42}$$

Image Quality for Scanning 243

where, for example, $Z = 1.96$ for 95% confidence and 1.28 for 80%. A few cases are illustrated in Table 9 to give the reader perspective on the precision of such experiments and the number of observers required. The first column shows the value of p, the fraction of observers finding the attribute of interest. The second column gives the confidence desired in % where 95 is common in many experiments, and 80 is about the lowest confidence cited in many texts and statistical tables. The numbers reported in the table are the deviations about the fraction in column one that constitute the confidence interval for the population of all observers that would detect the attribute. The unbolded italic numbers correspond to values of p that cannot be realized by straightforward means for an observer population as small as indicated (e.g., a "p" value of 0.99 could not be observed with only four people – it would take 100!). As an example, for a sample with an attribute that was seen 90% of the time by a sample of 20 observers, one can be 80% confident that 82 to 98% (0.90 ± 0.08) of all observers would see this attribute. One would also be 95% confident that between 77 and 100% (numerically 103%, which here is equivalent to 100%) would see it.

9 REFERENCE DATA AND CHARTS

The following pages are a collection of charts, graphs, nomograms and reference tables, which, along with several earlier ones, the authors find useful in applying first-order analyses to many image quality engineering problems. Needless to say, a small library of computer tools covering the same material would provide a useful package. In addition to those in this section, there are a few graphs, charts, and tables of value to engineering projects included in the text where their tutorial value was considered more important.

Table 9 Confidence Intervals Around p for Attribute Data from Statistics of Proportions Where $p =$ fraction favoring one of two choices, $n =$ number of respondents/observers

p	% Confidence	$n = 4$	$n = 8$	$n = 20$	$n = 100$	$n = 500$
0.99	95	*0.10*	*0.07*	*0.04*	**0.02**	**0.01**
	80	*0.06*	*0.05*	*0.03*	**0.01**	**0.005**
0.95	95	*0.21*	*0.15*	**0.10**	**0.04**	**0.02**
	80	*0.13*	*0.09*	**0.06**	**0.03**	**0.01**
0.90	95	*0.29*	**0.21**	**0.13**	**0.06**	**0.03**
	80	*0.17*	**0.12**	**0.08**	**0.04**	**0.02**
0.80	95	**0.39**	**0.28**	**0.18**	**0.08**	**0.03**
	80	**0.23**	**0.17**	**0.11**	**0.05**	**0.02**
0.60	95	**0.48**	**0.34**	**0.22**	**0.10**	**0.04**
	80	**0.31**	**0.21**	**0.13**	**0.06**	**0.02**
0.50	95	**0.49**	**0.35**	**0.22**	**0.10**	**0.04**
	80	**0.32**	**0.23**	**0.14**	**0.06**	**0.03**

Statistical uncertainties in experimental results for proportion data (e.g., percentages of "yes" or "no" answers). Table entries gives 80 and 95% confidence as one-sided confidence intervals, i.e., positive or negative deviation from the p value in column 1, at a few percentages of positive responses "p" (row headings) and a few numbers of respondents "n" (i.e., sizes of groups interviewed) as column headings. Italic unbolded entries are for p values that can not be realized or closely approximated with the associated n values.

These include Fig. 17 on CIE standard observer color matching function, Fig. 20 on scan frequency effects, Fig. 22 on nonuniformity guidelines, Figs 27 and 28 on MTF, and, finally, Figs 36 and 37 on edge noise calculations. The tables include Table 1 on halftone calculations, Table 3, which serves as a directory to Figs 52 to 56 in this section, and Table 9, giving confidence intervals for proportions.

In this section additional graphs on basic colorimetry are provided as Fig. 47 for a more precise x, y chromaticity diagram with a dominant wavelength example and some standard

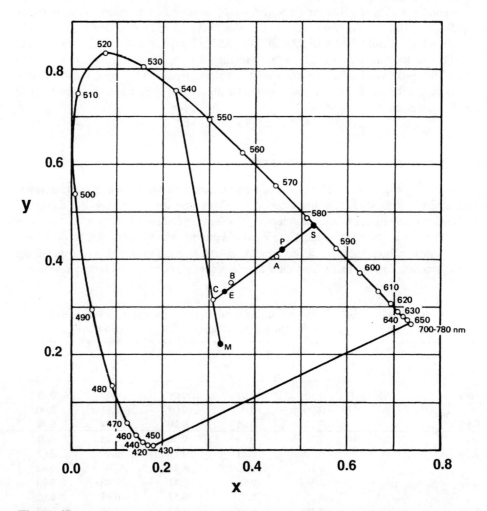

Figure 47 Dominant wavelength and purity plotted on the CIE x,y chromaticity diagram. The dominant wavelength for point P under illuminant C is found by drawing a straight line from the illuminant C point through P to the spectrum locus, where it intersects at 582 nm, the dominant wavelength. Excitation purity is the percentage defined by CP/CS, the percentage the distance from illuminant C to P is of the total distance from illuminant C to spectrum locus. Standard illuminants A, B, and E are also shown. See Fig. 49 for the relative spectral power distributions of A, B, and C. E has equal amounts of radiation in equal intervals of wavelength throughout the spectrum (see Ref. 67, R. Hunter, R. Harold, *The Measurement of Appearance*, 1987; reproduced with permission of John Wiley & Sons, Inc.).

Image Quality for Scanning 245

light sources, Fig. 48 for a uniform u' v' chromaticity diagram with Planckian radiator and some of the same standard light sources, Fig. 49 for spectral characteristic of four standard light sources, Fig. 50 for a conversion tool to enable transfers between points in x, y space and u', v' spaces showing a sample conversion. Figures 51 and 52 show, through annotations, the important structures in useful industry standard test patterns, two for monochrome in Fig. 52 and one for color in Fig. 51. Next are some useful MTF equations and their corresponding graphs (plotted in log–log form for easy graphical cascading).

Figure 53 is the MTF of two uniform, sharply bounded spread functions. The MTF of a uniform disc point spread function is defined as

$$T(N) = \frac{2J_1(Z)}{Z} \tag{43}$$

where $Z = \pi DN$, $N = cycles/mm$, and $D = diameter\ of\ disk\ in\ mm$.

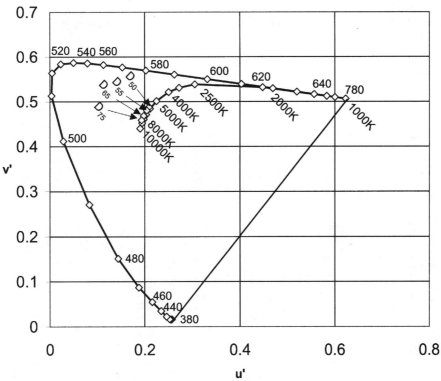

Figure 48 An alternative (to Figs. 18 and 47) and very commonly encountered form of chromaticity diagram, the u', v' diagram. This representation is also commonly chosen to show the loci of the chromaticities of Planckian (black body) radiators, noted here by their Kelvin temperature and the CIE - D Illuminants, that is, the daylight locus.

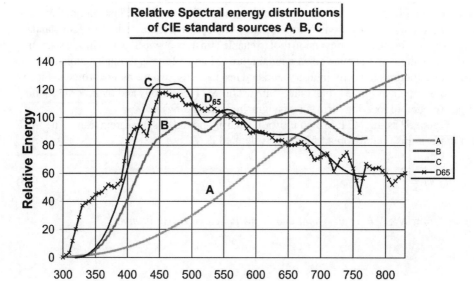

Figure 49 Wavelength is in nm. Standard illuminants A, B, C, and D_{65} showing relative spectral energy distribution.

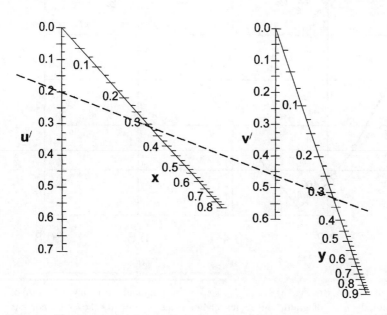

Figure 50 Nomogram for transferring data from the CIE x, y diagram to the CIE u', v' diagram and vice versa. A straight-edge placed across the four scales gives the values of u' and v' corresponding to those of x and y and vice versa (from Ref. 65).

Image Quality for Scanning

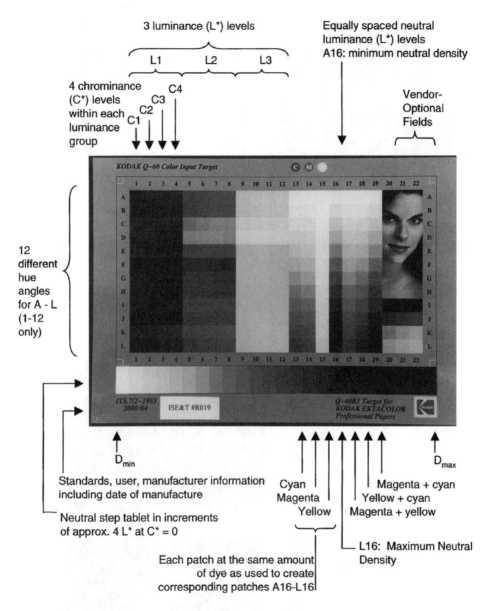

Figure 51 Layout of the IT8.7/1 (transmissive) and IT8.7/2 (reflective) scanner characterization targets. Details of colors are described in Table 5-1, 5-2, and 5-3 of Ref. 16, or ISO IT8.7/1 and 2-1993 (from Ref. 16). (*Note*: Do not attempt to use this reproduction as a test pattern.)

The MTF of a *uniform slit* or *uniform image motion* is defined as

$$T_{\text{slit}}(N) = \frac{\sin \pi DN}{\pi DN} \tag{44}$$

where D = width of slit in mm (or width of rectangular aperture or length of motion during image time) and N = cycles/mm.

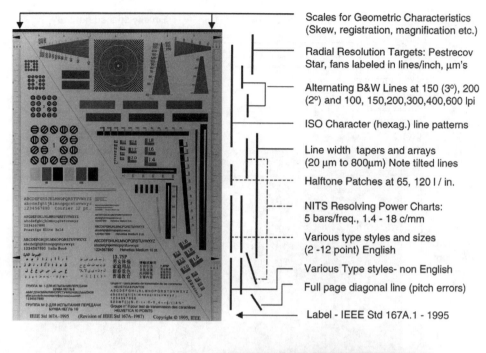

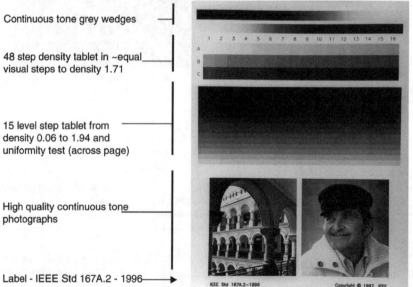

Figure 52 Images of two IEEE Standard Facsimile Test Charts which contain many elements valuable in assessing performance of scanning systems; (a) (top) IEEE Std. 167A.1-1995 – Bi-Level (black and white) chart, (b) (lower) IEEE Std. 167A.2 – 1996, High Contrast (gray scale) chart printed on glossy photographic paper. To identify what test pattern element each annotation refers to, project the relative vertical position of the bar in the specific annotation horizontally across the image of the test pattern. The bars are arranged from left to right in sequence. A full explanation of each is available on the IEEE web site which, at the time of this writing, was http://standard.ieee.org/catalog/167A.1-1995.htm See Fig. 31 for other resolving power targets and Table 10 for pointers to other standard test patterns. (*Note*: Do not attempt to use these reproductions as test patterns.)

Image Quality for Scanning

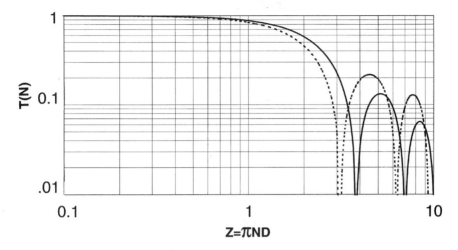

Figure 53 MTF of a uniform disk (solid line) and slit (dashed line) spread functions where N = frequency in cycles/mm and D = diameter of uniform disk, width of slit or rectangular aperture, or length of motion.

Figure 54 is the MTF of a Gaussian spread function $S(r)$

$$T(N) = e^{-a^2 N^2} \tag{45}$$

where $a = \pi/c$, and c = width of Gaussian spread function $S(r)$ of the form

$$S(r) = 2c^2 e^{-c^2 r^2} = 2c^2 e^{-c^2(x^2+y^2)} \tag{46}$$

where r = radius such that $r^2 = x^2 + y^2$; all are in mm^2.

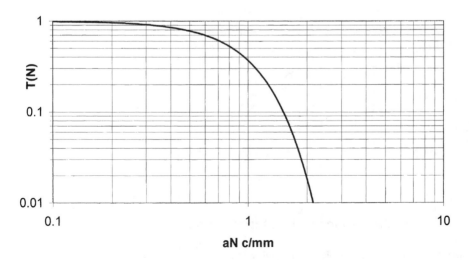

Figure 54 MTF for imaging system with gaussian spread function.

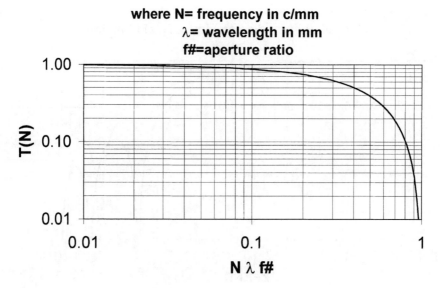

Figure 55 MTF of a diffraction-limited lens where $N =$ frequency in cycles/mm, $\lambda =$ wavelength in mm, and $f\# =$ aperture ratio.

Figure 55 is the MTF of a diffraction-limited lens, where

$$T(N) = \frac{2}{\pi}[\cos^{-1}\gamma - \gamma\sqrt{1-\gamma^2}] \tag{47}$$

where $\gamma = N\lambda f$ (object at ∞), $N =$ cycles/mm, $\lambda =$ wavelength of light in mm, and $f =$ aperture ratio $=$ [focal length]/[aperture diameter]

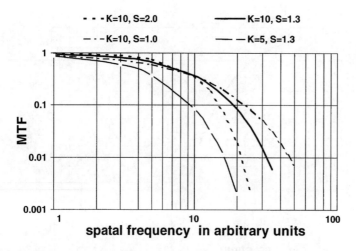

Figure 56 Examples of double exponential MTF.

Image Quality for Scanning

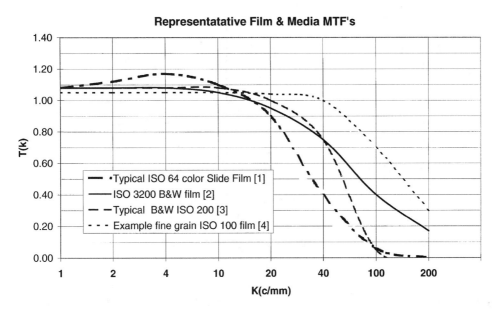

Figure 57 Data on four representative modern films.

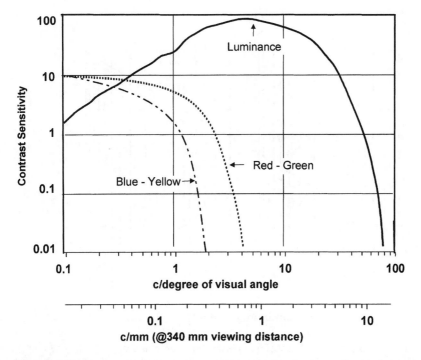

Figure 58 Typical visual spatial contrast sensitivity functions for luminance and indicated chromatic contrasts at constant luminance (From Ref. 3.)

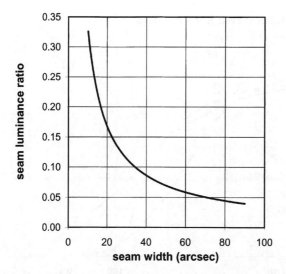

Figure 59 Line luminance visibility threshold as a function of line width for a black or white line on the opposite background derived from seam visibility for CRT displays. (From Ref. 122.).

Figure 56 is the double exponential MTF[193] that has been found (personal communication from Joe Kirkenaer, 1984) to fit data for many practical optical systems. Shown here are four specific combinations of the adjustable parameters S and K

$$T(N) = e^{-(N/K)^S} \tag{48}$$

where N is spatial frequency in arbitrary units.

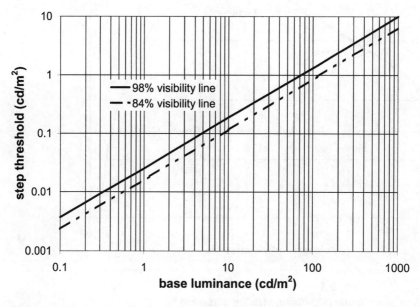

Figure 60 Thresholds for the visibility of a luminance difference at a step edge in a 17 by 5.25 degree CRT display where 84% detection = $dL_{85} = 0.01667 L^{0.8502}$. (From Ref. 122.)

Table 10 Standards of Interest in Scanned Imaging

Principal Standards Group	Subgroup	Sampling of Areas of Work or Example Standards	Website(s) and/or Address(es)
ANSI – American National Standards Institute (those assigned to NPES)	CGATS: Committee for Graphic Arts Technologies Standards	CGATS.4-1993 Graphic Technology – Graphic Arts Reflection Densitometry Measurements – Terminology, Equations, Image Elements and Procedures	www.npes.org/standards/cgats.html NPES Assoc. for Suppliers of Printing and Publishing Technologies, 1899 Preston White Dr., Reston, VA 22091
	IT8: committee on digital data exchange and color definition (assigned to CGATS in '94)	IT8.1 Exchange of Color Picture Data IT8.7/2-1993 Color Reflection Target for Input Scanner Calibration (also ISO 12641)	American National Standards Inst., 11 W. 42nd St. New York, NY 10036
Graphic Communications Association	GRACoL	General Requirements for Applications in Commercial Offset Lithography	www.gracol.org/index.html IDEAlliance, 100 Dangerfield Rd., Alexandria, VA
ICC-International Color Consortium	NA	ICC Profile Specification, ICC.1:2003-09 ver 4.1.0	www.color.org see NPES above
ISO/IEC	JTC 1/SC 28 (office equipment)	ISO 13660-2001 image quality measurement for hard copy output: large area density attributes and character and line attributes. Includes bitmaps for compliance testing	www.iso.ch www.iec.ch ISO Sectretariat International Organization for Standardization, la Rue de Varembé, Case postale 56, CH-1211 Geneva 20, Switzerland
	JTC 1/SC 29. (coding of multimedia information)	Coding of audio, picture, multimedia and hypermedia information includes bilevel and limited bits-per-pixel still pictures	
	ISO-TC42-WG18 (electronic still picture imaging) see Sec. 4.2	Photography-final draft 16067-2: Electronic scanners for photographic images – Spatial Resolution Measurements – Part 1: Scanners for reflective media;	

(continued)

Table 10 Continued

Principal Standards Group	Subgroup	Sampling of Areas of Work or Example Standards	Website(s) and/or Address(es)
NCITS (International Committee for Information Technology Standards); formerly NCITS '97–01 X3 '61–'96	W1 (office equipment)	Includes copiers, multifunction, fax machines, page printers, scanners and other office equipment.[79] Collaborates. w JTC1/SC 28 – for example on ISO 13660, above.	http://ncits.org/ incits/standards/ htm INCITS Secretariat C/O Information Technology Industry Council, 1250 Eye St. NW 200, Washington, DC 20005
ITU-T (International Telecommunication Union-Telecommunication Standardization Sector)	Joint Photographic Expert Group	ITU-T Rec. T.800 and ISO/ IEC 15444, Information Technology – Digital compression and coding of continuous tone still images (JPEG 2000) [see Ref.: 82, 158, 158a, 169 pp. 248–267]	www.itu.int/ITU-T/ International Telecommunications Union (ITU), Place des Nations, 1211 Geneva 20, Switzerland, ATTN. ITU-T
	Joint BiLevel Working Group	JBIG2 ITU-T Rec.T.88	
IEEE	Transmission Systems Committee of IEEE Communications Society	Facsimile imaging, which is of value to many general scanning areas of interest. See Fig. 52 for example test patterns	www.ieee.com The Institute of Electrical and Electronics Engineers, Inc., 345 East 47th St., New York, NY 10017-2394

Figure 57 presents data on four representative modem films, plotted here to provide perspective on the range of practical photographic characteristics. They are shown here to set scanning performance in perspective. These are not intended to be performance specifications of specific films.

Lastly we finish the reference curves with visual performance relationships. Figure 58 Illustrates recently developed visual contrast sensitivity curves (related to MTF of linear systems) including color components of vision, after Fairchild,[3] drawn with scales relating to the earlier published visual frequency response characteristics shown in Fig. 28. Figure 59 shows the line luminance visibility threshold as a function of line width, originally described as display "seam visibility" from display experiments after Alphonse and Lubin.[122] Figure 60 shows the edge contrast threshold visibility from display experiments of Lubin and Pica.[122]

The closing reference, Table 10, is a chart showing a *very* sparse cross-section of the standards that intercept the digital and scanning image quality technical world. These enable an engineer to get some orientation and pointers to important standards organizations.

ACKNOWLEDGMENTS

Memorial Remarks

This chapter is dedicated to the memory of Dr. John C. Urbach who died in his home, in Portola Valley, California, in February 2002, after several months of illness. He was a brilliant, dedicated, and prolific contributor to the field of optics and scanning. His outstanding contributions to Xerox research and the careers and ideas of many he worked with, both at Xerox and in the general technical community, are widely regarded with the highest esteem. This chapter would not have been completed without his efforts, as he continued to help in its editing, even during his last days. We will all miss his learned advice, special humor, and profound insights.

Contributions

We are indebted to Cherie Wright, John Moore, David Lieberman, and others on the staff of the Imaging Sciences Engineering and Technology Center of the Strategic Programs Development Unit at Xerox Corporation for their helpful participation in the preparation of various parts of this chapter, and to Xerox Corporation for the use of their resources. All illustrations, except as noted, are original drawings created for purposes of this chapter. However, those inspired by another author's way of illustrating a complex topic or providing a collection of useful data, reference his or her contribution with the note "(From Ref. X)". We are grateful for these authors' ideas or data, which we could build on here. We also wish to thank our reviewers Martin Banton, Guarav Sharma, Robert Loce, and Keith Knox for their time and many valuable suggestions, and our wives Jane Lehmbeck and Mary Urbach for their support and encouragement.

REFERENCES

1. Mertz, P.; Gray, F.A. A theory of scanning and its relation to the characteristics of the transmitted signal in telephotography and television. Bell System Tech. J. 1934, *13*, 464–515.
2. Dougherty, E.R. *Digital Image Processing Methods*; Marcel Dekker: New York, 1994.
3. Fairchild, M.D. *Color Appearance Models*; Addison-Wesley: Reading, MA, 1998.
4. Eschbach, R.; Braun, K. Eds. *Recent Progress in Color Science*; Society for Imaging Science & Technology: Springfield, VA, 1997.
5. Sharma, G. Ed. *Digital Color Imaging Handbook*; CRC Press: Boca Raton, Fl, 2003.
6. Eschbach, R. Ed. *Recent Progress in Digital Halftoning I and II*; Society for Imaging Science & Technology: Springfield, VA, 1994, 1999.
7. Kang, H. *Color Technology for Electronic Imaging Devices*; SPIE: Bellingham, WA, 1997.
8. Dougherty, E.R. Ed. *Electronic Imaging Technology*; SPIE: Bellingham, WA, 1999.
9. MacAdam, D.L. Ed. *Selected Papers on Colorimetry – Fundamentals, MS 77*; SPIE: Bellingham, WA, 1993.
10. Pennebaker, W.; Mitchell, J. *JPEG Still Image Data Compression Standard*; VanNostrand Reinhold: New York, 1993.
11. Engeldrum, P.G. *Psychometric Scaling: A Toolkit for Imaging Systems Development*; Imcotek Press: Winchester, MA, 2000.
12. Vollmerhausen, R.H.; Driggers, R.G. *Analysis of Sampled Imaging Systems Vol. TT39*; SPIE Press: Bellingham, WA, 2000.
13. Giorgianni, E.J.; Madden, T.E. *Digital Color Management Encoding Solutions*; Addison-Wesley: Reading, MA, 1998.
14. Watson, A.B. Ed. *Digital Images & Human Vision*; MIT Press: Cambridge, MA, 1993.

15. Cost, F. *Pocket Guide to Digital Printing*; Delmar Publishers: Albany, NY, 1997.
16. Adams, R.M.; Weisberg, J.B. *The GATF Practical Guide to Color Management*; GATF Press, Graphic Arts Technical Foundation: Pittsburgh, PA, 2000.
17. Sharma, G.; Wang, S.; Sidavanahalli, D.; Knox, K. The impact of UCR on scanner calibration. Proceedings of IS&T Image Processing, Image Quality, Image Capture, Systems Conference, Portland, OR, IS&T, 1998; 121–124.
18. Knox, K.T. Integrating cavity effect in scanners. Proceedings IS&T/OSA Optics and Imaging in the Information Age, Rochester, NY, 1996; 156–158.
19. Sharma, G.; Knox, K.T. Influence of resolution on scanner noise perceptibility. Proceedings of IS&T 54th Annual and Image Processing, Image Quality, Image Capture, Systems Conference, Montreal, Quebec, Canada, 2001; 137–141.
20. Loce, R.; Roetling, P.; Lin, Y. Digital halftoning for display and printing of electronic images. In *Electronic Imaging Technology*; Dougherty, E.R., Ed.; SPIE: Bellingham, WA, 1999.
21. Lieberman, D.J.; Allebach, J.P. On the relation between DBS and void and cluster. Proceedings of IS&T's NIP 14: International Conference on Digital Printing Technologies, Toronto, Ontario, Canada, 1998; 290–293.
22. Sharma, G.; Trussell, H.J. Digital color imaging. IEEE Trans. on Image Proc. 1997, *6*, 901–932.
23. Sharma, G.; Vrhel, M.; Trussell, H.J. Color imaging for multimedia. Proc. IEEE 1998, *86*, 1088–1108.
24. Jin, E.W.; Feng, X.F.; Newell, J. The development of a color visual difference model (CVDM). Proceedings of IS&T Image Processing, Image Quality, Image Capture, Systems Conference, Portland, OR, 1998; 154–158.
25. Sharma, G.; Trussell, H.J. Figures of merit for color scanners. IEEE Trans on Image Processing 1997; *6*, 990–1001.
26. Shaw, R. Quantum efficiency considerations in the comparison of analog and digital photography. Proceedings of IS&T Image Processing, Image Quality, Image Capture, Systems Conference, Portland, OR, 1998; 165–168.
27. Loce, R.; Lama, W.; Maltz, M. Vibration/banding. In *Electronic Imaging Technology*; Dougherty, E.R. Ed.; SPIE: Bellingham, WA, 1999.
28. Dalal, E.N.; Rasmussen, D.R.; Nakaya, F.; Crean, P.; Sato, M. Evaluating the overall image quality of hardcopy output. Proceedings of IS&T Image Processing, Image Quality, Image Capture, Systems Conference, Portland, OR, 1998; 169–173.
29. Rasmussen, D.R.; Crean, P.; Nakaya, F.; Sato, M.; Dalal, E.N. Image quality metrics: applications and requirements. Proceedings of IS&T Image Processing, Image Quality, Image Capture, Systems Conference, Portland, OR, 1998; 174–178.
30. Loce, R.; Dougherty, E. Enhancement of digital documents. In *Electronic Imaging Technology*; Dougherty, E.R., Ed.; SPIE: Bellingham, WA, 1999.
31. Lieberman, D.J.; Allebach, J.P. Image sharpening with reduced sensitivity to noise: a perceptually based approach. Proceedings of IS&T's NIP 14: International Conference on Digital Printing Technologies, Toronto, Ontario, Canada, 1998; 294–297.
32. Eschbach, R. Digital copying of medium frequency halftones. Proceedings of SPIE conference on Color Imaging: Device-Independent Color, Color Hardcopy, and Graphic Arts III, San Jose, CA, 1998; 398–404.
33. Engeldrum, P.G. A new approach to image quality. Proceedings of the 42nd Annual Meeting of IS&T, 1989; 461–464.
34. Engeldrum, P.G. A framework for image quality models. Imaging Sci. Technol. 1995; *39*, 312–323.
35. Engeldrum, P.G. *Chapter 2 Psychometric Scaling: A Toolkit for Imaging Systems Development*; IMCOTEK Press: Winchester, MA, 2000; 5–17.

36. Shannon, C.E. A mathematical theory of communication. Bell System Tech. J. 1948, *27*, 379, 623.
37. Roetling, P.G. Visual performance and image coding. Proceedings of the Society of Photo-Optical Instrumentation Engineers on Image Processing, Vol. 74, 1976; 195–199.
38. Roetling, P.G.; Loce, R.P. Digital halftoning. In *Digital Image Processing Methods*; Dougherty, E.R., Ed.; Marcel Dekker: New York, 1994; 363–413.
39. Eyer, J.A. The influence of emulsion granularity on quantitative photographic radiometry. Photog. Sci. Eng. 1962, *6*, 71–74.
40. Dainty, J.C.; Shaw, R. *Image Science: Principles, Analysis and Evaluation of Photographic-Type Imaging Processes*; Academic Press: New York, 1974.
41. Selwyn, E.W.H. A theory of graininess. Photog. J. 1935, *75*, 571–589.
42. Siedentopf, H. Concerning granularity, resolution, and the enlargement of photographic negatives. Physik Zeit. 1937, *38*, 454.
43. Altman, J.H.; Zweig, H.J. Effect of spread function on the storage of information on photographic emulsions. Photog. Sci. Eng. 1963, *7*, 173–177.
44. Vaysman, A.; Fairchild, M.D. Degree of quantization and spatial addressability tradeoffs in the perceived quality of color images. Proc SPIE on Color Imaging. III 1998, *3300*, 250.
45. Lehmbeck, D.R. Experimental study of the information storing properties of extended range film. Photog. Sci. Eng. 1967, *11*, 270–278.
46. Bryngdahl, O. J. Opt. Soc. Am. 1976, *66*, 87–98.
47. Jorgensen, G.W. Preferred tone reproduction for black and white halftones. In *Advances in Printing Science and Technology*; Banks, W.H., Ed.; Pentech Press: London, 1977; 109–142.
48. Jones, L.A.; Nelson, C.N. The control of photographic printing by measured characteristics of the negative. J. Opt. Soc. Am. 1942, *32*, 558–619.
49. Jones, L.A. Recent developments in the theory and practice of tone reproduction. Photogr. J. Sect. B 1949, *89B*, 126–151.
50. Bartleson, C.J.; Breneman, E.J. Brightness perception in complex fields. J Opt. Soc. Am. 1967, *57*, 953–957.
51. Nelson, C.N. Tone reproduction. In *The Theory of Photographic Process*; 4th Ed.; James, T.H., Ed.; Macmillan: New York, 1977; 536–560.
52. Nelson, C.N. The reproduction of tone. In *Neblette's Handbook of Photography and Reprography: Materials, Processes and Systems*, 7th Ed.; Sturge, J.M., Ed.; Van Nostrand Reinhold: New York, 1977; 234–246.
53. Holladay, T.M. An optimum algorithm for halftone generation for displays and hard copies. Proceedings of the SID 1980, *21*, 185–192.
54. Roetling, P.G.; Loce, R.P. Digital halftoning. In *Digital Image Processing Methods*; Dougherty, E.R., Ed.; Marcel Dekker: New York, 1994; 392–395.
55. Roetling, P.G. Analysis of detail and spurious signals in halftone images. J. Appl. Phot. Eng. 1977, *3*, 12–17.
56. Stoffel, J.C. *Graphical and Binary Image Processing and Applications*; Artech House: Norwood, MA, 1982; 285–350.
57. Stoffel, J.C.; Moreland, J.F. A survey of electronic techniques for pictorial image reproduction. IEEE Trans. Comm. 1981, *29*, 1898–1925.
58. Ulichney, R. *Digital Halftoning*; The MIT Press: Cambridge, MA, 1987.
59. Clapper, R.; Yule, J.A.C. The effect of multiple internal reflections on the densities of halftone prints on paper. J. Opt. Soc. Am., *43*, 600–603, 1953, as explained in Yule, J.A.C. *Principles of Color Reproduction*; John Wiley and Sons: New York, 1967, p. 214.
60. Yule, J.A.C.; Nielson, W.J. The penetration of light into paper and its effect on halftone reproduction. In *Research Laboratories Communication No. 416*; Kodak Research Laboratories: Rochester, NY, 1951 and in TAGA Proceedings, 1951, *3*, 65–76.

61. Lehmbeck, D.R. Light scattering model for predicting density relationships in reflection images. Proceedings of 28th Annual Conference of SPSE, Denver CO, Soc. of Phot. Sci. and Eng. 1975; 155–156.
62. Maltz, M. Light-scattering in xerographic images. J. Appl. Phot. Eng. 1983, 9, 83–89.
63. Kofender, J.L. *The Optical Spread Functions and Noise Characteristics of Selected Paper Substrates Measured in Typical Reflection Optical System Configurations*; Rochester Institute of Technology: Rochester, NY, 1987; *MS Thesis*.
64. Klees, K.J.; Holmes, J. *Subjective Evaluation of Noise Filters Applied to Bi-Level Images*. 25th Fall Symposia of Imaging (papers in summary form only). Springfield, VA, Soc. Phot. Sci. & Eng., 1985.
65. Hunt, R.W.G. *Reproduction of Colour in Photography, Printing & Television*, 5th Ed.; The Fountain Press: Tolworth, England, 1995.
66. Hunt, R.W.G. *Measuring Colour*; Ellis Horwood Limited, Halstead Press, John Wiley & Sons: NY, 1987.
67. Hunter, R.S.; Harold, R.W. *The Measurement of Appearance*, 2nd Ed.; John Wiley and Sons: New York, 1987; 191 pp.
68. Bestenreiner, F.; Greis, U.; Helmberger, J.; Stadler, K. Visibility and correction of periodic interference structures in line-by-line recorded images. J. Appl. Phot. Eng. 1976, 2, 86–92.
69. Sonnenberg, H. Laser-scanning parameters and latitudes in laser xerography. Appl. Opt. 1982, 21, 1745–1751.
70. Firth, R.R.; Kessler, D.; Muka, E.; Naor, K.; Owens, J.C. A continuous-tone laser color printer. J. Imaging Technol. 1988, 14, 78–89.
71. Goodman, N.B. Perception of spatial color variation caused by mass variations about single separations. Proceedings of IS&T's NIP14: International Conference on Digital Printing Technologies, Toronto, Ontario, Canada, 1998; 556–559.
72. Shade, O. Image reproduction by a line raster process. In *Perception of Displayed Information*; Biberman, L.M., Ed.; Plenum Press: New York, 1976; 233–277.
73. Shade, O. Image gradation, graininess and sharpness in TV and motion picture systems. J. SMPTE 1953, 61, 97–164.
74. Biberman, L.M. Ed. *Perception of Displayed Information*; Plenum Press: New York, 1976.
75. Lehmbeck, D.R.; Urbach, J.C. *Scanned Image Quality*, Xerox Internal Report X8800370; Xerox Corporation: Webster, NY, 1988.
76. Kipman, Y. Imagexpert Home Page, www.imagexpert.com; Nashua NH, 2003 (describes several scanning-based image quality tools).
77. Wolin, D.; Johnson, K.; Kipman, Y. *Importantance of Objective Analysis in IQ Evaluation*. IS&T's NIP14: International Conference on Digital Print Technologies, Toronto Ontario, Canada, 1998; 603.
78. Briggs, J.C.; Tse, M.K. Beyond density and color: print quality measurement using a new handheld instrument. Proceedings of ICIS 02: International Congress of Imaging Science, Tokyo, Japan, May 13–17, 2002, and describes other scanning-based image quality tools at QEA Inc., http://www.qea.com (accessed 2003).
79. Yuasa, M.; Spencer, P. NCITS-W1: developing standards for copiers and printers. Proceedings of IS&T Image Processing, Image Quality, Image Capture Systems (PICS) Conference, Savannah GA, 1999; 270.
80. Williams, D. Debunking of specsmanship: progress on ISO/TC42 standards for digital capture imaging performance. Proceedings of IS&T Processing Images, Image Quality, Capturing Images Systems Conference (PICS), Rochester, NY, 2003; 77–81.
81. Williams, D. Benchmarking of the ISO 12233 slanted edge spatial frequency response plug-in. Proceedings of IS&T Image Processing, Image Quality, Image Capture Systems (PICS) Conference, Portland, OR, 1998; 133–136.
82. Hubel, P.M. Color IQ in digital cameras. Proceedings of IS&T Image Processing, Image Quality, Image Capture Systems (PICS) Conference, 1999; 153.

83. Swing, R.E. *Selected Papers on Microdensitometry*; SPIE Optical Eng Press: Bellingham, WA, 1995.
84. Swing, R.E. *An Introduction to Microdensitometry*; SPIE Optical Eng Press: Bellingham, WA, 1997.
85. Lehmbeck, D.R.; Jakubowski, J.J. Optical-principles and practical considerations for reflection microdensitometry. J. Appl. Phot. Eng. 1979, *5*, 63–77.
86. Miller, M.; Segur, R. Perceived IQ and acceptability of photographic prints originating from different resolution digital capture devices. Proceedings of IS&T Image Processing, Image Quality, Image Capture Systems (PICS) Conference, Savannah, GA, 1999; 131–137.
87. Ptucha, R. IQ assessment of digital scanners and electronic still cameras. Proceedings of IS&T Image Processing, Image Quality, Image Capture Systems (PICS) Conference, Savannah, GA, 1999; 125.
88. Gonzalez, R.C.; Wintz, P. *Digital Image Processing*; Addison, Wesley: Reading, MA, 1977; 36–114.
89. Jakubowski, J.J. Methodology for quantifying flare in a microdensitometer. Opt. Eng. 1980, *19*, 122–131.
90. Knox, K.T. Integrating cavity effect in scanners. Proceedings of IS&T/OSA Optics and Imaging in the Information Age, Rochester, NY, 1996; 156–158.
91. Knox, K.T. US Patent #5,790,281, August 4, 1998.
92. Smith, W.J. *Modern Optical Engineering*; McGraw Hill: New York, 1966; 308–324.
93. Perrin, F.H. Methods of appraising photographic systems. J. SMPTE 1960, *69*, 151–156, 239–249.
94. Shade, O. *Image Quality, a Comparison of Photographic and Television Systems*; RCA Laboratories: Princeton, NJ, 1975.
95. Newell, J.T.; Triplett, R.L. An MTF analysis metric for digital scanners. Proceedings of IS&T 47th Annual Conference /ICPS, Rochester, NY, 1994; 451–455.
96. Lamberts, R.L. The prediction and use of variable transmittance sinusoidal test objects. Appl. Opt. 1963, *2*, 273–276.
97. Lamberts, R.L. *Engineering Notes – Use of Sinusoidal Test Patterns for MTF Evaluation*; Sine Patterns LLC: Penfield, NY, 1990.
98. Scott, F.; Scott, R.M.; Shack, R.V. The use of edge gradients in determining modulation transfer functions. Photog. Sci. Eng. 1963, *7*, 345–356.
99. Campbell, F.W. Proc. Australian Physiol. Soc. 1979, *10*, 1.
100. Gorog, I.; Carlson, C.R.; Cohen, R.W. Luminance perception – some new results. In Proceedings, SPSE Conference on Image Analysis and Evaluation, Shaw, R. Ed., Toronto, Ontario, Canada, 1976; 382–388.
101. Bryngdahl, O. Characterists of the visual system: psychophysical measurements of the response to spatial sine-wave stimuli in the photopic region. J. Opt. Soc. Am. 1966, *56*, 811–821.
102. Watanabe, H.A.; Mori, T.; Nagata, S.; Hiwatoshi, K. Vision Res. 1968, *8*, 1245–1254.
103. Glenn, W.E.; Glenn, G.; Bastian, C.J. Imaging system design based on psychophysical data. Proceedings of the SID 1985, *26*, 71–78.
104. Dooley, R.P.; Shaw, R. A statistical model of image noise perception. In *Image Science Mathematics Symposium*; Wilde, C.O., Barrett, E., Eds.; Western Periodicals: Hollywood, CA, 1977; 10–14.
105. Patterson, M. Proceedings of the SID 1986, *27*, 4.
106. Blakemore, C.; Campbell, F.W. J. Physio. 1969, *203*, 237–260.
107. Hufnagel, R. In *Perception of Displayed Information*; Biberman, L., Ed.; Plenum Press: New York, 1973; 48 pp.
108. Rogowitz, B.E. Proceedings of the SID 1983, *24*, 235–252.
109. Oppenheim, A.V.; Schafer, R. *Digital Signal Processing*; Prentice-Hall: Englewood Cliffs, NJ, 1975; 413–418.

110. Jones, R.C. New method of describing and measuring the granularity of photographic materials. J. Opt. Soc. Am. 1955, *45*, 799–808.
111. Lehmbeck, D.R. *Imaging Performance Measurement Methods for Scanners that Genereate Binary Output*. 43rd Annual Conference of SPSE, Rochester, NY, 1990; 202–203.
112. Shaw, R. The statistical analysis of detector limitations. In *Image Science Mathematics Symposium*; Wilde, C.O., Barrett, E., Ed.; Western Periodicals: Hollywood, CA, 1977; 1–9.
113. Kriss, M. Image structure. In *The Theory of Photographic Process*, 4th Ed.; James, T.H., Ed.; Plenum Press: New York, 1977; Chap. 21, 592–635.
114. Scott, F. Three-bar target modulation detectability. J. Photog. Sci. Eng. 1966, *10*, 49–52.
115. Charman, W.N.; Olin, A. Image quality criteria for aerial camera systems. J. Photogr. Sci. Eng. 1965, *9*, 385–397.
116. Burroughs, H.C.; Fallis, R.F.; Warnock, T.H.; Brit, J.H. *Quantitative Determination of Image Quality*, Boeing Corporation Report D2: 114058-1, 1967.
117. Snyder, H.L. Display image quality and the eye of the beholder. Proceedings of SPSE Conference on Image Analysis and Evaluation, Shaw, R., Ed.; Toronto, Ontario, Canada, 1976; 341–352.
118. Carlson, C.R.; Cohen, R.W. A simple psychophysical model for predicting the visibility of displayed information. Proc. of SID 1980, *21*, 229–246.
119. Barten, P.G.J. The Square Root Integral (SQRI): A new metric to describe the effect of various display parameters on perceived image quality. Proceedings of SPIE conference on Human Vision, Visual Processing, and Digital Display, Los Angeles, CA, 1989; Vol. 1077, 73–82.
120. Barten, P.G.J. The SQRI method: a new method for the evaluation of visible resolution on a display. Proc. SID 1987, *28*, 253–262.
121. Barten, P.G.J. Physical model for the contrast sensitivity of the human eye. Proceedings of the SPIE on Human Vision, Visual Processing, and Digital Display III, San Jose, CA, 1992; Vol. 1666, 57–72.
122. Lubin, J. The use of psychophysical data and models in the analysis of display system performance. In *Digital Images and Human Vision*; Watson, A.B., Ed.; MIT Press: Cambridge, MA, 1993; 163–178.
123. Daly, S. The visible differences predictor: an algorithm for the assessment of image fidelity. In *Digital Images and Human Vision*; Watson, A.B., Ed.; MIT Press: Cambridge, MA, 1993; 179–206.
124. Daly, S. The visible differences predictor: an algorithm for the assessment of image fidelity. Proceedings of the SPIE on Human Vision, Visual Processing, and Digital Display III, San Jose, CA, 1992; Vol. 1666, 2–15.
125. Frieser, H.; Biederman, K. Experiments on image quality in relation to modulation transfer function and graininess of photographs. J. Phot. Sci. Eng. 1963, *7*, 28–46.
126. Biederman, K. J. Photog. Korresp. 1967, *103*, 41–49.
127. Granger, E.M.; Cupery, K.N. An optical merit function (SQF) which correlates with subjective image judgements. J. Phot. Sci. Eng. 1972, *16*, 221–230.
128. Kriss, M.; O'Toole, J.; Kinard, J. Information capacity as a measure of image structure quality of the photographic image. Proceedings of SPSE Conference on Image Analysis and Evaluation, Toronto, Ontario, Canada, 1976; 122–133.
129. Miyake, Y.; Seidel, K.; Tomamichel, F. Color and tone corrections of digitized color pictures. J. Photogr. Sci. 1981, *29*, 111–118.
130. Crane, E.M. J. SMPTE 1964, *73*, 643.
131. Gendron, R.G. J. SMPTE 1973, *82*, 1009.
132. Granger, E.M. Visual limits to image quality. J. Proc. Soc. Photo-Opt. Instr. Engrs 1985, *528*, 95–102.
133. Natale-Hoffman, K.; Dalal, E.; Rasmussen, R.; Sato, M. Proceedings of IS&T Image Processing, Image Quality, Image Capture Systems (PICS) Conference, Savannah, GA, 1999; 266–273.

134. Dalal, E.; Rasmussen, R.; Nakaya, F.; Crean, P.; Sato, M. Evaluating the overall image quality of hardcopy output. Proceedings of IS&T Image Processing, Image Quality, Image Capture Systems (PICS) Conference, Portland, OR, 1998; 169–173.
135. Inagaki, T.; Miyagi, T.; Sasahara, S.; Matsuzaki, T.; Gotoh, T. Color image quality prediction models for color hard copy. Proceedings of SPIE 1997, *2171*, 253–257.
136. Goldman, S. *Information Theory*; Prentice Hall: New York, 1953; 1–63.
137. Felgett, P.B.; Linfoot, E.H. J. Philos. Trans. R. Soc. London 1955, *247*, 369–387.
138. McCamy, C.S. On the information in a photomicrograph. J. Appl. Opt. 1965, *4*, 405–411.
139. Huck, F.O.; Park, S.K. Optical-mechanical line-scan image process – its information capacity and efficiency. J. Appl. Opt. 1975, *14*, 2508–2520.
140. Huck, F.O.; Park, S.K.; Speray, D.E.; Halyo, N. Information density and efficiency of 2-dimensional (2-D) sampled imagery. Proc. Soc. Photo-Optical Instrum. Engrs 1981, *310*, 36–42.
141. Burke, J.J.; Snyder, H.L. Quality metrics of digitally derived imagery and their relation to interpreter performance. SPIE 1981, *310*, 16–23.
142. Sachs, M.B.; Nachmias, J.; Robson, J.G. J. Opt. Soc. Am. 1971, *61*, 1176.
143. Stromeyer, C.F.; Julesz, B. Spatial frequency masking in vision: critical bands and spread of masking. J. Opt. Soc. Am. 1972, *62*, 1221.
144. Miyake, Y.; Inoue, S.; Inui, M.; Kubo, S. An evaluation of image quality for quantized continuous tone image. J. Imag. Technol. 1986, *12*, 25–34.
145. Metz, J.H.; Ruchti, S.; Seidel, K. Comparison of image quality and information capacity for different model imaging systems. J. Photogr. Sci. 1978, *26*, 229.
146. Hunter, R.; Robinson, A.H. International digital facsimile coding standards. Proc. IEEE 1980, *68* (7), 854–867.
147. Rabbani, M. *Image Compression. Fundamentals and International Standards, Short Course Notes*; SPIE: Bellingham, WA, 1995.
148. Rabbani, M.; Jones, P.W. *Digital Image Compression Techniques, TT7*; SPIE Optical Engineering Press: Bellingham, WA, 1991.
149. Loce, R.P.; Dougherty, E.R. *Enhancement and Restoration of Digital Documents*; SPIE Optical Engineering Press: Bellingham, WA, 1997.
150. Joint BiLevel Working Group. *ITU-T Rec. T.82 and T.85*; Telecommunication Standardization Sector of the International Telecommunication Union, March 1995, August 1995.
151. Buckley, R.; Venable, D.; McIntyre, L. New developments in color facsimile and internet fax. Proceedings of IS&T 5th Annual Color Imaging Conference, Scottsdale, AZ, 1997; 296–300.
152. Huffman, D. A method for the construction of minimum redundancy codes. Proc. IRE 1962, *40*, 1098–1101.
153. Lempel, A.; Ziv, J. Compression of 2 dimensional data. IEEE Trans Info. Theory 1986, *IT-32* (1), 8–19.
154. Lempel, A.; Ziv, J. Compression of 2 dimensional data. IEEE Trans Info. Theory 1977, *IT-23*, 337–343.
155. Lempel, A.; Ziv, J. Compression of 2 dimensional data. IEEE Trans Info. Theory 1978, *IT-24*, 530–536.
156. Welch, T. A technique for high performance data compression. IEEE Trans Comput. 1984, *17*(6), 8–19.
157. Beretta, G. Compressing images for the internet. Proc. SPIE, Color Imaging, III 1998, *3300*, 405–409.
158. Lee, D.T. *Intro to Color Facsimile: Hardware, Software, Standards*. Proc. SPIE 1996, *2658*, 8–19.
158a. Marcellin, M.W.; Gornish, M.J.; Bilgin, A.; Boliek, M.P. An overview of JPEG 2000. SPIE Proceedings of 2000 Data Compression Conference, Snowbird, Utah, 2000, *2658*, 8–30.

159. Sharpe II, L.H.; Buckley, R. JPEG 2000.jpm file format: a layered imaging architecture for document imaging and basic animation on the web. Proceedings SPIE 45th Annual Meeting, San Diego, CA, 2001; 4115, 47.
160. Dougherty, E.R. Ed. *An Introduction to Morphological Image Processing*; SPIE Optical Engineering Press: Bellingham, WA, 1992.
161. Hamerly, J.R. An analysis of edge raggedness and blur. J. Appl. Phot. Eng. 1981, 7, 148–151.
162. Tung, C. Resolution enhacement in laser printers. Proceedings of SPIE Conference on Color Imaging: Device-Independent Color, Color Hardcopy, and Graphic Arts II, San Jose, CA, 1997.
163. Tung, C. Piece Wise Print Enhancement. US Patent 4,847,641, July 11, 1989, US Patent 5,005,139, April 2, 1991.
164. Walsh, B.F.; Halpert, D.E. Low Resolution Raster Images, US Patent 4,437,122, March 13, 1984.
165. Bassetti, L.W. Fine Line Enhancement, US Patent 4,544,264 October 1, 1985, Interacting Print Enhancement, US Patent 4,625,222, November 25, 1986.
166. Lung, C.Y. Edge Enhancement Method and Apparatus for Dot Matrix Devices. US Patent 5,029,108, July 2, 1991.
167. Frazier, A.L.; Pierson, J.S. Resolution transforming raster based imaging system, US Patent 5,134,495, July 28, 1992, Interleaving vertical pixels in raster-based laser printers, US Patent 5,193,008, March 9, 1993.
168. Has, M. Color management – current approaches, standards and future perspectives. IS&T, 11NIP Proceedings, Hilton Head, SC, 1995; 441.
169. Buckley, R. *Recent Progress in Color Management and Communication*; Society for Imaging Science and Technology (IS&T): Springfield, VA, 1998.
170. Newman, T. Making color plug and play. Proceedings IS&T/SID 5th Color Imaging Conference, Scottsdale, AZ, 1997; 284.
171. Tuijn, W.; Cliquet, C. Today's image capturing needs:going beyond color management. Proceedings IS&T/SID 5th Color Imaging Conference, Scottsdale, AZ, 1997; 203.
172. Gonzalez, G.; Hecht, T.; Ritzer, A.; Paul, A.; LeNest, J.F.; Has, M. Color management – how accurate need it be. Proceedings IS&T/SID 5th Color Imaging Conference, Scottsdale, AZ, 1997; 270.
173. Chung, R.; Kuo, S. Colormatching with ICC Profiles—Take One. Proc. IS&T/SID 4th Color Imaging Conference, Scottsdale, AZ, 1996, p. 10.
174. Rickmers, A.D.; Todd, H.N. *Statistics, an Introduction*; McGraw Hill: New York, 1967.
175. Dvorak, C.; Hamerly, J. Just noticeable differences for text quality components. J. Appl. Phot. Eng. 1983, 9, 97–100.
176. Hamerly, J. Just noticeable differences for solid area. J. Appl. Phot. Eng. 1983, 9, 14–17.
177. Bartleson, C.J.; Woodbury, W.W. Psychophysical methods for evaluating the quality of color transparencies III. Effect of number of categories, anchors and types of instructions on quality ratings. J. Photg. Sci. Eng. 1965, 9, 323–338.
178. Stevens, S.S. *Psychophysics: Introduction to Its Perceptual, Neural and Social Prospects*; John Wiley and Sons: New York, 1975; Reprinted: Transactions Inc.: New Brunswick, NJ, 1986.
179. Thurstone, L.L. Rank order as a psychophysical method. J. Exper. Psychol. 1931, 14, 187–195.
180. Stevens, S.S. On the theory of scales of measurement. J. Sci. 1946, 103, 677–687.
181. Kress, G. *Marketing Research*, 2nd Ed.; Reston Publishing Co. Inc.: a Prentice Hall Co.: Reston, VA, 1982.
182. Morrissey, J.H. New method for the assignment of psychometric scale values from incomplete paired comparisons. JOSA 1955, 45, 373–389.
183. Bartleson, C.J.; Breneman, E.J. Brightness perception in complex fields. JOSA 1967, 57, 953–960.
184. Bartleson, C.J. The combined influence of sharpness and graininess on the quality of color prints. J. Photogr. Sci. 1982, 30, 33–45.

185. Bartleson, C.J.; Grum, F. Eds. Vol. 5, Visual Measurements. In *Optical Radiation Measurements*, Academic Press: Orlando, FL, 1984.
186. Gescheider, G.A. *Psychophysics: The Fundamentals*, 3rd Ed.; Lawrence Erlbaum Assoc. Inc.: Mahwah, NJ, 1997.
187. Guilford, J.P. *Psychometric Methods*; McGraw Hill Book Co.: New York, 1954.
188. Malone, D. Psychometric methods. In *SPSE Handbook of Photographic Science and Engineering*, Chapter 19.4; A Wiley Interscience Publication, John Wiley &Sons: New York, 1973; 1113–1128.
189. Nunnally, J.C.; Bernstein, I.R. *Psychometric Theory*, 3rd ed.; McGraw Hill Book Co.: New York, 1994.
190. Torgerson, W.S. *Theory and Methods of Scaling*; J. Wiley &Sons: New York, 1958.
191. Scheff'e, H. An analysis of variance for paired comparisons. J. Am. Statist. Assoc. 1952, *47*, 381–395.
192. Daniels, C.M.; Ptucha, R.W.; Schaefer, L. The necessary resolution to zoom and crop hardcopy images. Proceedings of IS&T Image Processing, Image Quality, Image Capture Systems (PICS) Conference, Savannah, Georgia, 1999; 143.
193. Johnson, C.B. Image transfer properties of photoelectronic imaging devices, SPSE Conference on Image Analysis and Evaluation, Toronto, Ontario, Canada, 1976, pp. 103–113 and A method for characterizing electro-optical device modulation transfer functions. J. Phot. Sci. Eng., 1970, *14*, 413–415.

4

Polygonal Scanners: Components, Performance, and Design

GLENN STUTZ

Lincoln Laser Company, Phoenix, Arizona, U.S.A.

1 INTRODUCTION

Polygonal scanners have found a role in a wide range of applications including inspection, laser printing, medical imaging, laser marking, barcode scanning, and displays, to name a few. Ever since the laser was first discovered, engineers have needed a means to move the laser output in a repetitive fashion or scan passive scenes such as used in earlier military infrared systems.

The term polygonal scanner refers to a category of scanners that incorporate a rotating optical element with three or more reflective facets. The optical element in a polygonal scanner is usually a metal mirror. In addition to the polygonal scanner other scanners can have as few as one facet such as a pentaprism, cube beam splitter or "monogon." This section will concentrate on scanners that use a reflective mirror as the optical element.

Polygonal scanners share the beam steering market with other technologies including galvanometers, micromirrors, hologons, piezo mirrors and acousto-optic deflectors. Each technology has a niche where it excels. Polygonal scanners excel in applications requiring unidirectional scans, high scan rates, large apertures, large scan angles or high throughputs. The polygonal scanner in most applications is paired with another means for beam steering or object motion to produce a second axis. This creates a raster image with the polygonal scanner producing the fast scan axis of motion.

This chapter will provide information on types of scan mirrors, fabrication techniques to create these mirrors, and how to specify these mirrors. The motor and bearing systems used with the mirror to build a scanner are covered. A section on properly specifying a polygonal scanner as well as the cost drivers in the scanner design is included.

The incorporation of the scanner into a scan system including system level specifications and design approaches is reviewed. The final section covers system image defects to be aware of and methods used to compensate for these defects in a scanning system.

2 TYPES OF SCANNING MIRRORS

There are many types of scan mirrors, but most can be included in the following categories:

1. Prismatic polygonal scanning mirrors;
2. Pyramidal polygonal scanning mirrors;
3. "Monogons";
4. Irregular polygonal scanning mirrors.

2.1 Prismatic Polygonal Scanning Mirrors

A regular prismatic polygon is defined as one having a number of plane mirror facets that are parallel to, equidistant from, and face away from a central rotational axis (Fig. 1). This type of scan mirror is used to produce repetitive scans over the same image plane. It is the most cost-effective to manufacture and therefore finds its way into the vast majority of applications including barcode scanning and laser printing. An illustration of why the manufacturing cost can be lower than other types of scan mirrors is shown in Fig. 2. Here we see a stack of mirrors that can be moved through the manufacturing process as a single piece resulting in less handling, more consistency, and less machining time.

Figure 1 Regular prismatic polygonal scanning mirror.

Polygonal Scanners

Figure 2 Mirror stack reduces fabrication costs.

2.2 Pyramidal Polygonal Scanning Mirrors

A regular pyramidal polygon is defined as one having a number of facets inclined at the same angle, usually 45°, to the rotational axis (Fig. 3). This type of polygon is expensive to manufacture since one cannot stack mirrors together to process at the same time as is done with regular prismatic polygons.

A significant feature of the 45° pyramidal polygon is that it can produce half the output scan angle of a prismatic polygon for the same amount of shaft rotation. Prismatic

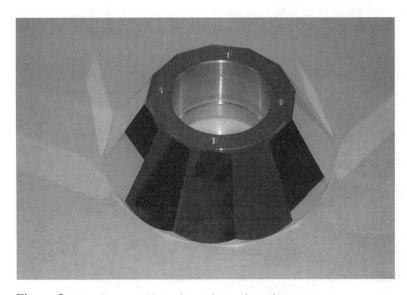

Figure 3 Regular pyramidal polygonal scanning mirror.

polygons are used primarily with the input beam perpendicular to the rotation axis whereas pyramidal polygons are used primarily with the input beam parallel to the rotation axis (Fig. 4). This feature can be used to the system designer's advantage by reducing data rates for a given polygon rotation speed.

2.3 "Monogons"

"Monogons" are scan mirrors where there is only one facet centered on the rotational axis. Because there is only one facet, a "monogon" is not a true polygon but they are an important subset of the scan mirror family. "Monogons" are also referred to as truncated mirrors and find application in internal drum scanning. In a typical system employing a monogon, the laser is directed toward the monogon along the rotation axis and the output sweeps a circle on an internal drum as the scanner rotates. This type of scan system can produce very accurate spot placement and very high resolution and finds application in the pre-press market. An example of a monogon scan mirror is shown in Fig. 5.

2.4 Irregular Polygonal Scanning Mirrors

An irregular polygonal scanning mirror is defined as one having a number of plane facets that are at a variety of angles with respect to, and face away from, the rotational axis (Fig. 6). The unique feature of this type of scan mirror is that it can produce a raster output

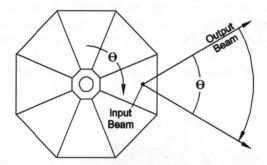

REGULAR 45° PYRAMIDAL POLYGON

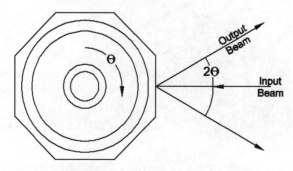

REGULAR PRISMATIC POLYGON

Figure 4 Scan angle vs. rotation angle.

Figure 5 "Monogon."

without a second axis of motion. The resulting output scans are nonsuperimposing if the facets are at different angles. This type of scanner finds its way into coarse scanning applications such as:

- point-of-sale barcode readers;
- laser heat-treating systems;
- intrusion alarm scanning systems.

These polygons typically cost significantly more than regular polygons because their asymmetry prevents any cost savings from stacking. Another disadvantage of these

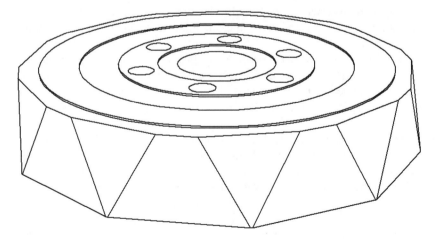

Figure 6 Irregular polygonal scanning mirror.

scanners is the inherent dynamic imbalance of the polygon during rotation. This limits their use to low-speed applications. A special case where equal and opposing facets are used on each side of the polygon helps with the balance problem. The result is the scan pattern is generated twice each revolution.

Now that the types of scan mirrors have been covered, a logical next step is to consider the materials used to fabricate the mirrors. The following section addresses the most common materials in use today.

3 MATERIALS

Material selection for polygonal mirrors is driven by considerations of performance and cost. The most common materials for polygonal mirrors are aluminum, plastic, glass, and beryllium. Facet distortion/flatness is a key performance consideration when choosing a material. It is measured as a fraction of a wavelength (λ).

Aluminum represents a good trade-off between cost and performance. This material has good stiffness, is relatively light, and lends itself to low-cost fabrication methods. The upper limit for the use of aluminum mirrors without the risk of facet distortion beyond $\lambda/10$ is on the order of a tip velocity of 76 m/s. Above this speed the size of the facet, the disc shape, the mounting method all play a role in the distortion of the facet. It is recommended that a finite element analysis be performed if you intend to operate above this level. An example of the shape change due to high-speed rotation, for a six-faceted polygon, is shown in Fig. 7.

Plastic is used in applications where cost is the primary concern and performance is good enough for the application. An example is in the hand-held barcode market and other short-range, low-resolution scanning applications. Injection molding techniques have come far in the past few years but it is still difficult to reliably produce plastic mirrors larger than 25 mm diameter with facets flat to better than 1 wave.

Glass polygons are used in a few applications, but few manufacturers like to produce scanners using this material. Glass mirror facets are adhered to a substrate to form the polygon. This requires precise alignment fixtures and good control of the curing process. This type of polygon is being replaced by aluminum in most applications. Glass can still find application in very short wavelength applications (deep ultraviolet) where its ability to be polished to very smooth surfaces is of benefit. This type of polygon construction does have speed limitations and is susceptible to microfractures that can lead to catastrophic destruction.

Beryllium has been used very successfully in applications where high speed and low distortion are required. It is a very expensive substrate and hazardous when machining, requiring specialized extraction and filtration equipment. Therefore it does not find wide usage and is a very expensive solution. Beryllium is typically nickel plated prior to polishing.

In some high-speed applications, distortions in facet flatness can be tolerated. In these cases the structural integrity of the polygonal mirror must be considered. The speed at which the dynamic stress will reach the yield strength (causing permanent distortion and dangerously close to the breaking speed) is found using the formula below.[1]

$$B = \sqrt{\frac{S}{(7.1e-6)w[(3+m)R^2 + (1-m)r^2]}} \qquad (1)$$

Polygonal Scanners

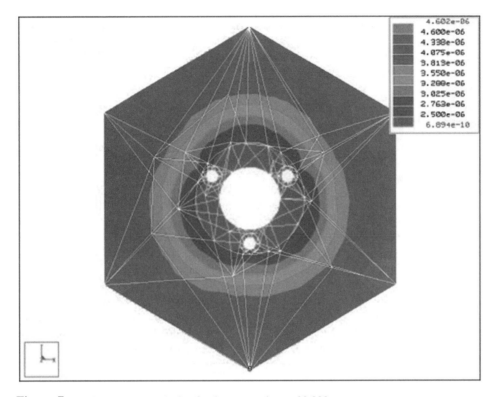

Figure 7 Finite element analysis of polygon rotating at 30,000 rpm.

where B = maximum safe speed (rpm), S = yield strength (lb/in.^{2}), w = weight of material (lb/in.^{3}), R = outer tip radius (in.), r = inner bore radius (in.), and m = Poisson's ratio. This formula does not have a margin of safety, so it would be wise to consider this and back off the results by an appropriate margin.

4 POLYGONAL MIRROR FABRICATION TECHNIQUES

Aluminum is the most common substrate for the fabrication of polygonal mirrors. There are two techniques for fabricating aluminum polygons that are widely used. These techniques are conventional polishing and single point diamond turning. Each technique has its advantages and the application will usually dictate the technique to be used.

4.1 Conventional Polishing

Conventional polishing in this context is pitch lapping in much the same manner as glass lenses and prisms are polished. A polishing tool is covered with a layer of pitch and a polishing compound is used that is a slurry composed of iron oxide and water. The pitch lap rubs against the optic using the polishing compound to remove material. Pitch lapping can be used to produce high-quality surfaces on a number of materials. Unfortunately, aluminum is not one of them. The aluminum surface is too susceptible to scratches during the polishing process. New techniques have been developed, but these rely on minimal abrasive mixtures and therefore material removal rates that are too slow to be cost-effective.

Because one cannot polish the aluminum directly, plating must be applied prior to polishing. Electro-less nickel is the most common plating applied. This combination provides the low cost and ease of machining of the aluminum as the structural material with the superior polishing properties and durability of nickel. The mirror facets are polished individually, blocked up in a surround as shown in Fig. 8. If the polygonal mirror is regular then a stack of polygons can be polished in one setup.

4.2 Single Point Diamond Turning

Single point diamond machining is a process of material removal using a finely sharpened single-crystal diamond-cutting tool. Diamond machining centers are available in the form of lathes and mills. The use of ultra-precise air-bearing spindles and table ways, coupled with vibration isolating mounting pads, enable machining to optical quality surface specifications. Figure 9 shows a diamond-machining center with a polygon in process.

Diamond machining has proven to be an efficient process for generating optical surfaces since it can be automated and the process time is a small fraction of the time required for conventional polishing. The diamond machined mirror is typically fabricated from aluminum, but satisfactory results have been obtained on other substrates. The diamond machined mirror face appears to be a perfect mirror, but upon close inspection the residual tool marks on the surface are apparent. These tool marks create a grating

Figure 8 Conventional polishing of polygonal mirrors.

Polygonal Scanners

pattern on the surface. This grating pattern can increase the scatter coming from the surface, particularly at wavelengths below 500 nm.

4.3 Polishing vs. Diamond Turning

Diamond turned aluminum scan mirrors are by far used in the highest volumes. This is due to the low manufacturing cost and good performance characteristics. Polished mirrors, however, have found a niche where they outperform diamond turned mirrors and justify the higher cost. These applications tend to be very scatter sensitive, such as writing on film. A polished mirror can approach surface roughness levels of 10 Å rms whereas diamond turned mirrors are limited to achieving roughness levels down to about 40 Å rms. Short-wavelength applications may also require the lower scatter of a polished mirror surface. Applications below 400 nm frequently need the lower scatter level of polished mirrors and the scatter can be a problem in applications up to about 500 nm.

5 POLYGON SPECIFICATIONS

In addition to selecting the type of polygon, the material to use, and the fabrication technique, several mechanical specifications need to be established. In a perfect world the polygon would have exactly the dimensions and angles that we specify on a print. Real-world manufacturing limitations cause us to have to add in a practical set of tolerances on

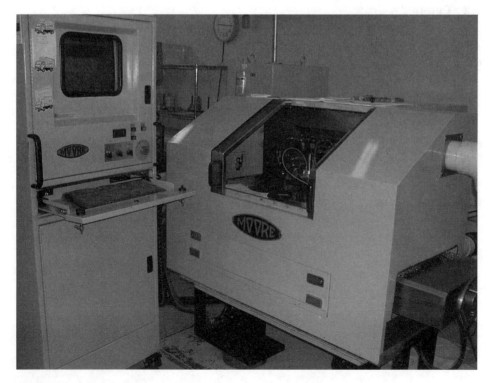

Figure 9 Diamond turning center.

the polygon and evaluate how these imperfections would affect system performance. Some of the items that need to be specified on a polygonal mirror include:

- facet-to-facet angle variance;
- pyramidal error;
- facet-to-axis variance (total and adjacent facet);
- facet radius:
  - nominal,
  - variation;
- surface figure (composed of power and irregularity);
- surface quality.

5.1 Facet-to-Facet Angle Variance

The definition of facet-to-facet angle variance (Δ) is the variation in the angle between the normals (Ψ) of the adjacent facets on the polygon (Fig. 10). This variation in angle causes timing errors from one facet to the next as the polygon rotates. Typical values for this angle range from ± 10 arc seconds to ± 30 arc seconds. Most scanning systems are not sensitive to errors in this range because of the use of start of scan sensors and/or encoders.

5.2 Pyramidal Error

Pyramidal error is defined as the average variation (Ω) from the desired angle between the facet and the mirror datum (Fig. 11). This variation results in a pointing error of the output beam and can also cause scan line bow. Typical values for this specification are ± 1 arc minute.

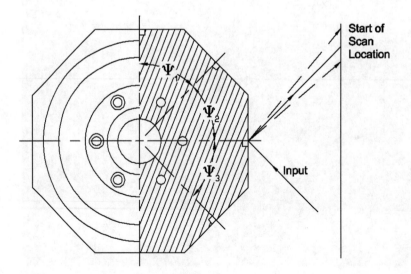

Figure 10 Facet-to-facet angle variance.

Polygonal Scanners

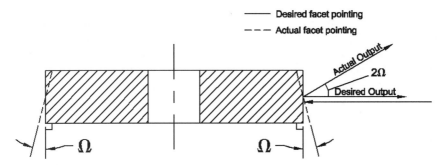

Figure 11 Pyramidal error.

5.3 Facet-to-Axis Variance

This is defined as the total variation of the pyramidal error from all the facets within one polygon (Fig. 12). This is a critical specification for the mirror and contributes to a scanner specification of dynamic track, discussed later. Typical values for this specification range from 2 arc seconds to 60 arc seconds.

Another parameter related to this is the adjacent facet-to-axis variance. This is defined as the largest step in the pyramidal angle from one facet to the next within a polygon. This is important to control in order to reduce banding artifacts in the final system. Typical values for this specification are in the range of 1–30 arc seconds.

Optical scanning systems may employ correction devices that allow this value to be reduced. System resolution plays a large part in determining the actual value required. Film writing applications tend to have the tightest requirements and passive reading systems tend to have the loosest requirements.

5.4 Facet Radius

The facet radius (referred to as facet height by some manufacturers) is the distance from the center of the polygon to the facet. The variation in this radius within the polygon

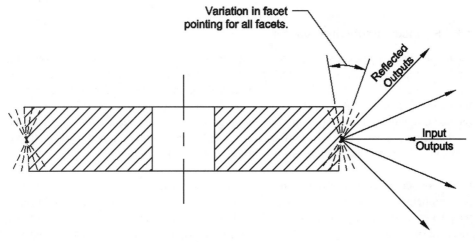

Figure 12 Facet-to-axis variance.

and the tolerance on the average radius are important to specify. The average facet radius is important because it locates the facet in the optical system. The variation of this radius within a polygon causes errors in the focal plane location from one facet to the next. It also causes linear speed variations within the scan line, which are usually small and show up as stability errors. Typical values for these parameters are ± 60 microns for the facet radius average position and ± 25 microns for the facet radius variation within a polygon.

5.5 Surface Figure

Surface figure is the macro shape of the polygon facet and is measured as the deviation from an ideal flat surface. The flatness of polygon facets will have an impact both on the aberrations in the beam as well as the pointing of the beam. The aberrations can affect the final focused spot size in the scan system. The pointing error results in velocity variations across the scan.

Several factors influence the flatness of polygon facets:

- initial fabrication tolerances;
- distortion due to mounting stresses;
- distortion due to forces induced when rotating at high speeds;
- distortion due to long-term stress relief.

Interferometers are commonly used to measure static flatness. The flatness is specified in wavelengths, λ, (or fractions thereof) of light. A typical flatness specification will read: $\lambda/8$ at 633 nm. Departure from flatness can have a variety of forms, depending on how the surface was fabricated. For example, conventionally polished mirror surfaces tend to depart from flat in a regular spherical form, either convex or concave. Diamond machined surfaces usually depart from flat in a regular cylindrical form, either convex or concave. A polygon will typically have two specifications related to flatness, a surface figure specification and irregularity. The irregularity is defined as the deviation from a best-fit sphere. Another common way of specifying the optical surface is in terms of power and pv-power (peak to valley error minus power), which separate the regular and irregular shapes. Most polygons used in reprographic applications are specified in the $\lambda/8$ to $\lambda/10$ range at the wavelength of interest.

5.6 Surface Quality and Scatter

Ideally a reflective optical surface will reflect all of the incident light without introducing any scattered components. In reality an optical surface has multiple defects of various sizes. The U.S. military developed a scratch and dig specification for surface defects, which is included in MIL-O-13830A and is in broad use within the optics industry. This method of quality determination involves close examination of a surface and identifying a scratch and dig level in a given unit area. A typical high-quality conventionally polished polygon will have a quality level of 40–20 scratch and dig.

Machined optical surfaces on the other hand, are made up of a precise regular pattern of scratches (machine tool marks), which are sufficiently high in frequency and low in height errors as to behave as a plane mirror at most visible and infrared wavelengths. The scratch and dig specification must be supplemented with an additional measure of surface quality here. A more representative definition for the overall surface quality is the rms surface roughness. The rms surface roughness can be measured directly by mechanical or optical profilometry means or indirectly by measuring the scatter from the surface.

A special test system is required to measure the scatter from the surface and correlate this to an rms roughness value. Different tests must be used for the measurement of scatter of diamond turned and conventionally polished mirrors. Diamond turned surfaces produce a significant fraction of scattered energy in a narrow cone around the reflected beam. Conventionally polished mirrors have the majority of their scattered energy in a cone significantly greater than the divergence angle of the reflected beam.

Conventionally polished mirrors can be tested using an integrating sphere that gathers a wide cone angle (Fig. 13). Scattered light in the cone of 4–180° is gathered with this test method. The diamond turned mirrors can be tested using a combination of the integrating sphere and a near angle test (Fig. 14). Scattered light in the cone of 0.4–4° is gathered using this test.

A correlation has been developed between the rms surface roughness and the total integrated scattered incident light:[2]

$$\text{rms roughness} = \frac{\lambda \sqrt{\ln(1-TIS)}}{4\pi} \qquad (2)$$

where *TIS* is the total integrated scatter.

The combination of scratch and dig along with rms surface roughness provide a good description of the surface structure higher in frequency than surface figure.

6 THIN FILM COATINGS

There are two major functions of optical coatings on polygons: to improve the reflectance of the surface and/or to improve durability. In the case of diamond machined polygons, the substrate is usually aluminum (in itself a good reflector over most of the visible spectrum). This aluminum surface is too soft without a coating. It is easily scratched during even a light cleaning. A thin layer of silicon monoxide, a dielectric material, is used

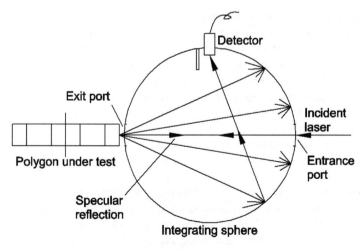

Figure 13 Test for wide angle scatter.

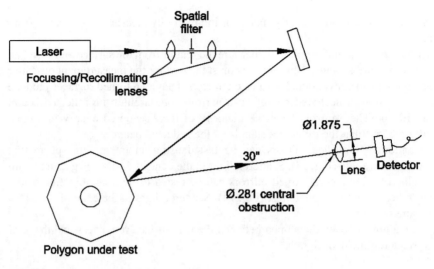

Figure 14 Test for near angle scatter.

as a surface protector. The optical thickness is usually about $\frac{1}{2}$ wavelength at the wavelength of interest. This material is more durable than the base aluminum and can be readily cleaned. The coating just described is typically referred to as a protected aluminum coating. This coating has a reflectivity of >88% across the 450–650 nm range.

The protected aluminum coating is fine for many applications and because of its simplicity is relatively inexpensive. Many applications, however, require enhancement coatings due to needs for higher reflectivity or performance at different wavelengths.

The first layer deposited in most applications is a metal, such as aluminum, silver, or gold. The layer or layers above the metal are composed of dielectric materials. The metal is selected based on the wavelengths of interest. As mentioned earlier, aluminum is a good choice in the visible region. It is also selected for ultraviolet applications since its reflectivity can be enhanced in this region with a dielectric stack. Gold is often selected as the base metal in applications above 600 nm. It has good reflectivity at 600 nm (90%) and very good reflectivity from 1 micron, out past 10.6 microns (>98%).

Silver exhibits very desirable reflectance characteristics over a broad spectrum and is frequently considered as a material for polygon coating. In practice, however, it is frequently a disappointing choice for the long term. The slightest pinhole (or minute scratch from cleaning) will expose the silver to reactive contaminants from the atmosphere, which over time (several days or weeks) will diffuse into the silver, producing an expanding blemish. Aluminum, which initially exhibits somewhat lower values of reflectance than silver, is far superior in terms of durability.

Common dielectric materials used to protect the surface and enhance reflectivity are silicon monoxide, silicon dioxide, and titanium dioxide. A quarter wave stack combining high refractive index and low refractive index materials is used to enhance the reflectivity in the wavelength region of interest. The term quarter wave stack refers to an alternating series of high- and low-index materials that are one-quarter of an optical wavelength thick at the wavelength of interest. The design of this quarter wave stack can be used to raise the reflectivity of the base metal significantly in various regions of interest. Most companies will offer a variety of standard reflectivity enhancing coatings for different wavelengths.

Polygonal Scanners

Thin film coatings are applied in vacuum deposition chambers such as the one shown in Fig. 15. The tooling to coat a polygon is specialized because the polygon has optical surfaces around its periphery. The polygons are stacked onto coating arbors that are placed in the chamber above the evaporant sources. The arbors are rotated with a drive mechanism during the deposition process at a constant rate so that all facets will see about the same thickness of coating material. The rotation rate must be fast enough that the time for one revolution is a small fraction of the deposition time for a single layer. Otherwise there will be significant variation around the polygon depending on where the shutter is opened and closed.

Reflectance uniformity both within a facet and facet-to-facet is an important polygon specification. The reflectance uniformity can impact the accuracy of written or read images using the polygon. In practice, the reflectance uniformity of rather large polygon facets (e.g., a few inches square) is more difficult to achieve than if the facets are small (e.g., $\frac{1}{2}$ inch square). This has to do with the consistency of the cleaning of the surface prior to coating and the variations in deposition rates with location and time within the coating chamber.

Aluminum polygons cannot be heated up to high temperatures during a coating run as one would typically do for good adhesion and layer density. The shape of the polygons makes them susceptible to slight stress changes during this heating cycle. This results in changes to the facet flatness. The coating process should be designed to keep the polygons below a temperature of 225°C to maintain the flatness.

Figure 15 Thin film deposition chamber.

Crucial to all of the desired characteristics of the coating of a polygon is its cleanliness prior to and during coating. Irrespective of the fabrication methods, polygons will be handled prior to coating during the inspection processes, transport, and installation into the coating chamber. The polygon must be cleaned thoroughly to remove foreign material that will degrade surface quality, prevent good coating adhesion, or outgas in the coating system.

Several tools are available to measure the optical performance of the coating. Common measuring tools to determine reflectances are spectrophotometers and laser reflectometers. Spectrophotometers are used to provide information on the reflectance vs. wavelength. The majority of spectrophotometers with a reflectance measuring attachment are limited to small sample sizes on the order of one to two inches in diameter. This fact usually precludes measuring the polygon itself. A witness sample is coated at the same time as the polygon and can be used to represent the actual part performance. This can be a reliable method of ascertaining the performance of the polygon as long as the witness sample has a similar surface preparation and quality level to the polygon.

Laser reflectometers compare the reflected beam to the incident beam at a specific wavelength and can be designed to test over a range of angles with either of the S or P polarizations. Reflectometers are useful for determining performance at one specific wavelength but cannot provide broadband information.

7 MOTORS AND BEARING SYSTEMS

The polygonal mirror requires a bearing system and a drive mechanism to turn it into a functional scanner. The drive mechanisms include pneumatic, AC hysteresis synchronous, and brushless DC. Bearing systems used in most applications are ball bearing, aerostatic air bearings, or aerodynamic air bearings.

7.1 Pneumatic Drives

Much of today's scan mirror technology has evolved from the development of ultra-high-speed polygon/turbine motors for the high-speed photography industry. Compressed air turbines continue to offer an attractive method of rotating a polygonal mirror at speeds beyond the capability of electric motors. The advantages of turbine drives are:

- Substantial horsepower can be delivered to the scan mirror to produce rapid acceleration and very high speed (up to 1,000,000 rpm).
- They are compact in size and low in weight in proportion to delivered power.
- They can be equipped with shaft seals so that the scan mirror can be used in a partial vacuum.

The disadvantages of turbine drives are:

- They require a compressed air source.
- They are asynchronous devices.
- They are relatively high in cost.
- They have a relatively short total running life.

Pneumatic drives are only recommended for short duty cycles and where ultra-high speed is essential.

7.2 Hysteresis Synchronous Motors

The rotor of a hysteresis synchronous motor is usually fabricated from a single piece of hardened steel selected out of a group (predominantly alloyed with cobalt) that exhibits substantial hysteresis loss. This resistance to the movement of magnetic flux in the material imparts torque to a rotor out of sync with the drive current. This torque is responsible for the motor's ability to start rotation. When the rotor approaches the speed of the stator flux, it becomes permanently magnetized and "locks in" to synchronism with the drive. If the motor is turned off and restarted, the stator flux demagnetizes the rotor and hysteresis takes over again. The synchronous mode of operation is more efficient than the hysteresis or startup mode, and in many systems a sync detector is used to reduce drive current after the motor is locked in to save energy and reduce heating.

AC hysteresis synchronous motors exhibit a characteristic called phase jitter (hunting). The rotors behave as though they were coupled to the drive waveform by a spring. Within synchronism the rotor springs forward and back in phase at a rate determined by the spring rate (flux density) and the torque/inertia ratio of the system. Typically, the frequency of this phase jitter is in the range of 0.5–2 or 3 Hz, at an amplitude of a few degrees (1–6° peak to peak). Under perfect conditions this jitter damps to zero values of amplitude. However, perfection is seldom seen and continual recurrence of jitter may be expected, caused by electrical transients on the input, mechanical shock to the assembly, variable resistance torque of the motor bearings, and so on. For many systems the 0.5–0.01% velocity error contribution of phase jitter is acceptably small. If this is not the case then a feedback loop is needed to reduce this level.

7.3 Brushless DC Motors

Brushless DC motors are by far the most common motor used to drive polygonal scanners. These motors use a permanent motor magnet and a stator that supplies the varying magnetic force. Motor magnets are composed of various materials including neodymium and ferrite depending on the application. Stators can be iron-based or ironless, with or without teeth. The number of magnetic poles is usually determined by the operating speed. Low-speed motors tend to have higher pole counts (8–12) while higher speed motors (>10,000 rpm) tend to have lower pole counts (4–6). The reason for the large number of poles at low speed is to achieve smoother rotation. At higher speeds this is not required and the lower pole count motors have less losses because the stator flux speed is lower.

Brushless motors do not have the hunting problem associated with AC hysteresis synchronous motors. The motor controls used to drive these motors can hold a tighter control loop. These motors can exhibit more high-frequency variations due to the torque available to rapidly change speed. This high-frequency velocity change is referred to as jitter. The amount of jitter is related to the rotor inertia and the number of feedback pulses per revolution. At higher speeds the inertia smoothes out the rotation and limits the amount of jitter. At lower speeds the number of feedback pulses helps keep the control loop errors small and therefore less velocity jitter when the motor has a correction torque applied. Hall effect devices are used at higher speeds to provide magnetic position feedback to the controller. Hall effect devices at lower speeds, where inertia is lower, can induce jitter as the controller chases the positional and triggering errors of the Hall effect devices. At lower speeds an encoder on the rotor may be required to achieve low jitter levels. Even incremental encoders can induce jitter errors due to disc quality, alignment, and component quality.

7.4 Bearing Types

Polygonal scanners require a bearing support system to allow the rotor to rotate. The most common bearings used in scanners are:

- ball bearings;
- aerostatic air bearings;
- aerodynamic air bearings.

These three types of bearing systems are discussed in detail in other chapters in this book. Ball bearings are used where possible due to their low cost. Applications requiring speeds less than 20,000 rpm and that can tolerate the bearing nonrepeatable errors, both in scan and cross-scan, are candidates for ball bearings.

Aerodynamic air bearings have made large inroads in laser scanning since the 1980s. An aerodynamic bearing generates its own air pressure as it rotates. It is commonly designed with two close-fitting cylinders for the radial bearing. The axial bearing can be either an air thrust bearing or magnetic. These systems have many advantages over both conventional ball bearing systems and aerostatic air bearings. The speed range for aerodynamic bearings is from approx. 4000 rpm up to over 100,000 rpm. These bearings are only slightly more costly than an equivalent ball bearing system. They have no wear while operating and require no external pressure support equipment. These bearings have been developed to withstand over 20,000 start/stop cycles. Aerodynamic bearings do have some limitations that limit their application. They are not well suited to dirty environments. Many designs exchange outside air frequently during operation, thereby ingesting the outside debris. Most designs cannot withstand high shock loads because the bearing stiffness is limited. The mass of the optic is limited in many applications due to both the lack of support and the need to withstand constant starting and stopping. Additional mass causes added wear to the bearing during startup and shutdown.

Aerostatic air bearings provide the ultimate in performance at a high cost. An aerostatic bearing uses pressurized air and closely spaced axial and radial bearing surfaces to float the rotor. When pressurized, the bearing has no contacting parts, resulting in extremely long life. These bearings are very stiff and have wobble errors less than 1 arc second. They are capable of supporting heavy loads and do not suffer from wear at startup and shutdown. They do require external components to supply the pressure to the bearing. This increases system complexity as well as cost.

8 SCANNER SPECIFICATIONS

Once the polygon, motor, and bearing system have been decided on, the packaging of the assembly becomes the next concern. One of the key elements in attaining high scanner performance is the mounting of the scan mirror to the rotating spindle.

To preserve the facet flatness achieved during initial polygon fabrication, it is necessary to fasten the polygon to its drive spindle with care, particularly if $\lambda/8$ or better flatness is required. The interface between the mirror and the rotor must not induce stress in the mirror that is translated out to the facets.

Polygonal Scanners

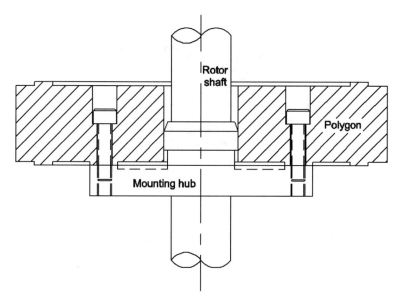

Figure 16 Mirror/rotor interface.

A typical mounting scheme is shown in Fig. 16. In this case the datum surface of the polygon and the locating annulus of the mounting hub are lapped to optical quality so that when the two are firmly held together, distortions are minimized.

Equally important to the accurate mounting of mirror datum and rotor hub surfaces is cleanliness at assembly and the appropriate torque levels of the fastening screws. Polygons can be attached in the manner described in low- and medium-speed applications. When tip velocities approach 76 m/s, other methods of mounting need to be considered.

In many applications the facets can be allowed to distort as long as they all change by the same amount. A symmetrical mounting method with screws aligned with every apex will work in this type of application. Other applications cannot stand significant shape change on the facets and require a true radially symmetric mounting method such as clamping. Clamping has been used successfully but this also requires a radial attachment means that may consist of an elastic material or aluminum shaped to have a spring force.

Once the polygonal mirror is integrated with the motor and the bearing system it can be referred to as a polygonal scanner. The scanner assembly has performance specifications that include:

- dynamic track;
- jitter;
- speed stability;
- balance;
- perpendicularity;
- time to sync.

8.1 Dynamic Track

Dynamic track is defined as the total mechanical angular variation from facet to facet perpendicular to the scanning direction. This is illustrated in Fig. 17. An optical beam

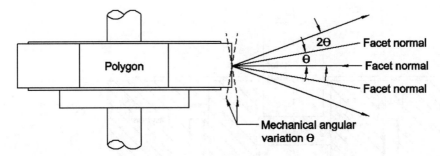

Figure 17 Dynamic track errors.

illuminating a polygon with a dynamic track of 10 arc seconds will have a scan envelope from all the facets of 20 arc seconds perpendicular to the scan direction. This is caused by an angle doubling effect on reflection from the rotating mirror.

There are three significant contributors to dynamic track error. The first is the polygon itself, which has a variation in the angle of each facet and can have a residual pyramidal (squareness) error. The second contribution comes from the mounting of the polygon to the rotating shaft. If the polygon is not perfectly perpendicular to the rotating shaft, then the facets will change their pointing in a sinusoidal manner with a period of one revolution. These first two contributions are fixed and repeatable. The third contribution is a random nonrepeatable error caused by the bearing support system. The random component of the dynamic track error will be 1–2 arc seconds for a ball bearing assembly and less than 1 arc second for air bearing assemblies. A final possible contribution to dynamic track is a repeatable wobble (conical orbit) and a cylindrical orbit from some air bearing systems.

The repeatable component of dynamic track (which tends to be larger) will show up in a laser writing system as a banding artifact. The line spacing will not be uniform and will repeat the pattern on each revolution of the polygon. Dynamic track errors can be reduced, if needed, through either active or passive correction means. These are methods discussed in Sec. 12.1.

8.2 Jitter and Speed Stability

Velocity errors from a polygonal scanner are important to minimize because they affect the pixel placement in a writing application and the receiving angle in a reading application. Velocity errors have both repeatable and nonrepeatable components. The repeatable components are easier to deal with than the nonrepeatable errors.

Specifications for velocity errors are broken into both high-frequency (jitter) and low-frequency (speed stability) components. The high-frequency components range from pixel-to-pixel to once per revolution. The low-frequency components are over multiple revolutions. There are many elements of the scanner system that contribute to either jitter or speed stability errors. These contributing elements are shown in Table 1.

This is a long, but certainly not exhaustive, list of causative elements contributing or potentially contributing to velocity errors. It becomes obvious that the entire scanner optical system is involved and influences the speed stability measurement and result.

Table 1 Elements Contributing to Jitter or Speed Stability Errors

Primary causes
- Optical system
    Fixed geometric errors of the scan lens
- Electronic driver stability
    Frequency and phase stability
    Voltage stability
    Noise
- Motor characteristics
    AC motor hunting (low frequency)
    Cogging (high frequency)
- Bearing behavior
    Varying resistance torque from lube migration
    Roughness from wear and or dirt
    Bearing pre-load
- Polygonal mirror characteristics
    Flatness
    Facet radius uniformity (distance from center of rotation)
Environmental (external shocks and vibrations)

Secondary causes
- Reflectance uniformity
- SOS detector/amplifier noise
- Facet (polygon) surface roughness
- Air turbulence in the optical path (high-speed systems)
- Polygon/motor tracking accuracy
- Laser pointing errors (dynamic)

8.3 Balance

Polygonal scanners are rotational devices that can operate at high speeds. As such, they need to be properly balanced to reduce the amount of imbalance forces generated during operation. This includes compensating for both static and dynamic imbalance. This requires that a two-plane balancing system is used. In a two-plane balancing system sensors are located at two separated planes where correction weights are to be applied. The sensors record the magnitude and phase of the imbalance.

Various methods of either adding or removing weight are used to balance scanners. The most common techniques are:

- drill balancing;
- epoxy balancing;
- screw balancing;
- grind balancing.

The preferred approach for high-speed operation is either grinding or drilling to remove material. The addition of material always brings risk of improper attachment and slinging of bonding agents.

Unbalance is typically measured in mg-mm, a mass multiplied by the distance from the rotational axis. An unbalance of 100 mgmm, for example, indicates one side of the rotor has an excess mass equivalent to 100 mg at a 1 mm radius. Typical values for small high-speed scanners range from 10–100 mgmm. The impact of unbalance on a scanner is vibration. This vibration can be measured and from this the actual scanner unbalance can be calculated.

8.4 Perpendicularity

Another important scanner parameter is the perpendicularity of the rotation axis to the mounting datum. This is important to ensure proper pointing of the beam after reflection from the polygon and to minimize the bow that can be created by striking the polygon out of the rotation plane.

8.5 Time to Synchronization

The time that it takes for the scanner to reach operating speed from a stopped condition can be important in some applications. This is a function of the motor/winding and the available current as well as the rotor inertia and the windage that must be overcome as the scanner approaches operating speed. Typical values range from 3–60 s.

9 SCANNER COST DRIVERS

Polygonal scanners can range from low-cost, easy-to-manufacture units, to high-cost state-of-the-art devices. It is important when designing a scan system to understand the cost drivers. One should try to minimize the overall cost through system level trade-offs. The scanner assembly has many cost drivers including:

- polygon shape;
- number of facets;
- fabrication method, conventionally polished or diamond turned;
- optical specifications including surface figure, surface roughness, and scratch/dig;
- coating requirements;
- polygon size;
- type of bearing system;
- speed;
- velocity stability;
- dynamic track specification.

In an earlier section the various shapes of polygons were discussed. In order to reduce costs it is advisable when possible to select either a regular polygon or a monogon. The other polygon shapes have cost penalties that may or may not be justified based on the application. While polygons can be manufactured with any number of facets, fewer facets results in lower cost. This is not a large cost component in a diamond turned mirror but has a large impact on the cost of a polished mirror.

The selection of diamond turned or polished mirror has a major impact on scanner cost. Diamond turned mirrors are the lowest cost and have surface roughness values greater than 40 Å rms. Conventionally polished mirrors are more costly but can bring the surface roughness down to 10 Å rms. All but the most scatter-sensitive short-wavelength systems can use diamond turned mirrors.

The optical specification of surface figure can also have a large influence on cost. Optical surface figure values of $\lambda/4$ per inch at 633 nm are common but surface figure values down to $\lambda/20$ can be achieved at additional cost. A scratch/dig specification of 80/50 is a typical standard, but specifications down to 10/5 can be achieved at significantly higher cost.

The optical coating chosen for the polygon can have a minor impact on the cost. The lowest cost option is a simple gold or aluminum coating with a silicon monoxide overcoat. As the reflectivity specifications get higher, more dielectric layers are needed to enhance the reflectivity, which can increase chamber time and therefore costs.

Bearing selection can have a significant effect on cost. In the speed range of 500–4000 rpm, the choice is between ball bearings and aerostatic air bearings. Ball bearing scanners are relatively low in cost and are the appropriate solution for many applications, but are susceptible to damage, generate many vibration frequencies, and can create motor speed instability. The aerostatic scanners are costly and require support equipment, but offer the ultimate in scanning performance.

The bearing choice in the speed range of 4000–20,000 rpm includes ball bearings, aerodynamic air bearings, and aerostatic air bearings. The selection is based on cost and performance criteria such as velocity stability and dynamic track. Above 20,000 rpm, aerodynamic air bearings are usually the best solution. These bearings are relatively low in cost and have long life operating at this speed. Ball bearings start to have life issues above 20,000 rpm and aerostatic bearings usually are not cost-effective.

Velocity stability standard specifications are a function of speed and mirror load. If speeds are too low or mirror loads too small then an encoder is required to achieve velocity stability. Velocity stability in this context is a measurement of the variation in the time for a beam reflected from the same facet of a scanner to cross two stationary detectors in an image plane over 500–1000 revolutions. Scanners operating faster than 4000 rpm can easily achieve 0.02% velocity stability. On most units this can be improved upon down to 0.002% at additional cost. Below 4000 rpm the mirror load becomes very important. The lighter the mirror and slower the speed, the more difficult it is to achieve tight velocity stability.

A final significant cost driver is the track specification placed on the assembly. Mechanical track values of 45 arc seconds results in low-cost assemblies, but specifications as tight as 1 arc second can be achieved by some vendors at much higher costs. This specification is a serious cost driver, so it is recommended that you review your actual needs carefully to obtain the most cost-effective design.

10 SYSTEM DESIGN CONSIDERATIONS

Laser scanning systems based on polygon technology can take on a variety of forms. Systems can range from very simple to extremely complex based on the performance level required. The first system consideration is whether it will be a reading or writing system. Writing systems tend to have much tighter performance requirements than reading systems. This is due to the fact that writing system errors tend to be visible, whereas the same level of error in a reading system will not be great enough to impact data integrity. A reading system, however, has the additional complexity of collecting the scattered light back from the target.

Reading systems will either use an external collection system that is separate from the scan system or an internal collection system where the scattered light passes back through the scan system and is de-rotated by the polygonal scanner. The internal collection system places increased demands on the scan system by requiring less backscattered light and reduced ghost images. For laser radars, one often has a scanning system for transmitting the laser, and a separate receiver, with a synchronized scanner to avoid this problem. This, however, is a very expensive solution. Another approach taken with laser radar is to increase the facet width and separate the transmission and receive apertures. Care must be taken in the design to ensure that the receiver instantaneous field of view encompasses the transmitter output over the distance range desired.

Beyond having knowledge of the basic system configurations it is important to develop a thorough list of performance specifications when starting the system design process. A list of key parameters and typical values are shown in Table 2.

The list in Table 2 covers the majority of specifications that are placed on a scanning system. Some scan systems will require additional specifications based on the unique nature of the writing or reading application.

The optical system used in laser scanners can be separated into two generic types: pre-objective and post-objective. Pre-objective is a term used to describe the use of a polygon to deflect a ray bundle, which after deflection is imaged by a lens or curved mirror (Fig. 18). This method of scanning places the function of focal plane definition on the lens, referred to as a scan lens, rather than on the scanning facet. Several desirable characteristics can be designed into the scan lens when employed in pre-objective scanning. An example is a lens design referred to as F-Theta. An F-Theta lens has the following characteristics:

- a flat focal plane;
- uniform spot diameter over the entire scan;
- linear spot velocity at the scan plane (assuming constant angular velocity of the polygon).

Usually it is desirable to have the scanning spot move with a highly accurate and constant velocity in the scan plane. Polygonal mirror deflectors provide angular velocity stability in the range 0.002–0.05%, depending on the speed and inertia of the scanner.

Table 2 List of Key Parameters

Wavelength	350–10,600 nm
Number of resolvable points	100–50,000
Spot size	1 micron–25 mm
Spot size variation across scan	$\leqslant 5$–15%
Scan length	1 mm–2 m
Telecentricity	0.5–30°
Bow	$\leqslant 0.001\%$ of scan line length
Scan efficiency	30–90%
Intensity nonuniformity	$\leqslant 2\%$ to $\leqslant 10\%$
Pixel placement accuracy	
• Jitter	$\leqslant 0.002\%$ to $<0.02\%$
• Cross scan error	$\leqslant 1\%$ to $\leqslant 25\%$ of line spacing
Scatter	$\leqslant 0.2\%$ to $\leqslant 5\%$
Data rate	
Laser noise levels	
Environmental factors and system interfaces	

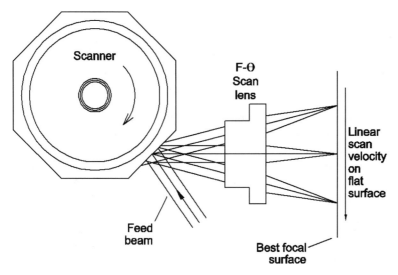

Figure 18 Pre-objective scanning system.

Without the aid of an F-Theta lens, however, the spot velocity variation on a flat focal surface will be proportional to the tangent of the scan angle, which for systems involving several degrees of scan means several percent variation.

Post-objective scanning is a term used to describe the use of a polygon to deflect a focusing ray bundle over a focal surface (Fig. 19). This method places the function of focal plane definition on the polygonal mirror, and the imaging (spot forming) lens is a relatively simple component located prior to the polygon.

The focal surface of a post-objective scanner is curved. The center of curvature is the center of the polygon facet. This type of scan system is typically used when the scan plane

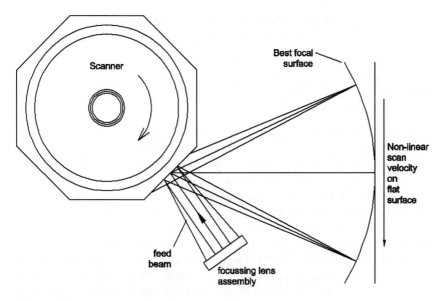

Figure 19 Post-objective scanning system.

can be curved to match the focal surface. Otherwise there are problems with spot size and velocity variations across the scan. The most popular system design incorporating post-objective scanning is a drum scanner. A drum scanner uses a monogon mirror, usually at 45°, with the source on the scan axis. As the monogon rotates, the focal surface is generated on the inside of a drum. Film or other flexible medium is located on this drum for image generation.

Post-objective scanning finds application in very high-resolution systems requiring greater than 25,000 points across the scan. Scanners designed for the pre-press industry use this design technique quite often.

Another factor to consider when designing a scan system is the degree of telecentricity required. A system is considered to be telecentric if the output from the scan system strikes the image plane at 90° for all points across the scan line. A post-objective scanner can be telecentric if the image plane can be curved to intercept the output from the scanner. If a flat image plane is required, a pre-objective scan system will need to have a scan lens that is slightly larger than the scan plane to meet the telecentric requirement. This can drive up the scan lens costs and result in a prohibitively expensive system. Normally, some level of deviation from telecentricity is given in a system specification.

In a writing application a decision as to how to use the available polygon facet is needed. Systems can either be under- or over-filled. Under-filled designs are the most common and do not waste available laser energy because the facet is sized such that the beam footprint on the facet never crosses over the edges of the facet during the full system scan angle. On the other hand, in an over-filled design the polygon facet is sized such that the beam completely fills the polygon facet over the entire full scan angle. Under-filled designs are preferred in many applications because there is less wasted energy and there is minimal diffraction from the facet edges. Over-filled designs have the one advantage that the system duty cycle can approach 100%. The duty cycle is the ratio of the active scan time to the full facet time.

11 POLYGON SIZE CALCULATION

Once a system concept is chosen, and the optical design completed, the polygon size needs to be calculated. A few key parameters must be known in order to size the polygon:

- scan angle, θ;
- beam feed angle, α;
- wavelength, λ;
- desired duty cycle, C.

θ is the full extent of the active scan measured in degrees as illustrated in Fig. 20. This value is usually in the range of 5–70°. α is the beam feed angle measured in degrees between the input beam to the polygon and the center of the scan exiting the polygon. It will be cost-effective to keep this angle as small as possible in order to reduce polygon size. In certain scanner applications the beam feed angle is zero. The beam is brought in through a beamsplitter in the center of scan or at a slight angle relative to the exiting scanned beam. λ is the operating wavelength expressed in microns and to be used in the calculation of the beam size on the polygon with a known desired spot size in the scan plane. C is the duty cycle, which is the ratio of active scan time to total time. Duty cycles in the range of 30–90% are common. However, the greater the duty cycle, the larger and more costly the polygon. With all conventional scan systems with the exception of monogon drum scanners some portion of the time will be spent transitioning from one

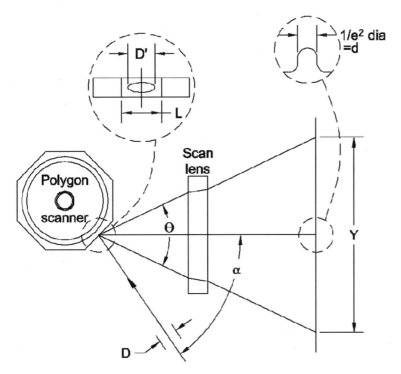

Figure 20 Illustration of scan angles.

facet to the next. We will assume that the design being considered is under-filled. This means that only one facet is being used to scan the image plane at any given time.

The number of facets, n, to be used is a trade-off that needs to be addressed. The formula for the number of facets is given by:

$$n = 720C/\theta \tag{3}$$

If this equation produces a noninteger answer, this means that there is no exact solution to provide the duty cycle desired at the same time as the optical scan angle requirement is satisfied. A next logical step is to fix the number of facets to an integer value near the result from the previous calculation and fix either the scan angle or the duty cycle and solve for the remaining variable.

$$C = n\theta/720 \tag{4}$$

For a writing application, once the duty cycle, scan angle, and number of facets is determined, the beam diameter D incident to the facet can be calculated. The following formulas assume a gaussian beam profile and the beam size defined at the $1/e^2$ intensity points.

$$D(\text{mm}) = \frac{1.27\lambda F}{d} \tag{5}$$

where F is the focal length of the scan lens in mm, and d is the $1/e^2$ beam diameter in the scan plane in microns.

The polygon can be sized without a scan lens by using the following formula:

$$D(\text{mm}) = \frac{1.27\lambda T}{d} \tag{6}$$

where T is the distance from the polygon to the focal surface in mm, and d is the $1/e^2$ beam diameter on the focal surface in microns.

For a reading system, D is a selected value based on the system-limiting aperture. The intensity profile across the diameter is no longer gaussian but top hat instead.

Since the size of the facet depends on the actual beam footprint on the facet, the feed angle effect on D must be taken into account. The value D' is the projected footprint on the polygon facet. It takes into account the truncation diameter and the cosine growth of the beam on the facet due to the beam feed angle. The formula for calculating the beam footprint is:

$$D' = 1.5D/\cos(\alpha/2) \tag{7}$$

The calculations assume a TEM00 gaussian beam that is truncated at the $1.5 \times 1/e^2$ diameter. If the application can tolerate more clipping at the start and end of scan the polygon size can be reduced.

The length of the facet (L) can be approximated from the beam footprint using the following [3]:

$$L(\text{mm}) = D'/(1 - C) \tag{8}$$

The polygon diameter can now be approximated as follows:

$$\text{Diam}_{\text{inscribed}} = L/[\tan(180/n)] \tag{9}$$

If the polygon diameter is too large then there are three options. The first is to reduce the duty cycle and suffer a higher speed and burst data rate. The second is to reduce the beam feed angle. The third is to allow more intensity variation across the scan by reducing the 1.5 multiplier. This in turn, reduces the facet length.

12 MINIMIZING IMAGE DEFECTS IN SCANNING SYSTEMS

In order to design a scanning system that accurately reproduces information, knowledge of the types of artifacts that the scan system can produce and visibility thresholds of these artifacts is needed. The specifications required to reduce the artifacts to acceptable levels vary by application; for example, a pre-press imager has different requirements from a laser printer.

12.1 Banding

Banding is one of the most common scan artifacts that will show up in scanning systems. Banding is a periodic variation in the line-to-line separation or density of the output. The human eye is very sensitive to periodic errors. The sensitivity is frequency dependent and great care must be taken to ensure that scan errors in the peak frequency range are minimized.[4]

Polygonal Scanners

Polygon reflectivity variations can be easily eliminated by properly specifying the polygon such that these errors will not be visible. A specification of less than 1% variation on all facets will be adequate for all but the most demanding applications.

In the case of a polygon-based scanning system, dynamic track errors or reflectivity variations between facets most often cause the banding. In continuous tone and halftone printers the line-to-line placement errors need to be reduced to less than 0.5% of the line spacing. In other applications this can be as large as 10–20% before banding becomes visible.

Either improving the polygon itself or compensating for the error can reduce dynamic track errors. Active correction techniques will compensate for repeatable errors, but not errors that vary throughout the scan line. Passive techniques will compensate for both repeatable and nonrepeatable errors. These passive methods will provide a significant reduction, but not perfect compensation due to pupil shifting. The pupil shift is due to the fact that the polygon rotates about its center rather than rotating about the facet. The facet vertex changes during rotation so the object point moves in and out as the facet rotates.

Active correction techniques are usually based on sampling the beam position errors perpendicular to the scan direction (cross-scan) between scans and applying a beam steering correction in the system prior to the polygon to change the beam pointing. These techniques are used primarily in low-speed systems due to the frequency response limitations of the beam steering components. Active correction systems are rare because there is added mechanical complexity, higher cost, and the lack of correction for changes that occur during scan.

Passive correction techniques are quite common and the basic concept is illustrated in Fig. 21. The polygon facet is re-imaged with some magnification to the scan plane in the cross-scan axis. A cylindrical lens element is typically used to create a line focus on the polygon facet. The re-imaging of this line in the cross-scan axis can be accomplished using a variety of components. Common methods include using a toroidal element near the polygon, or a cylindrical lens near the scan plane, or a cylindrical mirror near the scan plane.[5,6]

Banding does not necessarily result from optical effects. Other sources such as vibration or electrical noise can contribute to banding. Mechanical vibrations can be introduced by the rotating device and amplified by the scan system platform. If the platform is not rigidly coupled to the image plane, then relative motion between the scanner and the image can result in a banding artifact.

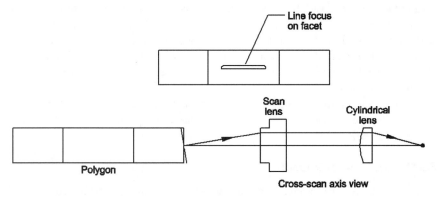

Figure 21 Passive cross-scan correction.

Electrical noise can be generated by lasers or laser power supplies. It can modulate the laser output directly or it can affect the performance of an external modulation device such as an acousto-optic modulator. Continuous tone applications can be particularly sensitive to electrical noise. Repeatable noise on the order of 0.5% of peak power can be visible. The electrical noise can appear to be banding if the frequency is near to or a multiple of the revolution rate.

In most scan systems the second axis of the image is controlled by a mechanical device, such as a translation stage, direct drive rollers, or a belt drive. The velocity stability of this second axis must be specified to the same level of requirements as the scanner device. Velocity errors in this second axis directly impact the banding in the image.

This section has shown that there are a variety of sources of banding. Care must be taken in the design phase to properly specify all components that contribute to this problem since it can be difficult to isolate the root cause when this defect appears in a scan system.

12.2 Jitter

Jitter is the high-frequency variation in the pixel placement along the scan direction. Various systems can tolerate different levels of jitter before artifacts become visible. Output scanners that place a premium on pixel placement will typically require 0.1-pixel accuracy whereas visual image outputs can tolerate up to 1 pixel in many applications. Jitter has both random and repeatable components. Random jitter is visually less objectionable than periodic jitter with a fixed pattern.

Random jitter errors can be produced by the ball bearings used in most low-speed scanning systems. The magnitude of these errors is dependent on the inertia of the rotor, the ball bearings chosen, and the bearing mounting method. Errors are usually small enough not to be of concern. Aerostatic air bearings offer an alternative if the system is sensitive to the ball bearing errors.

Motor cogging with brushless DC motors can also create jitter errors that repeat once per revolution. The motor controller can reduce these errors with proper feedback rates (encoders or start of scan feedback), but they cannot be eliminated. One method to overcome these errors in low-speed applications is to re-time the output data, based on actual scanner position information provided by an encoder. The only way to eliminate these errors is to find a motor with zero cogging torque. There is a new class of motors called thin gap motors that have close to zero cogging torque. They are expensive, but may become more affordable as they further penetrate the market.

Polygon facet flatness variations will result in a periodic jitter with a frequency of once per revolution or higher. The curvature causes small deviations in the angle of reflection from the facet. If the curvature of each facet varies, this causes the time between start of scan and end of scan to vary. A special case exists where there is no contribution to jitter if all facets have the same curvature. A facet flatness specification on the order of $\lambda/8$ is adequate for most applications.

Facet radius variations in systems using post-objective scanning result in beam displacement in the scan plane.[7] A facet radius variation specification of less than 25 microns is acceptable for most scanning applications.

12.3 Scatter and Ghost Images

There are many sources of scattered light and ghost images in an optical system. The majority of ghost images can be controlled through proper coatings and the placement of baffles. For example, if the strays are out of the plane of the scanned image an exit slit does a good job of eliminating them. Scan lenses can create problems with ghost images. The interior surfaces in these lens systems set up ghost images that are difficult if not impossible to eliminate with baffles. The anti-reflection coatings need to be high quality, reducing reflections to near zero.

As mentioned earlier in the chapter, the polygon surface can contribute to scatter. In extremely sensitive applications a diamond turned surface may produce too much scatter. This type of surface also produces a large percentage of near angle scatter that is difficult to baffle. A conventionally polished polygon produces wide-angle scatter and a much lower magnitude of total integrated scatter.

If the system contains an exit window or has a final lens close to the scan plane, then the cleanliness of this element is important. Dust particles on this element can cause localized scatter in the scan plane since the spot is typically small at this point in the optical system. If repeated each scan, this will result in a line being produced down the image.

Adjacent polygon facets tend to be problematic in many systems where there is significant reflection from the target surface. The beam can find its way back through the system to the next facet. The problem with this type of stray light is that it will be on axis. The best solution is to tilt the scan plane a few degrees relative to the scan system so scan plane reflections are out of plane. Another solution is to mask the polygon sufficiently to leave only the active scan aperture open.

The time between scans when the beam is passing over the tips of the polygon is another source for scatter. Light will scatter from the tips of the polygon and from the side of scan lens mounts. Turning off the beam between scans and using a time interval counter to turn the beam on just prior to the start of scan sensor will eliminate this possible source of problems.

Acousto-optic modulators can produce several undesirable effects. Scatter from the crystal can limit the extinction ratio. Long decay times may result in tails when transitioning from black to gray in a continuous tone application. The crystals used can also suffer from sound field reflections that show up as ghost images. Working with an application engineer at the modulator supplier is the best way to avoid these issues from affecting a scan system.

12.4 Intensity Variation

Variations in laser intensity can produce a variety of image artifacts depending on the frequency of the variation. A slowly varying fluctuation is much less objectionable than a high-frequency variation. Whereas intensity variation on the order of a few percent may be tolerable over an entire image, local intensity variations may need to be controlled to less than 0.5%.

Scan lens coatings and the coatings on any other elements located after the polygon can cause variations in the scan plane intensity across the scan. A transmission or reflection uniformity specification is needed to control this variable.

12.5 Distortion

Scanning systems typically employ a scan lens that has an F-theta characteristic. The lens distortion is controlled to produce image height that is proportional to the scan angle. This F-theta characteristic ensures linear scans with constant velocity. These lenses are not perfect but they do reduce the nonlinearity down to 0.01–0.1% range. This is adequate for all but the most critical applications. These residual errors are repeatable, therefore; intensity compensation for dwell time differences or variable clocking schemes for pixel placement differences can be employed to remove the residual error.

12.6 Bow

Bow is defined as the variation from straightness of a scan line. Bow is usually a slowly varying function across the scan. A considerable amount of bow can be tolerated before becoming visually objectionable. In most applications 0.05% of the scan line length of bow is an adequate specification. If the system is designed to have the beam brought in on the same axis as the scan then bow is caused by the errors in beam alignment. This can normally be adjusted to very fine levels and is therefore usually not a serious problem in a polygonal scan system. An equation for bow is given by Ref. 8:

$$E = F \sin \beta \left[\frac{1}{\cos \theta} - 1 \right] \tag{10}$$

where F is the focal length of the scan lens, E is the spot displacement as a function of field angle, θ is the field angle, and β is the angle between the incoming beam and the plane that is perpendicular to the rotation axis.

13 SUMMARY

This chapter has covered the components, performance characteristics, and design approaches for polygonal scanners and systems based on these scanners. This technology continues to evolve and thrives among increasing competition from other technologies both in writing and reading applications. I fully expect that the performance values that are stated in this chapter will be significantly improved on in the near future. However, the system level artifacts that a system designer must be careful to avoid tend to remain a constant. An in-depth knowledge of these artifacts and their root causes will help reduce development time for new systems.

ACKNOWLEDGMENTS

The author would like to thank Randy Sherman for his contributions to this chapter. Sections of this chapter were extracted from his chapter in *Optical Scanning* (1991) and updated. The assistance of Steve Lock with Westwind Air Bearings and Jim Oschmann with The National Solar Observatory with technical reviews of the chapter is greatly appreciated. The author would also like to thank Luis Gomez of Lincoln Laser Company for providing the illustrations. Photographs are courtesy of Lincoln Laser Company.

REFERENCES

1. Oberg, E. *Machinery's Handbook*, 23rd Ed; Industrial Press: New York, 1988; 196 pp.
2. Bennett, J.M.; Mattsson, L. *Introduction to Surface Roughness and Scattering*; Optical Society of America: Washington, DC, 1989; 50–52.
3. Beiser, L. Design equations for a polygon laser scanner. In *Beam Deflection and Scanning Technologies*; Marshall, G.F., Beiser, L., Eds; Proc. SPIE 1454; 1991; 60–65.

4. Bestenreiner, F.; Greis, U.; Helmberger, J.; Stadler, K. Visibility and corrections of periodic interference structures in line-by-line recorded images. J. Appl. Phot. Eng. **1976**, *2*, 86–92.
5. Fleischer, J. Light Scanning and Printing Systems. US Patent 3,750,189, July 1973.
6. Brueggemann, H. Scanner with Reflective Pyramid Error Compensation. US Patent 4,247,160, January 1981.
7. Horikawa, H.; Sugisaki, I.; Tashiro, M. Relationship between fluctuation in mirror radius (within polygon) and the jitter. In *Beam Deflection and Scanning Technologies*; Marshall, G.F., Beiser, L., Eds; Proc. SPIE 1454; 1991; 46–59.
8. Hopkins, R.; Stephenson, D. Optical systems for laser scanners. In *Optical Scanning*; Marcel Dekker: New York, 1991; 46 pp.

5

Motors and Controllers (Drivers) for High-Performance Polygonal Scanners

EMERY ERDELYI

Axsys Technologies, Inc., San Diego, California, U.S.A.

GERALD A. RYNKOWSKI

Axsys Technologies, Inc., Rochester Hills, Michigan, U.S.A.

1 INTRODUCTION

This chapter updates and expands upon the material covered by Gerald A. Rynkowski in *Optical Scanning*,[1] with greater emphasis being placed on brushless DC motors and the associated control electronics designed specifically for rotary scanning applications. Some background topics related to polygon scanning have been carried over to this chapter in order to clarify or illustrate control concepts. Other topics not directly related to motors and controllers, such as the discussion of air bearing design, have been omitted since these areas are covered in greater detail elsewhere in this book.

The availability of low-cost brushless DC motors and the continuing improvement in scanner control, as well as miniaturization of the drive electronics, have contributed greatly to the viability of opto-mechanical scanning in many new applications. Opto-mechanical scanning continues to be a cost-effective alternative to competing solid-state technologies. Several new application examples have been added that highlight the trend toward brushless DC motors and compact integrated control systems designed for military and commercial use.

Polygonal scanners have been designed, developed, and manufactured in all shapes and configurations during the past 30 years. These devices have been employed in military reconnaissance and earth resources studies, thermal imaging systems, film recorders, laser printers, flight simulators, and optical inspection systems, to name a few of the well-known applications. A common characteristic of all of these scanners is the requirement for precise control of polygon rotation, and consequently, the control of the beam scan. Recent advances in motor and control technologies have greatly improved the performance and efficiency of these scanners while simultaneously reducing the cost and size of the system.

This chapter explores the trends in motor and control technologies that are being utilized in today's polygonal scanners through the discussion of specific applications. In addition, motor characteristics, control techniques, and system models are presented to aid the opto-mechanical engineer in understanding these critical areas of scanning system design.

2 POLYGONAL SCANNER BASICS

Although a more thorough discussion of polygon geometry and scanning optics can be found elsewhere in this book, it will be useful to review some of the basic polygon configurations and optical designs as they influence the selection of the scan motor and control system. Also, a film recording system is presented in some detail in the next section in order to illustrate the influence of the scanner motor and controller characteristics on the overall system performance.

2.1 Polygon Configurations

In general, three types of scanner mirror configurations are popularly utilized in collimated or convergent, passive, or laser scanning optical systems. These rotating mirror spinners (polygons) are at the center of the electro-optical system. They direct incoming optical signals to a detector or steer outgoing modulated laser beams by virtue of their geometry and rotation about an axis.

The three most utilized scanner beam deflector configurations are the regular polygon, pyramidal, and the single-faceted cantilever design. The regular polygonal scanner is generally the most popular with system designers and can be utilized with either the collimated-beam or the convergent-beam scanning configurations. Figures 1 and 2 illustrate the two configurations using six-sided polygons having the spin axis projected into the page. In Fig. 1 the facets are illuminated with a collimated beam and reflected to a concave mirror that focuses at a curved focal plane.

Figure 2 shows a lens system that focuses the collimated beam prior to being reflected at the scanner facets, and then converging at a focus. Comparing the two configurations, it is obvious that both focal image surfaces are curved. This presents a problem to the system designer, but is usually corrected optically with a suitable field-flattening correction lens system, or perhaps the recording surface is curved to conform to the focal surface.

Another difference with regard to the polygon is that the convergent-beam bundle (Fig. 2) uses less area of the facet than the collimated configuration. However, maximum utilization of the facet area is desirable because the facet surface flatness irregularities tend to be averaged out, and therefore, minimize modulation of the exiting-beam scanning

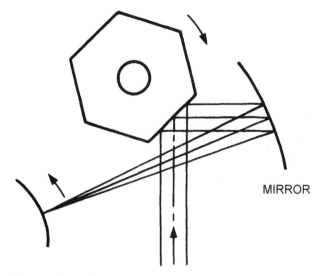

Figure 1 Collimated beam scanning (from Ref. 2).

angles. Most precision polygonal scanning systems use the collimated beam scanning configuration, which utilizes a larger proportion of the area of the facets.

Figure 3 illustrates a pyramidal mirror scanner commonly used with the rotational and optical axes parallel, but not coincident. Note that either the collimated or convergent configuration can be utilized with this design.

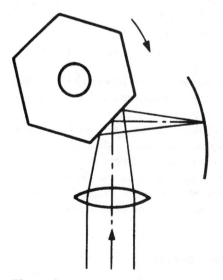

Figure 2 Convergent beam scanning (from Ref. 1).

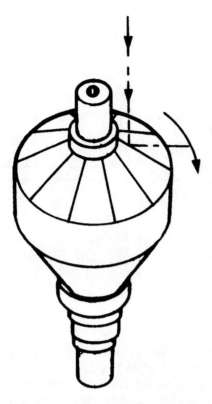

Figure 3 Parallel, but not coincident, optical and rotational axes (from Ref. 2).

Figure 4 illustrates a regular polygonal scanner in which the optical axis is normal to the rotational axis, or where the angle is acute to normal. Note that when the two axes, optical and rotational, are normal, the beam can reflect back upon itself.

Shown in Fig. 5 is a single-faceted cantilevered scanner with the beam and rotational axes coincident and reflecting from a 45° facet, and thereby generating a continuous 360° scan angle and a circular focused scan line.

This configuration has been used in passive infrared scanning systems having long focal length and requiring a large aperture. Nine-inch, clear-aperture scanners have been manufactured in this configuration for high collection efficiency.

This type of beam deflector design is also popular in scanning systems that are used in the image setting machines sold to the printing market. Many of these image setting machines have an "internal drum" design which involves the placement of a large piece of film on the inside surface of a cylinder. The single faceted "monogon" beam deflector rotates at high speed and scans a laser spot across the width of the film as it travels the length of the drum. Figure 6 illustrates a monogon beam deflector designed for rotational speeds exceeding 30,000 rpm.

2.2 Polygon Rotation and Scan Angle Relationship

At this point, it is noteworthy to realize the relation between the facet angle and scan angle. The facet angle is defined as $360°/N$ ($N =$ number of facets). The optical scan angle may

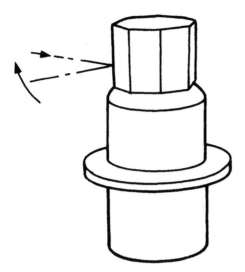

Figure 4 Optical and rotational axes normal (from Ref. 1).

be expressed as:

$$\text{Scan angle(degrees)} = 720/N$$

for $N \geq 2$. Observe that for $N \geq 2$, the optical scan angle is two times the shaft angle. This angle-doubling effect must obviously be considered when relating the shaft and facet parameters and their effects on the angular position of the focused spot at the focal plane. The optical angle-doubling effect places an even greater demand on the control of the polygon rotational velocity and must be carefully considered if the desired system accuracy is to be achieved.

The scanner rotational speed may be expressed as:

$$\text{rpm} = \frac{60\,W}{N}$$

where W = line scans/s, and N = number of facets.

An increase in the number of facets reduces the motor speed requirements as well as the maximum scan angle. However, the usable scan angle may in some cases also be limited by aperture size and the allowable vignette effect. In practice, the optical design will usually dictate the number of facets and consequently the motor and controller will have to be selected to accommodate the optical system designer. Figure 7 illustrates a small 12-faceted polygon mirror.

2.3 Polygon Speed Considerations

When a range of polygon rotation speeds are allowed by the optical design, it is best to avoid configurations that require speeds that are very low or very high. Problems with

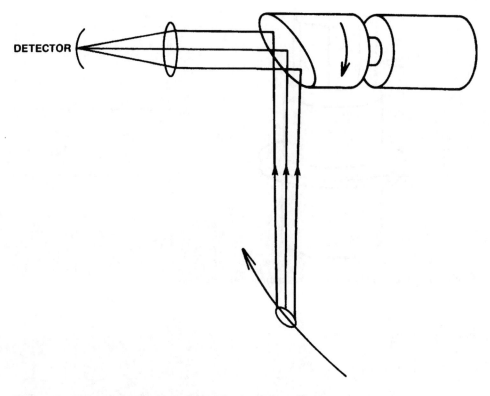

Figure 5 Parallel and coincident optical and rotational axis (from Ref. 1).

motor controllability may occur at speeds below about 60 rpm and motor efficiency may suffer when speeds exceeding 60,000 rpm are specified. To achieve good speed regulation at very low polygon rotation speeds, special motor and control designs are often required that will lead to increased system cost. A polygon rotating at 60 rpm and specified for 10 ppm (10 parts per million, or 0.001%) speed regulation will likely require a sinusoidally driven slotless brushless DC motor and complex controller design to achieve this level of performance.

Another complication associated with low-speed designs involves the selection of the velocity feedback device. Often the feedback device will be an optical encoder (Fig. 8), which functions both as a tachometer to monitor the polygon speed, and in some systems, as a position sensor for reporting the true position of the polygon to the scan processing electronics. Today's high-performance speed control systems operate using phase-lock loop techniques that provide excellent short-term as well as long-term speed regulation. These systems operate by comparing the frequency and phase of a stable reference signal with that of the polygon encoder, thereby generating an error signal, which is used to adjust the polygon speed. In order to achieve good speed control at low speeds, a high-resolution/high-accuracy encoder is necessary.

Depending on the polygon inertia, bearing friction, and the level of disturbances present, an encoder line density greater than 10,000 lines (counts) per revolution may be

Motors and Controllers (Drivers)

Figure 6 Monogon beam deflector.

required for a 60 rpm scanner. Optical encoders having greater than a few thousand line counts per revolution will also contribute to increased system cost since the disc pattern and the edge detection functions must be more precise.

As the polygon operating speed increases, fewer pulses per revolution from the encoder are required to achieve the same level of performance, all other factors being equal. This is due to the fact that at higher speeds, lower encoder resolutions can still produce an adequate number of pulses or "speed updates" per second from the encoder and allow for a reasonably fast control loop bandwidth. Higher control system bandwidth is desirable because the control loop can more readily react to short duration disturbances in speed, which may not be adequately attenuated by the polygon inertia.

The control system will receive some average polygon speed between encoder pulses from the phase detector, but it is essentially operating open loop until the next pulse arrives and updates the speed based on the encoder pulse phase relative to the reference clock. During this period between speed updates the polygon speed will generally decrease from the set point (and the last update) until the next encoder pulse arrives and the control system makes a correction. Therefore to achieve a specified update rate or bandwidth for the control system, the slower polygon will require a higher resolution encoder with correspondingly better pattern accuracy.

The actual speed change between encoder updates is a function of many factors, including the operating speed, total rotating inertia of the scanner, the amplitude and frequency of disturbances, the control system gain, and the friction. In general, higher

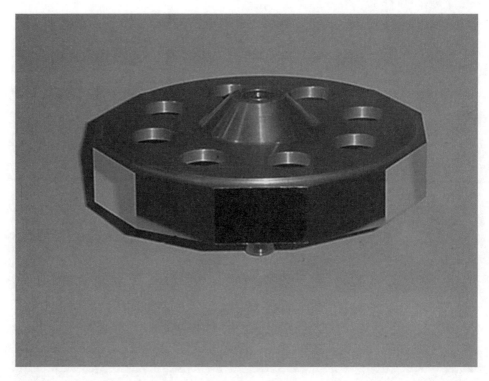

Figure 7 Twelve-faceted polygon.

polygon/motor inertia and lower bearing and windage friction is beneficial and will reduce the short-term speed variations present in the scanner. This will be discussed in greater detail in Sec. 5, which deals with the control system design.

At the other extreme, high operating speeds require that the polygon and the entire rotating assembly exhibit exceptionally low imbalance, ideally less than 10 μin-oz. Often this requires that the scanner must be balanced in two planes: at the polygon and at the motor/encoder location. This is necessary not only for maintaining the mechanical integrity and life of the rotating components, but also for achieving precise speed control within one revolution of the scanner. Any imbalance that causes concentricity error at the encoder will result in a sinusoidal scanner speed variation. This is especially true for air bearing scanners where the bearing stiffness is lower and allows for larger concentricity errors to be produced.

Also, the motor efficiency is adversely affected at high speeds, especially when the commutation frequency exceeds 1000 cycles/s. Special motor designs are required for efficient high-speed operation and these designs usually require expensive development time and have higher unit cost.

3 CASE STUDY: A FILM RECORDING SYSTEM

A film recording scanning system has been selected as a reference subsystem for purposes of discussing the scanner parameters as well as the dynamic performance requirements

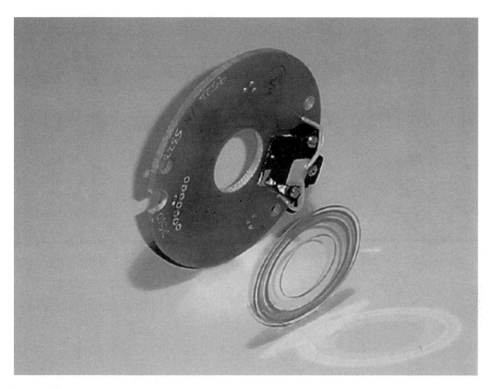

Figure 8 Optical encoder, disc, and readout electronics.

involving the motor and control system design. Figure 9 depicts a laser recording system capable of recording high-resolution video or digital data on film. The rotating polygon (spinner) generates a line scan at the film plane using a focused, intensity-modulated laser beam. The laser exposes the film in proportion to the intensity modulation in the video or digital signal. Line-to-line scan is accomplished by moving the photographic film at constant velocity and recording a continuous corridor of data limited only by the length of film. The film controller provides precise control of film velocity, which is locked to the polygon rotation. The expanded and collimated laser beam is intensity-modulated with video or digital data and then scanned by the facets of the spinner to the film plane. A field correction lens (F-θ) is used to focus the beam, to linearize the scan line with respect to the scan angle, and thereby provide a uniform spot size along the line at the film plane.

The scanner assembly contains a 12-faceted polygon required to perform the optical scan function. The rotating mirror, its drive motor rotor, and a precision optical tachometer are supported by externally pressurized gas bearings. The electronic controller provides precise motor speed control and synchronization between the reference frequency sync generator and the high-density data track of the optical encoder. The encoder also supplies an index pulse used for facet identification and derivation of the synchronized field frequency that is required for some raster scanning systems, as well as pixel registration and control of the film drive motor.

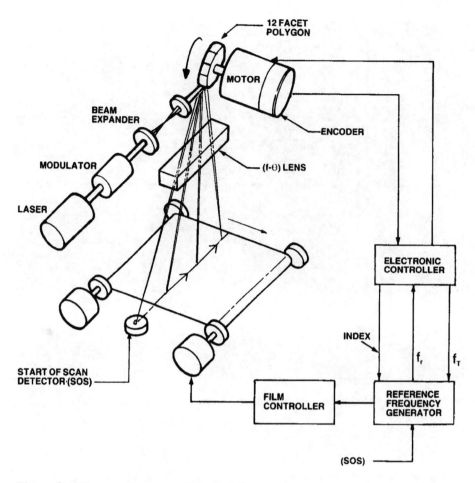

Figure 9 Film recording system (from Ref. 1).

3.1 System Performance Requirements

The system performance parameters for our example digital film recorder are discussed in the following paragraphs. These parameters are summarized in Table 1 and are typical and representative of a recently manufactured film recording system.

Table 1 Film Recorder System Requirements (Source: Ref. 1)

Line resolution (both directions)	10,000 pixels/line
Line scan length	13.97 cm
Scan rate	1200 lines/s
Film speed	1.6764 cm/s
Pixel frequency (clock)	12 MHz
Pixel diameter	10 μm at $1/e^2$
Pixel-to-pixel spacing	13.97 μm

Motors and Controllers (Drivers)

The system resolution requirements are defined at the film plane since all optical, scanning, and film transport errors will become evident on the recording media. The application requires 10,000 pixels of digital data per line at the scan plane using a 10 μm spot diameter measured at the $1/e^2$ irradiance level. One can calculate that for a 13.97-cm line scan length the pixel spacing, center to center, must be 13.97 μm, as must the spacing between scan lines. The line spacing variance, or film transport jitter, is conservatively specified to be less than ± 5 ppm (half the nominal spot size). For a reasonable throughput, the application dictates a scan rate requirement of 1200 lines/s. From this information the calculated pixel frequency is found to be 12 MHz (1200 lines/s × 10,000 pixels/line), and the film speed needed is 1.6764 cm/s (1200 lines/s × 13.97 μm/line).

3.2 Spinner Parameters

The polygon specification is determined and driven by the form, fit, and functional performances set by the optical requirements of the system. At this point, the optical engineer must optimize the design of the optical elements, which includes specifying the polygon type, facet number, facet width and height, inscribed diameter of the polygon, facet flatness and reflectance, and rotating speed. Owing to the interactions of the specifications and the high-accuracy requirements, the spinner is addressed as a scanner subsystem to allow the scanner designer to make trade-off decisions within the limits imposed by the optical and system performance requirements.

The scanner subsystem in our film recorder example consists of a one-piece beryllium scan mirror and shaft, suspended on hydrostatic gas bearings, and driven by a servo-controlled AC synchronous motor. The F-θ lens is designed to function with a 60° optical scan entrance angle which dictates a 30° facet angle specification for the polygon. The number of facets is therefore calculated to be 12, and a motor speed of 100 rev/s, or 6000 rpm, generates 1200 scan lines per second. Depicted in Table 2 is a summary of the scanner polygon requirements.

3.3 Scanner Specification Tolerances

Scanner specification tolerances are determined by the permissible static and dynamic pixel position errors acceptable at the film plane. These worst-case errors are referenced back through the optical system and scanner subsystem to be distributed and budgeted between the operational elements and reference datum. The acceptable variances are often

Table 2 Film Recorder Polygon Requirements (Source: Ref. 1)

Number of facets	12
Facet angle	30°
Inscribed diameter	4.0 in.
Facet height	0.5 in.
Facet reflectance	89–95%
Facet flatness	$\lambda/20$
Facet quality	MIL-F-48616
Scan rate	1200 scans/s
Rotational speed	6000 rpm

Table 3 Film Recorder Scanner Characteristics and Tolerances (Source: Ref. 1)

Characteristic	Tolerance	Comments
Polygon data		
Number of facets	N.A.	Determined by scan angle
Facet angle	± 10 arc sec	One pixel–pixel angle
Diameter	N.A.	Controlled by facet width dimension
Facet width	1.035 in mm	0.020-in. roll-off
Facet height	0.5 in mm	0.020-in. roll-off
Flatness	$\lambda/20$ max	Spot control
Reflectance	± 3%	Tolerance
Apex angle	1.00 arc sec	Total variation -10% of line–line angle
Speed regulation		
1 revolution	± 10 ppm	± 1.08 arc sec/line
long term	± 50 ppm	± 5.40 arc sec/line

Note: Scan error for any 12 scans $\leq \pm 12.96$ arc sec. N.A., not applicable.

specified as a percentage of pixel-to-pixel angle, pixel diameter, pixel-to-pixel spacing, or the motor speed regulation (stability) over one or more revolutions. The conversion of these variances to meaningful and quantifiable units is necessary for manufacturing, measurement, inspection, and testing of the scanning system components.

Our primary concern here is to relate the scanning system specifications, which ultimately determine the quality of the scanned image, to the performance requirements that are imposed on the motor and control system. It is apparent (Table 3) that the polygon rotation regulation plays a major role in the scanning system performance and that the design of the motor and control system must be given careful consideration in order to achieve the precise pixel placement accuracy.

Precision closed-loop control of the motor speed is essential if the performance targets outlined in Table 1 are to be met. To achieve the level of speed regulation required by the 13.97 μm pixel spacing specification, a phase-lock loop control system with a quartz oscillator frequency reference will be required. The 13.97 μm pixel spacing translates to a polygon speed regulation requirement of 0.001% (10 ppm) within one rotation of the scanner and over the time required to write the full page of the image onto the film. Speed variations within one turn of the polygon will produce an uneven scan line length, which will be visible as a variation along the edge of the film opposite the start of scan. Slower speed variations that occur over many revolutions but still within the same film frame may produce other undesirable image artifacts and distortions.

Subtle variations in the printed image such as shading and banding can also result from short-term speed changes within the scanner, which may occur over a few revolutions, and the human eye has a remarkable ability to detect these otherwise minor variations within the image.

3.4 High-Performance, Defined

Table 4 depicts the performance of the film recorder reference system in comparison to a state-of-the-art recording system considered by many as the highest resolution and fastest

Table 4 Scanner Performance Comparison (Source: Ref. 1)

Characteristics	Reference system	State-of-art system
Facet number	12	20
Facet tolerance	± 10 arc sec	± 1 arc sec
Apex angle error	± 0.4 arc sec	± 0.2 arc sec
Speed	6000 rpm	28,800 rpm
Scan rate	1200 scans/s	9600 scans/s
Speed regulation/rev	<10 ppm	<1 ppm
Pixels/scan	10,000	50,000
Pixel/jitter/rev	<± 25 ns	<± 2 ns
Pixel clock	12 MHz	480 MHz

system manufactured to date. Note that the polygon speed regulation required in the state-of-the-art system is less than one part per million, or 0.0001%.

4 MOTOR CONSIDERATIONS

4.1 Motor Requirements

In the most demanding scanning applications, which require high speed and exceptional accuracy, the precision of the integral polygon, shaft, and bearing assembly must not be degraded by the introduction of the motor rotor. This places a heavy burden on motor rotor selection.

Any motor rotor attached to the shaft must have very stable and predictable characteristics with regard to strength and temperature, and if possible, the rotor material should be homogeneous. If the rotor is of a complex mechanical configuration and consists of laminations and windings, the assembly may not maintain a precision balance (less than 20 μin-oz) when operating at high speeds. Additionally, thermal expansion and high centrifugal forces may shift and reposition the rotor and perhaps cause a catastrophic failure in an air bearing scanner.

Two motor designs are considered for use with high- and low-speed air bearing and ball bearing scanners, respectively; they are the hysteresis synchronous and the DC brushless. Pictured in Fig. 10 is a high-speed DC brushless motor with integral Hall-effect sensors for commutation and the associated rotor magnet mounted on a brass encoder hub assembly. Figure 11 depicts the two main components of a hysteresis synchronous motor: the stator and the hysteresis ring rotor.

4.2 Hysteresis Synchronous Motor

The difficult rotor mechanical stability requirements for high-speed scanner operation are easily achieved with the use of a hysteresis synchronous motor (Fig. 11). The hysteresis rotor is uncommonly simple in design and consists of a cylinder of hardened cobalt steel that is heat-shrunk onto the rotor shaft assembly. Careful calculations are required with regard to the centrifugal forces and thermal expansion influences on the rotor and shaft for safe and reliable operation. This type of motor is well suited for operation at speeds ranging from 1000 rpm to 120,000 rpm. Motors having output power as large as

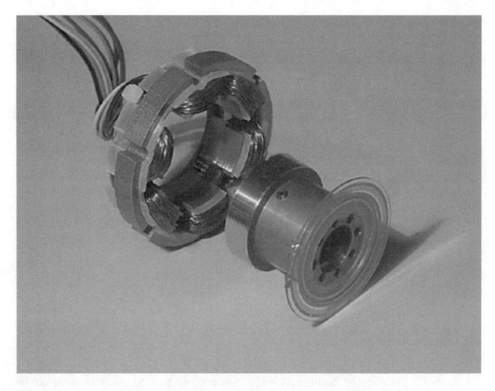

Figure 10 Brushless DC motor: stator and rotor.

2.2 kw have been successfully used on large-aperture IR scanning systems operating at 6000 rpm.

The operation of a hysteresis synchronous motor relies on the magnetic hysteresis characteristics of the rotor material. As the magnetizing force from a suitably powered stator (not unlike that used with reluctance-type motors) is applied to a cobalt steel rotor ring or cylinder, the induced rotor magnetic flux density will follow the stator coil current, as illustrated in Fig. 12.

The sinusoidal current is shown to increase from zero, along the initial magnetization curve, to point (a), thereby magnetizing the material to a corresponding flux level at the peak of the sinusoid. As the current decreases to zero, the rotor remains magnetized at point (b). If the current at this point in time were to remain at zero, the rotor would be permanently magnetized at the point (b) flux level. However, as the current reverses direction, the flux reduces to zero at some negative value of current as shown at point (c). Further decreases in current (negative direction) reverse the direction of flux as shown at point (d), corresponding with the negative peak of the current. The process continues to point (e) and back to point (a), completing the loop for one cycle of current. The figure generated is called a magnetic hysteresis loop. In physics, hysteresis is defined as a lag in the magnetization behind a varying magnetizing force.

By analogy, as the axis of the magnetizing force rotates, the axis of the lagging force of the rotor will accelerate the rotor in the same direction as the rotating field. As the rotor accelerates, its speed will increase until it reaches the synchronous rotating frequency of

Figure 11 Hysteresis synchronous motor: stator and rotor (from Ref. 3).

the field. At this point, the rotor becomes permanently magnetized and follows the rotating field in synchronism.

The synchronous speed of the rotor can be calculated with the following expression:

rpm = $120f/N$

where f = line frequency (Hz) and N = number of poles.

Figure 13 depicts a typical speed vs. torque characteristic curve for a hysteresis synchronous motor.

If a fixed line voltage and frequency is applied to the stator winding of the motor, an accelerating torque is developed equal to the starting torque, T_s, shown at point A. As the speed of the rotor increases, the operating point on the curve moves through the maximum torque developed at point B, and continues through point C, at which time synchronous speed is reached. The final operating point, D, is determined by the operating load torque presented to the shaft at torque level T_0. Note that if the operating load torque is greater than the in-sync torque, synchronous speed will not be reached.

Figure 14 is a vector representation of the rotating magnetizing field and the magnetized rotor field while in synchronism. Note that the rotor field vector lags the magnetizing field by an angle α. The operating torque (as developed by the motor in synchronism) is in proportion to the sine of the angle α in electrical degrees. If the load torque and stator frequency are absolutely constant, their frequencies will be precisely equal.

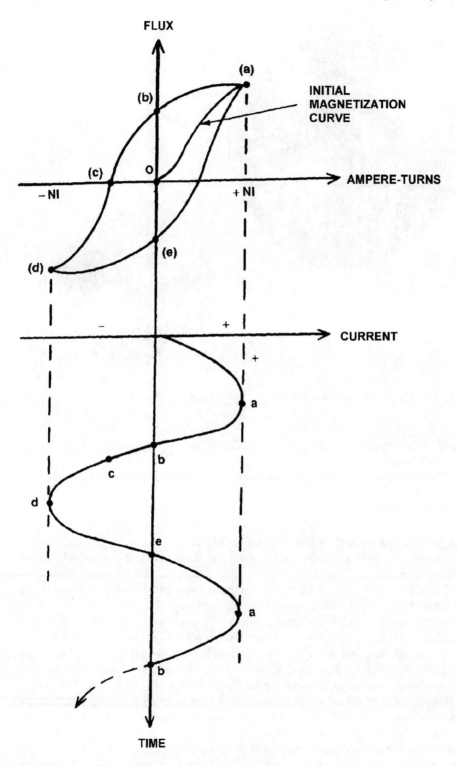

Figure 12 Magnetic hysteresis curve (from Ref. 4).

Motors and Controllers (Drivers)

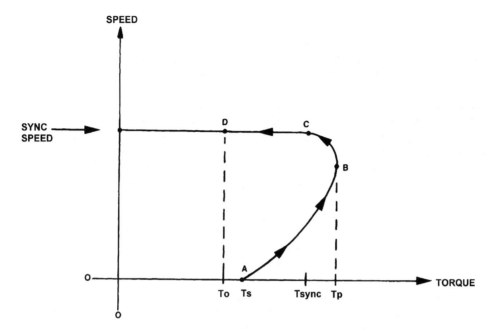

Figure 13 Speed/torque performance curve (from Ref. 4).

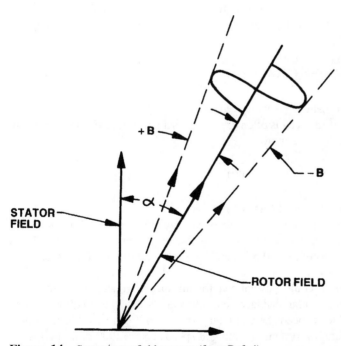

Figure 14 Stator/rotor field vectors (from Ref. 4).

However, should the torque angle be modulated sinusoidally, as indicated in Fig. 14 (with a torque variance of $\pm \beta$), the rotor vector will advance and retard as indicated about the average angle α. The long-term average speed will be as constant as the applied stator source frequency, but the instantaneous speed will follow the derivative of the sine wave on a one-to-one basis. The effect of torque perturbations that are not attenuated by system inertia will also modulate the shaft speed accordingly.

A characteristic of hysteresis synchronous motors (and other second-order devices and systems) called "hunting" may be observed when operating the motor in a system having low losses and damping factor. The motor rotor will oscillate in a sine wave fashion, not unlike that depicted in Fig. 14, if perturbed by applied forces internal or external to the system. If the perturbations are sustained, the oscillations will also be sustained.

However, if the perturbations are not sustained, the oscillations will diminish in amplitude to essentially zero. The internal damping factor of the motor can be influenced by the rotor resistivity, rotor-to-stator coupling coefficient, and driver source and stator impedance. In a typical open-loop operation (no external velocity or position feedback) the oscillations may not be predictable and, therefore, may suddenly appear due to an unknown source or sources of perturbing forces. The amplitude of the oscillations in shaft degrees can typically range from $1°$ to $10°$, and, as a practical matter, is very difficult to calculate; however, the rotor oscillating (hunting) frequency (W_n) can be estimated as

$$W_n = \sqrt{K/I}$$

where W_n = natural resonant frequency (rad/s), K = motor stiffness (in-oz/rad, or $\Delta T/\Delta\alpha$), and I = shaft moment of inertia (in-oz s^2).

The maximum instantaneous speed is determined by setting the derivative of the sine function to zero, and then calculating the maximum positional rate of change in radians per second:

$$\text{Change in speed(rad/s)} = \pm A_p W_n$$

where $A_p = \pm \beta$ and W_n = natural resonant frequency (rad/s).

The maximum change in speed is often expressed as a percentage change relative to the nominal operating speed of the motor:

$$\text{Velocity regulation}(\%) = 100(A_p W_n / W_s)$$

where W_s = nominal operating speed (rad/s).

Figure 15 shows two curves of percent velocity regulation vs. speed in rpm, for a typical open-loop scanning system having peak angular displacements of $1°$ and $5°$. A four-pole motor with a peak torque of 10 oz-in and having a total inertia of 0.076 oz-in s^2 was used for the calculations in this figure. Note that for peak angular displacements of $1°$, velocity regulation of 0.05% could be claimed for all speeds greater than 3000 rpm. However, should the peak angular displacements increase to $5°$, then 0.05% velocity regulation is only obtainable at speeds greater than 14,000 rpm. Figure 15 illustrates that, for a typical open-loop scanning system operating at speeds above 3000 rpm, the system designer can expect variance in velocity ranging from less than 0.05% to as high as 0.25%.

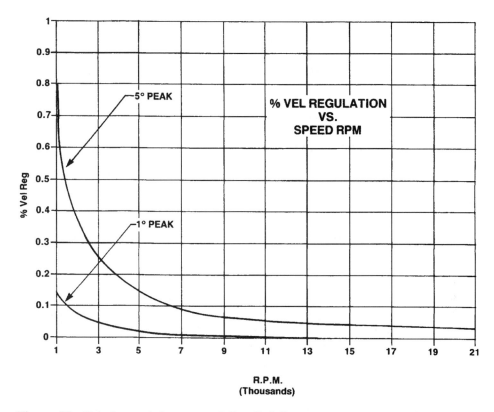

Figure 15 Velocity regulation vs. speed (from Ref. 1).

In conclusion, if system speed regulation requirements must be guaranteed to be less than 0.05% (500 ppm), closed-loop control of the motor speed using phase position feedback must be incorporated.

Another advantage of hysteresis synchronous motors in precision scanners is the near absence of rotor eddy currents when the rotor is synchronized with the rotating stator field. These currents can produce increased I^2R rotor losses that can cause rotor/shaft distortions due to generated temperature gradients and produce adverse effects, especially in air bearing designs. The primary source of rotor eddy current losses is from the spurious flux changes that occur as the rotor passes the stator slots. These parasitic losses are often referred to as "slot effect losses" and can be very significant at high speeds, rendering the device very inefficient as is often noted in older designs. Careful stator design can minimize these losses to the extent that the primary source of rotor/shaft heating is through the air bearing gap from the stator or due to air friction. Any residual heat generated by stator losses may be further reduced and diverted away from the bearing/rotor system by water or air cooling of the stator housing to minimize the rotor/shaft temperature rise.

In summary, AC hysteresis-synchronous motors were a natural choice in early polygon scanners, especially for high-speed applications. Simplicity of construction and reliable, maintenance free operation were key advantages. Also, since the long-term shaft speed was precisely determined by the excitation frequency, no speed control system was required in order to produce acceptable results in low-cost systems. However, as scanner speed control requirements became more critical, feedback devices such as optical encoders were added in order to control the short-term speed variation or "hunting" found in this type of motor. This required increase in control complexity paved the way for the entry of DC brushless motors for precision scanning applications.

4.3 Brushless DC Motor Characteristics

The brushless DC motor (Fig. 10) is well suited for speeds ranging from near zero to as high as 80,000 rpm. These motors exhibit the same characteristic as brush commutated types and can therefore be used in the same applications. They are also suited for velocity and position servo applications since they have a near ideal, linear control characteristic, meaning that the torque produced by the motor is in direct proportion to the applied current.

The elimination of brushes and commutating bars provides reduced electromagnetic interference, higher operating speeds and reliability, with no brush material debris from brush wear. The commutating switching function is accomplished by using magnetic or optical rotor position sensors that control the electronic commutating logic switching sequence. In actual operation, a DC current is applied to the stator windings, which generates a magnetic field that attracts the permanent magnets of the rotor, causing rotation. As the rotor magnetic field aligns with the stator field, the field currents are switched, thereby rotating the stator field and the rotor magnets follow accordingly.

The rotor will continue to accelerate until the motor output torque is equal to the load torque. Under no load conditions, the motor speed will increase until the back electromotive force (BEMF) generates a voltage equal to the stator supply voltage minus the DC winding resistance voltage drop. At this point, the rotor speed reaches an equilibrium level as determined by the BEMF motor constant.

The open-loop speed stability and regulation under controlled power supply and temperature conditions is usually 1–5%, so the device is typically used with closed-loop feedback control. In the closed-loop mode of operation, particularly in the phase-lock loop configuration, short-term speed stability of 1 ppm is obtainable. However, on a long-term basis, the speed stability and accuracy is only as good as the reference source, which is routinely specified at 50 ppm or less for quartz crystal oscillator references, which are used in phase-lock loop speed control systems.

The brushless DC motor, when properly commutated, will exhibit the same performance characteristics as a brush commutated DC motor, and for servo analysis the two may be considered equivalent devices. Both motor types may be characterized by the same set of parameters as described in the following discussion.

4.3.1 Torque and Winding Characteristics

The basic torque waveform of a brushless DC motor has a sinusoidal or trapezoidal shape. It is the result of the interaction between the rotor and stator magnetic fields, and is defined as the output torque generated relative to rotor position when a constant DC current is applied between two motor leads. With constant current drive, the torque waveform follows the shape of the back-EMF voltage waveform (BEMF) generated at any two motor

Motors and Controllers (Drivers)

winding leads. The frequency of the BEMF voltage waveform is equal to the number of pole pairs in the motor times the speed in revolutions per second. The BEMF waveform is easily observed by rotating the motor at a constant speed and is in fact often used to characterize the motor during testing.

The brushless DC motor exhibits torque/speed characteristics similar to a conventional brush-type DC motor. The stator excitation currents may be square wave or sinusoidal and should be applied in a sequence that provides a constant output torque with shaft rotation. Square wave excitation results in a small ripple in the output torque due to the finite commutation angle.

The commutation angle is defined as the angle the rotor must rotate through before the windings are switched. Ripple torque is typically expressed as a percentage of average-to-peak torque ratio and is present whenever the windings are switched by a step function either electrically via solid-state switches or mechanically via brushes. In brushless DC motors designed for square wave excitation, the ripple torque can be minimized by reducing the commutation angle through the use of a larger number of phases, which also improves motor efficiency. For a two-phase brushless motor, the commutation angle is 90 electrical degrees, which yields the largest ripple torque of about 17% average-to-peak. A three-phase delta-wound motor is shown in Fig. 16. The commutation angle is $60°$ and the ripple torque is approximately 7% average-to-peak. Since two-thirds of the available windings are used at any one time, compared to one-half for the two-phase motor, the three-phase system is more efficient.

The torque waveforms shown in Fig. 16 have a sinusoidal shape. For square wave excitation of the motor a trapezoidal torque waveform will produce improved torque uniformity. A trapezoidal torque waveform can be obtained by using a salient pole structure in conjunction with the necessary lamination/winding configuration. In practice, the trapezoidal torque waveform does not have a perfectly flat top and the benefit to torque ripple reduction may be small.

The commutation points and output torque for a three-phase brushless motor are shown in Fig. 16. Each phase (winding) is energized in the proper sequence and polarity to produce the sum torque shown at the bottom of the figure and is calculated to be equivalent to the current times the torque sensitivity of the motor (IK_T).

4.3.2 Brushless Motor Circuit Model

The equivalent electrical circuit model for a DC brushless motor is shown in Fig. 17. This model can be used to develop the electrical and speed–torque characteristic equations, which are used to predict the motor performance in a specific application.

The electrical equation is:

$$V_T = IR + L\,dI/dt + K_B \omega \tag{1}$$

where V_T = the terminal voltage across the active commutated phase, I = sum of the phase currents into the motor, R = equivalent input resistance of the active commutated phase, L = equivalent input inductance of the active commutate phase, K_B = back EMF constant of the active commutated phase, and ω = angular velocity of the rotor.

If the electrical time constant of the brushless DC motor is substantially less than the period of commutation, the steady-state equation describing the voltage across the motor

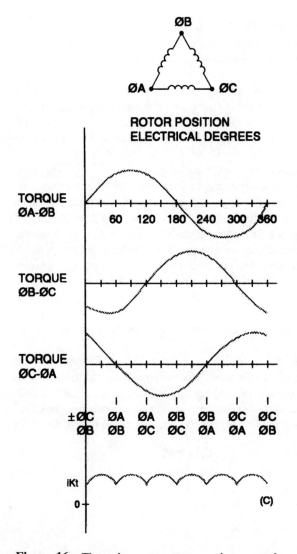

Figure 16 Three-phase motor torque and commutation points (from Ref. 5).

may be written as:

$$V_T = IR + K_B \omega \qquad (2)$$

The torque developed by the brushless DC motor is proportional to the input current such that:

$$T = IK_T$$

where $K_T =$ the torque sensitivity (oz-in/A).

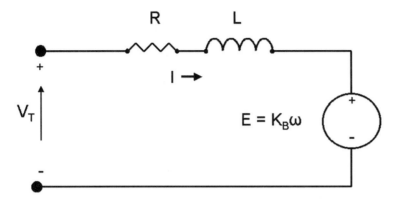

Figure 17 DC brushless motor equivalent circuit (from Ref. 5).

Solving for I and substituting into Eq. (2) yields:

$$V_T = T/K_T R + K_B \omega \qquad (3)$$

The first term represents the voltage required to produce the desired torque, and the second term represents the voltage required to overcome the back-EMF of the winding at the desired operating speed. If we solve Eq. (3) for rotor speed we obtain:

$$\omega = V_T/K_B - TR/K_B K_T \qquad (4)$$

Equation (4) is the speed–torque relationship for a permanent magnet DC motor. A family of speed–torque curves represented by Eq. (4) is shown in Fig. 18. The no-load speed may be obtained by substituting $T = 0$ into Eq. (4):

$$\omega(\text{no load}) = V_T/K_B$$

Stall torque can be found by substituting $\omega = 0$ into Eq. (4):

$$T(\text{stall}) = K_T V_T/R = I K_T$$

The slopes of the parallel line speed curves of Fig. 18 may be expressed by:

$$R/K_B K_T = \omega(\text{no load}) T(\text{stall})$$

Since the speed–torque curves are linear, their construction is not required for predicting motor performance. The system designer can calculate the required information for servo performance from the basic motor parameters given by the manufacturer.

4.3.3 Winding Configurations

Almost all of the brushless motors and drives produced today will be of the three-phase configuration, although two-phase motors have unique advantages, which are exploited in

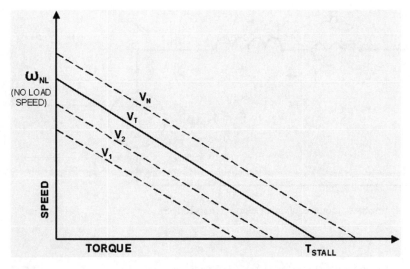

Figure 18 DC motor characteristic curves (from Ref. 5).

very low-speed applications. Three-phase windings can be connected in either a "delta" or "Y" configuration as shown in Fig. 19.

Excitation currents into the windings can be switched full-on, full-off, or be applied as a sinusoidal function depending on the application. The switch mode drive is the most commonly used system because it results in the most efficient use of the electronics. Two switches per phase (winding) terminal are required for the switch mode drive system. Therefore, only six switches are required for either the "Y" or "delta" configuration.

The delta windings form a continuous loop, so current flows through all three windings regardless of which pair of terminals are connected to the power supply. Since

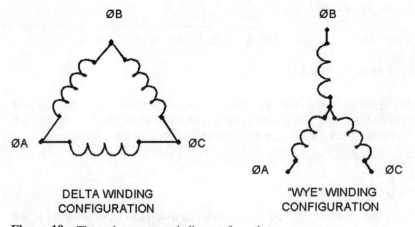

Figure 19 Three-phase motor winding configurations.

the internal resistance of each phase is equal, the current divides unequally, with two-thirds flowing through the one winding connected directly between the switched terminals, and one-third flowing through the two series-connected windings appearing in parallel with it. This results in switching only one-third of the total current from one winding to another as the windings are commutated.

For the "Y" connection, current flows through the two windings between the switched terminals. The third winding is isolated and carries no current. As the windings are commutated, the full load current must be switched from terminal to terminal. Owing to the electrical time constant of the windings, it takes a finite amount of time for the current to reach full value. At high motor speeds, the electrical time constant ($T_E = L/R$) may limit the switched current from reaching full load value during the commutation interval, and thus limits the generated torque. This is one of the reasons the delta configuration is preferred for applications requiring high operating speed. Other considerations are manufacturing factors, which permit the delta configuration to be fabricated with lower back-EMF constant, resistance, and inductance. A lower back-EMF constant allows the use of more common low-voltage power supplies, and the solid-state switches will not be required to switch high voltage. For other than high-speed applications, the "Y" connection is preferred because it provides greater motor efficiency when used in conjunction with brushless motors designed to generate a trapezoidal torque waveform.

4.3.4 Commutation Sensor Timing and Alignment

A brushless DC motor duplicates the performance characteristics of a DC motor only when its windings are properly commutated, which means that the winding currents are applied at the proper time, polarity, and in the correct order. Proper commutation involves the timing and sequence of stator winding excitation. Winding excitation must be timed to coincide with the rotor position that will produce optimum torque. The excitation sequence controls the polarity of generated torque, and therefore the direction of rotation. Rotor position sensors provide the information necessary for proper commutation. Sensor outputs are decoded by the commutation logic electronics and are fed to the power drive circuit, and activate the solid-state switches that control the winding current.

A useful method to achieve correct commutation timing is to align the position sensor to the back-EMF waveform. Since the back-EMF waveform is qualitatively equivalent to the torque waveform, the test motor can be driven at a constant speed by another motor, and the position sensors aligned to the generated back-EMF waveform. The sensor transition points relative to the corresponding back-EMF waveforms should be coincident when correct commutation has been achieved. For critical applications that require the commutation points to be optimized, the motor should be operated at its rated load point, and then the position sensors should be adjusted until the average winding current is at its minimum value.

The commutation points and output torque for a three-phase brushless motor are shown in Fig. 16. The commutation angle is 60 electrical degrees. The windings are switched "on" at 30 electrical degrees before the peak torque position, and switched "off" at 30 electrical degrees after the peak torque position. The current polarity must be reversed for negative torque peaks to produce continuous rotation. To identify each of the six commutation points, a minimum of three logic signals are required, as shown in Fig. 20. The logic signals are generated by three sensors which are spaced 60 electrical degrees apart and produce a 50% duty cycle.

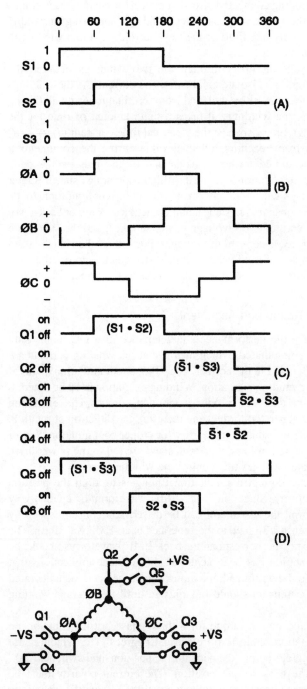

Figure 20 Three-phase motor commutation logic and excitation (from Ref. 5).

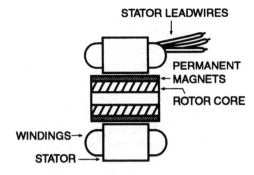

A) INNER ROTOR BRUSHLESS DC MOTOR

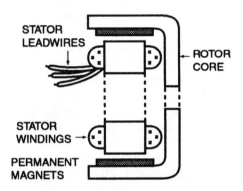

B) OUTER ROTOR BRUSHLESS DC MOTOR

Figure 21 DC brushless motor configurations (from Ref. 5).

As indicated in Figs. 16 and 20, sensor S1 can be readily aligned to the $\Phi A - \Phi B$ zero torque position. This can be accomplished by applying a constant current to the $\Phi A - \Phi B$ terminals. The rotor will rotate to the $\Phi A - \Phi B$ zero torque position and then stop. Sensor S1 should then be positioned so that its output just switches from a low to high logic state. Sensors S2 and S3 may then be positioned 120 and 240 electrical degrees respectively from S1 (either CW or CCW direction depending on the direction of rotation) and basic commutation will be established for the motor.

4.3.5 Rotor Configurations

The brushless DC motor rotor configurations (Fig. 21) most often used for scanners consist of rare-earth samarium–cobalt permanent magnets that are contained with a rigid ring or cup for the outer rotor configurations, and are usually bonded to a machined hub for the inner rotor configurations. The inner rotor configuration is generally used at the higher speeds because of the lower centrifugal forces resulting from a reduced rotor diameter.

Owing to the elastic characteristics of epoxy and other adhesives, and the need for stable and reliable precision balancing, the operating speed of DC brushless spinner rotors is generally less than what can be expected for the hysteresis motors. However, DC brushless motor technology is steadily displacing other motor types in many high-speed applications. Stainless steel sleeves have been used to aid in the retention of the rotor magnets and improve the rotor mechanical characteristics at high speeds. In addition, ring magnet rotors have been developed for high-speed designs, allowing DC brushless motors to operate in excess of 80,000 rpm. In this rotor configuration, a ring of the rotor material is magnetized by the pole pieces of a powerful electromagnet, which imprints the pole locations and polarities into the ring. This type of rotor design has in fact been employed to drive precision polygonal scanners for laser projection systems which operate at 81,000 rpm.

Pictured in Fig. 10 is a low-cost, high-speed DC brushless motor, which has been successfully used in several scanner designs. The simple stator design, which allows for machine winding of the coils, and the inexpensive ring magnet rotor, account for the low manufacturing cost of this rugged and reliable motor. The rotor magnet is further strengthened by an outer stainless steel sleeve.

The commutation sensors (Hall effect devices) are placed within the stator slots and do not require any timing adjustments. This motor is capable of delivering at least 50 W on a continuous duty basis. Also visible in Fig. 10 is the rotor hub assembly onto which the ring magnet and the optical encoder disc are mounted.

Pictured in Fig. 22 is a miniature eight-pole brushless DC motor, which is suitable for many low-power scanner drive applications. The stator configuration is noticeably more complex than the low-cost motor pictured in Fig. 10 and is intended for lower speed applications where cogging torque must be minimized. Commutation of the windings in this design is performed by dedicated channels of a shaft-mounted optical encoder rather than Hall effect sensors.

Commutation timing adjustment is made possible by rotating the encoder pattern relative to the rotor magnet position. Optimum timing at the desired operating speed is required in order to produce the lowest torque ripple and power consumption. The rotor assembly is of a more conventional design, where individual magnet pieces are bonded to a machined hub, which is then ground to the final dimensions.

For low-speed applications it is desirable to construct the motor using the highest pole count possible, that is consistent with the physical size and the electrical parameters specified. The smaller commutation angle and the resulting higher ripple torque frequency may be more readily attenuated by the rotating inertia in the system and produce a more constant speed within one revolution of the rotor.

5 CONTROL SYSTEM DESIGN

The basic control requirements for a precision polygonal spinner are to provide synchronization and velocity control for precise scan registration, whether on a film plane or detector array, or at a distant target being illuminated. To this end, the principles of feedback control are utilized for synchronization, velocity, and phase position (shaft angle) control.

In our film recorder example (Fig. 9), speed control is required for accurate pixel positioning, repeatability, and linearity, as well as line-to-line pixel registration and synchronization. In order to accurately position data pixels in a line at the film plane, the system must generate precision pulses spatially related to the facet angles. These pulses,

Motors and Controllers (Drivers)

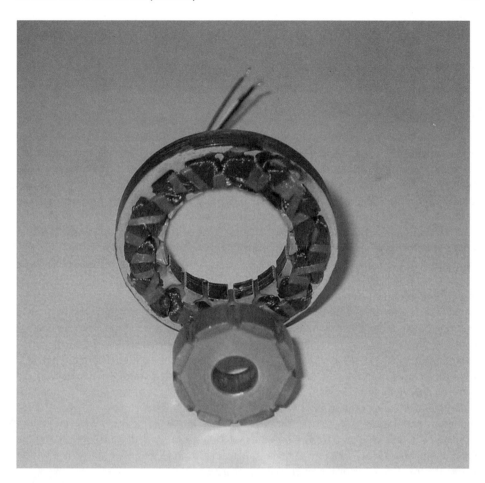

Figure 22 Low-speed DC brushless motor.

occurring on a one per pixel basis, are used to gate in and turn off the intensity-modulated video being projected to the light-sensitive film surface. Because there are 120,000 pixels in the film recorder example (12 times 10,000) per revolution, one clock pulse would be required for every 10 arc seconds of shaft rotation.

Optical encoders are well suited to the task of generating accurately timed and positioned pulses. However, incremental, high-density data track encoders are expensive, large in diameter, and difficult to mechanically interface with an integral spinner/motor/shaft assembly.

To overcome this problem, a smaller and inexpensive low-density optical encoder having 6000 pulses per revolution (PPR) was designed into the system. The required 120,000 PPR pixel clock pulses are obtained by electronically multiplying the encoder data track frequency (6000 PPR) by a factor of 20. Scanner speed control is accomplished by frequency/phase locking the encoder data track (600 KHz) to an accurate and stable crystal oscillator. An index pulse is accurately positioned at the normal of a facet on a second encoder track, thereby providing start of scan (refer to Fig. 9, SOS detector) synchronization and pixel registration.

This system of generating pixel clock pulses places a heavy burden on the performance accuracy of the rotating shaft assembly, the encoder design and adjustment, and the stability of the crystal oscillator. Nevertheless, the net speed control and jitter performance has been optically measured to be less than ± 10 ppm for one revolution of the scanner.

5.1 AC Synchronous Motor Control

Velocity control of the hysteresis synchronous motor is intrinsic to its design, that is, the long-term speed is as accurate as the applied frequency. With reference to Fig. 14, the stator and rotor fields rotate together (at an integer submultiple of the applied frequency) with the rotor lagging by the torque angle α and with possible modulations of $\pm \beta$, as was previously discussed. The control systems' task is to fix the rotor vector position, and therefore eliminate hunting and other speed variances.

To implement phase lock control, the shaft rotational frequency and phase position is measured with a shaft-mounted incremental encoder. The encoder pulses are frequency and phase compared with a stable reference frequency using a frequency/phase comparator, which has the transfer characteristics as shown in Fig. 25. The frequency/phase comparator has a unique transfer characteristic that allows the device to produce an output that is in saturation until the two input frequencies are equal. This is the frequency detection mode of the phase comparator. The saturation level is either positive or negative in value, and is useful in determining whether the motor speed is too high or too low. Assuming that the motor has reached synchronous speed, the tachometer frequency will equal the reference frequency, and the frequency/phase comparator will operate in the phase comparator mode. In this mode of operation, the output of the phase comparator is an analog voltage proportional to the phase difference between f_T and f_R. At zero frequency and phase differences, the two signals are edge locked, and the phase comparator output voltage is zero. Should the shaft advance or retard for any reason, the phase comparator error voltage will be in direct proportion to the phase difference within ± 360 electrical degrees of the reference frequency.

With reference to Fig. 23, the phase comparator error is processed through a proportional-integral-derivative (PID) controller compensation scheme and fed to the control input of the phase modulator.

The phase error-corrected phase modulator output frequency, f_M is applied to the motor, thereby completing the position control loop from the encoder to the frequency/phase comparator. The open-loop DC gain of the system is primarily determined by the product of the encoder, frequency/phase comparator, integrator, and phase modulator gains. The high DC gain of the integrator (100 dB) reduces the phase error between f_R and f_T to zero, resulting in near perfect synchronization. The differentiation gain constant provides sufficient damping to eliminate "hunting" and improve the overall dynamic performance and speed regulation to less than 1 ppm.

5.2 DC Brushless Motor Control

The velocity control of a brushless DC motor differs from that of hysteresis synchronous in that the brushless motor speed is a function of applied motor voltage, as opposed to the frequency/phase as the driving function of the latter. The same principles of feedback velocity/position control are utilized in a similar fashion and are depicted in Fig. 24. The elements within the closed-loop block diagram are essentially the same, with the exception of the motor transfer function and the addition of a DC power amplifier. The DC brushless

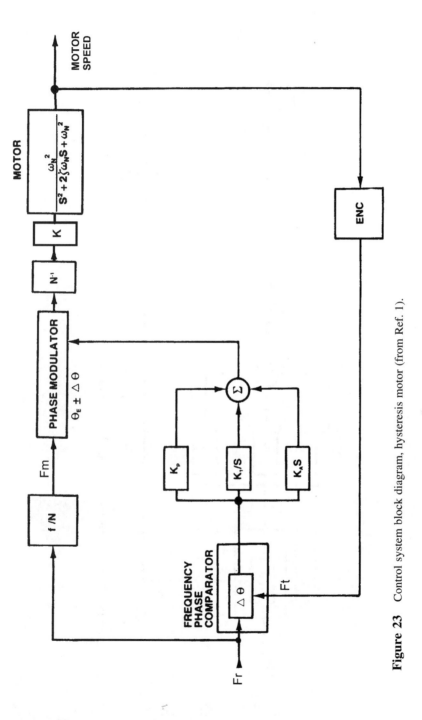

Figure 23 Control system block diagram, hysteresis motor (from Ref. 1).

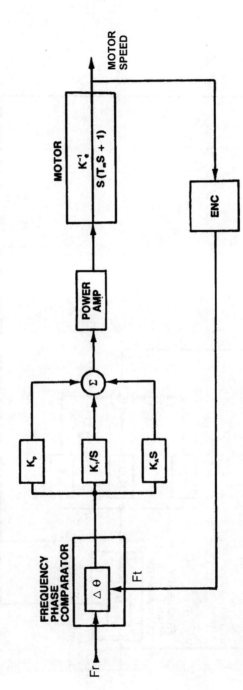

Figure 24 Control system block diagram, brushless DC motor.

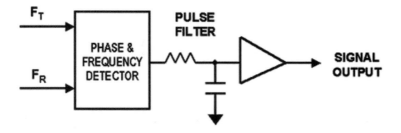

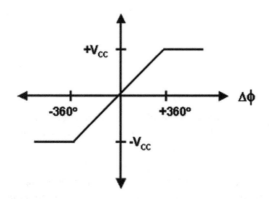

Figure 25 Phase/frequency detector (comparator) characteristics.

motor transfer function is shown in detail in Fig. 26. For simplification purposes, the commutating and pulsewidth modulation circuits have been omitted, but will be covered in a later discussion.

As before, to implement phase lock control, the shaft frequency and phase position are measured with a shaft-mounted incremental encoder. The encoder pulses are frequency and phase compared with the reference frequency using a frequency/phase comparator, which has the transfer characteristics shown in Fig. 25. The frequency/phase comparator has a unique transfer characteristic that allows the device to produce an output that is in saturation until the two input frequencies are equal. The saturation level is either positive or negative in value, and is useful in determining whether the motor speed is too high or too low.

The controller will accelerate or decelerate the motor until there is no frequency difference between the reference and the encoder signal. At this point a phase measurement is made by the comparator with every reference pulse cycle and an output voltage is generated that is proportional to the phase error.

The phase error signal is then processed through a PID controller and compensator, similar to that used in the hysteresis motor control system (Fig. 23).

The detailed DC motor transfer function shown in Fig. 26 relates the motor angular velocity ω_s to the applied terminal voltage V_T. In English units the constants are defined as

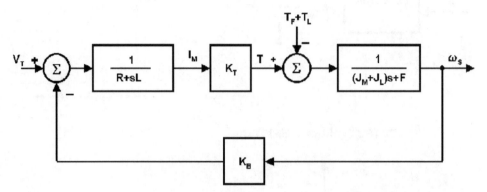

Figure 26 DC motor transfer function block diagram.

(across any two leads for a three-phase delta, or three-phase Y motor): $R =$ motor winding resistance (Ohms); $L =$ motor inductance (Henry); $I_M =$ motor current (Amperes); $K_T =$ motor torque sensitivity (oz-in/Amp); $K_B =$ motor back-EMF constant (Volts/radian/s); $J_M =$ motor moment of inertia (oz-in/s^2); $J_L =$ load moment of inertia (oz-in/s^2); $T_F + T_L =$ sum of the friction and load torque (oz-in); and $F =$ motor damping coefficient.

As the control system is turned on, the error signal is power amplified, causing the motor to accelerate to a speed at which f_T exceeds (overshoots) f_R. At this point, the error signal reverses polarity, reducing the motor speed until f_T equals f_R. Ultimately, a point of equilibrium is reached at which time the frequency/phase comparator error voltage is zero, and the integrator output voltage regulates the speed of the motor. Furthermore, the high DC gain of the integrator maintains a zero phase difference between f_T and f_R resulting in edge lock synchronization.

To ensure stable and accurate speed control, and to determine the gain coefficients K_P, K_I, and K_A, the motor and load characteristics should be modeled. Several very useful simulation programs such as *SIMULINK* (The Math Works, Inc., Natick, MA) are available for the systems designer which greatly reduce development time by allowing the rapid testing of various control configurations.

In conclusion, the DC brushless motor, when properly designed for the scanning application at hand, is capable of meeting or exceeding the performance of AC hysteresis motors in all but the highest speed applications. All of the successful scanning products that are discussed at the end of this chapter use brushless DC motor technology.

6 APPLICATION EXAMPLES

The following sections describe a few of the many scanner designs and control systems that have been developed over the last few years. The overall industry trend reflects the unrelenting drive of the end use market to improve performance, reduce power

Motors and Controllers (Drivers)

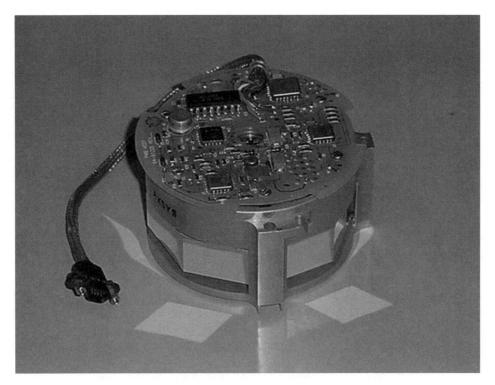

Figure 27 Military vehicle thermal imager scanner.

consumption and size of the supporting electronics, and to reduce the cost of the scanner subsystem.

Fortunately, the consumer electronics and automotive markets have provided many of the electronic components that have proven to be invaluable in the quest to reduce the size and cost of the scanning system. Progress in miniaturization and power reduction of the scanner control electronics is also largely due to the advancements in brushless DC motor technology. As the cost and complexity of the drive electronics for brushless motors approach those of conventional DC motors, brushless motor technology will likely displace all other motor types in scanning subsystems, as it offers high efficiency along with the high reliability, as previously found only in AC motors.

6.1 Military Vehicle Thermal Imager Scanner

Pictured in Fig. 27 is a small 12-facet polygon scanner that is designed for a military vehicle thermal camera system. This compact ball bearing scanner operates at 600 rpm and serves to generate the field scan function in conjunction with an infrared detector array. The motor and control system is designed to maintain polygon speed regulation to within 15 ppm (0.0015%). In addition, the control system must maintain the specified speed regulation in the presence of base disturbances, which are passed to the scanner as a result of vehicle motion. These challenges have been met in this compact scanner by the use of a high-resolution encoder and lightweight polygon design in conjunction with an agile control system.

The scanner employs a small low-voltage brushless DC motor, which is optimized for low cogging torque and smooth operation at low speed. Motor commutation is derived from three dedicated commutation tracks on the optical encoder disk, which also includes a high resolution 3000 count tachometer track as well as an index.

At the center of the control system is a single chip motor driver, which decodes the commutation information provided by the encoder and produces the correct three-phase motor current waveforms necessary for proper operation of the motor. Motor driver ICs of this type are commonplace devices found in computer disk drive and CD player applications. The motor driver produces a current through the motor windings that is proportional to a command voltage at its input and also incorporates a brake feature that is used to decelerate the motor for better control of the polygon speed under the influence of disturbances.

Tight speed control is accomplished by the use of a phase-lock loop regulation method as described in detail in the previous sections. At the operating speed of 600 rpm, the 3000 line optical encoder produces a tachometer frequency of 30 KHz, which is compared against an externally generated reference frequency in the phase/frequency comparator circuit. Any resulting phase error is amplified and filtered, and then fed to the motor driver, which increases or reduces the motor current to maintain the speed and minimize the phase error. If the motor control voltage falls below a predetermined level indicating that the scanner is operating above the reference speed, then the control system applies dynamic braking to the motor, which quickly decelerates the motor and the polygon.

A block diagram of the controller/driver is shown in Fig. 28.

6.2 Battery-Powered Thermal Imager Scanner

The polygon scanning system pictured in Fig. 29 was designed for a compact, low-cost, military thermal sighting device primarily intended for small arms applications. Many unique and difficult requirements are imposed by the system specifications on this relatively low-cost device. A wide operating temperature range, precise scanning speed, low cross-scan error, and low power consumption must simultaneously be met. In addition, the scanner must meet the demanding performance specifications for a high-resolution imaging system while being subjected to severe levels of shock and vibration. Particular attention was given to ensure the mechanical stability of the polygon under severe environmental conditions. Some of the scanner requirements are as shown in Table 5.

In order to maximize battery life, the total power consumption of the scanner was limited by the specification to less than 0.4 W while operating in synchronism at the low-temperature extreme.

A compact low-speed brushless DC motor similar to that in Fig. 22 was designed specifically to address the low power and low cogging torque requirements necessary for this application. A new, single chip motor driver, which employs PWM control, was selected to improve efficiency in the drive and help meet the power consumption requirement. The driver PWM efficiency benefit becomes more apparent as the motor load increases and the drive must supply more current to maintain speed regulation. With a linear motor driver such as the one used in the military vehicle scanner, additional power is lost in the form of heat dissipated in the driver power section.

Motors and Controllers (Drivers)

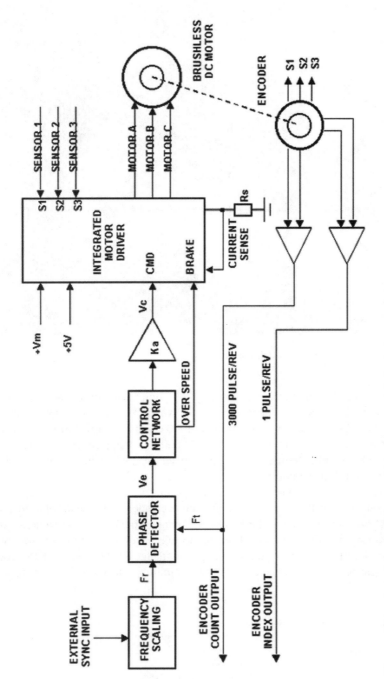

Figure 28 Block diagram, military vehicle scanner controller/driver.

Figure 29 Battery-powered thermal imager scanner.

Effective speed regulation is achieved by the implementation of the same control scheme used in the military vehicle scanner shown in Fig. 28. The optical encoder tachometer track line density was reduced to 1500 counts per revolution to meet the interface requirements of the imaging system. At 600 rpm the encoder tachometer track produces a 15 KHz pulse frequency, which is at an adequately high rate to maintain precise speed control of the polygon.

The phase-lock loop control system was optimized to maintain phase lock of the encoder tachometer with the synchronization reference under the influence of base motion disturbances and thereby prevent the loss of the image produced by the camera system during movement or under vibration. The need to reject base motion disturbances while providing precise speed control poses a unique challenge in the design of the control system. The scanner total rotational inertia must be minimized to allow the motor torque to

Table 5 Battery-Powered Scanner Characteristics

Operating temperature	-40 to $+75°C$
Polygon speed	600 rpm
Speed jitter (one rev)	15 ppm or less (0.0015%)
Encoder outputs	1500 ppr and index
Input voltage	$+10$ VDC and $+5$ VDC
Peak shock level	500 g (0.5 ms)

Motors and Controllers (Drivers)

Figure 30 Low-cost scanner controller.

accelerate and decelerate the polygon in order to overcome disturbance-induced speed changes. On the other hand, the rotational inertia within the scanner acts to reduce the effect of bearing and motor torque fluctuations, which tend to degrade speed stability. A compromise design was reached that adequately addresses both needs and meets the performance targets set for the scanner.

In order to reduce cost and update the electronic design of the control system, a new controller/driver circuit board was developed, which is shown in Fig. 30. Significant cost savings were realized as older, ceramic packaged military grade integrated circuits were replaced with industrial quality devices. Performance and environmental specifications for the scanner were met with the plastic-packaged industrial ICs, and the part obsolescence issues were solved that appear with ever greater frequency in the electronic components world.

6.3 High-Speed Single-Faceted Scanner

The successful scanner design shown in Fig. 31 was developed for the publishing and printing industry for use in the image setting machines that are the output devices leading to the manufacture of printing plates. The single faceted mirror, or "monogon," rotates at

Figure 31 High-speed scanner and controller/driver.

high speed and scans an intensity modulated laser spot along a sheet of film not unlike that shown in the film recorder system of Fig. 9. The exceptions are that the film sheet usually lies on the inside surface of a cylindrical drum and the F-θ lens is omitted. Also, the film sheet is stationary and the scanner rides the length of the drum on precision linear bearings driven by a ball-screw mechanism.

For improved speed stability and cross-scan accuracy, as well as greatly improved bearing life, the scanner rotating elements are supported by self-pumping conical air bearings. This type of bearing provides excellent high-speed stability and low friction with the benefit of virtually unlimited operating life. The self-pumping action of the air bearing design is a major factor in reducing the cost of the scanner because an external air supply is not required. Some versions of this scanner operate at 60,000 rpm and utilize the high-speed motor design shown in Fig. 10. The low-cost optical encoder shown in Fig. 9 provides up to 2000 pulses per revolution for effective speed control and also sends mirror position information to the imaging system. The success of this scanner design is partly due to the development of a low-cost single card controller, which is described in greater detail in the next section. The scanner and control system block diagram is shown in Fig. 33.

6.4 Versatile Single Board Controller and Driver

The need for a versatile, compact, and inexpensive motor driver and speed controller led to the development of a successful circuit capable of delivering up to 100 W of power

Figure 32 Single board scanner controller and motor driver.

to a three-phase brushless motor scanner. All of the functions necessary to achieve precise scanner speed control have been integrated into one unit. The reference frequency generator, phase-lock loop controller, and motor driver are combined on a single low-cost circuit card, which measures 4 × 8 in. A great deal of flexibility has been incorporated into this low-cost controller design so that many scanning applications can be readily accommodated without the need for circuit modifications. Various encoder resolutions and reference frequencies may be accommodated by setting jumpers that reconfigure the digital logic in order to present compatible frequencies to the phase comparator circuit.

This single board controller has been utilized to drive single-faceted as well as polygon scanners operating between 3000 and 81,000 rpm for a variety of laser scanning applications, and has demonstrated excellent speed regulation capabilities. The rotational speed jitter in many of the air bearing scanners driven by this controller was measured to be only a few parts per million within one rotation. This successful design has been incorporated into thousands of scanning systems sold to the printing and publishing industry worldwide. The controller is shown in Fig. 32.

As shown in Fig. 33, the circuit functions can be divided into the following main categories:

1. reference frequency generator and external sync processing circuit;
2. phase detector and PWM synchronization circuit;

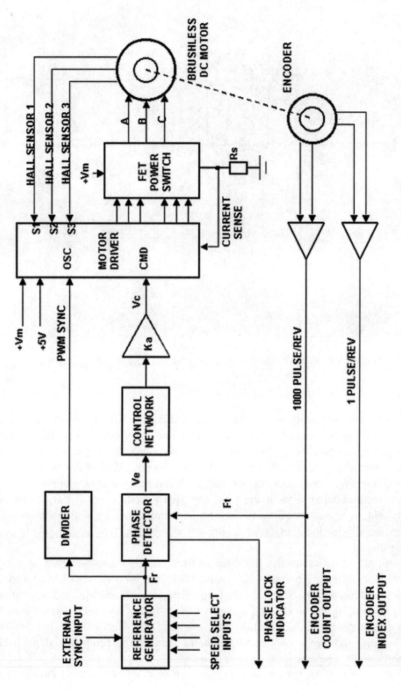

Figure 33 High-speed scanner and controller functional diagram.

3. control loop PID circuit; and
4. brushless motor controller and FET power section.

The controller reference frequency generator and external sync processing circuits function to provide a precise and stable frequency reference to the phase comparator. An on-board quartz reference oscillator and programmable divider provide for the selection of up to 16 preset operating speeds for the scanner. The speed may also be continuously varied within a wide band with the application of an external reference frequency from the imaging system controller. In this way, fine adjustments to the scanning speed may be made to trim the system optical parameters.

The phase detector and frequency comparator circuits produce the speed error voltage, which is then amplified and sent to the servo compensation network. The phase detector exhibits high gain since the output is in saturation until the two frequencies f_R and f_T are exactly the same, as discussed previously and shown in Fig. 25. This feature is responsible for the precise nature of phase-lock loop speed control systems.

Another useful innovation developed in the quest to provide the best possible speed regulation is the synchronization of the PWM oscillator with the phase detector reference frequency. The synchronization of these two frequencies ensures that the noise generated by the beat frequencies do not interfere with the scanner speed control circuits. When synchronized, the two frequencies produce a sum or difference (beat frequency), which will be constant. Stationary beat frequencies may be filtered or appear only as DC offsets, which may be subtracted from the speed control signal. The PID servo controller is similar to the arrangement shown in Fig. 24 and described in earlier sections. Because most of the scanning systems targeted for this controller are for stable and controlled environment applications, the servo control loop is optimized for speed control regulation rather than response time. In these applications, high rotating inertia in the scanner rotor and polygon is a benefit to speed regulation.

The motor driver and power output section consists of a monolithic (single chip) controller and discrete FET power switches. With a modest amount of forced air cooling, the FET power section is capable of delivering up to 4 A continuously and 6 A for several seconds to deliver higher motor starting current. The motor current is regulated by the driver IC using PWM control, which is effective in delivering power to the motor with minimum heat generation in the controller.

For some high-speed polygonal scanning applications it may not be possible to include an optical encoder within the scanner housing. In this instance, the polygon facet frequency may be used as the speed feedback sensor as depicted in Fig. 34. The optical pulse frequency should be at or above 1 KHz in order to provide precise speed regulation in this configuration.

7 CONCLUSIONS

The availability of low-cost brushless DC motors and drivers has made a significant impact on rotary scanner designs over the past ten years. Advances in the motor driver area have been especially important in that the size and cost of these devices have been reduced remarkably. The power efficiency of brushless DC motors as well as the motor driver allows for high mechanical power output with minimal temperature rise. For many polygon-scanning applications, it is now practical to integrate the drive and control functions directly on the scanner or motor body.

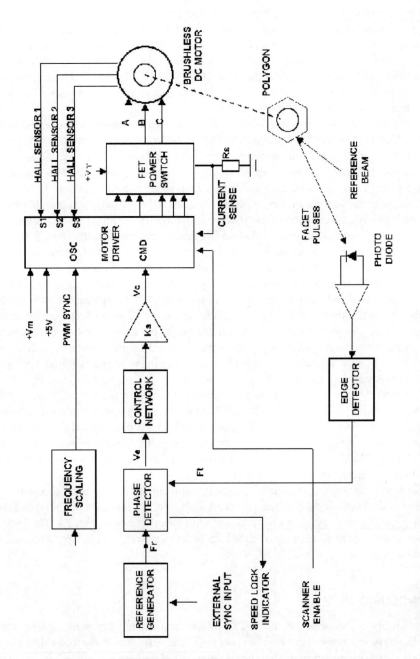

Figure 34 Scanner controller configured for high-speed polygon application.

ACKNOWLEDGMENTS

My thanks to Mr Gerald Rynkowski for compiling the groundwork material on which this chapter is based. His many years of experience and broad knowledge of control systems and opto-mechanical scanning design have contributed to the successful development of many commercial and military scanning systems.

Many thanks also to Mr David Fleming of S-Domain, Inc. (San Diego, CA) and Mr. Qunshan Du of Buehler Motor, Inc. (Cary, NC) for their review and valuable input in verifying the accuracy of the material in this chapter.

REFERENCES

1. Marshall, G.F.; Rynkowski, G.A. Eds. *Optical Scanning*; Marcel Dekker, Inc.: New York, 1991.
2. Speedring Systems Group, Rochester Hills, MI. Technical Bulletins: *Ultra Precise Bearings for High Speed Use 102-1, Gas Bearing Design Considerations 102-2, Rotating Mirror Scanners 101-1,101-2, 101-3.*
3. Rotors, H.C. *The Hysteresis Motor – Advances Which Permit Economical Fractional Horsepower Ratings*; AIEE Technical Paper 47-218, 1947.
4. Lloyd, T.C. *Electric Motors and Their Applications*; Wiley: New York, 1969.
5. Axsys Technologies Motion Control Products Division, San Diego, CA, *Brushless Motor Sourcebook*; Axsys: San Diego, CA, 1998.

6

Bearings for Rotary Scanners

CHRIS GERRARD

Westwind Air Bearings Ltd., Poole, Dorset, United Kingdom

1 INTRODUCTION

Although rotary scanners can take a variety of forms today, the basic concept of smoothly rotating a reflective, or holographic, optic remains the same. The optic must be rotated around a defined axis with a high degree of repeatability, and within a specified speed stability. These requirements will define, in the broadest sense, the type of bearings to be selected within the scanner assembly. Other considerations will include package price, maximum speed, thermal and environmental issues, and lifetime.

Owing to the design interactions of the different components within the scanner assembly, discrete parts can rarely be designed in isolation from one another. It is necessary to understand how the dynamics of the shaft interact with the bearing system, together with the effects of additional parts such as the motor, encoder, and optic.

The object of this chapter is to examine many of the compromises and trade-offs necessary to specify, rather than design, the correct bearing/shaft system, for the machine designer with limited experience of such systems. Owing to recent advances in the design of gas bearings and the ever increasing demands on performance, this chapter will focus more on this technology rather than on ball bearing designs, although all the alternatives will be discussed. A more detailed analysis of ball bearing design criteria can be reviewed in Ref. 1.

2 BEARING TYPES FOR ROTARY SCANNERS

When designing a new rotating product, the traditional first choice for most designers will be some form of rolling element bearing. Readily available, easy to incorporate, and usually relatively inexpensive, it appears the ideal solution for a rotary scanner, and indeed many successful designs were used in early internal drum and flatbed image setters as well as laser printers, plotters, faxes, and photocopiers.

However, with the advent of higher resolution and productivity, machine designers have had to find alternative solutions to the traditional ball bearing assembly. The main types of bearings that can be considered will now be examined briefly.

2.1 Gas-Lubricated Bearings

Regarded by most now as the industry standard for high-quality scanning devices, the use of gas-lubricated, and more specifically air-lubricated, bearings have become widespread across the image setting and laser printing industries. These bearings take the form primarily of self-acting, or aerodynamic, design, generating their own internal air film between the shaft and bearing, but there are also larger designs using externally pressurized bearings requiring some form of compressor. Each type has its own advantages and these will be examined in detail later in this chapter.

2.2 Oil-Lubricated Bearings

The hydrodynamic oil bearing, being self-acting, could be utilized in a rotary scanner, but its main disadvantage of static oil leakage limits it to special applications only.

2.3 Magnetic Bearings

The use of active magnetic bearings where the shaft is supported by a strong magnetic field, is becoming a practical alternative due to enormous improvements in the electronic controls to maintain the position of the shaft within the magnetic fields. A hybrid combination of passive magnetic bearings and air-lubricated bearings is also now in use.

2.4 Ball Bearings

For certain less demanding applications, angular contact ball bearings are still the ideal choice, especially with the recent advances in hybrid bearings utilizing ceramic ball bearing technology, and improved grease lubricants. These products can be found in certain desktop laser printers, fax machines, and most barcode scanner systems.

3 BEARING SELECTION

Before examining each bearing type in some technical detail, it is proposed to compare the relative benefits and disadvantages of each technology to allow the designer to focus quickly on the correct selection and move on to the relevant section of this chapter. Figure 1 shows a simple comparison chart of the most likely bearing systems to be considered for a rotary scanner based on rotational accuracy, maximum speed capability, relative price, and lifetime.

From Fig. 1, it is clear that for most high-accuracy scanner applications, air bearing technology is the most suitable, while for lower specification products, the ball bearing design would be the most cost-effective. However, a more detailed comparison may be necessary if the scanner is to be operated under special conditions, which could demand other bearing systems. Table 1 gives a more detailed comparison, and compares the merits of self-acting against aerostatic air bearings.

Bearings for Rotary Scanners

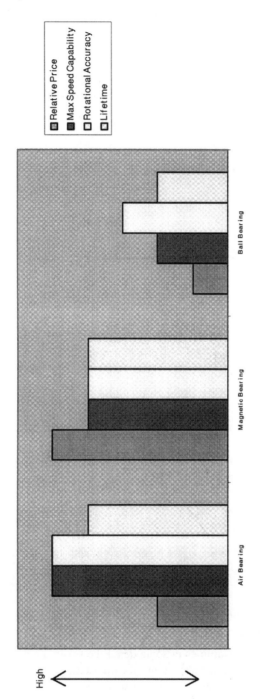

Figure 1 General comparison of bearing systems.

Table 1 Detailed Comparison of Bearing Systems

Parameter	Air bearing		Oil-bearing hydrodynamic	Mag bearing active	Ball bearing ang. contact
	Self-acting	Aerostatic			
Accuracy of rotation	Excellent	Excellent	Good	Good	Fair
Speed:					
<1000 rpm	Poor	Excellent	Excellent	Good	Fair
<1000–30,000 rpm	Excellent	Excellent	Fair	Good	Good
>30,000 rpm	Excellent	Excellent	Poor	Excellent	Poor
Low vibration	Excellent	Excellent	Excellent	Excellent	Fair
Shock resistance	Fair	Good	Excellent	Good	Good
Frequent stop/starts	Good	Excellent	Excellent	Excellent	Good
Low starting torque	Fair	Excellent	Good	Excellent	Good
Long lifetime (>20,000 hours)	Good	Excellent	Excellent	Excellent	Poor
Wide temperature range	Good	Excellent	Fair	Excellent	Fair
Contamination to surroundings	Excellent	Good	Poor	Excellent	Poor
Resistance to dust ingress	Fair	Excellent	Good	Good	Good
High axial/radial loads	Fair	Good	Excellent	Good	Excellent
High axial/radial stiffness	Fair	Excellent	Excellent	Good	Good
Small space envelope	Good	Fair	Good	Fair	Excellent
Low heat generation	Good	Good	Poor	Excellent	Good
Run in partial vacuum	Poor	Fair	Fair	Excellent	Fair
Low running costs	Excellent	Fair	Good	Fair	Good

4 GAS BEARINGS

This section of the chapter examines in some detail the two main types of gas bearings; namely self acting, aerodynamic and pressure-fed aerostatic designs. Typical mechanical construction of both types will also be investigated.

4.1 Background

The concept of using a gas as a lubricant is a logical derivative of the study of hydrodynamic fluid film bearings. Analytical work on the characteristics of a gas-lubricated bearing can be traced back as far as 1897 with the work of Kingsbury.[2] This was followed up in 1913 by Harrison,[3] who developed an approximate theory governing the performance of a gas-lubricated bearing, which allowed for the effects of compressibility.

At this early stage, the theory made clear that extreme accuracy would be necessary in the manufacture of a gas bearing. For this reason the concept lay dormant for the next 40 years. During the 1950s, many new fields of research were developing, most notably that of atomic energy. Nuclear reactors were being created, and the study of the radioactive environment necessitated the circulation of gas through the atomic pile. The demands on the circulators were considerable, with some power requirements being in excess of 100 HP. The original circulators designed for the purpose used conventional lubricants. Unfortunately, however, it was soon discovered that the radioactive environment caused the bearing lubricants to solidify, resulting in bearing seizure.

The failure of the gas circulators seriously hampered atomic research, and sparked an extensive search for a solution. It became clear that the only lubricant available for the circulators was the radioactive gas itself, and necessity became the mother of invention, as so often in the past.

Early work on research into gas bearings was carried out at Harwell, and the task was soon passed on to a number of major manufacturers of aero engines, these being the only companies at the time having facilities to achieve the level of accuracy required. Early results were encouraging, but following bearing seizures in many research establishments, it became clear that a bearing instability termed half speed whirl presented a major obstacle, which had to be surmounted before desired speeds could be achieved. A number of solutions were finally devised, and circulators were built capable of handling radioactive gases at temperatures of up to 500°C at pressures of 350 psig. One of the largest circulators was constructed by Societe Rateau for the Dragon reactor project. The pump ran at 12,000 rpm at 120 HP circulating helium at 289 psig at 350°C.

A further need for gas lubrication during this period was in the field of inertial navigation, where the replacement of miniature ball bearings resulted in a remarkable advance in accuracy of the instrument.

In the early stages of the above developments, theoretical performance characteristics gave little more than a guide, and the demands stimulated intensive theoretical and experimental research. Of the many vital theoretical contributions made, perhaps Raimondi should be singled out for his informative paper in 1961 entitled "A numerical solution for the gas lubricated full journal bearing of finite length".[4] By the use of computer-generated design charts in this paper, it proved possible to obtain excellent agreement between theory and practice for the aerodynamic bearing. Unfortunately, at this stage, half speed whirl defied accurate prediction, and solutions relied heavily on practical experience.

In parallel with studies on aerodynamic bearings, work was also proceeding on the theoretical performance of pressure-fed bearings. This type of bearing tended to be more amenable to prediction, and again many valuable contributions were made to assist engineers in their efforts to create practical gas-lubricated bearings.

One of the earliest practical applications of the pressure-fed air bearing was in the realm of dentistry. In the 1960s a dental drill produced by Westwind Air Bearings proved very successful in the field, operating at 500,000 rpm with minimal vibration. Other applications included precision grinding and drilling spindles for the machine tool industry.

It is only within recent times that the aerodynamic bearing has come to the fore once again, this time in the field of laser scanning. The characteristics of the aerodynamic bearing are ideally suited to this particular application, which demands high-speed rotation with very low levels of vibration, and zero contamination of the environment. Figure 2 shows typical aerodynamic scanners and items used in the internal drum laser scanning market.

4.2 Fundamentals

There are certain fundamental characteristics of gases that explain why gas bearings are particularly suitable for high-speed rotary scanner designs.

4.2.1 Low Heat Generation

Compared with even the lightest instrument bearing oil, dynamic viscosity of the common gases used in gas-bearing spindles are several orders of magnitude lower (Table 2). The

Figure 2 High-speed internal drum aerodynamic scanners and associated parts.

Table 2 Viscosity in cP, Oil vs. Several Gases

Gas/fluid	Temperature	
	27°C	100°C
Instrument oil	70	5.5
Argon	0.022	0.027
Air/helium	0.018	0.021
Nitrogen	0.017	0.021

main benefit of low viscosity can be seen from the equation below showing the power loss in a journal bearing.[5]

$$P_{Loss} = \frac{\pi \mu D^3 L \omega^2}{4c} \qquad (1)$$

where μ = viscosity of the fluid/gas, D = shaft diameter, L = length of journal, ω = angular velocity of shaft, and c = radial clearance between the shaft and journal.

It can be readily seen how the power consumed is proportional to the lubricant viscosity, and this allows the gas bearing to be run at much higher speeds for the same shaft diameter as in an oil bearing. Figure 3 shows typical journal heat generation figures for several common air bearing shaft sizes. Also, the shaft diameter has a critical effect on the heat generation, due to the cubic function in Eq. (1).

Similarly, the power loss in a thrust bearing is given by Eq. (2):[5]

$$P_{Loss} = \frac{\pi \mu \omega^2}{2h}(b^4 - a^4) \qquad (2)$$

where b = outside radius of the thrust bearing, a = inside radius of the thrust bearing, and h = axial clearance between the shaft and bearing surface.

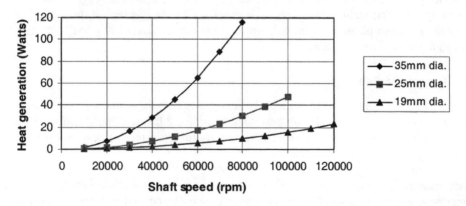

Figure 3 Air bearing journal heat generation vs. shaft speed for various shaft diameters (c = 12.7 microns, $L/D = 1$).

4.2.2 Wide Temperature Range

Another important factor is that the variation of viscosity with temperature is small for all the gases in Table 2. This allows gas bearings a very wide operating temperature range, and it is the mechanical properties of the bearing and shaft materials that will usually limit the maximum temperature of operation, not the bearing itself. This limit could be the differential thermal expansion between the shaft and journal changing the clearance to an unacceptable amount or even the maximum thermal conductivity of the bearing material to transmit heat out of the bearing system.

4.2.3 Noncontamination of Environment

With aerodynamic bearings, the gas surrounding the bearing is used as the lubricant, usually air for common rotary scanners, and no contamination of the gas will occur (provided the bearing materials do not chemically react with the gas). With aerostatic bearings, the pressurized gas supply (again usually air in normal scanners) will purge out of the bearing, mixing harmlessly with the environment. This has the added benefit of preventing the ingress of dust or other particles that could eventually cause damage to the shaft/bearing assembly by blocking the gap between the shaft and bearing.

4.2.4 Repeatability of Smoothness

As there is no physical contact between the shaft journal and the bearing during rotation, the axis of rotation, or the orbit of the shaft, will not degrade over the lifetime of the spindle, ensuring repeatable optic performance. The smoothness of rotation within one revolution of the shaft ensures minimal cross-scan errors off the optic.

4.2.5 Accuracy of Rotation

The shaft journal will find the average centerline of the bearing, as it is surrounded by the gaseous lubricant, which will conform to any local irregularities created during the manufacturing process, and as a general rule the shaft orbit will be an order of magnitude better than the measured roundness of the bearing in which it is revolving.

4.2.6 Noise and Vibration

Particularly with reference to aerodynamic bearings, audible noise is negligible from the bearing system. The main source of noise is generated by the windage of the optic. The damping properties of the gas film help ensure that transmission of any shaft vibration through to the bearing is reduced.

4.3 Aerostatic Bearings

A constant supply of pressurized gas must be supplied to both the radial and axial bearing gaps to support the shaft load with aerostatic bearings. Although the lift-off, or float pressure, will be at a very low pressure, to achieve useful loads and stiffness a pressure in the order of 3–6 bar is normally used.

This requires the use of an external compressor, which is a big disadvantage in many rotary scanner applications as the rest of the scanning system will not usually require compressed air, and additional noise, vibration, and cost associated with the compressor could be prohibitive. However, particularly when rotating large, overhung optics, the benefits of high radial and axial loads may justify the use of aerostatic bearings, particularly if very low speed operation is required (such as a few hundred revolutions per minute).

Another benefit with aerostatic bearings is that the shaft can be rotated in either direction with identical performance, something not normally possible with aerodynamic bearings. This feature could be of use where one aerostatic spindle is to be used in a variety of scanning products, which could well run in different directions. Finally, although the exhausting of gas through the end of the bearings creates additional noise, the self-cleaning effect could be useful to prevent particles created in the scanning machine (particularly paper and carbon) from being deposited inside the scanner.

The general design principles will now be examined, followed by some general notes on construction.

4.3.1 Aerostatic Journal Bearing

The general principle of operation can be explained by reference to Fig. 4. The bearing consists of an annular cylinder containing two sets of orifices, or jets, one row towards each end of the bearing.

The jets are supplied with gas at pressure P_s and the gas exhausts at P_a to atmosphere. With no load on the shaft (and ignoring its own mass) the downstream pressure in the gap between the shaft and the bearing is equal all round any circumference as shown in the cross-section of the bearing through one of the jet planes. The associated pressure profile diagram along the bearing shows how the discharge pressure P_d slowly drops as the gas flows towards the ends of the bearing until it exhausts at the atmospheric pressure P_a. In other words, there is a constant flow of gas between the jet plane and the end of the bearing, while the area between the jet planes remains at a constant pressure.

When a radial load is applied to the shaft, it will be displaced in the direction of the force, reducing the gap between the shaft and the bearing. The localized gas flow will reduce, causing an increase in pressure $(P_{d1} - P_a)$, with a similar reduction in pressure $(P_{d5} - P_a)$ due to an increase in flow on the other side of the shaft. This resultant pressure difference across the shaft will cause it to resist the applied load, preventing surface contact between the two parts. When the load is removed, the pressure distribution will return the shaft to the central position again.

In practice, due to the relatively small number of jets per row, typically 8 to 12, dispersion effects reduce the effective pressure zone between the rows of jets. This will reduce the load capacity slightly, as will circumferential flow around the bearing from the high pressure to the low pressure zone. Although there are other methods of feeding the gas into the bearings, such as slot feeds or the use of porous materials, the discrete jet orifice method has become the favorite for this market.

Load Capacity

The standard equation for expressing the radial load capacity of an aerostatic journal bearing is:

$$\text{Load } W = C_L(P_s - P_a)L \times D$$

where P_s = supply pressure, P_a = ambient pressure, L = bearing length, D = bearing diameter, and C_L = dimensionless load coefficient. The load coefficient C_L is affected by several different parameters, including the eccentricity ratio, the downstream pressure P_d,

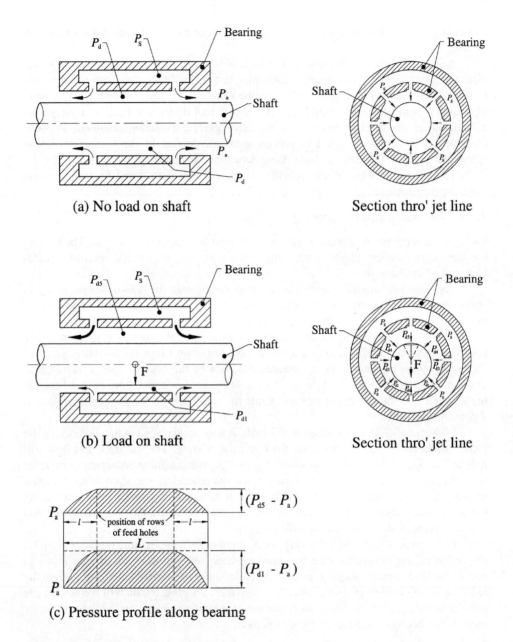

Figure 4 Aerostatic journal bearing.

the number of jets (the dispersion effect), and the jet position in relation to the end of the bearing. There are ways of estimating these effects as shown by Shires.[6]

To be able to estimate the actual maximum load capacity, the designer must decide how close to the bearing surface the shaft can be moved before local irregularities in the bearing or shaft cause actual contact; that is, balance, roundness of the shaft, ovality of the bearing, and squareness of the shaft to the bearing all have an effect on this decision.

Bearings for Rotary Scanners

This displacement is usually referred to as the eccentricity ratio ε of the shaft in the bearing and a typical maximum figure while rotating would be 0.5. That is, the shaft has been displaced by half the total radial clearance. Figure 5 shows typical load capacities for shaft diameters used in rotary scanners, with two different bearing length-to-diameter ratios: using $\varepsilon = 0.5$, jet position $= 0.25 \times L$, $P_s = 5.5$ bar g.

Radial Stiffness

Radial bearing stiffness is constant at low eccentricity ratios, and can readily be derived from the following equation.

$$\text{Stiffness } K = \frac{W_e}{\varepsilon \times C_0}$$

where W_e is the load capacity at $\varepsilon = 0.1$, $\varepsilon = 0.1$, and $C_0 =$ radial clearance between the shaft and bearing. Figure 6 shows the effect of clearance variation on radial stiffness for various shaft sizes.

Heat Generation

At first glance the designer would want to keep the radial clearance as small as possible to ensure maximum stiffness, but the trade-off is that bearing heat generation is inversely proportional to the clearance as shown in Eq. (1), and a compromise has to be reached between these two factors. Figure 7 shows bearing heat generation plotted against clearance for the common sizes of shaft.

Dependant upon the construction of the spindle, there will be a critical limit on the heat generation per square centimeter of bearing surface above which liquid cooling will be necessary to maintain the correct clearance between the shaft and bearing (due to thermal expansion).

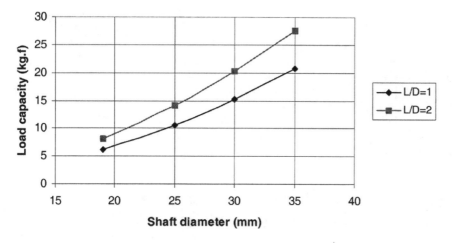

Figure 5 Radial load capacity vs. shaft diameter ($P_s = 5.5$ bar g, $\varepsilon = 0.5$).

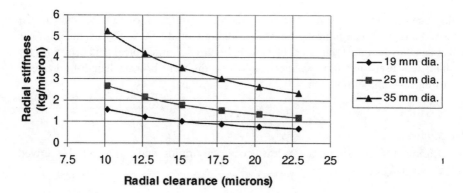

Figure 6 Radial stiffness vs. clearance for various shaft diameters ($\varepsilon = 0.1$, $L/D = 1$, $P_s = 5.5$ bar g).

Bearing Gas Flow

Before the air flow can be calculated, another design factor that must be considered is the shape of the jet orifice itself. There are two common forms of discrete jet: the plain jet and the pocketed jet. Figure 8 shows a simplified form of both types.

With the plain jet the smallest flow area is controlled by the bearing radial clearance c and hence the area used for flow calculations is the surface area of a hollow cylinder with the length equal to the radial clearance c.

$$A = \pi d c$$

However, with the pocketed jet (or simple orifice) the smallest flow area is controlled by the jet diameter itself, hence the area for flow calculations is the cross-sectional area of the jet itself.

$$A = \frac{\pi d^2}{4}$$

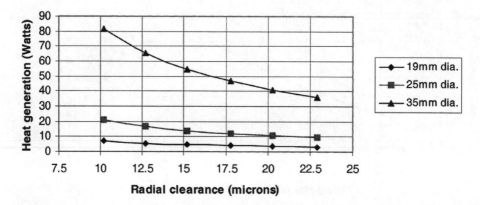

Figure 7 Journal bearing heat generation vs. radial clearance (60,000 rpm, $L/D = 1$).

Bearings for Rotary Scanners

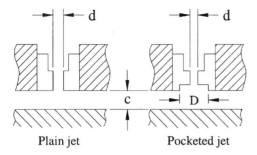

Figure 8 Common type of jet orifice.

Obviously, if very large clearances are used, which is rare, even a plain jet will run as a simple orifice.

For better control of flow, pocketed jets are preferable and yield higher stiffness due to reduced dispersion effects, but plain jets provide greater damping reducing the likelihood of instability. This instability or resonance can occur if the pocket volume is too large for the bearing design and a self-induced pneumatic hammer can be heard.

Figure 9 shows typical air flow for a 25 mm shaft for both pocketed jets and plain jets with a diameter of 0.16 mm.

4.3.2 Aerostatic Thrust Bearing

An aerostatic thrust, or axial bearing system consists of two opposed circular thrust plates sandwiching the shaft axial runner with the gap between the surfaces being controlled by a spacer, slightly thicker than the shaft runner, which is located around the outside diameter of the shaft.

Figure 10 shows a single thrust plate fitted with an annular row of discrete jets, linked by a narrow groove. The purpose of the groove is to create a pressure ring around the jet pitch circle diameter (PCD) for optimum performance, particularly when the shaft runner is almost at touchdown condition on the thrust face. The associated pressure profile is also shown.

For stability, two thrust plates are used in opposition, trapping the shaft runner between. In a similar way to the journal bearing mechanism, when a load is applied to the

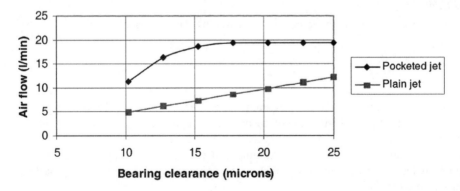

Figure 9 Air flow vs. bearing clearance for a 25 mm diameter bearing ($L/D = 1$, $P_s = 5.5$ bar g, 16 jets, $d = 0.16$ mm dia).

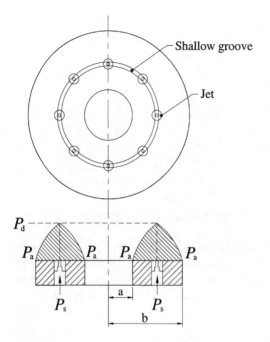

Figure 10 Aerostatic thrust bearing.

shaft axially the shaft will move towards one face of the axial bearing and the flow through the jets will drop, causing an increase in pressure over that bearing face. This will create an opposing force on the shaft runner, preventing it from moving closer to the bearing surface. Meanwhile, the opposite happens on the other bearing face, reducing the force on the other shaft runner face.

This mechanism can be seen more clearly with reference to Fig. 11(a), which shows the load capacity lines of both faces and where they cross will become the equilibrium position of the shaft runner. Other forms of axial bearing such as the center fed or journal fed are possible but the annular ring of jets is the most suitable for this market.

Load Capacity

The load capacity from one plate can be expressed by the following equation.[5]

$$W = \frac{(P_d - P_a)\pi(b - a)^2}{\ln(b/a)}$$

where P_d = downstream pressure, P_a = ambient pressure, b = outside radius of plate, and a = inside radius of plate. Estimation of the downstream pressure P_d is based on a number of factors, which are beyond the scope of this chapter (see Ref. 5). Again, with reference to Fig. 11(a), it can be seen that the maximum load capacity of an opposed thrust plate assembly is the summation of $(W_2 - W_1)$, at the point where the runner is approaching contact with thrust face 2.

Figure 11(b) shows some typical load values used in axial bearings for scanner applications, with a constant ratio between the outer radius and the inner radius of 1.6.

Bearings for Rotary Scanners

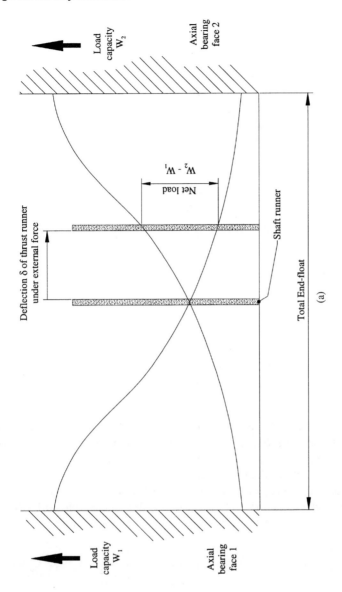

Figure 11 (a) Axial bearing diagram.

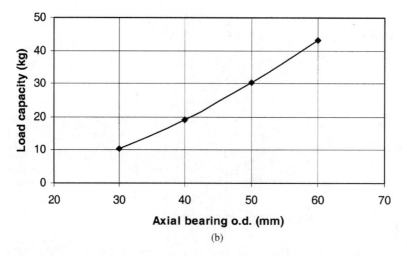

Figure 11 (b) Axial load capacity vs. bearing outside diameter ($b/a = 1.6$, $P_s = 5.5$ bar g, 8 jets, $d = 0.27$ mm dia).

Axial Stiffness

Axial stiffness is calculated using the equation

$$K = \frac{W_2 - W_1}{\delta}$$

where W_2 = load capacity of face 2 for the position of (equilibrium position $-\delta$), W_1 = load capacity of face 2 for the position of (equilibrium position $+\delta$), and δ = distance moved for applied external load. Note, for maximum stiffness, K is normally calculated for a δ of <10% of the total endfloat.

It can be seen from Fig. 12, a plot of axial stiffness against axial clearance for various bearing diameters, that the centerline (or equilibrium position) stiffness is highly dependant upon the clearance between the shaft runner and the bearing.

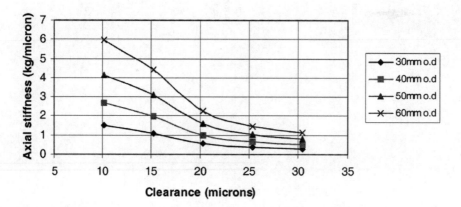

Figure 12 Axial stiffness vs. clearance for various axial bearing outside diameters ($a/b = 1.6$, $P_s = 5.5$ bar g).

Bearings for Rotary Scanners

To achieve optimum stiffness, the number of jets, the radius of the jets, and the dimension of the groove must all be considered. For minimum air flow, and highest stiffness, a small clearance is desirable, but again heat generation will be highest at this clearance.

Heat Generation

Using Eq. (2), the effect of small clearances on axial bearing heat generation are shown in Fig. 13.

4.3.3 Aerostatic Scanner Construction

A typical aerostatic bearing polygon scanner is shown in Fig. 14. The polygon is attached to a removable threaded mount in the front of the shaft and to help counterbalance the mass of the polygon the axial bearing system is located at the rear of the shaft, adjacent to the brushless DC motor. To minimize air flow, a long single radial bearing design has been used with only two rows of jets. For maximum radial stiffness and load capacity, a four-jet row, twin bearing system could be used at the expense of higher air flow. An optical encoder system is fitted at the rear of shaft to guarantee high accuracy speed control. The bearing materials would typically be bronze or gunmetal, while the shaft itself would be stainless steel.

4.4 Aerodynamic Bearings

In recent years, the use of aerodynamic or self-acting bearings has become more widespread, replacing ball bearing or aerostatic bearing assemblies in this market. In general, machining tolerances are much smaller for aerodynamic bearings and it is only with the recent advent of higher precision computer numerical control (CNC) machines that it has become more cost-effective to introduce this technology for high-volume manufacturing.

In its simplest form, the aerodynamic bearing consists of a plain circular tube in which the shaft rotates as shown in Fig. 15. If a load W is imposed onto the shaft as shown, causing the shaft to move off center by an amount εh_0, the pressure will rise in the reduced gap due to viscous shear of the gas, creating a "wedge" similar to the mechanism occurring in a hydrodynamic oil bearing. This high-pressure zone allows the shaft to float so it can rotate without contact in the bearing.

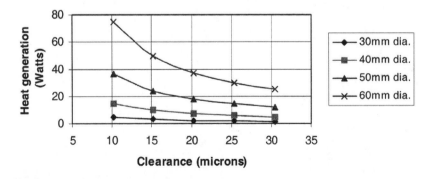

Figure 13 Heat generation vs. clearance for various axial bearing outside diameters (60,000 rpm $b/a = 1.6$).

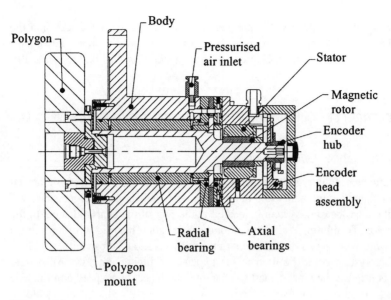

Figure 14 Aerostatic bearing polygon scanner with speed 5,000–7,500 rpm, shaft diameter 32 mm, $P_s = 5$ bar g, and air consumption 15 L/min.

Owing to the low viscosity of gases, the clearance between the shaft journal must be very small, typically a few microns for this effect to be useful. Obviously, when the shaft is stationary, there is no viscous shear, or supporting pressure, and the shaft journal and bearing will be in contact. To avoid damage to both surfaces when starting to rotate the shaft a high-torque motor is required to accelerate the shaft to floating speed in a very short time. For shaft diameters utilized in scanner assemblies, this speed is typically several hundred rpm.

Obviously gases are compressible, and this will reduce the effect of the pressure wedge compared with a liquid, and the resulting load calculations can be complex.

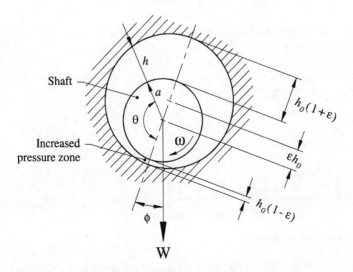

Figure 15 Aerodynamic journal bearing.

However, it can be shown that the compressibility number Λ can be used as a guidance of bearing performance.[6]

$$\frac{\Lambda}{6} = \frac{\mu\omega[r]^2}{P_a[c]^2} \tag{3}$$

where μ = viscosity, ω = shaft angular speed, P_a = ambient pressure, r = shaft radius, and c = bearing clearance.

Figure 16 shows how the load capacity at constant eccentricity ε for compressible and incompressible fluids varies with increasing compressibility number Λ. It can be seen that for gases at high compressibility numbers, the load capacity becomes independent of this number. Practically speaking, this means that there comes a point when the radial load capacity (and stiffness) become independent of the speed of the shaft.

Also from Fig. 15, it can be seen that when load W is applied to the shaft, the closest approach between the journal and bearing is not in direct opposition to W, but at an attitude angle ϕ. This angle can theoretically vary between zero and 90° and is mainly dependant on the compressibility factor.

Load Capacity

Estimating the load capacity of an aerodynamic bearing is more complex than for the aerostatic version, due to the compressibility effects. However, Raimondi[4] managed to compute numerical solutions and create design charts using the dimensionless group:

$$\text{load ratio } \frac{P}{P_a} = \frac{W}{2rLP_a}$$

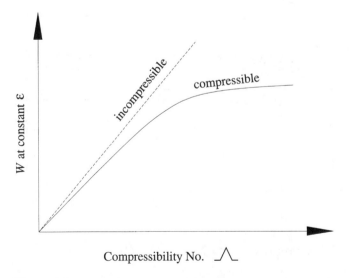

Figure 16 Variation of load capacity with compressibility number.

where P = bearing pressure, P_a = ambient pressure, W = load capacity, r = shaft radius, and L = bearing length, and the compressibility number Λ (Eq. 3) for various eccentricity ratios. Figure 17 shows the Raimondi graph[4] for a bearing L/D ratio of 2, probably the minimum number for practical aerodynamic bearings. Having calculated the compressibility number, and decided what is the maximum practical eccentricity ratio, the load capacity can be derived.

Unfortunately, from the explanation above, it is obvious that if no load is applied to the shaft in an aerodynamic bearing, then there is no wedge action and without the associated pressure effect, the shaft is essentially unstable. To overcome this instability, the designer must incorporate some kind of surface form into the bearing of the shaft that creates a wedge effect without the use of a load. Obviously, the shaft mass itself creates a small eccentricity if the shaft is running horizontally, but often it is not. Also with the correct surface form design, the effect of load at constant eccentricity becoming independent of speed at high compressibility numbers can be dramatically reduced.

There are two main surface forms used in aerodynamic scanner bearings; spiral grooves and lobing.

4.4.1 Spiral Groove Bearings

A series of shallow spiral grooves can be machined into either surface, but usually in the shaft journal, which are open to the atmosphere at one end. Gas is drawn into these grooves by viscous shear during rotation and creates a pressurized zone towards the closed end (Fig. 18). The important journal groove geometric parameters are: groove angle, α; no.

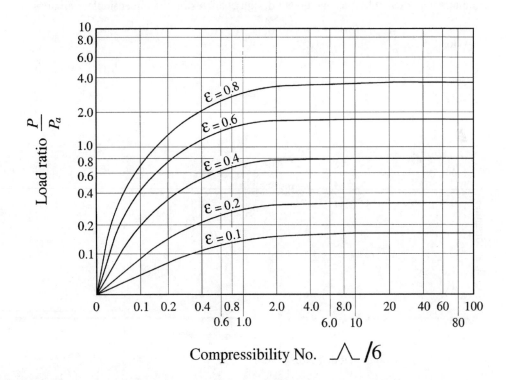

Figure 17 Load capacity vs. compressibility number for a bearing of $L/D = 2$.

of grooves, N; groove depth ratio, h_0/h; groove length ratio, L_g/L; and groove width ratio, $W_1/(W_1 + W_2)$.

The compressibility number from Eq. (3) can still be used and from using the appropriate graphs[7] the optimized geometry can be deduced.

The same concept can be applied to the axial annular bearings with a series of shallow grooves machined into one of the surfaces, which spiral in towards the center, again creating a pressurized zone at the closed inner ends. Figure 19 shows a typical example with the pressure profile. The important thrust groove geometric parameters are: radius ratio, $(r_b - r_i)/(r_b - r_a)$; width ratio, W_1/W_2; depth ratio, h_0/h; and groove angle, θ. Practical values of these parameters have been established by Whitley and Williams[8] and from these an estimation of the load capacity and stiffness can be made.

There are three basic bearing configurations that can be utilized with spiral groove technology: separate parallel radial and axial bearings (as above), conical bearings, or spherical bearings, as Fig. 20 shows.

Both the conic and the spherical bearing designs have the advantage of only one set of grooves, which makes for a compact design, especially if the motor can be placed between the bearings, but creating the bearing surfaces precisely is quite a production challenge, particularly for the hemispherical type.

4.4.2 Lobed Bearings/Shaft

Another form of aerodynamic geometry is the lobed bearing or shaft, which has an out-of-round surface that generates an axial increased pressure zone, or zones, along the length of the bearing as it rotates. Figure 21 shows a typical bearing form, which has three lobes and also a stabilizing groove that creates a small shaft eccentricity for stability (Westwind Patent No. EP0705393[9]).

This design has the advantage of simple manufacture in production volumes with stable performance over a wide speed range, and is ideally suited to supporting overhung optics due to the large bearing centerline separation.

Alternatively, the shaft could be lobed, and with the latest CNC cam grinding machines this form can be machined relatively easily. However, some form of spiral groove axial bearing will still be required on either design.

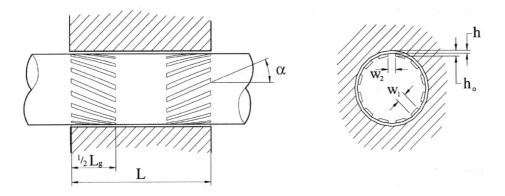

Figure 18 Spiral grooved shaft.

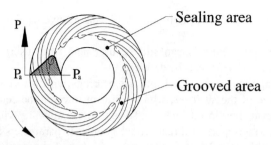

Direction of shaft to drag gas into groove

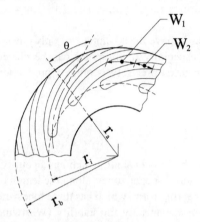

Figure 19 Axial bearing plate.

4.4.3 Spindle Construction

A typical overhung polygon scanner unit is shown in Fig. 22 using lobed radial bearing and spiral groove axial bearing technologies. To help balance out the mass of the polygon on the front of the shaft, the axial bearings, together with the motor and encoder, have been located at the rear. For vertical operation, additional repulsing magnets are contained at the rear of the spindle, one in the shaft and one in the housing, to provide additional upwards force to reduce starting frictional torque on the lower thrust plate.

To allow many thousands of stop/start cycles, bearing surfaces need to be coated with some form of antiscuff, low-friction material, probably containing PTFE (Teflon®). The choice of shaft and bearing materials is important to ensure that the correct bearing clearances are maintained over a wide range of temperatures.

For center-mounted polygon scanners, a conical bearing design with spiral groove technology is probably the most practical due to its compact shape, with the motor mounted between the bearings as shown in Fig. 23. In this case, the rotor rotates around the static central stator.

For many applications today, a monogon optic design is required, which by its very nature must be mounted onto the front of the shaft. A typical scanner cross-section is shown in Fig. 24 using a lobed radial bearing and spiral groove axial bearing technology.

Bearings for Rotary Scanners

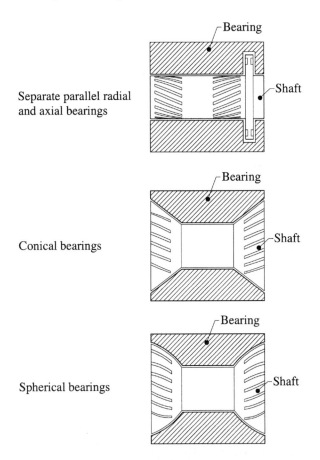

Figure 20 Spiral groove bearing types.

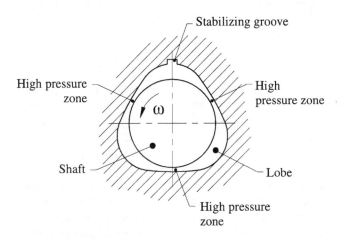

Figure 21 Lobed bearing.

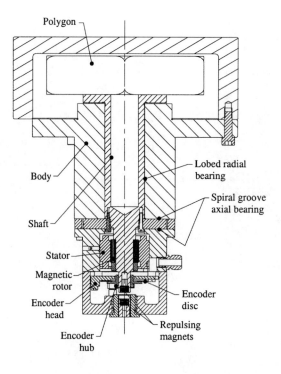

Speed : 5,000 rpm
Polygon size : 126mm a/f
Shaft dia : 32mm

Figure 22 Aerodynamic bearing polygon scanner.

Note that the axial bearing is now at the front of the spindle, close to the optic. This minimizes the forward axial growth of the shaft due to heat generation within the radial bearing system as monogon scanning systems are sensitive to axial movements (which can misplace the output beam).

4.5 Hybrid Gas Bearings

There is a special form of radial bearing that combines the technologies of both aerostatic and aerodynamic design. If the spacing between the two jet rows is increased significantly in an aerostatic design, a large increase in stiffness and load capacity will be achieved at high shaft rotational speed. Typically a length-to-diameter ratio of 2 would be considered ideal, with the jets spaced about 1/8th of the bearing lengths from the bearing ends, but to achieve significant hybrid performance, the bearing clearance must be kept to a minimum.

By utilizing the Raimondi curves, as mentioned previously for aerodynamic performance, the additional load capacity can be calculated and summated into the aerostatic load capacity calculations. From Fig. 25, the improvement can be seen in the radial bearing stiffness with three different bearing lengths as the speed increase (assuming constant bearing clearance over the speed range). This has the great benefit of increasing

Bearings for Rotary Scanners

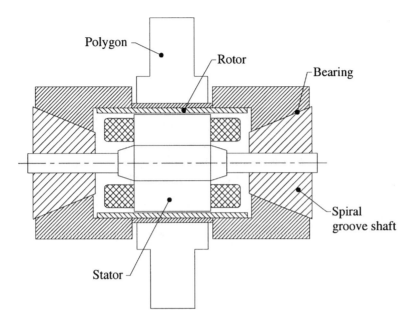

Figure 23 Scanner with conical gas bearing.

the gas film critical speeds and hence the maximum operating speed of the spindle. This will be discussed under Sec. 4.6 "Bearing and Shaft Dynamics."

4.6 Bearing and Shaft Dynamics

Whichever gas-bearing system is selected, certain bearing and shaft dynamics have to be considered to allow high-speed operation, and specifically the synchronous whirls, the half speed whirl, and the shaft natural frequency.

4.6.1 Synchronous Whirls

Synchronous whirls occur at the shaft rotation speed and can be due to an inherent imbalance in the shaft itself, which will increase quickly with speed as the out-of-balance forces are proportional to the square of the speed. Therefore careful dynamic balancing is necessary to minimize these forces. Typically a balance standard of better than G0.4 (International Standard ISO 1940) should be achieved for acceptable performance, involving a two-plane balancing methodology.

However, another phenomena is encountered in aerostatic bearings, which is due to the natural resonant frequencies of the gas film system. As these frequencies are approached, any out-of-balance force is magnified dramatically due to the almost total lack of damping. However, the shaft can be run through these frequencies and operate very comfortably in a "supercritical" mode, with the shaft now rotating around its mass center, and not its geometric center.

The speeds at which the natural resonant frequencies occur can be calculated as shown below, ω_1 being defined as a cylindrical whirl and ω_2 as a conical whirl mode.

$$\omega_1^2 = \frac{2k}{m}$$

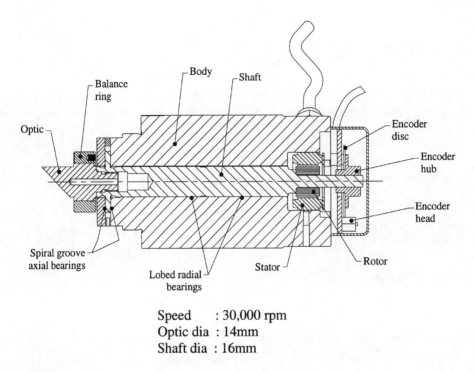

Figure 24 Aerodynamic bearing monogon scanner.

where k = gas film stiffness, and m = mass of the shaft.

$$\omega_2^2 = \frac{2kJ^2}{I - I_0}$$

where J = (distance between bearing centers/2), I = transverse moment of inertia of the shaft, and I_0 = polar moment of inertia of the shaft. Obviously, for maximum shaft speed,

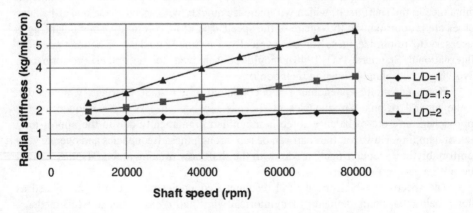

Figure 25 Hybrid radial bearing stiffness vs. shaft speed for a 25 mm shaft dia ($\varepsilon = 0.1$, P_s = 5.5 bar g, c = 12.7 microns).

Bearings for Rotary Scanners

the designer needs to achieve the highest radial stiffness possible and keep the shaft mass at a minimum.

4.6.2 Half Speed Whirl

This highly destructive phenomenon is encountered in both aerostatic and aerodynamic designs, and is usually the limiting speed parameter. In aerostatic designs, it occurs in practice at a speed somewhat below twice the lowest of the gas film resonant frequencies, typically at about 1.8 times. In aerodynamic designs it is much harder to predict the speed at which it will occur, but in general, spiral groove bearing designs do not suffer from this problem nearly as much. Half-speed whirl occurs when the rotor is orbiting the bearing centerline at a frequency equal to half its rotational speed. The shaft increases its orbit without a further increase in speed, and quickly contacts the bearing surface, causing seizure.

4.6.3 Shaft Natural Frequency

Like any other bearing system, the shaft natural frequency must be calculated using the normal processes. It must include the mass effect of additional items like the rotor, encoder disc, and, most importantly, the optic and its mount.

In small shaft scanner designs, the natural frequency is usually well above the operating speed range, maybe by a factor of 2 to 3, but with large shafts this frequency must be allowed for. In general, no shaft should be run beyond 80% of the shaft natural frequency.

The shaft critical speed can be calculated from the general formula for a uniform shaft (simply supported on short bearings):

$$\text{Angular velocity } \omega_{\text{crit}} = \frac{\pi^2}{l^2} \sqrt{EI/m}$$

where l = distance between bearings, E = modulus of elasticity, I = movement of inertia, and m = mass of the shaft. From this equation it can be seen that to keep the shaft mass and the bearing separation distance to a minimum is really important for high-speed operation.

4.7 Shaft Assembly

Although this chapter is primarily concerned with the different bearing types for use with scanners, it is important to realize that all the extra components mounted onto the shaft will have an impact on the performance of the scanner. As shown in the previous sections, the total mass of the rotor assembly affects both shaft natural frequency and the cylindrical whirl frequency, hence any additional masses built in the rotor must be minimized. Another consideration has to be the stresses that the shaft is imposing on these additional components when rotating at high speed.

4.7.1 Optics and Holders

The types of optics that can be fitted to rotary scanners fall into three basic groups: polygons, monogons, and holographic disks. All these optics are selected for specific applications and are discussed in much greater depth elsewhere in this book. However, the bearing design type of the scanner will in general be dictated by this optic selection.

Polygons

As shown in Sec. 4.4.3, polygons can be mounted on the front of the spindle (Fig. 22) or in the center of the bearing system (Fig. 23). From a bearing stability viewpoint, the center-mounted conic bearing design is better suited for small, high-speed polygons (<100 mm across flats) although the large bore size required to fit over the scanner body may cause optical distortion problems.

The polygon can also suffer from thermal distortion due to the close proximity of the motor rotor. Its compact space envelope is probably its greatest advantage and has widespread use across the scanning market.

Large diameter (>100 mm across flats) and also thick-faceted polygons are better suited to mounting on the front of the spindle. These tend to operate at lower speeds (<30,000 rpm), but require a substantial bearing system to support the overhung mass. There is a point where even a large, aerodynamic spindle is not really suitable due to high starting torque and insufficient load capacity and at this point an aerostatic, or hybrid, bearing system will have to be selected (as shown in Fig. 14).

Whichever system is selected, due design consideration must be given to the polygon mount design. The method of polygon attachment, be it using screws or bonding, must not affect the optical properties of the facets, either statically or dynamically. The choice of material for both the polygon and its mount must take into account the total additional mass being supported by the bearing system, the rotational stresses induced, and thermal growth characteristics between the hub and mounts. Most scanner polygons and mounts are manufactured from high-grade aluminum.

Finally, trapezoidal polygons, that is, polygons with facets not parallel to the axis of rotation, in general will be mounted on the front of the spindle to allow access to the incoming axial laser beam.

Monogons

Monogons, or single-faceted reflective optics, tend to be used when the output beam from the optic is to strike a circular, rather than a flat surface, such as in a cylindrical drum image setter, although if the output beam is passed through an F-theta lens it can then be used on flat surfaces too. The optics used can be of a simple open reflective surface form or a more complex glass form such as a prism. The input laser beam is usually co-incident and parallel to the shaft rotational axis and hence all monogons need to be mounted on the front of the spindle.

Particularly with reference to glass optics, but even with aluminum and beryllium, great care is required in the design of the optics holder. Unlike polygons, which normally have some form of bore for location, most glass optics are designed for their optical quality, not their ease of mounting. Figures 26(a) and (b) show two forms of housing into which glass prisms have been bonded. Figure 26(a) is a cube prism manufactured from BK7 optical quality glass with the two nonactive sides bonded to the extended cheeks of the housing. As can be seen, there are several square edges to the housing, which will cause noise and windage.

An improved version is shown in Fig. 26(b), where the optic bonded in the housing has had all the nonactive exposed corners ground to a spherical shape. This is called a ball prism. The housing is also more aerodynamically shaped for reduced turbulence. In both cases, the optics holder is manufactured in high-strength stainless steel. Although this adds extra weight to the shaft assembly over using aluminum, only steel can survive the very high forces exerted by the cheeks at the speeds used; typically 30,000–60,000 rpm.

Bearings for Rotary Scanners

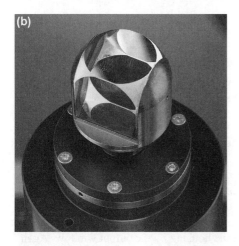

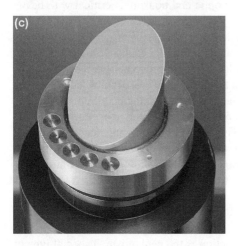

Figure 26 (a) Cube prism; (b) ball prism; (c) open face mirror; and (d) spherical housing containing 45° mirror.

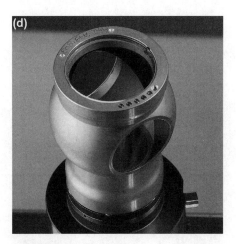

Figure 26 *Continued.*

Figure 26(c) shows a simple version of the open aluminum mirror. Although the mounting technique is easier, with a thread machined onto the back of the mirror stub, there is now the problem of imbalance to correct for, due to the asymmetric form of the mirror. An aluminum ring has to be added to the assembly as shown with heavy metal pins added in the appropriate position. In addition, the open mirror acts as a form of pump, which not only increases the turbulence around the mirror, but it actively attracts dust particles to the mirror surface.

For higher speed application, an improved version of the angled mirror design can be used as shown in Fig. 26(d). The entire mirror is now enclosed in a sphere-shaped housing with input and output windows to stop the pumping action. The stress analysis of this housing, especially around the output window, is of paramount importance due to the large centrifugal force exerted onto the edges of the housing by the output window. Again, balancing the asymmetric form of the combined optic and housing is critical with heavy metal pins added to the end face of the assembly as necessary.

With large clear apertures, up to 30 mm in diameter, optical distortion due to rotational centrifugal forces across the face of the mirror becomes a further challenge to contend with. The use of beryllium, rather than aluminum, reduces the deflection considerably due to its greater modulus of elasticity-to-density ratio, but there is a large cost penalty and processing can be a problem due to the health issues associated with machining beryllium.

Another solution that can be used if the application is designed to operate at only one fixed speed is to bias the surface of the optic. This involves machining a concave surface across the optic face, which becomes flat at the required operating speed. Although the extra process will add additional costs, this might allow the use of an aluminum substrate rather than beryllium, much reducing the overall cost of the optic assembly.

Obviously, the additional mass created by the spherical housing will reduce the top speed of the shaft assembly and Fig. 27 shows the typical maximum speed for each optic type and size.

One optic not already mentioned in this section is the pentaprism, a special type of prism with two internal reflectance faces. It has the unique ability to correct for a shaft error termed "wobble." This is a random conic motion of the shaft about its longitudinal axis, which occurs much more in ball bearing scanners, but can occur at a low level even in an air bearing. Owing to its relatively large mass, expensive manufacturing processes, and difficult mounting shape, it is rarely used in modern scanner designs.

Holographic Disc Optics

Certain scanning applications require the use of a holographic disc to be rotated at a relatively low speed. With a large outside diameter usually, and a small bore, this design is again best suited to mounting on the front of the spindle. As the disc is usually made of a glass sandwich, the mass is quite high, so either an aerodynamic or an aerostatic scanner may be used, depending on the minimum and maximum speeds at which the disc must run.

Owing to the principle of the holographic disc, this scanning process is not as optically demanding as rotating monogons and polygons, so many disc scanners in this market are still run on ball bearings.

4.7.2 Motors

For most scanning applications, some form of synchronous motor is required to ensure predictable speed control. The two usual designs employed are the hysteresis motor and the brushless DC motor. The lack of brushes and a commutation ring are a definite advantage in gas-bearing systems due to there being no wearing parts. In both cases the rotor is simple in mechanical construction, and ideal for high-speed rotation due to its composite nature.

The brushless DC motor consists of a wound laminated stator with integral "Hall effect" devices for commutation wound in. The rotor, in its simplest form, is a cylindrical hollow tube manufactured from sintered samarium cobalt material. This is then magnetized in a powerful magnetic fixture to the number of poles required. To withstand the high centrifugal forces, this material is usually contained inside another thin steel tube. Some typical brushless DC motors are shown in Fig. 28.

The rotor assembly can then be bonded directly onto the scanner shaft. In larger motor designs, discrete magnet pairs are used rather than the sintered material, but a steel or carbon fiber containment ring is still required to prevent individual magnets from separating from the shaft at high speed.

The brushless DC motor has become more widely used in recent years and is ideal for aerodynamic scanners, which require a very fast acceleration to reach floating speed

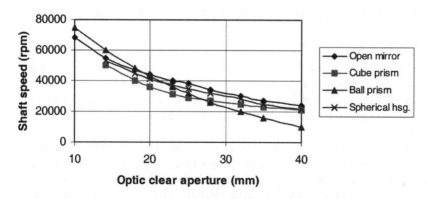

Figure 27 Speed vs. optic size for various monogon types.

Figure 28 Brushless DC motor parts.

before surface damage occurs within the bearings. This typically requires a high starting torque, which this motor type can deliver. Also, owing to the use of rare earth magnetic materials, the power density is very high, minimizing the amount of magnetic material and therefore keeping the weight of the rotor down. This is particularly important in overhung rotor designs, which are more difficult to design for high-speed operation than center motorized shafts.

The hysteresis motor has a wound, laminated stator, without "Hall effect" devices, and the rotor consists of a thin cylinder of hardened cobalt steel. This rotor can be shrunk directly onto the scanner shaft. Instead of the torque being created by the permanent magnets in the brushless DC motor, the magnetizing force from the stator induces magnetic fields in the cobalt steel due to hysteresis effect. This type of rotor produces very little heat due to the near absence of rotor eddy currents, and is therefore particularly suited to center motorized polygon spindles where low heat output from the rotor is critical.

4.7.3 Encoders

Many high-quality scanning systems require very accurate speed control, typically less than 10 ppm. A typical brushless DC motor can be speed controlled to about 2% using an open loop control system, but if an incremental encoder is fitted, the position of the rotor and shaft can be accurately measured during each revolution and controlled by a phase lock loop controller, yielding results better than 5 ppm with careful optimization.

Some typical encoder disc/hub assemblies and head assemblies are shown in Fig. 29. The head contains a photo diode emitter and receiver assembly, which is focused onto a fine grating engraved into a precision glass disc. Typical gratings vary from 200 to 1400 lines per revolution. In addition, there is a second track that provides a once per revolution index pulse, often used as a trigger for the start-of-scan process.

Bearings for Rotary Scanners

The glass disc is mounted on an aluminum hub very accurately, and the disc/hub assembly can then be fitted to the shaft end, usually after the rotor and stator have been fitted. This allows access for final adjustment of the head assembly but from the shaft dynamics viewpoint, this is yet another additional mass that must be allowed for in the design calculations, albeit a relatively light structure.

The strength of the glass disc can also be a limiting factor to the maximum speed of the scanner and for special very high-speed applications, a metal grating disc must be used. However, due to constructional reasons, high line counts with a metal disc are not possible.

For successful operation in a high-accuracy machine, the disc must be centered to a runout value of <5 microns and would typically exhibit an electronic jitter level of <2 ns.

5 BALL BEARINGS

Over the last 50 years or so, the quality of ball bearings has improved immensely and the designs have become well refined. The major manufacturers supply detailed applications design rules together with comprehensive ball bearing characteristic data. Hence it is not intended to revisit this information in detail within this chapter. However, a brief review of their application to rotary scanners is useful to understand the advantages and disadvantages of this bearing technology.

5.1 Bearing Design

For precision, high-speed ball bearing scanners, angular contact bearings are normally selected as shown in Fig. 30. Typically, the contact angle used will be in the range of 12–25°, with the greater the angle, the larger the thrust capacity available.

To ensure an accurate rotational axis, the bearings need to be used in pairs, with an axial preload, which will remove all the play in the shaft/bearing system. The preload can

Figure 29 Optical encoder assembly parts.

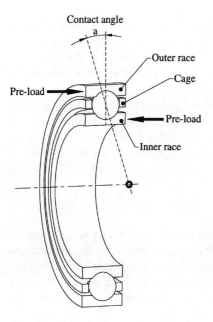

Figure 30 Angular contact ball bearing.

be ground into the bearing assembly by using different length spacers between the outer and inner races, or created by fitting disc springs between the outer face and the bearing housing.

ABEC 5 or ABEC 7 quality bearings are usually specified to ensure that the critical mechanical tolerances are controlled to known standards, particularly parameters such as the radial runout, which will have a big effect on the wobble characteristic of the shaft. Lubrication normally uses some form of light grease, with seals or shields on the bearings to help prevent any leakage. However, evaporation of the lubricant can be a major problem and will affect the life of the bearings.

In recent years one of the main improvements in ball bearing technology has been the introduction of new materials to improve overall performance. Ceramic balls, usually made from silicon nitride, can replace the steel balls in the ring and this design is known as the hybrid bearing. It is widely available from most manufacturers and gives the following advantages:

- significantly extended bearing life due to the improved running behavior between the ceramic and steel materials;
- higher speed rating due to the reduced density of the ceramic balls and hence lower centrifugal forces;
- lower thermal expansion coefficient of the balls reducing the variation in bearing preload; and
- higher bearing rigidity caused by the higher modulus of elasticity of the ceramics.

The improvement in the maximum speed capability of a hybrid bearing, over a standard precision steel ball bearing, for several common inner bore diameters, can be seen in

Fig. 31. These ratings are based on grease lubrication, ignoring preloads and other constructional constraints, and are extracted from typical manufacturers' available data.

A further improvement can be introduced by changing the material of the races themselves, from 440c typically to a finer structure steel, such as a high nitrogen content stainless steel. This allows for cooler operation and higher allowable contact pressure, again increasing the maximum speed rating still further, as shown in Fig. 31, depicted by the term "hybrid-ultra."

The improved running life between ceramics and steels is indicated in Fig. 32, which shows the expected service life for high-speed grease for the two different materials at various DN numbers (mean bearing diameter × speed). This graph can only be taken as an indication of the improvement in grease life, as the individual application, environment, and duty will have an effect on the numbers. The high DN values shown also indicate the recent advancements in synthetic grease technology to allow bearings to run above a DN of 1,000,000 at all without having to resort to oil mist lubrication. Information is based on manufacturers' available data.

5.2 Scanner Construction

Owing to the compact design of ball bearings, a variety of bearing configurations can be used in scanner designs. Polygons can be center mounted, overhung, or contained, as shown in Fig. 33 with both the polygon and the motor assembly sandwiched between the bearings. To keep the surface speed of the balls to a minimum for the longest life, the shafts for ball bearing designs tend to be small where they fit into the bearing bores, whichever configuration is chosen. This can lead to problems with the shaft critical speeds due to its low stiffness. For very compact designs, for laser printers for example, special motor technology is incorporated such as a "pancake" or radial wound motor.

6 MAGNETIC BEARINGS

Over the last ten years, active bearing spindles have been developed commercially for the machine tool industry for high-speed aluminum routing and more recently for

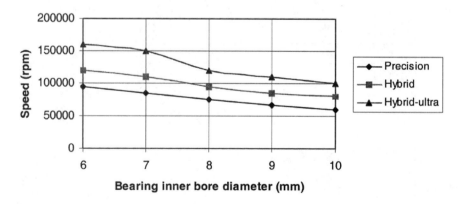

Figure 31 Speed vs. bore diameter for different ball bearing types (grease lubrication).

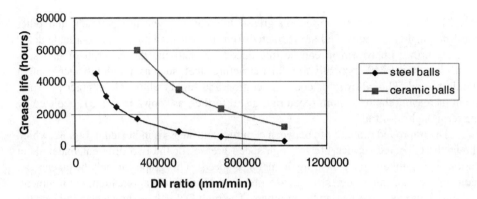

Figure 32 Grease life vs. bearing DN ratio for steel and ceramic balls.

turbo-molecular vacuum pumps in the semiconductor processing industry. Prior to this, magnetic bearing technology had been confined to special applications in the aerospace and satellite industry. The success of the commercialization lies mostly in the improvements in the electronic control system for the bearings, requiring powerful, very fast processor chips. With the cost of such chips plummeting in recent years, the largest drawback of the magnetic bearing, that is, the cost of the complete control system, has reduced dramatically.

These systems are now being applied to other lower cost applications within the electronics process industry, and it will not be long before magnetic bearing systems will appear in special applications in the scanning industry, particularly in vacuum conditions.

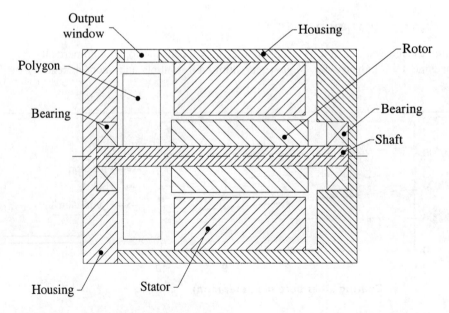

Figure 33 Scanner with ball bearings.

Bearings for Rotary Scanners 381

6.1 Bearing Design Principle

The principle of the active magnetic bearing is relatively simple. The journal bearing consists of a laminated core that is fitted onto the shaft, and is surrounded by a wound static stator, which, once energized, holds the shaft in its magnetic center. A displacement transducer close to the bearing constantly monitors the position of the shaft, and if any external force moves the shaft off center, then the control feedback system connected to the transducer adjusts the current in the stator coils to move the shaft back to the magnetic centerline. The feedback system can typically correct the shaft position every 100 μs. Obviously, this system is inherently unstable and in the case of transducer, or even electrical power failure, a set of catch bearings are necessary to prevent the shaft contacting the bearing coils and causing instant damage.

The main technical advantages of this are:

- very large clearances between the shaft and the stator coils minimizing heat generation within the bearings and reducing the motor power required to run the spindle;
- active damping control, which allows the shaft to be driven through shaft criticals that other bearing types could not accommodate; and
- operation in total vacuum conditions without contaminating the environment.

6.2 Scanner Construction

Figure 34 shows a diagrammatic cross-section of a magnetic bearing spindle. The design contains two radial and one bidirectional axial magnetic bearings with the high-frequency motor between the axial bearing and the rear bearing. In front of the front bearing there are position sensors looking on the shaft at 45° to the axis to provide radial and axial

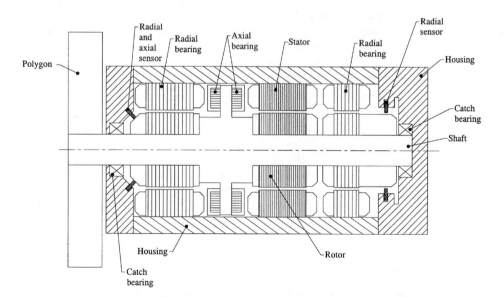

Figure 34 Scanner with magnetic bearings.

displacement information to the controller, and further radial sensors are placed next to the outer end of the rear bearing.

For overload conditions, or a power failure, two small angular contact "catch" bearings are located one at each end of the shaft, with about 0.1–0.2 mm clearance with the shaft under normal running conditions. Typically, the system controller can maintain the radial runout of the shaft to less than one micron.

7 OPTICAL SCANNING ERRORS

In any scanning system there will always be errors associated with the rotation of the optic that will lead to the output beam displaying certain distinguishable errors that can be traced back to either bearing-related or optic-related issues.

7.1 Bearing-Related Errors

Errors due to the rotation of the shaft can be subdivided into two categories: synchronous (repeatable) and asynchronous (nonrepeatable) motions. With synchronous errors, the shaft is usually performing some kind of regular conic (or wobble) motion around its axis, which could be caused by one of the following issues:

- poor shaft assembly balance;
- running at or close to an axial or radial bearing resonant speed, which will magnify any residual shaft imbalance;
- manufacturing errors within the bearings or shaft, such as ovality or misalignment; and
- magnetic pulsations from the rotor as it passes the windings.

Asynchronous errors are more difficult to locate by their very nature and the shaft may describe quite an irregular motion depending on the cause and the type of bearing.

With gas bearings, the onset of half speed whirl for whatever reason will produce strange shaft motion before normally leading to bearing failure. This could be due to an excessive bearing clearance caused by thermal effects within the scanner, or an over speeding of the shaft, or in the case of an aerostatic scanner, a sudden large drop in the supply pressure level. Also, pneumatic hammer occurring in the aerostatic scanner will create asynchronous errors in the shaft.

With ball bearing scanners, asynchronous errors can be caused by manufacturing errors within the ball set beating with the synchronous errors in the track races.

In either bearing system, the ingress of dirt into the bearing will cause intermittent motion errors, which may eventually result in premature failure.

Finally, the motor can cause its own form of asynchronous error, which is commonly called "jitter." It is the result of the feedback system controlling the motion trying to correct for speed variations, caused mainly by optic windage, which will always tend to over or undershoot the actual target speed. This can be minimized by using a closed-loop system with encoder feedback and reducing the effects of turbulence around the optic.

7.2 Optic-Related Errors

Although both polygons and monogons suffer from many of the same errors, their effects on the scanning system need to be examined separately.

Bearings for Rotary Scanners

7.2.1 Polygons

- *Mounting.* Misalignment of the polygon axis to the spin axis is likely to be the largest part of the total tracking error. Very accurate machining of the polygon bore and the hub on the shaft is required to minimize this tilting effect. *Effect*: Repeating weave pattern.
- *Manufacture.* Errors include pyramidal (facet-to-datum), facet-to-facet, dividing angle and facet flatness. *Effect*: Repeatable positional errors.
- *Dynamic distortion.* Loss of geometry due to thermal growth or mechanical stresses. *Effect*: Positional tracking errors varying with speed and time.

7.2.2 Monogons

- *Mounting.* Will only cause a slight permanent change of facet angle, which will occur on every revolution and will not usually affect the scan process. *Effect*: Small, permanent positional change of beam.
- *Manufacture.* Errors include facet flatness, deflection angle, wavefront distortion, and astigmatism. *Effect*: Spot quality and focus issues with speed and time.
- *Dynamic distortion.* Change in flatness and astigmatism due to thermal growth or mechanical stresses. *Effect*: Change in spot quality and focus with speed and time.

7.3 Error Correction

7.3.1 Polygons

Mounting errors can be minimized by machining the polygon hub *in situ* on the shaft running in its own bearings. This helps to correct for many of the synchronous bearings related errors as well as the mechanical errors. Alternatively, an adjustable mount can be used to fine tune the tilt of the polygon to bring it on spin axis. The mount can be manufactured from a thermally insulating material to stop thermal effect reaching the polygon.

With synchronous errors, an active correction system can be employed to slightly modify the beam path prior to striking the polygon facet to compensate for the error about to be put into the beam. This can be permanently preprogrammed in, or in more complicated systems, a facet error detector must be incorporated to constantly update the error compensation system.

7.3.2 Monogons

Error correction is more limited in monogon optic systems. To correct for dynamic mechanical optic distortion, biased optics can be used that will deform to the correct shape over a small specific running speed range, but this only usually refers to open facet mirrors, not prisms. However, many of the synchronous errors are not so noticeable as they occur on every scan line and in general will not cause banding.

As mentioned earlier in this chapter, the use of a pentaprism can dramatically reduce wobble errors generated in the bearings and is ideally suited for ball bearing scanners where wobble is a major problem in higher accuracy designs.

8 SUMMARY

Throughout this chapter, it has been the intention to provide enough theoretical and practical information for the designer to be able to understand, and therefore correctly specify, the bearing system most suitable for the scanning device under consideration. If the conclusion is that a gas bearing is required, then it is anticipated that in most cases, the reader will be intending to buy rather than make the scanning unit due to the complexity of design and manufacture. The graphs contained within this chapter, together with the theory, should provide valuable data regarding the critical parameters such as loads, stiffness, and heat generation. These parameters will have effects on the surrounding parts of the machine and must be taken into account during the machine design process.

Should the designer opt for ball bearing technology, then the option to make rather than buy is more realistic, as both design data and components are readily available. However, the design of the optic and the mount can be the most challenging part of the whole scanner and should not be treated lightly.

ACKNOWLEDGMENTS

The author wishes to acknowledge the assistance of many of his colleagues at Westwind Air Bearings, and Mike Tempest (former Chief Engineer, retired) for his help and support. In addition, the author's thanks go to Mike Tempest and Ron Woolley (Managing Director of Fluid Film Devices of Romsey, UK) for reviewing this document prior to publication.

REFERENCES

1. Shepherd, J. Bearings for rotary scanners. In Marshall, G. *Optical Scanning*; Marcel Dekker, Inc.: New York, 1991; Chap. 9.
2. Kingsbury, A. Experiments with an air lubricated journal. J. Am. Soc. Nav. Eng. 1897, *9*, 267–292.
3. Harrison, W.J. The hydrodynamic theory of lubrication with special reference to air as a lubricant. Trans. Camb. Phil. Soc. 1913, 22, 39.
4. Raimondi, A.A. A numerical solution for the gas lubricated full journal bearing of finite length. Trans. A.S.L.E. 1961, *4*(1).
5. Powell, J.W. *The Design of Aerostatic Bearings*; The Machinery Publishing Co.: Brighton, UK, 1970.
6. Grassam, N.S.; Powell, J.W. Eds. *Gas Lubricated Bearings*; Butterworth: London, 1964.
7. Hamrock, B.J. *Fundamentals of Fluid Film Lubrication*; McGraw-Hill, 1994.
8. Whitley, S.; Williams, L.G. The gas lubricated spiral-groove thrust bearing. U.K.A.E.A. I.G. Rep. 28 RD/CA, 1959.
9. Westwind Air Bearings Ltd. A improved bearing. (European EP) Patent No. 0705393, May 1994. Corresponding U.S. Patent No. 5593230.

7

Preobjective Polygonal Scanning

GERALD F. MARSHALL

Consultant in Optics, Niles, Michigan, U.S.A.

1 INTRODUCTION

Design equations for regular prismatic polygonal scanning systems have been analyzed and described by Kessler[1] and Beiser.[2,3] Beiser's analytical treatment is comprehensive in that the performance in terms of resolution is the key criterion used for the system designs and analyses. Henceforth, throughout this chapter, the term polygonal scanner shall infer a regular prismatic polygonal scanner.

The prime objective of this chapter is to provide a comprehensive visual understanding of the effects of changing the incident beam width (diameter) D, the incident beam offset angle 2β, which is the angle the incident beam is offset from the x-axis, and the number N of mirror facets on a polygonal scanner without regard to performance in terms of resolution. Diagrams, equations, and coordinates bring to light these insights.

Cartesian rectilinear coordinate axes Ox and Oy are chosen for the equations of lines, loci, and the coordinates of significant points. The origin coincides with the axis of rotation O of the regular prismatic polygonal scanner. The x-axis (Ox) is parallel to the optical axis of the objective lens.

There are three distinct topics associated with preobjective polygonal scanning systems that are covered in this chapter by three separate sections, 2, 3, and 4. To assist a reader interested in only one of the topics certain definitions are repeated for continuity of a topic in a section so that the reader does not have to cross reference back and forth to different sections. The topics are: equations and coordinates of a polygonal scanning system; instantaneous center-of-scan; and stationary ghost images outside the image format.

1.1 Equations and Coordinates of a Polygonal Scanning System

The midposition orientation of the polygonal scanner facets is such that the reflected collimated incident beam is parallel to the x-axis and defines the *scan axis*, both of which are chosen to be parallel to the optical axis of the objective lens (Fig. 1).

For a given incident beam offset angle 2β of the collimated incident beam, the equations for the scan axis, the incident beam, the mirror facet plane, and the objective lens optical axis scanner are expressed with respect to the rotation axis O of the polygonal scanner; likewise are the precise coordinates of significant points.[4]

1.2 Instantaneous Center-of-Scan (ICS)

Presented is the derivation of the parametric equations for the loci of the instantaneous center-of-scan (ICS) for six- and twelve-facet prismatic polygonal scanners.[5] Depicted are figures that show the changes in the loci characteristics for different incident beam offset angles 2β.

1.3 Stationary Ghost Images Outside the Image Format

Presented is a pictorial display of diagrams illustrating the permissible angular ranges of the incident beam offset angle 2β to ensure that the ghost images lie outside the image format of the scanned field image format.[6]

2 EQUATIONS AND COORDINATES OF A POLYGONAL SCANNING SYSTEM

The origin of the Cartesian rectilinear coordinate axes can be chosen to be either at the rotation axis O of the polygonal scanner, or at the point of incidence P on a mirror facet; each approach has an advantage for giving insights. In this section the rotation axis O has been chosen to be the origin.[4]

Consider Fig. 1. The diagram depicts a single facet ST of a regular prismatic polygonal scanner with N facets and its circumscribed circle of radius r. The facet ST is oriented so that the collimated incident beam, which is offset at an angle 2β from the x-axis, is reflected parallel to the x-axis. The beam has a finite width D (see Sec. 2.5).

2.1 Objective

The goal is to present the precise coordinates of significant points, the distances between these points, and the equations of three axes (incident beam, scan, and objective lens) with respect to the rotation axis O of the polygonal scanner, thereby eliminating manual or computer-aided iterative techniques. Furthermore, it is to provide unexpected interesting insights into the limitations of the optomechanical design layouts of polygonal scanning systems.

2.2 Midposition and Scan Axis

Shown in Fig. 1 is a single facet ST of a polygonal scanner oriented in a midposition such that a collimated incident beam is reflected parallel to the x-axis. The reflected beam axis PU in this midposition defines the *scan axis*.

2.3 Mirror Facet Angle A

From this midposition the reflected beam angularly scans symmetrically about the scan axis through an angle of $\pm A$. Character symbol A is the facet angle, which is the angle that

Preobjective Polygonal Scanning

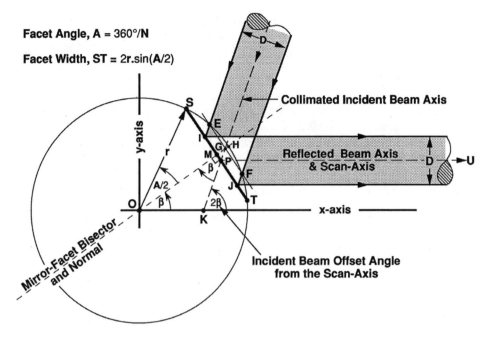

Figure 1 A single facet ST of a polygonal scanner oriented in the midposition. See Fig. 2 for greater clarity around the point of incidence P.

the facet ST subtends at the rotation axis O of the polygonal scanner.

$$A = 360°/N \tag{1}$$

2.4 Mirror Facet Width

The tangential width of the facet ST of a regular prismatic polygonal scanner is:

$$ST = 2r \sin(A/2) = 2r \sin(180/N) \tag{2}$$

2.5 Beam Width (Diameter) D

The gaussian laser beam width represented by D is the standard "$1/e^2$-beam width"* plus a margin of safety chosen by the system designer to minimize the imaged spot defects when a guassian beam is one-sidedly truncated by a facet edge as the polygonal scanner rotates. The margin of safety is inextricably linked to the desired *scan duty cycle* (scan efficiency) η of the scanning system for a given 2β, N, and r. For a one-sided truncation, optimally D is 40% greater than the $1/e^2$-beam width.[2,3]

---

*The $1/e^2$-beam width that is symbolized by D_{1/e^2} is the beam width (diameter) beyond which the residual laser beam power is $1/e^2$ of the total power of a laser beam that has a Gaussian distribution. For a Gaussian distribution, it uniquely and directly also corresponds to the laser beam width (diameter) at which the beam irradiance has dropped to $1/e^2$ of the axial peak laser beam irradiance (see Chapter 1).

In Fig. 1, in which the polygonal scanner is in the midposition:

1. The boundaries of the incident beam width D are designed to cut through the circumscribed circle at E and F such that the arcs SE and FT are equal. This ensures that the useful angular scan of the reflected beam is symmetrical about the scan axis.
2. The axis of the incident beam passes through the mid point G of the chord EF to impinge on a facet at the point P. Point G also lies on the facet sag MH. Henceforth, the point G, together with the point H, become points on the axis of the incident beam (Figs. 1 and 2).

2.6 Scan Duty Cycle (Scan Efficiency)

The *scan duty cycle* η of a polygonal scanner is the ratio of the useful scan angle during which the beam width D is unvignetted by the edges of the facets, to the full scan angle $\pm A$ of a beam with an infinitesimal beam width. One assumes that the tangential width of the footprint of the incident beam is less than the tangential width of the facet, when the polygonal scanner is in its midposition (see Chapter 2 of this volume). Henceforth, widths refer to tangential widths.

It can be shown[4] that

$$\eta = 1 - \frac{\arcsin[D/(2r\cos\beta)]}{180/N} \tag{3}$$

Knowing or selecting suitable values for r, N, β, and D will determine η. Alternatively, choosing suitable values for r, N, β, and η will determine the required incident beam width

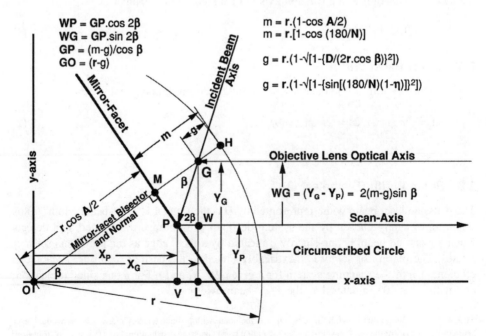

Figure 2 A geometrical diagram for determining the coordinates of G and P, namely, X_G, Y_G and X_P, Y_P.

D as follows. Transposing Eq. (3) gives

$$D = (2r \cos \beta) \sin[(180/N)(1 - \eta)] \tag{4}$$

More simply, if W represents the tangential facet width, this expression approximates to[2,3]

$$D_{approx} = W(\cos \beta)(1 - \eta) \tag{5}$$

Dividing Eq. (5) by Eq. (4) gives

$$D_{approx}/D = \frac{[\sin(180/N)](1 - \eta)}{\sin[(180/N)(1 - \eta)]} \tag{6}$$

The closeness of D_{approx} to D is illustrated in Table 1.

Let the polygonal scanner be in its midposition, then:

1. If the incident beam has an infinitesimal width ($D = 0$), the circumscribed circle of the polygonal scanner intersects the axis of the beam at the point H, which coincides with the top of the facet sag MH. The scan duty cycle is 100% ($\eta = 1$), ignoring the inevitable facet edge manufacturing roll-off (Fig. 2).
2. If the incident beam has a finite width D with a footprint that just covers the facet's tangential width ($D = 2r \sin(A/2)$), the beam axis is directed at M, at the base of the sag MH. The scan duty cycle is 0% ($\eta = 0$). Simultaneously, point P coincides with M, which is the midpoint of the facet chord ST.
3. For all finite incident beam widths D with a footprint width that is within the facet width ($D < 2r \sin(A/2)$), the beam axis passes through the midpoint G of the chord EF, and which lies on the facet sag MH, to impinge on a facet at the point P. The scan duty cycle is finite ($1 > \eta > 0$) [Eq. (3), Figs. 1 and 2].

2.7 Sag Dimensions

It can be shown[4] and with reference to the geometry in Fig. 2 that when sag MH = m

$$m = r[1 - \cos(A/2)] = r[1 - \cos(180/N)] \tag{7}$$

Table 1 Ratio [D_{approx}/D] for Scan Duty Cycle η vs. Number of Facets N

	η				
	0.00	0.25	0.50	0.75	1.00
$N = 3$	1.00	0.92	0.87	0.84	0.83
$N = 6$	1.00	0.98	0.97	0.96	0.95
$N = 12$	1.00	0.99	0.99	0.99	0.97
$N = 18$	1.00	1.00	1.00	1.00	1.00
$N = 24$	1.00	1.00	1.00	1.00	1.00

If the sag GH = g, then in terms of r, D, and β

$$g = r(1 - \sqrt{[1 - \{D/(2r\cos\beta)\}^2]}) \tag{8}$$

Or from Eqs. (7) and (8)

$$(m - g) = r(-\cos(180/N) + \sqrt{[1 - \{D/(2r\cos\beta)\}^2]}) \tag{9}$$

$$(r - g) = \sqrt{[1 - \{D/(2r\cos\beta)\}^2]} \tag{10}$$

Likewise, in terms of r, N, and η

$$g = r(1 - \sqrt{[1 - \{\sin[(180/N)(1 - \eta)]\}^2]}) \tag{11}$$

$$(m - g) = r(-\cos(180/N) + \sqrt{[1 - \{\sin[(180/N)(1 - \eta)]\}^2]}) \tag{12}$$

$$(r - g) = \sqrt{[1 - \{\sin[(180/N)(1 - \eta)]\}^2]} \tag{13}$$

(see Figs. 2 and 4).

2.8 Coordinates of G

From the geometry in Fig. 2:

$$X_G = (r - g)\cos\beta \tag{14}$$

and

$$Y_G = (r - g)\sin\beta \tag{15}$$

Substituting for $(r - g)$ from Eq. (10), and expressing X_G, Y_G in the terms of r, D, and β gives

$$X_G = r(\sqrt{[1 - \{D/(2r\cos\beta)\}^2]})\cos\beta \tag{16}$$

and

$$Y_G = r(\sqrt{[1 - \{D/(2r\cos\beta)\}^2]})\sin\beta \tag{17}$$

Likewise, substituting for $(r - g)$ from Eq. (10), and expressing X_G, Y_G in terms of r, N, and η gives

$$X_G = r(\sqrt{[1 - \{\sin[(180/N)(1 - \eta)]\}^2]})\cos\beta \tag{18}$$

and

$$Y_G = r(\sqrt{[1 - \{\sin[(180/N)(1 - \eta)]\}^2]})\sin\beta \tag{19}$$

2.9 Coordinates of P

Again from the geometry in Fig. 2:

$$X_P = X_G - (m-g)[(\cos 2\beta)/\cos \beta] \tag{20}$$

and

$$Y_P = Y_G - 2(m-g)\sin \beta \tag{21}$$

Substituting for X_G and Y_G from Eqs. (14) and (15) gives

$$X_P = (r-g)\cos \beta - (m-g)[(\cos 2\beta)/\cos \beta] \tag{22}$$

and

$$Y_P = (r-g)\sin \beta - 2(m-g)\sin \beta \tag{23}$$

By substituting for $(m-g)$ and $(r-g)$ from Eqs. (9) and (10) into Eqs. (22) and (23) one obtains the coordinates of X_P, Y_P expressed in terms of r, D, and β.

$$X_P = (r/\cos \beta)[\cos (180/N)\cos 2\beta + \sin 2\beta \sqrt{(1 - [D/(2r\cos \beta)]^2)}] \tag{24}$$

and

$$Y_P = (r\sin \beta)[2\cos (180/N) - \sqrt{(1 - [D/(2r\cos \beta)]^2)}] \tag{25}$$

By substituting for $(m-g)$ and $(r-g)$ from Eqs. (12) and (13) into Eqs. (22) and (23) one obtains the coordinates of X_P, Y_P expressed in terms r, N, and η.

$$X_P = (r/\cos \beta)[\cos (180/N) \cos 2\beta + \sin 2\beta \sqrt{(1 - \{\sin[(180/N)(1-\eta)]\}^2)}] \tag{26}$$
$$Y_P = (r\sin \beta)[2\cos (180/N) - \sqrt{(1 - \{\sin[(180/N)(1-\eta)]\}^2)}] \tag{27}$$

2.10 Optical Axis of the Objective Lens

The objective lens optical axis, which is parallel to both the x-axis and the scan axis, is directed through the point G to ensure that the scanning beam width D scans symmetrically across the aperture of the objective lens (Figs. 1, 2, and 3).

The separation between the objective lens optical axis and the scan axis is given by

$$WG = 2(m-g)\sin \beta \tag{28}$$

If the incident beam has an infinitesimal width, G coincides with H ($g=0$), and the separation WG between the objective lens optical axis and the scan axis is a maximum

$$WG_{max} = 2m(\sin \beta) \tag{29}$$

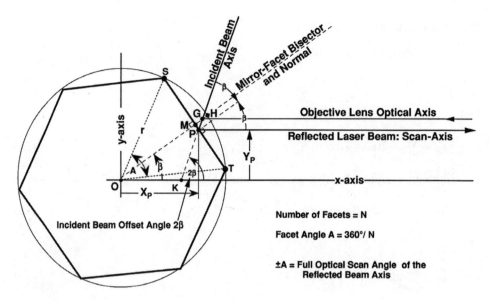

Figure 3 The six facets of the hexagonal scanner head oriented in the midposition in which the reflected incident beam axis lies along the scan-axis. The boundaries of the beam, which has a width D, are omitted to avoid overcrowding of the diagram.

Substituting for m from Eq. (7) leads to

$$WG_{max} = 2r[1 - \cos(180/N)]\sin\beta \tag{30}$$

As G and P simultaneously approach M the objective lens optical axis and the scan axis move toward each other at the same rate in the y-axis direction until they both coincide at M.

When the incident beam has a finite width D and the beam width of the footprint just covers the facet chord ST, G and P coincide with M, $m = g$, $(m - g) = 0$.

Substituting $(m - g) = 0$ into Eq. (28) gives

$$WG_{min} = 0 \tag{31}$$

Thus, the objective lens axis is coincident with the scan axis.

2.11 Equations

Except for the incident beam, the scan axis and objective lens optical axis are parallel to the x-axis and, therefore, have equations independent of x.

2.11.1 Scan Axis PU

The equation to the scan axis corresponds to Y_P, given in Eq. (23), namely

$$Y_P = (r - g)\sin\beta - 2(m - g)\sin\beta \tag{32}$$

(see Eqs. (25) and (27).)

In a reverse sense Y_P also represents the offset distance of the rotation axis O from the scan axis PU for a given offset angle β of the incident beam (see Sec. 2.12 and 4.9).

2.11.2 Objective Lens Optical Axis

The equation to the objective lens optical axis corresponds to Y_G given in Eq. (15), namely

$$Y_G = (r - g)\sin\beta \tag{33}$$

From Eq. (17) Expressing Y_G in terms of r, D, and β gives

$$Y_G = r\left(\sqrt{[1 - \{D/(2r\cos\beta)\}^2]}\right)\sin\beta \tag{34}$$

From Eq. (19), expressing Y_G in terms of r, N and η gives

$$Y_G = r\left(\sqrt{[1 - \{\sin[(180/N)(1 - \eta)]\}^2]}\right)\sin\beta \tag{35}$$

2.11.3 Incident Beam Axis Through GP

$$y = (\tan 2\beta)x - (r - g)[(\tan 2\beta)(\cos\beta) + \sin\beta] \tag{36}$$

where from Eq. (10) $(r - g)$ is expressed in terms of r, D, and β, namely

$$(r - g) = \sqrt{[1 - \{D/(2r\cos\beta)\}^2]} \tag{37}$$

Alternatively, where from Eq. (13) $(r - g)$ is expressed in terms of r, N, and η, namely

$$(r - g) = \left[\sqrt{1 - \{\sin[(180/N)(1 - \eta)]\}^2}\right] \tag{38}$$

2.11.4 Mirror Facet Bisector and Normal

The linear equation to the bisector of the mirror facet and normal has a slope of $\tan\beta$ with no intercept and passes through the rotation axis O, which is the origin of the coordinate system. Thus

$$y = (\tan\beta)x \tag{39}$$

2.12 Insights from an Alternative Analytical Approach

An alternative perspective for the analysis is to set the Cartesian rectilinear coordinate axes to be Px and Py with the origin at the point of incidence P on the facet for when the polygonal scanner is in a midposition (Fig. 4). In this approach the scan axis is collinear with the abscissa, the x-axis (Px), while the ordinate is the y-axis (Py). See the first paragraph at the beginning of this Sec. 2.

The immediate advantage is that equations for the scan axis, the incident beam, and the facet plane all pass through the origin P. The goal of this second approach is to determine the coordinates (X_O, Y_O) of the rotation axis O of the polygonal scanner with respect to the point of incidence P and the scan axis Px.

The approach presents a diagrammatic visualization of the existence of a finite areal zone, with respect to the fixed point P, of a set of loci for the rotation axis O_N of a

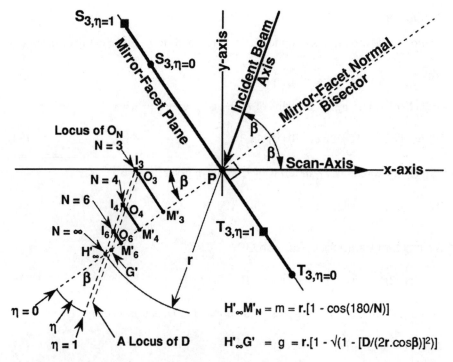

Figure 4 Displayed are the loci of the rotation axes O_N relative to the point of incidence P on a mirror-facet plane of a polygonal scanner oriented in a midposition.

polygonal scanner that results from changes in the number N of facets ($3 \leq N < \infty$), the laser beam width D, and the scan duty cycle η, ($0 \leq \eta \leq 1$), for a given circumscribed circle of radius r of the polygonal scanner (Fig. 4).[4]

All these coordinates and equations can be obtained from those already given in Sec. 2.8 to 2.11 by the transformation of the origin at O to an origin at P.

2.13 Features of Fig. 4

In Fig. 4, the loci for $N = 5$ and for $N > 6$ are omitted to avoid overcrowding the diagram. Certain character symbols are primed because of a direct, but not obvious, relationship to those corresponding unprimed symbols in Fig. 2.

The set of loci for the position of the rotation axis O_N are confined to the series of parallel base lines $I_N M_N$ of a nest of right triangles $H'_\infty M'_N I_N$ within the triangle $H'_\infty M'_3 I_3$. These base lines are parallel to the facet plane ST (Table 2).

$$I_N M_N = m \tan \beta = r[1 - \cos(180/N)] \tan \beta \qquad (40)$$

The base lines $I_N M_N$ are spaced at ever diminishing distances toward the apex H'_∞ of the triangle as the number N of facets increases. The spacing between every sixth base line is given by

$$[m_N - m_{N+6}] = r[\cos(180/\{N+6\}) - \cos(180/N)] \qquad (41)$$

Table 2 Facet Width W_N and Locus Length $I_N M_N$ vs. Number of Facets N for $r = 50$ mm

	\multicolumn{4}{c}{N}			
	6	12	18	24
$W_N = S_N T_N$ (mm)	50.00	25.9	17.4	13.1
$I_N M_N$ (mm)	3.87	0.46	0.13	0.06
$[S_N T_N]/[I_N M_N]$	12.9	56.7	130	232
$[m_N - m_{N+6}]$ (mm)	5.00	0.94	0.33	0.15

Simultaneously at a lesser rate, the facet widths $S_N T_N$ shorten as the number N of facets increases

$$S_N T_N = 2r \sin(180/N) \tag{42}$$

From Eqs. (42) and (40) the ratio of the facet width $S_N T_N$ to length of the locus $I_N M_N$ is expressed by

$$[S_N T_N]/[I_N M_N] = [\cotan(90/N)][(\cotan(180/N)] \tag{43}$$

for which the incident beam offset angle $2\beta = +A$.

The position of the rotation axis on a locus $I_N M_N$ depends on the scan duty cycle $(0 < \eta < 1)$. A fan of straight lines emanating from H'_∞ toward $I_N M_N$ represents a set of values for constant scan duty cycle η. The rotation axis O_N lies at the intersection of one of these fan lines of constant η with a base line $I_N M_N$.

A set of straight lines parallel to $H'_\infty I_3$ represents a set of values for constant beam width D. Similarly, the rotation axis O_N lies at the intersection of one of these parallel lines of constant D with a base line $I_N M_N$. The rotation axes O_N may not lie beyond M_N where the incident beam width footprint matches the facet width.

All facet widths $S_N T_N$ lie between the points $S_{3,\eta=1}$ and $T_{3,\eta=0}$ according to the values of N and η. The positional range of facet $S_N T_N$ directly corresponds to the range of the locus O_N, that is, the length of the baseline $I_N M_N$. Uniquely, the rotation axis O lies on the scan axis only when $N = 3$ and $D = 0$, that is, an incident beam of infinitesimal width.

2.14 Conclusion

The visualization of the effects of changing the controlling parameters N, β, D, η, and r of an optical scanning system helps in its design, while, in particular, the explicit coordinates and equations eliminate manual or computer-aided iterative techniques.

3 INSTANTANEOUS CENTER-OF-SCAN

Reflective scanning devices, resonant, galvanometric, and polygonal, have plane mirrors that oscillate or rotate about an axis. The rotation of the reflecting mirror deflects an incident light beam. When (1) the axis of rotation O is coincident with the mirror surface,

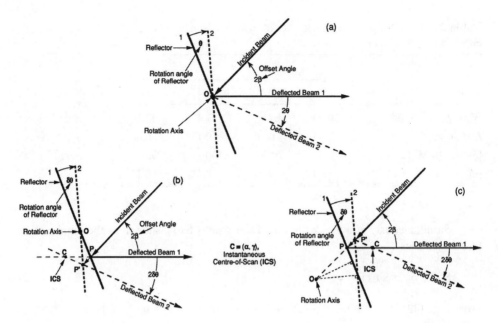

Figure 5 (a) The axis of rotation lies in the reflecting surface of a mirror (reflector) and the incident beam is directed at the axis of rotation O of the mirror. (b) The axis of rotation lies in the reflecting surface of a mirror but the incident beam is directed to one side of rotation axis O. (c) The incident beam is directed at the axis of rotation but the rotation axis O is displaced from the reflecting surface of the mirror.

and (2) the incident beam is directed at the axis of rotation, the instantaneous center-of-scan (ICS) is a single stationary point on, and at, the axis of rotation O for all angular positions of the mirror. These two conditions are difficult to achieve and are rarely met; as a result, the ICS moves with respect to the rotation axis O, and, therefore, is a locus (Fig. 5).[5]

3.1 Objective

This section explores and illustrates the characteristic form of the ICS locus for polygonal scanners with respect to the incident beam offset angle 2β, that is, the angle between the incident beam and the scan axis. The analysis, study, and depiction of the ICS loci for several incident beam offset angles for regular prismatic polygonal scanners of six and twelve facets give a visual appreciation of the asymmetry in the optical path lengths of the deflected beam as it sweeps through the full scan angle $\pm A$. These characteristics provide interesting insights for consideration when undertaking the design of a polygonal scanning system.

3.2 Origin of the Instantaneous Centers-of-Scan Locus

In Fig. 5(a) the center-of-scan of the reflected beam is a stationary point on the rotation axis O. This is because two conditions are met: (1) the axis of rotation lies in the reflecting surface of a mirror (reflector), and (2) the incident beam is directed at the axis of rotation O of the mirror.

Preobjective Polygonal Scanning

In Fig. 5(b) the center-of-scan is not a stationary point, because, although the axis of rotation lies in the reflecting surface of a mirror, the incident beam is directed to one side of the rotation axis O. However, there is an ICS at point $C \equiv (\alpha, \gamma)$ that has a locus.

Again, in Fig. 5(c) the center-of-scan is not a stationary point, because, although the incident beam is directed at the axis of rotation, the rotation axis O is displaced from the reflecting surface of the mirror. Similarly, there is an ICS at point $C \equiv (\alpha, \gamma)$ that has a locus.

3.3 Midposition and Scan Axis

Figure 6 depicts a cross-section of a hexagonal polygonal scanner set in a midposition with an incident beam offset angle 2β of 70°. The midposition is defined by two requirements: (1) the polygonal scanner is oriented such that the reflected incident beam from one of the facets is parallel to the x-axis (this reflected incident beam defines the scan axis); (2) the rotation axis O is offset from the scan axis to a position such that, as the polygonal scanner rotates, the reflected incident beam angularly scans symmetrically $\pm A$ about the scan axis.

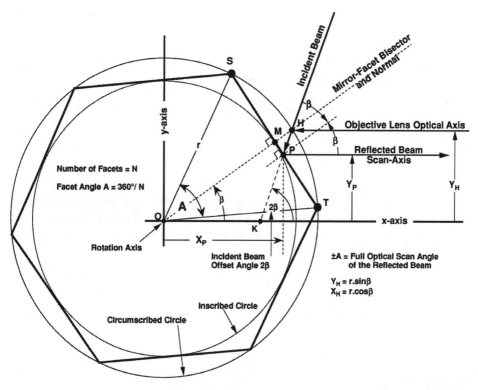

Figure 6 A scaled cross-section of a six-facet polygonal scanner in the midposition from which the incident beam at an offset angle of 2β is reflected parallel to both the objective lens optical axis and the x-axis. This reflected incident beam defines the scan-axis.

3.4 Derivation of the Instantaneous Center-of-Scan Coordinates

Consider a regular prismatic polygonal scanner with N facets and a circumscribed circle of radius r (Fig. 6). Cartesian rectilinear coordinate axes Ox and Oy are chosen for the equations of lines, loci, and the coordinates of significant points. The origin O coincides with the axis of rotation of the polygonal scanner. The x-axis (Ox) is parallel to the optical axis of the objective lens. The facet angle A, that is, the angle that the facet subtends at the axis of rotation O, is given by $360°/N$. For simplicity it is assumed that the beam width (diameter) is infinitesimal, such that a single ray represents the incident beam. Instantaneous Center-of-Scan loci for finite beam widths D are considered in Sec. 3.10.

Consider now an incident beam directed at the facet of a polygonal scanner in a midposition at an offset angle 2β (70°) (Fig. 6). Point H on the incident beam is where the circumscribed circle of the polygonal scanner and a facet bisector OM scanner intersects the incident beam.

Figure 7 depicts the position of one facet of the polygon after it has been rotated counterclockwise through an angle θ, and the resultant position and direction of the reflected incident beam that has been deflected through an angle 2θ.

The linear equation of the reflected beam passes through the ICS coordinates (α, γ) and is represented by

$$(y - \gamma) = [\tan(2\theta)](x - \alpha) \tag{44}$$

The incident beam linear equation expressed in the intercept form is

$$\frac{x}{(r/2)/\cos\beta} + \frac{y}{-(r/2)[\tan(2\beta)]/\cos\beta} = 1 \tag{45}$$

The linear equation for the line of intersection of the facet plane and the plane of incidence expressed in the intercept form is

$$\frac{x}{r\cos A/\cos(\beta + \theta)} + \frac{y}{r\cos A/\sin(\beta + \theta)} = 1 \tag{46}$$

From the three Eqs. (44), (45), and (46) the coordinates of α and γ may be determined. The technique is to differentiate Eqs. (44) and (45) with respect to θ remembering that α and γ are not variables, but constants at any instant.

Thus the derivative of Eq. (44) is

$$(x - \alpha) = (\cos 2\theta)^2 (y' - x' \tan 2\theta)/2 \tag{47}$$

And the derivative of Eq. (45) is

$$y' = x' \tan 2\beta \tag{48}$$

3.5 Solutions

Solving for $(x - \alpha)$ and $(y - \gamma)$: eliminating y' between Eqs. (47) and (48) gives

$$(x - \alpha) = x'(\cos 2\theta)^2 (\tan 2\beta - \tan 2\theta)/2 \tag{49}$$

Preobjective Polygonal Scanning

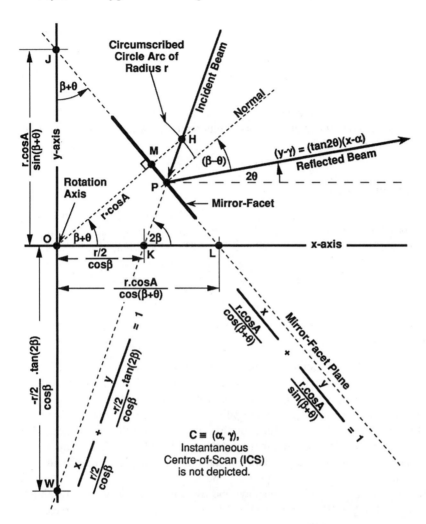

Figure 7 A trigonometrical diagram depicts the three key analytical equations of one facet of the polygonal scanner illustrated in Figure 6 which has been rotated counter-clockwise through an angle θ.

Substituting for $(x - \alpha)$ from Eq. (44) into Eq. (49) leads to

$$(y - \gamma) = x'(\sin 2\theta)(\cos 2\theta)(\tan 2\beta - \tan 2\theta)/2 \tag{50}$$

Inspection of Eqs. (49) and (50) shows the need to solve and substitute for x and y, and x' with expressions containing only r, A, β, and θ.

Solving for x and y, and x'

Note that simultaneous Eqs. (45) and (46) do not contain the ICS coordinates α and γ; thus solving for x and y gives the following parametric equations in terms of r, θ, A, and β:

$$x = \frac{r[\cos A/\tan 2\beta + \sin(\beta + \theta)/2\cos\beta]}{[\sin(\beta + \theta) + \cos(\beta + \theta)/\tan 2\beta]} \tag{51}$$

Likewise

$$y = \frac{r[\cos A - \sin(\beta + \theta)/2 \cos \beta]}{[\sin(\beta + \theta) + \cos(\beta + \theta)/\tan 2\beta]} \qquad (52)$$

Equations (51) and (52) also represent the locus of the point of incidence P, $(X_{P_\theta}, Y_{P_\theta})$, as θ varies and as the reflected incident beam scans. P inherently lies along the segment HP of the incident beam (Figs. 6 and 7).

3.6 Spreadsheet Program

The derivative x' is obtained by differentiating Eq. (51) with respect to θ. An explicit expression is possible but unnecessary when using a computer spreadsheet program. Tabulating data against θ, obtained by using a spreadsheet program, are the values of $(x - \alpha)$ and $(y - \gamma)$ from Eqs. (49) and (50); the values of x and y from Eqs. (51) and (52), and the values of the derivative x'; thence the coordinates α and γ are deduced and plotted (Figs. 8 to 12).

3.7 Instantaneous Center-of-Scan

Figures 8 to 11 display the ICS loci for four incident beam β offset angles, namely, $0°$, $70°$, $100°$, and $140°$. The data plots on the ICS loci correspond to the mechanical rotation angle θ of the polygonal scanner at two-degree intervals from its midposition.

It should be noted that a tangent at any point on the ICS locus is the position and direction of the reflected incident beam for a rotation angle θ. When the facet edges, S and T, on the circumscribed circle pass through the fixed point H on the incident beam, so also on the ICS locus do the tangents that represent the reflected incident beam at the full optical scan angles $\pm A$ ($\theta = \pm A/2$) (Figs. 6, 7, and 8).

The ICS locus shown in Fig. 8 displays the expected symmetry for an unlikely incident beam offset angle 2β of zero degrees. The peak of the ICS cusp characteristic touches the inscribed circle of the polygonal scanner and the locus extends beyond the circumscribed circle.

Figure 9 shows the asymmetry of the ICS locus for a realistic incident beam offset angle 2β of $70°$. The peak of the ICS cusp characteristic lies within the inscribed circle of the polygonal scanner, while one extremity lies beyond the circumscribed circle and the other lies between the two circles. The tangent on the ICS locus at the data point $\theta = 0°$ corresponds to the scan axis.

Figure 10 shows the asymmetry of the ICS locus for an incident beam offset angle 2β of $100°$. The peak of the ICS cusp characteristic lies within the inscribed circle of the polygonal scanner, as does one extremity $\theta = +30°$, while the other extremity $\theta = -30°$, lies between the inscribed and the circumscribed circles. The tangent on the ICS locus at the data point $\theta = 0°$ corresponds to the scan axis.

Figure 11 shows a more extreme asymmetry of the ICS locus for an incident beam offset angle 2β of $140°$ for a 12-facet polygonal scanner, $N = 12$. The peak of the ICS cusp has disappeared because, in part, the range of the full mechanical scan angle θ has been reduced from $\pm 30°$ to $\pm 15°$ by virtue of the increased number of facets from six to twelve. The ICS locus extremities range from $\theta = +15°$ within the inscribed circle of the

Preobjective Polygonal Scanning

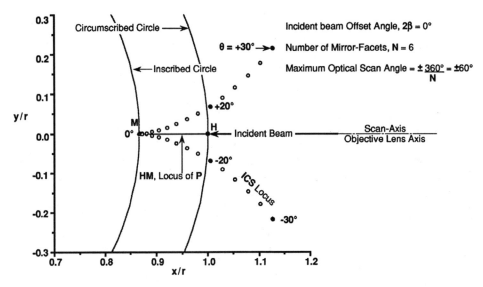

Figure 8 The ICS locus for an incident beam offset angle 2β of $0°$ for a six-facet polygonal scanner.

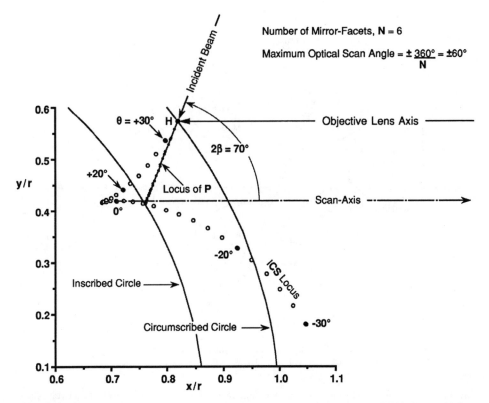

Figure 9 The ICS locus for an acute incident beam offset angle 2β of $70°$ for a six-facet polygonal scanner.

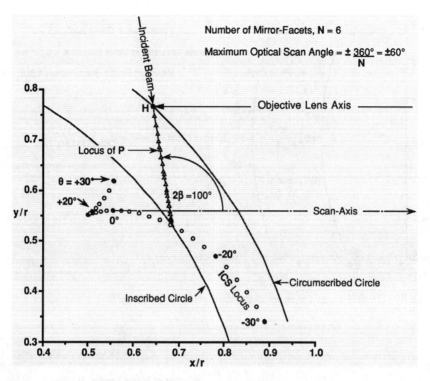

Figure 10 The ICS locus for an obtuse incident beam offset angle 2β of $100°$ for a six-facet polygonal scanner.

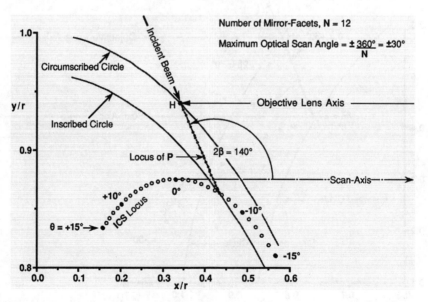

Figure 11 The ICS locus for an obtuse incident beam offset angle 2β of $140°$ for a twelve-facet polygonal scanner. A tangent to the ICS locus represents the position and direction of the reflected incident beam.

Preobjective Polygonal Scanning

polygonal scanner to $\theta = -15°$ between the inscribed and the circumscribed circles. The tangent on the ICS locus at the data point $\theta = 0°$ corresponds to the scan axis.

3.8 Locus of P

Axiomatically the locus of the point of incidence P lies along the incident beam line segment HP. Point P runs back and forth from the fixed point H on the incident beam, at which the circumference of the circumscribed circle of the polygonal scanner intersects. As the reflected incident beam scans through the full angular range $\pm A$ the locus of P overlaps itself.

The locus of P is inherently a straight line along the incident beam but doubles back on itself from and to the fixed point H. To provide visibility of this locus the data ordinate y/r values of Fig. 10 have been mathematically and linearly stretched in Fig. 12. For clarity the scale of the x/r axis is significantly magnified about tenfold. The markedly different spacing between data plots at two-degree intervals is indicative of a rapid acceleration and slow deceleration as the point of incidence P traverses the facet.

3.9 Offset Angle Limits

The incident beam is not likely to lie within the scan angle when the plane of incidence is normal to the axis of rotation; therefore, the smallest offset angle $[2\beta]_{min}$ for a beam with

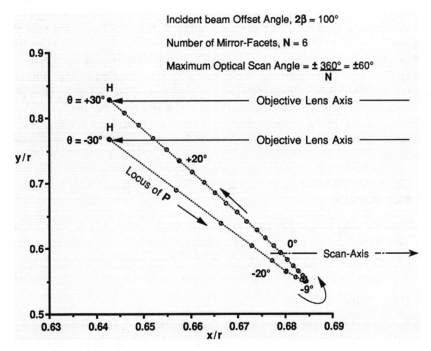

Figure 12 The locus of P for an incident beam offset angle 2β of $100°$ of Figure 10. For visibility and clarity the ordinate data and scale of the abscissa have been adjusted.

an infinitesimal diameter will always be equal to, or greater than, the semi full optical scan angle $+A$.

The maximum offset angle $[2\beta]_{max}$ of the incident beam with an infinitesimal diameter occurs when the incident beam is at grazing incidence for the semi full optical scan angle $-A$. Therefore, the upper limit to the offset angle will always be equal to, or less than $(180° - A)$.

Thus 2β lies in the range

$$360°/N \leq 2\beta \leq 180°(1 - 2/N), \qquad N \geq 4 \tag{53}$$

For real incident beams with a finite width (diameter) the minimum limit $360°/N$ will increase and the maximum limit $180°(1 - 2/N)$ will decrease (see scan duty cycle η in Sec. 2.6 and 4.8 of this chapter).

The expressions for the limits of the offset angles provide a useful guideline in the design of a scanning system. Although one will endeavor to design the incident beam to have an offset angle 2β to lie close to the semi full scan angle $+A$ of the reflected incident beam, there are occasions for reasons of packaging where this is not possible.

3.10 Finite Beam Width D

For simplicity the width of the beam has been assumed to be infinitesimal (Sec. 3.4) such that the objective lens optical axis is directed through the fixed point H on the incident beam where it is intersected by the circumscribed circle of the polygonal scanner in the midposition. If the incident beam has a finite width D, the radius r in Eq. (45) would be replaced by $r(1 - g/r)$ because the incident beam shifts to the left to pass through the point G. The dimensional symbol "g" is that shown in Fig. 2 (Sec. 2). The r in Eq. (46) remains unchanged.

For a finite beam width D the basic cusp-shape characteristic of the ICS loci that is shown in Figs. 8 to 11 remains the same, but, with the exception of Fig. 8 because of symmetry, it will be slightly displaced in an upward direction parallel to the facet plane by an amount $(g \tan \beta)$ (Fig. 2). The scan axis is raised and the objective lens axis is lowered, each by an amount $(g \sin \beta)$ with respect to the coordinate axes x/r and y/r. These displacement amounts are derived from Fig. 2, Eqs. (21) and (15) in Sec. 2.9 and 2.8, respectively.

3.11 Commentary

The ICS curves with offset angles greater than zero display interesting asymmetrical cusp-shaped characteristics that offer potential insights into pupil movement that give rise to asymmetric aberrations in the image plane of preobjective scanning systems.

3.12 Conclusion

An analysis of real beams of finite width (diameter) is fully expected to produce the same basic ICS characteristics as shown in the above documentation. The bottom line is that the instantaneous center-of-scan locus is of interest because it can give insight to the asymmetric wandering of the entrance pupil for the optical system lens designer who optimizes a design by minimizing the aberrations in the image plane regardless of the ICS locus.

4 STATIONARY GHOST IMAGES OUTSIDE THE IMAGE FORMAT

Ghost images are caused by both specular and scattered reflected rays from optical surfaces and are always unwanted, especially within the image format of the scanned field image plane. Various design innovations have been invented to minimize the effects or the presence of ghost images in the image format. Notable are those given in Refs. 7 and 8 in which a limited angular range for the incident beam offset angle (2β) from the scan axis is given so that stationary ghost images are formed outside the image format of the scanned field image plane of regular prismatic polygonal scanning systems.[6]

4.1 Objective

This section explores and illustrates the formation of stationary ghost images that are produced only by the reflected rays from the scanned field image plane itself. The goal is to determine the angular ranges and limits of the incident beam offset angles 2β (beyond that given in Refs. 7 and 8 mentioned above), with visual insights that ensure that the stationary ghost images lie outside the image format.

4.2 Stationary Ghost Images

Ghost images in the image plane from nonmoving optical components may be expected, but, at first thought, not from a rotating optical component such as a polygonal scanner, and if so, certainly not stationary. However, the rotating polygonal scanner itself synchronously derotates (descans) these unwanted diffusely reflected rays from the image plane itself, and they are then specularly re-reflected at the mirror facets. If these secondary specularly reflected rays are transmitted through the optics of the preobjective optical scanning system, stationary ghost images will be formed in the image plane.

4.3 Facet Angle A

The facet angle A is the angle that the facet subtends at the rotation axis O:

$$A = 360/N \tag{54}$$

where N represents the number of facets. For this section let $N = 10$. Then,

$$A = 36°, \quad \text{and} \quad 2A = 72° \tag{55}$$

4.4 Facet-to-Facet Tangential Angle

The mirror facet-to-facet tangential angle is the angle between successive facet normals in a plane perpendicular to the rotation axis. This angle is also denoted by the symbol A, because for a regular prismatic polygonal scanner the facet angle and the facet-to-facet tangential angle are geometrically identical.

4.5 Scan Axis

The scan axis is the axis about which the beam angularly scans symmetrically, $\pm A$ (see Sec. 2.2 and 3.3).

4.6 Offset Angle 2β

The incident beam offset angle 2β is the angle that the incident beam makes with the scan axis.

4.7 Midposition

The midposition of a scanner is that orientation of the polygonal scanner for which a facet reflects the incident beam collinearly with the scan axis, which is parallel to the objective lens optical axis (see Sec. 2.2, 3.3, and Fig. 1).

4.8 Scan Duty Cycle (Scan Efficiency) η

The maximum potential scan duty cycle η of a polygonal scanner is the ratio of the useful scan angle, during which the beam width D is effectively unvignetted by the edges of the facets, to the full scan angle $\pm A$ of a beam with an infinitesimal width ($D = 0$). We shall assume that the footprint of the beam's tangential width is less than the facet's tangential width in the midposition of a polygonal scanner.

$$\eta = 1 - \frac{\arcsin(D/[2r\cos\beta])}{180/N} \tag{56}$$

in which r represents the radius of the circumscribed circle of the polygonal scanner (see Sec. 2.6). It can be seen from Eq. (56) that for a given beam of finite width D the scan duty cycle η decreases with an increase in the offset angle 2β, or with an increase in the number of facets N.

4.9 Rotation Axis Offset Distance

The rotation axis offset distance is that distance $(-Y_P)$ of the rotation axis from the scan axis when the polygonal scanner is set in its midposition (Figs. 2 and 3).

The rotation axis offset distance $(-Y_P)$ depends on the number of facets N, the incident beam offset angle 2β, and beam width D. Replicating Eq. (25) with a negative sign from Sec. 2 gives

$$(-Y_P) = -r\sin\beta\left[2\cos(180/N) - \sqrt{(1 - [D/(2r\cos\beta)]^2)}\right] \tag{57}$$

in which r again represents the radius of the circumscribed circle of the polygonal scanner.

Alternatively, replicating Eq. (27) embodying in the scan duty cycle η from Sec. 2 leads to

$$(-Y_P) = -r\sin\beta\left[2\cos(180/N) - \sqrt{(1 - \{\sin[(180/N)(1 - \eta)]\}^2)}\right] \tag{58}$$

Table 3 Incident Beam Offset Angle 2β vs. Maximum Potential Scan Duty Cycle η

Incident beam offset angle, 2β	Maximum potential scan duty cycle, η	Rotation axis offset distance, Y_P, distance from the scan axis	Figure
27°	93.5%	$0.211r$	13
52°	92.9%	$0.395r$	14
92°	90.8%	$0.649r$	15
124°	86.4%	$0.797r$	16
164°	54.1%	$0.904r$	—

For an infinitesimal beam width ($D = 0$) or a 100% scan duty cycle ($\eta = 1$), Eqs. (57) and (58) both reduce to

$$(-Y_P) = -r \sin \beta [2 \cos(180/N) - 1], \quad N \geq 3 \tag{59}$$

It can be seen from Eq. (59), Table 3, and supported by Figs. 13 to 16, that as the incident beam offset angle 2β increases and/or the number of facets N, so also does the rotation axis offset distance $(-Y_P)$.

When $N = 3$, Eq. (59) leads to $(-Y_P) = 0$, which means the rotation axis lies on the scan axis (Fig. 4 and Sec. 2.13).

4.10 Choosing an Incident Beam Offset Angle 2β

Ideally, for symmetry of design, the incident beam should be directed along the scan axis, but this would obstruct the reflected scanning beam. Therefore, if the image format field angle is 2ω, then the incident beam offset angle 2β must at least be slightly greater than the semi-image format field angle ω to avoid this physical interference (Fig. 13). That is,

$$2\beta > \omega \tag{60}$$

Using Eqs. (56) and (57) with $N = 10$, $r = 25$ mm, and $D = 1$ mm, leads to Table 3.

4.11 Ghost Beams gh and Images GH

Pencils of scattered light rays from the incident scanning spot on the scanned surface plane are returned through the objective lens. The objective lens recollimates them as they proceed back to the polygonal scanner's facets, at which they are specularly reflected to produce what are known as ghost beams. In the figures these ghost beams are symbolized by the letters gh, with a subscript that identifies the facet whence they came.

Only if the ghost beams gh are reflected from a facet of the polygonal scanner at angles numerically much less than 90°, that is, towards the objective lens, is there a chance that they can traverse back through the objective lens, which will focus them onto the image plane to form stationary point ghost images GH. A subscript to GH refers to the respective facet whence the ghost beam gh came.

If these ghost beams are reflected from a facet of the polygonal scanner at angles numerically greater than 90°, that is, away from the objective lens, there is no chance of them traversing back through the objective lens to produce stationary ghost images GH.

4.12 Ghost Beam Field Angles ϕ

The field angle ϕ of all ghost beams gh, whether they produce stationary ghost images in the image format plane or not, is always at an angle that is a multiple of $2A$ away from, and on either side of, the incident beam offset angle 2β. This is because the mirror facet-to-facet tangential angle of a regular prismatic polygonal scanner is A. Thus

$$\phi = 2\beta \pm n(2A), \qquad |2\beta \pm n(2A)| < 90° \qquad (61)$$

in which n is an integer.

In Sec. 4.3, $N = 10$, therefore $2A = 72°$; hence all ghost beams in Figs. 13 to 16 occur at intervals of $72°$ from the incident beam offset angle 2β.

When $n = 0$, Eq. (61) represents the retroreflective ghost beam field angle ϕ_1 that is collinear with the incident beam. Ghost beam gh_1 is not relevant in this discussion and, to avoid confusion, it is not depicted in the figures.

If one increases the incident beam offset angle 2β by say $25°$ counterclockwise, and repositions the polygonal scanner to a midposition, all the field angles ϕ of the ghost beams will have also rotated by $25°$ counterclockwise, while the polygonal scanner will have only rotated $12.5°$ counterclockwise; and vice versa, if clockwise.

4.13 Incident Beam Location

Figures 13 to 16, which shall be discussed in turn, depict the first significant four incidence beam offset angular positions given in Table 3, namely, $27°$, $52°$, $92°$, and $124°$. The figures depict the respective orientations of the polygonal scanner in the midposition and the rotation axis offset distances Y_P, such that the reflected incident beam is collinear with the scan axis to focus it to the central point C in the image field format. Pencils of light rays, gh_2, gh_3, gh_4, and gh_{10}, scattered from point C, are shown passing back through the objective lens to meet the facets of the polygonal scanner, whence they are again reflected. The subscripts correspond to the facet number.

In Fig. 13 one such pencil, gh_2, passes back through the objective lens to produce the point image GH_2 below the image field format. There is one pencil, gh_1, which is reflected from facet S_1 that is not displayed so as not to overcrowd the diagram. Pencil gh_1 is the retroreflective pencil that returns collinearly with the path of the incident beam.

As predicted by Eq. (61) the angle between successive pencils of rays of ghost beams gh reflected from the five facets s_{10}, s_1, s_2, s_3, and s_4 is $2A$ (Figs. 13 to 16).

One should notice that in Figs. 13 to 16 the vertices of the fan depicting the full scan angle $\pm A$ and the image field format scan angle $\pm \omega$ do not coincide, nor do they touch the surface of the facet. They lie at two distinct locations. The first lies on the incident beam axis at its intersection with the circumscribed circle of the polygonal scanner; the second lies below, within the circumscribed circle and above the scan axis. The difference is best observed in Fig. 16 (see also Sec. 3).

4.14 Image Format Scan Duty Cycle η_ω

The image format scan duty cycle η_ω, is the ratio of the image field format angle 2ω to the full scan angle $2A$ of the polygonal scanner. It must not be confused with the maximum potential scan duty cycle (scan efficiency) η. The image format scan duty cycle η_ω

GHOST IMAGE FOR 2β = 27°

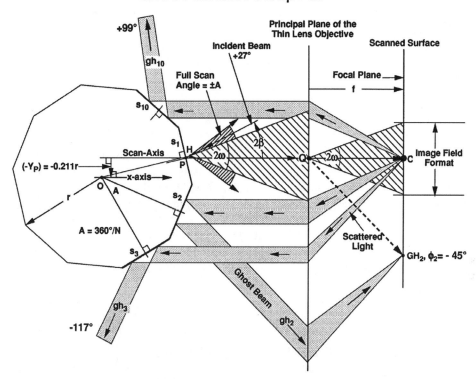

Figure 13 Formation of stationary ghost image GH_2 is produced by a pencil of scattered rays originating at C and re-reflected from facet S_2 at a field angle $\phi_2 = -45°$.

depends directly on the image format angle 2ω.

$$\eta_\omega = 2\omega/2A = \omega/A \tag{62}$$

In Figs. 13 to 16, $2\omega = \pm 20°$ (40°). Substituting for ω and A into Eq. (62) leads to

$$\eta_\omega = 20°/36° = 55.6\% \tag{63}$$

Since the above image field format scan duty cycle η_ω of 55.6% is greater than the maximum potential scan duty cycle η of 54.1% presented in Table 3 for an incident beam offset angle 2β of 164°, this offset angle is not relevant and no figure is provided. The image field format scan duty cycle η_ω must be less than the scan duty cycle η.

4.15 Incident Beam Offset Angle 27°

The incident beam offset angle of 27° in Fig. 13 is comfortably outside the semi-image format angle ω of $+20°$ to avoid physical obstruction of the scanning beam, but at an angle less than the half scan angle $A = +36°$, of the ten-facet polygonal scanner.

If the ghost beam gh_2 traverses the objective lens, there is only one stationary ghost image, GH_2, at a field angle of $\phi_2 = -45°$; and it lies outside and 25° below the image field format.

From Eq. (61) the field angles of ghost beams gh_{10} and gh_3 from face s_{10} and s_3, are $\phi_{10} = +99°$ and $\phi_3 = -117°$, respectively. These ghost beams are harmless ($|\phi| > 90°$).

4.16 Incident Beam Offset Angle 52°

In Fig. 13 the incident beam offset angle is 27°. Let the incident beam offset angle 2β with its accompanying ghost beams gh and ghost images GH be rotated counterclockwise through a positive angle of +25°. If the ghost beam gh_2 traverses the objective lens, the ghost image GH_2, $\phi_2 = -45°$, of Fig. 13 will move up to lie on the lower edge $(-\omega = -20°)$ of the image format field, $\phi_2 = (-45° + 25°) = -20°$. The incident beam offset angle increases to $2\beta = (+27° + 25°) = +52°$ (Fig. 14).

From Eq. (61) the field angles of ghost beams gh_{10} and gh_3 from facets s_{10} and s_3 become $\phi_{10} = +124°$ and $\phi_3 = -92°$, respectively. These ghost beams are harmless ($|\phi| > 90°$).

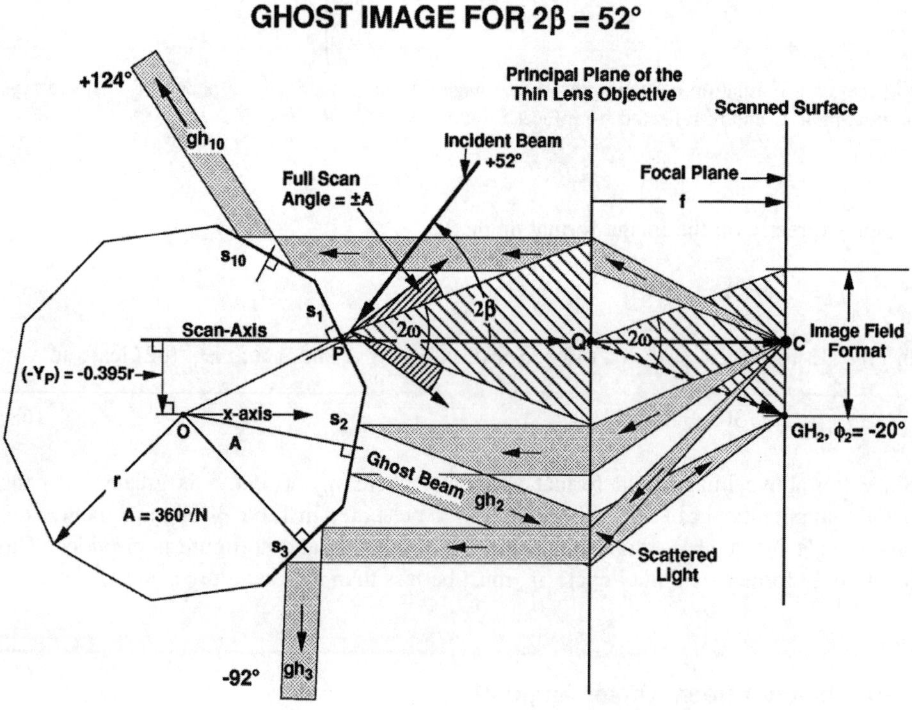

Figure 14 The stationary ghost image GH_2 on the lower edge of the image field format at a field angle $\phi_2 = -20°$ is produced by a pencil of scattered rays originating at C and re-reflected from facet s_2.

4.17 Incident Beam Offset Angle 92°

In Fig. 14 the incident beam offset angle is 52°. Let the incident beam offset angle 2β with its accompanying ghost beams gh and ghost images GH be rotated counterclockwise through a positive angle of $+40°$. If the ghost beam gh_2 traverses the objective lens, the ghost image GH_2, $\phi_2 = -\omega = -20°$, of Fig. 13 will move up to lie on the upper edge $(+\omega = +20°)$ of the image format field, $\phi_2 = (-\omega + 40°) = (-20° + 40°) = +20°$. The incident beam offset angle increases to $2\beta = (+52° + 40°) = 92°$ (Fig. 15).

For incident beam offset angles 27° and 52° there is only one stationary ghost image in the image format, namely GH_2. As ghost image GH_2 moves to the upper edge of the image format, a second ghost image GH_3 appears in the image format well below at a field angle, $\phi_3 = -52°$. Note that $(\phi_2 - \phi_3) = 2A = 72°$, as expected.

From Eq. (61) the field angle of the remaining ghost beam gh_{10} from facet s_{10} is $\phi_{10} = +164°$, and is harmless $(|\phi| > 90°)$.

4.18 Incident Beam Offset Angle 124°

In Fig. 15 the incident beam offset angle is 92°. Let the incident beam offset angle 2β with its accompanying ghost beams gh and ghost images GH be rotated counterclockwise through a positive angle of $+32°$. If the ghost beam gh_3 traverses the objective lens, the

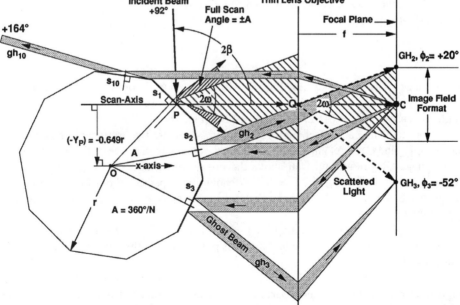

Figure 15 If the ghost beams gh_2 and gh_3 traverse the objective lens, there is a stationary ghost image GH_2 on the upper edge of the image field format $\phi_2 = +20°$ and a stationary ghost image GH_3 below it, $\phi_3 = -52°$.

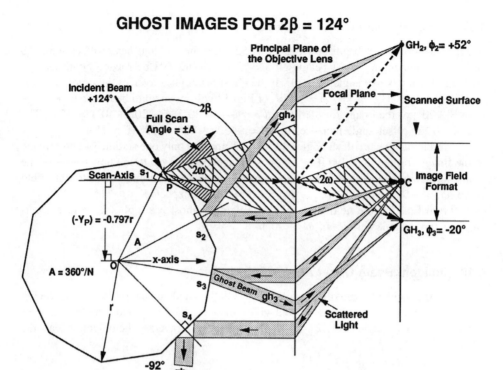

Figure 16 If the ghost beams gh$_2$ and gh$_3$ traverse the objective lens, there is a stationary ghost image GH$_3$ on the lower edge of the image field format $\phi_3 = -\omega = -20°$ and a stationary ghost image GH$_2$ above, $\phi_2 = +52°$.

ghost image GH$_3$, $\phi_3 = -52°$, of Fig. 15 will move up to lie on the lower edge ($-\omega = -20°$) of the image format field, $\phi_3 = (-52° + 32°) = -20°$. The incident beam offset angle increases to $2\beta = (+92° + 32°) = +124°$ (Fig. 16).

For incident beam offset angles 52° and 92° there are two stationary ghost images, namely GH$_2$ and GH$_3$. As ghost image GH$_3$ moves to the lower edge of the image format, ghost image GH$_2$ moves up well above the image format at a field angle, $\phi_2 = +52°$. Again note that $(\phi_2 - \phi_3) = 2A = 72°$, as should be expected. From Eq. (61) the field angle of the remaining ghost beam gh$_4$ from facet s$_4$ is $\phi_4 = -92°$, and is harmless ($|\phi| > 90°$).

A simple calculation of adding 72° to the field angle $\phi_{10} = 164°$ of ghost beam gh$_{10}$ in Fig. 15 produces a reflex angle of 236°, thus predicting that the ghost beam gh$_{10}$ can no longer exist.

A close inspection of Figs. 13 to 16 shows the center-of-scan of the total angular scan 2A of the scanner progressively becomes displaced from the center-of-scan of the image format scan angle 2ω with an increase in the incident beam offset angle β. This is valid because the instantaneous center-of-scan (ICS) is a locus (see Sec. 3 and Ref. 5).

Preobjective Polygonal Scanning

4.19 Ghost Images Inside the Image Format

A study of Figs. 14 and 15 shows that, if ghost image GH_2 is set on the lower and upper edges of the image format, the required incident beam offset angles are given by

$$2\beta = 2A + \omega \tag{64}$$

Thus, from Eq. (61) the range of 2β for ghost images to exist inside the image format is expressed by

$$n(2A) - \omega < 2\beta < n(2A) + \omega \tag{65}$$

in which n is zero or a positive integer, $A \geq \omega$, and $2\beta < 180°$.
Substituting $\omega = 20°$ and $2A = 72°$ leads to

when $n = 0$ $\quad -20° < 2\beta < +20°$ $\hfill(66)$
$\quad\quad n = 1$ $\quad +52° < 2\beta < +92°$ $\hfill(67)$
$\quad\quad n = 2$ $\quad +124° < 2\beta < +164°$ $\hfill(68)$

Each has a range of $40°$, which, not surprisingly, equates to 2ω.

A figure showing $2\beta = 164°$ is not relevant or depicted, because it has a scan duty cycle η less than the required image format duty cycle η_ω (Table 3).

4.20 Ghost Images Outside the Image Format

A study of expressions (66), (67), and (68) shows that when the incident beam offset angle lies between $+20°$ and $+52°$ no ghost image will appear in the image format. Likewise, when the incident beam offset angle lies between $+92°$ and $+124°$, each has a range of $32°$.[2]

Thus, to ensure ghost images lie outside the image format the condition is as follows:

$$n(2A) + \omega < 2\beta < (n+1)(2A) - \omega \tag{69}$$

in which n is zero or a positive integer, $A \geq \omega$, and $2\beta < 180°$.

Let ρ represent the angular range of 2β for ghost images outside the image format, then

$$\rho = 2A - 2\omega = 2(A - \omega) = 2(180/N - \omega) \tag{70}$$

and is independent of n.

4.21 Number of Facets

Subject to $A \geq \omega$, as the number of facets N increases, so also do the number of ghost beams gh and, therefore, there is a greater possibility of multiple ghost images GH in the scanned image plane. A critical case, in this example, occurs when $N = 18$. Then $A = +\omega = +20°$.

Substituting these values for A and ω into the inequalities (69) and (70) leads to Fig. 17.

$$\text{For } n = 0, \quad 20° < 2\beta < 20° \tag{71}$$
$$n = 1, \quad 60° < 2\beta < 60° \tag{72}$$
$$n = 2, \quad 100° < 2\beta < 100° \tag{73}$$

and the range

$$\rho = 0°$$

Hence, the positioning tolerance for the incident beam offset angle 2β is zero. For an adequate positioning tolerance for the incident beam

$$A > \omega \tag{74}$$

Substituting $A = 360°/N$ leads to the general condition

$$(360°/N) > \omega \tag{75}$$

In Fig. 17 P represents the point of incidence on the scanner facet. The image format field angle shown is $2\omega = 40°$. For the 18-facet polygon the angular range is zero, but theoretically available, and would simultaneously produce ghost images GH_2 and GH_{18} = at the upper, $\phi_2 = +\omega$, and lower, $\phi_{18} = -\omega$, edges of the image format, respectively, when the incident beam offset angle is $2\beta = +\omega, +3\omega, +5\omega, +7\omega$, and so on, subject to $2\beta < 180°$.

4.22 Diameters of Scanner and Objective Lens

No mention has been made with respect to the diameters of the objective lens, the scanner, nor the apertures near the scanner, or performance. These topics are out of the scope of this section, but all are important issues.[2]

However, the smaller the diameter of the scanner relative to the objective lens diameter, the greater the chance of a ghost beam returning to produce a ghost image in the scanned image plane. Likewise, the closer the scanner is to the objective lens, the greater the chance of a ghost beam returning to produce a ghost image in the scanned image plane.

4.23 Commentary

There is more than one angular zone for the incident beam offset angle to avoid ghost images appearing within the image format. These zones have acceptable scan duty cycles η, depending on beam width D, the diameter $2r$ of the polygonal scanner and the number of facets N (Fig. 17, Table 3).[2]

Preobjective Polygonal Scanning

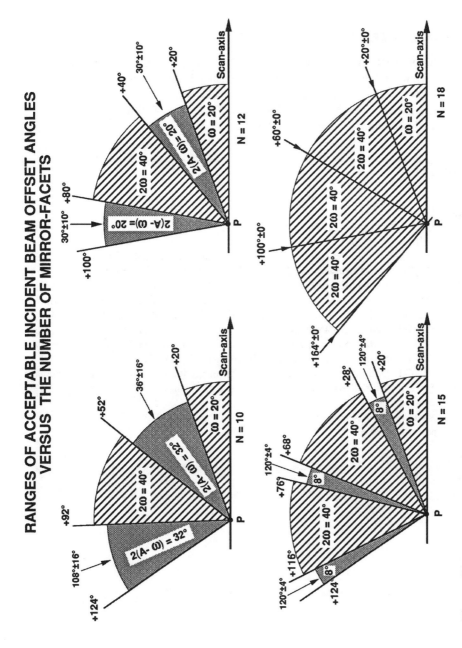

Figure 17 To ensure that stationary ghost images GH are outside the image format, the angular ranges ρ of the incident beam offset angle 2β are $32°$, $20°$, $8°$, and $0°$ are shown for 10-, 12-, 15-, and 18-facet polygonal scanners.

4.24 Conclusion

It behooves one to consider the possibility and the whereabouts of stationary ghost images in the image format plane during the initial optical system design stage.

ACKNOWLEDGMENTS

The author appreciates the time and expertise that Leo Beiser and Stephen Sagan have given in reviewing this chapter and providing many helpful suggestions. I thank the optomechanical design engineers of CSIRO, Australia, who encouraged me to solve and provide the explicit "Coordinates and Equations of a Polygonal Scanning System" for a beam with a finite width as presented in Sec. 2.

REFERENCES

1. Kessler, D.; DeJaeger, D.; Noethen, M. High resolution laser writer. Proc. SPIE *1079*, 1989, 27–35.
2. Beiser, L. *Unified Optical Scanning Technology*; IEEE Press, Wiley-Interscience, John Wiley & Sons: New York, 2003.
3. Beiser, L. Design equations for a polygon laser scanner. In Marshall, G.F.; Beiser, L.; Eds. *Beam Deflection and Scanning Technologies*, Proc. SPIE 1991, *1454*, 60–66.
4. Marshall, G.F. Geometrical determination of the positional relationship between the incident beam, the scan-axis, and the rotation axis of a prismatic polygonal scanner. In Sagan, S.F.; Marshall, G.F.; Beiser, L.; Eds. *Optical Scanning 2002*, Proc. SPIE 2002, *4773*, 38–51.
5. Marshall, G.F. Center-of-scan locus of an oscillating or rotating mirror. In Beiser, L.; Lenz, R.K.; Eds. *Recording Systems: High-Resolution Cameras and Recording Devices; Laser Scanning and Recording Systems*, Proc. SPIE 1993, *1987*, 221–232.
6. Marshall, G.F. Stationary ghost images outside the image format of the scanned image plane. In Sagan, S.F.; Marshall, G.F.; Beiser, L.; Eds. *Optical Scanning 2002*, Proc. SPIE *4773*, 132–140.
7. U.S. patent no. 5,191,463, 1990.
8. U.S. patent no. 4,993,792, 1986.

8

Galvanometric and Resonant Scanners

JEAN MONTAGU

Clinical MicroArrays, Inc., Natick, Massachusetts, U.S.A.

1 INTRODUCTION

The goal of this section is to offer the reader a comprehension of the parameters that shape the design and subsequently the applications of current oscillating optical scanners. Hopefully this will explain their design and possibly guide system engineers to reach the most desirable compromise between the numerous variables available to them.

It is also my hope that this may stimulate designers to extend the technology or pursue different technologies as they appreciate the constraints and limitations of current oscillating optical scanners and their applications.

This text is the third of a series edited by Gerald Marshall[1,2] covering the evolving field of optical scanning. Since oscillating scanners are developed to meet the needs of specific technical and scientific applications, it is constructive to review some of these applications. Applications are the stimulus that underlies past and future developments.

It is evident that the material presented here is evolutionary and a broader treatment can be found in the references. The reader is frequently referred to the previous texts.[1,2] Only Sec. 2.1.4, "Mirrors," and Sec. 2.23, "Induced Moving Coil Scanner," are reproduced here in toto. These are important subjects that are frequently disregarded by system designers, and no meaningful advances have taken place. On occasion, some material is taken from the previous editions in order to present the new material in a consistent manner. Section 2.1.3, "Bearings," as well as Sec. 2.16, "Dynamic Performances," contain some material from the previous edition as well as new material.

In addition, this edition is inversely organized when compared with its precursors. The technology underlying the components of scanners is reviewed up front and new applications are at the end of the section. Important evolutions of older applications are presented generally, while referring the reader to earlier texts for basic descriptions of the subject.

The past decade has seen extensive technology evolutions that have brought major changes in the market and manner of use of scanners as well as unexpected performances and designs of oscillating scanners. Improved performances of competing technologies have attracted applications previously the domain of oscillating mechanical scanners. Linear and two-dimensional solid-state arrays now dominate the vision and the night vision market, both military and commercial. Digital micromirror devices (DMDs) and liquid crystal displays (LCDs) have also captured the field of image projection away from oscillating scanners.

On the other hand, the advances of computer control in industry have benefited the laser micro-machining industry, which requires a high degree of flexibility and is well adapted to digital control. This has benefited galvanometric scanner manufacturers, who have responded by improving their product.

Improved scanner performances as well as greater choice and more economical associated technologies have broadened the market for scanners and stimulated new applications. This in turn has offered opportunities for new sources of supplies.

1.1 Historical Developments

The galvanometer is named after the French biologist and physicist Jacques d'Arsonval, who devised the first practical galvanometer in 1880. Initially it was used as a static measuring instrument. Its dynamic and optical scanning potential were recognized early on when galvanometers were employed to write sound tracks on the talking movies. Miniature galvanometers with bandwidths as high as 20 kHz were used for waveform recording on UV-sensitive photographic paper as late as 1960.

The invention of the laser broadened the applications of galvanometers in the graphic industry during the late 1960s. The first designs were open loop scanners, but very early in the 1970s, the position-servoed, better known as the closed loop scanner, came to reign in meeting the desire for more bandwidth and increased accurate positioning.

The servoed scanner enabled the accuracy of the device, relegated to the position transducer, to be dissociated from the torque motor. The next challenge was to minimize inertia and optimize rigidity. Cross-talk perturbations were mostly solved with the use of moving magnet torque motors and the practice of balancing the load and armature.

Demand for higher speed and greater accuracy forced the design of all the building blocks of scanning systems to be refined. The performance of scanners evolved along the evolution of its constituents: torque motor, transducers, amplifiers, and computers.

The first milestone in the early 1960s was the development of moving iron scanners, as they offered a compact magnetic torque motor. The compact, efficient, and economical design offered scanners beyond the capabilities of moving coils at that time.

The second milestone in the late 1980s came as a consequence of the commercialization of high-energy permanent magnets. Moving magnet torque motors were developed with much greater peak torque. In the same period, a new design of transducers appeared, driven by the availability of much improved electronic elements.

The third milestone that came into being in the last decade of the century is marked by the presence of computer power to mitigate the shortcomings of even the best galvanometers. The clock rate of ordinary PCs has reached the megahertz range and can compensate in real time for position encoder imperfections as well as optimize dynamic behavior of periodic and aperiodic armature motions. The PC has also simplified full system integration.

High-energy permanent magnets were also developed to power resonant scanners, but innovative new designs and the use of PCs form the underpinning of present-day devices.

At this writing, all high-performance optical scanners share a common architecture: a moving magnet torque motor, a position transducer built along a variable capacitor ceramic butterfly for high-precision work and optical sensors for less demanding applications. The performance of the galvanometric scanner is limited by the following parameters, which shall be covered in more detail in the following sections:

- The thermal impedance of the magnetic structure and specifically the drive coil. This in turn limits the available torque of the magnetic motor and induces unpredictable thermal drift of the position transducer.
- The thermal stability of the position transducer.
- The mechanical resonances of the armature and the load as they prevent the system from achieving a step response expected from available torque. An expert servo designer can appreciably optimize system performances if the elements are stable.

The ability to integrate all these disciplines, as well as optimum frame configuration, mirror design and mounting, drift compensation, and software has become a specialty so that a fully integrated subsystem is frequently selected rather than just the scanner. Figure 1 illustrates all these elements.

The progress of scanners is application-driven; consequently more recent applications will be reviewed in the last section.

2 COMPONENT AND DESIGN ISSUES

Scanners are like old soldiers; they fade away but never die. Moving coil and moving iron scanners, as well as some four-pole stepper motors built with laminated stators, are still available and priced competitively as their tooling has been amortized. Their design and performances have not evolved since they were evaluated previously by this author[1,2] and will not be reviewed here. Their desirable features do not compensate for their shortcomings such as iron saturation, nonlinearity, and strong unpredictable radial forces, the major cause of wobble, for the iron-based scanners and flexible armature and poor thermal properties for the moving coil units as well as high cost of entry and manufacturing.

2.1 Galvanometric Scanners

All modern high-performance oscillating galvanometer scanners are built with a moving magnet torque motor and all high-performance position transducers employ a two- or four-lobe ceramic variable capacitor.

All oscillating scanners designed in the last decade are built using the NdFeB family of permanent magnets. These alloys can have as much as five times the energy product of the best ALNICO magnet and certain other benefits, but they have a low Curie temperature, possibly as low as 310°C, such as for the 45 MGO material from Ugimag.[3]

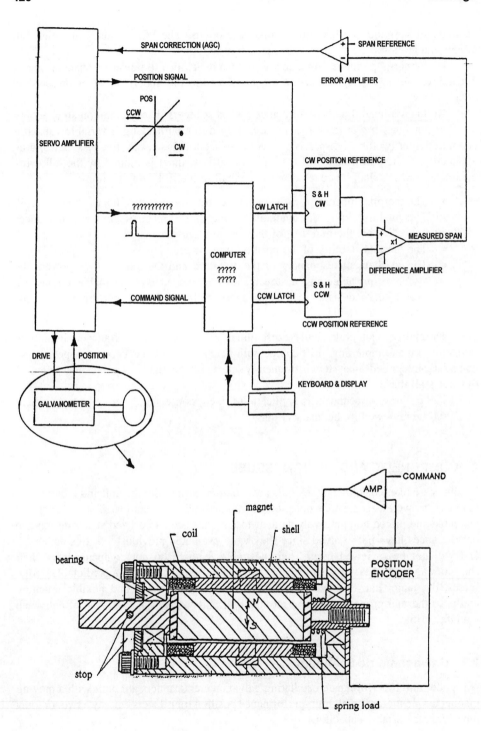

Figure 1 Galvanometric scanner and system management.

The higher the energy product, the lower the Curie temperature.[3] This has two important consequences:

- a typical magnetic strength temperature coefficient of $-0.8\%/°C$ in the range 22–85°C;
- an irreversible flux loss will take place each time the material is heated above 80–100°C. The range reflects the particular alloy selected and the magnetic design.

The coil design and its thermal conductivity are critical features of the galvanometric scanner because they are the major cause of transducer thermal drift.

2.1.1 Moving Magnet Torque Motor

The torque motor is selected for its ability to integrate with the other elements of the scanner, the mirror, the position sensor, and the electronic driver/controller. It must also support the dynamic performance requirements and those caused by environmental changes and perturbations.

The list of features of an ideal torque motor is long and frequently a compromise is reached where some necessary system properties are obtained through other means. Environmental control and electronic compensation schemes have become standard features of high-performance scanners.

The ideal galvanometric or resonant scanner driver would have the following properties:

- high torque-to-inertia ratio;
- low electrical time constant: inductance/resistance;
- linear relationship between torque, current, and angular position;
- freedom from cross-axis forces or excitations;
- no hysteresis or discontinuities;
- no elastic restraint;
- some mechanical damping, constant and uniform is acceptable;
- very high rigidity in torsion and bending;
- a balanced armature;
- low power consumption [figure of merit; torque/(inertia*Watt$^{1/2}$)];
- immunity to thermal expansion constraints;
- good heat dissipation;
- demagnetization protection;
- simplicity of installation and use;
- insensitivity to radio frequency (RF) and other environmental perturbations;
- absence of sensitivity to external perturbations;
- freedom from self-induced perturbations;
- infinite life with stable parameters;
- small, light, and cheap.

Additionally, for a resonant scanner, damping properties should be minimum along the axis of rotation and high for all other degrees of freedom.

Moving Magnet Torque Motor
The high energy of rare earth magnets are practically free of radial forces, which makes them attractive torque motors. They are the choice of all devices listed in Tables 1, 2,

Table 1 Comparative Performances of Moving Magnet Scanning Galvanometers: Inertia <1 gm·cm²

	Model						
	6200	6210	6220	RZ-15	6860	TGV-1	6230
Torque motor							
Rotor inertia, gm·cm²	0.012	0.02	0.14	0.34	0.6	0.65	1
Torque constant, gm·cm/A	10.8	25	57	40	93	123	114
Resistance, Ω	2.4	4.1	3.4	1.3	1.5	1.4	1.4
Thermal conductivity, °C/W	7.5	4	2		1.5		1
Figure of merit of torque motor							
Torque/inertia (watt)$^{1/2}$	580	625	221	103	126	160	96.3
Transducer	Opt.	Opt.	Opt.	Cap.	Cap.	Cap.	Cap.
Sensitivity, μA/o Opt.	24	24	22.8	100	29	50	23.4
Gain drift, ppm/°C	75	75	75	50	50	b	25
Null drift, μRad/°C	50	50	50	25	30	b	150
Repeatability, uRadian	30	30	30	6	16	4	30
Dynamic performance							
Small angle step response, ms	0.175	0.175	0.25	0.25	0.5	0.18	0.3

[a]Angular excursion: All scanners are rated 60° optical pick to pick (ptp), mechanical motion, minimum.
[b]Not applicable, scanner has internal feducial references.
[c]All angles are in degree optical.
[d]All optical detector have linearity >98% and all capacitive detectors have linearity >99.5%.

and 3. They are large-air-gap devices with comparatively low inductance, and simple to interface with electrically.

The driving stage is a conventional inside-out d'Arsonval movement, as shown in Figure 2. The torque can be calculated as the interaction of two fields or the effect of a field on a current. We shall follow the latter method.

Equation (1) is derived for coils with a total number of turns N and with conductors at 45° from the plane of symmetry. Equation (2) gives the torque generated by a device built with a coil as shown in Fig. 3 having a uniform conductor density distribution ±45° from the plane of symmetry. This assumes that D is the average diameter of the coil. Also the coil needs to be of tightly wound coils and facing the region of highest magnetic field of the rotor:

$$T = 0.90 K B_r L N I D \cos \gamma \tag{1}$$

The outer shell closes the magnetic circuit of the permanent magnet as well as that of the drive coil. It is preferably made of sintered high density 50/50 nickel–iron alloy. Low-carbon cold rolled steel such as C1020 and other steels have similar magnetic properties, but are rarely more economical as the finished part. Its radial thickness is recommended to be about one-quarter the diameter of the rotor when a rare earth magnet is used as the rotor.

Table 2 Comparative Performances of Moving Magnet Scanning Galvanometers: Inertia >1 gm·cm^2

	Model									
	M2	VM2000	6870	TGV-2	6450	M3	RZ-30	6880	TGV-3	TGV-4
Torque motor										
Rotor inertia, gm·cm^2	1.7	1.7	2	2.3	2.4 Note 5	4	5.9	6.4	7.4	14
Torque constant, gm·cm/A	230		180	335	450	500	278	254	550	650
Resistance, Ω	4.5		1.4	2.6	4	4.8	3	1	2.6	2.7
Thermal conductivity, °C/W	2.5		1		5	1.4		0.75	0.7	0.7
Figure of merit of torque motor[e]										
Torque/inertia*(Watts)$^{1/2}$	63.8		76	90	93.7	57	27.2	39.7	46.1	28.3
Transducer										
Sensitivity, μA/° Opt.	11		29	100	43	11	150	44	100	100
Gain drift, ppm/°C	−60	100	50	b	50	60	30	50	b	b
Null drift, μRad/°C	18	30	30	b	30	18	10	20	b	b
Repeatability, μRadian	12		16	2	4	2	2	16	2	2
Dynamic performance										
Small angle step response, ms		0.3	0.7	0.3			0.6			

[a]Angular excursion: All scanners are rated 60° optical ptp, mechanical motion, minimum.
[b]Not applicable, scanner has internal feducial references.
[c]All angles are in degree optical.
[d]All optical detectors have linearity >98% and all capacitive detectors have linearity >99.5%.
[e]Moving coil torque motor.
[f]All transducers are capacitive detectors.

Table 3 Performances of Moving Magnet Galvanometers with Flexure Bearings

	Model		
	Harmonicscan	FM200	Slowscan
Torque motor			
Rotor inertia, gm·cm^2	0.3	2.5	8.25
Torque constant, gm·cm/A	120	230	278
Resistance, Ω	1.3	4.5	5.5
Transducer			
Sensitivity, μA/° Opt.	70		90
Gain drift, ppm/°C	50	100	50
Null drift, μRad/°C	25	30	25
Repeatability, μRadian	5	1	2
Suspension			
Jitter, μRadian	4	1	1.7
Wobble, μRadian	1	0.5	0
Performances			
Small angle step response, ms	0.2	0.6	1.3

[a]Angular excursion: All scanners are rated 60° optical pick to pick (ptp), mechanical motion.
[b]All capacitive detectors have linearity >99.5%.
[c]Al angles are in degree optical.

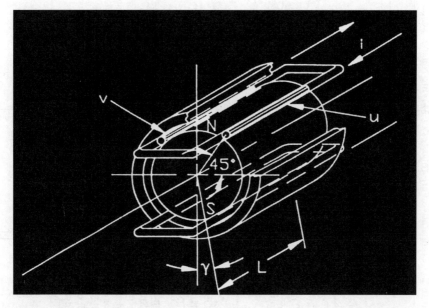

Figure 2 Inverted d'Arsonval movement.

Galvanometric and Resonant Scanners

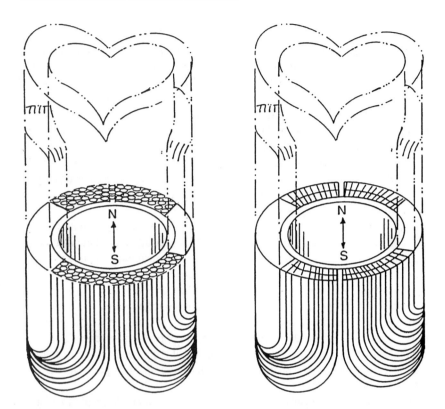

Figure 3 (*Left*) Coil wound with round wire conductors; (*right*) coil wound with ribbon wire conductors.

The constant K in Eq. (1) takes into consideration the space allocated for the coil with respect to the dimension of the magnet. It is expressed as

$$K = \frac{1}{1 + 2g \cdot B_r/\mu H_c \cdot d} \qquad (2)$$

where for rare earth magnets

$$B_r/\mu H_c = 1.1 \qquad (3)$$

B_r is the magnetic intrinsic induction, remanence of the rotor material; H_c is the magnetic demagnetization force, coercive force, of the rotor material; μ is the permeability of air; g is the radial gap between the rotor and the outer shell; and d is the diameter of the magnet. In practice, $\frac{1}{2} < K < 1$. It is evident that for a given number of coil windings and a given resistance, it is advantageous to minimize the radial gap.

Rare earth magnets have a very high intrinsic coercive force and they are practically impervious to operational demagnetization so that extremely high torques can be safely generated. The torque is only limited by the coil design and construction as it relates to the cooling capability to prevent either catastrophic failure or excessive thermal drift of the position transducer. Coil design and transducer designs are the critical features of galvanometric scanners as well as armature rigidity. These elements shall be reviewed in the following sections.

Coil Construction

The use of the space allocated to the coil windings in electromagnetic devices has been the subject of numerous texts and studies since electromagnetic devices have been made. Roters[4] gives a general description of the subject and early patents[5] demonstrate the appreciation that efficient packing density of coils as well as their thermal construction are critical features. Hodges[6] gives a detailed description of the benefits that can be derived from pressing a coil of round conductors into a minimum cross-section. He reports that thermal conductivity can be increased by a factor of 3 after compression.

Roters[4] shows that "resistance density" and "thermal conductivity" of a coil is nearly proportional to the "copper density" of the coil. Optimizing the copper density of a coil without prejudicing its reliability or cost has been the pursuit of optical scanner manufacturers as these devices are regularly driven extremely hard.

The thermal impedance of a coil is mostly defined by the electrical insulation of the conductor, the copper density as well as the encapsulation compound. The volume of insulation and encapsulation needs to be minimized. Single insulation of magnet wire occupies about 20% of the volume of the typical conductor used for these devices.

Packing efficiency has been recognized as another critical factor. The highest possible local packing density of large coils with round conductor wound in quincunx is 90.69% and layer winding reaches a maximum of 78.5%. Most common windings have a packing density under 60%, which consequently yields a copper density under 50%.

Two technologies have been developed to improve the copper packing density as well as the thermal impedance of galvanometer coils. One technique compacts the coil and is best described by Hodges[6] and Houtman.[7]

They recommend first preforming a coil with conventional single insulation round wire plus self-adhesive. This preformed coil is then compressed to conform to its final shape and achieve a higher packing density. A suitable choice of insulation and ductile conductor can improve the copper density by 20%, comparable to a quincunx winding. That indicates that when the base coil has a random winding, an increase in copper density from 50% to 70% copper density is possible.[7]

A slightly higher copper density can be achieved with low aspect ratio ribbon conductor. This construction in addition offers the best thermal impedance and a simpler construction. Figure 3 compares conventional coil and ribbon coil constructions with about equal resistance. This construction can yield a coil with thermal conductivity four times higher than can be obtained with a conventional random wound coil and 50% higher than can be achieved with a compressed coil construction.

Heat Dissipation

The heat dissipation constant is a critical parameter as the temperature rise of the scanner has numerous consequences with possible thermal runaway and catastrophic consequences as described earlier in this section. The elements directly affected by a temperature rise are:

- The coil temperature as its resistance increases with temperature rise as:

$$R_T = R_{25}(1 + 0.0039\Delta T) \quad (4)$$

where ΔT represents the temperature change from the 25°C resistance value.

- The thermal conductivity of the coil has major consequences for the temperature of the magnet as it is located at the center of the coil. The average coil temperature rise may be modest, but it may not be representative of the temperature at the center of the coil and the temperature of the magnet.
- The temperature stability of a hard magnetic material is inversely proportional to its energy product. The ALNICOs are the most stable, followed by the Samarium compounds and finally the NdFeB alloys. Most scanners are built with the most energetic NdFeB material with a high negative temperature sensitivity such that, as a first approximation, the magnetic field can be derived, from Ref. 3 data, for 22–85°C from:

$$B_T = B_{22}(1 - 0.008\Delta T) \tag{5}$$

- The thermal coupling of the position transducer and the torque motor. Symmetry and mounting designs are critical to minimize transducer drift. One manufacturer effectively separates the torque motor from the transducer with the suspension.

Manufacturers give particular attention to the heat dissipation coefficient of scanners for competitive reasons. It is advisable to obtain from the vendors the mode of measurement and judge if it applies to the application at hand.

The dissipation coefficient is frequently specified as "coil to case" because the best numbers are obtained when the change in coil temperature is derived from its change in resistance, as the device is being powered and held in a large fixed-temperature heat sink. Treating the galvanometer stator as an oven where a thermistor is located within a simulated armature yields the most meaningful values. The heat dissipation coefficient obtained in this manner may be half that of the coil to case values.

2.1.2 Position Transducer

The simplest and most economical position transducer is a torsion bar. The torque motor pushes against the torsion bar and positions the mirror. This forms a second-order system where bandwidth and position accuracy must be traded off. These units are best built as moving iron devices typically with very stable ALNICO magnets. They typically exhibit good temperature stability, around 150 ppm/°C.

Most high-performance scanners are closed loop servoed systems. They also must deliver bandwidth and positioning accuracy. High-energy magnet torque motors are very powerful and can deliver the bandwidth but the magnetic material, NdFeB, is temperature sensitive and they depend upon the position transducer to deliver the accuracy.

Gain and Pointing Stability Considerations

Galvanometric scanners applications can be divided into two groups: image/position acquisition and pointing/designation/laser-micro-machining. In both cases, systems need to be calibrated and drift needs to be compensated. This becomes evident in the tolerance budget of some advanced positioning systems demanding absolute beam positioning.

For example, as the technology advances, the density of the biology carriers in GeneChips is increasing from the present 4000 pixels per scan line to 10,000 and the number of scans per chip from 4000 to 10,000. Each one of the pixels must be correctly located.

High-precision laser micro-machining systems used in the manufacturing of flat panel displays or silicon devices such as DRAMs or the trimming of batches of MEMs (such as air bag accelerometers) or trimming of resistors on circuits on silicon need highly accurate and stable gain and pointing performance. These applications commonly demand addressability errors to be of the order of 1/20,000 or 50 ppm of the field of view and special application can be two or four times more demanding.

The addressability error has a number of sources:

- It may not be possible to accurately locate the work in the field of view with respect to the scanners' axis of reference. It is common practice to imbed fiducial marks for optical alignment and calibration.
- The environment and application may not permit a structural design sufficiently rigid.
- Scanner/transducer drifts may exceed acceptable tolerances for certain applications. In such cases it may be necessary to close the loop around the drift.

The two applications listed above cannot be satisfied with scanners built with only "open loop drift transducers." This becomes apparent upon the analysis of the temperature regulation required to meet the necessary pointing accuracy. The following case illustrates this point.

Let us consider an application where the optical scan angle is 0.4 radian, ptp, and the scanner exhibits a gain drift of 50 ppm/°C as well as a null (or zero) drift of 50 ppm/°C (20 μradian/°C over 0.4 radian excursion) or a total uncertainty for each measurement of 100 ppm/°C. Because two measurements are required to point the beam, one to define the system's references and one to point at the work, the uncertainty is 200 ppm/°C.

The DRAM application would demand that the scanner be held in an environment controlled to less than 0.25°C and the BioChip example could tolerate twice that. These are extremely demanding conditions, specially for aperiodic operation. For these reasons, closed loop drift compensation or closed loop drift position transducers are required.

Transducer Drift

Most high-accuracy capacitive position transducers are derivatives of Rohr's[8] design and built with a ceramic vane mounted on the armature of the scanner, which rotates between two stationary conductive plates as depicted in Fig. 6. A circuit creates a drive signal and detects changes in capacitance. The combined demands of high system bandwidth (0.5–5 kHz) and resolution (1–10 μradian) as well as stability (5–50 ppm/°C) are extremely difficult to achieve in an analog environment and, so far, no manufacturer offers a digital system able to approach this suite of specifications.

These transducers are complex analog devices with multiple sources of drift, both mechanical and electrical, induced by small changes of thermal or other origins including aging and hysteresis. A typical transducer with a 2 cm diameter ceramic butterfly vane can resolve better than 1 μradian in less than 1 ms. Under these conditions the tip of the ceramic vane has moved less than 0.01 μm. It is quite an achievement.

Unfortunately the stability of the same transducer is two orders of magnitude lower per °C. Scanner manufacturers specify short-term temperature gain drift as well as null drift and occasionally uncorrelated drift of the position transducer. Those quantities are difficult to quantify and it is valuable to know under which condition they were derived, as these may not be representative of the conditions where the scanners may be used.

Normally only gain drift is measured and the null or pointing drift value is derived from gain values measured at two extreme representative positions. The null drift quoted may be more representative of asymmetry in the gain drift.

Measurements are made with the scanner held and stabilized in a constan temperature oven as the controller is kept at room temperature. This is rarely representative of the operating environment of optical scanners. More common is a burst of energy to rapidly reposition the mirror then followed by rest time when part changes take place.

It is recommended to structure a representative drift test within the environment and process of the scanner application. If this is impractical, drift performances should be derated from the quoted values by at least a factor of 2.

Closed Loop Drift Transducers

It is frequently necessary that the beam be located with similar or greater precision than the resolution required for the task the instrument is to perform. To define location or correct for drift, reference points, known as fiducial marks, are commonly used. This approach, familiar in astronomy, is also frequently applied to recalibrate gain and null of galvanometric scanners.

It is important to keep in mind that most scanning systems are designed to the limit of the resolution capabilities achievable with the assembled elements: scanners, lasers, mirror size and flatness, focal length, and so on. Commonly, the optical elements performing the task are optimized for dynamic performances as well as optical performances. The addition of other scanner mounted optical sensors or the purpose of absolute position reconnaissance can substantially limit over all performances.

A number of optical techniques have been adopted to recalibrate gain and drift in process. Most of them use split cells – or fiducial marks – located in the work plane, beyond the angular reach of the work area. Weiss and colleagues[9,10] and others report the analyses of these techniques.

A different technique[11] to correct for drift is described by Montagu and colleagues. It takes advantage of the high-resolution capability of the capacitive transducer to incorporate capacitive fiducial features within the transducer proper. The leading edge of a step or a pulse is used for periodic recalibration. In this method, the fiducial features may be positioned on the inside of the operating range of the transducer or beyond. In most applications, this technique offers performances comparable to that obtained with fiducial marks in the work plane and can reduces drift errors by more than an order of magnitude.

Optical Transducers

Optical position transducers offer the attraction of low cost, low inertia, and small volume, as well as low power consumption and an analog signal in an analog environment. Unfortunately the same features are at the root of the difficulties to reach high temperature stability, good signal-to-noise ratio, and good linearity simultaneously.

Optical transducers have recently dramatically improved and the data listed in Table 1 lets us compare recent and older designs. Still, the performances of the best optical transducers fall short of those of the best capacitive units. From Tables 1 and 2 we can see that typical capacitive detectors have stability 50% better than the best optical transducers and their repeatability is about an order of magnitude better. The capacitive transducer also exhibits better linearity: one-quarter the nonlinearity.

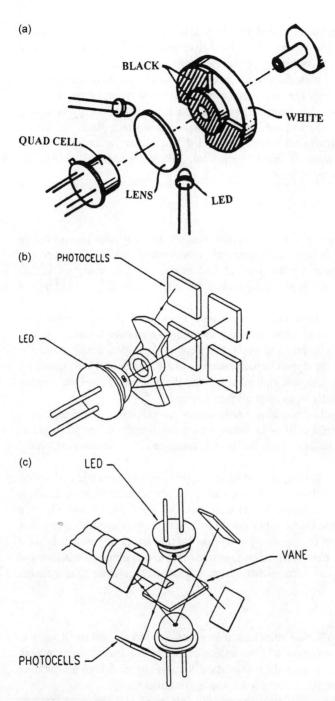

Figure 4 Construction of three optical scanners: (a) advanced optical position detector (General Scanning U.S. patent no. 5,235,180); (b) advanced optical position detector (Cambridge Technology U.S. patent no. 5,844,673); (c) radial optical position detector (Cambridge Technology U.S. patent no. 5,671,043).

Figure 4 shows the conceptual construction of three optical transducers from Refs. 12–24.

Capacitive Transducers

All capacitive transducers have a common concept where a movable ceramic element rotates between a driving plate and a pair of sensor plates wired in series according to the design of Robert Abbe.[15] The moving element can have two lobes ("butterfly" type) or four lobes ("iron cross" type). Figure 6 depicts the basic design mounted at the front end of a flexure-mounted scanner. This symmetrical construction is preferably located between the mirror and the torque motor to simplify the servo loop if the torque motor and the mirror lack a rigid coupling.

The previous sections have addressed the critical issue of stability. Numerous attempts to bring drift within the range of the resolution[16–18] have yielded limited commercial success. The use of fiducial marks in the field of regard has been the only reliable solution for the system designer. Although galvanometers with integrated fiducial marks have been commonly in use for military application, only recently have economical designs become commercially available.

2.1.3 Bearings

An entire chapter of this book (Chapter 5) is devoted to the design of bearing suspension for rotary scanners and the material addressed is applicable here. The design or selection of bearing suspension is critical for both galvanometric and resonant scanners. Galvanometric scanners are built with either ball bearing suspension or flexure bearing suspension. Resonant scanners are built with a variety of suspensions such as cross flexures or torsion bars.

Armatures as well as bearing tolerances of galvanometer scanners are similar to those encountered in the manufacturing of rotating polygons. Oscillating scanners benefit from the periodic motion and a suitable preload should ensure that all elements of the bearing will retrace their path. This should limits wobble to 2 or 3 optical µradian.

Moving magnet torque motors, common to galvanometric scanners, are built with large air gaps – where the drive coil is housed – and consequently impart negligible radial forces. In addition, a properly balanced load should not induce any radial forces. It is therefore practical to preload the ball bearing axially, which is compatible with conventional bearing design. In addition it is possible to use cross flexure bearings, which inherently have low radial rigidity. Torsion bar suspension of resonant scanners may also be employed for the same reasons.

Ball Bearings

Ball bearings should preferably be selected for their ability to operate at high speed. It is critical to prevent the balls from skidding. To that effect, the choices of lubricant as well as the magnitude of the preload are the major consideration. These are well within conventional bearing standards for scanners operating over a few degrees of motion. Periodic high-frequency continuous small motions, typically under 1 or 2 degrees, are known to cause rapid catastrophic failure, a condition known as "false brinelling" or "fretting corrosion."

BEARING WOBBLE. A typical ball bearing supported spindle, as for a polygon, exhibits 20–50 µradian of wobble normal to scan. This represents spindle-only errors and does not include any sagittal errors of the polygon reflective surface. A typical galvanometric scanner using the same ball bearings on the same spacing will exhibit wobble an order of magnitude lower, and typically under 2 μradian.

The components of a bearing, the inner and outer races as well as the balls, have specified tolerances. The same goes for the bearing seats and the shaft. Errors in excess of 1 μm are associated with each interface and the accumulation of all imperfections defines spindle wobble. Polygon inaccuracies must be added to these.

The armature construction of a well-designed galvanometric scanner has the same tolerances and imperfections as any spindle. The one mirror surface, however, is forced to keep a constant periodic relationship to all the bearing and other components, so the mirror repeats its sagittal behavior scan after scan and virtually no wobble is present.

The design of galvanometric scanners and mirror systems considers this periodicity as critical to their performances. The sections "Dynamic Imbalance" and "Mechanical Resonances" provide information for analysis of dynamic radial imbalances and confirm the importance of armature mirror balancing.

BEARING PRELOAD. The conventional method of achieving radial rigidity of a spindle is to have axial preload for the bearings following bearing manufacturers' recommendations. Moving magnet scanners exhibit extremely low radial forces and are normally preloaded axially. Scanner suspensions have benefited from the improvement of bearings, lubricant, and mounting technology developed for reliable spindles for torque motors that index magnetic heads of CD-ROMs. Ball bearing selection and installation are now well understood and must be tailored to the application and address the following considerations to assure proper life:

- pre-load forces;
- radial play and ABEC tolerance number;
- lubricant;
- acceleration (acceleration greater than 500g results in sliding ball condition and early failure);
- bearing rigidity;
- material selection. The choice of ceramic ball is preferred for high speed or high acceleration applications or to minimize fretting corrosion encountered when scanning very small excursion at high frequency.

Cross-Flexure Bearings

Cross-flexure bearings were incorporated in oscillating scanners quite early. Their low radial rigidity to angular rotation ratio has limited their use to angles of only 1 or 2° as they were built with moving iron architecture. The torque motor redesign with high-energy magnets has very much expanded their use. A thorough study of properties and designs of flexure bearings has been conducted by Wittrick[19] and Siddall.[20] It should be noted that radial motion of the axis of rotation is eliminated when the axis is located at one-third of the length of the flexure, as shown in Fig. 7.

Figure 5 is a section of a commercial cross-flexure bearing unit built inside a cylindrical housing. These units come close to the size of a ball bearing. Most scanners are built from photo-etched flexure assembled on a box-like armature designed for minimum inertia and maximum rigidity. Flexures must be stress-free at assembly to avoid catastrophic stress failure.

Galvanometric and Resonant Scanners

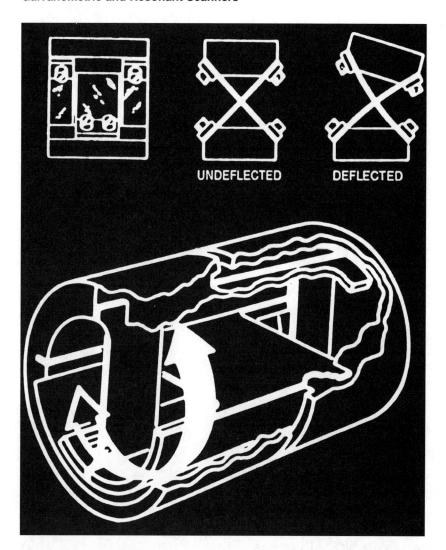

Figure 5 Free-flex flexural pivot.

Figure 6 shows a scanner with an armature supported on cross-flexure bearings. Cross-flexure pivot suspensions are totally free of jitter. Flexures are for oscillating scanners what air bearings are for polygons, but at a cost lower by at least one order of magnitude. Both flexures and air bearings have a common weakness however: low radial stiffness.

The advantages of flexure bearings are:

- freedom from wobble and jitter;
- nearly unlimited life in noncorrosive atmospheres;
- operation in a vacuum (non-out-gassing);
- no contamination of optics (vs. lubricated bearings);
- wide temperature range;

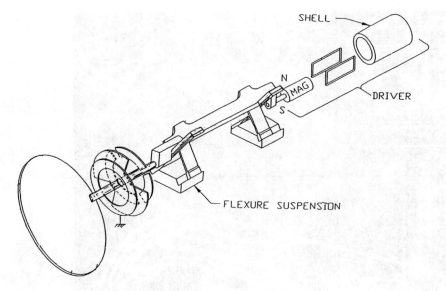

Figure 6 Scanner with cross-flexure bearings.

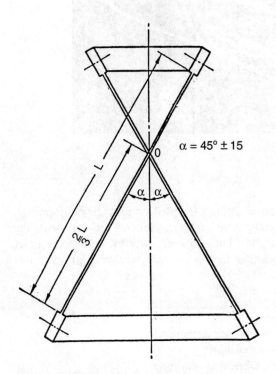

Figure 7 No center shift cross-flexure bearing.

- very low noise;
- very low damping losses.

The disadvantages are:

- bulkier than ball bearings;
- more expensive than ball bearings;
- limited angular excursion;
- low radial stiffness;
- rotation axis shift with angle for wide angle units;
- coupling of torsional and radial stiffness;
- multiple elastic modes with low damping;
- difficult installation;
- intolerance of axial loading;
- amplitude-dependent rigidity;
- temperature-dependent rigidity.

Fortunately, most of the shortcomings of flexure pivots can be circumvented by trade-offs in scanner design and installation procedures.

Cross-flexure bearings are the preferred suspension for applications with a need for extremely critical repeatability and low wobble. They are also the only solution for small angle excursions, under 4° optical. Small motion does not permit proper ball bearing lubrication, leading to "fretting corrosion" and catastrophic failure in short order. Clearing moves every few seconds in order to assist lubrication may mitigate the damages.

2.1.4 Mirrors

Many scanning problems can be attributed to mirror problems. If mirrors could be made infinitely rigid, flat, and reflective, and have negligible inertia, the design and operation of scanners would be extremely simple and there would be less demand for this text. No material able to fulfill these ideal requirements comes to mind; however, available materials do offer practical design solutions.

The facets of a polygon scanner are often viewed as the last link in the chain of components of a scanning system and so burdened with all the system's faults. The mirror of a galvanometric or resonant scanner, however, is rarely perceived in the same fashion and receives comparatively little attention. Actually, they both must meet the same requirements for reflectivity, balance, thermal and dynamic deformation, mounting, and so on.

The condition of the mirror of an oscillating scanner is of concern for that portion of its movement in which optical data is transmitted. Control of dynamic deformations is commonly associated with resonant scanners. Fabrication and installation requirements are commonly associated with galvanometric scanners. Thermal conductivity becomes an issue with the use of high-power lasers (and galvanometric scanners), which tend to induce temperature gradients.

In oscillating scanning systems, the design and installation of the mirror must be able to preserve the essential system features under all operating and storage conditions. Low wobble, low jitter, precise pointing or scanning accuracy, and long life are typical requirements that have to be addressed.

Mirror Construction and Mounting

Several guidelines for mirror design and mounting associated with scanning performances have been mentioned previously. These are:

- the mirror mass must be a minimum; and
- the mirror inertia must be a minimum.

The mirror must be mounted as close as possible to the front bearing of the scanner in order to lower cross-axis resonances. All moments of inertia with respect to the axis of rotation must be balanced in order to minimize wobble induced by angular acceleration and by environmental perturbations. Balancing is most imperative for resonant scanners with torsion bar suspension. Three other performance issues associated with mirror design and installation – alignment, mirror bonding, and mirror clamping/mounting – must be addressed.

ALIGNMENT. One consequence of mirror cross-axis misalignment is beam-positioning error, frequently expressed as a "smile" or a "frown." Limited compensation is possible for a line scan, but a precise area scanner, or designator, may require accurate mirror angular positioning. In addition, mounts may need to be electrically as well as thermally isolated. Alignment or balancing must be verified for each mirror along both axes to prevent imbalance and wobble.

An effective way to minimize problems with both alignment and bonding is to precision machine both the mirror's reflective surface and its shaft mounting hole from a single piece of metal to make an integral mirror. Mirror design criteria for high-resonance frequency and mounting stress isolation must be considered.

MIRROR BONDING. It is extremely difficult to bond a mirror to a mount and have it aligned with an accuracy of 1 mrad. For mirrors smaller than 1 cm the alignment tolerance can be as high as 5 mrad unless optical autocollimation methods and great care in bonding are used.

An improper bond process or mount design can cause mirror deformation when the adhesive cures or temperature changes during shipping or use with higher power lasers. These thermal stresses may also cause the mirror to break or become unbonded.

The elasticity of the bond can cause dynamic pointing errors as well as undesirable resonances that could reduce the system's bandwidth. The rigidity of the bond can cause mirror stresses and deformation when mounting to the shaft of the scanner.

MIRROR CLAMPING AND MOUNTING. Reiss[21] gives a brief overview of recommended mounting procedures. Clamp-like mirror mounts allows for repositioning and possible removal for replacement. By contrast, both integral and shaft bonded mirror mounting methods are semi-permanent conditions.

The most successful removable clamps are a form of collet that provides isolation of the mirror substrate from clamping stress. A disadvantage of the collet approach is potential loosening of the collet clamping forces, with a resulting drift of catastrophic failure. Collet clamps, if not overtightened, induce only compressive forces that produce no bending movement and thus no distortions in the mirror facesheet. One collet clamp technique is to mechanically isolate the shaft by relieved regions so that the distortions imposed by the clamping screw are not transmitted.

Fastening a mirror mount onto a shaft with setscrews is not recommended, as they can deform the shaft. When setscrews are properly fitted, removing them is nearly impossible. When not fitted properly, they act as a hinge, allowing wobble excitations. Set screws can fatigue and loosen.

DYNAMIC DEFORMATIONS. The accelerating torques imparted to a scanning mirror can produce significant mirror surface distortions. This is particularly true in scanners driven with a sawtooth waveform and in high-frequency resonant scanners. Brosens'[22] analysis of the deformations induced by accelerating torques yields the approximate formula

$$f = 0.065(s^2 T/Elh^3) \tag{6}$$

where f is the maximum deflection from the original mirror shape; s the width measured across the axis of rotation; E the Young's modulus of the mirror material; h the thickness; l the length in the axial direction, and T the applied torque. The torque is related to the angular acceleration a by

$$T = hs^3 lda/12 \tag{7}$$

with d the density of the mirror material. Combining the two equations, we obtain

$$f = 0.0055a(ds^5/Eh^2) \tag{8}$$

This expression points to the desirability of keeping mirrors as narrow as possible. For a glass mirror 1 cm in diameter and 1 mm thick, the resulting deflection at an acceleration of 10^6 rad/s^2 is about 1/25 of the sodium D-line wavelength. Since the inertia of such a mirror is 0.011 g·cm^2, the above acceleration corresponds to a torque of only 11,000 dyne·cm. The actual mirror deformation may be smaller when a substantial portion of the width of the mirror is cemented to a mount.

THERMAL DEFORMATIONS. When a scanning mirror is exposed to radiation, a substantial part of the radiation that the reflecting surface absorbs is transferred in the form of heat to the rear surface. This heat is discharged by conduction to the mount and by radiation and convection to the surrounding atmosphere. The conduction of heat to the rear surface causes differential expansion and, if the incident radiation is particularly intense, significant distortions can occur. Such distortions can be estimated by assuming one-dimensional heat transfer to the rear surface.

The radius of curvature R caused by a uniform temperature gradient is

$$R = \{a(du/dx)\}^{-1/2} \tag{9}$$

where a is the coefficient of linear thermal expansion, and du/dx is the temperature gradient in the material. From Fourier's law of conduction,

$$du/dx = q/kA \tag{10}$$

where q is the heat transfer rate, k is the thermal conductivity, and A is the cross-sectional area. The camber assumed by a plate of width s, when it curves with a radius of curvature R, is given to a first-order approximation by

$$e = s^2/2R \tag{11}$$

Combining this equation with the previous expressions, we obtain

$$e = aqs^2/2kA \qquad (12)$$

For a glass mirror of width 1 cm conducting 0.1 W/cm² to the back surface, the resulting camber is 0.5 μm, or about 1 wavelength.

EROSION. When a scanner mirror is moved through air at high speed, the collision of dust particles with its surface can cause gradual erosion of its reflective coating. Experience shows that for any coating the process of erosion does not occur below a critical impact velocity. It is believed that surface erosion occurs when the stress developed at the impact interface between the coating and the dust particle exceeds a value that is characteristic of the coating.

The stress developed by the impact of a rigid body against an elastic mass was analyzed by Timoshenko and Goodier.[23] The stress wave generated by impact is given by the formula

$$S = E(V/c) \qquad (13)$$

where E is the Young's modulus of the substrate, V is the relative speed at impact, and c is the velocity of wave propagation (sound velocity) in the substrate.

Experimental evidence shows that AIS1O coatings on fused silica degrade through erosion at all points where the speed of motion exceeds 3 m/s at any instant during the scan cycle.

Users of high-speed scanners should take precautions to minimize the presence of suspended dust particles near scanning mirrors. Where such protection cannot be provided, hard coatings should be used in combination with substrates of low Young's modulus.

MATERIAL SELECTION. All scanner applications do not have the same performance requirements, so there is no optimum mirror material. The selection of a substrate is application dependent and has to satisfy some or all the performance requirements reviewed earlier.

Table 4 lists the properties of materials suitable for mirror substrate as well as mounts. The figure of merit for resonant scanners, E/d^3, has been derived by Brosens and Vudler.[22] This is to be used as a comparative guide for a given geometry. One should keep in mind that design, construction, heat treatment, coating, and installation can each have a dominant influence on the performance of a mirror. The figure of merit for galvanometric scanner mirror design is E/d. This is based on fabrication requirements only (for discussion, see Ref. 25).

Cost, ease of fabrication, stability with time and environmental conditions (such as temperature and cyclic stresses), bonding capability, and mirror surface finishing are extremely important to the selection of a substrate material. Fatigue and yield strength are normally irrelevant.

MIRROR SURFACE FINISH. Definitions and specifications of available surface finishes and coatings for glass mirrors can be obtained from numerous sources, including government specifications, and will not be reviewed here.

For metal mirrors, difficulties begin with the form of the metal stock and the machining process. Each case is different and presents its own problems. After a blank has

Table 4 Mechanical and Thermal Properties of Substrates

Material	Density (g/cc)	Coef. T-exp. ($10^{-6}/°C$)	Therm. cond. (W/cm-°C)	E, Young's modulus (kg/cm² × 10^5)	Fig/merit ($E/d^3 \times 10^5$)	Fig/merit ($E/d \times 10^5$)
BK7	2.53	8.9	0.010	8.22	0.50	3.2
Fused silica	2.20	0.51	0.014	7.10	0.66	3.2
Fused quartz	2.20	0.51	0.014	7.10	0.66	3.2
Pyrex	2.23	3.3	0.011	6.67	0.54	3.0
Silicon	2.32	3.0	0.835	11.2	0.89	5.0
Aluminum	2.7	25	2.37	7.03	0.35	2.6
Iron alloys	7.86	0–20	0.1–0.8	13–21	0.03–0.04	<2.5
Al oxide	3.88	7.0	0.08	36.0	0.61	9.3
Titanium	4.3	8.5	0.20	11.2	0.14	2.6
Beryllium	1.8	12.0	2.10	30.8	5.2	17.0
Magnesium	1.7	26.0	1.59	4.2	0.80	2.5
Diamond	3.5	0.7	10–25	120.0	2.6	34.0
Silicon carbide	2.92	2.6	1.56	31.5	1.4	11.1
SXA	2.96	10.8	1.2	14.5	0.56	4.9
Tungsten carbide	15.3	5.94	0.5	68.5	0.02	4.5
Miralloy	2.10	6.3	1.1	20	2.1	9.5

been machined to finished dimensions, it has to be stress relieved and stabilized. Dynamic stressing can also be used. Thermal stabilization can be achieved with three or four cycles of processing from liquid nitrogen to boiling water.

The following processes are used for surface finishing when polishing the substrate is not acceptable.

PLATING AND POLISHING. All the surfaces of the mirror can be plated with equal thickness of hard nickel (typically 0.002–0.005 in.) to avoid thermal deformation. Any material removal during polishing may need to be allowed for when plating, and additional thermal stabilization may be required. Nickel can be ground and polished to a high-quality surface and then finished by conventional means.

REPLICATION. This is a process in which a reflective surface and one or more coatings are formed by successive evaporation in reverse order onto a master, and then transferred together onto the substrate and bonded. The bonding agent is typically an epoxy layer with a viscosity of 100 centipoise and a thickness of a few micrometers. If the thickness approaches 25 μm, the process introduces alignment errors of 0.1–1 mrad or greater. There is limited usable temperature range due to "bimetallic deformation." Weissman[24] shows that a 1 fringe curl per 25°C is to be expected for a disc with a 25 : 1 aspect ratio.

Additional limitations may be introduced by the power to be reflected from the mirror. Brosens and Vudler[22] calculate that the heat transfer coefficient of epoxy could limit replicated optics to energy pulsed under 0.0017 J/cm^2 or four orders of magnitude lower than polished copper optics at YAG and CO_2 wavelengths.

DIAMOND MACHINING. This technology has proven to be exceptionally successful in the manufacture of high-volume low-cost aluminum polygon mirrors with sagittal tolerance <50 μrad. Diamond machining is less attractive for beryllium and coated steel substrates, or when reflectivity requirements necessitate additional processing.

Mechanical Protection of Mirror Substrates

Table 4 lists the material and thermal properties of various mirror substrate material. It should be noted (last column) that beryllium has the highest figure of merit of all materials but for diamond.

2.1.5 Image Distortions

This section reviews the most common scanner-related image distortions. While it is important to identify the source of any error, only the most common imperfections deriving from scanner sources will be discussed here.

Cosine Fourth Law

Off-axis image distortions have been described by Smith.[26] They occur even when there is no vignetting. The illumination is usually lower than for the point on the axis. Figure 8 is a schematic drawing showing the relationship between exit pupil and image plane for point A on axis and point H off axis. The illumination at an image point is proportional to the solid angle that the exit pupil subtends from the point. It is apparent that for small values of ϕ, $\phi' = \phi \cos^2\phi$, and that $OA = OH \cos\theta$. Thus, the solid angle subtended by the pupil from H is reduced by a factor of $\cos^3\theta$ from that subtended at A. Now the illumination so far has been considered in a plane normal to the direction of propagation; it is apparent that at H the energy is spread over an area that is proportionately larger than at A because the cone strikes the surface at an angle θ from the normal; thus, a fourth $\cos\theta$ factor must be

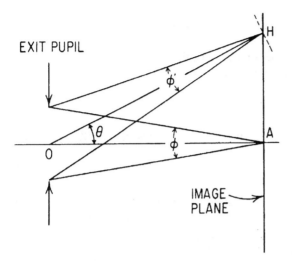

Figure 8 Cosine fourth image distortion.

added, and we find that:

$$\text{(illumination at H)} = \cos^4 \theta \text{ (illumination at A)} \tag{14}$$

Index of Refraction of Air

The index of refraction n for air is strongly dependent on the local pressure p, density d, and absolute temperature T. The pertinent relations are

$$p/d = RT \quad \text{and} \quad (n-1)/d = K \tag{15}$$

where R is the gas constant and K is the Gladstone–Dale constant, an empirical value. One should keep in mind that the local air density is proportional to its local velocity. In practice this forces the designer to enclose the optical bench and prevent air motion.

Excessive wobble of single-axis or bending of two-axis scanning systems operating at a slow speed (a few scans per second) is frequently due to air motion and can be eliminated by proper baffling and occasional vigorous air stirring, such as with a fan.

Microphonic perturbations are due to the variability of the index of refraction of air.

Air Dynamics

As we have seen, air is not the ideal medium for light to travel through. Air also adds to scanning difficulties due to the damping effect of its viscosity and the buffeting perturbations caused by its turbulence. These disturbances are pertinent for systems of high speed and high precision. At present they are frequently encountered with high-performance resonant scanners that are selected for their high frequency and large mirror capabilities. Many advanced systems being contemplated operate the scanners in partial vacuum or helium in order to minimize these disturbances.

There is no literature for low inertia scanners equivalent to that of Lawler and Shepherd[27] for polygons. Aerodynamic effects for low inertia scanners are complex due to the extremely low inertia and stored energy of the moving element compared to the effects of the aerodynamic forces generated. The Reynolds number has been used by Brosens for the purpose of evaluation.[28]

The Reynolds number Re is a dimensionless quantity function of the fluid density d, viscosity v, velocity V, and the mirror radius r:

$$Re = dVr/v \tag{16}$$

A practical expression for air at standard atmospheric conditions in the MKS system is

$$Re = (Vr)6.7 \times 10^{-4} \tag{17}$$

At Reynolds numbers above 2000 the pressure forces proportional to mirror tip velocity add to the viscous losses and are the dominant cause of low Q for resonant scanners. This is the region where laminar flow changes to turbulent.

It is also these turbulences that induce jitter in resonant scanners, which can exceed 5 μrad. This, more than any other effect, limits the dimensions and operating frequency of resonant scanners in air.

Mirror Surface Off Axis

For dynamic reasons, the reflecting surface of a scanner mirror is normally offset from the axis of rotation. This offset T causes an additional scan nonlinearity error. In order to minimize this effect, the beam should be centered below the axis of rotation by an amount K as shown in Fig. 8. This error E function is

$$E = (T - K \sin \alpha)/ \cos \alpha$$

where α is the angle of the mirror to the normal of the work plane.

Figure 9(a) and (b) show a graph of typical single-axis, flat-field scan error for a mirror with a unity surface offset. The scan angles are beam rotations. The beam scan angle is the rotation added to the reference angles of 37° and 45°, respectively.

Beam Path Distortions

Beam path distortion (BPD) may come from the scan head or imaging system. It is caused by path length variations for different portions of the beam. Portions of the image may be blurred or may focus before or after the image plane. Alternately, portions of the image beam may be directed to an incorrect position in the image plane.

In focused laser systems, BPD frequently appears as elongated or distorted spots. These defects appear in different axes in the near and far focus about the optimum focus or as "lobes" of energy projecting from the focused spot.

In vision systems, the image is not always diffraction limited, particularly with the larger apertures. As much as or even 1 wave of distortion may be acceptable.

Mirror nonflatness in imaging systems is a common cause of BPD and is likely to impair the image astigmatically.

Vertical and horizontal axes do not focus at the same plane. This is seen in the two views of the image beam in Fig. 10. The incorrect imaging position of certain bundles of light typically reduces contrast in the image.

Figure 10 shows the front and side views of an image beam reflected from a cylindrically deformed mirror. In the front view, because the beam's convergence is

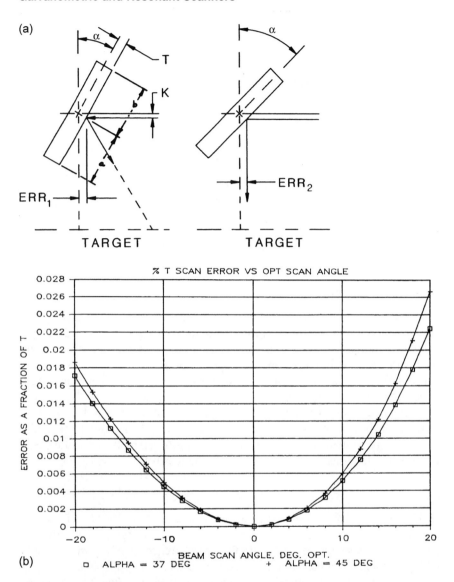

Figure 9 Single-axis, flat-field scan error: (a) compensation for mirror surface off-axis error; (b) scan error vs. scan angle.

reduced, it focuses at a farther point than the undisturbed focus cone in the side view. The spot diagrams show that, in this case, the spot takes on an oval or enlarged form; it never attains the correct size and shape, shown on the right.

Lens system defects, such as decentering, stress, and other manufacturing-related problems, can also cause image deformities. If image quality does not meet expectations, test the lens system without the scanner. Figure 10 shows common distortions.

F-θ lenses commonly exhibit more than 0.5% nonlinearity. Field-flattening lenses also have performance tolerances that need to be specified.

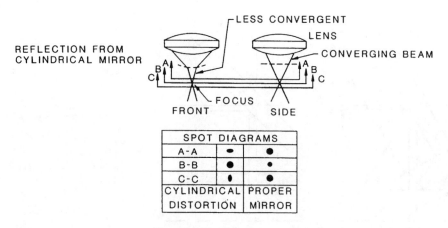

Figure 10 Astigmatism caused by cylindrical mirror surface.

2.1.6 Dynamic Performances

Galvanometric scanners are servo-controlled devices and must obey all standard closed loop servo control requirements for stability. Servo system theory is a well-developed discipline and this section shall only review the scanner design features affecting scanner performances and the effect of some common drive profile signals.

As the cost of high-speed computers has come down, they have become an integral element of scanner systems. Programs can be implemented that alter, on the fly, the drive signal to the power amplifier to optimize random addressing of scanners or laser beam in two or three dimensions. This may be used to circumvent amplifier saturation and high frequency excitations. System performances are strongly affected by the armature design as well as the driver amplifier and the algorithm chosen to drive the scanner.

Resonances

All mechanical elements have resonances. Galvanometric scanners should preferably be designed such that all elements and subelements be as rigid as possible. Ideally their lowest natural frequency should be higher than the highest frequency of the drive signal or any perturbing frequency that can be transmitted to it. This is rarely possible.

It is customary to identify offending resonances and exclude them from the drive signal. It should be kept in mind that scanners mounted on nonrigid lossy material, for isolation, may lose their registration to the work surface. Mirror installation is a frequent source of imbalance and should be executed with that in mind.

Dynamic Imbalance

A rotating body can be balanced, but never perfectly. It is necessary to qualify and quantify resulting imbalance forces in order to assess their consequences and judge if they are acceptable for the application.

Bearings are built with some degree of radial play. When they are subjected to periodic eccentric forces that exceed the constraints of their preload, radial or axial, damage follows that can result in catastrophic failure.

Galvanometric and Resonant Scanners

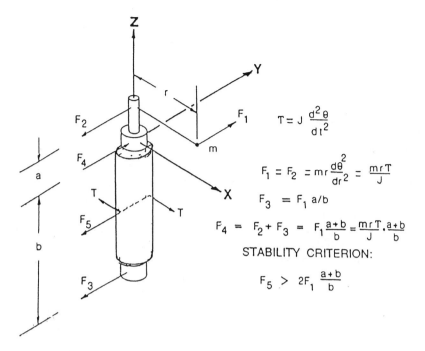

Figure 11 Dynamic forces on rotor.

Figure 11 is a schematic representation of a ball bearing mounted rotor from a galvanometer with an unbalanced load m. The total system inertia is J and a drive torque T imparts an acceleration $d^2\theta/dt^2$. The middle of the rotor is also subjected to a radial force F_5, which is the bearing preload. The derivation uses the symbols of the figure.

All the torques and forces on the rotor must balance; consequently, the following conditions must be satisfied. This will cause perturbations in the scan axis. Periodic compression and decompression of data will occur. If such a scanner is used to generate gray tone images, they will show lighter and darker waves normal to the raster lines.

$$T = mr^2 \, d^2\theta/dt^2 \tag{18}$$

$$F_1 = F_2 = mr \, d^2\theta/dt^2 \tag{19}$$

$$F_3 = F_1 a/b \tag{20}$$

$$F_4 = F_1 + F_3 = F_1(a+b)/b \tag{21}$$

The stability criterion is for the preload to be greater than the period eccentric forces. If it is assumed, according to the figure, that the eccentric mass represents the effect of the mirror, the front bearing is the most vulnerable. The following relationship must be satisfied:

$$F_4 = (a+b)/b * mr \, d^2\theta/dt^2 < F_5/2 \tag{22}$$

The most common mirror mounting technique is a mass balanced assembly with the reflecting surface forward of the axis of rotation. Lateral mass balance is equally imperative.

Armature imbalance of resonant scanners causes wobble and excites the instruments' chassis. Unacceptable audio coupling to the chassis can be minimized with massive construction or soft mounting of the scanner. Both are costly and undesirable as compared to a properly balanced armature.

Mechanical Resonances

A perfectly balanced armature mirror assembly can still produce unacceptable oscillations. These are caused by the excitation of any possible natural frequency of one or more of the elements of the armature or occasionally the stator. These structures commonly have no damping and can be excited by a magnetic imbalance or external shocks and vibrations. Such oscillations are usually sensed by the control circuitry and amplified to cause system instabilities; it is necessary to design the armature so that its first resonance in any mode is substantially beyond the cutoff frequency of the amplifier or is properly damped.

This is the most common limiting factor to the speed of response of a small mirror galvanometric scanner. Two familiar modes of oscillation are reviewed here.

MIRROR ON A LIMB. The mirror is overhung at the end of the shaft and behaves like a freely supported beam. With reference to Fig. 12 it has a deflection angle θ expressed by

$$\theta = Mb/3EI \tag{23}$$

where E is Young's modulus of the shaft's material and I its moment of inertia. For a heavy mirror of mass m the first cross-axis resonant frequency ω is expressed by:

$$\omega = (3EI/ma^3)^{1/2} \tag{24}$$

For a small mirror, the first cross-axis resonant frequency is the rotor resonance. This has been analyzed by Den Hartog.[29] If the rotor can be represented by an iron cylinder, its resonance can be expressed by

$$\omega = 500{,}000\, d/b^2 \tag{25}$$

where ω is in radians per second, d is an approximate value of the diameter of the rotor, and b is its length between bearings, as shown in Fig. 12 (d and b are both in inches here). Exciting this resonance will cause the mirror to wobble and/or render the servo unstable.

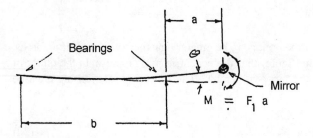

Figure 12 Bending forces on rotor.

TORSIONAL RESONANCES. The rotor and mirror are two freely supported inertias connected by a shaft. The resonant frequency of such a system, with reference to Fig. 13, is

$$\omega = \{K(J_1 + J_2)/J_1 J_2\}^{1/2} \qquad (26)$$

This will cause perturbations in the scan axis. Periodic compression and decompression of data will occur. If such a scanner is used to generate gray tone images, they will show lighter and darker waves normal to the raster lines.

Armature Construction

The armature of galvanometric scanners is a mass-spring system. The scanner's dynamic performance – step response – is limited by the first uncontrollable resonance. Most attempts for compensation beyond the first uncontrolled mechanical resonance have been so far mostly ineffective. The armature is made as rigid as possible and consequently exhibits very high Q resonances. These resonances are known to shift slightly with changes in temperature of the device or as a function of its mode of operation. This renders analog compensation extremely difficult and digital compensation is limited to slow operating systems. It is imperative to have very rigid constructions to minimize the number of possible resonances and raise all resonances preferably one order of magnitude beyond the desired system bandwidth.

Two armature architectures are commonly encountered and schematically exemplified in Figs 14(a) and 14(b) where the bearings are omitted. Both designs can have the same components but produce very different servo system responses.

The construction of Fig. 14(a) is more common as it is simpler to build. The mirror and the transducer are located at either end of the torque motor, beyond each bearing. Unfortunately, in order to satisfy the need for a high torque to inertia, the magnet is long and thin and adds one spring to the servo loop. This may also add a low cross resonance that may couple into the transducer signal. This may be inconsequential for a raster scanning application operating at a single frequency where these resonances may be excluded from the drive signal.

The construction of Fig. 14(b) offers a simpler servo system and consequently potentially a better response, but it is more complex to assemble because one shaft carries both the transducer armature and the mirror. This construction offers an efficient mounting where the scanner is held near the mirror and also serves as an efficient heat sink. It may be noticed in Tables 1 and 2 that scanners that have adopted this architecture also claim better thermal stability.

Drive Signals

It is imperative that the frequency content of the drive signal and the magnetic forces is kept away from exciting secondary resonances of the armature. For example, when designing a vector scanning micro-machining system with a 0.2 ms small step response, one would desirably see that all secondary mechanical resonances be beyond 50 kHz.

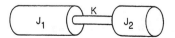

Figure 13 Angular rotor mirror presentation.

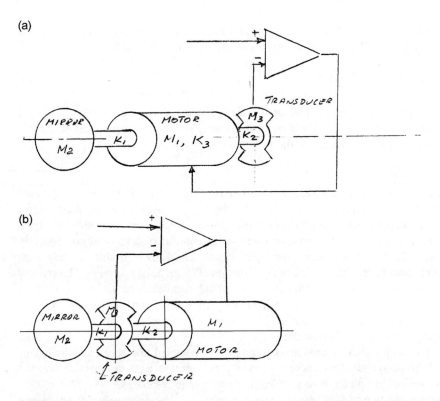

Figure 14 Armature architectures: (a) galvanometric scanner construction with mirror load and transducer at either end of the armature; (b) galvanometric scanner construction with mirror load and transducer at the same end of the armature.

Using Eq. (25) it can be shown that the resonance of a magnet 0.5 in. in diameter and 2 in. long, held on bearings at both ends, has a cross-axis resonance of about 10 kHz, which should be excluded from any drive signal.

VECTOR SCANNING. The shortest step response for a given excursion can be derived from the equation of motion of a second-order system under the assumption that current, voltage, resonances, and electrical time constants are negligible. This is an idealized model and should be used with that understanding. Experience tells us that small steps, 1°, can be expected to approximate these theoretical values. In most cases full torque is limited by the power supply and the driver and is delayed by the electrical time constant of the circuits and the torque motor. These limiting conditions are most experienced when large angular jumps are made.

Small angle stepping time can approximate the condition derived from Newton's law and depicted in Fig. 15(a). The minimum stepping time is derived as follows:

$$T = I d^2\theta/dt^2 \tag{27}$$

where T is the torque required to impart the angular acceleration $d^2\theta/dt^2$ and I is the moment of inertia. To optimize a reciprocating motion system, equal energy and time must be allowed for both acceleration and deceleration of the mirror and armature. The

Galvanometric and Resonant Scanners

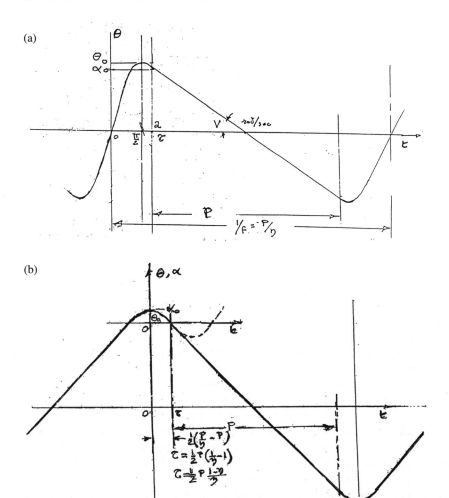

Figure 15 Derivation of (a) saw tooth motion; (b) triangular motion.

maximum potential mechanical energy W that can be given to a rotating system is expressed as

$$W = 1/2 T\beta \tag{28}$$

where β is the total angular displacement (peak-to-peak). This energy can be expressed as equivalent dynamic energy by the kinetic energy equation as follows:

$$W = 1/2 I (d\theta/dt)^{1/2} \tag{29}$$

Solving the above three equations for time yields the expression for the minimum stepping time or step response:

$$t = 2(I\beta/T)^{1/2} \tag{30}$$

In actual cases it is imperative that the drive signal excludes known resonances of the armature. In practice, structured waveforms are created that approximate a section of a sine wave joining the two endpoints of the scan angle. This is known as "cycloidal waveform" and minimizes high-frequency components.

The acceleration capability of moving magnet torque motors is extremely high and in some condition may cause the ball in the bearings to slip rather than rotate. This may lead to rapid catastrophic failure. For this reason, it is advisable to consider a drive signal other than the maximum acceleration condition described above. As expressed above, this condition can be very detrimental with a scanner operating at small amplitude, under 1 or 2°, which does not allow proper ball lubrication and leads to a condition known as false brinelling. A flexure bearing suspension is recommended for these applications.

RASTER SCANNING. The critical points to be attended to for raster scanning are scanner overheating and drive signal-induced vibrations as well as bearing damage as described in the above paragraph. A judicious drive signal design can avoid most of these difficulties.

Two types of raster modes are commonly used: saw tooth and triangular tooth signals.

SAW TOOTH DRIVE SIGNAL. Saw tooth drive signal yields a simpler overall architecture but makes much more demands on the scanner in the turn-around critical points. The fastest possible fly back time can be derived from Eq. (30) and is often referred to the "constant acceleration" drive signal. Frequently, amplifier and power supply saturation as well as inductance and other phase delays will cause the turn-around time to be longer than calculated. Also, caution must be taken to avoid exciting on-axis or cross-axis resonances that may couple through the transducer and the drive amplifier with positive feedback. Cross-axis vibration can couple into the elements of the transducer to induce an erroneous output. If these resonances are known they should be excluded from the frequency spectrum of the drive signal. The most desirable fly back signal, and frequently the fastest, is shaped as a segment of a sine wave where both the start and ending of the sine wave segment match the slope of the linear part of the signal as described in Fig. 15(a).

The relevant parameters of a saw tooth drive is illustrate in Fig. 15(a). The linear segment of the signal has duration p and slope V. The efficiency η can be expressed as a function of the frequency f of the signal in Hz according to

$$\eta = pf = p/p + 2\tau \tag{31}$$

The fly back signal is a sine wave in order to minimize the system's bandwidth. The sine wave and the linear segment join at an equal slope at time τ somewhat longer than $1/4$ period of the sinusoidal signal at time $\pi/2$. The angular position θ of the shaft is therefore expressed as

$$\theta = \theta_0 \sin \omega t \tag{32}$$

At time τ the angular position is α_0 such that

$$\alpha_0 = \theta_0 \sin \omega t \tag{33}$$

Galvanometric and Resonant Scanners

We match the slope of the sinusoidal motion to the slope of the linear motion such that

$$V = -2\alpha_0/p = \theta_0 \omega \cos \omega \tau \tag{34}$$

The ratio of these equations yields

$$\tan \omega \tau = -\theta p \omega / 2\alpha_0 \tag{35}$$

At time $t = \tau$, $\theta = \alpha_0$ and the equation simplifies to

$$\tan \omega \tau = -p\omega/2 \tag{36}$$

As both τ and p are defined by the application, ω can be derived by iteration. As a starting value for ω the value derived by setting $\theta_0 = \alpha_0$ yields $\omega = \pi/2\tau$

When ω is known, the first equation can be used to derive the value of θ_0 as both α_0 and τ are defined parameters.

The pick acceleration a can than be calculated as

$$a = \theta_0 \omega^2 \tag{37}$$

It is interesting to note that the natural frequency of the fly back is a function of only the scan efficiency $\eta = p/p + 2\tau$.

TRIANGULAR TOOTH SCANNING. Triangular tooth scanning normally yields a system running at more than twice the repetition rate with the same power consumption. The coding/decoding software must take into consideration phase shifts and possible lack of symmetry or linearity of the position transducer. Again the turn-around signal should be structured to minimize power consumption, to limit heating and undesirable frequencies. Again, the preferred drive signal is a segment of sine wave that matches the slope of the linear portion of the signal as described in Fig. 15(b), yielding a more robust system than the "constant acceleration" signal. The analysis of this model given below can readily be adapted for saw tooth drive.

If p is the linear segment of the signal with slope V and the efficiency is the magnitude of the overshoot, α can be derived for a sinusoidal turn-around waveform signal as outlined below according to the symbols of Fig. 15(b).

$$V = 2\theta_0/p \tag{38}$$
$$\alpha = \alpha_0 \cos \omega t \tag{39}$$

where

$$\omega = 2\pi\eta/2p(\eta - 1) \tag{40}$$

and therefore

$$\alpha = \alpha_0 \cos 2\pi\eta t/2p(\eta - 1) \tag{41}$$

We match the slope of the sinusoidal motion to the slope of the linear motion such that

$$d\alpha/dt = 2\theta_0/p = \alpha_0\{\pi\eta/p(1-\eta)\}\sin\pi\eta t/p(\eta-1) \qquad (42)$$

At time

$$\tau = p(1-\eta)/2\eta \qquad (43)$$

the angle of the sinusoid is $\pi/2$ and as $\sin \pi/2 = 1$ and the magnitude of the overshoot is derived from

$$\alpha_0 = 2\theta_0(1-\eta)/\pi\eta \qquad (44)$$

The second derivative of α is the angular acceleration a that shall be imparted to the armature during the turn-around function and it is maximum at $t = 0$. With Eq. (44) it yields

$$a = 2\theta_0\pi\eta/p^2(1-\eta) \qquad (45)$$

It can be noticed that the overshoot and the galvanometer acceleration, and therefore the torque requirement, are inversely affected by the scanning efficiency.

2.1.7 Evaluation Parameters

The accompanying tables list only scanners built as inside-out d'Arsonval movements. These are moving magnet designs with a large air gap in the magnetic circuit. This architecture is preferred for optical scanners as it best meets the list of desirable features listed in Sec. 2.1.1. Earlier optical scanners were built with torque motors incorporating iron in the stator or with moving coil armatures. These devices are reported in Ref. 1. To the knowledge of the author no advanced scanner has come from any further development of these two technologies.

As all these units are of similar torque motor design, the figure of merit of the torque motor reflects the coil copper packing density as well as its thermal conductivity. This feature may prompt the unit choice when the scanner is worked hard, such as in raster scanning, or if thermal drift is a critical element.

The choice of position transducers, capacitive and optical, is normally guided by the tolerable error in repeatability dictated by the application as well as by the duty cycle of the application. A low duty cycle with high peak power demand shall not permit the transducer to reach a stable temperature and therefore the actual drift parameters may need to be verified experimentally in the application or the use of fiducial marks should be considered.

The dynamic performances listed are to be used as references only as different standards are used for the various parameters. Dynamic performances also depend on the armature construction, the load, the load attachment, as well as the sophistication of the driver amplifier and often also the drive signal that is used.

The accompanying tables list commercial scanners according to their armature inertia. All the data is derived from published material from the following manufacturers:

- Cambridge Technology, Cambridge, MA: models 62xx, 64xx, and 68xx.
- GSI Lumonics (General Scanning Inc), Bedford, MA: models Mx and VM2000.
- Nutfield Technology Inc., Windham, NH: models RZ-xx.
- GalvoScan LLC, South Royalton, VT: models TGV-x.

Galvanometric and Resonant Scanners

Other commercial manufacturers of galvanometric scanners, whose products are not tabulated here are:

- EOS, Munich, Germany: moving iron scanners.
- Lasesys Corp., Santa Rosa, CA: stepper motor with an optical encoder.
- Laserwork, Orange, CA: entertainment products.

2.2 Resonant Scanners

The armatures of resonant scanners are low-mass, high-rigidity, and high-Q structures. Large excursion can be achieved with a low-torque simple drive motor. The major advantage of the resonant scanner is its simplicity, small size, long life, and in particular its low cost. Its major disadvantage has been its sinusoidal motion, and sensitivity to external as well as self-induced perturbations. This deficiency corroborates the scarcity of low-frequency resonant scanners. The induced moving coil of Fig. 16 is an older but simple low-frequency design that is commercially available.

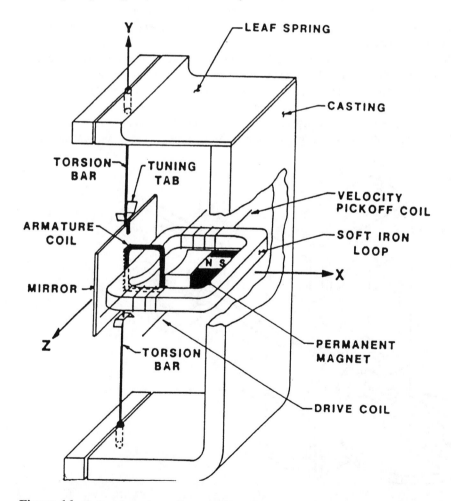

Figure 16 Induced moving coil resonant scanner.

2.2.1 New Designs

Taking advantage of new materials, Dean Paulsen[30-32] developed new concepts for the design of resonant scanners:

- the use of a high-energy permanent magnet, minimizing radial forces;
- the location of anchor points at vibration nodes of the armature;
- the use of highly lossy material, such as "Sorbutane" at the anchor points in order to damp out external as well as self-induced perturbations.

2.2.2 Suspension

Figures 17 and 18 exemplify these concepts as counter rotating resonant systems. The anchor point or points are located at nodes of the resonant armature that do not experience any movement. This is analogous to holding a musical tuning fork at its base. No energy is lost. High-frequency devices can be built with one support. Low-frequency devices need two supports to be insensitive to orientation or external perturbation. One can also note in Fig. 17 two orthogonal coils. One is the drive coil and the other the tachometer coil. As they are orthogonal, their fields do not interact, but each coil interacts with the permanent magnet of the armature.

Resonant scanners can also be built with cross-flexure suspensions. Dean Paulsen has also designed the ISX family of resonant scanner at General Scanning.[33]

Tunable resonant scanners have also been built. This author, in Refs. 1 and 34, evaluates two designs. Low-frequency large-aperture resonant scanners are frequently built with cross-flexure suspensions. A common design can be found in Ref. 1.

Table 5 presents performances of a number of commercial resonant scanners.

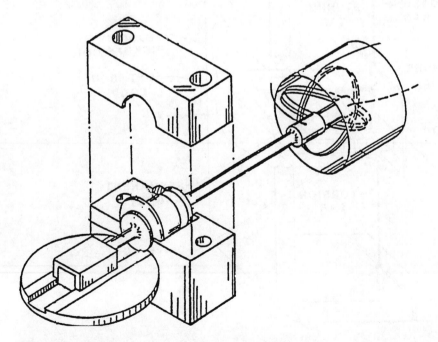

Figure 17 High-frequency tuned torsion bar resonant scanner.

Galvanometric and Resonant Scanners

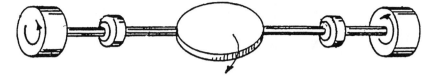

Figure 18 Low-frequency tuned torsion bar resonant scanner.

2.2.3 Induced Moving Coil

Two of the deficiencies of moving coil devices, lack of rigidity and moving electrical connections, can be bypassed by having a single turn coil energized by induction.[35,36] This technology is applicable to resonant scanners. A resonant scanner implementation is illustrated in Fig. 16. One of the most novel resonant low-inertia scanners to have been introduced is the balanced torsion bar design with an induction torque driver. The mirror is suspended by two torsion bars in a fully symmetrical arrangement designed to be mass balanced, so that accelerations along the three principal axes of translation will not cause torsional excitation of the mirror.

The drive coil and the armature are magnetically linked in a transformer-like fashion by a soft iron core. The armature is a single-turn, rectangular drive loop with an edge colinear with the torsion current in the drive loop. This current interacts with the return path of the flux of a permanent magnet to create the drive torque:

$$T = \mu ANIBlr/LR \tag{46}$$

where μ is the permeability of the iron core, A is the cross-section of the iron core, N is the number of turns of the drive coils, I is the driver coil current amplitude, B is the field of the permanent magnet, l is the length of the drive loop in the magnetic field, r is the drive loop acting radius, L is the length of the iron core path, and R is the resistance of the drive loop. In this manner a moving coil driver is obtained without having to provide leads or brush contacts to the moving armature. The motion of the armature induces a voltage and current, which are sensed by a pick-off coil. The actual voltage measured is the sum of this induced voltage, which is proportional to velocity, and of the portion of the drive coil voltage induced by transformer couplings. It is expressed as:

$$E = (\mu AIN^2/L)\omega \cos \omega t + NBLr\delta\theta/\delta t \tag{47}$$

where ω is the resonant frequency. The transformer coupling component is easily subtracted electronically. This velocity sensor permits extremely good, simple, and external amplitude control with drift below 100 ppm/°C of the peak-to-peak excursion. The resonant frequency is stable to $\sim$160 ppm/°C. A derivation of Eqs. (46) and (47) can be found in the appendix of Chapter 5 of Marshall.[1]

3 SCANNING SYSTEMS

Scanning systems can be treated analytically, but they involve a number of disciplines, most of which are treated superficially in this chapter. In laser machining applications the system performance also frequently requires a good understanding of the interaction of the

Table 5 Comparative Performances of Commercial Resonant Scanners

	Model						
	GRS	IMX200	IMX350	TRS	CRS4	CRS8	IDS	URS
Mirror/beam diameter, mm	<36	28	28	<20	12.7	7.8	9	<30
Flatness, wave@633	1/10	1/6	1/6	1/2	1/2	1/4	1/2	
Suspension, type	X-Flexure	X-Flexure	X-Flexure	X-Flexure	Counter Rot.	T Bar	2 T Bars	S-Flexure
Resonant frequency, Hz	<250	200	350	<10,000	4,000	8,000	<1,200	<500
Beam rotation max., degrees	72	60	30	60	20	26	60	90
Beam wobble, max., micro rad.	2	2	2	2	100	150	100	NA

NOTES:
- All units consume less than 1 W.
- All units incorporate a tachometer for self-excitation and amplitude control.
- Frequency stability 100 ppm/°C, typ.
- Line straightness, frequency, and amplitude jitter specifications should be requested.
- Large mirror units may impart considerable system vibration.
- Manufacturers: GSI Lumonics – IMX 200, IMX 350, CRS 4, CRS8, IDS
  Lasesys Corp. – GRS, TRS, URS.

work and the wavelength, radiation duration, and power of the laser, as well as possible damage to the optical elements of the scanning system. A number of manufacturers offer predesigned scanning packages also known as "scan heads." These are typically two- or three-axis vector scanning systems that include all scanning functions less the laser and the work surface. One of the valuable aspects of these packages is the matched driver-amplifier matched to the inertia of the mirrors. This is frequently the most economical and expeditious manner to construct a vector scanning system such as used for micro-machining. The goal of this section is to present the numerous available choices.

The scanning applications may be divided into two major classes: raster and vector scanning. The former frequently involves a galvanometric scanner or a resonant scanner sweeping the fast axis and a stepper motor driving the slow axis – frequently moving the workpiece as this minimizes the size of the lens of a preobjective scanner configuration. The latter demands that both axes have preferably identical dynamic behavior and is covered in Sec. 3.2. The former is exemplified in Sec. 5.2, "Microscopy", which looks at three raster optical scanning architectures, fixed objective, preobjective scanning, and flying objective scanning.

3.1 Scanning Architectures

Scanning systems can be split into two categories, namely the beam moving category and the moving objective lens category. The two are interdependent but can be described separately. First we will explore the fixed objectives choices.

3.1.1 Postobjective Scanning

In this configuration the scanners are located between the objective and the work. This demands an objective with a long focal length to accommodate the scanning mirrors; the focal point will approximately paint a sphere. If the work area is flat, this restricts the scan to a small angle where the depth of field can approximately intersect a plane. If large angular excursions are required, the objective lens is normally translated on a linear stage to accommodate the need. The bandwidth of the translation mechanism needs to be twice that of the fastest scanner, but less accurate position control is required. A galvanometer is frequently the best choice to drive the translation stage[37] and a number of scanner manufacturers offer such an assembly.

When used in microscopy with comparatively fast optics, this arrangement operates at small angles and consequently cosine fourth law aberrations[26] are negligible. Also this construction benefits for a comparatively small and low-cost objective where chromatic aberrations can be minimized.

This configuration is best applicable to long focal length objectives such as found in laser radars, range finders, and designators.

3.1.2 Preobjective Scanning

This architecture locates the objective lens between the scanners and the work. It is the most common choice for laser micro-machining as it permits a comparatively small spot size with high energy density. It is also the primary choice for conventional laser scanning confocal microscopes and Sec. 5.2.2 depicts the original design of Minsky.[38] The *Handbook of Biological Confocal Microscopy*[39] carries numerous examples of this concept.

A number of lens manufacturers offer off-the-shelf lenses for this application configured for YAG or CO_2 lasers. Multiple wavelength telecentric field-flattening objectives of this type, capable of yielding a diffraction limited small spot focus, are frequently the most expensive component of the entire scanning system, including the laser.

All the beam steering systems described in Sec. 3.2 are suitable for use as pre-objective systems.

3.1.3 Flying Objective Scanning

This scanning architecture demands that the work be moved in one axis while the beam is moved in the other axis. It is extremely advantageous for microscope raster scanning and is the preferred choice for biochip scanners as described in Sec. 5.2.3. It is also used for random axis scanning in the semiconductor industry for DRAM repairs. It offers a very low-cost system with multiple wavelength telecentric flat-field, diffraction limited small spot performance.

3.2 Two-Axis Beam Steering Systems

Oscillatory scanners are best suited for large excursion systems, over a few hundredths of a radian. Single-mirror or small motion two-axis beam steering (TABS) systems have been hotly pursued for SDI applications, and the *SPIE Proceedings*, vol. 1543 as well as in the *IR/EO Handbook* [40] examine a number of them. Conventional gimbal systems, where the torque motor for the second axis is transported on the structure mounted on the first axis are reviewed in Sec. 5.2.

Below are some of the most common architectures of two-dimensional scanning systems. They fall into two major classes: vector scanning and raster scanning. Vector scanning applications require that both axes have equal properties and normally they have two galvanometric scanners. Raster systems frequently have a more diversified architecture. Some use a polygon or resonant scanner for the fast or raster scan and a galvanometer, motor, or a linear transport for the other motion. The *LR/EO Handbook* [40] describes a number of such systems.

3.2.1 Single-Mirror TABS

The device symbolically described in Fig. 19 is capable of 1 radian motion in both axis. The drivers and encoders for both axes are stationary. The torque capabilities, range, and angular resolution are dissociated from the inertial load. The optical system behaves like a true point source with double pincushion distortions. The central block is on the origin of all three coordinates. The floating "L" bracket is necessary to prevent bearing lock.

3.2.2 Relay Lens TABS

This construction guarantees that both mirrors and scanners have equal performances. It is penalized by the added two optical elements with their associated cost and/or distortions. The distortions are again in the familiar pincushion pattern and can be compensated for in the design of the objective lens or the computer program. Transmission optics are frequently used, but when chromatic aberrations are critical, reflective optical elements are preferred, as shown in Fig. 20. It should be noted that the two axes need not be perpendicular.

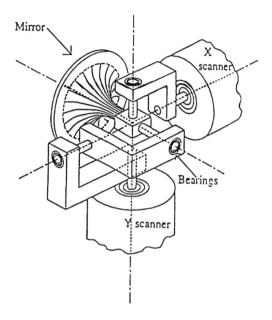

Figure 19 Single-mirror TABS.

3.2.3 Classic Two-Mirror Construction

Figure 21 is a generic model of this construction and is used to derive image distortions and focus variations. In this configuration the two mirrors have different inertia. In order to minimize their difference, the X galvanometer can be mounted at 15–20° from its

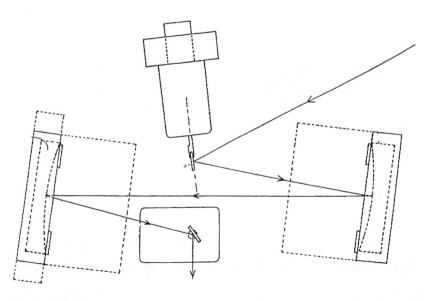

Figure 20 Relay lens TABS with reflective optics.

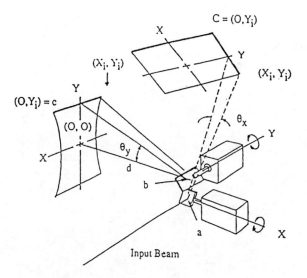

Figure 21 Two-mirror, two-axis flat-field assembly.

perpendicular axis. This allows a closer packing construction and a smaller Y mirror. The incoming beam must be parallel to the axis of the Y scanner.

In this configuration, a is the center of the X mirror, b is the center of the Y mirror, and c is the point at coordinates $(0, Y_i)$; d is the length from h to $(0, 0)$ and e is the length from a to b. The optical scanner angles are θ_x and θ_y and the coordinate (X_i, Y_i) is any point on the target field. It can be seen that when $X_i = Y_i = 0$, then $\theta_x = \theta_y = 0$. The equation that relates Y_i to θ_y is derived from the triangle of points $(0, 0)$, $(0, Y_i)$, and d. Solving for the length $(0, 0)$ to $(0, Y_i)$, which equals Y_i, we obtain

$$Y_i = d \tan \theta_y \tag{48}$$

The determination of the X equation is somewhat more complex and is best illustrated by projecting the target image onto the virtual image position of the Y mirror, as shown by the phantom lines and phantom coordinates $(0, Y_i)$, (X_i, Y_i), and a in Fig. 21. By solving the triangle of points a, $(0, Y_i)$, (X_i, Y_i), for the length $(0, 0)$ to $(0, Y_i)$, which equals X_i, we have

$$X_i = ac \tan \theta_x \tag{49}$$

Since $ac = (d^2 + Y_i^2)^{1/2} + e$ where $e = ab$, the solution is

$$X_i = \{(d^2 + Y_i^2)^{1/2} + e\} \tan \theta_x \tag{50}$$

If we solve for the length from a to (X_i, Y_i) we obtain the equation for the focus length:

$$f_i = [\{(d^2 + Y_i^2)^{1/2} + e\} + X_i^2]^{1/2} \tag{51}$$

Galvanometric and Resonant Scanners

The resulting change in focus length for (X_i, Y_i) is

$$\Delta f_i = [\{(d^2 + Y_i^2)^{1/2} + e\}^2 + X_i^2]^{1/2} - (d + e) \tag{52}$$

In looking at pincushion errors in simple two-mirror systems, we see that X_i (Eq. 50) can be combined with Y_i (Eq. 48) to yield

$$X_i = \left(\frac{d}{\cos\theta_y} + e\right)\tan\theta_x \tag{53}$$

The pincushion error ε is the ratio of the change in the value of X_i as θ_y changes from zero to a specified value, to the peak-to-peak amplitude $2X_i$ at $\theta_y = 0$:

$$\varepsilon = \frac{X_{i\theta y} - X_{i0}}{2X_{i0}} = (1 - \cos\theta_y)/2(1 + e/d)\cos\theta_y \tag{54}$$

3.2.4 Paddle Scanner Two-Mirror Configuration

In applications that need large excursion angles, high speed, and high precision, the single-mirror two-dimension scanner techniques described above exhibit handicaps. Paddle scanners are of great interest because they simulate a two-dimensional fulcrum. Figure 22 gives a symbolic representation.

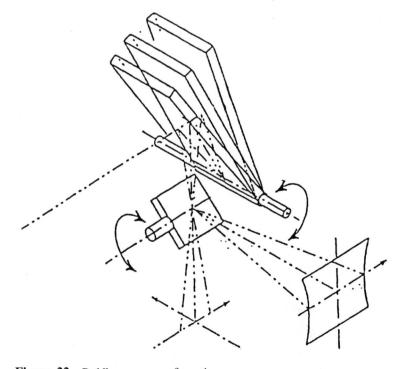

Figure 22 Paddle scanner configuration.

The inertia of the paddle is greater than the inertia of the other mirror. This configuration adapts well to raster scanning. The second scanner can be either a resonant scanner with sinusoidal, triangular, or saw tooth motion, or a rotating polygon. It should be remembered that at all high angular speeds, the loads should be statically and dynamically balanced in all axes.

The first mirror is mounted on a paddle-like arm and intersects the incoming beam at 45°. The mirror rotation, the size of the beam, and other geometric constraints determine the magnitude of this motion. The magnitude of lateral motion of the reflected beam at its pupil is small and is analyzed below from Fig. 23.

Note that the axis of rotation of the X scanner, the frame scanner, is within the plane of the mirror. In the drawing, the rest position of mirror OM is at 45° of the incoming beam. Radius ON is normal to incoming beam AN. Figure 23 also shows mirror OM

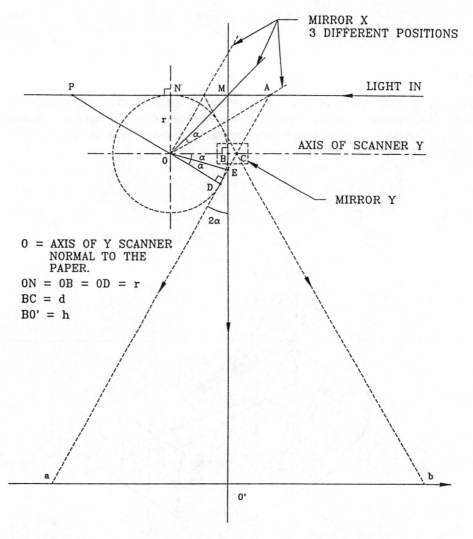

Figure 23 Paddle scanner pupil motion.

Galvanometric and Resonant Scanners

rotated by an angle α and the reflected beam OM' is rotated by the angle 2α to line Aa. Lines OB and OD are normal to the reflected beams at the two positions of the mirror. It is evident that point B is on the circle of radius ON. We shall show that point D is also on that circle.

From the above we can write:

$$\text{MEA} = 2\alpha \tag{55}$$

$$\text{NOM} = \text{NOA} - \alpha = \pi/4 \tag{56}$$

$$\text{NOA} + \text{NAO} = \pi/2$$

Consequently angle

$$\text{NAO} = \pi/4 - \alpha \tag{57}$$

Considering that angles APD and aEO' are equal as both their sides are normal, therefore

$$\text{angles APD} = \text{aEO}' = 2\alpha$$

As triangle PAD is rectangular

$$\text{angle NAD} = \pi/2 - \text{APD} = \pi/2 - 2\alpha = 2\text{NAO} \tag{58}$$

As angle NAD = NAO + OAD = 2NAO, we can conclude that angles NAO = OAD. The two rectangular triangles ONA and ODA have three equal angles and one common side, therefore they are equal and segments ON = OD and point D is on the circle centered at O and passing through point N and B.

It is therefore possible to derive the amount of translation the beam has on mirror Y as mirror X rotates. It is represented by the segment BC. Considering the triangle OCD, we can write

$$\text{BC} = \text{OD}(1/\cos 2\alpha - 1)$$

or, using symbols d and r,

$$d = r(1/\cos 2\alpha - 1) \tag{59}$$

It should be noted that the cosine does not change sign when the angle does and that d is always positive.

For an optical system where the distance between the mirrors is 10 mm and the angle of rotation of the frame mirror on the Y scanner is ± 0.15 radian, the beam walk d on the Y mirror is 0.33 mm.

3.2.5 Golf Club Two-Mirror Configuration

The diagram of a two-dimensional scanner known as a golf club scanner in Fig. 24 has similar features to the paddle scanner. The inertia of the page mirror and its mount is

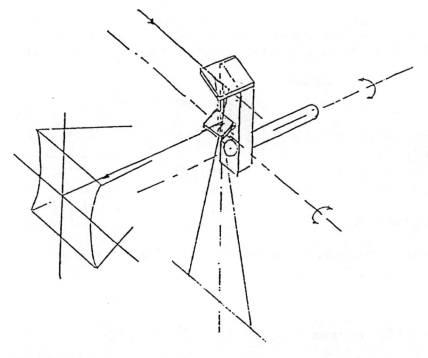

Figure 24 Golf club TABS.

typically double that of an equivalent paddle for the same mirror distance. Since some rotation and geometric constraints make this geometry desirable, it is reviewed here.

The defining feature of this arrangement can be seen in Fig. 24. In a raster scan application, the beam strikes the frame mirror first, the Y-axis, which is held at 45° from the incoming beam. This mirror is mounted on an arm whose axis of rotation intersects and is normal to the reflected beam at the rest position. The axis of rotation of the second mirror (the X mirror with reference to Fig. 25) is in the plane formed by these two lines. It is located approximately halfway between the point of incidence of the beam on the Y mirror at rest and its axis of rotation.

We shall show now that as the Y mirror oscillates, the reflected beam passes through the pupil location X in Fig. 25. We shall also derive the magnitude of "walk" st of the beam on the X mirror as a fraction of the radius of rotation $P_1Y = r$ of the Y mirror.

As an approximation we have

$$st = \frac{mn}{2^* \tan 2\theta} \tag{60}$$

As

$$mn = P_1 m - P_2 n \tag{61}$$

it can be seen that

$$P_1 m = \frac{P_1 P_2}{\tan 2\theta} \tag{62}$$

Galvanometric and Resonant Scanners

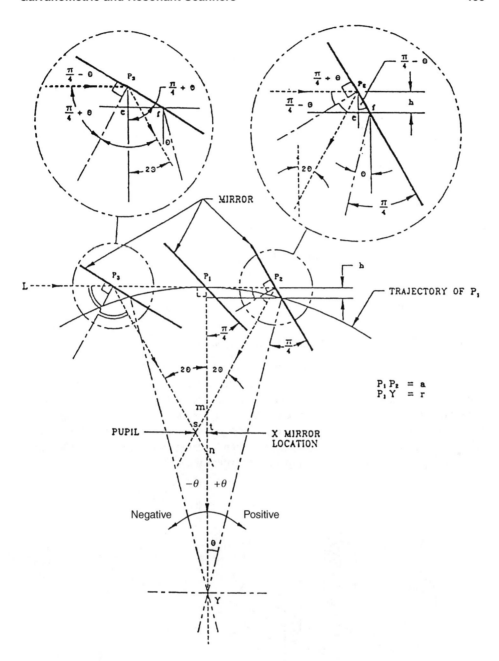

Figure 25 Golf club scanner.

Also

$$PrP_2 = r^* \sin\theta - ef \tag{63}$$

and segment

$$ef = h^* \tan(\pi/4 - \theta) \tag{64}$$

where

$$h = r^*(1 - \cos \theta) \tag{65}$$

The above simplifies to yield

$$P_1 m = \{\sin \theta - [1 - \cos \theta] \tan (\pi/4 - \theta)\} r / \tan 2\theta \tag{66}$$

Values for $P_2 n$ are derived similarly for the angle $-\theta$

$$P_2 m = \{\sin \theta + [1 - \cos \theta] \tan (\pi/4 - \theta)\} r / \tan 2\theta \tag{67}$$

and going back to Eq. (60) and (61) again, as a first approximation we have

$$st = \{\tan (\pi/4 + \theta) + \tan (\pi/4)\}[1 - \cos \theta] r / 2 \tag{68}$$

Comparing this geometry to the paddle, for a rotation of ± 0.15 radian and a 10 mm mirror spacing, the beam walk $st = 0.24$ mm.

3.2.6 TABS with Three Moving Optical Elements

The dominant feature of the design in Fig. 26 is the emulation of a perfect fulcrum with both axes having approximately the same dynamic capability. The "conditioner" scanner carries a glass wedge and is synchronized to the Y scanner. It translates the incoming beam onto the Y mirror so that the reflected beam always strikes the X mirror in its center. Despite the added cost, this is the preferred design for achieving high speed with large-beam wide-angle vector scanners, especially when focusing optics are needed. The details of this design can be found in Goodman, U.S. patent no. 4,685,775.

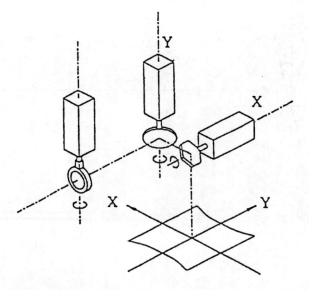

Figure 26 Three-optical-element TABS.

Galvanometric and Resonant Scanners

4 DRIVER AMPLIFIER

High-performance servo amplifiers for galvanometric or resonant scanners are very special products. They are critical to extract the performance of the galvanometer. They are as critical a system component as the galvanometer or the resonant scanner themselves. Commercial drivers are offered by scanner manufacturers and quoted performances are normally guaranteed only when the associated driver is used.

The design of these amplifiers is beyond the scope of this chapter. The list of elements to be addressed is also beyond the scope of this chapter and this author strongly suggest that anyone considering such a task should first refer to Ref. 41, where the parameters that need to be addressed to design a high-performance analog servo amplifier are reviewed.

At this time, all commercial galvanometer driver amplifiers are analog systems. Digital servos are commercially available to drive servo motor-based systems. The numerous attempts to design a high-performance digital equivalent have failed to meet the comparable bandwidth resolution product demanded by optical galvanometer scanners. The potential benefit of a full digital system continues to tempt and elude designers. Hope springs eternal.

Two critical features of these servo amplifiers are:

- *Low noise*. It is common to demand system resolution of 1 part in 100,000. Consequently each component in the system needs to perform 5 or 10 times better. The design of analog circuitry of that quality is an art in itself.
- *Frequency response compensation/filtering* for the galvanometer/mirror resonances.

5 SCANNING APPLICATIONS

The earlier edition of this text[2] carries numerous examples of applications that have shaped the development and performance of optical scanners, both galvanometric and resonant. Here are two categories that have seen vigorous development in the past decade. These are laser micro-machining and confocal microscopy.

5.1 Material Processing

The market for laser-based material processing has grown enormously in the last decade and has segmented itself. This is the result of a number of factors:

- new high-tech applications with no practical alternative solutions;
- penetration into the commercial market, pencils, and kitchenware;
- reliable lasers with a choice of wavelengths and sufficient power;
- scanners with sufficient precision;
- software packages complete with a powerful low-cost PC.

The general system architecture has not changed appreciably. Single lasers, multiple heads, and fiber optic laser light distribution are more common. In general, the market is divided into marking and micro-machining.

In the laser-marking field, the CO_2 laser dominates and as the need to ensure good marking visibility suggest large spot size and consequently small scanning mirrors as beam diameters between 10 and 20 mm cover most applications. High scanner speed is desirable to produce high throughput. Since the industrial waveguide laser was developed in the early 1980s – 10–50 W originally, and now as much as 200 W – its low cost and reliable design has stimulated this segment of the industry.

The CO_2 laser is also the preferred choice to mark packages on the fly or erode plastic packages to indicate and facilitate opening.

The YAG laser competes in marking applications for plastics, glasses, and paints that incorporate titanium oxide powder. This chemical exhibits very visible and permanent color changes when exposed to 1.06 nm wavelengths. It is frequently used for kitchenware.

The micro-machining segment of the industry addresses a variety of applications where different technologies compete. Scanning systems dominate applications where large areas of precisely located features need to be created or where a threshold of power/energy is required. Materials such as metals, polymers, and ceramics are commonly machined in this manner. Typical applications are:

- Nozzles for ink jet printing: 1 or 2 micron feature tolerances with submicron positioning and concentricity are required.
- Screens/sieves/membranes for filtering inks, biomedical fluids, gas separation, and so on. It is common to punch thousands of holes 15 microns in diameter, 20 micron on center, in 10 micron Kapton sheets for such applications;
- Via drilling for printed circuit boards or micro chip modules or green ceramic.
- Probe cards for circuit auto-testers may have 2000 to 3000 contact wires piercing a 2 in. square support.

In order to reach industrial applications, lasers need to be reliable and easy to maintain. A number of technologies have reached this stage. High-power diode pumped YAG lasers are now available at the following wavelengths: 1060, 532, 473, and 351 nm. The shorter wavelengths, pulsed or CW, compete efficiently with Excimer lasers for most polymer or organic micro-machining applications and are also used to stimulate opto-chemical reactions.

5.2 Microscopy

In the last decade scanners have participated in the expansion of microscopy. First came confocal microscopy and more recently large field-of-view microscopy. In these applications, images are assembled on a computer.

The development of confocal microscopy in the early 1950s by Marvin Minsky[38] has opened an array of applications for scanners. In this concept, the image of the work is constructed from single pixels acquired sequentially. This minimizes the demand on the optical design and offers major imaging benefits. Among these are sectioning capability, akin to tomography and the direct digitization of the image. In his original design, Minsky moved the work on a motorized *XY* stage under a conventional microscope. His invention gave images with greater resolution than had been available before, but the images were composed on a computer monitor and a number of decades had to pass before the technique was accepted.

Confocal microscopes are now commonly used either to improve resolution or to capture a large field in a single file without stitching a number of small fields of view. Scanners are most frequently incorporated in both classes of instruments. For simplicity, this review will address only the large field of view (centimeters square) model, as they encompass concepts found in all designs used to image arrays of biological material.

Arrays of biological material consist of large number of features such as dots or squares bonded to a glass substrate and labeled with fluorescent molecules. The role of

Galvanometric and Resonant Scanners

imaging systems is to view the entire array in a single image and present it for analysis to a specialized software program. High-density arrays may carry as many as one million separate 10 microns square. Low-density arrays have 400 spots per square centimeter; here each spot may be 150 micron in diameter, placed on a 200–300 micron grid. Large arrays are typically 22 × 66 mm square. Three basic technologies are used to expand the reach of conventional confocal fluorescent microscopes to meet the needs of biochip imaging and in all cases galvanometer scanners are the preferred actuator.

Scanning architecture	Manufacturers
XYZ stage: Minsky design	Packard, Brown Lab (website)
Preobjective scanning	Agilent, MD/Amersham
Flying objective scanning	Affymetrix, Virtek/BioRad, Axon

The target typically measures 1–2 cm square and the resolution needs to be between 2 and 10 microns. Fluorescence is an extremely inefficient phenomenon, so the image quality is extremely dependent upon the energy collection capability of the instrument. All of these instruments are epifluorescent pseudoconfocal laser scanning microscopes.

5.2.1 Preobjective Scanning

Conventional laser scanning epifluorescent microscopes are thoroughly described in the *Handbook of Confocal Microscopy* [39] and have been commercially available for the past 25 years. Zeiss, Nikon, BioRad, and Olympus are common names in that field. They are all built from conventional microscope objectives. Because fluorescence is a very low-energy phenomenon, a high numerical aperture is preferable. This yields a very small field of view, well under 1 mm square, so that multiple images are required to cover the large field of view associated with microarrays. By limiting the spectral range to 1 or 2 wavelengths and accepting a lower NA, it is possible to design, on this model, an instrument with a larger field of view (up to 10 mm). The Avalanche from Molecular Dynamics/Amersham, symbolically depicted in Fig. 27, is a scanner of this construction. It has a nine-element objective designed for a 10 mm field of view with a comparatively low NA, approximately 0.25. A similar instrument made by Agilent (formerly HP) has a 15 mm wide field of view. In order to make best use of the lens, the biochip or the microscope slide is translated under the objective and the beam scanned on the other coordinate.

5.2.2 The Marvin Minsky Confocal Microscope

The Minsky design shown in Fig. 28 can be built with a commercial microscope objective. The working clearance of high-power objectives is quite small and the slide needs to be held reliably to avoid catastrophic interference in a dynamic environment. The slide is mounted on a motorized XY stage. An optical scanner is preferably used to drive the high-speed scan stage as it outperforms a DC motor. The slide anchor mechanism must also be light enough to permit a reasonable scan rate without excessive vibration.

5.2.3 Flying Objective Scanning Microscope

The flying objective architecture transposes the difficult requirements posed in designing a large flat field of view lens by replacing it with a relatively simple objective lens, which is moved across the sample area. In other words, instead of moving the slide and its holding device in the fast axis, the beam of light is moved instead. This approach requires that

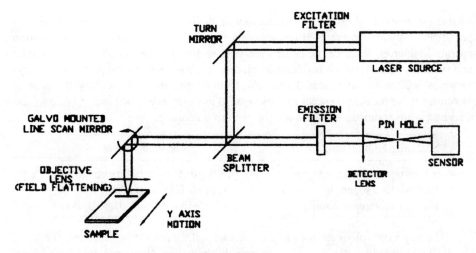

Figure 27 Preobjective line scan architecture.

good mechanical alignment be maintained throughout the entire range of motion, and carries the additional benefit from the use of a much larger numerical aperture objective lens. It also offers a high degree of measurement uniformity across the field of view as all pixels are acquired in an identical manner so that field flatness is not a problem.

Two basic designs are used: one design oscillates an objective lens mounted on a linear rail, and the other design oscillates an objective lens mounted at the end of a rotating arm.

5.2.4 Rectilinear Flying Objective Microscope

Hueton[42] describes this architecture. As shown in Fig. 29, the objective is translated on a linear stage and it is optically "tromboned" to the stationary optical elements. The

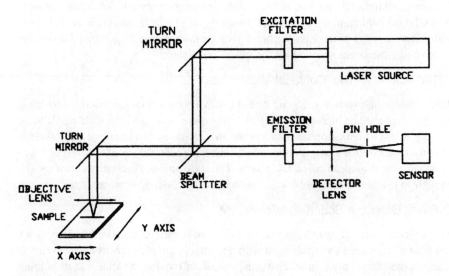

Figure 28 Moving stage architecture.

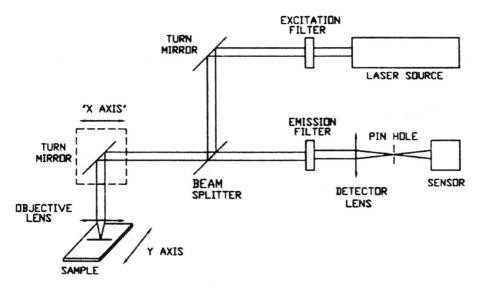

Figure 29 Rectilinear flying objective architecture.

objective can be driven by a linear motor, a voice coil, a stepper motor, but again, optical scanners offer the fastest drive mechanism. The scanning system offered by the Virtek Instrument oscillates a lens using a scanner to drive a stage in a similar construction.

5.2.5 Rotary Flying Objective Microscope

An alternative method to creat a rapidly scanning beam of light is the rotary architecture described by Overbeck and shown in Fig. 30.[46] It offers both speed and a constant optical path length. A periscope couples the scanning objective to the stationary elements of the microscope and thus keeps the light path length constant as it oscillates to scan the light beam. As the arc covering the width of the slide is in polar coordinates, the resultant image is instantly converted to Cartesian coordinates by the instrument's computer, so the image is then directly correlated with the sample spot pattern. A micro, aspheric, one-element

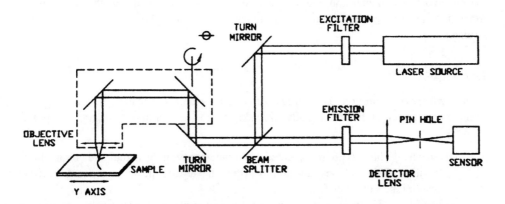

Figure 30 Rotary flying objective architecture.

lens is carried on a counterbalanced arm that extends perpendicular to the shaft of an optical scanner. The arm also carries mirrors that receive laser light on the axis of rotation via a periscope, send it out the length of the arm, and down through the lens. The optical axis is always normal to the biochip.

6 CONCLUSIONS

This chapter has described the current state of the art of optical scanning galvanometers as well as recent development of resonant scanners. These devices have developed over the past 40 years as part of the general electro-optic revolution.

All components of the electro-optic industry have witnessed a drastic price reduction in conjunction with a dramatic increase in demand. Galvanometric and resonant scanners may be a rare exception to that pattern of evolution. Prices have grown with demand.

The elements used to construct these oscillating devices are the same as those found in the head movers of every computer hard drive or in CD players or high-quality motors that can be purchased for a few hundred dollars.

High-volume applications have stimulated the development of other technologies: rotating polygones, MEMS, LIC scanners, and so on.

Presently the total worldwide annual sales of oscillating scanners lies between 20 and 30 million dollars, with an additional 15 or 20 millions dollars of annual sales of subsystems integrating scanners with their driver amplifiers and some optical elements.

ACKNOWLEDGMENTS

The author wishes to acknowledge the contributions made by Dr Jim Overbeck and Dr Miles Mace to the design of flying objective confocal scanning microscopes. The author also wishes to thanks Herman DeWeerd at GalvoScan for his thorough and painstaking review of this manuscript and the many discussions, suggestions, and corrections. The author would like to take this opportunity recognize the exceptional contributions that Dean/Valerie Paulsen made to the concepts and design of resonant scanners during the many years in residence at General Scanning Inc. It should be noted that the long list of patents under his name do not properly assess his contribution. Finally, the author would like to express his appreciation to Bruce Rohr, the founder of Cambridge Technology, for the review of this chapter and take the opportunity to remember him as the father of all modern "capacitive position transducers" found in current high-performance oscillating scanners.

GLOSSARY

This short list of definitions is offered so that it may help the reader. It is a subset of terminology recorded by Alan Ludwiszewski,[43] and is commonly accepted in this industry.

Accuracy. The maximum expected difference between the actual and commanded position. This includes any nonlinearities, hysteresis, noise, encountered drifts, resolution, and other factors.

Bandwidth. The maximum frequency for which a system can track a sinusoidal input with an output attenuated to no less than 0.7 (-3 dB point) of the command. For open loop frequency responses with a phase margin of 90°, the open loop crossover frequency will equal the closed loop bandwidth. For other phase margins the relationship is not as straightforward. A complete treatment of these relationships can be found in any control theory text.

Drift, Mechanical Null. The drift of the steady-state position of an unpowered scanner when the armature is restrained with a torsion bar. This drift can occur with time and with temperature. Drift is usually specified in terms of the change in optical angle per amount of correlated influence, such as time or temperature.

Drift, Position Detector. The change in relationship between the output of the position detector and the position of the output shaft. It consists of the sum of the gain drift, null drift, and others.

Drift, Position Detector Gain. The change in scale factor of the position detector. Since the absolute magnitude of this change is dependent upon angle, it is specified in terms of the ratio of the change in output over the output per unit time or temperature (that is, ppm/°C or %/1000 hours). This takes into account that the effect is seen most at extreme angles where the output is greatest.

Drift, Position Detector Null. The drift of the electrical zero of the position detector with time and temperature. Drift that occurs with temperature change is specified in units of angle per degree temperature (that is, urad/°C). Drift that occurs with time is specified in units of angle per units of time (that is, urad/1000 hours).

Drift, Uncorrelated. Drift that cannot be attributed to a change in a particular external condition, such as time or temperature. Often caused by mechanical ratchetting or noncatastrophic damage to the system due to overstress.

Jitter. Nonrepeatable position error fluctuations caused by velocity perturbation in a scanner. Generally described in units of optical scan angle and often expressed as the standard deviation of the maximum jitter error observed in each scan line of a large number of consecutive scans. Some applications may require specification of the frequency as well as the magnitude of acceptable jitter.

Nonlinearity, Best Fit Straight Line. This method of quantifying nonlinearity involves finding a first-order linear function that is the closest approximation to the measured data. The nonlinearity is then calculated as the maximum observed deviation from this line. This will result in the smallest measurement of nonlinearity.

Nonlinearity, Pinned Center. Pinned center nonlinearity uses a straight line that intersects a given datum point, such as the mechanical or electrical null of a scanner, and has a slope that best approximates that of the measured data. Sensor nonlinearity is then calculated from this reference.

Null, Electrical. The zero output point of the position transducer.

Null, Mechanical. The steady-state position of an unpowered scanner. This position is determined by the torsion spring, if any, and the magnetic spring of the scanner. In many scanners without a torsion spring, the magnetic spring is not of great enough magnitude to overcome frictional forces and make this an absolute position.

Repeatability. The inaccuracies in final position encountered while implementing a series of identical command inputs.

Repeatability, Bidirectional. The inaccuracies in final position encountered while attempting to return to a position from different directions.

Resolution. Resolution is the ability to discern individual spots in the target field. This is not to be confused with accuracy, which includes gain and offset drift, noise, resolution, and other factors. Dependant upon system design, the limit to resolution may be due to optical considerations, digital resolution, or position detector signal-to-noise ratio and drift.

Resolution, Optical. The optical resolution of a scanning system can be described as the number of separately resolvable spots that can be produced. For diffraction limited optics, this is dependant upon the aperture width in the scan direction, the aperture shape factor, the wavelength of the source, and the total scan angle. These factors are related through the scan equation, which can be found in may optical texts.

Resolution, Scanner. Scanner resolution is limited by the noise and drift of the position detector. The RMS signal-to-noise ratio will determine the statistical resolvability of a given level of command in a given frequency range. Filtering can improve low-frequency resolution, but drift factor will also come into effect.

Resonance, Cross-Axis. Structural resonances that cause motion perpendicular to the scan axis are referred to as cross-axis resonances. These resonances may be accentuated by poor mirror design and will cause periodic wobble, possible system instabilities, and may be a limit to achievable system bandwidth.

Ideal distribution of pixels along a scan line

Distribution of pixels with jitter error

Jitter: Error from average pixel position, measured in the scan direction.

Distribution of pixels with wobble error

Wobble: Error from average pixel position, measured in a direction perpendicular to the scan direction.

Resonance, Torsional. An on-axis resonance that appears in scanners due to the distributed masses on a compliant rotor shaft the flexible coil of a d'Arsonval system. These resonances can appear as periodic jitter and may cause controllability difficulties due to the resonant peak created in the scanners' transfer function. Mirror design and mounting will have a significant effect on torsional resonance.

Response Time. The response time of a scanning system is defined as the tracking error divided by slew rate. Owing to the characteristics of the controller, stage saturation, aerodynamics, and other nonlinearities, response time will not necessarily be a constant. Although not a constant, at least for vector tuned controllers it is approximately so, if slew rates are neither driving the tracking error to near zero nor approaching the maximum slew rate. Response time is an indication of the relative speed of a scanner and the ultimate performance obtainable with a given load and tuning.

REFERENCES

1. Marshall, G.F. *Optical Scanning*; Marcel Dekker, Inc.: New York, 1991.
2. Marshall, G.F. *Laser Beam Scanning*; Marcel Dekker, Inc.: New York, 1985.
3. *Permanent Magnet Data Sheet*; UGIMAG, Ugimag 45M2 material.
4. Roters, H.C. *Electromagnetic Devices*; John Wiley & Sons, Inc, 1955.
5. Keller. U.S. patent nos. 985,420 and 1,041,293.
6. Hodges. U.S. patent no. 3,348,183.
7. Houtman, J.A. U.S. patent no. 3,528,171.
8. Rohr, B. U.S. patent no. 4,864,295.
9. Proc. SPIE 1991, *1454*, 265–271.
10. Dillon, R.; Trepanier, P. U.S. patent no. 6,000,030.
11. Montagu, J.; Honkanen, P.; Weiner, N. U.S. patent no. 6,218,803.
12. Montagu, J.I. U.S. patent no. 5,235,180.
13. Ivers, R. U.S. patent no. 5,844,673.
14. Ivers, R. U.S. patent no. 5,671,043.
15. Abbe, R. U.S. patent no. 3,990,005.
16. Rohr, B. U.S. patent no. 41,864,295 and 4,142,144.
17. Stokes, B. U.S. patent no. 5,099,368.
18. Dowd, R. U.S. patent no. 5,537,109.
19. Wittrick, W.H. The Aeronautical Quaterly 1951, *II*.
20. Siddall, G.J. *The Design and Performance of Flexure Pivots for Instruments*; University of Aberdeen, Sept. 1970; M.Sc. Thesis.
21. Reiss, R.S. OE Reports, May 1989.
22. Brosens, P.J.; Vudler, V. Opt. Eng. 1989, *28*, 61–65.
23. Timoshenko, S.; Goodier, J.N. *Theory of Elasticity*; McGraw-Hill: New York, 1951.
24. Weissman, H. Opt. Eng. 1976, *5*, 435–441.
25. Yoder, P. *Opto-Mechanical System Design*; Marcel Dekker: New York, 1986; 71–77.
26. Smith, W.J. *Modern Optical Engineering*; McGraw-Hill: New York, 1966; 132–133.
27. Lawler, A.; Shepherd, J. *Laser Beam Scanning*; Marshall, G.F., Ed.; Marcel Dekker, Inc.: New York, 1985; 125–147.
28. Marshall, G.F. *Optical Scanning*; Marcel Dekker, Inc.: New York, 1991; 560 pp.
29. Den Hartog, J.P. *Mechanical Vibrations*; McGraw-Hill: New York, 1956; 396 pp.
30. Paulsen, D.R. U.S. patent no. 4,878,721.
31. Paulsen, D.R. U.S. patent no. 4,919,500.

32. Paulsen, D.R. U.S. patent no. 4,990,808.
33. J. Laser & Optronics 1998, *17* (2).
34. Proc. SPIE 1987, *817*.
35. Montagu, J.I. Electro Optics 1983, *May*, 51–56.
36. Montagu, J.I. U.S. patent no. 4,502,752, 1985.
37. Montagu, J.I.; Pelsuel, K. U.S. patent no. 4,525,030.
38. Minsky, M. Microscopy Apparatus. U.S. Patent no. 3,013,467.
39. Pawley, J.; Ed. *Handbook of Biological Confocal Microscopy*, 2nd Ed.; Plenum, 1995.
40. Roggatto, W.D., Ed., IR/EO Systems Handbook. SPIE Press, 1993.
41. Proc. SPIE 1991, *1454*, 185–195.
42. Huerton, I.; Van Gelder, E. High Speed Fluorescence Scanner. U.S. patent no. 5,459,325, 1995.
43. Proc. SPIE 1991, *1454*, 174–185.
44. Hueton, I. High Speed Florescent Scanner. U.S. patent no. 5,459,325.
45. Montagu, J. Positioner for Optical Elements. U.S. patent no. 4,525,030.
46. Overbeck, J. U.S. patent no. 6,335,824.

9

Flexure Pivots for Oscillatory Scanners

DAVID C. BROWN

GSI Lumonics, Inc., Billerica, Massachusetts, U.S.A.

1 INTRODUCTION

The memory is the knapsack of the mind. When setting out on a journey, particularly one that promises to be long and challenging, it is better to fill it with maps and recipes, fishhooks and twine, matches and a few candles rather than tinned goods. These things are more useful, and are sure to last far longer. For those of you who cannot resist the quick fix, we hope that there are a few chocolate bars scattered about as well.

Flexures are quite ancient, and their use as pivots is also ancient. Long before the use of the most primitive bearings, leather strap flexures were used as trunk lid hinges and the like. Early war engines, for example the ballista of the Romans, the limbs of technically advanced hand bows, such as those attributed to the Turks, and the crossbows of the 14th century all employ flexures as their enabling technology.

Many pendulum clocks suspend their pendula by means of a flexure. The mechanical metronome, a specialized inverted form of clock, relies heavily on the flexure suspension in its design. It is at least arguable that the tuning fork is a pair of coupled flexures, and the music box comb is a set of flexures tuned to self-resonate at desirable frequencies.

All of these examples, and of course there are many others, exploit the flexure pivot because of its simplicity, reliability, lack of internal clearance, long service life, ease of construction, and often, its high mechanical "Q". The flexure pivots used in scientific instruments, including optical and laser scanning equipment, exploit these very same attributes.

Given the great attention devoted to flexure pivots during the 1960s and 1970s as a result of the need for rugged, reliable, light-weight, unlubricated pivots and bearings for space exploration applications, and the fact that the best and brightest were naturally attracted to what was undoubtedly the biggest science of its time, so much progress was made in flexure pivots and suspensions that one might be tempted, like Charles H. Duell, Commissioner of U.S. Office of Patents, who urged President William McKinley to abolish his office in 1899, to conclude that everything useful had already been invented.

However, as the second section of this chapter suggests, flexure technology may yet again provide one of the enabling elements in mankind's next leap of technological advancement, universal connectivity by means of light. Of course, like the flexure itself, this is an old idea, widely practised on a small scale for many centuries. The Romans communicated between the watch towers along the borders of their far-flung empire in this way, and the indigenous people of North and South America did so as well.

The organization and content of this chapter are intended to reflect the contrast between the rather settled state of macro-scale flexure pivots, and the as yet entirely unexplored space of the microelectromechanical flexure scanners (MEMs) flexure pivot. In the case of the former, the primary author has chosen to dwell largely on the details of manufacturability, which are the fruit gained from many years spent extracting the stones from the fertile soil of the garden of flexures – the practicum, as it were, of the craft. This is the sort of information that is not generally gained through textbooks, and, in the end, differentiates a possible mechanism from a practicable, not to say elegant one. In the case of the latter, the primary author has endeavored to present the theoretical underpinnings of integrated benders, as well as the fabrication and characterization of high-speed, large-angle optical scanners constructed with them.

1.1 Introduction to Macro-Scale Flexure Pivots

First, let us begin by defining what we shall mean by the term "pivot." The Chambers 20th Century Dictionary defines a pivot as "A pin on which anything turns." We must begin by modifying the term "pin," because flexure pivots sometimes have virtual pins or axles. A good working definition of pivot in this context might be "A device which defines a virtual axis of rotation, over a limited angle, while fixing the other five degrees of freedom. A device which is able to resist moments except in one axis of rotation, and may be able to resist relatively small moments in that as well." In general, these pivots are of interest in instrument-quality or scientific applications, as opposed to things like lorry fifth wheel applications, where they would work, but would have no particular advantages over the more common commodity sorts of bearings, such as sleeve, ball, roller, and the like.

What then are the attributes that equip flexure pivots for instrument and scientific applications? In no particular order, they are:

- low mass;
- zero clearance;
- intrinsic restoring force;
- infinite life;
- no lubrication;
- low hysteresis;
- no viscous losses;
- no friction (at least none in the usual sense);
- no particulate generation;
- no outgassing;
- no intrinsic temperature limit;
- extremely high load capability;

Flexure Pivots for Oscillatory Scanners

- extremely high force linearity over small angles;
- flexibility in design configuration, in the sense that the pivot members may be pierced to permit transparency, or even to permit the passage of solid mechanical parts or objects (in contrast with conventional bearings, which are pretty much impenetrable three-dimensional solids);
- low design and tooling cost;
- predictability;
- the ability to withstand high shock and vibration loading;
- short lead time; and
- others for very specific applications.

The choice of material for a flexure is driven by conflicting requirements. For example, it is often the case that the precision of location of the pivot point is secondary to some other requirement, such as long life, low operating force, or high strength, or all the above. It may be that linearity of operating force over the desired angle of motion is also unimportant. There are often operating environment considerations. It may be that easy replacement of a damaged pivot is required. The choice of leather for trunk lid flexures results from meeting requirements such as these.

On the other hand, the designer of a bow limb has the opposite requirement that the operating force be large. Because, however, he must deal with stresses in one direction only, unlike the trunk lid flexure designer, he will exploit that difference to his advantage, and construct a flexure that is a composite. The front or tension side of his limb will be constructed from horn, which is quiet stiff in tension, while the belly, or compression side, will be constructed of sinew, which is stiff in compression. He will rightly conclude that the respective stresses in this structure are controlled by their separation from the neutral plane, and will provide a "filler" of wood properly shaped to maximize the stresses in the working "skins" of his bow without precipitating failure.

This pair of examples is designed to illustrate the broad range of potential applications of flexure pivots, and in particular the degree to which clever design is able to cope with many-order-of-magnitude differences in an application-specific parameter. Leather is able to store only a few tenths of a Joule per kilogram, whereas the best Turkish bows store over 750 Joules per kilogram, eclipsing steel and rivalling the best of the modern composites. The attribute of energy storage, related to the density, strength, and Young's modulus of the materials, may be of critical importance, required to be very high or very low, or of no importance at all in a particular application.

The absence of mechanical noise is a benefit of flexure pivots as well. Flexures are inherently low-noise pivots. Because they have no loose parts, it is unnecessary to preload them to remove clearances. As they are monolithic structures, they display very little "noise" in their force versus displacement curves. Other pivot types, comprised as they are of loose parts, make some mechanical noise during operation, and as they wear and the clearances increase, the noise level rises. Any particulates released during the wear process, of course, can cause sudden changes in the position and parasitic torque of the pivot, as well as mechanical noise, if they jam between the moving parts. Bearing "rumble" is widely recognized and even quantified as an inherent characteristic of ball bearings. In many cases, the presence of this bearing noise places an upper limit on the allowable noise bandwidth of any servo system associated with the pivot.

The possibility of a pivot design with inherently low mechanical loss is an area where flexure pivots shine. All other types of pivots known are lossy. Losses resulting from lubricant viscous friction and ball and race deformation energy not only place upper limits on the speeds at which these devices can operate, but also consume energy, limiting the mechanical efficiency of all known moving-element bearings to a low value.

Flexures, on the other hand, have no lubrication requirement, and are limited only by the internal energy of deformation. Mechanical "Q" above 5000 is regularly achieved, even in very high speed designs. This attribute of flexures makes possible resonant scanners operating above 10 kHz at very low power levels. Of course, the absence of lubrication is itself a benefit in applications in optical instrumentation, spectroscopy, space research, medicine, and semiconductor processing, where even minute contamination levels are an issue.

Flexures work extremely well for small-angle applications, where stiction, traction, surface finish limitations, mechanical tolerances, lubrication distribution requirements, and so forth put demands on other pivot types, which they cannot meet successfully. Flexures are, of course, free of any looseness or play, so there is no "backlash" whatsoever associated with their use; moreover, because they rely on molecular stretching, their intrinsic hysteresis over small angles near the neutral position is always less than the unbalanced forces resulting from unavoidable asymmetries in their mountings (which produce some hysteresis in realizable designs; hysteresis levels of less than 0.1% are difficult to achieve, but possible). On the other hand, unlike other pivot types, flexures are not capable of continuous rotation. While several hundred degrees of rotation seem plausible, the authors know of no application of flexure pivots designed to operate over 90°, and most designs are for much smaller angles.

Other precision pivot types, such as ball bearings and jewel ("watch") bearings are limited in their geometrical precision by the limits of accuracy that their respective manufacturing processes impose on them. For example, the best class of bearing balls, class 3, has an allowed out-of-roundness of 3×10^{-6} inches. If averaged in a bearing with nine balls, the best error in "wobble" of the whole assembly will be about 9×10^{-2} less, or about 1 micro-inch. If two such bearings 1 inch apart support an axle, then the axle will have a wobble of about 2 microradians, not including errors in the concentricity of the bearing rings, axle to ring mounting, and so on. This kind of error in pivots, that is, error that is associated with geometrical features on the parts, tends to be coherent with the motion of the pivot, and so is periodic. As a result, these errors give rise, in raster scanning systems in particular, to undesirable moiré patterns. Moirés are vastly more noticeable by human observers than are nonperiodic or random errors, and can sometimes lead to rejection of a scanner that actually meets its design requirements for wobble. This effect is particularly noticeable in systems designed for printing applications, where the eye is sensitive to moirés generated by periodic angular errors in the 1/10 microradian range. Flexure pivots have none of these sources of error. Production printing engines using flexure scanners in a raster mode built at the authors' factory regularly achieve a level of periodic error below 1/10 microradian, and were developed specifically to solve the moiré problem.

There is a tendency for conventional pivot bearings to become more fragile as the level of precision required of them is increased. This is not surprising, because the accuracy of the pivot is established by careful dimensional and geometrical control of several loose parts. The single 1 microinch dent in a bearing ring, which is undetectable in a 10 microradian application, becomes the limiting factor in a 1 microradian application. Flexures, on the other hand, are extremely robust, impervious to dust and other mechanical and most chemical contamination, and are very insensitive to shock and vibration. In most cases, a flexure pivot will survive indefinitely in a factory-floor or outdoor application in which other pivot types have very limited service life, if they operate at all.

Lastly, flexure pivots have inherently infinite life if operated at a peak stress level below the fatigue limit for the material from which they are made. Some care is required in eliminating any sort of stress riser in the design, and also in determining the effective fatigue limit for the material. GSI Lumonics (Billerica, MA) regularly warrants its flexure products for five years of operation in any environment, at any duty cycle, with no statistically significant failure rate. For an 8 kHz scanner operated continuously, this amounts to more than 10^{12} cycles.

Of course, flexure pivots are not perfect. In general, the transverse stiffness of the flexure pivot is inferior to that achievable with ball bearings, resulting in unintended cross-axis motions in the presence of significant environmental stimuli, or in gyroscopic reaction to very large accelerations of the axle. It is also more difficult to achieve very large scanning angles with flexure pivots, for the reason that the curl of the flexures at extreme angles reduces even further their transverse stiffness. The allowable stress limit at the extreme angle legislates in favor of quite thin flexures, further reducing their stiffness. While flexure scanners with scan angles up to 80° optical have been produced, their tolerance of environmental vibration and axle acceleration is low because of these effects, particularly when equipped with the very thin flexures required to obtain extremely long lifetimes.

Flexure pivots generally are designed to constrain the minimum number of degrees of freedom required by the application. Often, for example, a translation of the axle parallel to its axis is allowed, although even when the translation is not obnoxious optically, it has the potential to set up undesirable vibrations, and, in the limit, catastrophic mechanical positive feedback. With crossed-flexure pivots, it is possible to construct a geometry with zero axle shift, and practical to expect such a design to produce translation of the axis of less than a few microns over a small angle of rotation.

2 FLEXURE DESIGN

The possible combinations and permutations of flexures in a design approach for an arbitrary application is so vast that no attempt will be made here to cover any but the simplest form of flexure, the single leg. In the case of symmetrical designs, the easiest approach is to multiply the width of a single leg by the total number of flexures to model the flexure as a unit. In the case of asymmetric designs, other approaches to fit the circumstances will require invention. The formulas below assume fully reversed stress; that is, that the flexure is bent symmetrically through the same angle on both sides of the neutral position. As in the case of the Turkish bow, unreversed, or partially reversed stress provides an opportunity for clever design and perhaps nonmonolithic flexure materials. In fact, the definition of "fatigue limit" as it is used here, "That stress which represents the maximum stress which will allow the flexure to operate indefinitely without failure under a particular set of circumstances," is a hotly contested topic, and those who believe they have a suitable answer guard their secret with the most intense passion. The case of the partially reversed stress is an even more volatile subject. Obviously one of the corner stones of flexure design is an understanding of the safe upper limit of flexure stress. In the absence of reliable published data, and for the purpose of designing for a standard laboratory environment, it is the considered opinion of the authors that, for ferrous material, a value of 35% of the ultimate strength is safe under dry operating conditions.

2.1 Useful Formulas

The primary parameters of interest to a flexure pivot designer include the rotational spring rate of the pivot, the maximum scan angle achievable under some maximum stress level, the maximum stress applied to a flexure bent through some scan angle, the allowable thickness for a flexure whose fatigue limit, length, Young's modulus, and angle of operation are known, and the first resonant frequency of the axle–pivot assembly.

It will be obvious that these formulas are a first-order approximation. It has been found, however, that the errors produced are small enough so that minor variations in the density of the materials used, the actual effective length and thickness of the flexures, and so on, dominate the result, and any application that requires very precise control of a particular parameter, such as the frequency of a mechanical resonator, will require some "tweaking" mechanism for final tuning on an individual basis.

These formulas also assume that the flexure mountings are infinitely rigid, support the flexure completely and uniformly, introduce no stress risers, and allow no relative motion. Except for the rigidity, these conditions are usually met satisfactorily by careful design. The first-time designer would be well advised to do a careful finite element analysis (FEA) of the mounting rigidity, or allow for an iteration or two in the design schedule. The formulas are given as follows.

Rotational spring rate

$$K = EWT * 3/12L$$

Peak mechanical angle in radians

$$A = \frac{2LS}{ET}$$

Maximum stress

$$S = \frac{ETA}{2L}$$

Thickness

$$T = \frac{2LS}{EA}$$

Resonant frequency

$$F = 1/2\pi(K/J) * 1/2$$

where E is the Young's modulus of the flexure material, W is the width of the flexure, T is the thickness of the flexure, L is the effective active length of the flexure, A is the peak scan angle of the flexure in radians, S is the peak stress in the flexure, F is the first rotational resonant frequency of the flexure/axle system, and J is the combined moment of inertia of the rotating system. Of course, use of consistent units is required.

Flexure Pivots for Oscillatory Scanners

There is an extensive bibliography on the subject of detailed design of flexure pivots, of which several are listed in the reference section.

2.2 Flexure Materials

It will be noticed immediately from inspection of the relationships among the physical constants pertinent to flexure materials that the stress is directly proportional to Young's modulus. The allowable stress is proportional to the fatigue limit of the flexure material, which most workers agree is itself proportional to the ultimate strength of the material. (Since it is now generally agreed that the so-called proportional limit, or yield strength, is not a useful concept, we shall avoid using it here.) The natural conclusion from this train of reasoning is that one could rank flexure materials by a figure of merit that is the result of dividing the material's fatigue limit by its Young's modulus. Because the exact value of the fatigue limit is unknown, or at least unpublished for the materials of interest, we shall use the ultimate strength instead, since the two are believed to be directly proportional.

Leaving for those who follow us those exotic materials whose currently available form or lack of observable malleability or other defect make their use questionable, we can construct Table 1. One should, therefore, try to get material rolled as thoroughly as possible, so that the rise in modulus during use will be minimized.

There are some applications in which inertia, or minimizing it, are paramount. In this case, specific strength, rather than ultimate strength, can be used in the construction of a figure of merit list, which might look like Table 2. One should, therefore, try to get material rolled as thoroughly as possible, so that the rise in modulus during use will be minimized.

As one would expect, aluminum and titanium have moved up the list, but little else has changed. One could, of course, construct other lists for other specific purposes, such as

Table 1 Parameters for Various Materials, Including Ultimate Strength

Material	Young's modulus (psi)	Ultimate strength (psi)	Ratio
Carbon/graphite	$<2 \times 10^6$	375×10^3	0.19
Diamond	150×10^6	7.69×10^6	0.051
Glass reinf. epoxy	5×10^6	240×10^3	0.05[a]
Silicon	27.5×10^6	1.02×10^6	0.037
Be/Au	15×10^6	210×10^3	0.014
Spring gold	15×10^6	180×10^3	0.012
Be/Cu	18×10^6	180×10^3	0.01[b]
Ti^{-6}Al^{-4}V	19×10^6	205×10^3	0.01
7075 Al	10×10^6	98×10^3	0.009
Udeholm 718	30×10^6	265×10^3	0.009
17-7PH	30×10^6	235×10^3	0.008
Inconel	31×10^6	250×10^3	0.008
302SS	28×10^6	200×10^3	0.007[c]

[a] This material's lack of electrical conductivity may limit its usefulness.
[b] Be/Cu has published figures of fatigue limit of 40 ksi.[1]
[c] Interestingly, this material not only work hardens, but its Young's modulus rises with work as well. The range of moduli published are $24-28 \times 10^6$ psi.[1]

Table 2 Parameters for Various Materials, Including Specific Strength

Material	Young's modulus (psi)	Specific strength (psi)	Ratio
Carbon/graphite	$<2 \times 10^6$	3509×10^3	1.75
Glass reinf. epoxy	5×10^6	3200×10^3	0.64[a]
Silicon	27.5×10^6	12300×10^3	0.448
Diamond	150×10^6	61000×10^6	0.407
7075 Al	10×10^6	961×10^3	0.10
Ti^{-6}Al^{-4}V	18.5×10^6	1120×10^3	0.06
Be/Cu	18×10^6	557×10^3	0.03[b]
Udeholm 718	30×10^6	936×10^3	0.03
17-7PH	30×10^6	830×10^3	0.03
Be/Au	15×10^6	301×10^3	0.02
Spring gold	15×10^6	301×10^3	0.02
Inconel	31×10^6	749×10^3	0.02
302SS	28×10^6	697×10^3	0.02[c]

[a] This material's lack of electrical conductivity may limit its usefulness.
[b] Be/Cu has published figures of fatigue limit of 40 ksi.[1]
[c] Interestingly, this material not only work hardens, but its Young's modulus rises with work as well. The range of moduli published are $24-28 \times 10^6$ psi.[1]

minimization of weight. Dividing specific strength by specific stiffness would produce such a list.

However, there are two caveats. First, most engineering enterprises are constrained by commercial goals, and so the use of exotic materials is quite properly frowned upon unless there is robust technical or other justification for their use. To the knowledge of the authors, precious metal alloys such as BeAu and spring gold are used almost exclusively in the manufacture of very high quality fountain pen nibs. In this application, they are justified because the unique combination of fatigue resistance, corrosion resistance, wear resistance, appearance, and "feel" enhance the marketability of the end product. Secondly, the fact is that very little is actually known about the long-term fatigue resistance of nonferrous alloys. This is partly the result of the fact that ferrous alloys have been in use for a very long time, and their overall combination of characteristics have recommended them for so many demanding applications that they are both well studied and in ready supply. One could argue that copper alloys have been in use longer, and while that is true, these alloys were generally eclipsed by their ferrous brethren as soon as they became available. The stainless steels have replaced bronzes where corrosion resistance is important, and BeCu and phosphor bronze for springs except for applications where magnetic susceptibility or electrical or thermal conductivity are issues.

In fact, it seems from the published literature on the subject that nonferrous metals have disappointingly low fatigue limits.[1] Experimental work at GSI Lumonics with BeCu flexure materials reinforces this view.

However, the primary reason for their use is probably simply that ferrous materials make very good flexures, and except for very unusual circumstances, the apparent technical superiority of the "other" material, if actually realizable, would require the investment of such large development or other resources that return on the investment might well be negative. As a case in point, GSI Lumonics has abandoned the "stress-proof" steels, number 8 on the list, in favor of 302SS, which is last in both Tables 1 and 2. While the manufacturer of air compressors has available to him thicknesses of material appropriate to his flapper-valve needs, no one stocks the thin foils needed for flexures. This material cannot be re-rolled, and so a new thickness requirement calls for a mill run minimum at a cost of $25,000 or more and two years wait, and which yields a supply of material which is vastly in excess of the needs of any flexure product lifetime yet known, and must be stored carefully, inventoried, and so forth for many, many years against possible future demand, or else expensed and discarded.

Of course, not all pivots are required to last indefinitely. It is likely that an exploratory scientific mission to Mars, for example, might require only a few hundred thousand actuations. In circumstances such as these, since low weight and high reliability are much more important than life, it is probable that resources can be saved by designing a limited-life flexure. It is easy to think of other applications in which this may be true, and one of the authors (Brown) designed and flew successfully flexure systems required to operate only a few tens of cycles during atmospheric testing of atomic weapons.

Unfortunately, it appears that the borderline between indefinite life and finite life is clifflike. It is fairly easy to predict accurately life VS stress for small to moderate cycle lives, because testing to validate the model is practical. As an example, the Bendix Corporation (Utica, NY) published graphs[2] showing life VS cycles for their standard flex pivots under various conditions of loading. These graphs show load lines at 35,000 cycles, 220,000 cycles, and indefinite life. These curves cover a factor of 2 in angle, and 0–100% loading. Interpolation is possible, but as one approaches indefinite life, small errors in actual loading and angle can cause one to fall off the cliff, with disastrous results. Testing a device to demonstrate "indefinite" life takes a very long time, and various forms of accelerated testing should be approached with caution. That said, many manufacturers of flexures have developed accelerated testing methodology. They, like GSI Lumonics, protect these methods as trade secrets. A rule of thumb, in keeping with the most widely accepted theory of fatigue failure, is that a flexure design that passes 3×10^7 cycles without failure is destined to operate indefinitely in a dry environment at this or lower level of stress.

As the next section discusses, there are some hidden requirements for the qualities of the raw material form that should be understood before extravagant expense is lavished on the development of a material whose soundness of internal structure and finishability are questionable. That said, whenever an exotic, hard-to-work, expensive, material has significant technical justification, then it is worth exploring. The example of the folks who succeeded in bringing diamond phonograph styli to the mass market should be firmly held in mind.

2.3 Stress Risers

If the failures of flexures were ranked by cause, it would be found that, except for instances that resulted from exceeding the design stress, all failures resulted from either stress risers or corrosion or a combination. The next section deals with the germane corrosion issues. Here we will deal with the problem of stress risers, which generally arise from poor finish on the surface of the flexure foil, nicks in the edges of the flexure as manufactured, mounting defects, or inclusions in the flexure material itself, in about that order of importance. The sections on flexure manufacturing and flexure mounting deal in some detail with the avoidance of stress risers resulting respectively from manufacturing and from mounting.

A stress riser is anything that raises the stress in the flexure, either locally or globally, above the value that the designer intended that it have at that point. Because most flexures of interest to this discussion are made from rolled material, it is important to note that such materials are not really isotropic in their mechanical characteristics, the "grain" having been elongated more in one direction (the direction perpendicular to the axis of the rolls) than the other; in other words, the grain is longer in the length direction than it is in the width direction. As a result, the fatigue resistance of the material is anisotropic as well, being higher across the grain than it is parallel to the grain. It is therefore necessary to specify the direction of the grain when flexures are produced from the sheet of foil, and if the as-rolled coil is cut up into smaller sheets, the direction of rolling must be marked on each sheet in order to preserve the grain direction information. Since fatigue resistance is paramount in flexures, it is usual to specify that the flexure be cut out of the sheet in such a way as to cause the bending in operation to be across the grain.

The stress risers most prevalent in oriented flexures are the result of scratches in the sheet surfaces. The best quality rolled finish obtainable is none too good for flexure material, particularly when the flexure is thin. For example, a flexure 0.001 in. thick, made from material with a standard mill finish of 32 microinches roughness average (RA), will have scratches on both sides that are up to 0.0001 in. deep. If two of them lie on top of each other, then the local thickness of the flexure is 0.0008 in.

Because the stress is proportional to the cube of the thickness, the local stress will exceed the designer's intent by 95%, and is likely to cause failure. Of course, an inclusion of microscopic size will produce the same effect.

GSI Lumonics specifies 4RA or better finish for its rolled flexure strip. For critical applications, it is well to have the rolled sheet 100% x-rayed to find any inclusions or cracks or voids or other defects inside the material.

The last important item is not associated with the quality of the raw material, but is one of the most easily overlooked elements of a stress riser free design. Nearly every flexure has one or more radical changes in width, usually associated with attachment. It is absolutely critical that these section changes are made with a stress-distributing fillet or radius in the corner. Otherwise, failure cracks will inevitably originate at these locations.

Figure 1 is a drawing of one of the flexure assemblies in current production at GSI Lumonics. This part is actually two flexures joined together by a bar (the horizontal part along datum "B"). Making a pair of flexures this way ensures the utmost in symmetry, since they are made at the same time from nearby regions of the same piece of metal. Also, the "self-fixturing" feature of the bar eliminates the necessity for adjusting their parallelism and individual effective lengths during assembly.

Attention should be paid to the specification legend in the upper left-hand corner, as well as to the grain-defining arrows and the corner radii at every section change.

2.4 Corrosion

It is assumed here that the flexure designer will take into account the environment in which the equipment will live, and take suitable precautions against atmospheric or environment corrosion. We will discuss here only the forms of corrosion that result from the highly stressed conditions of the flexures during use, the effects of galvanic couples, and the hydrogen embrittlement associated with electroplating.

Flexure Pivots for Oscillatory Scanners

NOTES:
1. MATERIAL: STAINLESS STEEL, TYPE 301 OR 302, FULL HARD CONDITION. SURFACE QUALITY 4RA (RAHNS 2B, PHOTOETCH QUALITY).
2. FINISH: PASSIVATE, PER QQ-P-35.
3. PHOTOETCH ATTACHMENT POINTS ALONG DATUM B ONLY.

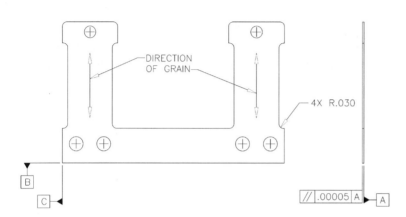

Figure 1 Typical flexure.

Corrosion in flexures is intimately connected with stress risers, because most of the corrosion effects encountered begin in cracks and crevices either caused by stress cracking or enhancing to stress cracking. However, there is one area of corrosion difficulty encountered that is not directly associated with stress, but is instead a form of electrolytic corrosion resulting from the contact between metals far enough apart on the electromotive series. This effect is not significant in very dry conditions, but the presence of an electrolyte can cause rapid erosion of the anodic member of the couple. Even atmospheric water held in cracks by capillary action can be a powerful electrolyte under conditions of stress. It is up to the designer to decide which member of the couple he would prefer to be dissolved; in general, it is preferable to protect both to the extent possible. Theoretic galvanic couples are presented in Table 3.

In general, MIL-STD-186 allows adjacent groups, or, in some cases, metals two groups away from each other to be coupled. This does not mean that galvanic action does not take place. It just means that for most purposes the rate of corrosion is so slow as to be unimportant.

At GSI Lumonics, standard product joins group 5 flexures with group 14 mounting materials using group 13 fasteners with satisfactory results. Because the aluminum is quite anodic to the flexures, it is slowly dissolved, but the more sensitive flexures are unaffected. For demanding applications, all the ferrous parts are tin-plated before assembly, and the aluminum is anodized.

In this context, it is useful to discuss a side effect of electroplating of high-strength steel, the so-called hydrogen embrittlement. During electroplating, the electrolyte, if aqueous, contains hydrogen ions, which are hydrogen atoms stripped of their electrons. These very small ions are able to squeeze into the grain boundary lattice structure of the

Table 3 Theoretical Galvanic Couples

Group	Material	EMF with respect to Calomel in sea water
1	Gold, platinum	+0.15 V most cathodic
2	Rhodium, graphite	+0.05
3	Silver	0.00
4	Nickel, monel, titanium	−0.15
5	Copper, Ni–chrome, austenitic SS	−0.20
6	Yellow brasses and bronzes	−0.05
7	High brasses and bronzes	−0.30
8	18% chromium SS	−0.35
9	12% chromium SS, chromium	−0.45
10	Tin, tin–lead solders	−0.50
11	Lead	−0.55
12	2000 series aluminum	−0.60
13	Iron and alloy steels	−0.70
14	Wrought aluminum other than 2000	−0.75
15	Nonsilicon cast aluminum	−0.80
16	Galvanized steel	−1.05
17	Zinc	−1.10
18	Magnesium	−1.60 most anodic

steel, where they can produce pressures approaching 13,000 atmospheres. Under conditions of stress, the combined forces of the hydrogen and the external stress rupture the metal. It is one of the cruel jokes of nature that the very high strength steels such as those used for flexures are severely afflicted by this process, and that the heat treatment required to drive out the hydrogen severely limits the strength that can be obtained in precipitation-hardened stainless steels such as 17-4 PH. This is one of the reasons for choosing an age-hardening grade of steel such as 302 for flexures that might require electoprocessing.

The theory of stress–corrosion cracking favored by the authors is the electrochemical theory.[3] According to electrochemical theory, galvanic cells are set up between metal grains and anodic paths are established by heterogeneous phases. For example, the precipitation of $CuAl_2$ from an Al-4Cu alloy along grain boundaries produces copper-depleted paths in the edges of the grains. When the alloy, stressed in tension, is exposed to a corrosive environment, the ensuing localized electrochemical dissolution of the metal, combined with plastic deformation, opens up a crack in the metal. Supporting this theory is the existence of a measurable potential in the metal at grain boundaries, which is negative with respect to the potential of the grains. Once the crack exists, capillary forces aid the delivery of electrolyte to the tip of the crack, and it propagates. Figure 2 is an illustration of this concept.

3 FLEXURE MANUFACTURING

3.1 Manufacturing the Material

Flexures have been manufactured by every conceivable process, but for the purposes of small precision pivots, carefully rolled sheet or foil produces the best densification,

Flexure Pivots for Oscillatory Scanners

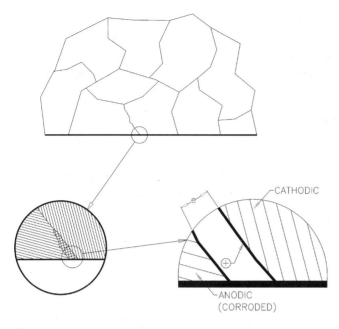

Figure 2 Stress–corrosion cracking.

uniformity of metallurgy, and surface finish. In addition, many work-hardenable materials come off the rolls with the proper temper for use. It cannot be overemphasized that surface finish is critical to the performance of thin flexures. No process now known is able to improve over the uniformity of thickness, flatness, freedom from scratches, and the consistency of temper of ferrous materials precision rolled on machines specially designed and constructed for the purpose. However tempting it may be to obtain experimental stock by grinding and polishing or etching down existing material, whether chemically or in combination with ion impact, these processes will change the performance values of the material. Unless one is prepared to make all the required flexures by an identical process stream, likely extremely uneconomical, the experimental flexures should be made from a material process exactly as the production flexures will be processed. In fact, considering that 100 pounds of flexure material such as type 302 stainless can be obtained as a minimum run from a reputable re-roller for under $3000, and 100 pounds of foil will make many tens of thousands of flexures, there is hardly any reason not to make the experimental parts from the production material, and there are several reasons why this is the best approach.

First, the process of arriving at the desired thickness, for the sake of argument 0.00500 in. thick ± 0.000020 in., at the required as-rolled temper, usually requires that the re-roller start with carefully annealed material whose thickness is the nearest standard thickness above 10 times the finished thickness, in this case, probably 1/16 in. Even so, the finished temper of the material will vary by a few percent run-to-run, the thickness will vary by 10 to 20 millionths run-to-run, and the cross-sheet crown will also vary by a few millionths run-to-run. As a result, even if the edges of the sheet are trimmed off the material used for flexures, which is good practise, for the best uniformity flexure-to-flexure, it is best to buy all the material needed for the product lifetime in one mill run. If this is not possible, or if product demand exceeds all expectations, it is best to re-qualify each succeeding flexure material run, and to make adjustments in the flexure width to compensate for any variations in modulus of elasticity or spring constant discovered.

It is also worth the expense, for critical applications at least, to 100% x-ray the material to find any inclusions, cracks, or voids.

3.2 Cutting Out the Flexures

Flexures may be cut out of sheet or foil by punching, conventional machining of stacked blanks, laser, water-jet, or E-beam cutting, or chemical etching of photolithographically produced patterns in photo resist. The last process is the standard process at GSI Lumonics, and is that preferred by most workers under most circumstances, because it is universally applicable, widely available, accurate to micro-inch tolerances, repeatable, fast, localized to the area to be cut, the photoresist protects the sensitive surfaces, and the process is inexpensive.

The other methods cause, or can cause, performance variation as the result of modified metallurgy in the zone near the cut. These potential modifications include heat effects, work hardening, and changes in the composition or phase of the alloy. Changes in the thickness of the flexure locally, and the introduction of stress risers, invariably accompany the relatively large-scale, relatively uncontrolled motions of mechanical machining and punching, including fine-blanking, and are also associated with jets of abrasive or abrasive in water, laser cutting, and E-beam cutting.

It is also difficult to provide adequate surface protection during processing by these methods.

Even photoetching produces artifacts that have the potential for undesirable effects. For example, the edge pattern left by double-sided etching is a double cusp that is irregular, sharp, and full of stress risers. Figure 3 shows such an edge.

Also, the position of the attachment points between the flexures to be and the parent sheet should be specified in such a place, usually on an unstressed tie bar or mounting tab, that the stresses associated with their disconnect are not obnoxious. The best way to remove the cusps and their associated stress risers is by tumbling in appropriate media, chosen to batter rather than to scratch, so that the cusps are beaten down, but the surfaces are not scratched. The edges of the flexures should be 100% inspected under suitable magnification, and any flexure with a nick or scratch visible at $20\times$ should be rejected.

3.3 Corrosion Protection

Most of the materials usually chosen for flexure material have intrinsic resistance to laboratory or office environments which is adequate protection. However, in shop floor, marine, outdoor, or specialized applications, further protection is often desirable. Such mechanisms include sealing and purging the entire mechanism. When such approaches are not practical, then the flexures themselves (and, if necessary, their mated parts) can be protected. Of course, with the flexures in particular, the protection should not interfere any more than necessary with the underlying functional performance of the part. At GSI Lumonics, hot-oil reflowed tin plating of the ferrous parts, including clamps and screws, is the protection method used. Tin has excellent protection in very thin layers, is anodic to the type 302 flexure material, has a very small Young's modulus, is a lubricating film, helps to distribute the clamping forces uniformly, and has low mass. It is, however, cathodic to 5, 6, and 7000 series aluminum, as well as type 355 and 356 cast aluminum. It

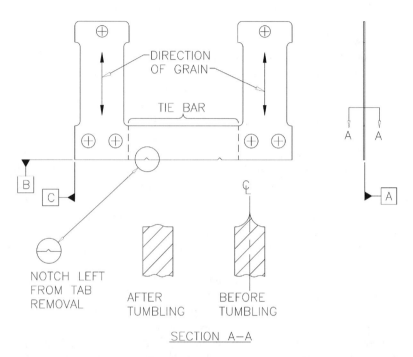

Figure 3 Edge stress risers.

is, of course, extremely cathodic to magnesium and its alloys. The answer here is anodizing of any aluminum or magnesium mated parts, brought to completion, and sealed with a suitable material such as sodium silicate. Although MIL-STD-186 allows tin in contact with magnesium, the best practise is anodization of the magnesium, or, in the case of fairly porous castings, vacuum impregnation of the casting with epoxy or equivalent followed by anodization.

4 FLEXURE MOUNTING

The careful mounting of flexures is essential if the full potential of careful flexure design is to be reached. There are three aspects of mounting that deserve thought. The first is the transition from the flexure itself to the mounting provision for that flexure, whatever it may be. For example, flexures may have integral mounting tabs intended to be supported with re-enforcing loose clamps, clamps attached by adhesive or solder to the flexure tab, flexures without tabs intended to be clamped or soldered or cemented into a slot, flexures with or without tabs intended to be welded in place, and so on. Whatever the method used, it is essential that the distribution of stress from the highly stressed operating region to the mount is gradual, controlled, and without stress risers. In general, neither the mounting area nor the body of the mount can be made to carry the concentrated stress loads that the flexure itself carries. Typically, the flexure is flared out in the region of the mounting so that the stressed region is increased in cross-sectional area, and the intensity of the stress is reduced thereby to a level low enough to prevent fracture of the mounting region before the main flexure. While any of the above methods may be used, the standard practise at GSI Lumonics is to produce a flared tab design similar to that in Figs 1 and 2. The corner radii are essential to achieving a transition without a stress raiser in the corner.

Of course, the presence of a radius in the corner raises the question of just where the effective region ends, and where the mounting region begins. It has been found by experiment that if the effective length of the flexure is taken to be the distance between the points of tangency between the constant-width section of the flexure and the radii at the ends, plus one radius, the error in the result will be dominated by tolerances on the flexure thickness.

Secondly, alignment of the flexure(s) to the stationary and moving members must be thought out. It is often the case that a multiplicity of flexure pairs or sets positioned along the axis of rotation is required in order to resist the cross-axis moments of the system. Thus, as many as four or six individual flexures may require individual alignment in a typical system. In this case, it is convenient to manufacture the flexures attached together by means of a tie bar, which may either be left attached or removed after assembly. This reduces the number of alignments required by at least two. Such a tie bar is illustrated in Fig. 3.

Thirdly, the designer must decide whether or not to try to make the flexure mountings on the fixed or on the moving or both elements indefinitely stiff. If he succeeds, then the design formulas given earlier will produce predicable results. If not, then some iteration may be required before satisfactory performance is obtained.

However, before leaping to a conclusion here, the designer should take into consideration the fact that the economical materials of choice for the rotor and/or stator are often not high fatigue strength materials, and may not even be highly qualified materials and so may have relatively loose, or unknown specifications of physical parameters such as Young's modulus and strength. In addition, it may well be that some other parameter is of more importance to their overall function than fatigue strength. One might posit thermal conductivity for the stator, and low inertia for the rotor. Whatever the reason, it is often desirable to allow the flexure-mated parts to be made of less intrinsically stiff material than is the material of the flexure itself. One then has the option of increasing the cross-sectional area of the mounting region appropriately. However, if the reason for using, say, an aluminum–magnesium alloy for the rotor had to do with inertial minimization, then bulking up the extreme-from-the-axis parts (where the flexure ends inevitably attach) to increase their stiffness is not likely to find favor. It may, in fact, be a better overall solution to allow the mountings to bend a little under load, and recalculate the flexures, if necessary, to compensate. It should be recognized that some deflection of the flexure mountings is required anyway, because if they are as stiff as or stiffer than the flexures, the clamping loads will be at least partially transferred to the flexures, potentially overloading them at the worst possible location, the mounting transition region. In this context, it is well to pay as much attention to the flatness and surface finish of the flexure-contact pads as to that of the flexures themselves. For this reason, slots are discouraged because good surface finish on the sides of slots is difficult to achieve, and inspecting them is next to impossible. A carefully machined and lapped pad, with a loose, lapped clamping plate under the screw heads to distribute their local loading is much better, and less costly in the long run. A layer of tin or indium foil or adhesive between the flexure and its mating surface on each side is advised as a method of filling microscopic voids and the valleys between the peaks on the surfaces, as well as providing corrosion protection at the interfacial joints. Care in mounting flexures cannot be overemphasized, particularly when attempts have been made to make the mating structures extremely stiff.

5 CROSSED-AXIS FLEXURE PIVOTS

5.1 General Introduction

Of course, there are many uses for flexures, and many flexure types. Flexure hinges, previously mentioned, are perhaps the most ancient. Straight-line motion mechanisms, arguably the most difficult bearing types to design, have reaped great benefit from flexure technology. One of the authors (Brown) spent 20 years designing and building bearings for scanning Michaelson interferometers, including diaphragm-flexure, "porch-swing" parallel-flexure, Bendix pivot porch-swing, and others. Because these straight-line motions are not generally considered pivots, they will not be discussed in detail here. The reader interested primarily in straight-line motions is recommended to Refs. 4 and 5. Flexures of the torsional type are discussed in detail elsewhere in this book.

There is, however, a very interesting flexure pivot type whose design is due to the late Niels Young, and is illustrated in Fig. 4. This pivot began life as a straight-line

Figure 4 Support diaphragm for straight-line mechanism.

mechanism for a scanning Michaelson interferometer. The diaphragms, instead of being corrugated or plane, were pierced by a number of quasi-radial curved slots. These slots increased the axial compliance of the diaphragm considerably without materially decreasing the radial stiffness, or introducing a departure from the straight-line motion. It was discovered, however, that if two of these diaphragms were mounted to their separating pillar with the slots aligned, a small rotation of the pillar accompanied translation. This was, of course, undesirable in the application, so the diaphragms were mounted reversed with respect to each other, and the rotation was restrained thereby.

Somewhat later one of the authors (Brown), having need of a very tiny precisely adjustable rotation in order to align an optical component, recalled the "defect" of the Young flexure, seized upon it as a virtue, and succeeded in putting it to good use.

Since the 1960s it has been generally recognized that the crossed-axis flexural pivot has the most widely applicable characteristics of any single-flexure pivot type. This wide adaptability was so compelling that the Bendix Corporation designed and manufactured a family of self-contained flexure pivots in various materials and sizes, and was so successful in its implementation that this type of pivot has come to be known worldwide as the Bendix pivot.

5.2 The Bendix Pivot

The Bendix "free-flex" pivot was introduced to the world in November 1962 in a seminal paper in Automatic Control magazine, entitled "Considerations in the Application of Flexural Pivots."[6]

Available still from Lucas Aerospace, these high-quality, standardized, well-quantified pivots should be considered whenever there is space to include them, and the product envisioned will be produced in modest quantities. Much time and effort, particularly in qualification testing, can be saved by using these devices. Made in both single-ended (cantilever) and double-ended types, Fig. 5 shows the general principle of construction.[7]

This general form of layout, the 90° symmetrical cross, has become the standard of construction, and the departure point for many specialized designs. However, this layout has the defect that translation of the axis of rotation takes place during angular motion of the pivot. This motion is neither linear nor small, and while many optical applications are quite tolerant of translation perpendicular to the plane of a mirror, the speed of the system may be limited by vibration produced by the translation of significant mass. As mentioned in the introduction to this chapter, it is possible to design a zero translation crossed-axis flexure pivot, and in principle possible to construct one, although the precision of assembly required is very high. The amount of translation, and the shape of the angle VS translation curve is dominated by the ratio of the leg lengths of the flexure. In the Bendix design, the leg lengths are equal, and the point of crossing, which lies on the axis of rotation, is initially central to the mounting tube. As the pivot is rotated, the axis of rotation departs from the center along a curve whose cusp is central and symmetric, but whose shape depends on the ratio of lengths of the flexure arms at intersection. Since the axis of rotation is usually a line fixed in space with respect to the rest of the mechanism, the axle is constrained by its load, and the flexures must cross on this axis, the only convenient way to change the ratio of the flexure arms is to extend the arm from the axis toward the stator (leaving the arm length between the axis and the rotor attachment point fixed) until the desired arm ratio is obtained. This inevitably has the effect of increasing the diameter of

Flexure Pivots for Oscillatory Scanners

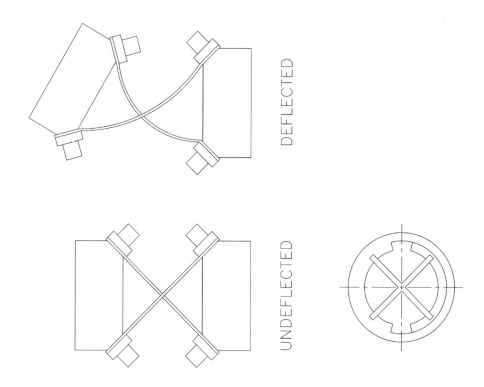

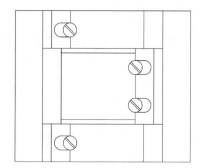

Figure 5 The Bendix pivot.

the stator flexure attachment point with respect to the axis of rotation. Since the ratio of arm lengths required for theoretical zero translation is 12.5 to 87.5%, this can become a big deal, so most designs find a compromise that is workable.

5.3 GSI Lumonics Design Example

Figures 6 through 9 show the successive stages of assembly of a flexure-pivot optical scanner mechanism, which represents the state of the art at GSI Lumonics.

This scanner, designed in 1995 and still in production, is used in a high-quality large-format printing engine to produce multicolor magazine illustrations. It produces 30° optical scans of an elliptical 30 mm aperture mirror at 160 Hz, with line straightness of a few microradians. Every feature of design and construction highlighted in this text as preferred is illustrated in the design. Of the several thousand of these scanners delivered, of which many have more than 20,000 hours of operation at 160 Hz (4×10^9 cycles), none has ever been returned for any reason.

Figure 6 shows the assembly of the flexures to the rotor. Note the radii in the mounting transition region of the flexures, the tie bar 5, and the lapped loose clamps 6 under the screw heads. These flexures are photo-etched, tumbled, 100% edge inspected under 20× magnification and tin plated. Figure 7 shows the assembly of the rotor and its flexures into the stator. There are registration notches on the mounting pads on both the stator and the rotor, which position the ends of the flexures and their clamps, assuring equal effective length of the flexures and parallelism between the axis of the rotor and the

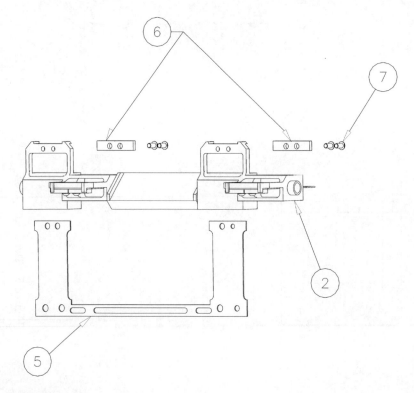

Figure 6 Rotor-flexure assembly.

Flexure Pivots for Oscillatory Scanners

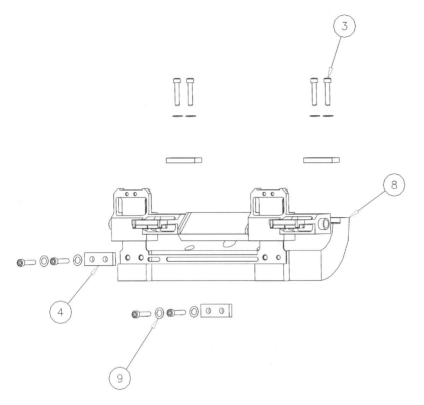

Figure 7 Rotor-stator assembly.

axis of the stator. Notice that the flexure mounting pads are islands, which can be machined, lapped, and inspected easily. Once assembled to the stator sector, the entire pivot mechanism is complete, and can be inspected and tested easily without unnecessary obstruction. Once qualified, it is mounted into the final housing, such as the one shown in Figs 8 and 9.

An interesting variation of this design is illustrated in Ref. 8. This scanner was required to operate in a mechanism that was quite sensitive to microscopic levels of vibration and other mechanical noise. Since crossed-axis pivots operated at high speeds produce a periodic translation of the axle, they have the potential to transmit vibration to the rest of the mechanism through their mounting means. In this case, the stator–rotor assembly, instead of being bolted directly to its housing, is supported on a set of flexures that permit an additional small rotation between the stator and the mounting. These flexures allow the pivot assembly to oscillate torsionally and transversely with respect to the housing. The second flexures therefore isolate the pivot assembly from the housing, and allow the housing to be bolted to the optical system without imparting an undesirable level of vibration in it. The second set of flexures allow the stator to rotate in response to the reaction torque from the rotor. The rotor and the stator counter-rotate, and the relative amplitude of the angular oscillations of the rotor and the stator is approximately in inverse proportion to their respective inertias. A typical ratio of inertias for the rotor and the stator is 1 : 150, so that the angular deflection of the second set of flexures is approximately 1/150 as great as that of the rotor flexures. The second set of flexures also allows the stator to

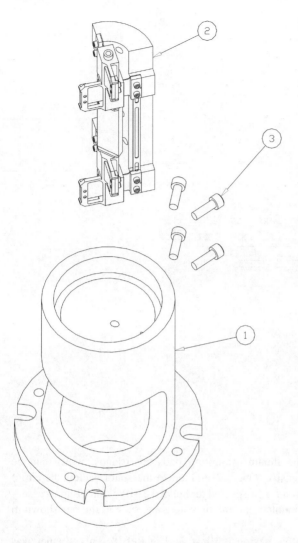

Figure 8 Stator-housing assembly, exploded view.

translate in response to the oscillatory translations of the rotor's center of gravity, which result from the translation of the rotor as it rotates. The forces required to dynamically accelerate the rotor and the stator largely balance each other. The amplitude of the residual translation force, compared with the translation force of the same pivot assembly bolted directly to the housing, is given approximately by the ratio of the pivot assembly mass to the housing mass, typically 1 : 15 or greater. One could, of course, apply ever more stages of isolation by this means of flexure pivots within flexure pivots if one had the need to do so.

6 MICROELECTROMECHANICAL FLEXURE SCANNERS

Microelectromechanical scanners (MEMS) integrate flexures and electromechanical actuators to yield small scanners of high performance and low cost. These are interesting in a number of applications but none more than optical switching in telephony.

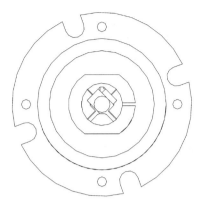

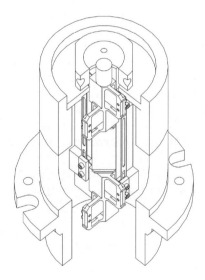

Figure 9 Final scanner assembly.

The scanning mirror may be formed direct onto the MEMS carrier, or it may be a separate component bonded to the mechanical assembly. Its aperture is typically between 0.1×0.1 mm and 3×3 mm. Currently, optical scan angles up to $20°$ are practical. Resonant frequencies are typically in the range 10–40 kHz and are dependent on system design, mirror size, and maximum scan angle.

There is a huge body of expertise in the manufacture of precise, small structures from silicon or on silicon substrates. It is fortuitous that silicon possesses attractive properties for small flexures, so we can exploit the existing semiconductor manufacturing infrastructure to make MEMS. This same manufacturing technology is well suited to the fabrication of the piezoelectric actuators.

6.1 MEMS Design

Piezoelectric bimorphs consist of two strips of piezoelectric elements joined over their long surfaces, provided with electrodes in such a manner that when an electric field is applied, one strip elongates and the other contracts. This results in a bending motion of both strips. The motion of the tip can be considerable, which is what makes these devices such useful tools as converters of electrical energy to mechanical energy and vice versa. They are used in different applications such as ultrasonic motors,[9] laser beam detectors,[10,11] fans for cooling electronics,[11] numeric displays,[12] filters,[13] accelerometers,[14–16] optical choppers,[17] and more recently as the legs of microrobots.[18] They are suitable as converters of electrical signals into sound (speakers) and similarly as pick-up elements for the detection of sound.[19] They are also used as the control element to reduce vibration in space-borne structures such as solar panels and in the walls of offices for the reduction of sound transmission.[20] While there is a huge variety of concepts that could be realized, we confine this discussion to a relatively simple layout as represented in Fig. 10.

The two piezoelectric actuators are of a form that has been used for the last 70 years or more.[9] This "bender," as we shall refer to it, has various names in the literature, including "piezoelectric bimorph," a term not strictly accurate in this case. It comprises two bars of piezoelectric material, typically PZT, having opposite polarities and sandwiched between two electrodes as indicated in Fig. 11. A voltage applied to the electrodes results in a field that causes the length of one bar to increase and that of the other to reduce. This produces curvature of the structure similar to the bending of a bimetal structure as a result of a temperature change. In our case, the benders are, strictly speaking, monomorphs; that is, there is only one active layer of piezoelectric material.

The free ends of the two benders are connected by torsion bars to the mirror substrate. The benders are driven by equal and opposite polarity so their motion is equal and opposite. The mirror substrate rotates about a central axis. This geometry has been studied in depth by Smits.

When using PZT material, an electric field is applied across two plates separated by poled ferroelectric material. PZT ceramics are isotropic and not piezoelectric prior to

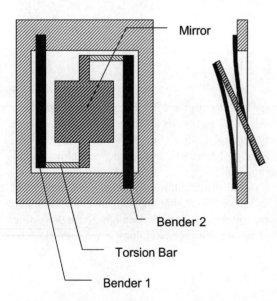

Figure 10 Simple piezoelectric "bender" scanner.

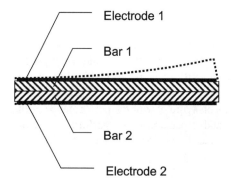

Figure 11 Piezoelectric "Bender" structure.

poling. Upon poling, they become anisotropic and display directionally dependent piezoelectric and mechanical properties. Poling is accomplished by heating the material to its Curie temperature and applying a DC field to the crystal, which then aligns its previously randomly oriented dipoles parallel to the field. Upon cooling, the dipoles maintain this preferred arrangement. As a result, there is a crystal distortion that causes growth in the dimension parallel with the field and also in the dimension perpendicular to the field. The axial strain resulting is typically small (0.2%) and is accompanied by hysteresis.

The displacement of the PZT material in the actuators when an electric field is supplied is the source of the bending moment of the "J" arms connected to the mirror platforms. When the electric field is applied, the material has displacement in two directions: parallel to the field and at a right angle to the field. The parallel and perpendicular displacements are opposite in sign so that when the film expands in parallel with the field it contracts its dimension perpendicular to the field and vice versa. The polarity of the parallel displacement is determined by the direction of the electric field with respect to the domains of the material. (The directions of the domains were established during the polarization process.) If the polarization and electric field are opposed, that is, the domains are in the opposite direction to the applied electric field, the material will expand parallel to the field and shrink perpendicular to the field. Opposed domains are stable in the sense that they are already arranged to cancel the applied electric field. If the domains and electric field are parallel, then the material will shrink as the electric field gets stronger until 50% of the domains have switched their direction to be opposed. This is effectively re-polarizing the material. When more than 50% of the domains have switched, the material will begin to expand again.

The amount of displacement caused in the case of the opposed situation can be estimated by the following formula:

$$D_3 = +d_{33} \times (V_3)$$

where D_3 is the displacement in the "3" direction, (the "3" direction is perpendicular to the capacitor plates), V_3 is the applied voltage between the plates of the capacitor, and d_{33} is the coefficient for displacement parallel to the field, typically 7×10^{-12} m/V^2.

6.2 MEMS Manufacture

Exploitation of the existing silicon fabrication technology leads naturally to the manufacturing process represented by Fig. 12. Each MEMS module is typically 4 × 4 mm. The modules are produced in arrays on a silicon wafer to yield 50 to several hundred modules per wafer.

The correct operation of the MEMS depends on symmetry of the performance of the two benders in the assembly. This symmetry and repeatability of performance from module to module are achieved only through precise dimensional, process, and material control.

The fabrication process is as follows:

1. A silicon wafer with a buried oxide layer is oxidized [Fig. 12(a)].
2. Front- and backside alignments are applied in photoresist (not shown).
3. The wafer is etched in BOE (buffered oxide etch), removing the oxide layer, leaving identical marks on front- and backside [Fig. 12(c)].
4. The wafer is etched in NaOH leaving a protruding pyramid at the backside and an inverted pyramid on the front side. [Fig. 12(c)].
5. The wafer is re-oxidized [Fig. 12(d)].
6. The backside is now etched to depth of 350 microns, which leaves a varying thickness of the membrane on the front side, according to the location of the point where the measurement is made. A Pt under electrode is deposited and PZT is applied as a sol-gel. A Pt top electrode is deposited [Fig. 12(f)].
7. On the front side the outline of the double monomorph optical scanner structure is etched out to the depth of the insulating buried oxide layer, while the backside is protected [Fig. 12(g)].
8. The wafer is re-oxidized (not shown).
9. 25 microns is removed from the backside, to etch free the double monomorph optical scanner structure. The etching stops at the buried oxide layer [Fig. 12(h)].

Figure 13 shows a schematic of a double monomorph optical scanner. Figure 14 shows an SEM photograph of a scanner at 50× magnification.

6.3 Operation of the Scanner

In Fig. 15 a cross-section is given, in which the eye is located in the plane of the wafer. The angle β is the angle the mirror makes with the plane of the wafer.

6.4 Material Properties

The properties required of the driving materials are a reasonably high coefficient of strain vs. applied field, low or predictable hysteresis, and high repeatability. The properties required of the support materials are ease of fabrication, very high stability, and a high fatigue limit. These are met satisfactorily in the silicon mechanical material chosen: it was in fact, these demonstrated characteristics of silicon that led to its choice as a basis material. On the other hand, while the PZT material has a satisfactory coefficient of strain, these constraints remain. Although these materials are excellent for submicron positioning, they have inherent hysteresis and creep, and so they lack the level of repeatability

Flexure Pivots for Oscillatory Scanners

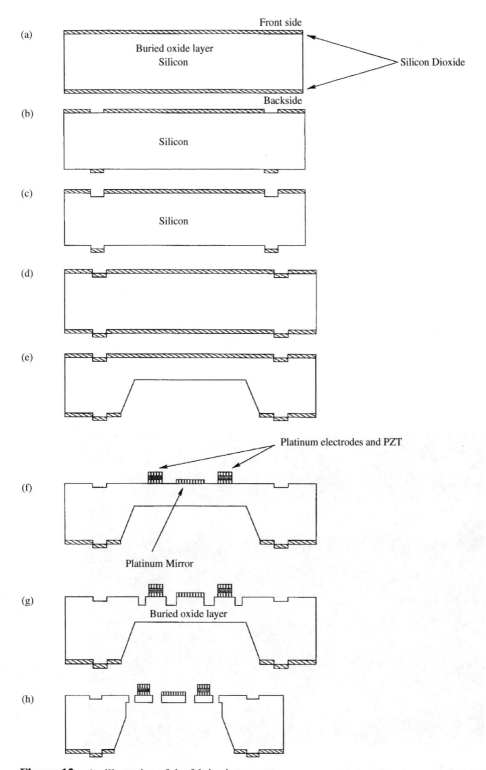

Figure 12 An illustration of the fabrication process.

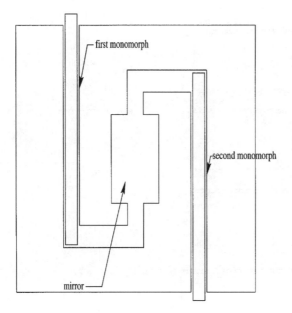

Figure 13 Schematic of double monomorph optical scanner.

Figure 14 50× SEM photograph of MEMS scanner.

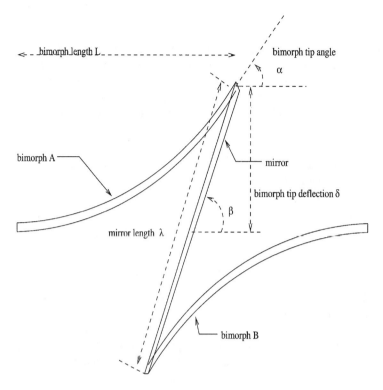

Figure 15 Cross-sectional schematic of double monomorph optical scanner.

required in a practical open-loop positioning device. Long-term creep can approach 15%, and hysteresis can approach 12%.

6.5 Static Performance

6.5.1 Hysteresis

Hysteresis appears as apparent "backlash" upon reversal of the direction of motion, but unlike the backlash of conventional mechanical systems and stiction, which can largely be predicted and compensated, hysteresis depends on the recent history of motions and is difficult to model, predict, and compensate. The creep is, at this time, completely unpredictable. As a result, open-loop piezoelectric devices are limited to applications in which repeatability of 10% is satisfactory. Of course, when combined with a position feedback system, this defect can largely be overcome, but the cost of this level of control is disproportionate to the economics of the production of MEMS.

6.5.2 Linearity

Linearity of motion over mirror angles of 10° mechanical is in the range of 2%, smoothly changing, monotonic, and predictable. This characteristic is considered satisfactory for most laboratory uses. However, PZT has a rather large strain sensitivity to temperature, so that linearity of motion can be compromised by changes in temperature.

6.5.3 Uniformity

Uniformity of performance scanner-to-scanner within a wafer is entirely a question of process control, and is not expected to pose long-term problems. Standard levels of production process control are also expected to address wafer-to-wafer performance uniformity issues satisfactorily. At the moment, we are achieving uniformity of characteristics of about 10% across the wafer and wafer to wafer in a pilot production phase.

6.5.4 Yield

Yield is an issue that falls into the process control purview as well. At the moment, we are achieving yields above 80%, with most of the dropouts the result of mirror defects.

6.6 Dynamic Performance

6.6.1 Dynamics

These scanners have a first torsional resonance typically well above 20 kHz with integral mirrors. As a result, the dynamics permit full-scale (10°) steps in well under 100 µs. With high-quality mirror flakes 150 microns thick cemented over the integral mirror, the resonant frequency falls to about 8 kHz, which is satisfactory for most microscanning purposes. The flatness of the present examples of integral mirrors is poor, as is the surface finish, but attached flakes achieve any required surface quality. Testing has verified that 150 micron thickness is adequate to preserve quarter wave or better flatness over the 1.8×2.6 aperture of the standard scanner because of the underlying silicon support ring.

6.6.2 Life

Life testing has revealed no observable damage after 10^{10} 60° optical scans when driven by an 8 kHz oscillator. Flexure stress is expected to be below the endurance limit for silicon at all angles of operation up to 30° mechanical.

6.6.3 Degradation Processes

No degradation process is known, other than excessive voltage breakdown inside the PZT crystal, and heating above the Curie point. High-speed operation (above 20 kHz) has the potential to encounter impact degradation of the tips of the bender arms and the edges of the mirror if operated in open air. The standard packaging presently envisioned for individual scanners is a hermetically sealed TO5 package.

6.7 Application Rules

6.7.1 When and When Not to Use MEMS

The current state of the art in PZT open-loop scanners has limited repeatability and "staring" positional stability because of creep and hysteresis. As a result, MEMS scanners should not be specified for applications in which positional precision better than 10% of full scale is required unless some form of active position feedback is anticipated. On the other hand, the device behaves as though it were a pair of capacitors of a few nanofarads capacity. Therefore, self-heating even during aggressive scan profiles is not an issue, and cooling is not required except in environments above 100°C. Because the thermal transport path from the mirror through the structure is long, the effect of power coupled from the scanned beam may be significant. In this case, it is desirable to provide a direct mirror cooling path through a suitable gas, such as helium, directly to the scanner case.

Vibration and inertial forces applied to the scanner are likely to cause deformation of the flexure structure, and, since the system is very poorly damped, are likely to produce long-lasting settling periods. Otherwise, these scanners have no particular susceptibility to environmental stimuli such as temperature, barometric pressure, humidity, magnetic and electric fields, and so forth.

6.8 Anticipated Developments

It is anticipated that electrostrictive actuators may replace piezoelectric actuators in MEMS in the near future. A typical electrostrictive material, such as lead–magnesium–niobate (PMN), should provide an order of magnitude improvement over PZT in positional stability.

For PMN materials, the change in length is proportional to the field voltage squared, so the coefficient is a factor of two higher than PZT. Unlike piezoelectric materials, PMN crystals are not poled. Positive or negative voltage changes result in an elongation in the direction of the applied field, regardless of its polarity. Because PMN is not poled, it is inherently more stable than PZT, resulting in a reduction of long-term creep from 15% to 3%. Also, PMN materials have better hysteresis than does PZT. While PZT exhibits 12% hysteresis, PMN displays only 2%.

Two properties of PMN contribute to its thermal stability. First is the strain sensitivity to temperature. The PMN is much more robust than PZT in this regard, especially over large temperature ranges. Secondly, the coefficient of expansion of PMN is twice that of PZT.

6.9 Conclusions

It is clear that MEMS have not yet grown from infancy. Unlike their mature brethren, the macroscale flexure scanners, the only conclusion one could legitimately draw at this time is that the future is bright for small scanning and pointing devices that draw little power, pack closely together, demonstrate extreme reliability, and offer a price–performance index beyond the capability of any macroscale device. It seems certain that these attributes of MEMS will be exploited. Just how MEMS will look in their maturity years hence is anyone's guess.

7 CONCLUSION

It seems that flexure pivots have come of age. They have been used in a great diversity of products where their attributes of sensitivity, accuracy, and repeatability are enabling. They are also low-cost, lubricant-free, low-mass, high "Q," and capable of storing considerable energy. They can be cascaded to provide vibration isolation. There are stand-alone commercial versions to suit many purposes. The published data on detail design rules are considerable, and no competent engineer should have unusual difficulty in producing a workable pivot on the first try. It is the hope of the authors that the information in this chapter will help neophytes to avoid most of the impediments to success that lurk in the mysteries of material selection, processing, and the mounting of flexures and that many more useful applications of flexure pivots will be found and pursued with success.

ACKNOWLEDGEMENTS

First of all, may I thank my friend and colleague, Felix Stukalin, without whose encouragement and forbearance this chapter would not have been written at all. Brian Stone performed the labors of Hercules in turning my sketches and doodles into the illustrations. Michael Nussbaum read the manuscript, and made many helpful suggestions. Dr. Tim Weedon and Reggie Tobias were both very perceptive as well as diplomatic in their criticisms. Last but not least, Professor Jan Smits did all the heavy lifting in the second part of the chapter. Assisting him in his laboratory at Boston University, Koji Fujimoto and Vladimir Kleptsyn did much of the MEMs construction and testing, and Steven Vargo and Dean Wibig of JPL, and Joe Evens and Gerry Velasquez of Radiant Technologies Inc. provided enabling processing services.

REFERENCES

1. Weinstein, W.D. Flexure pivot bearings. Machine Design 1965, *37*, 136–145.
2. Bendix flexural pivot. *Bendix Electric and Fluid Power Division*, Application Notes, Catalog, Bendix Corp., Utica, NY.
3. Boyer, H.E., Ed. Failure analysis and prevention. In *Metals Handbook*, 8th ed., Vol. 10, 208–249. American Society for Metals: Metals Park, OH.
4. Paros, J.M.; Weisbord, L. How to design flexure hinges. Machine Design 1965, *37*, 151–156.
5. Neugebauer, G.H. Designing springs for parallel motion. Machine Design 1980, *52*, 119–120.
6. Troeger, H. Considerations in the application of flexural pivots. Automatic Control 1962, 17 (4) 41–46.
7. Paulsen, D.R. Flexural Pivot. U.S. Patent 4,802,720, Feb. 7, 1989.
8. Brosens, P.J. Resonant Optical Scanner. U.S. Patent 5,521,740, May 28, 1996.
9. Sawyer, C.B. The use of Rochelle salt crystals for electrical reproducers and microphones. Proc. Inst. Radio Eng. 1931, *19* (11), 2020–2029.
10. Smits, J.G.; Dalke, S.I.; Cooney, T.K. The constituent equatons for piezoelectric bimorphs. Sensors and Actuators 1991, *28*, 41–61.
11. Kugel, V.D.; Xu, B.; Zhang, Q.M.; Cross, L.E. *Bimorph based piezoelectric air acoustic transducer: a model*; Sensors and Actuators.
12. Caliano, G.; Lamberti, N.; Iula, A.; Pappalardo, M. A piezoelectric bimorphstatic pressure sensor. Sensors and Actuators A 1995, *46* (1–3), 176–178.
13. Coughlin, M.F.; Stamenokic, D.; Smits, G. Determining spring stiffness by the resonance frequency of cantilevered piezoelectric bimorphs. IEEE Trans. Ultrasonics, Ferroelectrics and Frequency Control 1977, *44*, 730–733.
14. Kielczynski, P.; Pajenski, W.; Salewski, M. Piezoelectric sensors for the investigation of microstructures. Sensors and Actuators A 1998, *65* (1), 13–18.
15. Juan, I.; Roh, Y. Design and fabrication of piezoceramic bimorphvibration sensors. Sensors and Actuators A 1998, *69* (3), 259–266.
16. Van Mullem, C.J.; Blom, F.R.; Fluitman, J.H.J.; Elwenspock, M. Piezoelectrically driven silicon beam force sensor. Sensors and Actuators A 1991, *26* (1–3), 379–383.
17. Naber, A. The tuning fork as a sensor for dynamic force control in scanning near-field optical microscopy. J. Microscopy–Oxford 1999, *194* (Part 2–3), 307–331.
18. Yamada, H.; Itoh, H.; Watanabe, S.; Kobayashi, K.; Matsushige, K. Scanning near-field optical microscopy using piezoelectric cantilevers. Surface and Interface Analysis 1999, *27* (5–6), 503–506.
19. Kielczynski, P.; Pajewsli, W.; Sealcwski, M. Piezoelectric sensor applied in ultrasonic contact microscopy for the investigation of material surfaces. IEEE Trans Ultrasonics, Ferroelectrics and Frequency Control 1999, *46* (1), 233–238.
20. Edwards, H.; Taylor, L.; Duncan, W.; Melemed, A.J. Fast, high-resolution atomic force microscopy using a quartz tuning fork as actuator and sensor. J. Appl. Phys. 1997, *82* (3), 980–984.

10

Holographic Barcode Scanners: Applications, Performance, and Design

LEROY D. DICKSON

Wasatch Photonics, Inc., Logan, Utah, U.S.A.

TIMOTHY A. GOOD

Metrologic Instruments, Inc., Blackwood, New Jersey, U.S.A.

1 INTRODUCTION

Significant changes have occurred in the field of laser scanning since the first edition of this book was published over ten years ago. Specifically, visible laser diodes (VLDs) have become the laser light source of choice in the scanning industry, allowing scanners to become much smaller, in the form of hand-held and wearable scanners. Holographic scanning, however, does not yet have a very significant presence in these applications.

In the first edition, much attention was given to supermarket scanners and most of the examples were given in reference to such designs. Over the past decade, however, a great deal of growth has occurred in the industrial scanning market, and the adaptability of holography has helped it grow into this market more significantly. Accordingly, more examples are given in this edition with respect to the industrial scanning market.

The fundamentals of scanning, however, have not changed. Neither has the presence of several decades of barcode and scanner design specifications and laser standards been diminished. For example, printing specifications of barcodes are based on reflectance properties at the wavelength of helium neon lasers, predominantly used years ago but no longer in scanners today. As such we will begin with a basic discussion of barcodes.

A barcode is a sequence of dark bars on a light background or the equivalent of this with respect to the light reflecting properties of the surface. The coding is contained in the relative widths or spacings of the dark bars and light spaces. Perhaps the most familiar barcode is the universal product code (UPC), which appears on nearly all of the grocery items in supermarkets today. Figure 1 is an example of a UPC.

Figure 1 Typical UPC barcode.

A barcode scanner is an optical device that reads the code by scanning a focused beam of light, generally a laser beam, across the barcode and detecting the variations in reflected light. The scanner converts these light variations into electrical variations, which are subsequently digitized and fed into the decoding unit, which is programmed to convert the relative widths of the digitized dark/light spacings into numbers and/or letters.

The concept of barcode scanning for automatic identification purposes was first proposed by N.J. Woodland and B. Silver in a patent application filed in 1949. A patent, titled "Classifying Apparatus and Method," was granted in 1952 as U.S. Patent No. 2,612,994. This patent contained many of the concepts that would later appear in barcode scanning systems designed to read the UPC.

1.1 The UPC Code

In the early 1970s, the supermarket industry recognized a need for greater efficiency and productivity in their stores. Representatives of the various grocery manufacturers and supermarket chains formed a committee to investigate the possibility of applying a coded symbol to all grocery items to allow automatic identification of the product at the checkout counter. This committee, the Uniform Grocery Product Code Council, Inc., established a symbol standardization subcommittee, whose purpose was to solicit and review suggestions from vendors for a standard product code to be applied to all supermarket items.

On April 3, 1973, the Uniform Grocery Product Code Council announced their choice. The code chosen was a linear barcode similar to a design proposed by IBM. The characteristics of this barcode, the now familiar UPC, are described in detail in the article by Savir and Laurer.[1]

The UPC is a fixed-length numeric-only code. It consists of a pair of left guard bars, a pair of right guard bars, and, in the standard version A symbol shown in Fig. 1, a pair of center guard bars. Each character is represented by two dark bars and two light spaces. The version A symbol contains 12 characters, 6 in the left half and 6 in the right half. Thus, a version A UPC symbol will have 30 dark bars and 29 light spaces, counting the 6 guard bars – left, right, and center. The first character in the left half is always a number system character. For example, grocery items are given the number system 0, which often appears on the left of the symbol. The last character on the right is always a check character. This sometimes appears to the right of the barcode symbol.

The remaining five characters in the left half of the version A UPC symbol identify the manufacturer of the product. For example, the left-half number 20000 represents

Green Giant products. This left-half five-digit code is assigned to the various manufacturers by the Uniform Product Code Council.

The remaining five characters in the right half of the version A UPC symbol identify the particular product. This right-half five-digit code is assigned by the product manufacturer at his discretion. For example, Green Giant has assigned the right-half number 10473 to their 17 oz. can of corn. Therefore, the complete UPC code for the Green Giant 17 oz. can of corn, ignoring the number system character and the check character, is 20000–10473.

There are a number of other properties of the UPC code and symbol that are significant relative to the design and use of equipment for reading the code. First, the left and right halves of the version A symbol are independent. That is, each half can be read independently of the other half and then combined with the other half in the logic portion of the reader to yield the full UPC code. Furthermore, as shown in Fig. 2, each half of the UPC symbol is "over-square." That is, the symbol dimension parallel to the bars is greater than the symbol dimension perpendicular to the bars. The aspect ratio of the barcode is vital in the determination of a minimum scan pattern for reading the code, as we will see in a later discussion.

It should be noted that the original "over-square" design of the UPC code is not always adhered to by product manufacturers. Often they will truncate the height of the code. These truncated codes, while discouraged by the Uniform Product Code Council, are often used by manufacturers to maximize the space available for product information. That, in turn, has meant that scanner designs must be more complex and the decoding algorithms more sophisticated.

Each character of the code is represented by two dark bars and two light spaces. Individual bars and spaces can vary in width from one module wide to four modules wide. (Note, on all UPC codes, the guard bars are always one module wide and separated by a one-module-wide space.) The total number of modules in each character is always seven. The left-half characters are coded inversely from the right-half characters. As shown in Fig. 3, for example, if we let white = 0, and black = 1, then the code for the number 2 in the left half is 0010011 while in the right half it is 1101100.

The fact that each character is always seven modules wide leads to a second major property of the UPC code: it is self-clocking. Therefore, absolute time measurements are unimportant. What is measured is the time that is required to go from the leading edge of the first black bar to the leading edge of the third black bar (i.e., the first black bar of the next character). This time interval is then divided into seven equal intervals, and the relative widths of the two black bars and the two white spaces are then determined for

Figure 2 Two UPC half-symbols.

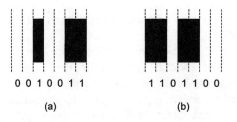

Figure 3 An example of character encoding for the number 2: (a) left-half character; (b) right-half character.

decoding purposes. Thus, the total width of a character – black-white-black-white – is measured, and the relative widths of the two black bars and the two white spaces are determined for decoding.

This self-clocking feature is very important in the design of scanners for reading the UPC code. It means that the velocity of a scanning light beam for reading the code does not have to be constant across the full width of the code. The velocity only needs to be reasonably constant across a single character. This means that the UPC code can be read by moderately nonlinear scan patterns, such as sinusoidal or Lissajous patterns. It also means that the code can be read on curved surfaces. Furthermore, the scan lines reading the code do not have to be perpendicular to the bars and spaces in the code. As originally designed, satisfactory reading of a UPC code can be obtained with scan lines that pass through the code at any angle relative to the bars and spaces so long as a single scan line passes completely through a full half symbol, including the center guard bars and one pair of edge guard bars. More sophisticated decoding algorithms now exist that can "stitch" together three smaller, individually scanned pieces of a UPC code (or other symbol), and, as such, the process is commonly referred to as stitching. The use of stitching allows a slightly less thorough scan pattern to do as good a job as a better pattern and makes a good pattern even better; however, the UPC code was designed with only the left and right sides in mind. Stitching was a software adaptation developed later on.[2] One of the factors driving its development was the occurrence of the aforementioned truncated codes, for which stitching is particularly useful.

A third property of the UPC symbol of significance to the scanner designer is the size of the symbol, which is allowed to vary from the nominal size [about 1.0 in. × 1.25 in. (25.4 mm × 31.75 mm) for the version A symbol] down to 0.8 × nominal and up to 2.0 × nominal. This size variation allows the use of small labels on small packages with good print quality and large labels on larger packages with poorer print quality. From the scanner designer's viewpoint, the small label will establish the minimum bar width to be read and the large label will set a lower limit on the size of the scan pattern.

The minimum bar width established by the UPC specification, including tolerances, is 0.008 in. (0.2 mm). This number establishes the maximum attainable depth of field for the optical reader. In practice, the depth of field of the typical laser scanner designed to read the UPC code will easily meet, and exceed, the 1 in. (25.4 mm) depth of field required by the early UPC guidelines. However, this 1 in. depth of field did not take into consideration the manner in which the scanners would eventually be used. Depths of field of several inches (100 mm+) are required for today's UPC barcode readers.

Finally, the contrast specification for the UPC symbol requires that the contrast be measured using a photomultiplier detector (PMT) with an S-4 photocathode response curve coupled with a Wratten 26 filter. This combination has a peak response at a wavelength of approximately 610 nm (24 μin.), falling to zero at approximately 590 nm (23.2 μin.) and 650 nm (25.6 μin.). This response includes, not coincidentally, the wavelength of the helium–neon laser, 632.8 nm (24.9 μin.), which was the preferred laser at the time that laser scanners were first being considered. While several of the inks used for printing UPC labels can provide acceptable contrast out to 700 nm (27.56 μin.) there are many other inks in use that do not provide acceptable contrast beyond 650 nm (25.6 μin.). These inks would preclude general use of longer-wavelength light sources. Today, nearly all UPC scanners use one or more diode lasers as their light source(s); however, the wavelengths of these lasers fall within the original UPC wavelength specification.

1.2 Other Barcodes

The UPC code is not widely used in the industrial environment (the manufacturing, warehouse, and distribution applications). Here, the requirements are different from those of the supermarket, so the codes used are different than the UPC code. The preferred codes for the industrial environment are Bar Code 39, Interleaved 2 of 5, and Codabar.

The most common barcode in the industrial environment is the so-called Bar Code 39, or 3 of 9 barcode. This code is fully alpha-numeric and is self-checking. For a full discussion of Bar Code 39, as well as several other codes, and the definition of such terms as "self-checking," see work by Allais.[3] The code shown in Fig. 4 is an example of Bar Code 39.

Bar Code 39 got its name from the fact that it originally encoded 39 characters: the 26 letters of the alphabet, the numbers from 0 through 9, and the symbols -, ., and SPACE, plus a unique start/stop character, the asterisk (*). Today it also encodes the four so-called special characters: $, /, +, and %, for a total of 43 characters. However, it is still referred to as Bar Code 39. It is also often called the 3 of 9 code, because each character in the code is represented by nine elements (five dark bars and four light spaces), and three of them are wide with the remaining six narrow. In the primary set of 39 characters, two of the wide elements are dark bars. In the four special characters, the wide elements are all light spaces.

The 2 of 5 code is a subset of the 3 of 9 code. In 2 of 5, only the bars are used for encodation. Two of the five bars are wide, just as in the original 3 of 9 code. The spaces are not used. This code is strictly numeric. The basic 2 of 5 code is not widely used in the industrial environment, but a variation of it, called the Interleaved 2 of 5 Code, is used extensively for manufacturing and distribution applications. This code uses the bars to encode one character in the standard 2 of 5 Code and then uses the interleaving spaces to encode a second character in 2 of 5 code. This allows more characters to be encoded in a fixed barcode length than either 3 of 9 Code or 2 of 5 Code. This code is also only numeric, but, due to the interleaving feature, it can encode nearly 80% more characters per unit

Figure 4 An example of Bar Code 39.

length than Bar Code 39, assuming both codes have the same minimum bar width. For this reason, the Interleaved 2 of 5 code is often used where space limitations will not permit the use of Bar Code 39.

A third code that is used extensively in medical institutions, and which was adopted as an early standard by the American Blood Commission for use in identifying blood bags, is Codabar. This code is also frequently seen in some transportation and distribution applications.

1.3 Barcode Properties

From the standpoint of the scanner, the important properties of any barcode are:

1. Minimum bar width: generally specified in millimeters or mils (thousandths of an inch). And often referred to as the "X" dimension.
2. Contrast: a measure of the reflectance of the bars and spaces. Contrast is generally expressed in terms of the print contrast signal (PCS), defined as

$$\text{PCS} = \frac{r_s - r_b}{r_s} \quad (1)$$

where r_s is the reflectance of a space and r_b is the reflectance of a bar. It should be noted that PCS is usually measured for one particular wavelength of light. In the majority of applications this wavelength is 633 nm (24.9 μin.), the wavelength of the helium–neon laser, which was the most common light source for most of the early laser scanners. [Some applications allow PCS to be measured at 900 nm (35.4 μin.), the wavelength of some infrared light sources used in some readers.] This is an important point to remember since PCS will vary drastically as a function of wavelength if colored inks or backgrounds are used. In practice, the most important reflectance property of the barcode is the absolute contrast, which is simply the space reflectance minus the bar reflectance [the numerator in Eq. (1)].
3. Code length: the physical length of a barcode is determined by the density of the code (which is determined by the minimum bar width) and the number of characters in the code. The physical length of the code determines how long the scan lines must be and, when combined with the code height, will determine how accurately the scan line must be oriented with respect to the barcode.
4. Code height: the height of the barcode (the dimension parallel to the bars) will determine the angular accuracy required in orienting the scan line relative to the barcode.
5. Barcode quality: this includes both the quality of the printing or etching of the code itself and the quality of the surface on which the code is printed. Obviously, the better the quality of both, the easier it will be for the scanner to successfully scan and decode the barcode.

There is a great deal more that could be said about the barcodes themselves. However, a more detailed analysis of the fundamental properties of the barcodes is beyond the scope of this review and is not really necessary for the purposes of our discussion of barcode scanning.

2 NONHOLOGRAPHIC UPC SCANNERS

A block diagram of a typical laser scanner system for reading the UPC code is shown in Fig. 5. The focused laser beam scans the UPC symbol on a package as the package passes over the read window of the scanner. The laser beam is reflected from the symbol as it passes over the dark bars and light spaces. The diffuse portion of the reflected light is modulated by the reflectivity variations of the symbol (bars and spaces). This light modulation is detected by the photodetector, which converts the light modulation into electrical modulation. The electrical "signal" is then amplified, digitized, and transmitted to the "candidate select" block. This block acts as a filter, allowing only valid UPC half-symbols to pass to the decoder. The decoder converts the signals for each half-symbol into characters and then combines the characters for the two half-symbols together to yield a complete UPC product identification code. The computer then searches its memory for a description and price of the item identified by this UPC code. This information is transmitted back to the checkout terminal where it appears on the display and the customer receipt. Simultaneously, the store inventory is updated to reflect the sale of the identified item. All of this takes place in a few milliseconds.

The focused spot size of the scanning laser beam must be about 0.2 mm in order to be able to read the labels with the smallest bar widths while still yielding adequate depth of field. This requires an optical f-number of approximately 250, which, when combined with scanner geometry, establishes the focusing optics' requirements.

A number of technologies are available for deflecting the focused laser beam in a conventional, nonholographic, UPC scanner. Cost and performance requirements limit the choice to mechanical deflectors – generally either rotating or oscillating mirrors, or a combination of these. The scan pattern created by the laser deflection mechanism must be capable of reading a full-size version A UPC symbol regardless of the orientation of the symbol in the scan window. In other words, the scanner must be omnidirectional. An omnidirectional scanner will allow maximum freedom for the scanner operator when bringing the item across the scan window.

We have already seen that the two halves of the UPC symbol can be read independently of each other. We have also seen that each half-symbol is over-square. Therefore, the minimum scan pattern that will allow omnidirectional scanning is a pair of perpendicular scan lines in the form of an X [see Fig. 6(a)]. As the UPC symbol passes over the scan window, at least one of the legs of the X will pass through the entire half-symbol at some point in the window. The figure shows two extreme orientations of the

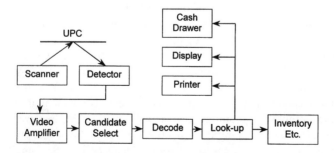

Figure 5 Block diagram of a UPC scanner.

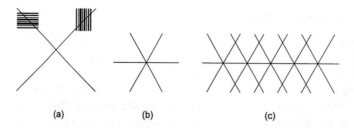

Figure 6 Omnidirectional scan patterns in the plane of the window: (a) the minimum scan pattern; (b) basic pattern of an optimum scan pattern; (c) optimum pattern for a rectangular window.

symbol as it is passed over the window. These are the worst-case orientations in that they allow the minimum time for scanning the symbol satisfactorily.

The amount that the half-symbol is over-square, when combined with the maximum item velocity of 2.54 m/s (100 in./s), determines the minimum pattern repetition rate to guarantee at least one good scan through the symbol as it passes across the scan window, regardless of its orientation. The pattern repetition rate, the total scan length, and the width of the smallest UPC module establishes the maximum video signal rate seen by the photodetector.

Although the pattern in Fig. 6(a) is the minimum scan pattern required to yield an omnidirectional scanner for the UPC symbol, it is not an "optimum" pattern. The pattern repetition rate required to guarantee one scan through a UPC half-symbol moving across this pattern at 2.54 m/s (100 in./s), at the worst-case orientation relative to the scan pattern, is very high. This results in high scanning spot velocities and subsequently high video signal rates. A "better" scan pattern can guarantee one good scan at lower pattern repetition rates and lower scan velocities. An optimum scan pattern will minimize scan velocity, thereby minimizing video signal rates.

If one could increase the amount that a symbol is over-square (i.e., improve the aspect ratio), then one could reduce the pattern repetition rate, and the scan velocity, and still guarantee one good scan through the UPC half-symbol at maximum symbol velocity and worst-case symbol orientation. While the symbol itself cannot be changed, one can effectively improve the aspect ratio of the symbol by using a scan pattern where the scan lines are separated by angles less than 90°. Thus, a scan pattern consisting of three scan lines, for example, instead of two, could be repeated less often and still be able to read the UPC symbol under the worst-case conditions mentioned above. Increasing the number of scan lines has the effect of increasing the total linear distance scanned by the scan pattern, which, by itself, would increase the scan velocity of the scanning spot. However, the reduction in the pattern repetition rate is greater than the increase in the scan length. The net result is a better scan pattern, in the sense described above.

Can we continue to improve the scan pattern by adding still more lines? Unfortunately, the answer is no. The reduction in the required pattern repetition rate realized by using four scan lines is nearly offset by the increased scan velocity resulting from the greater distance traveled by the scanning spot. The small amount of gain is not enough to justify the increased cost and complexity required to generate the four-line pattern. Beyond four lines, there is no gain when scanning UPC symbols. (We will, however, see later that in certain industrial applications four- and five-line patterns can be quite effective, especially when scanning symbologies with extreme aspect ratios.) It appears,

Holographic Barcode Scanners

then, that for the UPC code the optimum scan pattern in the plane of the window would be one based on the three-line pattern shown in Fig. 6(b). This fundamental three-line pattern, which is still the basic criterion used in the design of UPC scanners in 2004, formed the basis of the first scanner designed to read the UPC code, the IBM 3666 scanner. The linear equivalent of the Lissajous scan pattern used in the IBM 3666 scanner is shown in Fig. 6(c).

2.1 Forward-Looking Scanners

Initially, all UPC scanners were conceived as "bottom scanners." That is, the scanning laser beam pointed directly upward to read the UPC symbols on the bottoms of packages as they passed over the scan window. A major problem was encountered in the design of this type of scanner. UPC symbols printed on shiny surfaces were difficult to read because the specular reflection from the shiny surfaces contained no bar-space modulation in the specularly reflected light. In addition, the specular reflection created saturation problems in the photodetector because the specularly reflected light was so much more intense than the diffusely reflected light. In most scanners, the photodetector was located back along the general direction of the outgoing laser beam. Therefore, some solution had to be found that would keep the specularly reflected light from being directed back along the laser beam path.

One solution to this problem is shown in Fig. 7(a). The laser beam is tilted at an angle of approximately 45° relative to the scanner window. In this configuration, the specularly reflected light is reflected away from the photodetector, thereby eliminating the specular reflection problem.

A fringe benefit occurs when this scanner geometry is used. The tilted beam can be used to read UPC symbols on the front of packages without tilting the packages forward [see Fig. 7(b)]. Of course, this increases the depth of field required to read these upright labels, but the laser scanner has the capability to provide a depth of field of several inches (100 mm+), which is usually sufficient for side reading with the tilted beam geometry. Nearly all UPC scanners today employ some form of forward-looking, tilted-beam reading geometry.

2.2 Scan Pattern Wrap-Around

The next development in the evolution of the scan pattern was the introduction of scan pattern wrap-around. Several scan patterns were introduced that took the basic three-line

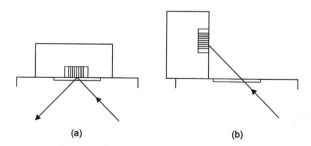

Figure 7 Tilted-beam scanning: (a) removing the specular reflection problem; (b) side reading with a forward-looking tilted beam.

optimum scan element shown above and created it in such a way that the scan lines were directed at the items from points within the scanner that were slightly off to the sides of the package. In these scanners, horizontal lines were projected forward from immediately in front of the item and from directions slightly off to each side. Vertical lines were also projected from slightly off to each side. The pattern projected from the two sides was essentially a cross pattern, while the scan pattern projected from immediately in front of the item was a horizontal line, as shown in Fig. 8. The overall pattern was created by using a rotating mirror deflector and an array of fixed folding mirrors.

This type of scan pattern was effective in reading the UPC symbol on the scan window because it employed the basic three-line optimum scan element. It was also effective in reading upright items because it projected a pattern of perpendicular horizontal and vertical lines on the front of the items. Such a pattern is effective for upright reading because the bars in the UPC symbol will usually be either vertical or horizontal when the package is presented to the scanner in this manner.

The major advantage of this type of scan pattern is that it "wraps around" to the sides of the packages to some degree. This means that the operator does not have to align the item as carefully when he brings the item across the scan window. The UPC symbol can be on the bottom of the package, on the front of the package, or on the side of the package and still be readable by the scanner. This has a positive effect on operator productivity.

This concept of wrap-around was further exploited in the development of "bi-optic" scanners, exemplified by the NCR 7875, the PSC Magellan SL, and the Metrologic Stratos. The bi-optic-type scanner is currently becoming the most used type of scanner in supermarkets and other large-volume, point-of-sale applications. The bi-optic nature of the scanner refers to the two separate scan windows it possesses, as shown in Fig. 9. In this configuration there is a horizontal scan window and a second window at or near vertical orientation, depending on the specific scanner in question.

Improved performance is gained by employing the wrap-around concept from both windows. In the preferred package-presentation orientation, two of the six faces of an item directly face a scanner window and are seen by the primary three-line pattern. Two other faces, typically the faces in the direction of item motion (see Fig. 9), are targeted by the wrap-around patterns of both scanner windows. The package surface that faces away from the vertical window still potentially sees the wrap-around pattern out of the horizontal window. Finally the top surface of the package, which has the least exposure to the scan

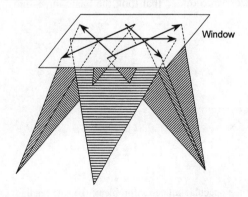

Figure 8 Projected wrap-around scan pattern.

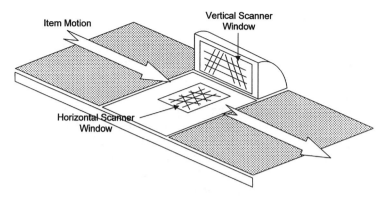

Figure 9 A "bi-optic" supermarket scanner.

pattern, may still have some chance of being scanned by the wrap-around pattern of the vertical window, depending on the package height and code orientation. The net result is effective scanning throughout the majority of the 360° horizontal orientation range and equally effective scanning through nearly 270° of vertical orientation. Such a wide range of acceptable presentation orientations means very little of an operator's time needs to be spent paying attention to the position and orientation of a code on an item.

2.3 Depth of Field

The multidirectional, three-line scan pattern forms the basis for nearly all present-day UPC scanners, holographic and nonholographic. Unfortunately, the forward-looking feature increases the depth of field requirement considerably. As much as 150 mm (6 in.) depth of field may be required to read some barcodes on upright items. Because the codes must, in many cases, be read by a tilted beam, the resultant spot ellipticity on the barcode will increase the effective scanning spot diameter. This will reduce the depth of field of the scanner.

Providing satisfactory scanning performance over such a large depth of field, with a tilted scanning beam, is a significant challenge to the scanner designer. Significant improvements in signal processing over the last decade have allowed smaller bar widths to be read without reducing the actual size of the spot, helping to solve this problem and increase depth of field. Another means of easing the problem, without relying on electronic improvements, is to design a scanner that can provide more than one focal plane. Such a scanner could focus some of its scan lines close to the scan window and some of its lines further from the scan window, thereby increasing the effective depth of field of the scanner.

Holographic scanning allows the scanner designer to add this additional degree of flexibility to the scanner. The holographic scanning element essentially allows each scan line to be optimally focused to provide increased depth of field and increased flexibility in the placement of beam-folding mirrors for the creation of the scan pattern. This also allows for a more complex, and more effective, scan pattern. The need for greater depth of field and the desire for a more effective scan pattern led to the development of the holographic barcode scanner.

3 HOLOGRAPHIC BARCODE SCANNERS

The concept of holographic scanning has been around for three decades,[4] and during this time many different applications have been suggested,[5,6] but few have been demonstrated and an even smaller number have made it into the marketplace. A general review of holographic scanning and various applications can be found in the book by Beiser[7] as well as in previous volumes of optical engineering.[8]

Holographic barcode scanners first appeared commercially in 1980 with the introduction of holographic UPC scanners by IBM and Fujitsu. Today, Metrologic manufactures holographic scanners primarily for industrial applications. Such applications range from large depth of field [greater than 1 m (40 in.)] overhead scanners, to high-density, high-resolution scanners for large aspect-ratio codes, to completely automated hands-off scanning tunnels for bulk-mail centers.

3.1 What is a Holographic Deflector?

Photography is a light-recording process in which a two-dimensional light-intensity distribution incident on a light-sensitive medium is recorded by that medium. In contrast, holography is a light-recording process in which both the amplitude and phase distribution of a complex wavefront incident on the recording medium can be recorded by that medium.

Holography, therefore, differs from photography in that it is able to record all of the information that is needed by the eye, or any other optical system, to interpret the full three-dimensional nature of the object.[9,10] This information is accessed when the recording (the "hologram") is illuminated by the proper light source – usually, but not always, a laser.

The most common form of hologram creates, when viewed, a three-dimensional image of a complex, three-dimensional object. As will be described below, the reduction of the three-dimensional object to a single point-source produces a special case of particular importance to deflection – a hologram that acts as a lens to focus an incident laser beam. This type of hologram is referred to as a holographic optical element (HOE).

The concept of holographic recording and reconstruction – more importantly, how it deflects light – can best be understood with reference to Figs. 10 and 11. In Fig. 10(a), two wavefronts of equal intensity created from a laser are directed to overlap in some region of space where the recording is to be made. If the optical path difference from the point of beam separation to the region of overlap is within the coherence length of the source, the resulting interference pattern will be stationary in both space and time and will have high fringe contrast. The intensity distribution in these fringes can be exposed onto, or more properly into, a suitable photosensitive medium such as a photographic emulsion. After processing, the recording contains a variation in optical density, refractive index, or optical thickness – sometimes a combination of all three – and is the hologram. When this recording is repositioned and illuminated by one of the wavefronts, such as the diverging wavefront in Fig. 10(b), the structure at each point within the hologram diffracts the illuminating light and creates a new wavefront that is identical to the original second wavefront. In the case of Fig. 10(b) the result is a converging and deflected wavefront. This simple HOE is the equivalent of a positive or converging lens in combination with a prism converting and deflecting light from a point object to a point image. The efficiency and quality with which this wavefront conversion occurs is directly related to the recording configuration and selection of recording material. In Fig. 10(c) we see that the equivalent

Holographic Barcode Scanners

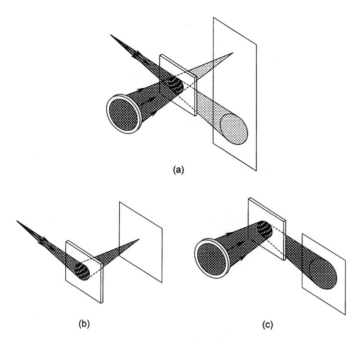

Figure 10 Simple holography: (a) recording the hologram; (b) reconstruction of the convergent wavefront (positive lens); (c) reconstruction of the divergent wavefront (negative lens).

of a negative lens is realized by illuminating the same HOE with the converging wavefront, thereby reproducing the original diverging wavefront. Holographic recording of complex multidimensional objects can be treated as the recording of a superposition of individual spherical waves from all the points in the object field.

By using a combination of small area illumination and HOE translation, the reconstruction geometry of Fig. 10 can be used to create simultaneous light deflection and focusing. This is shown in Fig. 11, where the HOE is initially located at position 1 and has a small subarea, on the right side, illuminated by a diverging wavefront that corresponds to the diverging reference-construction-beam of Fig. 10(a). The light is focused by this subarea of the HOE to the point in the image plane labeled "position 1" (Fig. 11). This

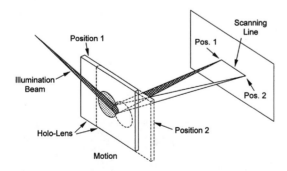

Figure 11 Principle of holographic deflection.

point in the image plane corresponds to the location, with respect to the displaced HOE, of the original convergence point of the converging construction wavefront in Fig. 10(a). As the HOE is translated, different subareas of the HOE are passed under the illuminating beam, and the reconstructed image point is caused to translate by the same distance, in this case to position 2. This is completely analogous to the deflection and focusing that would occur if a conventional lens were illuminated off axis with a collimated beam and the lens displaced normal to its optic axis. A continuous back and forth motion in either case produces the same motion in the focused spot.

In practice, however, a continuous rotary motion rather than a reciprocating motion is easier to implement. Higher scan speeds can be realized, and different holograms can be easily accessed. Consequently, most holographic deflectors consist of a number of unique HOEs placed circumferentially as sectors on a glass disc, as shown in Fig. 12. Other materials can be used, and other geometries besides the disc geometry can be used,[8] but, for simplicity, we will restrict our discussion to the glass disc, which is the most common medium and geometry used in holographic scanners today. It should be noted that plane linear gratings, producing prismatic deviation without focal power, must be rotated to generate scanning. Translation of a plane grating in one direction will not produce scanning.

A holographic deflector-disc, when properly illuminated by a laser and rotated about its axis of symmetry, can produce a complex variety of scanning laser beams. The optical and geometrical properties of each of these beams can be distinctly different from all of the others. This is the most important feature of holographic scanning. It is the major feature that distinguishes it from conventional laser scanning technology and, in a barcode scanner, allows the introduction of capabilities that could not be readily achieved with conventional technology.[11]

The holographic disc works in the following manner. Each sector, or facet, of the holographic disc is a unique HOE of the type previously described – the holographic equivalent of a prism and lens combined. When a facet is illuminated by a laser beam, the beam is diffracted, or bent, by the facet and focused to some point in space (see Fig. 13). The focal length and deflection angle are established during the holographic construction of the facet and may vary from facet to facet.

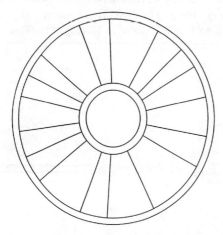

Figure 12 A holographic deflector disc.

Holographic Barcode Scanners

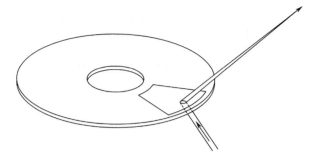

Figure 13 Light deflection and focusing by a holographic disc.

As the disc rotates, the deflected, focused laser beam scans. When the beam scans across a barcode, some of the diffusely reflected light will return to the facet that generated the scanning beam. The facet now acts as a light-collection lens, combined with a prism, to collect a portion of the reflected light and direct it toward a photodetector.[12]

3.2 Novel Properties of Holographic Barcode Scanning

The use of holography in barcode scanning allows the introduction of scanning concepts that are not available to the designer of conventional barcode scanners, at least not in any economically practical design. Such concepts as multiple focal planes, overlapping focal zones, variable light-collection aperture, facet identification, and scan-angle magnification allow holographic scanning to bring to barcode scanning some significant design and performance capabilities.

A conventional barcode scanner contains a lens for focusing the laser beam, a device for deflecting the laser beam, and some optics for collecting a portion of the laser light reflected from the barcode and focusing it onto the photodetector. In a holographic scanner, all of these properties – focusing, deflecting, and light collection – are contained in the holographic disc. As indicated earlier, these properties may be different in each sector of the holographic disc; thus, a 16-sector holographic disc, for example, would contain 16 unique optical systems. Each of these systems would have its own focal length, scan angle, and light-collection aperture. One revolution of such a holographic disc would produce the equivalent of scanning with 16 different scanners.

Because each facet of the holographic disc may be different, with its own combination of focal length, deflection angle, and facet area, then one complete rotation of the disc will create multiple scan lines with multiple deflection angles, multiple focal lengths and multiple light-collection systems. This enables the holographic scanner to introduce some novel operational characteristics.

One of the major advantages of using a holographic disc in a barcode scanner is that it can provide a much larger depth of field than would be attainable with a conventional, single-focal-length, barcode scanner. In order to understand this point, we need to briefly review the subject of depth of field.

3.3 Depth of Field for a Conventional Optics Barcode Scanner

In barcode scanning, depth of field is the distance along the laser beam, centered around the focal point of the scanner, over which the barcode can be successfully scanned. The spot size profile of a laser beam along its direction of propagation is established by the beam waist diameter and the wavelength of the laser light source. The depth of field of a barcode scanner employing such a beam depends on the size of the minimum bar width in the barcode being read and also on the resolving ability of the scanner electronics. For a given resolving ability there will be a normalized spot-size definition, or resolution criterion [C in Eqs. (3)–(5) below], based on the assumed Gaussian intensity profile of the laser beam. This resolution criterion simply defines the relative intensity level at which the beam diameter is measured. A commonly used criterion is the $1/e^2$ beam width (13.5% intensity level, $C = 0.135$). Typical resolving abilities of scanners range from the 50 to 70% intensity levels, or possibly even higher. Once all these beam, code, and electronic parameters have been established, there is little that can be done in a conventional barcode scanner to increase the depth of field.

Figure 14 illustrates the concept of depth of field. The lens in the scanner focuses the laser beam to a relatively small spot size at the focal point. The diameter of the beam at the focal point is determined by the focal length of the lens, the diameter of the beam at the lens, and the wavelength of the laser being used. When the optical system of the scanner is properly designed, the minimum spot size will be somewhat smaller than the minimum bar width and will therefore be able to successfully scan the barcodes. As one moves the barcode to either side of the focal point, toward or away from the scanner, the spot size increases as the beam becomes out of focus. Eventually, a point is reached where the out-of-focus spot size is larger than the minimum bar width in the barcode. When this occurs, the barcode can no longer be successfully scanned by the beam. The distance between the two points to either side of the focal point where the limit of scanning capability occurs is, by definition, the depth of field.

The major factors determining the depth of field are the spot size at the focal point, the wavelength of the laser and the minimum bar width. (For convenience, we will assume throughout this discussion that the minimum space width is the same as the minimum bar width.) Using the notation of Dickson[13] for the variation in beam radius of a propagating Gaussian beam, the $1/e^2$ beam radius, r, at a distance, z, from a beam waist of $1/e^2$ radius r_0 is given by

$$r = r_0 \left[1 + \left(\frac{\lambda z}{\pi r_0^2} \right)^2 \right]^{1/2} \qquad (2)$$

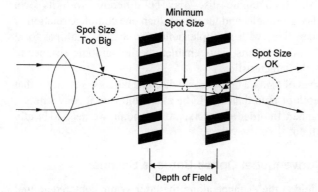

Figure 14 Depth of field for a conventional optical system.

Holographic Barcode Scanners

The beam radius, r_c, for a different resolution criterion, C, is given by

$$r_c = rK = r\sqrt{-\ln(C)/2} \qquad (3)$$

Similarly, the waist size at that resolution criterion is scaled down by $r_{0c} = r_0 K$. Eqs. (2) and (3) can be combined and rearranged to express the depth of field for a given beam radius, r_c, resolution criterion, C, and waist radius, r_{0c}, as

$$\text{DOF} = \Delta z = 2|z| = \frac{-4\pi}{\lambda \ln(C)}\sqrt{r_{0c}^2 r_c^2 - r_{0c}^4} \qquad (4)$$

It can be shown that, for any resolution criterion, the depth of field given by Eq. (4) is maximized for a given minimum bar width when the focused spot size, as measured by the particular resolution criterion being used, is equal to the minimum bar width divided by $\sqrt{2}$. Applying that condition requires the substitutions of $2r_{0c} = w_{\min}/\sqrt{2}$ and $2r_c = w_{\min}$ into Eq. (4), which then reduces to

$$\Delta z = \frac{-\pi w_{\min}^2}{2\lambda \ln(C)} \qquad (5)$$

At a wavelength of 650 nm, typical of VLDs being used in scanners today (2004), and assuming a reasonable resolution criterion of $C = 0.6$ (60%), Eq. (5) can be approximated as

$$\Delta z = \frac{w_{\min}^2}{8.3} \qquad (6)$$

where Δz is the depth of field in inches when the minimum bar width is in mils or as

$$\Delta z = \frac{w_{\min}^2}{210} \qquad (7)$$

where Δz is the depth of field in millimeters when the minimum bar width is in microns.

For example, if the scanner is optimally designed to read a barcode that has a minimum bar width of 8 mils (200 μm), the depth of field will be 7.7 in. (200 mm). Note that the above equation tells us that the depth of field is strongly dependent on the size of the minimum bar width to be read. Therefore, a small minimum bar width is always accompanied by a small depth of field.

Furthermore, if the scanner is not optimally designed for the minimum bar width to be read, the depth of field will be smaller than it could be. This will be true whether the scanner is optimally designed to read either a higher density barcode or a lower density barcode. In addition, if the scanner uses a laser with a different wavelength, the depth of field will be multiplied by the ratio of the wavelength for which the design is optimized to the wavelength of the laser being used. (This assumes that the new laser has its minimum spot size focused to the same size as the original design.)

There is very little that can be done in a conventional barcode scanner to increase the depth of field. It is possible to use an autofocus scanner. Autofocus scanners have been designed and built for the industrial scanning market by Accu-Sort Systems (Telford, PA). However, a problem with such scanners is that the reaction time of the auto focus system has to be very fast to accommodate fast-moving items on a conveyor system. Because all autofocus systems today require mechanical movement of some of the optics of the scanning system, the reaction time may not be fast enough, depending on the application. In addition, such systems will add cost and complexity to the scanner.

One could also add a supplemental optical element to the scanner that could move into position in the laser beam path to change the net focal length of the scanner. This moving element would allow, for example, two different focal lengths to be selected. In practice, this approach would allow only two or three different focal lengths to be selected, giving only a slight increase in the depth of field.

Furthermore, in either an autofocus system or a dual or triple-focal-length system, only the focal lengths can be easily changed. Ideally, one should also change the aperture of the scanner as the focal length is changed. This would maintain a constant level of light collection over the full range of focus, thereby optimizing the performance of the scanner over the full range of readability. However, rapid variation of the light-collection aperture is difficult to accomplish in a conventional barcode scanner.

3.4 Depth of Field for a Holographic Barcode Scanner

The use of holography in a barcode scanner would allow the introduction of a true multifocal-plane scanner with a variable light-collection aperture. The way this would be accomplished in a holographic scanner is illustrated in Fig. 15.

Figure 15 shows focusing of the laser beam by two consecutive facets on the holographic disc. Each facet will exhibit a conventional depth of field as established by the focal length of the facet, the beam diameter at the disc, and the wavelength of the laser. Notice, however, that the two facets are focused at different distances from the disc. Therefore, while each facet has only a conventional depth of field, the combined depth of field of the two facets is twice as great as for either facet alone, assuming that the focal

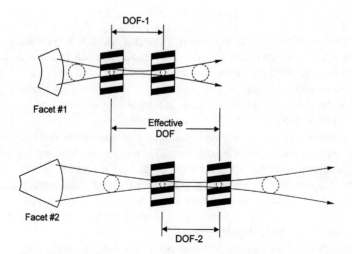

Figure 15 Combined depth of field for two holographic facets.

lengths are chosen so that the end of the depth of field for facet 1 coincides with the beginning of the depth of field for facet 2.

Thus, with only two focal planes, the holographic disc can double the depth of field of a conventional, nonholographic scanner.

If the disc is designed so that all of the facets are focused at different distances, then a much larger overall depth of field can be achieved. For example, if the minimum bar width to be scanned is 0.2 mm (8 mils), a conventional, single-focal-length scanner would have a depth of field of approximately 200 mm (8 in.). However, a properly designed holographic scanner could provide as much as 800 mm (32 in.) depth of field for the same 0.2 mm minimum bar width code by using only four facets.

Even greater depth of field can be achieved with more facets; however, there are diminishing returns to this route to greater depth of field. Ideally, we would like to be able to control the focused spot size in each focal plane; however, this would require automatic aperture adjustment of the outgoing beam (similar to autofocus), which would prove mechanically difficult and very costly. As a compromise, typically, the center-most focal plane is optimized for maximum depth of field, which causes the other focal planes to be close to optimum as well.

Another limit to the continuing growth of depth of field is the very large distances that are eventually encountered as facet focal lengths increase. While theoretically this is not a problem for the resolution of the outgoing beam profile, it does present a greater challenge to light collection, which will be discussed later.

4 OTHER FEATURES OF HOLOGRAPHIC SCANNING

There are other novel features of holographic scanning that are not as obvious as the ability to provide a large depth of field. The major features of holographic scanning are:

1. Multiple focal planes;
2. Overlapping focal zones;
3. Variable light-collection aperture;
4. Facet identification and scan tracking;
5. Scan-angle magnification.

We have already discussed the multiple-focal-plane feature and the large depth of field that it provides. Let us now examine the other features to see what they are, how they are produced by the holographic disc, and what capability they provide.

4.1 Overlapping Focal Zones

We showed in the previous section how holographic scanning can provide a large depth of field by designing the holographic disc so that the depth-of-field region of each successive facet was contiguous with the depth-of-field regions of the facets immediately preceding it and following it. This may not, in practice, be the best disc design. It may, in fact, be better to design the holographic disc so that the focal point of one facet coincides with the limit of the depth of field for the preceding and following facets, resulting in an overlapping focal zone design. The reason why this design may be superior is explained in the following paragraphs.

One of the major contributors to decoding problems in a barcode scanner is the existence, or creation, of noise in the so-called quiet zone, the white, or clear, region immediately preceding and following the barcode. One of the contributing factors to noise in this region, and throughout the barcode, is substrate noise, or paper noise. Paper noise occurs when the size of the focused spot of the scanning laser beam is about the same as the size of the granularity of the substrate material. Paper fibers, for example, can be as large as 0.1 mm (4 mils). For very coarse paper or cardboard, the fiber size can be even greater. For nonpaper substrates, such as for barcodes etched into plastic or metal, the granularity can be greater still.

If a barcode on a noisy substrate is scanned at the focal point of a scanning laser beam, the small, in-focus spot will "see" the granularity of the substrate material. This will introduce paper noise on the return light signal that will, in turn, lower the probability of achieving a successful read.

While noise could, in general, be reduced with low-pass electrical filtering, the filter properties would have to be altered for each facet to correct for the differences in spot velocity. That is, a low-pass filter designed to remove noise from the short-focal-length facets would, at the same time, filter out the barcode signals from long-focal-length facets. Electrical filtering does not appear to be a practical solution for large depth-of-field scanners.

The solution to this problem is to scan the noisy barcode with a slightly out-of-focus spot. This spot will be larger than the in-focus spot, but still small enough to read the barcode. This larger spot will, while scanning the barcode, act as a filter to smooth out the surface roughness, effectively lowering the paper noise and increasing the probability of achieving a successful read.

Figure 16 shows the analog photodetector signal for a noisy barcode when scanned (a) by an in-focus spot and (b) by a slightly out-of-focus spot. The noise on the signal from the in-focus scanning spot is apparent. The resultant reduction in the noise level due to scanning the same barcode with a slightly out-of-focus scanning spot is equally apparent.

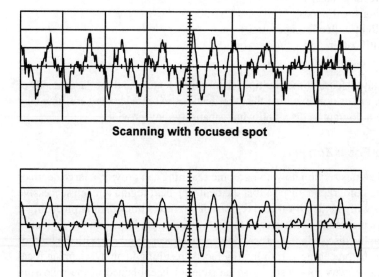

Figure 16 Photodetector signals for in-focus and out-of-focus scanning laser beams.

Holographic Barcode Scanners

By overlapping the focal zones of the individual holographic facets, as shown in Fig. 17, we can guarantee that all barcodes will be scanned by both an in-focus scanning spot and one or more slightly out-of-focus scanning spots. The slightly out-of-focus spots will be small enough to read the barcodes, but large enough to smooth out the substrate noise.

This in-focus/out-of-focus capability, which would be difficult to implement with conventional scanning technology, is relatively simple to introduce with holographic scanning. One merely selects, during the master holographic disc design phase, the focal length for each of the facets that guarantees the desired amount of focal zone overlap.

4.2 Variable Light-Collection Aperture

There is more to successfully achieving a large depth-of-field scanner than simply providing multiple focal planes. If, for example, one designs a scanner with a one meter depth of field where the optical throw (closest reading distance) is 200 mm (8 in.) and the range (farthest reading distance) is 1200 mm (47 in.), then the variation in the light level returned to the detector, for barcodes with identical reflection characteristics, will be 36 : 1, the square of the ratio of the far and near distances. This places a severe dynamic range requirement on the analog electronics in the scanner. The problem is worse in practice since other factors, such as label skew and label reflectivity variations, also affect the amount of light returned.

In order to reduce the variation in the light level of the return light in a multiple-focal-plane scanning system, it would be desirable to vary the light-collection aperture to compensate for changes in the distance to the barcode being scanned. One could then use a relatively small aperture for a near-focus scan line and a relatively larger aperture for a far-focus scan line.

Holographic scanning allows one to do exactly that. Figure 18 shows a holographic scanning disc designed for focus distances ranging from 1000 to 1680 mm (39.5 to 66 in.). Notice that the light collection area of each facet is different. The facet with the shortest

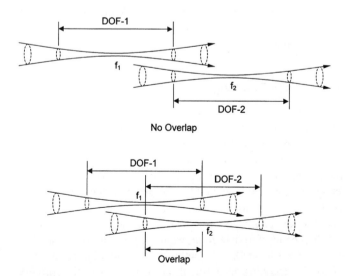

Figure 17 Overlapping focal zones of two holographic facets.

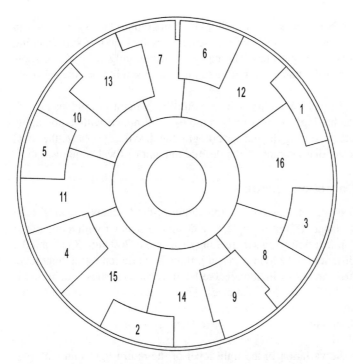

Figure 18 The Metrologic Penta holographic scanning disc showing large variation in facet areas.

focal length has the smallest light collection area while the facet with the longest focal length has the largest light collection area. The light collection area of each of the remaining intermediate facets is a direct function of its focal length.

This difference in light collection area for the near and far facets of the holographic disc allows the light collection to be approximately uniform over the total depth of field of the scanner. This is a major advantage in obtaining decoding accuracy over a large depth of field for a barcode scanning system.

4.3 Facet Identification and Scan Tracking

Note that the disc design shown in Fig. 18 incorporates a gap in the outer annulus between two of the holographic facets. This gap may be transparent or opaque, depending on the application, and is referred to as the home-pulse gap. Because the outgoing laser beam is incident on the disc in this outer section, a detector placed in the proper location above the disc can sense this gap by measuring the laser power incident on the detector. As the gap passes over the laser the change in power recognized by the detector generates a home pulse in the analog signal. With this information embedded in the signal we can determine the rotational speed of the disc and, thereby, where on the disc the laser beam is currently incident. This knowledge can then be used to determine what facet is currently scanning, and even where in that facet the beam is located. This method of facet identification can be used in several ways to improve the decoding accuracy of the scanner, as well as provide for additional features.

Holographic Barcode Scanners

If we knew, for example, that we were on a short-focus facet, we could decrease the electrical gain in the analog electronics. If we were on a long-focus facet, we could increase the gain of the analog electronics. This electronic automatic gain control (AGC) would add to the already existing optical AGC, introduced by the variable light-collection aperture, to further improve the decoding accuracy.

We could also vary the internal clock rate from facet to facet to improve resolution. Because the scanner is an angular scanner, the linear velocity of the scanning beam will vary directly with the distance from the scanner. The bit rate seen by the detector while scanning with a long-focus facet would be greater than the bit rate seen when scanning with a short-focus facet, assuming that the code density is the same in both cases. By making the clock rate vary from facet to facet to maintain the optimum clock rate for a given bit rate, we could, once again, improve the decoding accuracy of the scanner.

An additional capability made possible by this facet identification feature is scan tracking, for which several patents have been filed by Metrologic Instruments, Blackwood, NJ (U.S. patent numbers 6,382,515 B1; 6,457,642 B1; 6,517,004 B2; and 6,554,189 B1). Knowing precisely where the incident beam is striking each facet at each instant in time allows indirect determination of where the item being scanned is located in a given three-dimensional, spatial reference system. This knowledge is extremely useful in fully automated systems where little human interaction is desired, such as in scan tunnels in bulk-shipping centers. With the location of a scanned package identified, other automated mechanical systems can then redirect packages to their intended destination.

The reason holography lends itself so well to scan-tracking is that the facets on a holographic disc are easily repeatable in the manufacturing process with high precision. Obtaining good repeatability of the deflector with mirrored scanning systems is more difficult.

4.4 Scan-Angle Multiplication

Holographic scanning discs used in barcode scanners are frequently designed to be illuminated with a collimated beam incident normal to the surface of the holographic disc. This illumination geometry provides a scanning spot that is free from aberrations across the entire scan line because the relative incidence geometry with respect to the hologram does not change as the disc rotates. This is a special case of the more general aberration-free illumination geometry for rotationally symmetric systems in which the designed illumination beam is a converging spherical wavefront, converging toward a point located on the rotational axis of the disc. (A normally incident collimated beam has a point of convergence on the axis at infinity.) Under these conditions, the illuminating wavefront remains unchanged with respect to the hologram, always converging to the hologram's design convergence point even as the disc is rotated. Because the HOE was designed to produce an aberration-free diffracted beam when the incidence wavefront converges to that point, the diffracted beam remains aberration-free throughout the motion of the hologram. In essence, the playback (illumination) beam remains identical to the reference (recording) beam, which is the condition for zero aberrations.

If the holographic disc is designed to be illuminated with a collimated beam inclined at a non-normal angle, then some amount of aberrations will be introduced in the scanning beam. Each facet of the disc can be designed to still provide zero aberrations at the center of its corresponding scan line, but there will always be aberrations introduced as the disc rotates because of the resulting mismatch between the recording and playback wavefronts. The amount of the aberrations will be dependent on the amount of rotation away from the center of the facet and the amount of tilt of the collimated incident beam.

There is, however, one advantage to tilting the incident beam. It can be shown that a tilted, collimated reference beam will provide a greater scan-angle multiplication factor relative to an untilted, collimated reference beam geometry. A precise determination of the multiplication factor requires the use of a computer program because of the interdependence of the diffracted beam elevation angle (β in Fig. 19) and the rotation angle (ϕ_{rot}). A first-order approximation, accurate to a few percent for holograms whose construction geometry is not too extreme, can be obtained from the following simple relationship. The variable f represents the focal length of the holographic lens (facet). The other terms and the geometry are defined in Fig. 19.

$$\phi_{scan} = \phi_{rot}\left(\frac{r}{f}\cos\gamma + \cos\alpha + \cos\beta\right) = \phi_{rot}\left(\frac{r}{f}\cos\gamma + \frac{\lambda}{d}\right) \tag{8}$$

where d represents the grating spacing of the hologram and the angle γ is determined from Eq. (9):

$$\sin\gamma = \cos\beta \sin\theta_{skew} \tag{9}$$

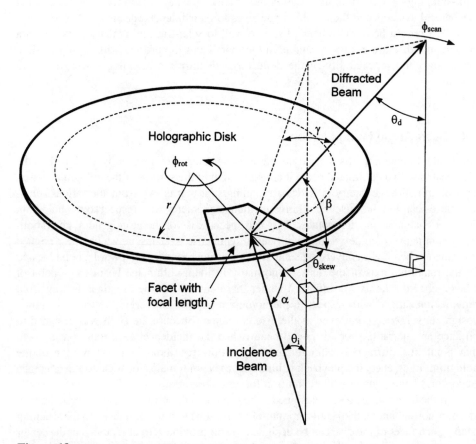

Figure 19 Scan-angle magnification parameters.

The multiplication effect of the tilted reference beam is due to the cos α term in Eq. (8); for normal incidence this term goes to zero.

As an example of the effect of the tilted reference beam, consider a holographic facet with $f = 350$ mm (13.8 in.), $\theta_{skew} = 40°$, $\beta = 66°$, and $r = 72$ mm (2.8 in.). For a normal reference (and incidence) beam ($\alpha = 90°$), $\phi_{scan} = 0.605\ \phi_{rot}$, while for a tilted reference resulting in an incident beam tilted at an angle of 22° relative to the normal ($\alpha = 68°$ in Fig. 19), $\phi_{scan} = 0.980\ \phi_{rot}$. The relative multiplication factor obtained by using the tilted incident beam is 1.62. This is a significant amount of scan-angle multiplication.

Scan-angle multiplication factors of this magnitude provide significant design flexibility. For a given scan length one could: make the disc smaller to produce a more compact unit; keep the disc the same size and add facets to provide more scan lines or to generate a more complex scan pattern; move the disc closer to the window to increase the light collection efficiency of the individual facets; or some combination of all three options. One may, of course, elect to use scan-angle magnification to just generate a longer scan line.

The incident-beam tilt angle, the aberrations, and the scan-angle multiplication factor are interrelated: the smaller the amount of tilt of a collimated incident beam, the smaller the aberrations; conversely, the larger the tilt, the larger the scan-angle multiplication factor. By careful selection of the tilt angle of the incident beam, one can get a significant amount of scan-angle multiplication while maintaining an acceptable amount of aberrations. The amount of aberrations introduced can be determined with a suitable ray tracing program.[11] Acceptable levels of aberrations are established by the individual application.

If, on the other hand, the incidence angle is dictated by some other mechanical constraint in the scanner system, the aberrations can be controlled, as alluded to before, by controlling the designed convergence of the incidence beam. The closer the convergence point is to lying on the axis of disc rotation, the less severe the aberrations will be. This, however, has the same trade-off as manipulating the incidence angle. As the convergence of the incidence beam becomes shorter, the focal length, f, of the facet will become longer, perhaps even reaching or "passing" infinity and becoming negative. As this happens, the first term in Eq. (8) gets smaller or, in the case of negative f, becomes negative, thereby canceling out some of the scan multiplication factor of the other two terms. Whether or not this is a disadvantage depends on what is more important to the application, scan-line length or spot quality.

5 HOLOGRAPHIC DEFLECTOR MEDIA FOR HOLOGRAPHIC BARCODE SCANNERS

All holographic barcode scanners today (2004) use a rotating circular disc as the substrate for the recording medium. Other geometries have been considered, but, for barcode reading applications, the disc geometry offers a number of manufacturing advantages and is generally less expensive.

All of today's holographic barcode readers also operate only in the transmission mode. It would not be impossible to develop a reflective holographic barcode scanner, but the transmission mode provides a simpler design, an easier manufacturing process, and less susceptibility to disc wobble.[8]

There are two general types of media suitable for recording holographic optical elements on a disc surface for use in a holographic barcode scanner: surface-relief phase media, such as photoresist, and volume-phase media, such as bleached silver halide and dichromated gelatin. There are advantages and disadvantages associated with both types of media. For a more general review of the wide variety of holographic recording materials, see work by Bartolini[14] and Smith.[15]

The major factors influencing the selection of the type of holographic medium are manufacturing cost, diffraction efficiency, and, as will be described, the scan pattern density.

5.1 Surface Relief Phase Media

There are only two significant surface-relief phase media presently being used for holographic scanning discs – photoresist, and plastic copies made either directly from photoresist or from intermediate copies of the photoresist. As will be discussed later, this latter type is probably the least expensive of all to manufacture in a high-volume process and, from a purely cost consideration, this medium would appear to be the best choice for a holographic deflector disc.

The major disadvantage of the low-cost surface-relief material is that when using a simple mechanical replication process the resulting diffraction efficiency will be relatively low, on the order of 30%. This is because a much larger grating aspect ratio (depth vs. spacing) is required to produce high efficiency. Mechanically releasing such a high-efficiency copy from the master can be very difficult, often damaging the master and the copy. This is a major drawback in barcode scanner applications. The low efficiency means that a higher power laser must be used to get sufficient laser power onto the barcode symbol to obtain a good reading. The greater cost of the higher power laser may offset the lower cost of the disc.

It is possible to get high diffraction efficiency using a surface-relief medium for light that is incident on the disk in an S-polarized orientation.[16,17] It is not possible at the present time, however, to mechanically replicate these high-efficiency surface-relief holograms because of the aforementioned aspect ratio of the relief profile. (An example of such a high aspect ratio is shown in Fig. 20.) This means that original holograms, not inexpensive copies, would have to be used. In some holographic deflector applications, this is acceptable. In general, however, this is not an acceptable alternative for a holographic barcode scanner due to the higher cost of the discs and their greater susceptibility to physical damage. (Surface relief holographic discs cannot be protected by a cover glass since an index-matching adhesive would effectively eliminate the surface relief structure.)

The low diffraction efficiency of the mechanically replicated surface-relief material also means that the collected light will be low in a system employing the holographic facets of the disc in a retroreflective mode. This loss in collected light cannot be compensated by increasing the laser power further because of the limitations established by the federal laser safety standards. The only means left to compensate for the low diffraction efficiency in the collected light is to increase the size of the facets on the holographic disc. However, this reduces the total number of facets, hence the total number of independent scan lines, and the subsequent scan pattern density. In some barcode scanning applications, this may be an acceptable trade-off. In other applications, such a trade-off may be unacceptable.

For example, in supermarket/retail barcode scanning applications, the depth of field requirement is relatively moderate, so that a holographic scanner with as few as two focal

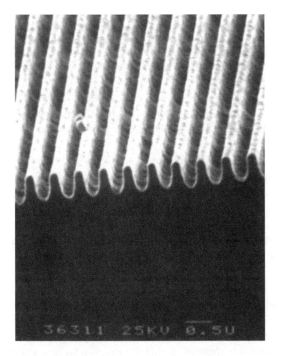

Figure 20 Surface-relief hologram in positive photoresist.

planes can provide adequate performance. However, in industrial barcode scanning applications, the required depth of field may be as large as 1 m (40 in.), or more, for medium density barcodes (barcodes with a minimum bar width on the order of 0.3 mm). This kind of depth-of-field requirement can only be met with a scanner that can provide a large number of focal planes. A holographic scanner can be designed to provide this capability, but the number of independent facets on the scanning disc must be as large as possible. Therefore, any reduction in the number of facets imposed by a low diffraction efficiency recording medium will reduce the depth of field.

For many retrocollection scanning applications even the "high-efficiency" holograms mentioned above do not have a high enough diffraction efficiency for good light collection. In order to eliminate specular reflection noise in a barcode scanner, a polarizer is often placed in front of the light detector. This is done so that one linear light polarization can be used for the outgoing, scanning beam while the other, perpendicular light polarization will return to the scanner, pass through the polarizer, and be detected. This presents a problem for surface-relief holograms in that an even higher aspect ratio relief profile is necessary to diffract P-polarized light than is necessary to diffract S-polarized light. As a result, while the outgoing efficiency could produce a strong scanning beam, the light collection ability would be relatively weak (or vice versa) requiring the same kind of trade-offs as discussed above.

Even so, despite the relatively low diffraction efficiency of the holographic discs produced by mechanical replication, which is a surface relief process, the low cost of such discs make them very attractive in supermarket and retail barcode scanners in which component cost is a major factor.

5.2 Volume Phase Media

Volume phase materials are capable of very high diffraction efficiencies, on the order of 90% or more. Such high efficiencies means that the individual facets on the disc can be relatively small, even when the disc is used in a retro reflective mode. This means that there can be more facets on the disc, which, in turn, means that the scanner can generate more independent scan lines, resulting in a larger depth of field and/or a more complex scan pattern. The higher diffraction efficiency also means that a lower power laser can be used to generate the scan lines.

There are a number of materials that are suitable for use as volume phase materials in holographic scanners. The first material that comes to mind is bleached silver halide. In this process, the absorptive structure in a photographic emulsion hologram is chemically converted from metallic silver to a material having a refractive index different than the surrounding gelatin matrix.[18-21] For example, the silver may be rehalogenated by exposure to bromine vapor. Holograms created with this material can have high diffraction efficiencies, on the order of 80% or more.[22,23] Processing is relatively simple, and the holograms are reasonably stable. There are a few bleaches, however, that leave reaction products behind in the emulsion. Some of these products are photosensitive and exhibit printout effects, particularly when subjected to intense ultraviolet irradiation. Nevertheless, moderately efficient holograms can be realized, and the advantages of photographic emulsions, such as extended spectral response and speed, may be exploited. One practical disadvantage associated with this material is that the discs must be coated and sensitized by one of the major companies producing general photographic materials. Such companies are usually reluctant to stock odd substrate shapes (like discs) and coat them to a user's specifications, in quantities that are, for them, relatively small. This creates a very real sourcing problem.

The next most attractive volume phase material is photopolymer.[24,25] Cross-linking in these materials is produced when they are exposed to light of relatively short wavelength, blue to ultraviolet. When a photopolymer is exposed to a holographic fringe pattern at the proper wavelength, the periodic variation in light intensity of the fringe pattern produces a corresponding periodic variation in cross-linking in the polymer. When developed, the photopolymer will exhibit a periodic variation in refractive index corresponding to this periodic variation in cross-linking. These materials are relatively stable when exposed to normal levels of ambient light, heat, and humidity.

The main drawback to these materials has been their relatively small change in refractive index, Δn, produced by exposure to light and subsequent processing. This means that, in order to get high diffraction efficiency, the thickness of the photopolymer coating has to be on the order of 50 microns (2 mils). Such a large thickness would make the holographic deflector disc very sensitive to the Bragg angle. That is, very slight deviations in the angle of incidence of the reconstruction beam in the scanner would cause severe reductions in the disc diffraction efficiency. Deviations on the order of $1/4°$ (4.4 mrad) could cut the diffraction efficiency in half.[26] This is generally unacceptable in a product where the total angular manufacturing tolerances could easily be this large. Furthermore, the anticipated mode of operation could cause the effective angle of incidence to vary by $1/4°$ (4.4 mrad) during disc rotation.

The DuPont[27] and Polaroid[28] photopolymers have exhibited refractive index changes that are much greater than those of previous photopolymers. Δn values approaching the values obtainable with dichromated gelatin (nearly ten times as great as earlier photopolymer Δn values) have been obtained. These materials have great potential for use as a recording medium for holographic deflectors used in barcode scanners, since high diffraction efficiency should be achievable in relatively thin coatings, on the order of 5 microns (200 μin.).

The volume phase material that has, up to now, been the most successful material for use in holographic deflectors for barcode scanners is dichromated gelatin (DCG).[29-31] The major advantage of this material, as a medium for holographic deflector discs, is that its diffraction efficiency can be very high (>90%) in a relatively thin (3-5 microns or 120-200 μin.) coating because of its high Δn (0.10-0.15 or greater). This means that dichromated gelatin can have, simultaneously, high diffraction efficiency and very low Bragg-angle sensitivity. This is a significant advantage from both a manufacturing and an application standpoint.

The major disadvantage of DCG is that it is extremely sensitive to moisture. Holograms made with DCG must be sealed to protect them from environmental moisture.

From the standpoint of the development of a barcode scanner, there is one other disadvantage to DCG. Although it has been around a long time, DCG is the least understood of all the holographic recording media. There are at least three theories that claim to explain the mechanism of image formation,[32-34] and there are as many recipes for processing DCG as there are authors writing on the subject. Many of them start with gelatin that is already coated on photographic plates,[35] a procedure which is unacceptable for the same reasons that bleached silver halide is unacceptable: the sourcing problem.

In most large corporations, one will also find considerable resistance to the use of dichromated gelatin. Most chemists feel comfortable with well-understood inorganic materials, such as silicon, and the more traditional organics, such as photoresist, photopolymer, and so on. Dichromated gelatin is an organic material whose properties are poorly understood and relatively unpredictable. Gelatin is, after all, made from the skins, bones, and connective tissues of animals. Its properties can vary depending on what the animals ate or where they were raised.

Nevertheless, because of its excellent holographic qualities, DCG is one of the best recording materials for holographic deflectors used in barcode scanners. It is relatively stable when exposed to normal ambient temperatures. However, it is extremely moisture sensitive and must be sealed to protect it from normal ambient humidity.

Dichromated gelatin is generally sensitive only to the short wavelength portion of the visible spectrum, $\lambda < 520$ nm (20.5 μin.), and although it is possible to sensitize it to the red end of the spectrum[36-40] only moderate success has been achieved. The primary problem has been removal of the residual sensitizing dye to give a complete phase structure. For barcode scanning applications, where the light source in the scanner is generally a visible laser diode (VLD) with a wavelength somewhere in the red region of the spectrum, unsensitized DCG cannot be used to make the master holograms at the operating wavelength. Because of this, the DCG holographic disc must be made as a copy of masters formed in one of two ways.

In the first method, the wavelength that will be used in the scanner is used to construct the masters with a material that is sensitive to that region of the spectrum, such as silver halide. This allows a relatively simple optical set-up that will produce an aberration-free hologram. Typically, DCG submasters (in the form of a submaster disc, or individual submasters) are made from the masters, providing greater efficiency to the production process.

In the second method, the masters are made directly in DCG using a wavelength within its spectral sensitivity range, such as 488 nm (19.2 μin.), one of the high-power lines of an argon laser. The difference of the exposure wavelength from that which the scanner will employ requires some kind of aberration-compensating optics in the master-exposure set-up. This makes the set-up more complicated, but the result can be a higher-quality hologram since the submaster step of the process is not needed, and the use of additional aberration-correcting optics will maintain the essentially aberration-free performance of the final hologram.

Whichever method is used, the DCG copy disc can then be made using any wavelength to which the DCG is sensitive. There will be no aberrations introduced in the copy process, regardless of the wavelength used in the copy process. We will have more to say about this in the section on disc fabrication.

Dichromated gelatin is processed in a sequence of alcohol/water baths of varying concentrations of alcohol and varying temperatures. Times, temperatures, and concentrations vary, depending on whose process is used.

Diffraction efficiencies obtained with DCG approach the theoretical limits for volume phase materials. Efficiencies greater than 90% can be readily obtained. The only things limiting the diffraction efficiencies of a sealed DCG holographic disc are reflections off the glass surfaces and absorption and scattering losses in the gelatin. If antireflection (AR) coatings are not employed, the Fresnel reflections at the air/glass interfaces will cause the primary losses. This may limit maximum efficiency to about 70%, depending on polarization. If, however, good AR coatings can be applied, there will still be some minor Fresnel losses at the internal gelatin/glass interface and some small amount of scattering and absorption in the gelatin. If the film properties can be controlled enough to keep scattering to a minimum, then efficiencies of 95% are achievable. This control, however, is the key, and is sometimes easier said than done. Even so, efficiencies in excess of 85% are relatively easy to maintain.

6 FABRICATION OF HOLOGRAPHIC DEFLECTORS

6.1 The DCG Holographic Disc

One possible DCG holographic disc fabrication process using DCG masters (no submasters) is shown schematically in Fig. 21. Each facet of the holographic master disc is recorded individually using an argon laser and a vibration-isolated table. The facets are recorded on rectangular DCG holographic plates, which are then processed in the water/alcohol baths. When a suitable efficiency is achieved, resulting in an optimum intensity-ratio of diffracted to transmitted light, the facets are "capped" (sealed with another piece of glass using an index-matching optical adhesive) for moisture protection then cut to a size appropriate to an automated copy-exposure machine. They are then masked to provide the designed size and shape when exposed by the copy beam.

The DCG copy disc fabrication process resembles, but is not identical to, a photographic contact-copy process. All of the DCG masters are placed on a computer-controlled wheel in the appropriate sequence. When the exposure cycle is started, a DCG disc is brought in under the master wheel with a slight air gap between the disc and the master. Because of this gap the process is not a true contact-copy process and some settling time is required after the motion stops before the exposure can begin. In situations where relative motion between the master and the disc is not required (such as with a submaster disc), index matching fluid can be used between the master and the disc. This limits relative motion and also greatly reduces reflections at the interface. For air gap copying,

Holographic Barcode Scanners

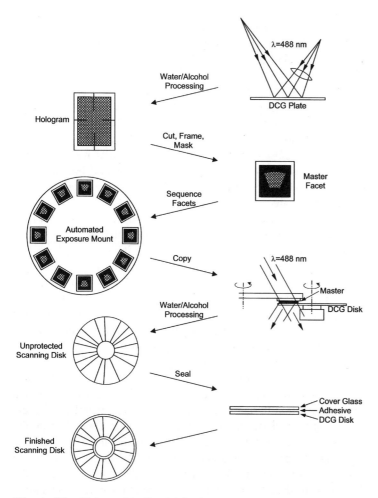

Figure 21 Holographic disc fabrication process.

AR-coated caps are recommended for the masters. During the exposure sequence each master facet swings into position and is then sequentially illuminated in a step-and-repeat exposure process, using an expanded beam from an argon laser. That beam may be collimated, divergent, or convergent, depending on the desired characteristics of the copy HOE. The angle of illumination of the laser beam is modified from facet to facet considering each facet's construction and the difference between the copy exposure wavelength and the scanner wavelength.

Exposure of each master holographic facet creates an optically identical holographic copy facet in the DCG through the interference of the diffracted beam with the undiffracted zero-order beam. As long as the copy process is reasonably close to a contact-copy process there will be no aberrations introduced by copying, regardless of copy wavelength or exposure angle. The reason for this is that the recorded interference pattern in the hologram will accept any configuration of incidence beam and produce a corresponding conjugate diffraction beam. Because those two beams were created from that interference pattern, those two beams are the exact beams that will recreate that same interference pattern, provided the recording medium is located in the same place as the master hologram. The more removed the recording medium is from the master hologram, the more the properties of the two will differ. For a very small space, the difference is insignificant.

When all of the facets of the DCG copy disc are exposed, the disc is processed in a sequence of water and alcohol baths. The process is essentially the same as the process for the DCG masters, although some of the timing may be different, and precise control and consistency are, ironically, more important in the production process than they are in the master process. This is because DCG can be reprocessed if the initial results are not satisfactory; however, reprocessing leads to inconsistency and inefficiency, both of which are undesirable in a production environment. The details of the baths, the relative times, and the temperatures of the liquids, are proprietary.

A major objective in any DCG disc manufacturing operation is to establish a total exposure and development process that provides consistent results and high yield. One of the more difficult problems encountered in attempting to do this is the problem of "gel swell" – the tendency for the exposed and processed gelatin to be thicker than the unexposed, unprocessed gelatin. This residual gelatin swell causes a shifting of the Bragg planes within the thickness of the gelatin so that the angle of the Bragg planes, relative to the surface of the gelatin, is not the same after processing as it was during exposure (see Fig. 22). This results in a decrease in diffraction efficiency when the reconstruction beam is at the designed incidence angle. Any attempt to increase the diffraction efficiency by changing the reconstruction beam angle will introduce undesirable aberrations.

A similar effect also occurs if the post-processing bulk refractive index is different from that anticipated. A different refractive index causes the incident reconstruction beam to refract into the gelatin at an angle different from that expected and, thereby, meet the Bragg planes at the wrong angle. Because the bulk index in DCG is dependent on the amount of microscopic air voids present in the processed gelatin, the bulk index is sensitive to changes in both the film preparation process and the water/alcohol process. In preparation, if the film is excessively hardened it will be more difficult to form voids and the bulk index will be higher. In general this also limits the range of the index modulation, Δn. During wet processing, if the hologram is left too long in hot water the gel can be oversoftened, creating excessive voids and lowering the bulk index. This also can cause

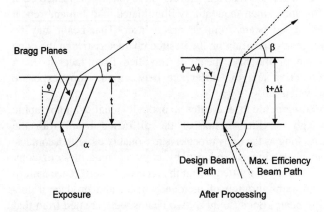

Figure 22 Effect of gel swell on the angle of the Bragg planes.

the problem of excess scattering losses since the voids can become larger and, therefore, make the gel appear less homogenous to the scanner wavelength.

Gel swell and bulk index changes are separate effects that can be separately measured; however, symptomatically they produce the same effect – reduction in efficiency due to "Bragg error." Several methods of eliminating these undesirable effects have been described in the literature.[35] Generally, these involve either some sort of post-processing chemical treatment or some form of post-processing baking of the hologram. None of these methods is predictable enough to be suitable in a manufacturing process. If, however, the gel swell is predictable and consistent, it can be compensated for in the copy process by reducing the calculated angle of incidence of the argon laser copy beam. This increases the Bragg plane tilt angle. After processing, the gel swell will raise the Bragg planes, decreasing the tilt angle until it equals the original value. This process is described in greater detail by Dickson.[42]

If the processing methods are well controlled, so that the gel swell and bulk index are both predictable and consistent, then one can eliminate the Bragg error problem by altering the copy beam angle. Altering the copy beam angle, incidentally, has no effect on the optical properties of the holographic copy disc since these are fixed in the surface fringe structure of the master. The copy process will always faithfully reproduce this fringe structure.

After the DCG disc is exposed and processed, several millimeters of the gelatin are stripped from the outer and inner edges to inhibit wicking-in of moisture in the sealed disc. The disc is then sealed with a glass cover disc for protection from moisture, using an index-matching optical adhesive. A metal hub is then bonded to the inner diameter of the disc and the disc is dynamically balanced.

The optical properties of each and every DCG copy disc will be identical to those of the master holographic disc. The optical characteristics of the holographic scanner are essentially established at the time of the construction of the holographic master. While it is possible to modify these characteristics somewhat through variations in pre-disc or post-disc optics, this is generally not done in holographic barcode scanners.

6.2 The Mechanically Replicated Surface-Relief Holographic Disc

The other primary holographic recording material, photoresist, has a diffracting structure in the form of surface deformation or relief as shown in Fig. 20.[16,17] Consideration can therefore be given to using mechanical replication for mass production. This is not to imply that optical copying techniques cannot be used with surface relief, because they certainly can, either as master or copy or both.

Mechanical replication of surface relief is not new, having been developed decades ago for low-cost replication of mechanically ruled gratings. Today, it has become one of the primary manufacturing techniques of the production of high-quality gratings from holographic masters.[43]

Although replication of a surface relief hologram can be performed directly from the photoresist master, there is some danger that the photoresist may not stand up to repeated mechanical pressures, elevated temperatures, and/or the copy-master release process. Considering the difficulty and potential expense associated with master fabrication, a replication process that permits maximum replication volume is required. This is accomplished by the fabrication of a more durable submaster and usually results in the master itself being sacrificed. A very early technique, borrowed from the audio recording industry uses a metal "stamper" to emboss or compression mold the relief into a vinyl thermoplastic.[44,45]

In the adaptation of this process,[46,47] a very fine grain layer of nickel or gold is deposited by evaporation or sputtering onto the relief to form a conductive conformal coating, typically a few hundred Angstroms thick. Nickel formation is then continued by other methods, such as electrochemical deposition, until a thickness of several hundred microns is achieved. At this point, the outer nickel surface has no significant relief and it can be attached to a rigid substrate. The sandwich is separated at the nickel/resist interface and residual photoresist dissolved away leaving behind a rigid metal replication of the relief. This structure is a negative of the original and can be used in hot pressing, injection molding, or epoxy replication processes, which will be discussed later.

An alternative method of submaster preparation is to transfer the resist relief downward into its own substrate by radio frequency (RF) sputter etching or reactive ion etching (RIE) techniques.[48,49] In these methods, the relief surface is removed at a uniform rate by bombarding the relief with accelerated ions or, in the case of RIE, with reactive atoms that react with the substrate molecules to form a volatile gas. The valleys of the resist pattern disappear first and the underlying substrate is exposed and etching occurs. By the time the resist peaks have disappeared, the valley areas are deeply cut into the substrate. Proper choice of photoresist, substrate materials (such as silicon or quartz), and plasma parameters allows the surface relief to be accurately transferred into the substrate and the cross-sectional shape to be preserved.[50] These processes result in a submaster having a positive replication of the original master in contrast to the negative shape of the previous nickel submaster.

Once fabricated, these more durable submasters may be used to generate multiple copies. One such method, thermal mechanical embossing or compression molding, is accomplished by pressing the relief into a heated and softened thermoplastic film such as polymethyl methacrylate (PMMA) or polyvinyl chloride (PVC). Bartolini and colleagues[46] rolled the submaster together with a vinyl strip between two heated cylinders. Gale et al.[47] used a conventional hot stamping press at 150°C and 3 atm. to emboss into PVC sheets. A similar pressing technique was used by Iwata and Tsujiuchi[51] with separation of the copy from the mold performed by sudden cooling and differential contraction.

Replication by pressing tends to introduce considerable strain and other inhomogeneities into the new substrate. These problems can be overcome by using injection molding techniques that have been developed for high-volume, high-quality fabrication of plastic lenses.[52] In this case, the submaster is one surface and an optically polished stainless steel flat is used as the facing, parallel surface. The appropriate polymer is plasticized to a more fluid state than used by compression molding and introduced into the temperature-controlled mold under high pressure.[53] Most of the materials in use are copolymers of PVC, polyvinyl acetate (PVA), and acrylic (PMMA) compounds. The acrylic material has an advantage over vinyl due to the lack of birefringence in the finished substrate, and the stability and ease with which it can be machined and polished.

The final alternative is to use a polymer that can be cross-linked by ultraviolet illumination.[54] This technique eliminates the need for high-temperature processing and reduces the possibility of induced stresses and dimensional changes upon cooling/curing. An injection mold apparatus can be adapted for these purposes as long as one plate is sufficiently transparent for the UV illumination. Depending on the use of release agents

Holographic Barcode Scanners

and relative adhesion, the replica can also be attached directly to a rigid substrate in the same operation.

7 AN EXAMPLE OF A HOLOGRAPHIC BARCODE SCANNER: THE METROLOGIC PENTA SCANNER

As an example of a holographic scanner that uses most of the design techniques and methods discussed thus far we will now discuss the Metrologic Penta Scanner, an industrial application scanner that exploits many of the advantages of holography. First we will discuss the design of the scan pattern and then the means by which it is produced.

7.1 The Penta Scan Pattern

The Penta scanner was designed as a large-scale "pass-through" scanner. In general, it was designed to create an aggressive scan volume at some distance from the scanner through which packages would pass in a roughly uniform manner. Ideally, a large range of package sizes with barcodes of moderate resolution would be able to pass through the scan volume, either manually or automatically, and as long as the package was roughly facing the scanner it would be successfully scanned throughout a large depth of field.

This broad application definition places several requirements on the design. A large depth of field will require multiple focal planes. With no specification given to package orientation the scanner will have to read essentially all orientations. Also, the variety of package sizes requires a large scan pattern size coming out of the scanner. All of these requirements can be met, quite easily, with holography.

The name of the "Penta" scanner comes from the pentagonal configuration of the scan pattern, shown in Fig. 23. The basic pattern is formed by combining five simple rasters (groups of parallel lines) of different angular orientation. The rasters are evenly spaced through an entire 360°, making this pattern omnidirectional. Omnidirectionality is

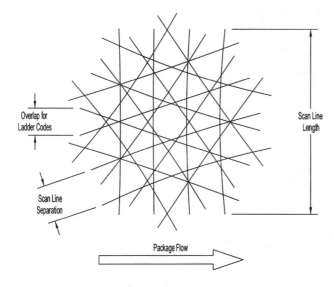

Figure 23 The two-dimensional representation of the Penta scan pattern.

an essential characteristic for a barcode scanner if it is desired that the operator not waste time trying to identify the proper orientation of the code for a good scan. In automated applications, where the highest efficiency is desired, packages often are presented in truly random orientations as they pass under the scanner on a conveyor belt.

Assuming a direction of package flow, as shown in Fig. 23, the primary parameters that define the pattern are the scan line length, the line separation, and the number of lines per orientation group. These parameters must be combined in such a way as to provide the desired total pattern omniwidth (width over which full omnidirectional scanning is possible). At the same time enough overlap in the near-vertical fields must be maintained such that codes in the "ladder" orientation (i.e., codes travelling in the direction perpendicular to its own bars and spaces) cannot slip through the pattern unscanned.

This pentagonal pattern not only provides for omnidirectional scanning, but it is also very aggressive on codes with higher aspect ratios, which are more difficult to scan than square codes like the original UPC code. The smaller angle between adjacent scan-line groups means each group has a smaller angular range that it must cover, and, therefore,

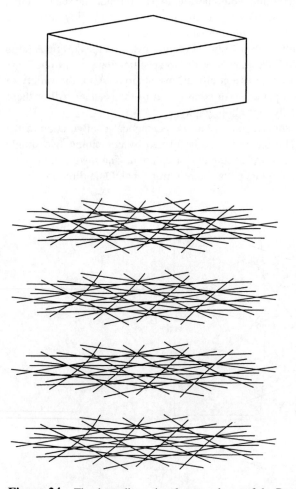

Figure 24 The three-dimensional scan volume of the Penta scanner.

Holographic Barcode Scanners

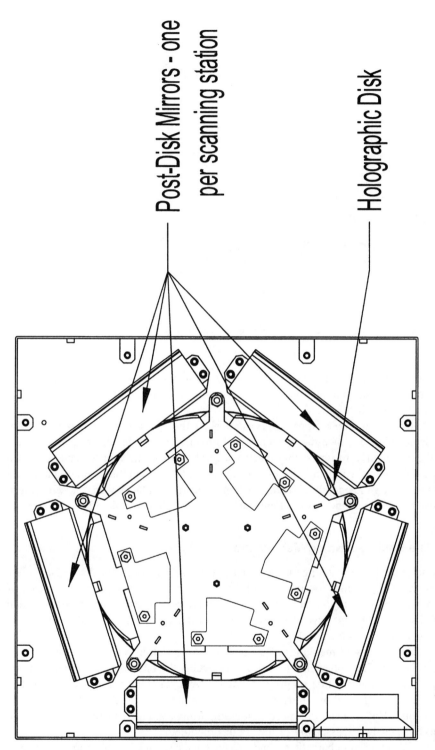

Figure 25 Penta scanner top view showing the different scanning stations.

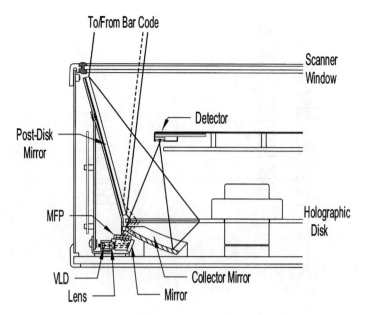

Figure 26 Penta scanner side view showing the scanning and light-collection optics.

less code height is required. This also means there will be less reliance on software stitching algorithms, but such algorithms can still be employed to make the scanner's performance even better.

Once the optimum scan pattern is established, the task still remains to provide for a large depth of field. Since the optimum performance pattern has been determined, the logical conclusion is to repeat that pattern several times at different distances from the scanner. Different focal planes are established providing acceptable overlap and producing a full, contiguous depth of field. For Penta four focal planes were chosen. A three-dimensional representation of the Penta scan pattern is shown in Fig. 24.

7.2 The Penta Scanning Mechanism

The heart of the Penta scanning mechanism is, of course, a holographic disc. Also included are five scanning stations located around the periphery of the disc, each comprised of a VLD module prior to the disc and a folding mirror after the disc to direct the beams out the scanner window. A top view of the scanner with the cover removed, showing the five scanning stations, is shown in Fig. 25.

A clearer view of the optics of a single scanning station can be seen from the side view in Fig. 26. The dotted line in Fig. 26 represents the outgoing beam path. The path starts at the VLD and is first roughly collimated by a conventional, aspheric lens. From there it reflects off a folding mirror, which directs the beam to the multifunction plate (MFP). The MFP is a multipurpose hologram, which, along with the VLD, lens, and mirror, finishes the subassembly of the "optics module."

The VLDs used in barcode scanners today have certain inherent properties, some of which are undesirable. With the use of the MFP, however, some of the undesirable effects can be alleviated. The functions performed by the MFP include beam aspect ratio

Holographic Barcode Scanners

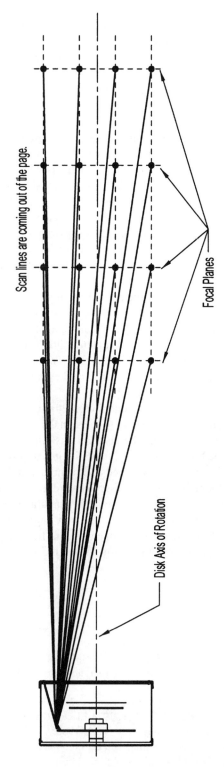

Figure 27 Side view of the scan lines produced by one scanning station.

modification, astigmatism reduction, and dispersion minimization. In fact, VLDs inherently produce beams with an elliptical shape and a characteristic astigmatism. Both of these properties of the beam can be manipulated by a single MFP (or by more than one if a greater range of control is desired) by simply choosing the incidence and diffraction angles of the MFP properly. At the same time, the dispersion produced by the facets of the disc (natural to all diffraction gratings) can be minimized by the same MFP by countering it with its own dispersion.

Following the MFP the beam heads directly to the holographic disc, incident on it at a specified angle. This is where the primary pattern formation occurs. Each of the five rasters in the pattern in Fig. 23 contains four lines, and that pattern is repeated in four focal planes. This requires 16 unique facets on the disc, which were shown previously in Fig. 18. Each one of these facets has a different diffraction angle, θ_d, and a different focal length, f, which results in the beam being focused at a different distance, s. The result of combining all the different focal lengths and diffraction angles is shown in Fig. 27. Choosing the right combination of focal lengths and diffraction angles results in scan lines that are laterally equidistant from the disc rotation axis, thereby reproducing the desired pattern at the different distances.

The solid lines in Fig. 26 represent the path of the light reflected off a barcode as it makes its way to the signal detector. The light that is collected returns essentially along the same path by which it left the scanner, making the scanner a retrocollective system. On the way back the light has diffusely spread out to completely fill the aperture of the facet, after first being reflected off the large pattern-folding mirror. The light is then diffracted back towards the module; however, a collector mirror is in the path everywhere except for a small hole through which the outgoing beam passed.

This collector mirror generally has a parabolic or elliptical shape. The light incident on the collector is focused and directed upward to the signal detector; however, in order to get there it must, once more, pass through the disc. On this third pass, however, the desire is for the disc to not affect the beam in any way. In reality it is impossible not to have some losses occur, but if the disc is correctly designed and manufactured those losses need not be much worse than that of a plain piece of glass. This is due to the angular sensitivity of the diffraction efficiency of the holograms. A properly manufactured disc will only have high efficiency at the designed incidence and diffraction angles. Because the rays proceeding from the collector to the detector are incident on the disc at angles far enough removed from the design incidence and diffraction angles, the resulting transmission of the disc is relatively high.

REFERENCES

1. Savir, D.; Laurer, D.J. IBM Systems J. 1975, *14*, 16.
2. Broockman, E. U.S. Patent 4,717,818, assigned to IBM, Jan. 5, 1988.
3. Allais, D.C. *Bar Code Symbology*; Intermec Corporation: Everett WA, 1984.
4. Cindrich, I. Appl. Opt. 1967, *6*, 531.
5. Beiser, L. Proc. 1975 Electro-Opt. Syst. Des. Conf. 1975, 333.
6. Beiser, L.; Darcey, E.; Kleinschmitt, D. Proc. 1973 Electro-Opt. Syst. Des. Conf. 1973, 75.
7. Beiser, L. *Holographic Scanning*; Wiley: New York, 1988.
8. Sincerbox, G.T. *Laser Beam Scanning*; Marshall, G., Ed.; Marcel Dekker: New York, 1985; 1.
9. Gabor, D. Nature 1948, *161*, 777.
10. Leith, E.; Upatnieks, J. J. Opt. Soc. Am. 1962, *52*, 1123.

11. Dickson, L.D.; Sincerbox, G.T.; Wolfheimer, A.D. IBM J. Res. Dev. 1982, 26, 228.
12. Pole, R.V.; Werlich, H.W.; Krusche, R. Appl. Opt. 1978, 17, 3294.
13. Dickson, L.D. Appl. Opt. 1970, 9, 1854.
14. Bartolini, R.A. Proc. SPIE 1977, 123, 2.
15. Smith, H.M., Ed. *Holographic Recording Materials*; Springer-Verlag: New York, 1977.
16. Werlich, H.; Sincerbox, G.; Yung, B. Dig. 1983 Conf. Lasers Electro-Opt. 1983, 224.
17. Werlich, H.; Sincerbox, G.; Yung, B. J. Imaging Tech. 1984, 10 (3); 105.
18. Rogers, G. J. Opt. Soc. Amer. 1965, 55, 1185.
19. Upatnieks, J.; Leonard, C. Appl. Opt. 1969, 8, 85.
20. Pennington, K.; Harper, J. Appl. Opt. 1970, 9, 1643.
21. Graube, A. Appl. Opt. 1974, 13, 2942.
22. Phillips, N.; Porter, D. J. Phys. E. 1976, 9, 631.
23. Phillips, N.; Cullen, R.; Ward, A.; Porter, D. Photogr. Sci. Eng. 1980, 24, 120.
24. Booth, B. J. Appl. Phot. Eng. 1977, 3, 24.
25. Chandross, E.; Tomlinson, W.; Aumiller, G. Appl. Opt. 1978, 17, 566.
26. Kogelnik, H. Bell. Sys. Tech. J. 1969, 48, 2909.
27. Gambogi, W.J.; Gerstadt, W.A.; Mackara, S.R.; Weber, A.M. Proc. SPIE 1991, 1555, 256.
28. Ingwall, R. Proc. SPIE 1986, 615, 81.
29. Shankoff, T. Appl. Opt. 1968, 7, 2101.
30. Lin, L. Appl. Opt. 1969, 8, 903.
31. Chang, B.J. Opt. Eng. 1980, 19, 642.
32. Meyerhofer, D. RCA Rev. 1972, 33, 111.
33. Samoilovich, D.; Zeichner, A.; Freisem, A. Photogr. Sci. Eng. 1980, 24, 161.
34. Sjolinder, S. Photogr. Sci. Eng. 1981, 25, 112.
35. Chang, B.J.; Leonard, C.D. Appl. Opt. 1979, 18, 2407.
36. Graube, A. Opt. Commun. 1973, 8, 251.
37. Graube, A. Photogr. Sci. Eng. 1978, 22, 37.
38. Kubota, T.; Ose, T. Appl. Opt. 1979, 18, 2538.
39. Akagi, M. Photogr. Sci. Eng. 1974, 18, 248.
40. Kubota, T.; Ose, T.; Sasaki, M.; Honda, M. Appl. Opt. 1976, 15, 556.
41. Lin, L.; Doherty, E. Appl. Opt. 1971, 10, 1314.
42. Dickson, L.D. U.S. Patent 4,416,505, assigned to IBM, Nov. 22, 1983.
43. Lerner, J.; Flamand, J.; Thevenon, A. Proc. SPIE 1982, 353, 68.
44. Ruda, J.C. J. Audio Eng. Soc. 1977, 25, 702.
45. Roys, W.E. Ed. *Disc Recording and Reproduction*; Dowden, Hutchinson & Ross: Stroudsburg, PA, 1978.
46. Bartolini, R.; Feldstein, N.; Ryan, R. J. J. Electrochem. Soc. 1973, 120, 1408.
47. Gale, M.T.; Kane, J.; Knop, K. J. Appl. Phot. Eng. 1978, 4, 41.
48. Hanak, J.J.; Russell, J.P. RCA Rev. 1971, 32, 319.
49. Lehman, H.W.; Widner, R. J. Vac. Sci. Tech. 1980, 17, 1177.
50. Matsui, S.; Moriwaki, K.; Aritome, H.; Namba, S.; Shin, S.; Suga, S. Appl. Opt. 1982, 21, 2787.
51. Iwata, F.; Tsujiuchi, J. Appl. Opt. 1974, 13, 1327.
52. Wolpert, H.D. Photonics spectra 1983, 17 (2–3), 68.
53. Ryan, R.J. RCA Rev. 1978, 39, 87.
54. Okino, Y.; Sano, K.; Kashihara, T. Proc. SPIE 1982, 329, 236.

11

Optical Disk Scanning Technology

TETSUO SAIMI

Matsushita Electric Industrial Co., Ltd., Kadoma, Osaka, Japan

1 INTRODUCTION

The aim of this chapter is to describe important aspects of optical disk recording and readout technologies, with a brief historical introduction and references for further study. The selected topics are based on the contemporary analysis and experimental results of general interest.

1.1 Progress in Optical Disk Technology

The fundamental concept of an optical disk dates back to 1961 when Stanford Research Laboratories developed a video disk using photographic technology. However, the low luminance of available light sources yielded reproduced images of low quality. Columbia Broadcasting System (CBS) announced the EVR (Electronic Video Recorder) system in 1967, but enormous costs ultimately forced them to discontinue development. The invention of the laser by T.H. Maiman *et al.* in 1960 provided the light source considered the most suitable for optical disks.

Lasers have good temporal and spatial coherence, which enables one to obtain the small, diffraction-limited beam spot necessary for high-quality information retrieval from optical disks. After many approaches were considered, the basic design of optical disks, the "bit-by-bit" recording method, was developed in the 1970s. The first optical video disk system for commercial use, the VLP (video long play), was released in 1973 by Philips of Holland and MCA (Music Corporation of America) of the United States. In early systems, the He–Ne laser was the preferred light source. The introduction of many new optical disk systems soon followed. The 12 cm diameter digital audio disk (DAD), later called the CD (compact disk), was announced in 1978. Standardized CD products from several manufacturers became available in December 1982. CD players use semiconductor lasers to allow the design of small and lightweight players. In 1996, the digital versatile disk (DVD) for players was released. These playback-only systems marked the inception of optical disk products. Write-once optical disk systems were first introduced by Philips in 1978.

Development of rewritable optical disk systems accelerated in the 1980s as the performance of reversible media progressed. Magneto-optical disks that utilize a magnetic field reversal for recording and the Kerr effect for playback were commercialized in 1988 by Sony. In 1989, the first phase-change rewritable disk, containing 470 megabytes user capacity and utilizing an amorphous-to-crystalline phase change[1] for recording and playback was commercialized by Matsushita. In 2000, rewritable DVDs (DVD-RAM,-RW) were released and the development for higher density DVD media started.

1.2 Characteristics of Optical Disks

Optical disks are now used in various applications, including audio, computer memory devices, picture files, document files and video files. The advantages of optical disks over other known memory devices are:

1. Large capacity/high information density. The information capacity of a 120 mm diameter DVD disk is 4.7 Gbytes for single-layer ROM and RAM, and 8.5 Gbytes for double-layer ROM. The recording density of commercial products is about 3.3 Gbits/in^2 for DVD-ROM and 4 Gbits/in^2 for DVD-RAM. Recent developments for next generation products show that an information density of more than 16 Gbits/in^2 can be achieved by using a blue laser and an objective lens of high numerical aperture.
2. Fast random-access library systems allow access to large mass memories. Changing mechanisms provide access within seconds to several petabytes of information.
3. Reliability. The information surface of an optical disk is covered with a protective layer, which ensures a long archival life. Information retrieval is achieved without physical contact between the optical pick-up and disk, which increases the reliability of stored information.
4. Replication. Mass production using injection molding or other high-volume techniques is possible. Replicated optical disks benefit from lower cost per bit than the rigid magnetic disk or tapes.
5. Removability/ROM-RAM compatibility. A large quantity of data can be handled easily by exchanging disks. Compatibility between replicated and recordable disks and interchangeability between standardized drives provide this capability. These advantages leads to the ubiquitous uses of optical disk products in consumer and computer applications.

1.3 Principles of Optical Read/Write[2-6]

In many optical disks, as in the normal audio disk, information is recorded in a spiral groove referred to as the "track." The information cells shown in Fig. 1 are called "pits," or "marks." They are discontinuous small depressions, differential reflectivity patterns or phase-shifting patterns, all showing differential reflectivity. Information signals are derived from changes in luminance caused by diffraction of the laser beam by the pits or marks (which are about 0.3 µm^2 diffraction cells). The laser beam emerging from the

Optical Disk Scanning Technology

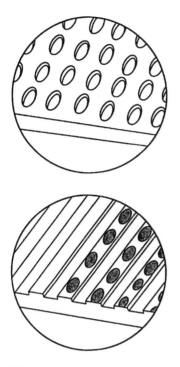

Figure 1 Pit patterns: (a) phase pit, (b) amplitude pit.

objective lens is focused to a spot on the disk. The spot size is proportional to the wavelength λ of the laser beam and inversely proportional to the numerical aperture (NA) of the objective lens.

The numerical aperture NA is given by the sine of the angle θ between the optical axis and the marginal rays:

$$NA = n \sin \theta \tag{1}$$

where n represents the refractive index of the medium in object space. The full-width-at-half-maximum (FWHM) intensity diameter of the beam spot (D_s) on the disk is expressed as

$$D_s = k \frac{\lambda}{NA} \tag{2}$$

where k represents a constant dependent upon the light amplitude distribution at the objective lens pupil. If the incident beam to the objective lens is plane wave, the value of k is 0.53. When the incident beam is Gaussian or contains some aberrations, k becomes larger. Since the information density of the disk is inversely proportional to the square of D_s, smaller k is more desirable. Supposing $k = 0.53$, $\lambda = 0.405$ μm (15.9 μin.), and $NA = 0.85$, we obtain the beam spot diameter $D_s = 0.25$ μm (9.9 μin.). More than 1.4×10^{11} information bits can be stored on one side of a 5.25 in. optical disk using a beam of this size.

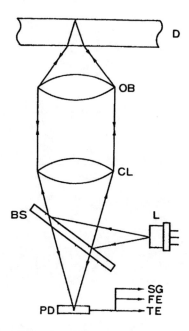

Figure 2 Playback optics.

Figure 2 shows the playback optics for a reflective optical disk. The laser emission from the semiconductor laser (L) is reflected by a beam splitter (BS) and is incident on the objective lens (OB) through a collimating lens (CL). The wavelength λ generally used for CD is in the range 780–800 nm and for DVD, 635–660 nm. The numerical aperture of the objective lens is generally 0.45 when used for CD and 0.6 for DVD. Next generation optical disks will have a wavelength of 405 nm and a numerical aperture of 0.85. The objective lens aperture limits the spatial frequency response of the optical system.

The laser beam reflected from the disk is intensity-modulated by the pits prior to a second pass through the objective lens. Part of the return beam is transmitted through the beam splitter and is incident on the photodetector (PD). The information signal (SG), focus error signal (FE), and tracking error signal (TE) are generated from the photodetectors.

Reflective or transmissive mode systems can be constructed, but the reflective mode is used in most optical disk systems. In the transmissive mode, a second optical pick-up with the photodetectors must be positioned on the other side of the disk, complicating the design of the drive. Another problem is that the pits must be very deep and replication becomes more difficult, which leads to degradation of signals during read. A third problem with the transmission mode is difficulty in obtaining a good focus error signal with a satisfactory S/N ratio. Simple focus error signal detection methods are easily achieved in reflective mode.

2 APPLICATIONS OF OPTICAL DISK SYSTEMS

2.1 Read-Only Optical Disk Systems

Four types of standardized players are available for read-only optical disks: video disk, audio disk (CD), data file disk (CD-ROM), and DVD. Among the advantages of read-only optical disks are (1) mass replication, (2) a relatively simple optical layout as compared to write-once or rewritable system, and (3) ease of commercialization due to its use as a stand-alone unit. Signals in read-only optical disk systems are generally encoded with pulse width modulation (PWM), resulting in high recording density.

2.1.1 Video Disk

Optical video disk systems have been commercially available for many years. They have the international standard name of LV (laser vision), and use analog signal recording. Two types of disks having diameters of 30 and 20 cm are utilized. Rotational speed is constant at 1800 rpm for the constant-angular-velocity (CAV) mode, and a variable rotational speed of 600–1800 rpm is used for the constant-linear-velocity (CLV) mode. Laser vision has relatively low recording density and is now being replaced by DVD.

2.1.2 CD/CD-ROM

The digital audio disk system has been standardized using the term compact disk. The diameter of the disk is 12 cm (4.7 in.), and the thickness of the polycarbonate protective layer is 1.2 mm. The linear velocity can vary from 1.2 to 1.4 m/s (3.9–4.6 ft/s) in the CLV mode. The maximum playback time is about 75 minutes, long enough to accommodate a fairly long classical music selection on a single disk. Audio signals are quantitized using 16 bits, allowing a dynamic range of 96 dB in playback. CD is now dominant in package media and is widely used in the music market. Using the ordinary data signal coding of the CD system, computer compatible data can be stored for use as a read-only memory of a personal computer (CD-ROM). More than 650 Mbytes of information can be stored on one side of a disk, enough to store the entire text of *Encyclopedia Britannica* on one side of a CD-ROM disk. Most personal computers are equipped with a CD-ROM, although DVD drives are rapidly replacing the CD drives in this application.

2.1.3 DVD

The successor to the CD is the integrated DVD optical disk system. The DVD specifications for Read-Only Disk were issued in 1996, followed by Rewritable (ver. 1.0) in 1997, Rewritable (ver. 2.0), and Re-recording (ver. 1.0) in 1999, and DVD-R for General (ver. 2.0) in 2000. These DVD systems are all integrated into DVD for Multi. The DVD disk has a storage capacity of 4.7 Gbytes, and more than 135 minutes of MPEG2 video signal can be stored on one side of a 12 cm disk. The DVD has overtaken almost all use of LD and CDV in video and music videos.

2.2 Write-Once Disk Systems[7–9]

Write-once disks have been commercially implemented in applications such as archival data memory devices for computers, document storage, and picture filing systems. Polycarbonate (PC) is being used to form injection molded disk substrates. The recording mechanism of the disk may be (1) phase-changing, (2) hole-burning, and (3) bubble-forming. In Fig. 3, the pits formed by these different recording methods are schematically shown.

Signal pits are recorded by irradiation with a semiconductor laser focused to a spot less than 0.3 up to 1 μm in diameter. This irradiation increases the temperature of the

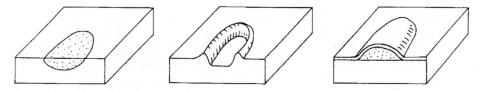

Figure 3 Pits formed by different recording methods.

recording medium to about 200–600°C (392–1272°F), and the recording takes place as the result of the consequent physical or chemical change of the medium.

2.2.1 CD-R

Write-once disk systems can be used for archival storage of large data files. The removability of optical disks and the standardization of products provides a broad range of application. CD-R is currently the most commonly used write-once disk. Typical specifications of CD-R disk systems are shown in Table 1.

2.3 Erasable Optical Disk Systems

Two major families of erasable media are available: phase-change rewritable (PCR) and magneto-optical (MO). Data recording on phase-change rewritable media is accomplished by inducing a transition from a crystalline phase to an amorphous phase. Differences in reflectivity of the two phases allow signal playback.

Magneto-optical recording is accomplished by establishing the magnetization of a mark by heating it in the presence of a magnetic field. Read back utilizes the polarization change of the laser beam induced by magnetic modulation according to the Kerr effect. The principal characteristics of erasable disks are shown in Table 2. The overwrite mechanism of a PCR disk is easy to design. However, reversibility is better in MO disks. The MO disk drive requires a complicated system for applying the write and erase magnetic fields, which have opposite polarities.

Table 1 Specifications of CD-R Disk Systems

Items	Unit	Specifications
User data capacity	Mbytes	~650
Disk diameter	mm	120
NA of objective lens		0.5
Wavelength	nm	775–795
Wavefront distortion of pick-up	λ	<0.050
Recording power	mW	$4 < P_o < 8$
Playback power	mW	<0.7
Thickness of disk substrate	mm	1.2
Rim intensities		Tangential 0.14 ± 0.04
		Radial 0.7 ± 0.10

Optical Disk Scanning Technology

Table 2 Characteristics of Erasable Optical Disks

	PCR	MO
Recording and erasing mode	Phase change	Change of magnetization
Read	Change of amplitude	Change of polarization
Material of medium	Te-Ge-Sb	Tb-Fe-Ni-Co
Overwrite mechanism	Simple	Complicated
Magnetic field	Not required	Required
Reversibility	Fair	Good
Required power	High	Medium

PCR, phase-change rewritable; MO, magneto-optical.

2.3.1 PCR Disk[7]

Figure 4 shows PCR optical data file drives (DVD-RAM). Figure 5 shows the principle of the direct overwriting mechanism for the PCR disk. The laser intensity at the disk is modulated, in correspondence with the pit pattern to be recorded, between the maximum level (A) and the intermediate level (B) as illustrated in Fig. 5(a). At exposure level (A), the material reaches a melting temperature of over 600°C. Rapid quenching forces the material to remain in the amorphous phase, giving low surface reflectivity. Exposure level (B) heats the material to about 400°C, allowing rapid crystallization to proceed, giving an increased reflectivity. Figure 5(b) is a schematic illustration of the overwriting operation.

Figure 4 Optical data file drives (DVD-RAM).

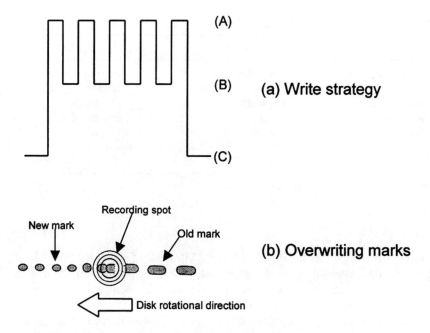

Figure 5 Principle of the direct overwriting mechanism for the PCR disk.

2.3.2 MO Disk[10,11]

In magneto-optical disks, a light beam is directed at a magnetic material to record or erase information. The underlying principle is the utilization of a temperature-dependent change of magnetic properties. There are several methods, including Curie point recording and compensation point recording, which can be used for recording. Figure 6 is an elementary illustration showing the principle of Curie point recording. In this example the initial magnetization of the recording layer is uniformly oriented in a given direction, as shown in Fig. 6(a). When a limited area of the recording layer is irradiated with light sufficient in intensity to heat it to a temperature above the Curie point T_c, the magnetization of the local area is lost, as shown in Fig. 6(b). When the exposure is discontinued, the temperature of the recording layer falls below T_c. The exposed area is remagnetized, but the direction of this magnetization coincides with the direction of the applied external magnetic field. Therefore, if the external magnetic field is applied in a direction opposite that of the original magnetization of the recording layer, as shown in Fig. 6(b), a magnetic domain different from the surrounding area remains, as shown in Fig. 6(c), enabling the recording of binary information. For reading the signal, the recording layer is irradiated with a laser light of low power. The polarization rotations of the reflected beam from the signal surface and the land surface are in opposite directions, as shown in Fig. 6(c). These beams are detected with a polarization analyzer to obtain the read signal. To erase the information, a selected area is again heated to a temperature above the Curie point, as shown in Fig. 6(d). The direction of the external magnetic field is reversed from that for recording.

In signal readout, the linear polarization angle of the incident beam is set at θ. The Kerr rotation angle is given by $\pm \Phi_k$. Referring to Fig. 7, the differential output ΔI of the

Optical Disk Scanning Technology

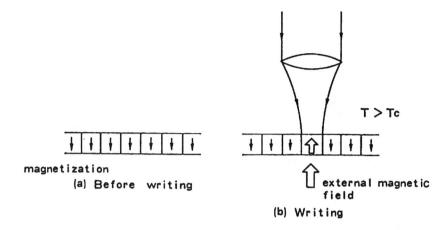

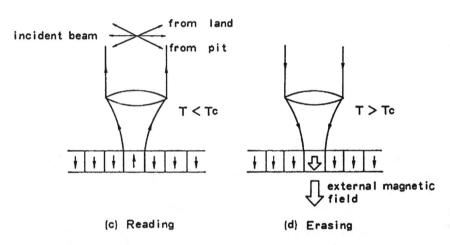

Figure 6 Magneto-optical disk method: (a) before writing, (b) writing, (c) reading, (d) erasing.

analyzer between the x and y directions is given by

$$\Delta I = I_0 R[\cos^2(\theta - \Phi_k) - \cos^2(\theta + \Phi_k)]$$
$$= (1/2)I_0 R \sin(2\theta) \sin(2\Phi_k) \tag{3}$$

where R is the disk reflectivity.

Since the linear polarization angle of the incident beam is $\pi/4$ and $\Phi_k \ll 1$

$$\Delta I \sim I_0 R \Phi_k \tag{4}$$

Thus, the playback signal level is proportional to the incident light intensity I_0, the disk reflectivity R, and the Kerr rotation angle Φ_k.

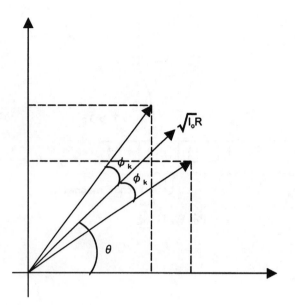

Figure 7 Readout of MO signal.

3 BASIC DESIGN OF OPTICAL DISK SYSTEMS

3.1 Pick-Up Optics

The many types of optical disks described in the preceding sections each have optimized optical pick-ups. The methods of design for the optics and mechanics of a writable optical pick-up will be described in this section.

The following factors determine the quality of read/write signals.

1. frequency characteristics of signals;
2. cross-talk from the adjacent tracks, which degrades read/write signals;
3. carrier-to-noise ratio (CNR) of read/write signals;
4. errors rate in read/write signals.

Factors 1 and 2 are mainly dependent on the wave aberrations of the optics. Factor 3 is as much associated with the characteristics of elements such as the semiconductor laser, detector and electronics as with wave aberrations, and factor 4 is mainly dependent on defects in the disk.

3.1.1 Optical Layout

The schematic construction of the optics for a writable optical pick-up is shown in Fig. 8. In this example, the astigmatic method is used for detecting the focusing signal and the push–pull method is used for detecting the tracking signal. The laser beam emitted from the semiconductor laser (LR) has a near-field pattern elongated in the direction of the active layer of the laser and is polarized in the same direction. The beam waist in this direction lies within the laser, and the beam waist in a direction perpendicular to the above direction is situated at the end facet of the laser active layer. The beam emergent from the laser is therefore anamorphic, and its far-field distribution is elliptical in cross-section with

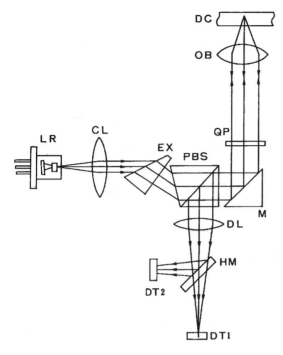

Figure 8 Schematic construction of a writable optical pick-up.

an ellipticity of 2 to 3. To correct this elliptical distribution, it is necessary to use a one-dimensional afocal system after the collimating lens (CL), consisting of two cylindrical lenses or a wedge prism. With a single wedge prism, the designed incident angle of the laser beam must be approximately 69–72°. Since the wedge prism has chromatic dispersion, a change of the wavelength of the laser results in an angular deviation of the beam. Taking this angular deviation as $\Delta\theta$ and the focal length of the objective lens (OB) as f_o, the beam spot moves approximately by $f_o \cdot \Delta\theta$ on the disk (DC). Using a single BK7 wedge prism with the incident angle 72°, an objective lens with focal length 4.5 mm and wavelength 0.78 μm, the beam spot displacement on the disk is approximately 0.073 μm for a change of 1 nm in wavelength. Therefore, the optics should be designed such that the direction of this movement will not cause a track offset. For this reason it is good practice to use two wedge prisms as illustrated in Fig. 8. In the case of the playback-only optical pick-up, the influence of the elliptical and astigmatic beam can be small at the cost of beam utilization efficiency.

In Fig. 8, the laser beam transmitted through a polarizing beam splitter (PBS) as a p-polarized beam passes through the $\lambda/4$ plate (QP) to become a circularly polarized beam, which is incident on the objective lens (OB). The beam emerging from the objective lens is incident on the disk (DC) to form a beam spot for recording and reproducing the signals. The beam reflected at the disk enters the objective lens and again passes through the $\lambda/4$ plate (QP) to become an s-polarized beam and is reflected to the detection lens (DL) by the polarizing beam splitter. The beam emergent from the detection lens is partially reflected by a half-mirror (HM) and incident on the detector (DT2) for push–pull tracking signal detection. Because the convergent beam passing the half-mirror is astigmatic, it is received by a quadrant detector (DTI) to give a focusing signal. The data signal is retrieved by sum of the output from both detectors (DT1 and DT2). The astigmatic focusing and push–pull tracking methods will be described in detail later.

3.1.2 Influence of Intensity Distribution

The intensity distribution of the beam incident on the objective lens is dependent on the beam divergence angle distribution of the semiconductor laser. With the objective lens aperture radius being standardized as unity and the intensity distribution of the incident beam assumed to be $\exp(-\alpha r^2)$, the amplitude distribution is given by Fourier–Bessel transform:

$$g(s) = \int \exp(-\alpha r^2) J_0(sr) r \, dr \tag{5}$$

with $s = 2\pi n R/\lambda f_0$, where f_0 is the focal length of the objective lens and R is the polar coordinate in the focal plane. Integration gives (Appendix A, Eq. A4):

$$g(s) = \sum_{n=0}^{\infty} 2^n \alpha^n e^{-\alpha} \left(\frac{2 J_{n+1}(s)}{s^{n+1}} \right) \tag{6}$$

Since $\alpha = 0$ for plane-wave incidence,

$$g(s)|_{\alpha=0} = \frac{2 J_0(s)}{s} \tag{7}$$

This is the well-known Airy distribution. When $\alpha = 1$, the beam intensity distribution around the objective lens aperture is $1/e^2$. Figure 9 shows the intensity distribution of the $|g(s)|^2$ beam spot with various values of α. It is apparent from Fig. 9 that when $\alpha = 1$, the FWHM of the beam spot is increased by about 10% (relative to $\alpha = 0$), and the peak of the side-lobe diffraction ring is made sufficiently small.

In order for the reproduced signal to have satisfactory frequency characteristics, the value of α in the signal direction must be in the range of $\alpha \ll 1$. On the other hand, in the direction perpendicular to the signal direction, the cross-talk from the adjacent track must be minimized. This cross-talk can be small by using α close to 1. Therefore, the spot on the disk need not be truly round, but an improved frequency characteristic is sometimes obtained when the beam spot is elliptical with an ellipticity of about 10%.

3.2 Wave Aberrations[12]

When root-mean-square wave aberration W exists, the on-axis energy density Strehl definition (SD) of the beam spot is expressed by

$$SD = 1 - k^2 W^2 \quad \text{with } k = 2\pi/\lambda \tag{8}$$

where λ is the wavelength.

Figure 10 shows the relation between rms wave aberration and on-axis energy density. The on-axis energy density SD is a factor directly associated with reproduced signal SNR or record/reproduced signal SNR. The allowable rms wave aberration for the

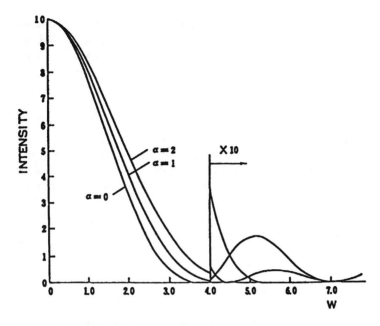

Figure 9 Intensity distributions of laser beam spot.

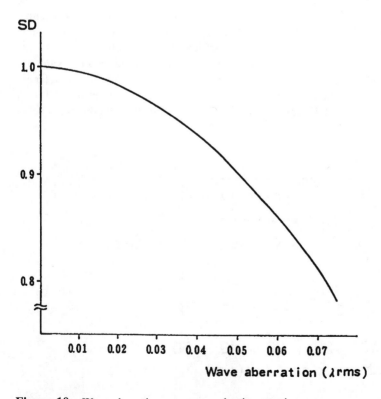

Figure 10 Wave aberrations vs. energy density on axis.

whole optical disk system is subject to Maréchal's criterion that the rms wave aberration is 0.070 λ when SD has decreased about 20% from the level at no aberration. The validity of the criterion has been endorsed by read/write experiments. This allowable wave aberration for the whole system must be allocated to disk thickness error and tilt error, initial optical aberration, and defocus value.

3.2.1 Aberration Derived from Disk Substrate

The aberration originating from the disk substrate is composed of the aberration W_{ST} due to the error Δt of substrate thickness t and the aberration W_{TL} due to the inclination θ of the substrate. These aberrations, when small, are expressed by the following equations (also Eqs. A11 and A14):

$$W_{ST} = \frac{\Delta t(n^2 - 1)(NA)^4}{8\sqrt{180}n^3} \tag{9}$$

$$W_{TL} = \frac{t(n^2 - 1)\theta(NA)^3}{2\sqrt{72}n^3} \tag{10}$$

where NA is the numerical aperture of the objective lens and n is the refractive index of the disk substrate. Figure 11 shows the relation between disk thickness error Δt and wave aberration W_{ST} with NA and wavelength λ as the parameters. Figure 12 shows the relation between disk tilt angle and wave aberration W_{TL} with NA and wavelength λ as the parameter. In the usual recordable CD, the practical values are $NA = 0.5$, $t = 1.2$ mm, $n = 1.51$, $\lambda = 780$ nm, $\Delta t = 40$ µm, and $\theta = 4$ mrad. Substituting these values, we obtain $W_{ST} = 0.011$ λ and $W_{TL} = 0.017$ λ. In the DVD optical disk, the values are $NA = 0.6$, $t = 0.6$ mm, $n = 1.51$, $\lambda = 650$ nm, $\Delta t = 16$ µm, and $\theta = 3.9$ mrad for the same values of $W_{ST} = 0.011$ λ and $W_{TL} = 0.0172$ λ. The plot for higher $NA = 0.85$ and short wavelength $\lambda = 405$ nm is shown as a reference. Thus the tolerance of the tilt angle for the DVD disk is almost the same as for the CD, as contrasted with the tolerance of the thickness being small.

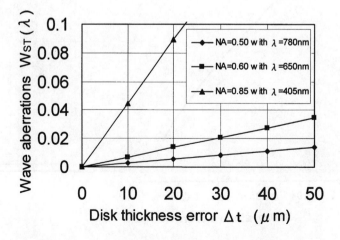

Figure 11 Wave aberrations vs. disk thickness error.

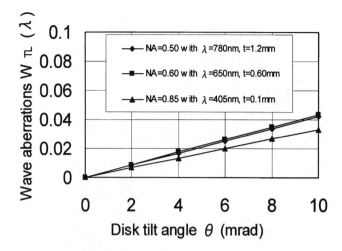

Figure 12 Wave aberrations vs. disk tilt angle.

3.2.2 Wave Aberrations of Optical Components

Because mass-produced items are used for the disk optical components, the influence of variations in wave aberrations cannot be disregarded. Of all the components of the optical pick-up, the objective lens and the collimating lens have the largest wave aberrations. Both the objective lens and the collimating lens usually are aspherical pressed glass (APG) available on a mass production basis. Figure 13 shows an example of a mass-produced aspherical objective lens. Figure 14 shows the wave aberrations of a typical APG objective lens as measured with a Fizeau interferometer. The wave aberrations of prism systems are generally small, but as the number of prisms increases, the allowance for the entire pick-up is consumed.

3.2.3 Aberration Due to the Semiconductor Laser

Semiconductor lasers are generally astigmatic, and as this astigmatism is propagated and focused on the disk, variations in frequency characteristics may occur according to relative directions on the disk, or there may be reduced focusing latitude. In the record-playback optics, the electromagnetic emission from a semiconductor laser is passed through a collimating lens to yield a beam of substantially parallel rays, which is then converted by an anamorphic beam expander to a beam having substantially isotropic distribution.

The correction for astigmatism is carried out concurrently in this stage. If stationary prism is used for correction, the astigmatism generated at an angle of 45° with the prism cannot be corrected. Therefore, when the semiconductor laser is mounted at an angle θ with the horizontal direction of the anamorphic expander prism, there occurs a residual astigmatism. The residual wave aberration W_{LA} due to this astigmatism is expressed as (Eq. A20)

$$W_{LA} = \frac{\tan \theta \Delta_L (NA_C)^2}{\sqrt{6} \cos^2 \theta} \tag{11}$$

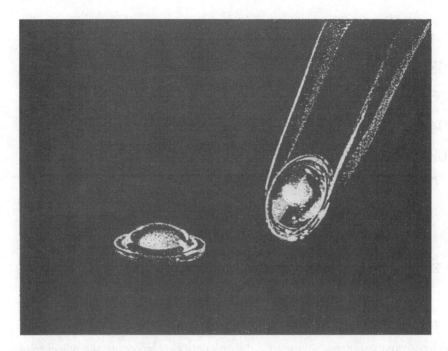

Figure 13 Mass-produced aspherical objective lens.

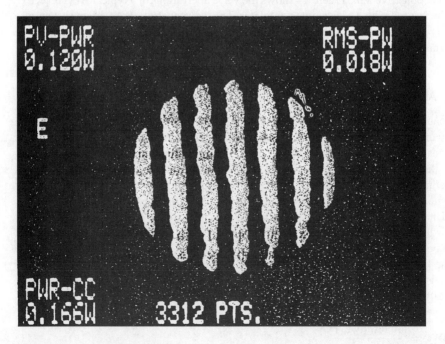

Figure 14 Wave aberrations of a typical APG objective lens.

Optical Disk Scanning Technology

where NA_C is the NA of the collimating lens and Δ_L is the astigmatism of the semiconductor laser. Assuming that the allowable wave aberration dependent on the semiconductor laser is 0.010 λ, the NA of the collimating lens is less than 0.25 and the astigmatism of the semiconductor laser is less than 8 μm (0.32 mil), and the allowable angle θ of the semiconductor laser is found from Eq. (11) to be

$$\theta \leq \pm 4° \quad (12)$$

3.2.4 Defocus

The factors responsible for defocus in an optical system may be classified as in Table 3. The relationship between the amount of defocus ε and the maximum optical path difference Δ_{DF} of the wavefront can be found from Fig. 15:

$$\Delta_{DF} = \varepsilon(NA)^2/2 \quad (13)$$

Using the relationship between maximum optical path difference Δ_{DF} and wave aberration W_{DF}, wave aberration can be expressed as

$$W_{DF} = \frac{\varepsilon(NA)^2}{4\sqrt{3}} \quad (14)$$

For initial focus setting, the use of a diffraction grating having a spatial frequency near one-half of the cutoff frequency $2NA/\lambda$ of the disk optics is advantageous, for the influence of defocus is then most pronounced. Because the track pitch of the optical disk is usually the space frequency in the vicinity, the position of best focus is where the modulation by the track is maximal. By this adjustment, the setting error can be reduced to less than ± 0.14 μm. Thus, with an optical disk of $NA = 0.6$ and $\lambda = 650$ nm and a defocus of ± 0.14 μm, the wave aberration is $W_{DF} = 0.011\ \lambda$.

3.2.5 Allowable Wave Aberration

Table 4 shows the typical wave aberration classified by causative factors. These wave aberrations can be integrated into the system allowance limit of 0.070 λ rms as a totality. Since most of the factors responsible for wave aberrations are independent by nature, it is possible, in the actual design of an optical pick-up, that the allowable aberration value of each optical component is fairly liberal, as shown in Table 4.

Table 3 Defocusing Factors

Defocus	Static defocus	Initial setting error
		Aging error
	Dynamic defocus	Servo residual error
		Temperature- and humidity-related error

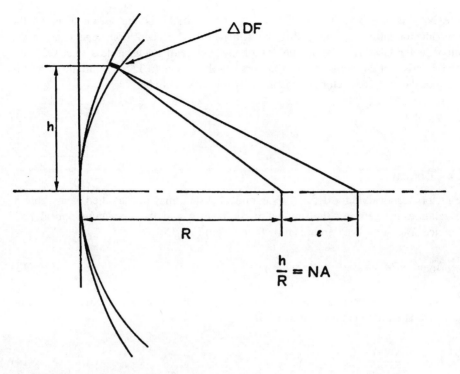

Figure 15 Optical path difference vs. defocus.

3.3 Optical Pick-Up Mechanism

3.3.1 Optical Pick-Up Construction[13,14]

The optical pick-up generally consists of an optical base forming the optics assembly and an actuator for allowing the objective lens to follow the disk plane and tracking groove.

Table 4 Factors Responsible for Wave Aberrations and Amounts of Aberrations for DVD

	System allowance limit 0.070λ	
Disk 0.028λ	Thickness error $< \pm 1.6$ μm	$\leqslant 0.011\lambda$
	Tilt $< \pm 4$ mrad	$\leqslant 0.017\lambda$
Head 0.054λ	Semiconductor laser	$\leqslant 0.010\lambda$
	Objective lens	$\leqslant 0.035\lambda$
	Collimating lens	$\leqslant 0.025\lambda$
	Wedge prism	$\leqslant 0.014\lambda$
	PBS	$\leqslant 0.020\lambda$
	λ/4 plate	$\leqslant 0.020\lambda$
Defocus 0.036λ	Perpetual change	$\leqslant 0.011\lambda$
	Initial setting error	$\leqslant 0.011\lambda$
	Servo residual error	$\leqslant 0.023\lambda$
	Temperature-dependent error	$\leqslant 0.023\lambda$

Optical Disk Scanning Technology

A typical optical pick-up construction for DVD is shown in Fig. 16. Laser and photodetectors are combined on one silicon substrate. The beam emitting from the laser is initially parallel to the silicon surface and is then reflected by an engraved mirror to become perpendicular to the silicon surface. A polarizing hologram, shown in Fig. 17, is used as a beam splitter. It is transparent for the p-polarized beam emitted from the laser and diffractive for the s-polarized beam reflected from the disk. The p-polarized beam becomes an s-polarized beam as a consequence of the double pass through the quarter-wave plate. Photodetectors formed on the silicon surface are located on both sides of the laser, and receive the beam diffracted by the polarizing hologram. The environmental resistance and reliability of the system are greatly enhanced when the number of reflective surfaces is minimized throughout the optical path from the laser to the objective lens. Thus the construction with the integrated laser detector unit (LDU) is advantageous for reliability. Figure 18 is a view showing a laser detector unit for a DVD player.

The optical base has a three-point support structure that enables a two-dimensional adjustment of the tilt angle of the optical pick-up. The axis of the objective lens is aligned perpendicular to the disk plane by mechanical adjustment, and the tilt of the optical base or the angle of inclination of the actuator must be adequately adjusted. The actuator and the optical base are provided with a convex and a concave spherical surface, respectively, whereby tilt correction can be made in two dimensions by means of couple of screws and springs. When the center of the sphere is aligned with the focal point of the objective lens, there is no transverse shift of the beam as it passes through the objective lens.

3.3.2 Actuator

The actuator has both a focusing drive mechanism for following the axial position of the disk and a tracking drive mechanism for following the track on the disk. The actuator must include a balanced combination of these two mechanisms, and must be designed in such a manner that there will be a minimum of interference between the mechanisms. The essential conditions that must be satisfied in the design of an actuator are:

1. satisfactory frequency characteristics;
2. high acceleration characteristics;
3. high current sensitivity; and
4. broad dynamic ranges for both focusing and tracking.

Two actuator constructions satisfying these criteria are shown in Fig. 19. The wire-suspended actuator[15] in Fig. 19(a) is quite simple in construction and can be moved in the focusing and tracking directions by driving the center of gravity of its movable segment. Moreover, reliability is high because the four wires can be utilized as leads to the coils. The rotational actuator in Fig. 19(b) is characterized by a small tilting angle of the critical axis and a large focusing dynamic range.

The first-order resonant frequency f_0 of the wire-suspended actuator is

$$f_0 = 2\pi\sqrt{\frac{K}{m}} \qquad (15)$$

where K is a spring constant and m is a movable mass. As a rule of thumb, the dynamic frequency range of an actuator is approximately from the level of the basic disk rotation

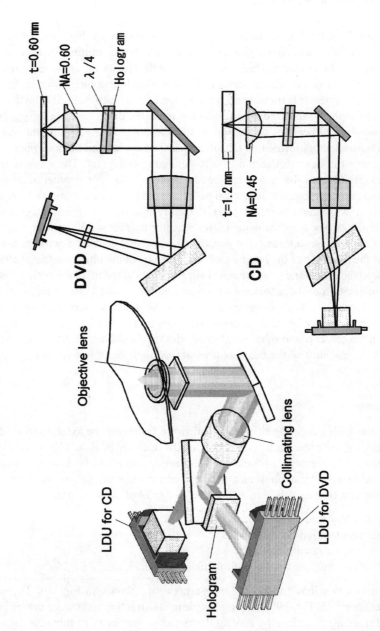

Figure 16 Optical pick-up for DVD.

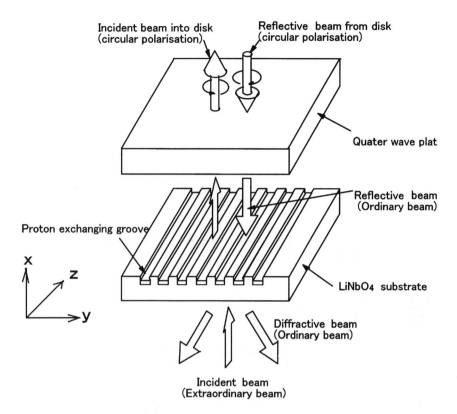

Figure 17 Polarizing hologram.

frequency to the peak level of high-order resonant frequency. The larger this dynamic range value is, the larger is the servo gain that can be obtained.

4 SEMICONDUCTOR LASER

4.1 Laser Structure

4.1.1 Operating Principles of an Al-Ga-As Double Hetero-Junction Laser[16]

The energy band diagram of a double hetero-junction semiconductor laser is shown in Fig. 20. This laser consists of three layers having dissimilar energy gaps Eg, with increased energy gaps for the n-type and p-type cladding layers, which are on both sides of the active layer. As a photon hvg_2 corresponding to the active layer energy gap Eg_2 ($Eg_2 = hvg_2$) passes through the active layer, the electrons in the conduction band drop into the positive holes in the valence band to trigger a stimulated emission in phase with the incident photon. As a current I_p in the normal direction is passed through this diode, the probability of the presence of electrons in the active layer 2 is increased by the energy barrier ΔE_c, the conduction band. On the other hand, in the valence band, the probability of the presence of positive holes in the active layer 2 is increased by the energy barrier ΔE_v to cause a population inversion in the active layer 2. In the active layer, therefore, the conduction band becomes full of electrons normally absent at thermal equilibrium, and the probability

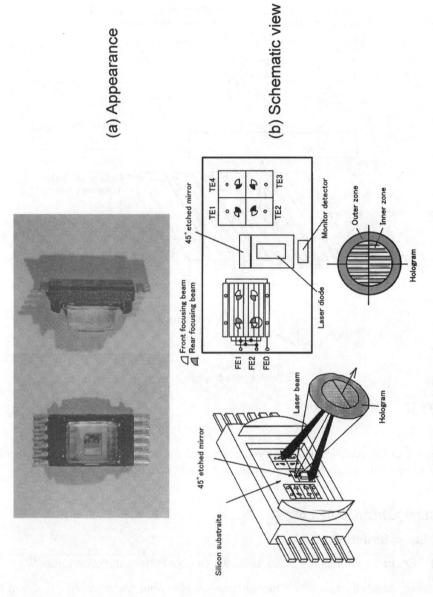

Figure 18 Integrated laser detector unit for DVD.

Optical Disk Scanning Technology

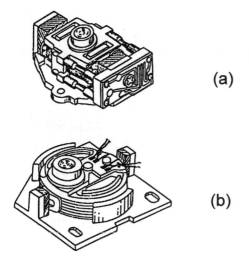

Figure 19 Two different types of actuator: (a) wire-suspended and (b) rotational.

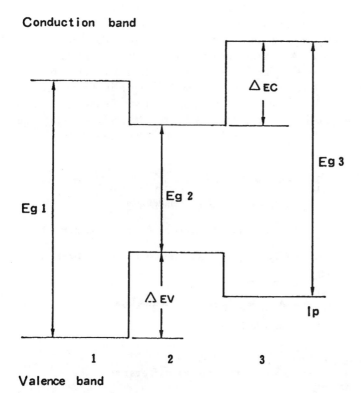

Figure 20 Energy band diagram.

of recombination of electron–hole pairs with stimulated emission is increased. An incoming photon into the active layer is thereby amplified. The feedback mirrors at both ends of the active layer constitute a resonant cavity, and as the amplification surpasses the losses within the resonance cavity laser emission takes place.

4.1.2 High-Power Laser Technology[17]

Low-current and high-temperature operating laser diode of 650 nm AlGaLnP with a real refractive index guided self-aligned (RISA) structure is schematically illustrated in Fig. 21. The RISA structure is characterized by an AlInP current blocking layer, which leads to small internal loss in the waveguide and substantially reduces operating carrier density. The resultant operating current for 950 mW continuous wave at 70°C is less than 100 mA. The RISA laser is produced by two steps of MOCVD growth. In the first MOCVD growth, an n-GaAs buffer layer, a cladding layer, the MQW active region, the optical confinement layer, the current blocking layer, and a nondoped GaAs(0.01 μm) are successively grown. Then, the stripe region for the current path is formed by chemical etching. In the second MOCVD growth, the cladding layer, buffer layer, and contact layer are grown. The cavity length is 500 μm. Front and rear facets are coated to obtain reflectivities of 4 and 90%, respectively.

4.2 Astigmatism of the Laser

There are two categories of semiconductor lasers: gain-guided and index-guided. In a gain-guided laser the direction of beam propagation is not perpendicular to the wavefront. This mismatch causes relatively large astigmatism. Some lasers classified as index-guided also have weak evanescent waves. As a result, the beam waist in the horizontal direction is situated inwardly by Δ_L from the plane of beam emergence and thereby produces astigmatism. Whereas this astigmatism Δ_L is as large as 10–50 μm in the gain-guided laser, it is about 5–10 μm (0.2–0.4 mil) in the index-guided laser. Figure 22 shows a typical distribution of astigmatism in index-guided lasers. Generally speaking, the astigmatism of these lasers tends to decrease as the laser output increases.

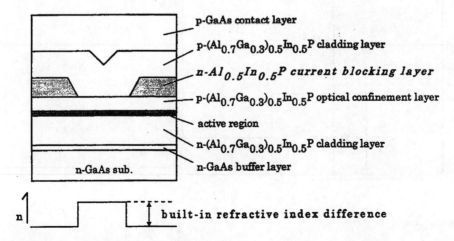

Figure 21 Schematic drawing of the RISA laser structure.

Optical Disk Scanning Technology

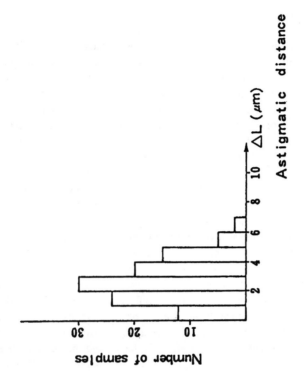

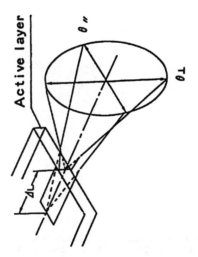

Figure 22 Distributions of astigmatic distance.

4.3 Laser Noise[18]

Semiconductor diode lasers heat up during operation because of power dissipation arising from the injection current. The temperature increase induces mode hopping, which is a small shift in the wavelength of the output light. Figure 23 shows the temperature dependence of a semiconductor laser. As the temperature of the laser increases, the longitudinal mode of the laser is shifted toward longer wavelengths. Substantial noise accompanies this type of mode hopping. Figure 24 shows the relative intensity of noise (RIN) vs. the laser heat sink temperature. Figure 24(a) shows the characteristic of the element itself, and the broken line represents the allowable noise level for an optical pick-up.

If a small amount of the output beam is directed back into the output aperture of the laser, the laser emission will become unstable, exhibiting both mode hopping and excessive amplitude noise. If the level of return light is about 0.5%, the relative intensity noise will be above the noise level allowable for an optical pick-up. Noise of this magnitude not only leads to a decrease in disk recording/reproduction SNR, but may also lead to instabilities in the focus and tracking servos.

An optical isolator consisting of a $\lambda/4$ plate and a polarizing beam splitter is typically used in the optical path to minimize light reflections back into the laser output operations. However, the return light cannot be completely eliminated due to the birefringence in the disk and variations in isolator performance. As a consequence, the generation of noise caused by return light is inevitable in semiconductor lasers of longitudinal single mode. Within the constraints of the basic aspects of the optical pick-up design, return light noise can be best controlled by broadening the emission spectral line width of the semiconductor laser to reduce the coherence of the light. Introducing the multiple longitudinal modes can broaden the emission spectral line width.

In index-guided lasers, the transverse mode behavior becomes single mode at an emission output of 1 mW or higher, due to a confining effect on the transverse mode. In

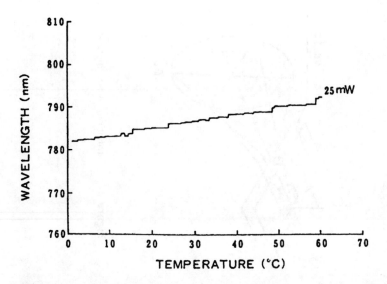

Figure 23 Temperature dependency of wavelength.

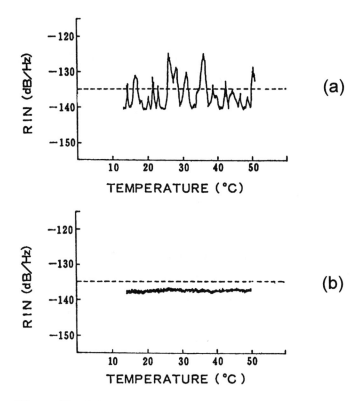

Figure 24 Noise characteristics of semiconductor laser.

gain-guided lasers, the transverse mode is confined by the gain corresponding to the carrier density and that generally results in multiple mode output. The influence of return light in a multimode laser is small and hence laser emission is unaffected. However, the inherent noise level is higher than in single-mode lasers. As shown in Fig. 24(b), when the index-guided laser operating single mode is modulated with high-frequency carrier, the longitudinal mode becomes multimode with the result that the noise level is lowered. Figure 25 shows the return-light noise levels of various lasers.

Noise level RIN is calculated as:

$$\text{RIN} = \frac{\langle \Delta P^2 \rangle}{P^2 \Delta f} \qquad (16)$$

where $\langle \Delta P^2 \rangle$ is the mean square of noise power, P is the output power, and Δf is the noise bandwidth. When the high-frequency oscillation is set at 300–600 MHz and the modulation level is set below the laser emission threshold, the light output becomes a pulse emission providing a multimode operation. Figure 26 shows an example of (upper curve) the read signal obtained without high-frequency oscillation and (lower curve) the read signal with high-frequency oscillation. The addition of the high-frequency oscillation improves the CNR of the carrier signal by about 5 dB.

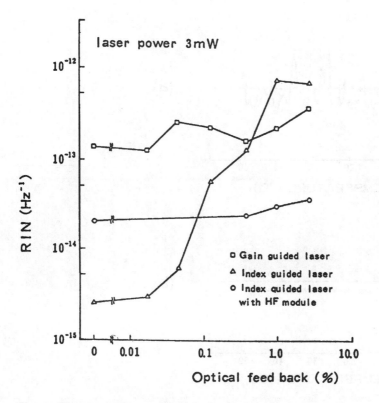

Figure 25 Laser noise vs. optical feedback.

5 FOCUSING AND TRACKING TECHNIQUES

5.1 Focusing Servo System and Method of Error Signal Detection

The laser beam in an optical disk system is focused on the disk surface while the disk rotates at a high speed. An optical disk spinning at a high speed typically exhibits motion in the axial direction of tens to hundreds of micrometers. It is necessary that the objective lens follows this motion to keep the focus position of the beam on the signal plane of the disk within the allowable limits of defocus of the optical system. The focusing mechanism generally used for this purpose is a moving-coil actuator employing a magnet and a coil. The required frequency response of the system is from several Hz to more than 10 kHz. Figure 27 shows a block diagram of the optical disk focusing system. The focusing servo loop comprises a focusing error signal detection unit, a circuit for amplification with phase correction of the detected error signal and an actuator for driving the objective lens. This actuator is designed to follow the axial motion of the disk in response to a servo signal in the presence of external noise associated with the movement of the actuator and the interference from the tracking signal. In designing an autofocusing mechanism for the optical disk, external light noise must be reduced as much as possible. A balanced design must be employed and take into consideration: (1) interference from the tracking signal that occurs when the beam traverses the tracks, (2) mutual interference from motion of the focusing and tracking actuators, and (3) false focusing error signals associated with movement of the beam on the detector in the course of tracking.

Optical Disk Scanning Technology

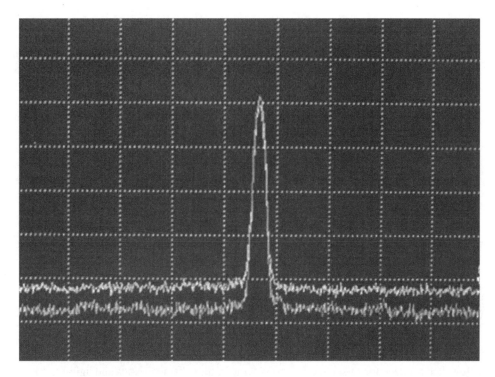

Figure 26 Reproducing carrier signals: upper baseline shows the noise level without high-frequency oscillation; lower baseline shows the noise level with high-frequency oscillation.

Focusing errors are introduced by the axial motion of the disk, vibrations of the device and other causes. The focusing error information contained in the laser light reflected from the disk can be transformed into intensity or phase differences to derive an error signal. Each of the following beam characteristics can be utilized to generate a focusing error signal:

1. change of beam shape;
2. movement of the beam position; and
3. phase of the modulated waveform of the beam.

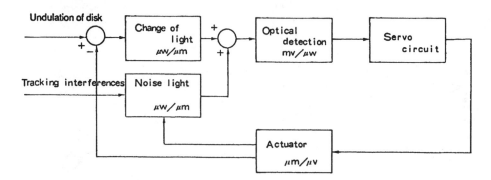

Figure 27 Block diagram of focusing servo.

5.1.1 Beam Shape Detection Method

Two separate techniques can be used to detect the beam shape to obtain a focusing error signal. These are the astigmatic focusing detection method and the spot size detection method.

Figure 28 shows a basic optical system for the astigmatic focusing detection method using a tilted parallel plate. Although the conventional implementation includes use of a cylindrical lens, the tilted plate method is advantageous in the simplicity of the optics. The sensitivity of focusing error signal detection is dependent on the thickness and refractive index of the parallel plate assuming that the magnification of the objective lens is constant. The greater the thickness of the plate and/or the larger the refractive index, the larger the astigmatism and, hence, detection sensitivity. Figure 29 shows a typical focusing error signal in the optimum design. When the detection sensitivity is relatively low, defocus becomes large due to false signals caused by dropouts in the disk or movement of the objective lens during tracking. When the detection sensitivity is too high, the dynamic range of the focusing servo is diminished and the stability of the servo is decreased.

5.1.2 Spot Size Detection Method

Figure 30 shows the operating principle of the spot size detection method. The central part of the beam is received in front of and beyond the focal point of the beam by two three-segment detectors. A focusing error signal is derived from the intensity difference on the central and outer segments. The beam shape detection method generally has a large allowance for detector offset and has good temperature characteristics and aging stability. Recent progress of holographic technology enables use of a holographic optical element (HOE) to detect focusing signals.[13]

5.1.3 Beam Position Detection Method

The beam position detection method converts the movement parallel to the optical axis – such as axial motion of the optical disk – to a beam movement in a plane perpendicular to the optical axis to obtain a focusing error signal. This detection method uses relatively simple hardware construction for detection and gives a broad focusing dynamic range.

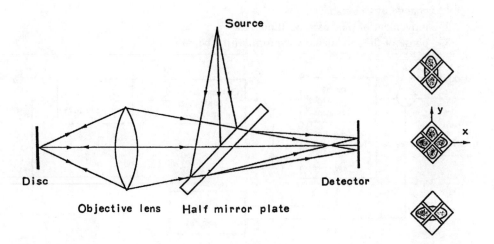

Figure 28 Astigmatic focusing method with plate.

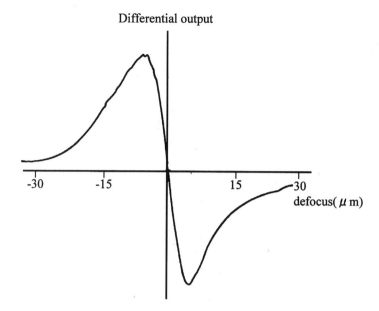

Figure 29 Typical astigmatic focusing error signal.

Figure 31 shows a focusing error signal detection system using a bi-prism. This is an example of the Foucault focusing detection method. As the distance between the disk and the objective lens decreases, the intensity on the inner side of the respective split detectors increases and an increasing distance between the disk and the objective lens results in increasing intensity on the outer sides of the split detectors.

Figure 32 shows a focusing detection system using the critical angle of a prism. The beam of rays reflected from the disk enters the prism as a divergent beam when the distance between the disk and the objective lens is small, or as a convergent beam when the distance is large. When the prism is set at the critical angle, and the rays are not parallel, the beam will be partially transmitted by the prism to establish differential intensities on the detectors. In the case of a divergent beam, the near detector receives less light; in the case of a convergent beam, the far detector receives less light.

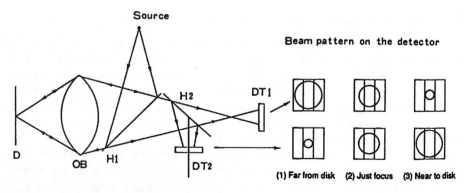

Figure 30 Spot size detection method.

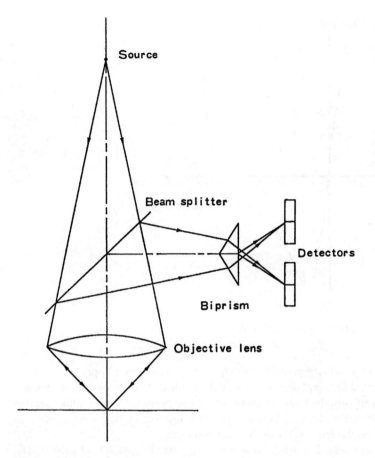

Figure 31 Foucault focusing method.

There are other methods for focusing detection, such as a skew beam focusing detection system, a system wherein the incident beam is eccentric with respect to the objective lens axis, a system using a single knife-edge, a beam rotation focusing detection system, and others.

5.1.4 Beam Phase Difference Detection

There are two methods to detect beam phase difference: the spatial phase difference detection method and the temporal phase difference detection method. In the spatial phase difference detection method, illustrated in Fig. 33, the phase of the beam located in the far-field pattern of the reflected beam diffracted by a given pattern in the optical disk (for example, the pregrooved track pattern) is detected. This method is dependent on beam wavelength, and the dynamic range of focusing error signal is narrow. The temporal phase difference detection method is also known as the wobbling method. In this method, the focal point of the beam irradiating the optical disk is modulated along the optical axis with a wobbler. The phase of the modulated signal obtained with a detector is compared with the phase of the modulated drive signal of the wobbler to obtain a focusing error signal proportional to the phase difference.

Optical Disk Scanning Technology

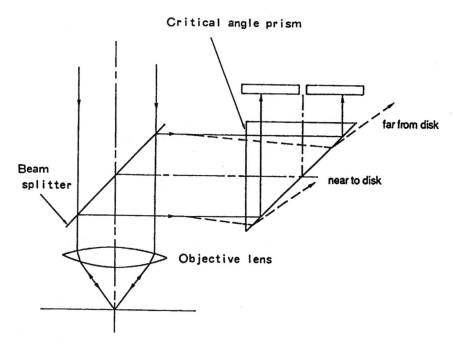

Figure 32 Critical angle focusing method.

5.2 Track Error Signal Detection Method

5.2.1 Detection Methods

The signal tracks on a DVD disk have a pitch of 0.74 μm, and the signal tracks on a conventional CD disk have a pitch of 1.6 μm. The beam spot for reproducing the signal must follow this track within an accuracy of 0.04–0.1 μm. This tracking performance is achieved by driving the objective lens in a lateral direction with a voice coil actuator. The following methods are commonly used for optical detection of the tracking error signal:

1. Detection using two auxiliary beams generated by grating (3-B or 3-beam method);

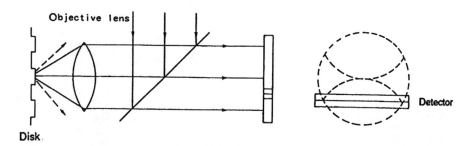

Figure 33 Spatial phase detection method.

2. Detection of the far-field distribution of the read/write beam reflected from the disk (PP or push–pull method);
3. Detection from the difference between two signal levels obtained with sample pits disposed at an offset of $\pm 1/4$ pitch from the track (SS or sampled servo method);
4. Detection of the phase difference between the differential output from diagonal playback signal of quadrant detector and the sum output signal of quadrant detector (DPD or differential phase detection method); and
5. Detection of the phase difference between the playback signal obtained by a slight induced displacement of the beam in the direction perpendicular to the track and the phase of the corresponding drive signal (wobbling method).

5.2.2 3-Beam Method

The 3-Beam method using auxiliary beams is shown in Fig. 34. The two first-order beams obtained by passing the laser beam through a diffractive grating are aligned to positions on the disk of about plus and minus one-quarter of the track pitch apart from the track center (B1, B2). The reflected two beams are received by two detectors (D1, D2) to obtain a track error signal. The fundamental beam (B0) is used for information signal detection by a central detector (D0). While this method is well suited to the read-only optical pick-up like CD, it must be carefully designed for use in read/write optics. In this case the beam intensity is increased during the writing mode, introducing the risk of harmful recording by the auxiliary beams.

5.2.3 Wobbling Method

In the wobbling method, the track error signal corresponds to the phase difference between a signal to the transducer that induces a slight displacement of the beam in the direction

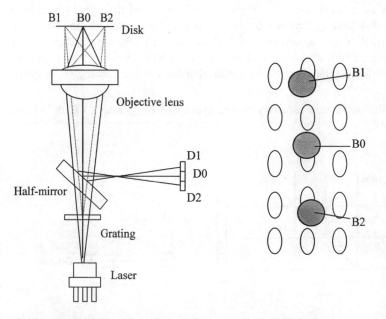

Figure 34 Tracking error signal detection with 3-beam method.

Optical Disk Scanning Technology

perpendicular to the track and a signal from the beam modulated by track edge diffraction. This method has only been implemented in certain limited applications. This is partly due to the poor stability of the wobbling frequency and partly due to the 0.1 μm wobble displacement required to obtain a tracking error signal with satisfactory S/N ratio. Wobble displacement of 0.1 μm is close to the maximum allowable tracking error value.

5.2.4 Differential Phase Detection (DPD) Method

This method uses the quadrant detectors to detect the phase difference between the differential output from the diagonal playback signal $(D1 + D4) - (D2 + D3)$ of the quadrant detector and the sum output signal $(D1 + D2 + D3 + D4)$ of the quadrant detector. The DPD tracking error detection is shown in Fig. 35. This is the method recommended for playback tracking error signal detection in the DVD specifications.[19]

5.2.5 Push–Pull Track Error Signal Detection Method

The simplest method for obtaining a track error signal in read/write is called the "push–pull method." A split detector is inserted in the far-field of the beam in such a manner that the line of division of the detector is lined up with the track.

Figure 36 shows a basic optical system for signal detection using the push–pull method. The intensity pattern incident on the split detector is a combination of the zero-order and first-order beams (due to diffraction caused by the track on the disk). The track error signal is derived from the difference in the signals from the two detectors. Figure 37 shows the far-field beam distributions according to the beam spot position on the track. The asymmetry of the far-field beam intensity distribution and the track error signal level

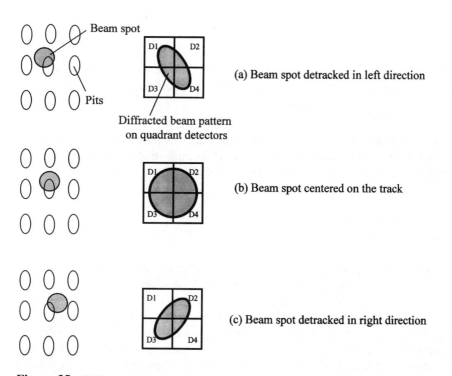

Figure 35 Differential phase detection method.

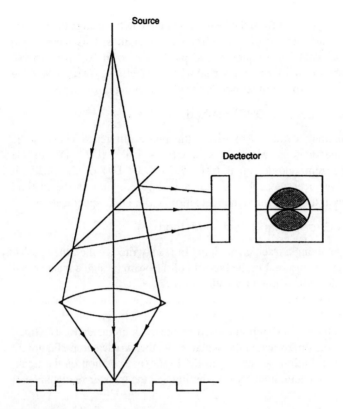

Figure 36 Push–pull method.

are maximum when the track groove depth is $\lambda/8$. (When the depth is $\lambda/4$ multiplied by an integer, asymmetry disappears and no track error signal is obtained.)

5.2.6 Slit Detection Method[20]

When the beam spot is located in the center of the track, the phase difference ψ between the zero-order beam and the two first-order beams in the push–pull system is dependent on the diffraction by the track and on the defocus ΔZ and can be expressed as[21]

$$\psi = \frac{\pi}{2} + \frac{2\pi}{\lambda}\left[\sqrt{1 - \left(\frac{\lambda}{p} - \sin\alpha\right)^2} - \cos(\alpha)\right]\Delta Z \qquad (17)$$

where α is an angle between the optical axis and an arbitrary point in a far-field image.

From Eq. (17), the phase difference is constant for $\alpha = \sin^{-1}(\lambda/2p)$ in the far-field image and independent of defocus but exclusively dependent on diffraction by the track. Figure 38 shows a typical far-field beam distribution in the presence of defocus. By utilizing this property in the far-field, the control range of tracking with respect to defocus can be expanded. Thus, the defocus characteristic of the track error signal can be improved by providing slits symmetrically in the centers of overlaps between the zero-order beam and the two first-order beams as illustrated in Fig. 39.

Optical Disk Scanning Technology

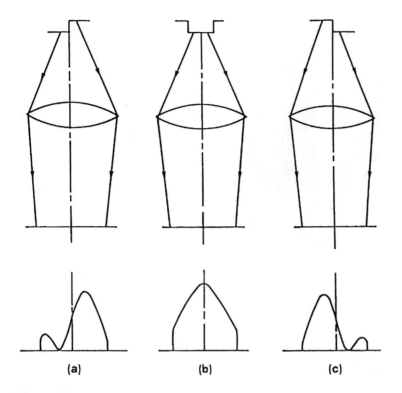

Figure 37 Far-field beam distribution.

Figure 40 shows the change in track error signal level according to defocus at various slit widths. If the slit width is about 20% of the far-field pattern, the proportion of change in the track error signal due to defocus is improved by about a factor of 2. If the slit width is made too narrow, the S/N of the track error signal will decrease. The optical parameters in these experiments and theoretical calculations are as follows:

- objective lens: $NA = 0.5$;
- laser wavelength: $L = 830$ nm;
- track pitch: $t = 1.6$ μm;
- slit width: $W = 0.8$ and 0.4 mm.

5.2.7 Sampled Tracking Method[22]

In a sampled tracking system, track error signal detection pits are periodically provided in lieu of the continuous groove in a conventional grooved disk. The sample pits consist of two pits displaced by about $\pm 1/4$ of the track pitch from the track center and one pit centered on the track. Figure 41 shows the principle of sampled track error signal detection. When the beam spot is (a) off-track upwardly from the track center, the first pit output is large and the second pit output is small. In the on-track condition, the first pit output level and the second pit output level are equal (b). If the beam spot is downwardly off-track, the first pit output is small and the second pit output is large (c). The off-track condition is diagnosed by comparing these pit output levels. The third pit is used for

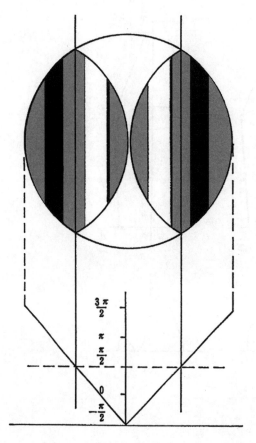

Figure 38 Far-field pattern when defocus occurred.

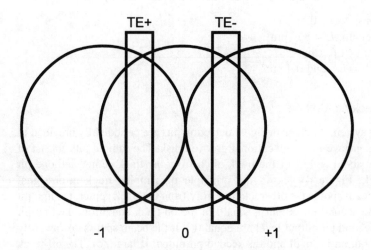

Figure 39 Slit detection method.

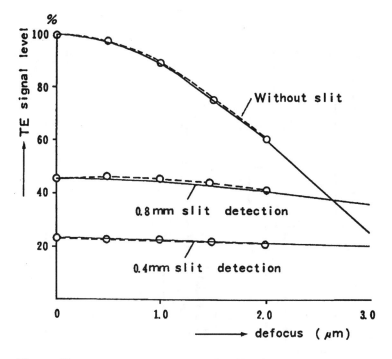

Figure 40 Defocus vs. tracking error signal level.

making a sampling clock, and the track error signal detection is constructed from these outputs. Figure 42 shows a block diagram of the detection circuit.

In the sampled tracking system, each set of pits provided for the track error signal detection uses a track length equivalent to that used to store 1 byte of information. The sampling frequency must be higher than about 10 times the cutoff frequency of the tracking servo. This reduces the size of the usable data area, but the overall system performance improves because there is less degradation of data signals and interference effects on the focusing servo caused by the track groove. Further, an inclination of the disk induces a track offset in the push–pull system, but not in the sampled tracking method. For

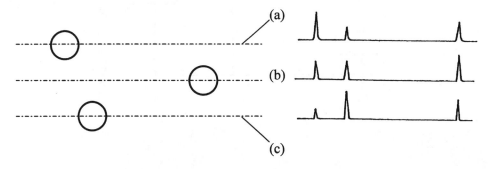

Figure 41 Tracking error signal detection from sampled pits.

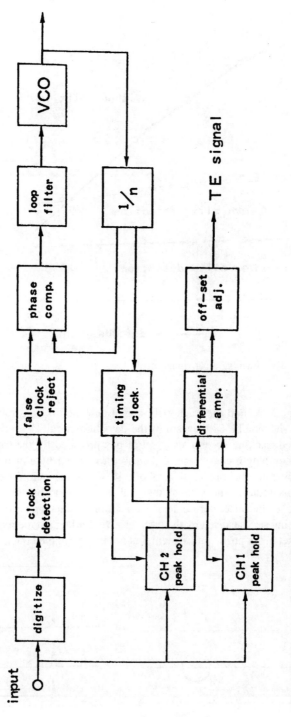

Figure 42 Block diagram of sampled servo tracking method.

example, a disk inclination of 0.7° of a 1.2 mm thick substrate with $NA = 0.5$ and $\lambda = 830$ nm causes a 0.1 μm lateral shift in the quiescent operating position of the push–pull tracking servo. This type of systematic track error is decreased by about a factor of 5 when the sampled tracking method is used.

6 RADIAL ACCESS AND DRIVING TECHNIQUE

6.1 Fast Random Access

A critical aspect of an optical disk memory system is fast random access to the stored information. This random access is accomplished via two mechanisms: optical pick-up motion for rough positioning and the tracking actuator for precise positioning. A linear actuator is used as the coarse positioning means. To minimize access time, it is necessary to (1) develop a small and lightweight optical pick-up, (2) increase the resonant frequency of the linear actuator, and (3) develop a transfer mechanism with a minimum of friction. In the typical linear actuator designed for video recording applications, the transfer segment weighs only 78 g and has a thrust of 3.0 N/A. This linear actuator gives an average access time of 75 ms or less.[9] The optical pick-up base is provided with roller bearings so that it freely moves on the guide rods. The low-frequency component of the track error signal is fed to the linear actuator so that the center of drive of the objective lens will lie at the center of the tracking mechanism motion range.

Figure 43 shows the modes of access by the linear actuator/tracking actuator combination.[23] First, the tracking actuator is disabled. Then the linear actuator is accelerated and decelerated at the maximum speed to position the optical pick-up in the vicinity of the correct track (A to B). The tracking actuator is reactivated and the track address is read (B to C). The number of tracks between the actual and desired address is calculated, and access is completed by executing a multitrack jump. Figure 44 is a block diagram of the tracking servo circuit and linear actuator circuitry. The track error signal obtained by the procedure described in Sec. 5.2 is fed to an amplification circuit, a switching circuit, and a drive circuit in succession to drive the tracking actuator.

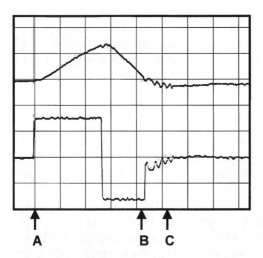

Figure 43 Modes of access by the linear actuator and tracking actuator.

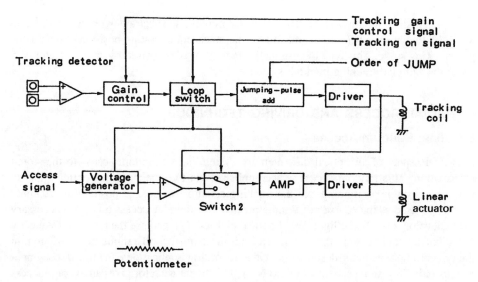

Figure 44 Block diagram of tracking and accessing servo.

The tracking drive signal is also used to drive the linear actuator. During random access, the drive signal is removed from the tracking actuator and a voltage corresponding to the access signal is generated and supplied to the linear actuator drive circuit. Two methods are available for ascertaining the actual position with respect to the target track. The first method involves calculating the number of tracks between the current and desired positions, then detecting and counting optically each pregroove as it is passed over in the radial scan. The scan is stopped when the correct number of tracks have been crossed.

Since this method counts the tracks themselves, the distance to the target track can be accurately computed. The track detection bandwidth must be broad enough to prevent miscounts of the tracks during the peak speed of the linear actuator. This method is less applicable when using sampled format disks; even track addresses can complicate the track-counting process in continuous format disks. A second method provides the optical pick-up with a position sensor for detecting the current position. This provides a stable position signal, and the access servo can be damped using the output signal from this sensor. Examples of position sensors include linear scale sensors, optical position sensors, and slide resistance sensors. Figure 45 shows a typical optical position sensor.

6.2 Optical Drive System

The optical disk system consists of hardware comprising the disk, optical pick-up, accessing circuit, signal processing circuit, error correction circuit, microcomputer, and so on, and software for processing the various signals. The height of the 5.25 in. optical disk drive is either full-height (82 mm), half-height (41 mm), 1 in. height or half-inch height, corresponding to standardized magnetic disk products. In the half-height drive, design goals include the use of a disk cartridge that is inserted and clamped and low profiles of the component parts for the access mechanism. The height of the optical pick-up must be less than 15–16 mm. For the thinner drive like those used in a notebook computer, the height

Optical Disk Scanning Technology

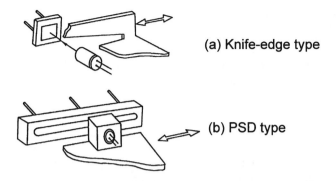

Figure 45 Optical position sensor: (a) knife-edge type; (b) PSD type.

of the drive is less than 12.7 mm. A height of the optical pick-up of around 7.5 mm is needed and an ingenious design is required. Figure 46 shows the thin optical disk drive for a DVD-ROM. The removability of the optical disk affects differences in the amounts of eccentricity and undulation each time a disk is mounted. Moreover, since the disk substrate is made of plastic, the amount of undulation increases with the age of the disk. The dynamic balance of the disk is affected by these factors, and vibrations are induced. It is necessary, therefore, to design the various actuators such that these vibrations do not degrade performance.

ACKNOWLEDGMENTS

I give my special thanks to Gerald F. Marshall, volume editor of "Optical Scanning," who has given me a chance to contribute to this book and is generous over my untrained English. I extend my profound thanks to the two following reviewers of my chapter, experts of optics, for their patient work in pointing out important directions and suggestions: Masud Mansuripur of the University of Arizona, and David Strand, of Energy Conversion Devices, Inc.

Figure 46 Thin optical disk drive for DVD-ROM.

APPENDIX

Appendix A

When the amplitude distribution in the pupil is given by $f(r^2)$ and the radius of the pupil is one unit, the integral of the Fourier–Bessel transform is written as

$$g(s) = \int_0^1 f(r^2) J_0(sr) \, d(r^2) \tag{A1}$$

A plurality of solutions exist for this integral. However, the solution given by A. Boivin[24] is easily understood. Thus, the amplitude of the Fourier spectrum $g(w)$ is written in the form of a Bessel series:

$$g(s) = \sum (-1)^n 2^{n+1} f_{(1)}^n \frac{J_{n+1}(s)}{s^{n+1}} \tag{A2}$$

where $f^n(r^2)$ denotes the nth differential of the function $f(r^2)$. For calculating the Fourier spectrum of a truncated gaussian, $f(r^2)$ is expressed by $\exp(-\alpha r^2)$. Then the integral of Fourier–Bessel transform is written as

$$g(s) = \int_0^1 \exp(-\alpha r^2) J_0(sr) r \, dr \tag{A3}$$

and the result becomes[24]

$$g(s) = \sum_{n=0}^\infty 2^n \alpha^n e^{-\alpha} \left[\frac{2 J_{n+1}(s)}{s^{n+1}} \right] \tag{A4}$$

Appendix B

Thickness variations, index changes, and tilts of the disk substrate all cause wavefront aberrations. Here, we calculate the optical path differences Δ_0 between two rays: the first is the on-axis ray, and the second is the outermost ray, which determines the numerical aperture of the objective lens. From Fig. 47, the following relations can be easily calculated:

$$\sin(\psi - \theta) = n \sin(r_1) \tag{A5}$$
$$\sin(\theta) = n \sin(r_0) \tag{A6}$$
$$\Delta_0 = nt\{1/\cos(r_1) - 1/\cos(r_0)\} + t\{\cos(\psi - \theta)/[\cos(r_1)\cos(\theta)] \\ - 1/\cos(r_1)\}/n \tag{A7}$$

Developing the power series of ψ and θ, the next quadratic terms are obtained:

$$\Delta = t(1 - n^2)\{\psi^4 - 4\psi^3\theta + 8\psi^2\theta^2 + 8\psi\theta^3\}/8n^3 \tag{A8}$$

Optical Disk Scanning Technology

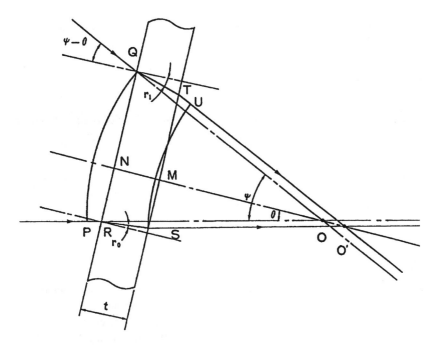

Figure 47 Optical path differences between two rays.

where ψ is the numerical aperture (NA) of the objective lens and t is the thickness of the disk substrate. Here, each term denotes the Siedel aberration. When only a thickness error Δt exists, the spherical aberration S_1 is generated:

$$S_1 = (n^2 - 1)\psi^4 \Delta t / 8n^3 \tag{A9}$$

The relation between the wave aberration W_{ST} and the spherical aberration S_1 can be calculated from Maréchal's equation:[25]

$$W_{ST}^2 = d^2/12 + dS_1/6 + 4S_1^2/45 = (d + S_1)^2/12 + S_1^2/180 \tag{A10}$$

where the wave aberration W_{ST} is minimal when the defocus d is equal to the third spherical aberration S_1. Thus, the wave aberration due to the spherical aberration becomes

$$W_{ST} = \frac{S_1}{\sqrt{180}} = \frac{\Delta t (n^2 - 1)(NA)^4}{8\sqrt{180}\, n^3} \tag{A11}$$

For a small tilt θ of the disk substrate, high orders can be disregarded and the only important aberration is the coma C_1:

$$C_1 = t(n^2 - 1)\psi^3 \theta / 2n^3 \tag{A12}$$

The relationship between the wave aberration W_{TL} and the coma C_1 can be calculated, again from Maréchal's equation[25]

$$W_{TL}^2 = K^2/12 - KC_1/6 + C_1^2/18 = (K - C_1)^2/12 + C_1^2/72 \tag{A13}$$

where the wave aberration W_{TL} is minimal when the tilt of wavefront K is equal to coma C_1. Then the wave aberration W_{TL} due to a tilt of the disk substrate becomes

$$W_{TL} = \frac{C_1}{\sqrt{72}} = \frac{t(n^2-1)\psi^3\theta}{2\sqrt{72}n^3} \tag{A14}$$

Appendix C

The allowable limit of laser-mounting angle is calculated. When the wavefront of the beam emerging from a laser is astigmatic and its axis is not in agreement with the x, y-axis of the optics, there exists a residual astigmatism. With the y-axis as a reference, the phase difference $\psi(x)$ along the x-axis within the pupil plane of the collimating lens with an astigmatic distance of Δ_L and a focal length of f_c is expressed by

$$\psi(x) = \Delta_L x^2/(2f_c^2) \tag{A15}$$

Assuming that the wavefront is inclined through an angle θ with respect to the y-axis, the phase difference $\psi(x, y)$ is expressed as

$$\begin{aligned}\psi(x, y) &= \Delta_L(x - y\tan\theta)^2/(2\cos^2\theta f_c^2) \\ &= \Delta_L(x^2 + y^2\tan^2\theta - 2xy\tan\theta)/(2\cos^2\theta f_c^2)\end{aligned} \tag{A16}$$

By focusing the optics, the term $x^2 + y^2\tan^2(\theta)$ can be zero. Therefore, the wavefront aberration assumes a maximum value in the direction of $x = y = h$:

$$\psi_o = \Delta_L \tan\theta h^2/(\cos^2\theta f_c^2) \tag{A17}$$

Since h/f_c is the NA of the collimating lens, the above equation may be rewritten as:

$$\psi_o = \Delta_L \tan\theta (NA_c)^2/\cos^2\theta \tag{A18}$$

Since Maréchal's equation gives the relationship between maximum astigmatism ψ_o and wavefront aberration W_{LA} as

$$W_{LA}^2 = \psi_o^2/6 \tag{A19}$$

we obtain

$$W_{LA} = \frac{\Delta_L \tan \theta (NA_c)^2}{\sqrt{6} \cos^2 \theta} \quad (A20)$$

REFERENCES

1. Feinleib, J.; de Neufville, J.; Moss, S.C.; Ovshinsky, S.R. Rapid reversible light-induced crystallization of amorphous semiconductors. Appl. Phys. Lett. 1971, *18*, 254.
2. Hopkins, H.H. Diffraction theory of laser readout systems for optical video disks. J. Opt. Soc. Am. 1979, *69*, 4–24.
3. Goodman, J.W. *Introduction to Fourier Optics*; McGraw Hill: New York, 1968; Chap. 6.3.
4. Braat, J. *Principles of Optical Disk System*; Adam Hilger Ltd.: New York, 1985; 7–85.
5. Firester, A.H.; Caroll, C.B.; Gorog, I.; Heller, M.E.; Russell, J.P.; Stewart, W.C. Optical read out of RCA video disk. RCA Review 1978, *39* (3), 392–407.
6. Mansuripur, M. Scanning optical microscopy part 1. Opt. & Photonics News 1998, *May*, 56–59.
7. Yoshida, T. Tellurium sub-oxide thin film disk. Proc. SPIE Optical Disks Systems and Applications 1983, *421*, 79–84.
8. Saimi, T. Compact optical pick-up for three dimensional recording and playing system. CLEO '82 Pheonix, April 1982.
9. Imanaka, R.; Saimi, T.; Okino, Y.; Tanji, T.; Yoshimatsu, T.; Yoshizumi, K.; Kamio, K. Recording and playing system having a compatibility with mass produced replica disk. IEEE Consumer Electonics 1983, *CE-29* (3), 135–140.
10. Hartmann, M.; Jacobs, B.A.J.; Braat, J.J.M. Erasable magneto-optical recording. Philips Tech. Rev. 1985, *42* (2), 37–47.
11. Deguchi, T.; Katayama, H.; Takahashi, A.; Ohta, K.; Kobayashi, S.; Okamoto, T. Digital magneto-optical disk drive. Appl. Opt. 1984, *23* (22), 3972–3978.
12. Born, M.; Wolf, E. *Principles of Optics*; Pergamon Press: Oxford, 1970.
13. Saimi, T. *PD Head for "PD" System*, National Technical Report, Dec. 1995; Vol. 41, No. 41.
14. Shih, Hsi-Fu. Holographic laser module with dual wavelength for digital versatile disk optical heads. Jpn. J. Appl. Phys. 1999, *38*, 1750–1754.
15. Nakamura, H. *Fine Focus 1-Beam Optical Pick-Up System*, National Technical Report, 1986; 72–80.
16. Finck, J.C.J.; van der Laak, H.J.M.; Schrama, J.T. A semiconductor laser for information readout. Philips Tech. Rev. 1980, *139* (2), 37–47.
17. Imafuji, O.; Fukuhisa, T.; Yuri, M.; Mannoh, M.; Yoshikawa, A.; Itoh, K. Low operating current and high-temperature operation of 650-nm AlGaInP high-power laser diode with real refractive index guided self-aligned structure. IEEE J. Selected Topics in Quantum Electronics 1999, *5* (3), 721–728.
18. Chinone, N.; Ojima, M.; Nakamura, M. A semiconductor laser below allowance of noise due to the optical feed-
back by adding the high frequency generating circuit. Nikkei Electronics 1983, *10* (10), 173–194.
19. ECMA Standardizing Information and Communication System Standard ECMA-267, December 1997.
20. Saimi, T.; Mizuno, S.; Itoh, N. Amelioration of tracking signals by using slit-detection method. Proc. Conference of Japan Society of Applied Physics 1987, *34,29a-ZL-7*, 743; Tokyo, March 1987.
21. Oudenhuysen, Ad.; Lee, Wai-Hon. Optical component inspection for data storage applications. Proc. SPIE Optical Mass Data Storage II 1986, *695*, 206–214.

22. Tsunoda, Y. On-land composite pregrove method for high tract density recording. Proc. SPIE Optical Mass Data Storage I 1986, *695*, 224–229.
23. Saito, A.; Maeda, T.; Tunoda, Y. Fast accessible optical pick-up, O plus E; Shingijyutsu Communications: Japan, 1986, *76*, 84–87.
24. Boivin, A. *Théorie et Calcul des Figures de Diffractions*; Press de l'Université Laval: Quebec, 1964; 118–122.
25. Maréchal, A.; Françon, M. *Diffraction Structure des Images*; Masson & Cie: Paris, 1970; 105–112.

12

Acousto-Optic Scanners and Modulators

REEDER N. WARD

Noah Industries, Inc., Melbourne, Florida, U.S.A.

MARK T. MONTGOMERY

Direct2Data Technologies, Melbourne, Florida, U.S.A.

MILTON GOTTLIEB

Consultant, Carnegie Mellon University, Pittsburgh, Pennsylvania, U.S.A.

1 INTRODUCTION

It will be apparent to the reader of this book that there are a great variety of applications of lasers for which scanning devices are required, and that these applications include a wide range of performance requirements on the scanner. The basic specifications include speed, resolution, and random access time, and the choice of a scanner will be determined by these parameters. Acousto-optic (AO) scanners are best suited to those systems that are of moderate cost, since the cost of AO Bragg cells and the associated drive electronics are by no means trivial, and for which the resolution requirement is about 1000 spots. In addition, AO technology is most appropriate where random access times on the order of 10 µs are needed, or where it may be desired to perform intensity modulation on the laser beam, as in image recording. There are currently many systems employing AO scanners, perhaps the most familiar being laser printers, in which the scanner capability is an excellent match to the system requirements. Large-area television display was one of the first applications considered for AO scanners, and it performs this function very well, although such display systems are relatively uncommon. These, as well as other applications of acousto-optic scanners, will be described in detail in a later section.

The interaction of light waves with sound waves has in recent years been the basis of a large number of devices related to various laser systems for display, information handling, optical signal processing, and numerous other applications requiring the spatial or temporal modulation of coherent light. The phenomena underlying these interactions were largely understood as long ago as the mid-1930s, but remained as scientific curiosities, having no practical significance, until the 1960s. During this period several technologies were developing rapidly, at the same time that many applications of the laser were being suggested that require high-speed, high-resolution scanning methods. These new technologies gave rise to high-efficiency, wideband acoustic transducers capable of operation to several gigahertz, high-power wideband solid-state amplifiers to drive such transducers, and the development of a number of new, synthetic acousto-optic crystals with very large figure of merit (low-drive-power requirements) and low acoustic losses at high frequencies. This combination of properties makes acousto-optics the method of choice for many systems, and is very often the only approach to satisfy demanding requirements. In this chapter, the underlying principles of acousto-optic interactions will be reviewed, and this will be followed by a description of the materials considerations and the relevant acoustic technology. Acousto-optic scanning devices will be described in some detail, including the important features of optical design for various types of systems.

2 ACOUSTO-OPTIC INTERACTIONS

2.1 The Photoelastic Effect

The underlying mechanism of all acousto-optic interactions is very simply the change induced in the refractive index of an optical medium due to the presence of an acoustic wave. An acoustic wave is a traveling pressure disturbance that produces regions of compression and rarefaction in the material, and the refractive index is related to the density, for the case of an ideal gas, by the Lorentz–Lorenz relation

$$\frac{n^2 - 1}{n^2 + 2} = \text{constant} \times \rho \tag{1}$$

where n is the refractive index and ρ is the density. In fact, this relation is adhered to remarkably well for most simple solid materials as well. The elasto-optic coefficient is obtained directly by differentiation of Eq. (1):

$$\rho \frac{\partial n}{\partial \rho} = \frac{(n^2 - 1)(n^2 + 2)}{6n} \tag{2}$$

where it is understood that the derivative is taken under isentropic conditions. This is generally the case for ultrasonic waves, in which the flow of energy by thermal conduction is slow compared with the rate at which density changes within a volume smaller than an acoustic wavelength. The fundamental quantity given by Eq. (2), also known as the photoelastic constant p, can be easily related to the pressure applied, with the result

$$p = \frac{1}{\beta} \frac{\partial n}{\partial P} \tag{3}$$

where P is the applied pressure and β is the compressibility of the material. The photoelastic constant of an ideal material with refractive index of 1.5 is 0.59. It will be seen later that the photoelastic constants of a wide variety of materials lie in the range from about 0.1 to 0.6, so that this simple theory gives a reasonably good approximation to measured values.

The relation in Eq. (3) follows from the usual definition of the photoelastic constant:

$$\Delta\left(\frac{1}{\epsilon}\right) = \Delta\left(\frac{1}{n^2}\right) = pe \tag{4}$$

where ϵ is the dielectric constant ($\epsilon = n^2$) and e is the strain amplitude produced by the acoustic wave. From Eq. (4) it is easily seen that the change in refractive index, Δn, produced by the strain is

$$\Delta n = -\frac{1}{2}n^3 pe \tag{5}$$

where e is of the form $e_0 \exp(i\Omega t)$ for an acoustic wave of frequency Ω. The magnitude of refractive index change typical for acousto-optic devices is not large. Strain amplitudes lie in the range 10^{-8} to 10^{-5}, so that using the above expressions for Δn and p gives for Δn about 10^{-8} to 10^{-5} (for $n = 1.5$). It may be somewhat surprising, then, that devices based upon such a small change in refractive index are capable of generating large effects, but it will be seen that this comes about because these devices are configured in a way that can produce large phase changes at optical wavelengths.

The relation defining the photoelastic interaction has been written in Eq. (5) as a scalar relation, in which the photoelastic constant is independent of the directional properties of the material. In fact, even for an isotropic material such as glass, longitudinal acoustic waves and transverse (shear) acoustic waves cause the photoelastic interaction to assume different parameters. A complete description of the interaction, particularly for anisotropic materials, requires a tensor relation between the dielectric properties, the elastic strain, and the photoelastic coefficient. This may be represented by the tensor equation

$$\Delta\left(\frac{1}{n^2}\right)_{ij} = \sum_{kl} p_{ijkl} e_{kl} \tag{6}$$

where $(1/n^2)_{ij}$ is a component of the optical index ellipsoid, e_{kl} are the cartesian strain components, and p_{ijkl} are the components of the photoelastic tensor. The crystal symmetry of any particular material determines which of the components of the photoelastic tensor may be nonzero, and also which components are related to others. This may be useful in determining whether some crystal, based only upon its symmetry, may even be considered for certain applications.

2.2 Isotropic AO Interaction

The most useful photoelastic effect is the ability of acoustic waves to diffract a light beam. There are several ways to understand how diffraction comes about; the acoustic wave may be thought of as a diffraction grating, made up of periodic changes in optical phase, rather than transparency, and moving at sonic velocity rather than being stationary. Thus, it is possible to analyze the diffraction as resulting from a moving phase grating. Alternatively, the light and sound may be thought of as particles, photons, and phonons, undergoing collisions in which energy and momentum are conserved. Either of these descriptions may be used to obtain all the important diffraction effects, but some are more easily understood on the basis of one or the other. It will be useful, then, to outline both of these approaches.

To examine the simplest case of plane acoustic waves interacting with plane light waves, consider Fig. 1. Suppose the light wave, of frequency ω and wavelength λ, is incident from the left into a delay line with an acoustic wave of frequency Ω and wavelength Λ. If the refractive index of the delay medium is $n + \Delta n$ in the presence of the acoustic wave, the phase of the optical wave will be changed by an amount

$$\Delta \phi = 2\pi \frac{L}{\lambda} \Delta n \tag{7}$$

if the length of the delay line is L. Some typical values of $\Delta \phi$ can be obtained by assuming $L = 2.5$ cm (1 in.) and $\lambda = 0.5$ μm, with Δn reaching a peak value of 10^{-5}. This yields a phase change of π rad, which is, of course, quite large. It is large because L/λ, the number of optical wavelengths, is 50,000, so that a very small Δn can still produce a sizable $\Delta \phi$. If the electric field incident on the delay line is represented by

$$E = E_0 e^{i\omega t} \tag{8}$$

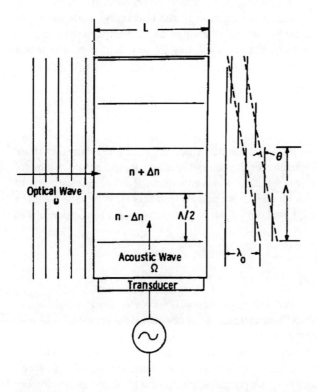

Figure 1 Tilting of optical wavefronts caused by upward-traveling acoustic wave.

then the field of the phase-modulated emerging light will be

$$E = E_0 e^{i(\omega t + \Delta\phi)} = e^{i\omega t} e^{i2\pi(L/\lambda)(a_0 \sin \Omega t)} \qquad (9)$$

We shall not give a detailed derivation of the resulting temporal and spatial distribution of the light field, but we can use intuition and analogy with radiowave modulation to arrive at the resultant fields. It is well known from radio frequency (RF) engineering that the spectrum of a phase-modulated carrier of frequency ω consists of components separated by multiples of the modulation frequency Ω, as shown in Fig. 2. There is a multiplicity of sidebands about the carrier frequency, such that the frequency of the nth sideband is $\omega + n\Omega$, where n is both positive and negative. The amplitude of each sideband is proportional to the Bessel function of order equal to the sideband number, and whose argument is the modulation index $\Delta\phi$. Although not shown by Fig. 2, note that the odd-numbered negative orders are 180° out of phase with the others. The light emerging from the delay line is composed of a number of light waves whose frequencies have been shifted by $n\Omega$ from the frequency ω of the incident light. The relative amplitudes will be determined by the peak change in the refractive index.

In order to understand the diffraction of the light by the acoustic wave, consider the optical wavefronts in Fig. 1. Because the velocity of light is about five orders of magnitude greater than the velocity of sound, it is a good approximation to assume that the acoustic wave is stationary in the time that it takes the optical wave to traverse the delay line. Suppose that during this instant the half-wavelength region labeled $n + \Delta n$ is under compression and $n - \Delta n$ is under rarefaction. Then the part of the optical plane wave passing through the compression will be slowed (relative to the undisturbed material of index n) while the part passing through the rarefaction will be speeded up. In this rough picture, the emerging wavefront will be "corrugated," so that if the corrugations are joined by a continuous plane its direction is tilted relative to that of the incident light wavefronts. Because the optical phase changes by 2π for each acoustic wavelength Λ along the acoustic beam direction, the tilt angle will be given by $\theta \cong \lambda/\Lambda$.

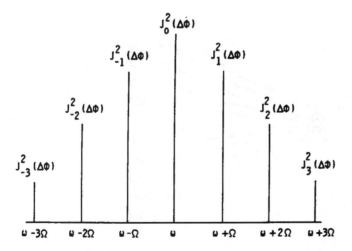

Figure 2 Intensity of diffracted orders due to Raman–Nath interaction.

The direction normal to the tilted plane is the direction of optical power flow and represents the diffracted light beam. Note that the corrugated wavefront could just as well have been connected by a tilted wavefront at an angle given by $\theta \cong -\lambda/\Lambda$. This corresponds to the first negative order, the other to the first positive order. At this point we will note that an important consideration in the operation of AO Bragg cells, or for that matter of most ultrasonic devices, is the ratio of the acoustic wavelength Λ to the transducer length L. The assumption that the acoustic energy propagates as a plane wave is valid when this ratio is very small or when there is little diffraction of the wave. However, when this ratio is not large, the acoustic propagation is more properly described in terms of the sum of plane waves, the angular spectrum of such plane waves increasing as the ratio increases. If we consider that partial wave which is propagating at an angle λ/Λ to the forward direction, then we see that the light that has been diffracted into the first order may be diffracted a second time by this partial wave into an angle $2\theta = 2\lambda/\Lambda$, and that the frequency of this light will once again be upshifted, for a total frequency shift of 2ω. If the angular spectrum of acoustic waves contains sufficient power of still higher orders, then this process can be repeated again, so that light will be multiply diffracted into higher order angles, $n\theta = n\lambda/\Lambda$ each with a frequency shift $n\omega$. A similar argument holds for the negative orders, so that a complete set of diffracted light beams will appear as shown in Fig. 3, where the angular deflection corresponding to the nth order is given by $\theta_n \cong \pm n\lambda/\Lambda$ and the frequency of the light deflected into the nth order is $\omega \pm n\Omega$. The intensity of the carrier wave, or zeroth order, will be zero when the modulation index $\Delta\phi$ is equal to 2.4. The generally important first order will have a maximum value of 34% of the input for $\Delta\phi = 1.8$, decreasing for higher modulation. These phenomena were described by Debye and Sears[1] and are often referred to as Debye–Sears diffraction. Similar observations were published almost simultaneously by Lucas and Biquard.[2] An extensive theoretical analysis of the effect was given by Raman and Nath,[3] and so it is alternatively referred to as Raman–Nath diffraction. A distinctive feature of this type of diffraction is that it is limited to low acoustic frequencies (or relatively long wavelengths). The origin of this limitation lies in the diffraction spreading of the light beam as it traverses apertures formed by the columns of compression and rarefaction in the acoustic beam. If the length of the

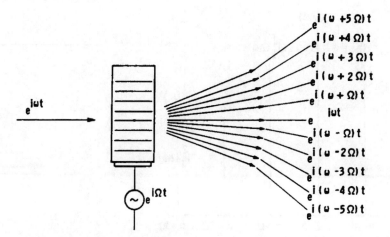

Figure 3 Raman–Nath diffraction of light into multiple orders.

acoustic beam along the light propagation direction is large enough, the diffraction spread of the light between adjacent compression and rarefaction regions will overlap, so that the Debye–Sears model is no longer valid. To estimate the characteristic length L_0 that bounds the Debye–Sears model, suppose the compression and rarefaction apertures are one-half an acoustic wavelength, $\Lambda/2$, so that the angular diffraction spread of the light is $\delta\phi \approx 2\lambda/\Lambda$. Then L_0 can be defined as that interaction length for which the aperture diffraction spreads the light by one-half an acoustic wavelength,

$$L_0 \delta\phi = \frac{\Lambda}{2} \tag{10}$$

or

$$L_0 = \frac{\Lambda^2}{4\lambda} \tag{11}$$

The interaction length is sometimes expressed in terms of a ratio known as the Raman–Nath parameter (often referred to as the Klein–Cook parameter) Q, as

$$Q \equiv \frac{2\pi L \lambda}{\Lambda^2}$$

Devices with interaction length $Q < \pi$ are said to operate in the Raman–Nath regime, while devices with $Q > 4\pi$ are said to operate in the Bragg regime. For typical values of $L = 1$ cm and $\lambda = 6.33 \times 10^{-5}$ cm, $Q = 1$ for $\Lambda = 0.0159$ cm (0.006 in.), which corresponds to a frequency of 31.4 MHz for a material whose acoustic velocity is 5×10^5 cm/s (2×10^5 in./s).

In the Bragg regime, the thin grating approximation no longer holds. If the incident light beam is normal to the sound beam propagation direction, the higher diffraction orders interfere destructively beyond L_0, eventually completely wiping out the diffraction pattern. In order for constructive interference to take place, the angle of incidence must be tilted with respect to the acoustic beam direction. To better understand what conditions must be satisfied for this, it is easier to think of the light and sound waves as colliding particles, photons, and phonons. In this description, the light and sound take on the attributes of particles, and the dynamics of their collisions are governed by the laws of conservation of energy and conservation of momentum. The magnitudes of the momenta of the light and sound waves are given by the well-known expressions

$$|k| = \frac{\omega n}{c} = \frac{2\pi n}{\lambda_0} \tag{12}$$

and

$$|K| = \frac{\Omega}{v} = \frac{2\pi}{\Lambda} \tag{13}$$

respectively. In the latter equation, v is the velocity of sound in the delay medium, $v = 2\pi\Omega\Lambda$. Conservation of momentum is expressed by the vector relation

$$k_i + K = k_d \tag{14}$$

the diagram for which is shown in Fig. 4(a), where k_i and k_d represent the momentum of the incident photon and the diffracted photon, respectively. The process may be thought of as one in which the acoustic phonon is absorbed by the incident photon to form the

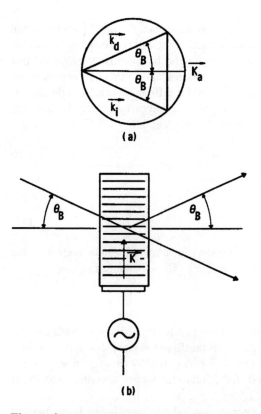

Figure 4 Bragg diffraction in isotropic medium.

diffracted photon. Thus, conservation of energy requires that

$$h\omega_0 = h\omega_1 + h\Omega \tag{15}$$

or

$$\omega_d = \omega_i + \Omega$$

in which h is Planck's constant.

Because ω_i lies in the optical frequency range, and Ω will typically lie in the RF or microwave range, $\Omega \ll \omega_i$ so that $\omega_d \approx \omega_i$. This results in the magnitudes of k_i and k_d being almost equal, so that the momentum triangle of Fig. 4(a) is isosceles, and the angle of incidence (with respect to the normal to K) is equal to the angle of diffraction. This angle of incidence is easily obtained from Fig. 4(a) as

$$\sin \theta_B = \frac{1}{2}\frac{K}{k} = \frac{1}{2}\frac{\lambda}{\Lambda} \tag{16}$$

It is called the Bragg angle because of its similarity to the angle of diffraction of x-rays from the regularly spaced planes of atoms in crystals. The configuration of these vectors in relation to the delay line is shown in Fig. 4(b). In order for diffraction to take place, the light must be incident at the angle θ_B, and the diffracted beam will appear only at this same angle. In contrast to the Debye–Sears regime, there are no higher-order diffracted beams. In the full mathematical treatment of the Bragg limit ($Q \gg 1$), light energy may appear at the higher orders, but the probability of its doing so is extremely small so that the intensity at higher orders is essentially zero. The diagrams in Fig. 4 show the interaction in which the diffracted photon is higher in energy than that of the incident photon, but the reverse can also take place. If the sense of the vector K is reversed with respect to k_i, then $\omega_d = \omega_i - \Omega$, and the diffracted negative first order results.

It is important to understand that the Debye–Sears effect and Bragg diffraction are not different phenomena, but are the limits of the same mechanism. The Raman–Nath parameter Q determines which is the appropriate limit for a given set of values λ, Λ, and L. Quite commonly in practice, these values will be chosen such that neither limit applies, and $Q \approx 1$. In this case, the mathematical treatment is quite complex, and experimentally it is found that one of the two first-order diffracted beams may be favored, but that higher orders will be present.

Having obtained the angular behavior of light diffracted by acoustic waves, the next most important characteristic is the intensity of the diffracted beam. Again, the full mathematical treatment is beyond the scope of this book, but a very good intuitive calculation leads to results that are useful. Referring to the spectrum of a phase-modulated wave shown in Fig. 2, we can see that the ratio of the intensity in the first order to that in the zero order is

$$\frac{I_1}{I_0} = \left[\frac{J_1(\Delta\phi)}{J_0(\Delta\phi)}\right]^2 \tag{17}$$

We shall now show in detail how this result comes about for acousto-optically diffracted light. The acoustic power flow is given by

$$P = \frac{1}{2}cve^2 \tag{18}$$

where c is the elastic stiffness constant. The elastic stiffness constant is related to the bulk modulus β and the density and acoustic velocity through the well-known expression

$$c = \frac{1}{\beta} = \rho v^2 \tag{19}$$

Thus, the acoustic power density is

$$P_A = \frac{1}{2}\rho v^3 e^2 \tag{20}$$

We can express the phase modulation depth in terms of the acoustic power density, using Eq. (5) for Δn and Eq. (7) for $\Delta\phi$, with the result

$$\Delta\phi = 2\pi\frac{L}{\lambda}\Delta n = -\pi\frac{L}{\lambda}n^3 p\left(\frac{2P_A}{\rho v^3}\right)^{1/2} \tag{21}$$

For small modulation index, the zero-order and first-order Bessel functions can be approximated by

$$J_0(\Delta\phi) \approx \cos(\Delta\phi) \approx 1 - \Delta\phi \tag{22}$$

and

$$J_1(\Delta\phi) \approx \sin(\Delta\phi) \approx \Delta\phi$$

so that the small-signal approximation to the diffracted light is, from Eq. (17),

$$\frac{I_1}{I_0} \approx (\Delta\phi)^2 = \frac{\pi^2}{2}\left(\frac{L}{\lambda}\right)^2 \left(\frac{n^6 p^2}{\rho v^3}\right) P_A \tag{23}$$

This efficiency may be expressed in terms of the total acoustic power P,

$$P = P_A(LH) \tag{24}$$

where H is the height of the transducer, and

$$\frac{I_1}{I_0} = \frac{\pi^2}{2}\frac{L}{H}\left(\frac{n^6 p^2}{\rho v^3}\right)\frac{P}{\lambda^2} \tag{25}$$

The quantity in parentheses depends only upon the intrinsic properties of the acousto-optic material, while the other parameters depend upon external factors. It is therefore defined as the figure of merit of the material,

$$M_2 = \left(\frac{n^6 p^2}{\rho v^3}\right) \tag{26}$$

from which it can be seen that, in general, the most important factors leading to high acousto-optic efficiency will be a high refractive index and a low acoustic velocity. This does not guarantee a large figure of merit, since the photoelastic constant may be very small, or even zero.

The other factors in Eq. (25) have the following effect on the diffraction efficiency. The efficiency decreases quadratically with increasing wavelength, so that the power requirements for operation in the infrared (IR) may be hundreds of times that required for the visible. For high efficiency, it will be desirable to have a large aspect ratio, L/H, leading to a configuration as shown in Fig. 5. It is difficult to make conventional bulk devices with H much less than 1 mm, so that aspect ratios up to about 50 can be achieved. Much higher aspect ratios can be reached in guided optical wave devices. A more exact calculation of the diffraction efficiency in the Bragg regime[4] yields the result

$$\frac{I_1}{I_0} = \sin^2\left[\frac{\pi^2}{2}\frac{L}{H}M_2\frac{P}{\lambda^2}\right]^{1/2} \tag{27}$$

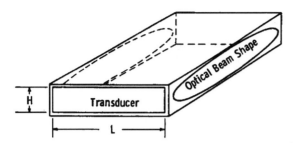

Figure 5 Transducer and optical beam shapes for optimization of acousto-optic diffraction.

For low signal levels Eq. (27) reduces to the same expression as in Eq. (25). To obtain an order of magnitude for the power requirements of an acousto-optic deflector, let us assume a material with $n = 1.5$, $\rho = 3$, $v = 5 \times 10^5$ cm/s, and a photoelastic constant calculated from the Lorentz–Lorenz expression, $p \cong 0.6$, so that $M \simeq 1.1 \times 10^{-17}\,\mathrm{s}^3/\mathrm{g}$. If the remaining parameters are $L = 1$ cm and $\lambda = 0.6$ μm, then by assuming a maximum acoustic power density for CW operation of 1 W/cm^2 (10^7 ergs/cm^2-s), the maximum obtainable efficiency is 15%. We shall see later, however, that materials and designs are available that are capable of realizing higher efficiencies with lower power levels.

2.3 Anisotropic Diffraction

Optical materials such as glass, or crystals with cubic structure, are isotropic with respect to their optical properties; that is, they do not vary with direction. Many crystals, on the other hand, are of such structure, or symmetry, that their optical properties depend on the direction of polarization of the light in relation to the crystal axes. They are birefringent; that is, the refractive index is different for different direction of light polarization.

The theory of diffraction of light thus far presented has assumed that the optical medium is isotropic, or at least that it is not birefringent. A number of important acousto-optic devices make use of the properties of birefringent materials, so a brief description of the important characteristics of anisotropic diffraction will be given here. The essential difference from diffraction in isotropic media is that the momentum of the light,

$$k = \frac{2\pi}{\lambda} = \frac{2\pi n}{\lambda_0} \tag{28}$$

will in general, be different for different light polarization directions. Thus, the vector diagram representing conservation of momentum will no longer be the simple isosceles triangle of Fig. 4(a). The momentum vectors for light that is ordinary polarized will terminate on a circle, as shown in Fig. 6, while those for light that is extraordinary polarized will terminate on the ellipse of Fig. 6.

To understand the effect of anisotropy on diffraction, it is necessary to mention another phenomenon that occurs when light interacts with shear acoustic waves, that is, waves in which the displacement of matter is perpendicular to the direction of propagation of the acoustic wave. A shear acoustic wave may cause the direction of polarization of the diffracted light to be rotated by 90°. The underlying reason for this is that the shear disturbance induces a birefringence which acts upon the incident light as does a

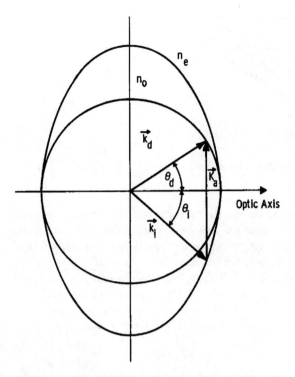

Figure 6 Vector diagram for diffraction in birefringent medium.

birefringent plate; in other words, it causes the plane of polarization to be rotated. This phenomenon occurs in isotropic materials as well as in anisotropic materials; however, in isotropic materials the momentum vector, $k = 2\pi n/\lambda_0$, will be the same for both polarizations, so there is no effect on the diffraction relations. Suppose, instead, that the interaction occurs in a birefringent crystal in a plane containing the optic axis. Let us choose the example as shown in the index surfaces in Fig. 6, in which the incident light is an extraordinary ray and the diffracted light is an ordinary ray. For this example,

$$k_i = \frac{2\pi n_e}{\lambda_0} \quad \text{and} \quad k_d = \frac{2\pi n_0}{\lambda_0} \qquad (29)$$

and the angles of incidence, θ_i, and diffraction, θ_d, are in general not equal. The theory of anisotropic diffraction was developed by Dixon,[5] in whose work the expressions for the anisotropic Bragg angles were derived as

$$\sin \theta_i = \frac{1}{2n_i} \frac{\lambda_0 f}{v} \left[1 + \left(\frac{v}{\lambda_0 f} \right)^2 (n_i^2 - n_d^2) \right] \qquad (30)$$

$$\sin \theta_d = \frac{1}{2n_d} \frac{\lambda_0 f}{v} \left[1 - \left(\frac{v}{\lambda_0 f} \right)^2 (n_i^2 - n_d^2) \right] \qquad (31)$$

where n_i and n_d are the refractive indices corresponding to the incident and the diffracted light polarizations, and f is the acoustic frequency,

$$f = \frac{v}{\Lambda} \tag{32}$$

These angles are plotted in Fig. 7 about that frequency f_m, for which there is a minimum in the angle of incidence. These curves, the general shapes of which are similar for all birefringent crystals, have a number of interesting characteristics that are useful for several types of acousto-optic devices. The minimum frequency for which an interaction may take place corresponds to $\theta_i = 90°$ and $\theta_d = -90°$, for which all three vectors will be collinear, as shown in Fig. 8. It is easily shown that for this case, since the vector equation for conservation of momentum can be written as a scalar equation,

$$|k_i| + |K| = |k_d| \tag{33}$$

the frequency for which collinear diffraction takes place is

$$f = \frac{v(n_i - n_d)}{\lambda_0} \tag{34}$$

Such collinear phase matching has been used as the basis of an important device, the electronically tunable acousto-optic filter.[6] Note that if the incident light had been chosen as ordinary rather than extraordinary polarized, the sense of the acoustic vector K would be reversed. In fact, the roles of the two curves in Fig. 7 would be reversed by interchanging n_i and n_d.

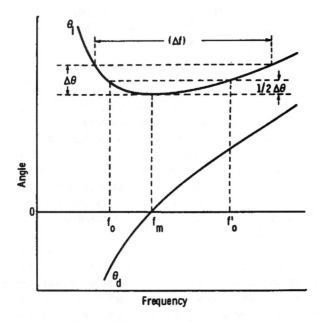

Figure 7 Angles of incidence and diffraction for anisotropic birefringent diffraction.

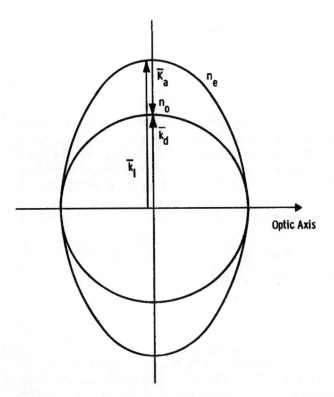

Figure 8 Vector diagram for collinear diffraction in birefringent medium.

Another interesting region of anisotropic diffraction occurs at the minimum value in the curve representing θ_i, at which frequency $\theta_d = 0$. This frequency, f_m, is obtained by setting the quantity in brackets in Eq. (31) equal to zero:

$$f = \frac{v}{\lambda_0}\sqrt{n_i^2 - n_d^2} \tag{35}$$

The significance of this point is that the angle of incidence of a scanned beam is relatively insensitive to change over a very broad range of frequencies. This frequency has important implications for the design of scanners because the bandwidth can be much greater than for a comparable isotropic scanner. Because the incident beam angle reaches a minimum value while the diffracted beam angle passes through zero at this point, and increases approximately linearly with frequency, the Bragg angle matching can be maintained over a large range, as will be described later. It will be seen that it is very difficult to achieve an interaction bandwidth this large by any other method.

The description of the interaction of light with sound we have given above is perhaps the simplest in terms of giving an intuitive understanding of the phenomena. Other descriptions, with totally different mathematical formalisms, have been carried out, and these lead to many details and subtleties in the behavior of acousto-optic systems that are beyond the scope of this book. Exact calculations have been carried out to extend the range of validity[7] from the limits allowed by the Raman–Nath theory,[8] and this has been experimentally investigated.[9] Other studies have also been carried out to give accurate numerical results for the intensity distribution of light in the various diffraction orders.[10] The diffraction process has been reviewed and analyzed by Klein and Cook,[11] using a coupled mode formulation, and there is continuing recent interest in refining the plane-wave scattering theory to give explicit results for intermediate cases.[12,13] Finally, the acousto-optic interaction can be viewed as a parametric process in which the incident optic wave mixes with the acoustic wave to generate polarization waves at sum and difference frequencies, leading to new optical frequencies; this approach has been reviewed by Chang.[14]

One of the most remarkable materials to have appeared recently for acousto-optic applications is paratellurite (TeO_2).[15] It has a unique combination of properties, which leads to an extraordinarily high figure of merit for a shear-wave interaction in a convenient RF range. It will be recalled that the anisotropic Bragg relations of Eqs. (30) and (31) led to a particular frequency, given by Eq. (35), for which the angle of incidence is a minimum and therefore satisfies the Bragg condition over a wide frequency range. However, typical values of birefringence place this frequency around 1 GHz or higher. Of particular interest in TeO_2 is its optical activity for light propagating along the c-axis, or (001) direction; the indices of refraction for left- and right-hand circularly polarized light are different, so that plane polarized light undergoes a rotation of its plane of polarization by an amount

$$R = \frac{2n_0}{\lambda} \delta \qquad (36)$$

where δ is the index splitting between left- and right-hand polarized light,

$$\delta = \frac{n_l - n_r}{2n_0} \qquad (37)$$

Just as acoustic shear waves can phase-match two linearly polarized light waves, they can also phase-match two oppositely circularly polarized light waves. Thus, shear waves propagating in the (110) direction, with shear polarization in the (110) direction, will diffract left- or right-hand polarized light propagating along the (001) direction, one into the other. The anisotropic Bragg relations apply to crystals with optical activity, where the birefringence is interpreted as

$$\Delta n = n_l - n_r = 2n_0 \delta \qquad (38)$$

and the value of δ obtained from specific rotation is wavelength-dependent. For the light and sound wave propagation directions described above, the acoustic velocity is 0.62×10^5 cm/s (0.24×10^5 in./s) and the figure of merit, M_2, is 515 relative to fused quartz. The frequency for which the Bragg angle of incidence is a minimum, as evaluated from Eq. (35) for $\lambda = 0.633$ μm, is $f = 42$ MHz, a very convenient frequency. For other important wavelengths, the minima occur at 36 MHz for $\lambda = 0.85$ μm and at 22 MHz for $\lambda = 1.15$ μm.

The application of the anisotropic Bragg equations to optically active crystals was discussed in detail by Warner et al.[16] They showed that near the optic axis the indices of refraction are approximated by the relations (for right-handed crystals, $n_r < n_l$)

$$\frac{n_r^2(\theta)\cos^2\theta}{n_0^2(1-\delta)^2} + \frac{n_r^2(\theta)\sin^2\theta}{n_1^2} = 1 \tag{39}$$

and

$$\frac{n_l^2(\theta)\cos^2\theta}{n_0^2(1+\delta)^2} + \frac{n_l^2(\theta)\sin^2\theta}{n_0^2} = 1 \tag{40}$$

For incident angles near zero with respect to the optic axis and for small values of δ,

$$n_r^2 = n_0^2\left(1 - 2\delta + \frac{n_1^2 - n_0^2}{n_1^2}\sin^2\theta\right) \tag{41}$$

and

$$n_l^2 = n_0^2(1 + 2\delta\cos^2\theta) \tag{42}$$

For light incident exactly along the optic axis the two refractive indices are simply

$$n_r = n_0(1 - \delta) \tag{43}$$

and

$$n_l = n_0(1 + \delta) \tag{44}$$

The anisotropic Bragg equations for optically active crystals are obtained by substitution of Eqs. (41) and (42) into Eqs. (30) and (31) for n_i and n_d. By ignoring the higher-order terms, this results in

$$\sin\theta_i \cong \frac{\lambda f}{2n_0 v}\left[1 + \frac{4n_0^2 v^2}{\lambda^2 f^2}\delta + \frac{\sin^2\theta_r n_0^2}{\lambda^2 f^2}\left(\frac{n_1^2 - n_0^2}{n_e^2}\right)\right] \tag{45}$$

and

$$\sin\theta_d = \frac{\lambda f}{2n_0 v}\left[1 - \frac{4n_0^2 v^2}{\lambda^2 f^2}\delta - \frac{\sin^2\theta_e n_0^2}{\lambda^2 f^2}\left(\frac{n_1^2 - n_0^2}{n_0^2}\right)\right] \tag{46}$$

The anisotropic Bragg angles (as measured external to the crystal) are shown in Fig. 9 for TeO$_2$ at $\lambda = 0.6328$ μm. It is obvious that for frequencies around the minimum, it will be possible to achieve a much larger bandwidth for a given interaction length than is possible with normal Bragg diffraction; a one-octave bandwidth corresponds to a variation in angle of incidence for perfect phase matching of only 0.16°. A useful advantage of such operation is that large bandwidths are compatible with large interaction lengths, which avoids higher diffraction orders from Raman–Nath effects. For normal Bragg diffraction,

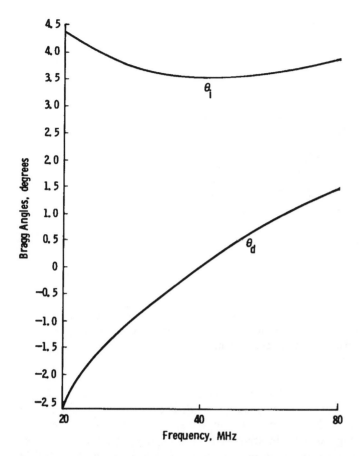

Figure 9 Bragg angles of incidence and diffraction (external) for anisotropic TeO_2 scanner, $\lambda = 0.6328$ μm.

on the other hand, large bandwidths can only be reached with interaction lengths that are so small that significant higher-order diffraction occurs. This decreases the efficiency with which light can be directed to the desired first order, and also limits the bandwidth to less than one octave in order to avoid overlapping low-frequency second-order with higher-frequency first-order diffracted light. However, tellurium dioxide operating in the anisotropic mode can always be made to diffract in the Bragg mode with no bandwidth limitation on length, so that this feature combined with the extraordinary high figure of merit leads to deflector operation with very low drive powers.

An important degeneracy occurs for anisotropic Bragg diffraction, which causes a pronounced dip in the diffracted light intensity at the midband frequency, where θ_i has its minimum. This degeneracy was explained by Warner et al.,[16] and is easily understood by referring to the diagram in Fig. 10. Two sets of curves are shown in the figure; the solid pair represent θ_i and θ_d when the incident light momentum vector has a positive component along the acoustic momentum vector, and the dotted pair represent these angles when the incident light momentum vector has a negative component along the acoustic vector. In the former case, the frequency of the diffracted light is upshifted, and in

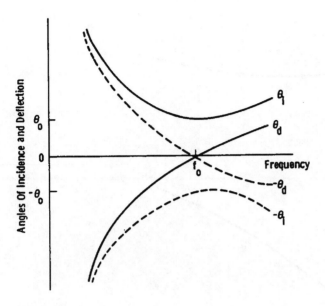

Figure 10 Angles of incidence and diffraction for anisotropic diffraction. Solid curves are for incident light having a component in the same direction as the acoustic wave, and dotted curves are for incident light having a component in the opposite direction.

the latter it is downshifted. The vector diagram for this process is shown in Fig. 11. Light is incident to the acoustic wave of frequency f_0 at an angle θ_0, and is diffracted as a frequency upshifted beam, $(\nu + f_0)$, normal to the acoustic wave. This light, in turn, may be rediffracted; referring to Fig. 10, it can be seen that for a frequency f_0 light that is incident at $\theta = 0°$ can be rediffracted to either θ_0 or $-\theta_0$. In the former case, the light will be downshifted to the original incident light frequency ν, and in the latter case it is upshifted to $\nu + 2f_0$. Note that this degeneracy can only occur at the frequency f_0 where light incident normal to the acoustic wave is phase-matched for diffraction into both θ_0 and $-\theta_0$. How the light is distributed in intensity between the three modes will depend upon the interaction length and the acoustic power level. The exact solution to this is found by setting up the coupled mode propagation equations under phase-matched conditions. The result of this is that maximum efficiency for deflection into the desired mode at f_0 is 50%. At low acoustic power, the deflection of light into the undesired mode is negligible; at high powers the unwanted deflection increases so that, for example, if the efficiency is 50% for

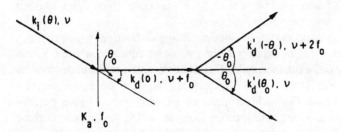

Figure 11 Vector diagram for midband degeneracy of Bragg diffraction in birefringent medium.

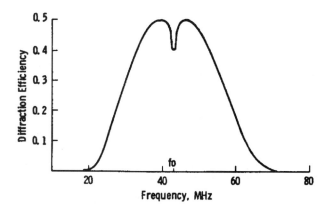

Figure 12 Effect of midband degeneracy on diffraction efficiency, for a maximum efficiency of 50%.

frequencies away from f_0, it will be 40% at f_0. The theoretical response of such a deflector is shown in Fig. 12, and is in excellent agreement with experimental results.

3 ACOUSTO-OPTIC MODULATOR AND DEFLECTOR DESIGN

3.1 Resolution and Bandwidth Considerations

Resolution, bandwidth, and speed are the important characteristics of acousto-optic scanners, shared by all types of scanning devices. Depending upon the application, only one, or all of these characteristics may have to be optimized; in this section, we will examine which acousto-optic design parameters are involved in the determination of resolution, bandwidth, and speed. Consider an acousto-optic scanner with a collimated incident beam of width D, diffracted to an angle θ_0 at the center of its bandwidth Δf. If the diffracted beam is focused onto a plane by a lens, or lens combination, at the scanner, the diffraction spread of the optical beam will be

$$\delta x = F \delta \phi \cong \frac{F\lambda}{D} \tag{47}$$

where F is the focal length of the lens. The light intensity will be distributed in the focal plane as illustrated in Fig. 13. As an example for diffraction-limited optics, the spot size for a 25 mm (1 in.) wide light beam of wavelength 6.33 μm at a distance of 30 cm (12 in.) from the delay line is 7.6 μm (3×10^{-4} in.). There are, however, aberrations that prevent this from being fully realized, as will be discussed later.

The number of resolvable spots will be the angular scan range divided by the angular diffraction spread,

$$N = \frac{\Delta \theta}{\delta \phi} \tag{48}$$

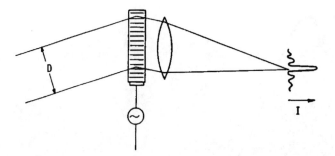

Figure 13 Distribution of light intensity due to diffraction by acoustic field.

where $\Delta\theta$ is the range of the angular scan. Differentiating the Bragg angle formula yields

$$\Delta\theta = \frac{\lambda}{v\cos\theta_0}\Delta f \tag{49}$$

and

$$N = (\Delta f)\left(\frac{D}{v\cos\theta_0}\right) = \Delta f \tau \tag{50}$$

where τ is the time that it takes the acoustic wave to cross the optical aperture. The resulting expression is the time–bandwidth product of the acousto-optic scanner, a concept applied to a variety of electronic devices as a measure of information capacity. The time–bandwidth product of an acousto-optic Bragg cell is equivalent to the number of bits of information that may be instantaneously processed by the system. For an acousto-optic modulator that is strictly a temporal modulator, a time–bandwidth product near unity is generally desired as the goal is often to modulate as fast as possible and therefore to minimize the aperture delay time. In contrast, for an acousto-optic deflector the time–bandwidth product is generally desired to be as large as possible to produce a large number of resolution elements.

There are two factors limiting the bandwidth of an acousto-optic device: the bandwidth of the transducer structure (discussed later) and the acoustic absorption in the delay medium. The acoustic absorption increases with increasing frequency; for high-purity single crystals the increase generally goes with the square of the frequency. For glassy materials, on the other hand, the attenuation will increase more slowly with frequency, often approaching a linear function. The maximum frequency is generally taken as that for which the attenuation of the acoustic wave across the optical aperture is equal to 3 dB. A reasonable approximation of the maximum attainable bandwidth is $\Delta f = 0.7 f_{max}$, so that we may derive some relationships for the maximum number of resolution elements.

For a material with a quadratic dependence of attenuation on frequency,

$$\alpha(f) = \Gamma f^2 \tag{51}$$

Acousto-Optic Scanners and Modulators

and the maximum aperture for 3 dB loss is

$$D = \frac{3}{\Gamma f^2} \tag{52}$$

Using these results, the maximum number of resolution elements is

$$N_{\max} \simeq \sqrt{\frac{1.5D}{v^2 \Gamma}} \tag{53}$$

from which it can be seen that, in principle, it is always advantageous to make the delay line as long as possible. In practice, the aperture will be limited by the largest crystals that can be prepared, or ultimately by the size of the optical system. For a glassy material for which the attenuation increases linearly with frequency,

$$\alpha(f) = \Gamma' f \tag{54}$$

and the maximum number of resolvable spots will be

$$N_{\max} \simeq \frac{2}{\Gamma' v} \tag{55}$$

which is independent of the size of the aperture, being determined only by the material attenuation constant and the acoustic velocity.

In the next section we will review material considerations in some detail, and see what the performance limits are of currently available acousto-optic materials. As a numerical example, however, the highest-quality fused quartz has an attenuation of about 3 dB/cm at 500 MHz and an acoustic velocity of 5.96×10^5 cm/s (2.35×10^5 in./s) (for longitudinal waves), leading to $N_{\max} = 560$.

3.2 Interaction Bandwidth

The number of resolution elements will be determined by the frequency bandwidth of the transducer and delay line, but a number of other bandwidth considerations are also of importance for the operation of a scanning system. While a large value of τ leads to a large value of N, the speed of the device is just equal to $1/\tau$. That is, the position of a spot cannot be changed randomly in a time less than τ. If the acoustic cell is being used to temporally modulate the light as well as to scan, then obviously the modulation bandwidth will similarly be limited by the travel time of the acoustic wave across the optical aperture. In order to increase the modulation bandwidth, the light beam must be focused to a small width, w, in the acoustic field. The 3 dB modulation bandwidth is approximately

$$\Delta f = \frac{0.75}{\tau} = \frac{0.75v}{w} \tag{56}$$

and the diffraction limited beam waist (the $1/e^2$ power points) of a gaussian beam is

$$w_0 = \frac{2\lambda_0 F}{\pi D} \qquad (57)$$

where D is the incident beam diameter and F is the focal length of the lens. With this value of beam waist, the maximum modulation bandwidth is

$$\Delta f = 0.36\pi \frac{vD}{\lambda_0 F} \qquad (58)$$

It can be seen from Eq. (58) that the modulation bandwidth for a diffraction-limited focused gaussian beam can be very high; for example, for a material of acoustic velocity 5×10^5 cm/s (2×10^5 in./s) the bandwidth of a 0.633 μm light beam focused with an $f/10$ lens is about 1 GHz. Such a system, however, is practically useless, because the diffraction efficiency would be extremely small.

In order for the Bragg interaction bandwidth to be large, there must be a large spread of either the acoustic or the optical beam directions, $\delta\theta_a$ and $\delta\theta_0$, respectively, or both. This spread may occur either by focusing, which in the case of the acoustic beam is achieved by curving the plane of the transducer, or it may be due simply to the aperture diffraction for both beams. It follows from fairly simple arguments that the optimum configuration for the most efficient utilization of optical and acoustic energy corresponds to approximately equal angular spreading, $\delta\theta_0 \cong \delta\theta_a$, as illustrated in Fig. 14.

For an acousto-optic deflector, the angular spread of the acoustic beam should be made large enough to match Bragg diffraction over the frequency range of the transducer-driving circuit bandwidth. As mentioned previously, this will result in some reduction in efficiency. To examine the relationship between bandwidth and the efficiency, we must first state another well-known result of acoustically diffracted light. This is, as shown by

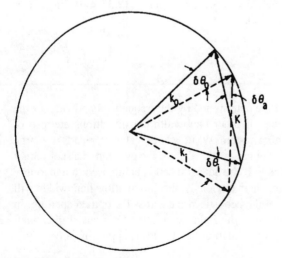

Figure 14 Vector diagram for Bragg diffraction in isotropic medium with angular spread of acoustic beam direction.

Cohen and Gordon,[17] that the angular distribution of the diffracted light will represent the Fourier transform of the spatial distribution of the acoustic beam. This Fourier transform pair is illustrated in Fig. 15 for the usual case of the rectangular acoustic beam profile. It seems intuitively obvious for this simple case, in which the diffraction spread of the incident optical beam is ignored, that there will be components in the diffracted light corresponding to the acoustic field side lobes. It is shown in Ref. 17 that the Fourier transform relationship holds for an arbitrary acoustic beam profile.

For the rectangular profile, the angular dependence of the diffracted light, illustrated in Fig. 15, is

$$\frac{I(\theta)}{I_0} \propto \left[\frac{\sin \frac{1}{2} KL(\theta - \theta_B)}{\frac{1}{2} KL(\theta - \theta_B)} \right]^2 \tag{59}$$

for which the -3 dB points occur at

$$\frac{1}{2} KL(\Delta\theta)_{1/2} \simeq \pm 0.45\pi \tag{60}$$

where $(\Delta\theta)_{1/2}$ is the value of $\theta - \theta_B$ at the half-power points. This yields a value for the angular width of the optical beam, just equal to the diffraction spread of the acoustic beam, namely

$$2(\Delta\theta)_{1/2} \simeq \frac{1.8\pi}{KL} \tag{61}$$

The frequency bandwidth is obtained by equating this result to the differential of the Bragg condition:

$$\delta\theta = \frac{\lambda_0 \Delta f}{nv \cos \theta_B} \tag{62}$$

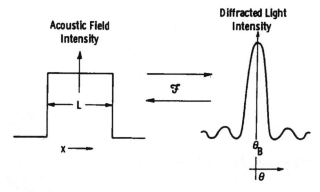

Figure 15 Fourier transform relationship between acoustic field intensity and diffracted light intensity.

This result is

$$\Delta f = \frac{1.8 n v^2 \cos \theta_B}{L f_0 \lambda_0} \qquad (63)$$

For acousto-optic scanning devices in which the bandwidth as well as the diffraction efficiency is of importance, a more relevant figure of merit may be the product of the bandwidth with the efficiency. By combining Eqs. (63) and (25), this product is

$$2 f_0 \Delta f \cdot \frac{I_1}{I_0} = \frac{1.8 \pi^2}{\lambda_0^3 H \cos \theta_B} \left[\frac{n^7 p^2}{\rho v} \right] P \qquad (64)$$

The quantity in brackets can be regarded as the figure of merit of the material when the efficiency–bandwidth product is the important criterion, and is designated as

$$M_1 = \frac{n^7 p^2}{\rho v} \qquad (65)$$

Other methods of achieving a large interaction bandwidth include transducer designs that steer the acoustic beam in direction in order to track the Bragg angle as it changes with frequency. A description of beam steering will be included in the section on transducers. Still another figure of merit was introduced by Dixon[18] in connection with wideband acousto-optic devices. Because the power requirements decrease as the transducer height H decreases, it is advantageous to make H as small as possible. If there are no limitations on the minimum size of H, it can be as small as the optical beam waist in the region of the interaction, h_{min}. The modulation bandwidth is determined by the travel time of the acoustic wave across this beam waist,

$$\tau \approx \frac{1}{\Delta f} = \frac{h_{min}}{v} \qquad (66)$$

so that

$$h_{min} = \frac{v}{\Delta f} \qquad (67)$$

Substitution of this value for H in Eq. (62) results in the relation

$$2 f_0 \frac{I_1}{I_0} = \frac{1.8 \pi^2}{\lambda_0^3 \cos \theta_B} \left[\frac{n^7 p^2}{\rho v^2} \right] \qquad (68)$$

and the appropriate figure of merit for this situation is the quantity in brackets:

$$M_3 = \left[\frac{n^7 p^2}{\rho v^2} \right] \qquad (69)$$

Acousto-Optic Scanners and Modulators

Note that the optical wavelength appears as λ_0^3 in both Eqs. (64) and (68), so that operation at long wavelengths is relatively more difficult in terms of power requirements for configurations optimizing bandwidth as well as efficiency.

3.3 Deflector Design Procedure

The useful optical aperture of a Bragg cell is usually considered to be that length across which the difference in acoustic attenuation between the highest and the lowest frequencies within the operating bandwidth of the cell is 3 dB. A particular application may dictate either a bandwidth or a resolution, that is, a time–bandwidth product. In general, an optimized Bragg cell design will maximize the number of resolvable spots, as well as other transducer structure parameters.

The number of resolvable spots, or the time–bandwidth product, will be determined by three key factors:[19] the acoustic attenuation Γ, the optical aperture of the acousto-optic crystal D, the angular beam spreading of the acoustic wave, which is determined by the transducer length L, and the acoustic wavelength. The constraints placed upon the number of resolvable spots N by these three factors is given by the relations[19]

$$N \leq \frac{1.5\Lambda_c}{\Gamma \Lambda_1^2} \tag{70}$$

$$N \leq \frac{D}{2\Lambda_c} \tag{71}$$

$$N \leq \left(\frac{L}{2\Lambda_c}\right)^2 \tag{72}$$

where Λ_c is the acoustic wavelength at the center frequency, Λ_1 is the acoustic wavelength at 1 GHz, and Γ is the acoustic attenuation in dB per unit length, normalized to 1 GHz (under the usual assumption that the attenuation increases quadratically with frequency). Note that Eq. (70) allows for a 3 dB attenuation. Once the center frequency and the bandwidth of the cell have been determined, the transducer structure must be designed. This will include the electrode length L and height H. The length must be chosen so that it is small enough to allow sufficient beam spread to satisfy the Bragg angle matching requirements over the desired bandwidth (for a fixed angle of incidence of the optical beam). At the same time, the diffraction efficiency will decrease as L decreases, so that we will want L to be as large as possible within the interaction bandwidth constraint.

3.4 Modulator Design Procedure

Acousto-optic deflectors and modulators have very similar design requirements and in some cases, one design may be suitable for either application or both. While the key design parameter for deflectors is typically the number of resolvable spots, the key design parameter for modulators is typically rise time or modulation bandwidth. These differing design parameters lead to the characteristic that, for deflectors, the optical aperture is typically made as large as possible, while for modulators, the optical beam is made as small as possible.

The rise time of an acousto-optic modulator (AOM) is fundamentally limited by the acoustic velocity of the modulator material. When an acoustic pulse is transmitted from the transducer, diffraction will begin when the leading edge of the pulse reaches the optical beam. Full, diffracted beam power will not be obtained until the acoustic wavefront reaches the opposite end of the optical beam. The shape of the rising optical pulse will depend on the shape of the optical beam.

For a gaussian optical beam, the time required for the acoustic wave to cross the $1/e^2$ beam diameter is

$$\tau = \frac{D_{1/e}^2}{V_a} \tag{73}$$

where V_a is the acoustic velocity. This beam width corresponds to a rise time from 2.3 to 97.7%. The more conventional rise time from 10 to 90% is calculated as

$$t_R = 0.64\tau \tag{74}$$

For video modulation applications, the rise time limits the frequency response of the modulator. The modulator bandwidth can be expressed as the frequency at which 3 dB roll-off occurs, which is estimated by the standard relation

$$f_0 = \frac{0.35}{t_R} \tag{75}$$

In the case of square pulse video modulation, the modulation speed may be defined by a specific dynamic extinction ratio requirement. For square wave modulation at frequency f_0, the dynamic extinction ratio is approximately 10:1. For high extinction ratios on the order of 1000:1, the maximum square wave modulation frequency is approximately $f_0/2$. For a given rise time, the design beam diameter can be calculated using the above relations. Note that the optical beam cannot be made arbitrarily narrow because the beam must remain relatively collimated over the acoustic interaction length L. If the beam waist is too small, the beam divergence over the length L will result in a longer rise time than predicted based on the beam waist. The minimum value of L is constrained by the need to stay in the Bragg regime (Sec. 2.2) and achieve a specified diffraction efficiency (Sec. 2.3).

The above discussion of dynamic extinction ratio assumes that the static extinction ratio is not limiting. The limit of static extinction ratio is determined by the ability to discriminate the diffracted beam from undiffracted or scattered light. Scattered light is a function of the quality of the material and the surface finish of the acousto-optic cell and is typically the limiting parameter for static extinction ratio when the diffracted beam separation is made large enough. The beam separation between the zero- and first-order beams is given by

$$\Delta\phi = \frac{\lambda}{\Lambda} \tag{76}$$

If the separation angle is made equal to the full divergence angle of the optical beam and a knife edge placed halfway between the zero- and first-order beams, approximately 2.3% of the blocked beam will pass the knife edge. This means the minimum static extinction ratio would be about 40:1. If the beam separation is increased to twice the beam angle, the optical power passing the knife edge is decreased to 0.003%. For most applications, this amount of beam separation is sufficient to make the extinction ratio limitation due to beam separation negligible. Using this condition with the formula of divergence of a gaussian beam gives

$$\Delta\phi > \frac{8\lambda}{\pi D_0} \qquad (77)$$

Combining the above two equations gives the following condition for beam separation

$$\Lambda > \frac{\pi D_0}{8} \qquad (78)$$

Another consideration is that the angular acceptance window of the acoustic field should be large enough to allow Bragg interaction over the optical beam. If the angular acoustic field is too narrow, the optical beam will be apodized in angle, resulting in output beam distortion and decreased diffraction efficiency. The angular acoustic intensity from a rectangular transducer is described by

$$I(\theta) = \text{sinc}^2\left[\frac{\theta}{(\Lambda/L)}\right] \qquad (79)$$

The null-to-null width is therefore $2\pi\Lambda/L$. This width should be much greater than the $1/e^2$ beam angle to maintain good diffraction efficiency and prevent distortion. Using the formula for beam divergence, the condition for L becomes

$$L \ll \frac{\pi^2 \Lambda D_0 n}{2\lambda} \qquad (80)$$

Remember that for Bragg interaction L cannot be made arbitrarily small. In practice, the length is chosen to be as short as tolerable based on requirements for efficiency and suppression of higher orders, typically such that Raman–Nath parameter $Q \approx 12$.

4 SPECIALIZED ACOUSTO-OPTIC DEVICES FOR SCANNING

4.1 Acoustic Traveling Wave Lens

Most acousto-optic applications are based on diffraction effects, requiring interaction over at least a few periods sinusoidal index change in the acoustic media. However, it is also possible to use the index change over a fraction of a period to act as a lens to focus light by refraction. In this case, the index change produced by a segment of an acoustic wave can be viewed as a gradient-index cylinder lens moving with the speed and direction of the acoustic wave.

Consider a conventional scanning system with a single axis scan device, in this case an acousto-optic Bragg deflector, followed by a scan lens. The total number of resolvable spots in the scan is determined by the aperture size and scan angle of deflector, and can be determined approximately as the product of the bandwidth and the acoustic transit time (Eq. 50). The scan lens can be altered to change the scan length and spot size, but the total number of spots remains the same. However, the number of spots may be increased by adding a traveling lens after the scan lens as shown in Fig. 16. In this case, the speed and timing of the deflector is synchronized with the speed and phase of the acoustic traveling wave such that the input beam tracks the acoustic lens as it propagates in the scan direction. The acoustic lens reduces the scanned spot size but has no effect on the scan length, thereby producing spot gain.

4.1.1 Design Considerations

The sinusoidal index variation is approximately parabolic near the index minimum and acts as a focusing (positive) lens. A quarter-wave aperture centered on the index minimum will provide the lens with near diffraction-limited performance. The focal length of the traveling lens was derived by Foster:[20]

$$F = \left(\frac{\Lambda}{4}\right)\left(\frac{n_0}{\Delta n}\right)^{1/2} \tag{81}$$

where Λ is the acoustic wavelength, n_0 is the index of refraction, and Δn is the peak refractive index variation. The focal length (F) in Eq. (81) refers to the focal distance internal to the lens. It can also be viewed as the lens thickness required to produce a quarter-pitch lens. If the thickness of the lens is less than F, the effective focal length will be longer.

Foster also derived an expression for the *f*-number of the lens by considering the path of the extreme rays at $\pm \Lambda/8$:

$$f\text{-}number = \frac{F}{D} = \frac{2}{\pi(n_0/\Delta n)^{1/2}} \tag{82}$$

Assuming a gaussian beam of diameter D_0 ($1/e^2$) is input to the traveling lens, the size of the focused spot is estimated as

$$D_1 = \frac{4\lambda F}{\pi n_0 D_0} \tag{83}$$

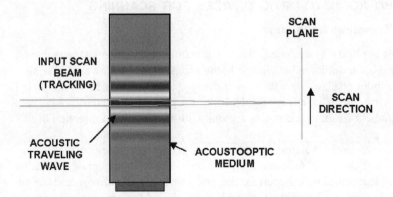

Figure 16 Application of an acoustic traveling-wave lens.

Combining Eqs. (81), (82), and (83) gives an expression for the output spot size from the traveling lens:

$$D_1 = \frac{8\lambda}{\pi^2 (n_0/\Delta n)^{1/2}} \tag{84}$$

The above derivation uses the assumptions that the lens thickness is approximately equal to F and the input beam diameter (D_0) is equal to $\Lambda/4$. In an application where the traveling lens is the final scan lens, F is typically made less than the lens thickness such that the focus occurs outside the lens and with a back focal distance sufficient to reach the scan plane.

As an example, assume that a prescanner produces a linear scan of 50 mm with a 0.5 mm diameter beam of wavelength 633 nm. A traveling acoustic lens device made from dense flint glass (SF-59) will be added to provide a final spot size of 0.05 mm or spot gain of 10. The required refractive index change (Δn) is determined to be 0.000157 from Eq. (84). The acoustic wavelength will be $4D_0$ or 2 mm, and therefore F is determined to be 55.7 mm from Eq. (81).

The length of the acoustic transducer (L), corresponding to the thickness of the lens, is chosen to be 45 mm, which is slightly less than F such that the focus can be outside the traveling lens device. The height of the transducer (H) is chosen to be 15 mm, such that the acoustic near field is longer than the scan length to prevent excessive acoustic loss along the scan direction due to diffraction spreading.

The amount of refractive index change is proportional to the square root of the acoustic intensity:

$$\Delta n = \left(\frac{M_2 P_A}{2}\right)^{1/2} \tag{85}$$

where M_2 is the acousto-optic figure of merit and P_A is the acoustic intensity. Using Eq. (85), the acoustic intensity required is 2.6 watts per mm^2. The instantaneous power required is therefore 1800 W.

While the power requirement is very large compared to those for typical Bragg cells, the average power required can be reduced significantly by pulsing the acoustic signal once per scan line. Even so, the large instantaneous and average power requirements for traveling acoustic wave devices are a significant challenge to practical implementation. One way to reduce the power requirement is to narrow the acousto-optic cell height to a fraction of an acoustic wavelength to form an acoustic slab-waveguide. The waveguide properties eliminate the acoustic diffraction spreading problem and associated need for a tall transducer.

4.2 Chirp Lens

In a typical acousto-optic deflector application, the transducer frequency is swept linearly over a range Δf to provide a linear angular scan ($\Delta \theta$) of the output beam. This frequency sweep is referred to as a chirp. The equivalent length of the acoustic chirp is equal to the product of the chirp time and the acoustic velocity. If the aperture were large enough to cover the entire acoustic chirp, then the diffracted angle would vary over the range $\Delta \theta$

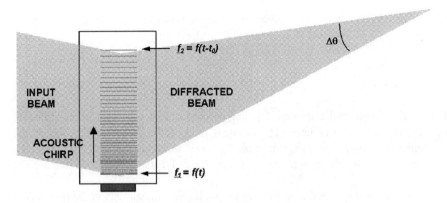

Figure 17 Focusing effect of an acoustic chirp.

along the chirp length as shown in Fig. 17. Using the small angle approximation, the acoustic chirp acts as a lens with f-number inversely proportional to $\Delta\theta$:

$$f\text{-number} \approx \frac{1}{\Delta\theta} \qquad \text{Acoustic aperture} \geq \text{Chirp length} \qquad (86)$$

If the aperture is smaller than the chirp length, as is typically the case, then the f-number will be inversely proportional to the fraction of the chirp length covered by the aperture:

$$f\text{-number} \approx \frac{1}{\Delta\theta} \frac{T_{\text{chirp}}}{T} \qquad \text{Acoustic aperture} < \text{Chirp length} \qquad (87)$$

For a typical beam deflector application, the chirp length T_{chirp} is much larger than the acoustic aperture time T. In this case, the focusing power of the chirp will be much less than that of an f–θ lens, and can be neglected. For very fast scanners, where T_{chirp} approaches T, the chirp focusing effect may contribute to the number of scan spots. Note that if $T_{\text{chirp}} = T$, the chirp focusing effect is equivalent to the power of an f–θ lens. However, this arrangement does not make a useful scanner, as the scan time is approximately the same as the access time.

By making the chirp smaller than the aperture length, the chirp can be used as a traveling lens. The chief application of a traveling chirp lens is to employ it as a post scan lens as described in the previous section. The traveling chirp lens is diffractive, and does not require the high instantaneous acoustic power needed to produce the refractive quarter-wave lens with comparable f-number. It also allows more flexibility with lens aperture, as the lens size is a function of chirp time and not a function of acoustic frequency. Unlike the refractive lens, the chirp lens is subject to diffractive losses and the diffraction efficiency may not be uniform across the aperture. Another disadvantage is that the quality of the lens is a function of the linearity of the chirp signal, such that phase error in the chirp signal will translate into aberrations in the lens.

4.3 Multichannel Acousto-Optic Modulator

Acousto-optic modulators are often used in conjunction with a scan beam to produce a raster scan. The scan line rate may be limited by the response time of the acousto-optic modulator, the exposure time, or the scan rate of the scan generating components. One way to increase throughput for a raster scan is to use multiple beams in a parallel scan arrangement. The beams may pass through a common set of scan elements, such that the final scans are identical except for position offset at the scan plane. Although scanned together, each beam must have its own modulation sequence. This can be accomplished by using a multichannel acousto-optic modulator.

Multichannel acousto-optic modulators can be fabricated using the same number of steps as single-channel modulators, with the main difference being that an array of electrodes is deposited on the transducer substrate instead of a single electrode. In application, a parallel array of beams is registered to the transducer array of the modulator as shown in Fig. 18.

The effect of one channel upon another is of particular concern for multichannel devices, especially as larger numbers of channels are integrated in a single device. Cross-talk between channels can occur through a number of mechanisms including electrical cross-talk in the feed circuitry or electrodes and acoustic cross-talk between adjacent channels. Another cross-talk mechanism is thermal, where the heat load added from turning on one or more channels causes a thermo-optic or strain-optic effect that alters the output of an independent channel. Electrical cross-talk is controlled by using good RF design practices including use of controlled impedance microstrip lines and providing good ground continuity from the feed circuitry to the transducer. Acoustic cross-talk is dependent on the location of the optical aperture and the spacing between electrodes. Because the acoustic field in isotropic materials will spread once it propagates beyond the near-field zone adjacent to the transducer, the amount of acoustic overlap is generally greater if the optical aperture is placed further from the transducer.

The degree of acoustic overlap between adjacent channels can also be manipulated through the electrode design. For an electrode with a simple rectangular shape, the acoustic intensity will have a sinc^2 angular distribution, which has side lobes 13 dB down from the main lobe. Other shapes, such as diamond or Gaussian envelope, can produce much lower side lobe values, although sometimes at the expense of faster spread of the main lobe.

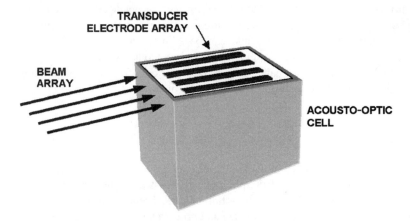

Figure 18 Multiple transducer channels on a monolithic acousto-optic device.

5 MATERIALS FOR ACOUSTO-OPTIC DEVICES

5.1 General Considerations

We have seen in the preceding section that two important criteria for choosing materials for acousto-optic scanning systems are the acousto-optic figure of merit and the high-frequency acoustic loss characteristics. Other properties that determine the usefulness of a material are its optical transmission range, optical quality, availability in suitable sizes, mechanical and handling characteristics as they may pertain to polishing and fabrication procedures, and chemical stability under normal conditions. As with most components, cost will be an important factor, even when all the other factors may be positive, if competing techniques are available.

One of the limitations on the use of acousto-optic scanners before the late 1960s was the availability of materials with reasonably high figure of merit. As we have seen, fused quartz, which is used as the standard for comparison, has a figure of merit so low that only a few percent diffraction efficiency can be obtained for scanners of typical dimensions, and with RF powers that can be applied without causing damage to transducer structures. Water is a fairly efficient material, with a figure of merit about 100 times larger than fused quartz and has actually found use in some scanning systems. As with most liquids, it cannot be used at frequencies higher than about 50 MHz, so that large numbers of resolution elements cannot be achieved. Since the late 1960s, many new materials have been synthesized and existing ones were found to have excellent properties. Materials can now be found for most scanning applications from the UV through the intermediate IR where high bandwidth is required.

The selection of a material for any particular device will be dictated by the type of operation under consideration. In general, it is desirable to select a material with low-drive-power requirements, suggesting those with large refractive index and low density and acoustic velocity. If, however, high-speed modulation is of paramount importance, then a low acoustic velocity may lead to slower than required speeds. In the following section, we will review the factors and trade-offs involved in the selection of materials for various acousto-optic applications. Whatever the particular material requirements may be, there are also a number of practical considerations that dictate several generally important material properties whatever the application: (1) the optical quality must be high so that not only absorption but scattering and large-scale inhomogeneities are small; (2) good chemical stability is required so that protective enclosures are not needed to maintain integrity; (3) good mechanical properties are required so that the device can be cut and polished without extraordinary procedures and can be adjusted and used with normal handling techniques; (4) the availability of crystal growth methods for obtaining suitably large, high-quality boules with reasonable cost is needed; and (5) a low-temperature coefficient of velocity is required to avoid drift of scan properties.

5.2 Theoretical Guidelines

There is no simple microscopic theory of the photoelastic effect in crystals. Therefore it is not possible to predict the magnitude of the photoelastic constants from first principles. However, Pinnow[21] has suggested the use of certain empirical relationships between the various physical properties in order to systematize and group acousto-optic materials. It is well known that such relations exist, for example, for the refractive index and the acoustic velocity for such groups as the alkali halides, the mineral oxides, and the III–IV compounds.

A large amount of data has been collected on the refractive indices of crystals, and generally good agreement is found with the Gladstone–Dale[22] equation

$$\frac{n-1}{\rho} = \sum_i q_i R_i \tag{88}$$

in which R_i is the specific refraction of the ith component and q_i is percentage by weight. Reliable values of R_i have been determined from mineralogical data over many years. From the expression for the acousto-optic figure of merit, it is apparent that a high value of refractive index is desirable for achieving high diffraction efficiency. It is not, however, possible simply to select for consideration those materials with high refractive index, as even a casual survey shows that such materials tend to be opaque at shorter wavelengths. This trend was examined in great detail by Wemple and DiDomenico,[23] who found that the refractive index is simply related to the energy band gap. The semiempirical relation for oxide materials is

$$n^2 = 1 + \frac{15}{E_g} \tag{89}$$

where E_g is the energy gap (expressed in electron volts). For other classes of materials the energy gap constant will be different, but the same form holds. It can be seen from Eq. (89) that the largest refractive index for an oxide material transparent over the entire visible range (cutoff wavelength at 0.4 μm) is 2.44. Higher refractive indices can be chosen only by sacrificing transparency at short wavelengths.

Pinnow[21] has found that a good approximation to the acoustic velocity for a wide range of materials is obtained with the relation

$$\log\left(\frac{v}{\rho}\right) = -b\bar{M} + d \tag{90}$$

where $\bar{M}$ is the mean atomic weight, defined as the total molecular weight divided by the number of atoms per molecule, and b and d are constants. Large values of d are generally associated with harder materials, while b does not vary greatly for oxides. Thus, in general, low acoustic velocities tend to be found in materials of high density, as is intuitively expected. Another useful velocity relationship has been pointed out by Uchida and Niizeki;[24] this is the Lindemann formula relating the melting temperature T_m and the mean acoustic velocity v_m,

$$v_m^2 = \frac{cT_m}{\bar{M}} \tag{91}$$

in which c is a constant dependent upon the material class. This relation suggests that high-efficiency materials would likely be found among those with large mean atomic weight and low melting temperature, that is, dense, soft materials.

In order for an acousto-optic material to be useful for wideband applications, the ultrasonic attenuation must be small at high frequencies. An attenuation that is often taken as an upper limit is 1 dB/μs (so that the useful aperture will depend upon the velocity). Many materials that might be highly efficient and otherwise suitable are excessively lossy at high frequency. A microscopic treatment of ultrasonic attenuation was carried out by Woodruff and Ehrenreich.[25] Their formula for the ultrasonic attenuation is

$$\alpha = \frac{\gamma^2 \Omega^2 \kappa T}{\rho v^5} \tag{92}$$

where Ω is the radian frequency, γ is the Grünneisen constant, κ is the thermal conductivity, and T is the absolute temperature. This formula would suggest that the requirement of low acoustic velocity and low attenuation conflict with each other, since $\alpha \sim v^{-5}$; it is quite unusual for materials with low acoustic velocity to not also have a high absorption, at least for the low-velocity modes.

The determination of the photoelastic constants of materials is essentially an empirical study, although a microscopic theory of Mueller,[26] developed for cubic and amorphous structures, is still referenced. For both ionic and covalent bonded materials the photoelastic effect derives from two mechanisms: the change of refractive index with density, and the change in index with polarizability under the strain. Both of these effects may have the same or opposite sign under a given strain, and one or the other may be the larger. It is for this reason that the magnitude or even the sign of the photoelastic constant cannot be predicted, since the effects may completely cancel each other. It is possible, however, to estimate the maximum constants for groups of materials. This has been done for three important groups with the result

$$|P_{max}| = \begin{cases} 0.21 & \text{water-insoluble oxides} \\ 0.35 & \text{water-soluble oxides} \\ 0.20 & \text{alkali halides} \end{cases}$$

In general, the photoelastic tensor components corresponding to shear strain will be less than those corresponding to compressional strain because there is no change, to first order, of density with shear; only the polarizability effect will be present. It is always possible that exceptionally large values of shear-related photoelastic coefficients may be found, but in no case could they be expected to be larger than the estimated value of $|P_{max}|$. The maximum values of photoelastic constant are shown in Table 1 for a number of important oxides and other materials.

5.3 Selected Materials for Acousto-Optic Scanners

Among older materials, those that have been shown useful for acousto-optic applications are fused quartz, because of its excellent optical quality and low cost for large sizes, and sapphire and lithium niobate, because of their exceptionally low acoustic losses at microwave frequencies. For infrared applications germanium[27] has proven very useful, as has arsenic trisulfide glass, where bandwidth requirements are not high. Among the newer crystal materials, very good acousto-optic performance has been obtained in the visible with GaP[28] and PbMoO$_4$.[29,30] One of the most interesting new materials to be developed within the past several years is TeO$_2$,[31] which along with PbMoO$_4$ has found

Table 1 Maximum Photoelastic Coefficients[a]

Material	$|P_{max}|_{measured}$
LiNbO$_3$	0.20
TiO$_2$	0.17
Al$_2$O$_3$	0.25
PbMoO$_4$	0.28
TeO$_2$	0.23
Sr$_{.5}$Ba$_{.5}$Nb$_2$O$_6$	0.23
SiO$_2$	0.27
YIG	0.07
Ba(NO$_3$)$_2$	0.35
α-HIO$_3$	0.50
Pb(NO$_3$)$_2$	0.60
ADP	0.30
CdS	0.14
GaAs	0.16
As$_2$S$_3$	0.30

[a]From Ref. 9.

wide use in commercially available acousto-optic scanners. More design details for devices employing this material will be given later. Among the new materials that have been developed for infrared applications, very high performance has been reached with several chalcogenide crystals.[32] Particularly important members of this group of materials include Tl$_3$AsS$_4$[33] and Tl$_3$PSe$_4$.[34] The compound Tl$_3$AsSe$_3$[35] is particularly interesting beyond its possible use as an infrared acousto-optic modulator material. Since Tl$_3$AsSe$_3$ belongs to the crystal class 3m, its symmetry permits it to possess a nonzero p_{41} photoelastic coefficient, and it is suitable for use as a collinear tunable acousto-optic filter, a device first realized by Harris,[36] using lithium niobate. Tables 2–4 summarize the properties of some of the materials that have been studied for acousto-optic applications. The acoustic attenuation constant in these tables is defined as

$$\Gamma = \frac{\alpha}{f^2} \tag{93}$$

which supposes that the attenuation increases quadratically with frequency. This will be the case for good-quality single crystals, but not for polycrystalline, highly impure, or amorphous materials. For the latter, the constant given in the tables is a rough estimate, based on measurements at the higher frequencies. The light polarization direction is designated as parallel or perpendicular according to whether the light polarization is parallel or perpendicular to the acoustic beam direction. Table 2 lists some of the more important amorphous materials, which may be useful if large sizes are desired or very low cost is required, but none of which can be used at frequencies much above 30 MHz. Table 3 lists the most important class of materials, crystals that are transparent throughout the visible with very low acoustic losses. Table 4 lists high-efficiency crystal materials that are transparent in the infrared and have reasonably low acoustic losses.

Table 2 Acousto-Optic Properties of Amorphous Materials

Material	Transmission range (μm)	Acoustic mode	v (cm/s × 10^5)	Γ (dB/cm-GHz^2)	Opt. Pol. Dir.	n (0.633 μm)	M_1 (cm^2-s/g × 10^{-7})	M_2 (s^3/g × 10^{-18})	M_3 (cm-s^2/g × 10^{-12})
Water	0.2–0.9	L	1.49	2400	∥ or ⊥	1.33	37.2	126	25
Fused quartz	0.2–4.5	L	5.96	12	⊥	1.46	8.05	1.56	1.35
SF-4	0.38–1.8	L	3.63	220	⊥	1.62	1.83	4.51	3.97
SF-59	0.46–2.5	L	3.20	1200	∥ or ⊥	1.95	39	19	12
SF-58		L	3.26	1200	∥ or ⊥	1.91	18.2	9	5.6
SF-57		L	3.41	500	∥	1.84	19.3	9	5.65
SF-6		L	3.51	500	∥ or ⊥	1.80	15.5	7	4.42
As_2S_3	0.6–11	L	2.6	170	∥	2.61	762	433	293
As_2S_5	0.5–10	L	2.22		∥	2.2	278	256 (est.)	125

Table 3 Acousto-Optic Properties of Crystals for the Visible

Material	Transmission range (μm)	Acoustic mode & prop. dir.	v (cm/s × 10^5)	Γ (dB/cm-GHz^2)	Opt. pol. dir.	n (0.633 μm)	M_1 (cm^2-s/g × 10^{-7})	M_2 (s^3/g × 10^{-18})	M_3 (cm-s^2/g × 10^{-12})
LiNbO$_3$	0.04–4.5	L[100]	6.57	0.15		2.20	66.5	7.0	10.1
		S[001]	3.59	2.6	$\perp$	2.29	9.2	2.92	2.4
Al$_2$O$_3$	0.15–6.5	L[100]	11.0	0.2	$\parallel$ = $\perp$	1.77	7.7	0.36	0.7
YAG	0.3–5.5	L[110]	8.60	0.25	$\parallel$ or $\perp$	1.83	0.98	0.073	0.114
		S[100]	5.03	1.1	$\parallel$ or $\perp$	1.83	1.1	0.25	0.23
TiO$_2$	0.45–6	L[001]	10.3	0.55	$\perp$	2.58	44	1.52	4
SiO$_2$	0.12–4.5	L[001]	6.32	2.1	$\perp$	1.54	9.11	1.48	1.44
		L[100]	5.72	3.0	[001]	1.55	12.1	2.38	2.11
α-HIO$_3$	0.3–1.8	L[001]	2.44	10	[100]	1.99	103	86	42
PbMoO$_4$		L[001]	3.63	15	$\parallel$ = $\perp$	2.62	108	36.3	29.8
TeO$_2$	0.35–5	L[001]	4.20	15	$\perp$	2.26	138	34.5	32.8
		S[110]	0.616	90	Circ [001]	2.26	68.0	793	110
Pb$_2$MoO$_5$	0.4–5	L a-axis	2.96	25	b-axis	2.183	242	127	82

Table 4 Acousto-Optic Properties of Infrared Crystals

Material	Transmission range (μm)	Acoustic mode & prop. div.	v (cm/s × 10^5)	Γ (dB/cm-GHz²)	Opt. pol. dir.	λ (μm)	n	M_1 (cm²-s/g × 10^{-7})	M_2 (s³/g × 10^{-18})	M_3 (cm-s²/g × 10^{-12})
Ge	2–20	L[111]	5.50	30	∥	10.6	4.00	10,200	840	1850
		S[100]	3.51	9	∥ or ⊥	10.6	4.00	1430	290	400
Tl₃AsS₄	0.6–12	L[001]	2.5	29	∥	1.15	2.63	620	510	290
GaAs	1–11	L[110]	5.15	30	∥ or ⊥	1.15	3.37	925	104	179
		S[100]	3.32		∥	1.15	3.37	155	46	49
Ag₃AsS₃	0.6–13.5	L[001]	2.65	800	⊥	.633	2.98	816	390	308
Tl₃AsSe₃	1.25–18	L[100]	2.15	314	∥	3.39	3.15	654	445	303
Tl₃PSe₄	0.85–9	L[100]	2.0	150	∥	1.15	2.9	2866	2069	1288
TlGaSe₂	0.6–20	L[001]	2.67	240	∥	.633	2.9	430	393	161
CdS	0.5–11	L[100]	4.17	90	∥	.633	2.44	52	12	12
ZnTe	0.55–20	L[110]	3.37	130	∥	1.15	2.77	75	18	19
GaP	0.6–10	L[110]	6.32	6.0	∥	.633	3.31	75	30	71
ZnS	0.4–12	L[001]	5.82	27	∥	.633	2.35	27	3.4	4.7
		S[001]	2.63	130	∥	.633	2.35	14	8.4	5.2
Te	5–20	L[100]	2.2	60	∥	10.6	4.8	10,200	4400	4640

Acousto-Optic Scanners and Modulators

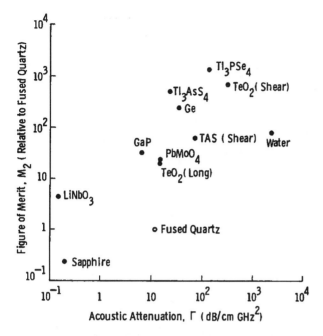

Figure 19 Figure of merit vs. acoustic attenuation.

An overall summary of a few outstanding (in one or another respect) selected acousto-optic materials presented in these tables is shown in Fig. 19. Using figure of merit and acoustic attenuation as criteria of quality, it is clear that a trade-off between these two parameters exists, and that the selection of the optimum material will be determined by the system requirements.

6 ACOUSTIC TRANSDUCER DESIGN

6.1 Transducer Characteristics

The second key component of the acousto-optic scanner after the optical medium, is the transducer structure, which includes the piezoelectric layer, bonding films, backing layers, and matching network. Recent advances in this area have made available a number of new piezoelectric materials of very high electromechanical conversion efficiency, and bonding techniques that permit this high conversion efficiency to be maintained over a large bandwidth. Furthermore, the design of high-performance transducer structures utilizing this new technology has been facilitated by new analytical tools[37,38] that lend themselves to computer programs for optimizing this performance.

The most elementary configuration of a thickness-driven transducer structure is shown in Fig. 20. It consists of the piezoelectric layer, thin film or plate, excited by metallic electrodes on both faces, and a bonding layer to acoustically couple the piezoelectric to the delay medium, or optical crystal. The backing is applied to mechanically load the transducer for bandwidth adjustment but may simply be left as air. The thickness of the transducer is about half an acoustic wavelength at the resonant frequency, and the thickness of the bonding layer is chosen to allow high, broadband

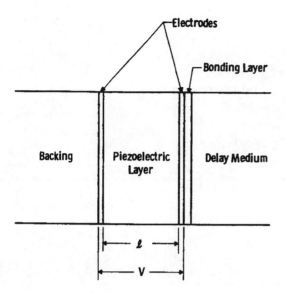

Figure 20 Transducer structure.

acoustic transmission. The most efficient operation of the transducer is obtained when the mechanical impedances of all the layers are equal. The mechanical impedance is

$$Z = \rho v \qquad (94)$$

and in general there is not sufficient choice of available materials to satisfy this condition. When the impedances are unequal, reflection occurs at the interfaces, reducing the efficiency of energy transfer. The reflection and transmission coefficients at the boundary between two media of impedances Z_1 and Z_2 are

$$R = \frac{(Z_1 - Z_2)^2}{(Z_1 + Z_2)^2} \qquad (95)$$

$$T = \frac{4Z_1 Z_2}{(Z_1 + Z_2)^2} \qquad (96)$$

The electromechanical analysis is generally carried out in terms of an equivalent circuit model, first proposed by Mason.[39] Several variations of the equivalent circuit have since been developed, but the one due to Mason is shown in Fig. 21. The fundamental constants of the transducer are permittivity s, acoustic velocity v, and electromechanical coupling factor k. The other parameters are transducer thickness l and area S. With these parameters, the circuit components shown in Fig. 21 are

$$C_0 = \epsilon \frac{S}{l} \qquad (97)$$

$$\phi = \kappa \left(\frac{1}{\pi} \omega_0 c_0 Z_0 \right) 1/2 \qquad (98)$$

$$Z_A = jZ_0 \tan \frac{\gamma}{2} \qquad (99)$$

$$Z_B = -j \frac{Z_0}{\sin \gamma} \qquad (100)$$

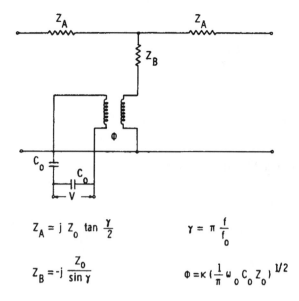

Figure 21 Equivalent circuit of Mason.

where

$$\omega_0 = \frac{\pi v}{l} \tag{101}$$

$$\gamma = \pi \frac{\omega}{\omega_0} \tag{102}$$

$$Z_0 = S\rho v \tag{103}$$

This equivalent circuit was used by Sittig[37] and Meitzler and Sittig[38] to analyze the propagation characteristics of acoustic energy between a piezoelectric and a delay medium. This was done in terms of a two-port electromechanical network, described by the chain matrix

$$\begin{pmatrix} A & B \\ C & D \end{pmatrix} = \prod_m \begin{pmatrix} A_m & B_m \\ C_m & D_m \end{pmatrix} \tag{104}$$

If the equivalent circuit of Fig. 21 is terminated at the input with a voltage source V_s and impedance Z_s, and at the output with a transmission medium of mechanical impedance Z_t, output voltage V_l, and load impedance Z_l, as shown in Fig. 22, then the insertion loss is

$$L = 20 \log \frac{V_s}{V_l} + 20 \log \left| \frac{Z_s + Z_l}{Z_l} \right| \text{ dB} \tag{105}$$

The impedances Z_s and Z_l are assumed to be purely resistive and

$$\frac{V_l}{V_s} = \frac{2Z_l Z_t}{\{AZ_t + B + Z_s(CZ_t + D)\}\{AZ_t + B + Z_l(CZ_t + D)\}} \tag{106}$$

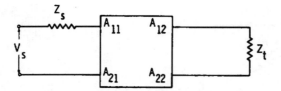

Figure 22 Terminated two-port transducer.

The two-port transfer matrix was obtained by Sittig,[40] with the result

$$A = \frac{1}{\phi H} \begin{vmatrix} A' & B' \\ C' & D' \end{vmatrix} \begin{vmatrix} \cos\gamma + jz_b \sin\gamma & Z_0(z_b \cos\gamma + z\sin\gamma) \\ \dfrac{j\sin\gamma}{Z_0} & 2(\cos\gamma - 1) + jz_b \sin\gamma \end{vmatrix} \quad (107)$$

where

$$z_b = \frac{Z_b}{Z_0}, \qquad H = \cos\gamma - 1 + jZ_b \sin\gamma \quad (108)$$

and

$$A' = 1, \qquad B' = j\frac{\phi^2}{\omega C_0}, \qquad C' = j\omega C_0, \qquad D' = 0 \quad (109)$$

The impedance Z_b represents the mechanical impedance of layers placed on the back surface of the transducer for loading, $Z_b = S\rho_b v_b$. In case the transducer is simply air-backed, $Z_b \simeq 0$. Electrical matching may be done at the input network by adding inductors either in parallel or in series in order to be electrically resonant with the transducer capacity C_0 at midband, $\omega = \omega_0$. If no inductances are added, the minimum loss condition is achieved for

$$R_s = \frac{1}{\omega_0 C_0} \quad (110)$$

where R_s is the source resistance. The inductance, if added, is chosen so that

$$L = \frac{1}{\omega_0^2 C_0} \quad (111)$$

A result of the matrix analysis shows that when piezoelectric materials with large values of the coupling constant κ are used, it is possible to achieve large fractional bandwidths without the necessity for electrical matching networks. As an example of the results obtained with this formalism, several plots of the frequency dependence of transducer loss for different values of the coupling constant are shown in Fig. 23.

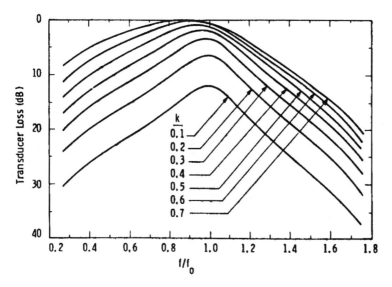

Figure 23 Transducer loss for various values of κ; $z_{0t} = 0.4$ and $R_s = (w_0 C_0)^{-1}$.

6.2 Transducer Materials

The piezoelectric material itself is perhaps the single most important factor governing the efficiency with which electrical energy can be converted to acoustic energy, this through the electromechanical coupling factor κ. The coupling efficiency is equal to κ. Prior to the discovery of lithium niobate, quartz was the most commonly used high-frequency transducer material, although its coupling factor, even for the most efficient crystal orientations, is rather small. The very-high-efficiency transducers were introduced with the discovery of various new ferroelectrics, such as lithium niobate, lithium tantalate, and the ceramic PZT materials, lead-titanate-zirconate. While the PZT transducers have among the highest values of κ, up to 0.7, they are not suitable for high-frequency applications since they cannot be polished to very thin plates. The most suitable piezoelectric transducer materials for high-frequency applications and their important properties are listed in Table 5, which is based on a compilation of Meitzler.[41] In order to produce transducers in the high-frequency range, say larger than 100 MHz, the piezoelectric crystal must be very thin (<20–30 μm). There are three well-established techniques for fabricating such thin transducers. In the first method, the piezoelectric plate is lapped to the desired thickness by the usual optical shop methods and then bonded to the delay medium. This method becomes impossibly difficult for transducers of even small area as their frequency increases, because such thin plates cannot be manipulated. A much more convenient technique is to bond the piezoelectric plates with a convenient thickness, say several tenths of a millimeter, to the delay medium and then lap the plate to the final thickness. In both methods, one electrode is first deposited on the delay medium, and in the case of thinning the piezoelectric after bonding, the second electrode and back layers are deposited as the final step. Care is required in lapping the bonded transducer so that the base electrode is not damaged by the polishing compound. If a chemically active compound, such as Cyton, is used, the delay medium as well as the electrode may be attacked and must be protected by some appropriate coating, such as photoresist. The final

Table 5 Properties of Transducer Materials

Material	Density	Mode	Orientation	K	ε_{rel}	v (cm/s)	Z (g/s-cm^2)
LiNbO$_3$	4.64	L	36° Y	0.49	38.6	7.4×10^5	34.3×10^5
		S	163° Y	0.62	42.9	4.56×10^5	21.2×10^5
		S	X	0.68	44.3	4.8×10^5	22.3×10^5
LiTaO$_3$	7.45	L	47° Y	0.29	42.7	7.4×10^5	55.2×10^5
		S	X	0.44	42.6	4.2×10^5	31.4×10^5
LiIO$_3$	4.5	L	Z	0.51	6	2.5×10^5	11.3×10^5
		S	Y	0.6	8	2.5×10^5	11.3×10^5
Ba$_2$NaNb$_5$O$_{15}$	5.41	L	Z	0.57	32	6.2×10^5	33.3×10^5
		S	Y	0.25	227	3.7×10^5	19.8×10^5
LiGeO$_2$	4.19	L	Z	0.30	8.5	6.3×10^5	26.2×10^5
LiGeO$_3$	3.50	L	Z	0.31	12.1	6.5×10^5	22.8×10^5
αSiO$_2$	2.65	L	X	0.098	4.58	5.7×10^5	15.2×10^5
		S	Y	0.137	4.58	3.8×10^5	10.2×10^5
ZnO	5.68	L	Z	0.27	8.8	6.4×10^5	36.2×10^5
		S	39° Y	0.35	8.6	3.2×10^5	18.4×10^5
		S	Y	0.31	8.3	2.9×10^5	16.4×10^5
CdS	4.82	L	Z	0.15	9.5	4.5×10^5	21.7×10^5
		S	40° Y	0.21	9.3	2.1×10^5	10.1×10^5
Bi$_{12}$GeO$_{20}$	9.22	L	(111)	0.19	38.6	3.3×10^5	30.4×10^5
		S	(110)	0.32	38.6	1.8×10^5	16.2×10^5
AlN	3.26	L	Z	0.20	8.5	10.4×10^5	34.0×10^5

electrical connection to the top electrode must be made in some fashion that does not mass-load the transducer and distort its bandpass characteristics, or be so small as to cause hot spots from high current densities. The usual method is to bond thin gold wire or ribbon onto electrode tabs, as is done for electronic circuit chips. The most successful method for fabricating very-high-frequency transducers for longitudinal wave generation is by deposition of thin films of piezoelectric materials by methods that yield a desired crystallographic orientation.[42,43] The materials used are CdS and ZnO, whose properties are shown in Table 5.

Such piezoelectric thin films generally cannot be grown with values of κ as high as that of the bulk material, but in the best circumstances κ may approach 90%. Thin-film transducers with band center frequencies up to 5 GHz can be prepared by these techniques.

A problem that arises with large-area transducers, or even with small-area transducers at very high frequencies, is that of matching the electrical impedance to the source impedance. It is especially true for the ferroelectric, piezoelectric materials of very high dielectric constant that the impedance of the transducer may be so low that it becomes difficult to efficiently couple electrical power from the source to the transducer. This problem can be largely overcome by dividing the transducer into a series connected mosaic, as reported by Weinert and deKlerk.[44] A schematic representation of such a mosaic transducer is shown in Fig. 24. If a transducer of given area is divided into N elements, which are connected in series, the capacity of the transducer will be reduced by a factor of N^2. As an example, a 1 GHz lithium niobate transducer of 0.25 cm^2 (0.4 in.^{2}) area would represent a capacitive impedance of only 0.038 Ω; if this area were divided

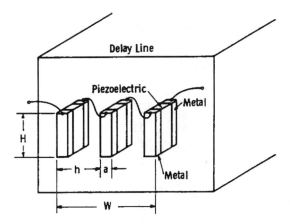

Figure 24 Schematic of mosaic transducer.

into a 16-element mosaic, the impedance would be increased to 10 Ω. A 40-element thin-film transducer is shown in Fig. 25. The same considerations will apply at lower frequencies for transducers with large areas, about 1 cm² or more. Because most ferroelectric transducer materials, such as the PZTs or lithium niobate, have high dielectric constants, the large areas will lead to very large capacitance values for frequencies far below 100 MHz. Thus, large-area transducers are usually divided into multiple elements, which are then wired in series to obtain the desired 50 Ω impedance to match to the RF driver. A large-area transducer that has been so wired is shown in Fig. 26.

6.3 Array Transducers

One of the serious limitations of normal (i.e., isotropic) Bragg acousto-optic deflectors is that imposed by the bandwidth as limited by the Bragg interaction. The most straightforward method of enlarging the interaction bandwidth is simply to shorten the interaction length in order to increase the acoustic beam diffraction spread. This is generally not a very desirable method to increase bandwidth for systems in which the light to the Bragg cells is collimated because it wastes acoustic power; only those momentum components of the acoustic beam that can be phase-matched to incident and diffracted light momentum components are useful. Furthermore, as the interaction length shrinks, the transducer becomes increasingly narrow, with a corresponding increase in power density. This increase in power density may produce heating at the transducer, which can cause thermal distortion in the deflector due to gradients in the acoustic velocity and refractive index.

An ideal solution to this difficulty would be one in which the acoustic beam changes in direction as the frequency is changed, so that for every frequency the Bragg angle is perfectly matched. The first approximation to such acoustic beam steering was carried out by Korpel[45] for a television display system. This transducer consisted of a stepped array, as shown in Fig. 27. The height of each step is one-half an acoustic wavelength at the band center $\Lambda_0/2$, and the spacing s between elements is chosen so as to optimize the tracking of the Bragg angle. Each element is driven π rad out of phase with respect to the adjacent

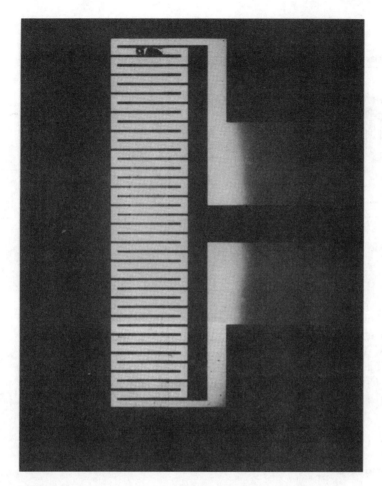

Figure 25 Forty-element thin-film mosaic transducer array.

elements, and the net effect of such a transducer is to generate an acoustic wave with corrugated wavefronts, which are tilted at an angle with respect to the transducer surfaces when the frequency differs from the band center frequency f_0. For this transducer configuration, the acoustic beam steers with frequency but matches the Bragg angle only imperfectly.

To understand the steering properties of such an acoustic array, which was analyzed in detail by Coquin et al.,[46] consider the somewhat simpler arrangement shown in Fig. 28, in which each transducer element is driven Ψ rad out of phase with respect to the next one, and Ψ may be electrically varied. This causes the effective wavefront to be tilted by an angle θ_e with respect to the piecewise wavefronts radiating from the individual elements. If θ_e is small, it can be approximated by

$$\theta_e \approx \tan \theta_e = \frac{\Psi}{2\pi} \frac{\Lambda}{s} = \frac{\Psi}{Ks} \qquad (112)$$

Figure 26 Four-element, series-connected lithium niobate transducer metal-bonded to Bragg cell.

If the incident light beam makes an angle θ_0 with the plane of the transducer and if the Bragg angle is $\theta_B = K/2k$, then the angular error from perfect matching is

$$\Delta\theta = (\theta_0 - \theta_e) - \theta_B = \left(\theta_0 - \frac{\Psi}{Ks}\right) - \frac{K}{2k} \tag{113}$$

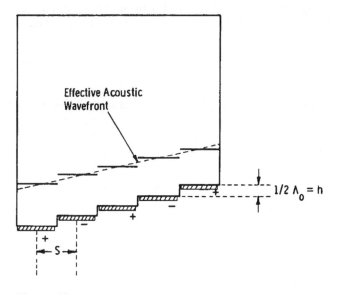

Figure 27 Stepped transducer array.

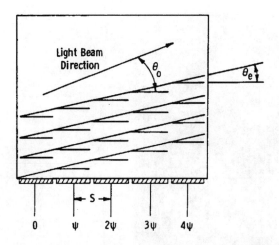

Figure 28 Steering of an acoustic beam by a phased array transducer.

The condition for perfect beam steering is that $\Delta\theta = 0$ for all values of K; setting $\Delta\theta = 0$, the required phase for perfect beam steering is

$$\Psi_p = \theta_0 Ks - \frac{K^2}{2k} s \tag{114}$$

from which it can be seen that the phase must be a quadratic function of the acoustic frequency.

Most of the work done on acoustic beam steering has involved making various approximations to this condition. One such approximation is obtained by making Ψ a linear function of frequency, with $\Psi = 0$ at f_0, the midband frequency. This was accomplished in the step transducer method of Fig. 27, as described in Ref. 45. For this case, the angle that the effective wavefronts make with respect to the transducer plane is

$$\theta_e \approx \frac{\pi}{Ks} - \frac{h}{s} \tag{115}$$

where h is the step height, and there is 180° phase shift between adjacent elements. The resulting beam steering error is

$$\Delta\theta = \left(\theta_0 - \frac{K}{2k}\right) + \left(\frac{h}{s} - \frac{\pi}{Ks}\right) \tag{116}$$

which can be made zero at the midband frequency f_0 by choosing

$$h = \frac{1}{2}\Lambda_0$$
$$s = \frac{\Lambda_0^2}{\lambda} \tag{117}$$

and

$$\theta_0 = \frac{1}{2}\frac{\lambda}{\Lambda_0}$$

A further improvement can be achieved by noting from Eq. (115) that θ_e varies as $1/f$, whereas perfect beam steering should lead to a linear variation of θ_e with f. Therefore, the constants h, s, and θ_e may be chosen to agree with the perfect beam steering case at two frequencies, rather than only one, as shown in Fig. 29. This first-order beam steering can yield substantial improvements in performance for systems requiring less than one octave bandwidth,[47] but bandwidths larger than this require a better approximation to the quadratic dependence of the phase on the acoustic frequency. The next higher approximation to perfect beam steering was carried out by Coquin et al.[46] for a 10-element array, as shown in Fig. 30. If the phase applied to each transducer corresponds to that for perfect steering, $\Psi_l = l\Psi_p$, and the element spacing is $s = \Lambda_0^2/\lambda$, the bandwidth extends from 0 to about $1.6 f_0$, the high-frequency drop-off being determined by the finite element spacing. Coquin pointed out that the deflector performance is very tolerant of errors in the individual phases; for example, if the phase applied to each transducer is within 45° of the perfect beam steering phase, there is a loss of only 0.8 dB in diffracted light intensity. If the phase error is increased to 90°, the loss increases to 3 dB. Thus, for

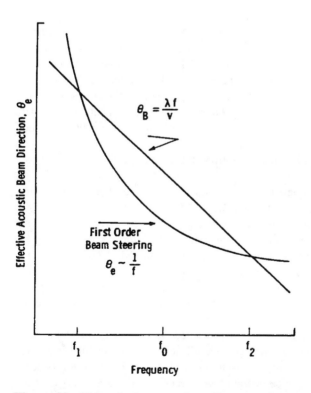

Figure 29 First-order beam steering with exact match at two frequencies.

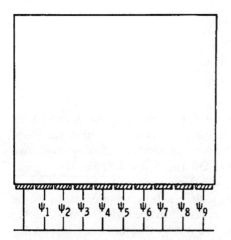

Figure 30 Ten-element phased array transducer, in which $\Psi = 0$, 90, 180, or 270°, leading to diffracted intensity less than 0.8 dB lower than for perfect beam steering.

deflectors, in which this degree of ripple is permissible, the transducer array may be driven by logic circuitry that sets digital phase shifters. This requires prior knowledge of the input frequency, or analog phase shifters, which accomplish the same function without the need for logic circuits.

An entirely different approach to broadband Bragg acousto-optic interaction matching is the use of the tilted transducer array, first reported by Eschler.[48] A tilted transducer array consists of two or more transducers electrically connected in parallel and tilted in angle with respect to each other, as illustrated in Fig. 31. Each transducer element in the array is designed to cover some fraction of the entire bandwidth, and its angle with respect to the incident light direction is chosen to match the Bragg angle at the center of its subband. For frequencies near the midband of any of the transducer elements, the incident light will interact strongly only with the sound wave emanating from that element; interaction will be weak from the other elements both because the angle of incidence will be mismatched and the frequency will be far from the resonance frequency of those elements.

On the other hand, for frequencies that are midway between the resonance frequencies of adjacent elements, that is, $(f_{01} + f_{02})/2$ or $(f_{02} + f_{03})/2$, the contributions to the acoustic fields from both elements are about equal and the effective wavefront direction lies midway between those of the components. Thus, the array behaves very much as if the acoustic wave were steering with frequency, although this is not true in a strict sense. The diffraction efficiency of the tilted transducer array is shown in Fig. 32, in which the solid curves represent the efficiency of the individual elements and the dotted curve represents the overall efficiency. There will typically be about 1 dB of ripple across the full band, which is acceptable for most applications.

There are two additional advantages to the tilted array transducer. First, it is obviously relatively simpler to design a larger overall acoustic bandwidth, since each element of the tilted array need be only about one-third of the total bandwidth. Second, tilted-array transducers can generally be operated more deeply in the Bragg mode, as the combined acoustic wavefronts from adjacent elements are twice the length of that from a

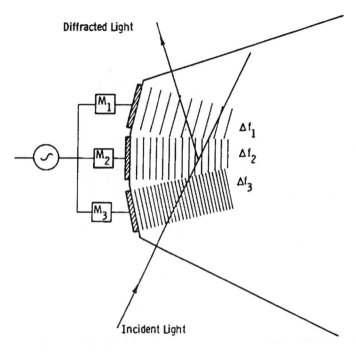

Figure 31 Tilted transducer array, in which each element is optimized for part of the entire frequency band.

single element. If the second-order diffracted light is sufficiently low, then operation over a frequency range larger than one octave is possible. The elements of the array can be connected in parallel since they will tend to behave as bandpass filters, the power being directed to the element with the closest frequency range. In practice, it is generally necessary to provide impedance-matching networks, as indicated in Fig. 31, because of the low reactance obtained with a parallel network.

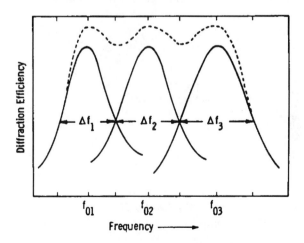

Figure 32 Diffraction efficiency of a three-element tilted transducer array. Solid curves represent efficiency of individual elements, and dotted curve represents efficiency of entire array.

7 ACOUSTO-OPTIC DEVICE FABRICATION

7.1 Cell Fabrication

The acousto-optic material is in most cases a crystal or an optical-grade glass. For crystalline materials, specific acoustic and optic propagation axes are typically required and the crystallographic orientation of the material must therefore be identified and marked. A typical cell configuration is shown in Fig. 33. The optical window surfaces are prepared on the faces perpendicular to the optical axis.

These surfaces are typically polished to a high-quality window surface with a flatness of $\lambda/20$ or better. Because the cells may be relatively thick (up to several centimeters) the homogeneity of the optical material may be a significant factor in the overall wavefront distortion. Therefore, the wavefront distortion may need to be specified for the cell in transmission.

Optical scatter is often a critical parameter for AO devices used in scanning systems. A portion of scatter from the zero-order beam will fall within the aperture of the deflected beam. This scattered light is unmodulated and may limit the extinction ratio in AOM applications. Scatter may occur at the surfaces due to contamination or imperfections or within the material due to defects or inhomogeneities.

Unlike most lenses and mirrors where the clear aperture is in the middle of the optic, the desired clear aperture of an acousto-optic device usually starts at the edge of the transducer face. For AOMs, the beams may be centered less than 0.5 mm from the edge of the cell. Therefore, special care must be taken to achieve the surface flatness, polish, and antireflection coating required across the entire clear aperture.

In particular with AOMs, the laser beam is focused to a small spot at the AOM to achieve the desired rise time and can lead to high optical intensity. Therefore, anti-reflection coatings often must be specified with high damage thresholds.

The acoustic transducer bonding surface is nominally parallel to the optic axis. This surface is prepared for bonding typically by polishing to an optical-quality surface. In most applications, the face of the cell opposite of the acoustic transducer surface is cut or ground at an off angle so that acoustic waves will not reflect directly back to the optical aperture or acoustic transducer. This is to avoid modulation or intermodulation from acoustic echo. Other techniques employed to reduce back reflection are to bond an acoustic absorbing

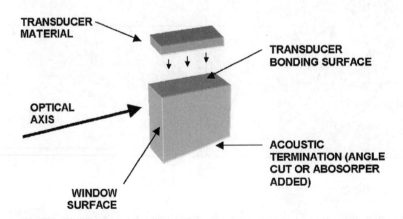

Figure 33 Bonding of acoustic transducer and acousto-optic medium

material to back surface or grinding a rough finish on the back surface to diffuse the back reflection.

7.2 Transducer Bonding

For transducers in the frequency range for which crystal plates are bonded to the delay medium, the bonding procedure is probably the most critical and most difficult step in fabricating the structure. The bonding layer can drastically modify the transmission of acoustic energy between the piezoelectric and the delay media; this is because the bond layer must provide molecular contact between the two surfaces, which will otherwise result in incomplete transfer, and because the mechanical impedance of the bond layer may produce a large acoustic mismatch with low transmission. In addition to these considerations, if the bond material is acoustically lossy, further decrease in transmission will result.

Because of the special properties required, there is only a very limited number of known bonding materials available. For temporary attachments, a commonly used agent is "salol," phenyl salicylate. It is easily applied as a liquid, which is crystallized by addition of a small seed. It is reliquified by gentle heating, and therefore is useful for various test measurements, but does not yield wide bandwidth or efficient coupling. A more satisfactory bond is made with epoxy resin, mixed to a very low viscosity, which may be compressed to a layer less than one micrometer thick before setting. Such thin layers require a high degree of cleanliness to avoid inclusion of any dust particles. Because of the low impedance of epoxy compared with such transducer materials as lithium niobate, thicker bonding layers would cause serious impedance mismatch problems around 100 MHz, where this technique has been successfully used.

Good results can also be obtained with a low-viscosity, ultraviolet, light-cured cement. For frequencies higher than about 100 MHz, other techniques, capable of yielding still thinner bond layers, which must be kept to a small fraction of an acoustic wavelength, must be used. Vacuum-deposited metallic layers are well suited for this purpose, because their thickness can be very accurately controlled down to the smallest dimensions, and impedances much closer to those of commonly used piezoelectric materials are available. Very good results were first obtained with indium bonds,[49] which are deposited to a thickness of several thousand angstroms on both surfaces, and without removal from the vacuum systems are mated under a pressure of about 100 psi. This technique yields a cold-welded bond, which has excellent mechanical properties with large acoustic bandwidth, if properly designed, and low insertion loss at frequencies of hundreds of megahertz.

The greatest fabrication difficulty is due to the necessity of maintaining the deposited films under vacuum to prevent oxidation. This requires a vacuum system with rather elaborate fixtures to bring the two surfaces together after film deposition and to apply the hydraulic pressure. The inside of a vacuum system in which this procedure is carried out is shown in Fig. 34. The substrates are held on either side of the evaporation filament sources during film deposition and are quickly brought into contact before contamination can occur. Using this technique, compression bonds of indium, tin, aluminum, gold, and silver are routinely made. It is essential for these procedures to be carried out in a dust-free atmosphere in order to avoid contaminating the interface with particulates. Even the smallest particles will prevent good acoustic contact between the transducer and the cell, so that fabrication must be carried out in a clean-room facility. A typical acousto-optic device clean room is shown in Fig. 35.

Figure 34 Vacuum compression bonding system. Metal films are deposited on transducer and delay line surfaces, which are then brought into contact.

In a modification of the indium compression bond[50], which allows the freshly deposited indium surfaces to be removed from the vacuum system for handling, the work is then placed in an oven under a pressure of several hundred psi, raised in temperature to slightly below the melting point of indium (156°C), and slowly cooled. This procedure forms a molecular bond in spite of the oxidation that may occur, and gives results similar to the vacuum bond. The principal drawback is that upon cooling, differential thermal expansion coefficients between the delay line material and the transducer material may set up unacceptable strains in the optical path. For some systems, this may not be a problem; for example, quartz or even lithium niobate transducers on fused or crystal quartz delay lines can be routinely made by this method. On the other hand, such crystals as tellurium dioxide require a great deal of care in handling, since they are extremely sensitive to thermal shock and strains. Differential contraction between the crystal and transducer for

Figure 35 A clean room for acousto-optic device fabrication.

a bond made in this fashion may easily be severe enough to fracture the crystal. Therefore, its applicability will depend upon the materials and sizes involved and upon the degree of freedom from residual strain required.

For frequencies approaching 1 GHz, the attenuation of indium layers may become excessive, and better results can be achieved with metals with lower acoustic loss constants. Among such metals are gold, silver, and aluminum.

Although these are made by the vacuum compression method, they generally require higher pressure. Still another method that has been used with these, as well as indium, is ultrasonic welding[51]. The chief advantage to be gained is that the procedure is carried out in normal atmosphere, since the ultrasonic energy breaks up the oxidation layer that forms on the surface. Some heating occurs as a result, but the temperature remains well below that required in the indium thermocompression method, with much lower residual strains. The technique requires the simultaneous application of pressures up to 3000 psi; this may be excessive for easily fractured or deformed materials or where odd-shaped samples are involved. A summary of the important properties of a few bonding materials, also used for electrodes and intermediate impedance-matching layers, is given in Table 6.

At lower frequencies, the effects of thin electrode and bonding layers on the performance of the transducer may be entirely negligible, but near 100 MHz, they become increasingly large, and even for layers less than 1 μm thick the effect may not be negzligible if the impedance mismatch to the rest of the structure is large. The effects of the electrode layer can be determined by setting $Z_b = 0$ in Eq. (107), and the entire effect of the back layers will be due to the impedance of the electrode z_{b1} of thickness t_{b1}, so the normalized impedance

$$z_b = jz_{b1} \tan(t_{b1}\gamma) \equiv j\tan\delta \qquad (118)$$

and the matrix of Eq. (107) becomes more complex.

Table 6 Acoustic Properties of Bond Layer Materials

Material	Longitudinal waves			Shear waves		
	Velocity (cm/s)	Impedance (g/s-cm^2)	Attenuation (dB/μm @ 1 GHz)	Velocity (cm/s)	Impedance (g/s-cm^2)	Attenuation (dB/μm @ 1 GHz)
Epoxy	2.6×10^5	2.86×10^5	Very large	1.22×10^5	1.34×10^5	Very large
Indium	2.25×10^5	16.4×10^5	8	0.19×10^5	6.4×10^5	16
Gold	3.24×10^5	62.5×10^5	0.02	1.2×10^5	23.2	0.1
Silver	5.65×10^5	38×10^5	0.025	1.61×10^5	16.7×10^5	
Aluminum	6.42×10^5	17.3×10^5	0.02	3.04×10^5	8.2×10^5	
Copper	5.01×10^5	40.6×10^5		2.11×10^5	18.3×10^5	

The effect of the bond layer and front electrode is even more complex, but an interesting illustrative example of varying the bond layer thickness is shown in Fig. 36. For this example, the normalized impedance of the bond layer is taken to be rather low, $z = 0.1$, and it can be seen that even for a fairly small thickness, the effect on the transducer loss is quite marked. Such a low value of impedance would correspond to the nonmetallic bond materials, but for the metallic bond materials the impedance mismatch would not be as severe, and the curve of transducer loss would be correspondingly less influenced. This influence of intermediate layers on the shape of the transducer loss curve can be used to determine the bandpass characteristics of the transducer structure. Such impedance transformers can be used, for example, to make the response symmetric about the band center f_0 by making the intermediate layer thickness one-quarter wavelength at f_0. By choosing other values for the thickness, the bandwidth can be enlarged, ripples

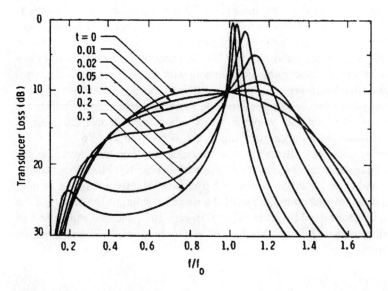

Figure 36 Transducer loss for various values of normalized transducer thickness t and intermediate layer normalized thickness 0.1. $R_s = (w_0 C_0)^{-1}$, $z_{0t} = 1$, $k = 0.2$.

smoothed, or various distortions introduced. In general, however, any such objectives are achieved at the expense of increased transducer loss.

7.3 Packaging

Acousto-optic device packaging must take into account needs for optical mounting and electrical connectivity and in many cases, thermal path management as well. Often it is desirable to attach the optical cell to a metal mount suitable for mechanical attachment in a larger optical system. This is achieved by bonding one or more of the surfaces to the mount with adhesive. Low shear strength adhesives are needed in many cases to alleviate temperature-induced strain between the metal and the optical material. This is because most metals have a thermal expansion coefficient significantly greater than that of glass or other optical crystals. For example, the linear expansion coefficient of aluminum is approximately 23 parts per million per °C compared to only 0.5 parts per million per °C for fused silica. If the cell is bonded with a thin layer of high strength epoxy, difference in expansion over temperature may cause significant strain birefringence in the optical cell.

For devices that use a watt or more of RF power, thermal management becomes important. Most of the power input to the device will end up as heat either from electrical resistive losses or acoustic attenuation loss. This heat must be transferred from the modulator efficiently enough to keep the operating temperature within desired limits. A significant fraction of the input power may be lost at the transducer surface of the AOM, including resistive losses from bond wires and electrodes, and acoustic losses in the transducer and bond layers. Localized heating from the transducer will form temperature gradients in the optical cell that can cause optical distortion due to the change in index with temperature. The effect may be reduced by attaching a heat sink to the back side of the transducer or by choosing an optical aperture further from the transducer. The remaining power is converted into acoustic power that dissipates inside the cell. Because most optic materials are poor thermal conductors, the cell may need to be intimately contacted with a good heat sink path on as much surface area as possible to keep temperature rise to a minimum.

For modulators operating with frequencies of hundreds of MHz or more, the transducer electrode is typically deposited as a thin film (less than 1 μm thickness) by vacuum deposition. Electrical connection to the electrodes is made by bond wiring. Direct soldering to the electrodes may cause damage to the electrode and cause excess loading on the acoustic transducer. A circuit card may be used to bring the input RF from an external connector up to the acoustic transducer. The feed circuit typically includes a passive matching circuit to optimize the impedance matching, nominally 50 Ω.

8 APPLICATIONS OF ACOUSTO-OPTIC SCANNERS

8.1 Multichannel Acousto-Optic Modulator for Polygonal Scanner

Acousto-optic modulators or deflectors can be fabricated with many independent channels on a single monolithic device. This approach allows multiple parallel beams to be modulated with a single device. This approach is used in the Etec Systems ALTA 3000 mask writing machine, which writes semiconductor photomasks using a rasterized simultaneous 32-beam scan.

The scanner architecture of the ALTA 3000 system is illustrated in Fig. 37. An argon-ion laser generates a single Gaussian beam at a wavelength of approximately

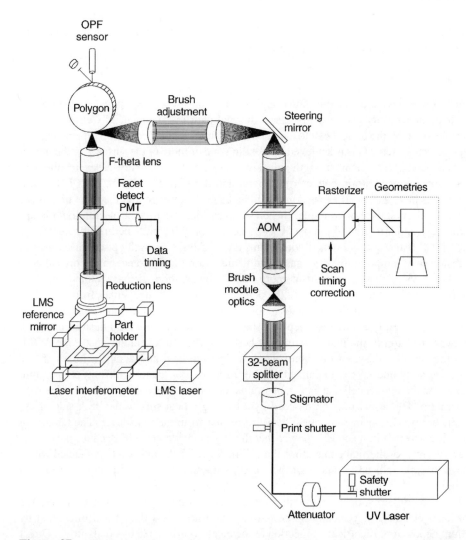

Figure 37 Use of a multichannel modulator in a precision semiconductor mask writing machine.

364 nm. A beam-splitter subassembly creates 32 separate beams, referred to as *the brush*, that pass through the AOM, which can independently turn on and off each of the beams. The modulation of the beams is controlled by the data path subsystem by varying the RF power to each channel of the AOM. The scan is created by a rotating polygonal mirror, and the scanned angle is converted to a spatial displacement by an $f-\theta$ lens. The final image of the brush at the photomask is obtained after transmitting it through a $20\times$, $0.6\,NA$ reduction lens. At the image plane, the FWHM diameter of the spot size is approximately 360 nm. During the scan, a translation stage moves the photomask perpendicular to the direction of scan.

The AOM and other optics are made from UV-grade fused silica, which has excellent transmittance and resistance to radiation darkening at UV wavelengths. Unfortunately, the M_2 value for fused silica is relatively low, making power requirements for the modulator an important design parameter. A nominal drive level of 500 mW per channel at 200 MHz is required to achieve a diffraction efficiency of 50%.

The brush used to print at the photomask is an image of the one first created at the AOM. Therefore, the arrangement of beams at the AOM is dictated by the requirements for the final print. In this case, this arrangement is two sets of 16 beams with each beam separated by approximately 3 $1/e^2$ beam diameters. The size of the beams at the AOM is set at 144 μm $1/e^2$ diameter based on the modulation bandwidth requirement of 50 megapixels per second, and the beam-to-beam spacing is 412 μm.

The design of the transducer electrode size and shape must consider the impacts on power efficiency, beam distortion, and channel-to-channel cross-talk. The limited aperture width and angular acceptance window of this design distort the output beam to a net far-field ellipticity of approximately 1.3:1, which is acceptable for this application. The modulator also introduces astigmatism, which would be detrimental to the system performance. This problem is corrected by precompensating optics referred to as the stigmator on Fig. 37.

The tight channel spacing requires careful attention to the prevention of electrical cross-talk in the feed network to the transducer electrodes. Radio frequency signals are fed to the modulator from the system electronics by an array of coaxial cables. The coaxial cables are connected to a printed circuit card with an individual trace for each of the 32 channels. Each trace ends with a land registering to one of the transducer electrodes. Bond wires are applied to make the final connection between the feed circuit and transducer electrodes.

When all 32 channels are on, this equates to 16 W of RF power. This creates localized heating in the AO cell near the transducer, also corresponding to the useful optical aperture of the AOM. Because the index of the fused silica is temperature-dependent, these localized temperature gradients correspond to index gradients in the glass that may cause several waves of distortion across the optical aperture. A temperature gradient across the aperture will have the effect of an optical wedge, causing the angular orientation of the beams to shift. To minimize these effects, a heat sink is applied to the back side of the transducer to keep the temperature rise at the transducer as small as possible. The AOM mount is also water cooled to carry the bulk heat away from the device and the surrounding optical system.

8.2 Infrared Laser Scanning

Acousto-optic beam scanners for use with infrared lasers have been under consideration in recent years by the aerospace industry in connection with laser radar and optical communications systems. Where system requirements place excessive demands on mechanical scanning methods, various electronic approaches become attractive. In general, the carbon dioxide laser, with wavelengths from 9 to 11 μm, is the most common one for long-wavelength operation. There are a number of electronic approaches to infrared beam scanning besides acousto-optic, and all of them are quite difficult to implement, usually for reasons related to the long interaction length needed to achieve large optical phase excursion. For acousto-optic diffraction, we have seen that the RF power needed for a given diffraction efficiency increases quadratically with wavelength. At 10.6 μm therefore, 280 times the power for 0.633 μm is necessary. Clearly, there will be severe constraints on the available materials for such devices, and on their performance. Referring to Table 4, we can see that only a few materials that transmit to 11 μm and have large acousto-optic figure of merit have been identified. The most common of these is germanium, which can be purchased in very large single crystals of excellent optical quality; germanium acousto-optic scanners have been commercially available for a number of years. They operate best in the isotropic mode, typically near 100 MHz RF. Another favorable infrared material for use in the 9 to 11 μm range is thallium arsenic selenide, which has very recently become available on a commercial basis. This crystal is best used in an anisotropic mode, as described in a previous section. The acousto-optic figure of merit is very high due to the low value of shear acoustic wave velocity. The low velocity produces another result that may simplify the design of scanner optics. The scan angles at infrared wavelengths are quite large; the angular dispersion for 10.6 μm carbon dioxide wavelength for this Bragg cell is shown in Fig. 38. For an RF bandwidth of 30% around 110 MHz center frequency, a scan-angle range of 16° is reached. For many applications, no magnification of the scan angle will be needed, as may be the case for acousto-optic scanners in the visible.

One of the major problems associated with carbon dioxide laser beam scanners is heating, due both to absorption of optical energy and RF power heating. If very high laser beam powers are used, then even a small absorption coefficient in the scanner may cause unacceptable heating. For germanium and thallium arsenic selenide, the absorption coefficients at 10.6 μm are 0.032 and 0.015 cm^{-1}, respectively. The thermal conductivity of germanium is much higher than that of thallium arsenic selenide, but design considerations may favor one or the other, depending upon the detailed effects of thermal gradients. High RF power operation is limited by heating at the transducer, which will eventually damage the transducer bond. Such thermal effects can be reduced by water- or air-cooling the RF mount and by heat-sinking the transducer; a photograph of a high-

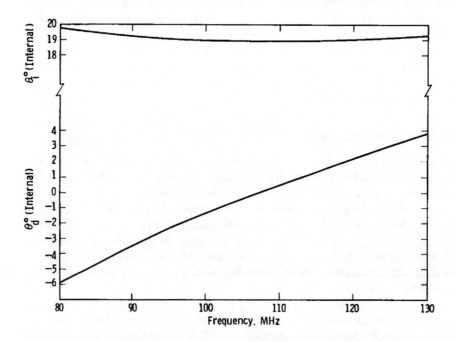

Figure 38 Anisotropic Bragg diffraction in TAS at λ = 10.6 μm.

power transducer with a matching sapphire heat sink, which also serves to make electrical contact, is shown in Fig. 39. The resolution of such infrared acousto-optic scanners will, in general, be limited by RF heating if high diffraction efficiency is to be obtained. This comes about because a large interaction bandwidth requires a small interaction length, so that high efficiency can only be achieved by high power. The relationship between resolution and acoustic power for the infrared scanner materials is shown in Fig. 40. If CW operation is required, it can be seen that no more than a few hundred spots can be obtained for an aperture of 1 or 2 cm.

8.3 Two-Stage Acousto-Optic Scanner

Acousto-optic deflection has been successfully applied to semiconductor photomask inspection. The KLA-Tencor 3000 series mask inspection machine checks photomasks for defects by scanning a focused spot across the mask and detecting the transmitted light (brightfield inspection). In this application, the feature sizes are very small (0.3 μm) compared to the size of the mask to be inspected (typically 6 in. diameter). To achieve desired throughput, a net scan rate of 50 megapixels per second is required. The number of spots is too great to cover the photomask with a single scan, therefore the approach is to use a high-speed optical scanner to produce a lineal subscan, and use x–y mechanical translation to pass the photomask beneath the scan location. Significant control is required to provide the precise translation required and register all the subscan data together such that the data for the entire photomask can be seamlessly reconstructed.

8.3.1 Scanner Optics

The lineal scan is produced by an acousto-optic scan pair and has no moving parts. A block diagram of the scan optical portion of the system is shown in Fig. 41. The source for the

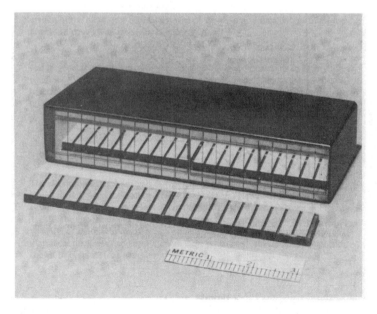

Figure 39 Transducer structure on TAS acousto-optic device, with matching heat sink for high RF power operation.

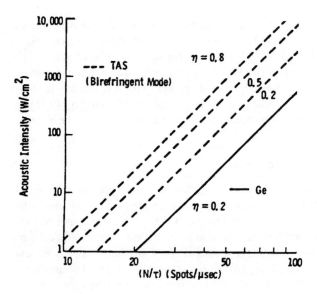

Figure 40 Acoustic intensity for various deflection efficiencies at the Bragg angle as a function of bandwidth using shear acoustic waves in TAS.

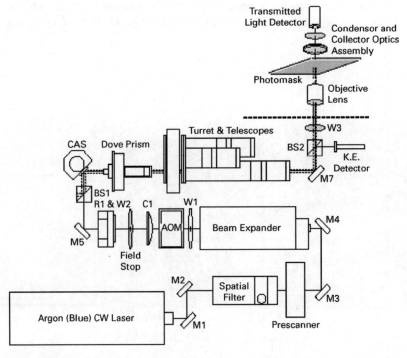

Figure 41 Use of an acousto-optic deflector (prescanner) and traveling chirp lens (AOM) in a semiconductor mask inspection machine. (Go to www.dekker.com to view this figure in color.)

scan beam is an argon-ion laser operating at 488 nm wavelength. The beam is passed through a spatial filter and then focused to a 400 μm beam diameter at the first acousto-optic device, the predeflector. This device is made from SF-6 glass and operates in the longitudinal acoustic mode. The center frequency for the device is 90 MHz, and the operating bandwidth is 14.4 MHz. This corresponds to a deflection angle in air of 12.5 mrads ± 1 mrads.

Following the predeflector, prescan optics transform the deflector output into a telecentric scan with magnification appropriate for the traveling chirp lens. The spot size at the input of the traveling chirp device is 12 mm $1/e^2$, and is designed to overfill the traveling chirp lens.

The chirp lens is produced from TeO_2, and uses the slow shear acoustic mode. Unlike the predeflector, the diffraction performance of a slow shear TeO_2 is very polarization dependent, and requires the input to be right circularly polarized. A wave plate is used to convert the linear polarization from the laser to circular polarization ahead of the TeO_2 device. The lens is formed from a linear chirp from 75 to 125 MHz over 7.5 μs. This makes the lens aperture in the scan direction 4.6 mm (7.5 × 0.616).

The aperture size in the cross-scan direction is controlled by the acoustic transducer height and is set to also be approximately 4.6 mm. The length of the scan is 14 μs or 8.6 mm. Therefore, the traveling chirp cell must have a clear aperture of at least 12.2 mm in the scan direction.

A cylinder lens is placed immediately after the chirp device to focus the spot on the cross-scan axis. A scan plane occurs at one focal length from the chirp lens. This is the object plane that is relayed and demagnified to produce the final scan at the photomask. Using the approximation $N = \tau\Delta f$, the number of resolvable spots from the predeflector is 1.6. While this performance would be of little use in direct scanning applications, its purpose in this application is to track the traveling chirp lens and maintain optimal illumination. The spot gain from the traveling chirp, also estimated from the time–bandwidth product, is 375. Therefore, the scan resolution is dominated by performance of the traveling chirp lens. The approximate scan size based on these approximations is 600 spots. However, due to limited aperture at the traveling chirp lens, the scanned spot is better approximated by an Airy disk function than a gaussian, and the scan size based on the null-to-null spacing approximately 1000 spots.

8.3.2 Driver

Both AO devices are driven by analog electronics that consist of a voltage-controlled oscillator followed by amplifier stages. Linear voltage ramps are generated by the system electronics and supplied to the drivers to produce the linear frequency chirps required by the AO devices. Both drive inputs are derived from the same clock to ensure synchronization is maintained between the two devices. Note that chirp linearity is not critical for the predeflector as it would be with an $f–\theta$ configuration. In this system, the scan linearity is controlled by the propagation of the traveling lens and the dominant concern is variation of acoustic velocity in the traveling lens cell due to changes in temperature. Chirp linearity is critical for the traveling lens, as the nonlinearities in the chirp signal will appear as aberrations in the lens. Therefore, precompensation is included in the voltage ramp fed to the traveling chirp voltage controlled oscillator to correct for inherent nonlinearities.

9 CONCLUSIONS

Acousto-optic devices have been utilized in a variety of scanning applications from direct spatial or temporal modulation to predeflection or post-scan lensing. For most applications of these devices there are significant design trades between modulation bandwidth, efficiency, and other performance parameters. Acousto-optic deflectors are often advantageous in applications where high precision is required over a relatively small angular scan. Acousto-optic modulators are effective for pixelation of raster scans or video modulation when it is impractical to directly modulate the optical source. The ability to modulate light without moving parts should continue to make acousto-optics an attractive technology in the future.

ACKNOWLEDGMENTS

We wish to thank Dr. Robert Montgomery for his review and assistance, and Damon Kvamme and Bryan Bolt for sharing their experiences with the application of acousto-optic devices.

REFERENCES

1. Debye, P.; Sears, F.W. Proc. Natl. Acad. Sci. 1932, *18*, 409.
2. Lucas, R.; Biquard, P. J. Phys. Rad. 1932, *3* (7), 464.
3. Raman, C.F.; Nath, N.S.N. Proc. Indian Acad. Sci. I 1935, *2*, 406.
4. Gordon, E.I. Proc. IEEE 1966, *54*, 1391.
5. Dixon, R.W. IEEE J. Quantum Electronics 1967, *QE-3*, 85.
6. Harris, S.E.; Nieh, S.T.R.; Winslow, D.K. Appl. Phys. Lett. 1969, *15*, 325.
7. Mertens, R. Meded. K. Vlaam. Acad. Wet. Lett. Schone Kunsten Relg., Kl. Wet. 1950, *12*, 1.
8. Exterman, R.; Wannier, G. Helv. Phys. Acta 1936, *9*, 520.
9. Klein, W.R.; Hiedemann, E.A. Physica 1963, *29*, 981.
10. Nomoto, O. Jpn. J. Appl. Phys. 1971, *10*, 611.
11. Klein, W.R.; Cook, B.D. IEEE Trans. Sonics Ultrason. 1967, *SU-14*, 723.
12. Korpel, A. J. Opt. Soc. Am. 1979, *69*, 678.
13. Korpel, A.; Poon, T. J. Opt. Soc. Am. 1980, *70*, 817.
14. Chang, I.C. IEEE Trans. Sonics Ultrason. 1976, *SU-23*, 2.
15. Uchida, N.; Ohmachi, Y. J. Appl. Phys. 1969, *40*, 4692.
16. Warner, A.W.; White, D.L.; Bonner, W.A. J. Appl. Phys. 1972, *43*, 4489.
17. Cohen, M.; Gordon, E.I. Bell Syst. Tech. J. 1965, *44*, 693.
18. Dixon, R.W. J. Appl. Phys. 1962, *38*, 5149.
19. Young, E.H.; Yao, S.K. Proc. IEEE 1981, *69*, 54.
20. Foster, L.C.; Crumly, C.B.; Cohoon, R.L. A high-resolution linear optical scanner using a traveling-wave acoustic lens. Appl. Opt. 1970, *9*, 2154–2160.
21. Pinnow, D.A. IEEE J. Quantum Electronics 1970, *QE-6*, 223.
22. Gladstone, J.H.; Dale, T.P. Phil. Trans. Roy. Soc. London 1964, *153*, 37.
23. Wemple, S.H.; DiDomenico, M. J. Appl. Phys. 1969, *40*, 735.
24. Uchida, N.; Niizeki, N. Proc. IEEE 1973, *61*, 1073.
25. Woodruff, T.O.; Ehrenreich, H. Phys. Rev. 1961, *123*, 1553.
26. Mueller, H. Phys. Rev. 1935, *47*, 947.
27. Abrams, R.L.; Pinnow, D.A. J. Appl. Phys. 1970, *41*, 2765.
28. Dixon, R.W. J. Appl. Phys. 1967, *38*, 5149.
29. Pinnow, D.A.; Van Uitert, L.G.; Warner, A.W.; Bonner, W.A. Appl. Phys. Lett. 1969, *15*, 83.

30. Coquin, G.A.; Pinnow, D.A.; Warner, A.W. J. Appl. Phys. 1971, *42*, 2162.
31. Ohmachi, Y.; Uchida, N. J. Appl. Phys. 1969, *40*, 4692.
32. Gottlieb, M.; Isaacs, T.J.; Feichtner, J.D.; Roland, G.W. J. Appl. Phys. 1969, *40*, 4692.
33. Roland, G.W.; Gottlieb, M.; Feichtner, J.D. Appl. Phys. Lett. 1972, *21*, 52.
34. Isaacs, T.J.; Gottlieb, M.; Feichtner, J.D. Appl. Phys. Lett. 1974, *24*, 107.
35. Feichtner, J.D.; Roland, G.W. Appl. Optics 1972, *11*, 993.
36. Harris, S.E.; Wallace, R.W. J. Opt. Soc. Am. 1969, *59*, 744.
37. Sittig, E.K. IEEE Trans. Sonics and Ultrasonics 1969, *SU-16*, 2.
38. Meitzler, A.H.; Sittig, E.K. J. Appl. Phys. 1969, *40*, 4341.
39. Mason, W.P. *Electromechanical Transducers and Wave Filters*; Van Nostrand Reinhold: Princeton, NJ, 1948.
40. Sittig, E.K. IEEE Trans. Sonics and Ultrasonics 1969, *16*, 2.
41. Meitzler, A.H. *Ultrasonic Transducer Materials*; Mattiat, O.E., Ed.; Plenum: New York, 1971.
42. deKlerk, J. *Physical Acoustics*; Mason, W.P., Ed.; Academic Press: New York, 1970; Vol. IV, Chap. 5.
43. deKlerk, J. IEEE Trans. on Sonics and Ultrasonics 1966, *SU-13*, 100.
44. Weinert, R.W.; deKlerk, J. IEEE Trans. on Sonics and Ultrasonics 1972, *SU-19*, 354.
45. Korpel, A. et al. Proc. IEEE 1966, *54*, 1429.
46. Coquin, G.; Griffin, J.; Anderson, L. IEEE Trans. on Sonics and Ultrasonics 1971, *SU-7*, 34.
47. Pinnow, D.A. IEEE Trans. on Sonics and Ultrasonics 1971, *SU-18*, 209.
48. Eschler, H. Optics Communications 1972, *6*, 230.
49. Sittig, E.K.; Cook, H.D. Proc. IEEE 1968, *56*, 1375.
50. Konog, W.F.; Lambert, L.B.; Schilling, D.L. IRE Int. Conv. Rec. 1961, *9* (6), 285.
51. Larson, J.D.; Winslow, D.K. IEEE Trans. Sonics and Ultrasonics 1971, *SU-18*, 142.

13

Electro-Optical Scanners

TIMOTHY K. DEIS

Consultant, Pittsburgh, Pennsylvania, U.S.A.

DANIEL D. STANCIL

Carnegie Mellon University, Pittsburgh, Pennsylvania, U.S.A.

CARL E. CONTI

Consultant, Hammondsport, New York, U.S.A.

1 INTRODUCTION

Electro-optic deflection systems are most typically considered in applications where high deflection speed is the paramount selection criteria. They can also have other attractive features, such as high optical efficiency and physical robustness, but these are usually secondary concerns after deflection speeds that can exceed 10^9 rad/s.

Requirements for great speeds are driven by advancements in two technical areas: lasers and computing. Progress in laser technology has resulted in reliable, low-cost compact sources with high power. Many applications, such as material marking, require a certain amount of energy to be delivered and are somewhat independent of the time period that the energy is delivered over. More powerful lasers can deliver the required energy dose in a short period of time, putting pressure on deflection system designers for higher operating speeds.

Advances in computing and communications have resulted in ever-higher data rates delivered to the laser deflection system. Data rates in applications such as displays now far surpass the control bandwidth of most mechanical systems such as galvanometers, piezoelectric deflectors, and microelectromechanical systems (MEMS) mirrors, which are limited by mechanical inertia effects. Relatively low bandwidth, usually less than 10 kHz, makes these systems a poor choice if a laser or other light source can deliver the required energy dose within a time period that matches the desired data rate, which often exceeds

MHz speeds. In the realm of electrical circuits, electro-optic deflectors are essentially capacitors, which can be charged very quickly with appropriate driver circuitry and can operate at MHz speeds with ease.

The most common technical competition to electro-optic (EO) deflection is acousto-optic (AO) deflection, covered elsewhere in this volume. Electro-optic systems are often faster and more optically efficient than AO systems. They often can handle higher beam powers than AO systems because there is no need to achieve a tight beam focus in the device for maximum speeds. The AO systems can be designed to have much higher resolutions, offering hundreds or thousands of resolvable spots compared to 5 to 100 that are more typical of EO systems. Both EO scanners and many AO scanners have polarization-dependent characteristics, which can limit their application. In general, the crystal and glass materials for AO devices are more readily available and have uniform high quality, unlike the materials for EO deflectors, which can be hard to procure or may have uncertain property characteristics. Electronic drivers for AO deflection systems usually require higher input power than those for EO systems.

Examples of current applications where the combination of high-power compact laser sources and high-speed data delivery are driving the evaluation or adoption of EO deflection systems include: encoding image data onto printing plates in plate setting machines, controlling optical paths in communications switching systems, creating jitter-free laser projection displays, and maintaining alignment in free space communications systems over long distances.

Electro-optic deflectors or scanners can be configured to meet a variety of requirements. Choices of the optical material, typically a crystal, the drive electronics, and optical path can be made to address all of the following functional requirements:

- Error correction: high-speed (10^6 rad/s) and low deflection (order of milliradian) characteristics can be used to provide facet-to-facet error correction of polygon scanners. The same characteristics can also be used to "debounce" a galvanometer-based display system.
- Switching: high speed (<100 ns steps) and low deflection (order of 10 resolution spots) can be used to switch collimated light in fiber optic switches, optical backplanes, and optical computers.
- Modulation: in cases where it is not possible to directly modulate a laser, as with some diode-pumped sources, a deflector can be used to modulate the beam by deflecting it toward or past a beam stop. Applications of this sort include displays, printing plate production, and marking.

There is a class of EO device that is sometimes confused with deflectors or scanners. This class uses the Kerr effect effectively to rotate the polarization of light traversing the device. Such devices can be used as modulators for polarized light sources by rotating the plane of polarization to be parallel or perpendicular to a polarizer downstream in the system. Such systems are commercially available, and have been used in some printing applications.

If a polarization-dependent mirror or beam splitter is used in place of the polarizer in the above system, the beam may be switched to one of two output positions by electro-optically controlling the polarization of the beam. These devices may be cascaded to produce many output positions. These and related systems are not the focus of the current text and will not be covered further.

Electro-Optical Scanners

The present effort builds on the previous edition, authored by Clive L.M. Ireland and John Martin Ley.[1] Much material has been carried over, especially that related to materials properties and basic physics. New material has been added to cover domain inverted scanner designs, which have recently been applied in commercial products. A new section was added to address some of the options and characteristics of electronic drivers suitable for commercial applications. Experience has shown that this part of the system can be as difficult to realize as the optical elements. As with the material relating to optical materials and designs, the electronic driver material is intended to serve as a general reference for guiding development efforts, not as a definitive treatise on the subject.

2 THEORY OF THE ELECTRO-OPTIC EFFECT

2.1 The Electro-Optic Effect

The electro-optic deflectors and scanners discussed below all rely on the electro-optic (EO) effect evidenced to some degree by all materials. Only a few materials exhibit property changes with an applied electric field large enough to be exploited for use in deflecting and switching applications. The index of refraction of these materials, typically crystals, changes when an electric field is applied to such a degree that it can be used to usefully deflect a beam according to the rules of normal refractive optics.

In crystals, the direction of the polarization induced by an applied electric field may differ in direction from that of the applied field. Mathematically, this means that the relative permittivity must be represented by a second-rank tensor:

$$D_i = \varepsilon_0 \kappa_{ij} E_j = \varepsilon_0 E_i + P_i \tag{1}$$

where ε_0 is the permittivity of free space, κ_{ij} is the relative dielectric tensor, and E_i and P_i are the ith components of the electric field and the induced polarization, respectively, and summation over repeated indices is assumed. We will restrict our discussion to crystals that are essentially neither magnetic nor optically active, and that exhibit negligible absorption. In this case, κ_{ij} is a real, symmetric tensor.

A convenient geometric representation of any symmetric second-rank tensor S_{ij} is an ellipsoidal or hyperboloidal surface defined by

$$S_{ij} x_i x_j = 1 \tag{2}$$

Thus we can construct such a surface for κ_{ij}, or its inverse $(1/\kappa)_{ij}$. In contrast, the refractive index – given by the square root of the relative permittivity in isotropic materials – does not transform as a second-rank tensor. Since $\kappa = n^2$ in an isotropic material, it is conventional to adopt the following notation:

$$\left(\frac{1}{n^2}\right)_{ij} x_i x_j = 1 \tag{3}$$

This ellipsoidal surface is called the index ellipsoid, or indicatrix. In the coordinate system in which $(1/n^2)_{ij}$ is diagonal, Eq. (3) reduces to

$$\frac{x^2}{n_x^2} + \frac{y^2}{n_y^2} + \frac{z^2}{n_z^2} = 1 \tag{4}$$

This surface has a simple geometric interpretation. The principal axes of the ellipsoid correspond to directions in the crystal for which D is parallel to E, and the refractive indices for waves polarized along these directions are n_x, n_y, and n_z.

2.2 The Linear Electro-Optic Effect

The crystal materials commonly used for EO devices do not possess inversion symmetry (see Stancil[2] for a complete explication of the tensor properties of crystals), meaning that the application of an electric field induces a small change in the refractive index that is proportional to the field, and reverses in sign when the field reverses. This is known as Pockel's Effect, or the Linear Electro-Optic Effect.

In the presence of a uniform electric field, the changes in the indices of refraction of such materials can be shown to be

$$\Delta\left(\frac{1}{n^2}\right)_{ij} = r_{ij,k} E_k \tag{5}$$

where $r_{ij,k}$ is the linear EO tensor, whose values are readily available from data published in the literature and from crystal suppliers.

As an example, consider the EO effect in KH_2PO_4, also known as KDP. In this crystal (and all crystals with symmetry $\bar{4}2m$) the only nonzero components of the EO tensor are $r_{41} = r_{52}$, and r_{63}. For simplicity, we will consider the case of a static electric field applied along the optic axis, E_3. For further simplicity, for small Δn, which is normally the case, $\Delta(1/n^2) = -2\Delta n/n^3$. Thus, the refractive index seen by an optical wave polarized along the $\langle 110 \rangle$ direction is approximately given by

$$n_1 = n_o - \frac{1}{2} n_o^3 r_{63} E_3 \tag{6}$$

where use has been made of the fact that $r_{63} E_3 \ll 1/n_o^2$. Similarly, the index along the $\langle 1\bar{1}0 \rangle$ direction is

$$n_2 = n_o + \frac{1}{2} n_o^3 r_{63} E_3 \tag{7}$$

The index along the optical axis $\langle 001 \rangle$ is unchanged ($n_3 = n_e$).

Note that the index change observed depends on the polarization of the light transiting the region with the applied field. This effect makes the performance of many EO scanners "polarization dependent," where the deflection achieved depends on the polarization of the beam. This can result in splitting randomly polarized beams and is the reason that most EO scanners are used with polarized beams only.

2.3 The Quadratic Electro-Optic Effect

Refractive index changes proportional to the square of the applied field are permitted by symmetry in all materials. Besides crystals such as KDP and $LiNbO_3$, liquids that are strongly polar are of particular EO interest since they can exhibit a high anisotropic, optic

Electro-Optical Scanners

polarizability. By applying a strong external field, the molecules of these substances partially align with the field, causing the bulk material to become birefringent.

The component of a beam polarized parallel to the main polarizability of the molecules, usually nearly parallel to the dipole moment of the molecule, sees an increase in refractive index relative to that of the orthogonal polarization. This effect, which was observed by Kerr in glass and other materials, is generally described by the following equation:

$$n_p - n_s = B\lambda E^2 \qquad (8)$$

Here λ is the vacuum wavelength of the beam, B is the Kerr constant for the material, and n_p and n_s are the parallel and orthogonal refractive index components, respectively, and E is the applied electric field.

A variety of Kerr materials and devices have been studied in the past, see Lee and Hauser[3] and Kruger et al..[4] At the time of this writing, late 2002, there do not appear to be any commercial devices or systems based on the quadratic EO effect in production.

3 PRINCIPAL TYPES OF ELECTRO-OPTIC DEFLECTORS

3.1 Basic Topologies

The EO effect can be utilized in a variety of basic ways, with details seemingly limited only by the imagination of very clever practitioners. The design problem is one of selecting a gross geometry that the beam can traverse, and then selecting the geometry and magnitude of desired index change. To provide some order, the following nomenclature is used:

- Shaping of gross geometry: at its simplest, an EO scanner can consist of a prismatic crystal element with electrodes covering both ends. As the potential across the element is changed, it acts like an electrically controlled prism.
- Electrically shaped fields: When a bulk prism would be too small for easy handling, or if the field strength desired would result in electrical breakdowns across the side surfaces, electrical fields may be shaped. This is normally achieved by using shaped electrodes of constant potential. A less common approach is to use electrodes with finite or graded conductivity so that a varying voltage profile is obtained when a current is applied.
- Poled structures: some EO materials, such as $LiNbO_3$ and $LiTaO_3$, may be "poled," a process that can result in geometrically precise crystalline domains within the bulk material. A uniform electric field applied to poled structures will result in equal and opposite changes in index within each domain according to its orientation.

3.2 Terminology for Describing Electro-Optic Scanners

Electro-optic scanners can be discussed using much of the same terminology as other types of scanners. The difficult trade-offs between size, voltage, and material properties creates some nuances that are important to consider.

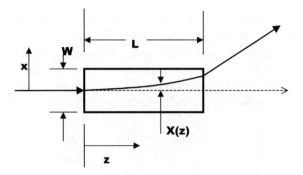

Figure 1 Schematic electro-optical scanner of width W and length L. An index of refraction profile of the form $n(x) = n_0 + kx$ is assumed. The optical beam to be deflected enters from the left.

3.2.1 Beam Displacement and Deflection Angle

A schematic EO scanner is shown in Fig. 1. In the presence of a linear index variation (constant gradient) across the width of the scanner, the trajectory of the center of the beam within the scanner is described by the parabolic relation[5]

$$X(z) = \frac{1}{n}\frac{dn}{dx}\frac{z^2}{2} \approx \frac{1}{2}\frac{\Delta n}{n}\frac{z^2}{W} \tag{9}$$

where $X(z)$ is the displacement of the beam center from the optical axis at position z, Δn is the total change in index across the scanner, n is the nominal index of refraction (in the absence of an EO shift), and W is the width of the scanner.

The deflection angle at a particular position z is given by the slope of the trajectory at that point, or the derivative of Eq. (9):[5,6]

$$\theta_{in}(z) = \frac{1}{n}\frac{dn}{dx}z \approx \frac{\Delta n}{n}\frac{z}{W} \tag{10}$$

When the beam exits the material, the angle is increased by the factor n, owing to the small-angle form of Snell's Law. The external deflection angle for the scanner is therefore obtained by evaluating Eq. (10) at $z = L$ and multiplying by n:

$$\theta_{def} = \frac{dn}{dx}L \approx \Delta n \frac{L}{W} \tag{11}$$

The displacement of the beam at the output facet of the scanner is finally given by

$$\delta = \frac{1}{2}\frac{\Delta n}{n}\frac{L^2}{W} \tag{12}$$

3.2.2 Pivot Point

Comparison of Eqs. (12) and (10) shows that the output displacement can also be expressed as

$$\delta = \frac{1}{2}\theta_{in}L \tag{13}$$

This suggests that although the actual trajectory is parabolic, the output angle and displacement is correctly given by assuming that the beam has an abrupt deflection of θ_{in} a

Electro-Optical Scanners

distance of $L/2$ from the output plane.[7] We call this the *pivot point*, and define it more generally as

$$L_{P,in} = \frac{X(L)}{\theta_{in}(L)} \tag{14}$$

A scanner has a well-defined pivot point when $L_{P,in}$ does not depend on the magnitude of the index variation Δn.

When the beam deflection is viewed outside the scanner, the deflection also appears to have a pivot point, although displaced, just as an object on the bottom of a swimming pool appears to be displaced when viewed from outside the pool. From outside the scanner, the pivot point appears to be a distance L_P from the output plane:

$$L_P = \frac{X(L)}{\theta_{def}(L)} = \frac{L_{P,in}}{n} \tag{15}$$

The existence of a pivot point is significant for the design of optical systems containing EO scanners. From the optical system point of view, the scanner can be represented simply by a mirror at a distance L_P from the output plane that introduces the deflection θ_{def}.

3.2.3 Resolvable Spots

Perhaps the best way to compare various scanner technologies is to use the number of "resolvable spots." The deflection angle can be magnified or reduced with other optical elements, but the number of resolvable spots will remain constant. The number of resolvable spots is given by the number of beam diameters corresponding to the lateral displacement at a certain distance, usually the far field. Clearly this number depends on the way that the beam diameter is defined. For our purposes, we will assume the beam is well described by a fundamental Gaussian beam whose $1/e^2$ intensity radius is given by the Gaussian beam waist $w(z)$:

$$w(z) = w_0 \sqrt{1 + \left(\frac{\lambda(z - z_0)}{\pi w_0^2}\right)^2} \tag{16}$$

where w_0 is the minimum radius at the beam waist, λ is the optical wavelength, and z_0 is the location of the minimum radius, or waist. The number of resolvable spots is given by

$$N_U(z) = \frac{S(z)}{2w(z)} + 1 \tag{17}$$

where N_U is the number of spots assuming unipolar deflection (deflection only to one side of the optical axis) and S is the displacement of the beam at the observation plane (see Fig. 2). Often scanners are used with bipolar drive voltages, resulting in a total beam displacement of $2S$. Thus the number of bipolar resolvable spots is

$$N_B(z) = \frac{S(z)}{w(z)} + 1 \tag{18}$$

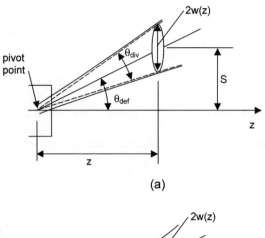

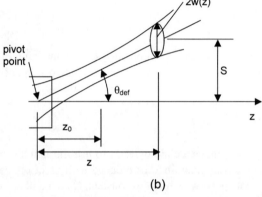

Figure 2 Geometry illustrating the concept of resolvable spots: (a) beam focus/waist near the pivot point of the scanner, and (b) beam focus/waist beyond the output.

The displacement S at a distance z from the pivot point is given by

$$S = \theta_{\text{def}} z \tag{19}$$

It is instructive to consider the case with the waist collocated with the pivot point ($z_0 = 0$) and the observation point arbitrarily far away. This limit results in the number of spots in the far-field:

$$N_{\text{U,FF}} \approx \frac{\theta_{\text{def}} \pi w_0}{2\lambda} + 1 = \frac{\theta_{\text{def}}}{\theta_{\text{div}}} + 1 \tag{20}$$

$$N_{\text{B,FF}} \approx \frac{2\theta_{\text{def}}}{\theta_{\text{div}}} + 1 \tag{21}$$

where

$$\theta_{\text{div}} = \frac{2\lambda}{\pi w_0} \tag{22}$$

is the far-field divergence (full-angle) of the Gaussian beam. The displacement clearly gets larger the further away the observation plane is, but this does not result in an increase in spots since the beam divergence causes the spot diameter to increase at the same rate in the far-field.

It is important to note that the maximum number of resolvable spots is achieved only in the far-field. Many practical EO systems, such as deflection-based modulators, do not require operation in the far-field. It is critically important to perform accurate simulations and ray tracings as part of the design process.

Other definitions of beam diameter are sometimes used, depending on the application. For instance, flat-top power profiles are useful for some laser machining operations. The effect of the diameter definition, and the correspondingly correct description of beam divergence, should be considered when discussing the number of resolvable spots that a particular system may exhibit.

3.3 Single Elements and Assemblies of Single Elements

One of the most basic optical elements is a prism. Accordingly, one of the most basic EO elements is an electrically controlled prism (Fig. 3).

In practice, the actual deflection about the mean is very small. For example, an equilateral triangle prism fabricated from lithium niobate and operated to ± 1 kV/mm will undergo an index change of approximately ± 0.0002, or 0.01%. This change is actually smaller than the accuracy bounds typically quoted by commercial crystal manufacturers for property values.

As discussed previously, the sign of the refractive index change reverses with the direction of polarization (usually coincident with the optic axis of the crystal). Thus a direct extension of the single prism implementation is to assemble several discrete prisms with alternating polarization as shown in Fig. 4. With proper choice of optical polarization, crystal, and orientation, a voltage applied between conducting electrodes on opposing transverse sides of the assembly increases and decreases the index in alternating prisms generating the Δn required for scanner operation. The major disadvantage of this type of construction is the labor-intensive cutting, polishing, and assembly of the prisms.

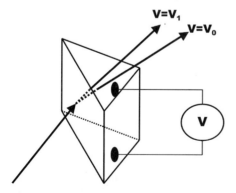

Figure 3 The simplest electro-optical scanner: a prism fabricated from an electro-optic material having electrodes at each end. As the voltage is varied, the angle of the exiting beam is changed.

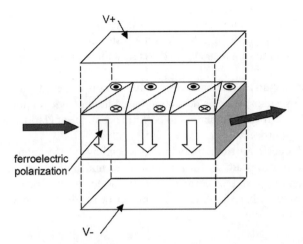

Figure 4 Electro-optic prism scanner made from alternating discrete prisms.[6]

3.4 Shaped Fields

An analysis of the energized multi-element assembly shows that it can be accurately represented by a band of material with linearly varying index of refraction.[5] This leads to two interesting design realizations: graded electrical fields with uniform crystal structure, covered in this section, and uniform electrical field and graded crystal structures created through domain inversion, covered in the following section.

3.4.1 Graded Index with Uniform Applied Voltage

One way to create a linearly varying spatial index profile is to apply a linearly varying electric field to an EO crystal. Most electrical circuits, especially those designed for low-power operation, are designed to provide a single voltage across an electrode. However, if the geometric spacing between electrodes is not fixed to a single value, the electrical field between the electrodes will exhibit a spatial variation. If this variation is linear where the beam transits the crystal, a good scanner can be produced. A quadrupole field as shown in Fig. 5 has the desired behavior near the origin.[8,9] If the electrodes are shaped to follow the hyperbolae

$$xy = \pm \frac{R_0^2}{2} \tag{23}$$

then the potential in the region between the electrodes is given by

$$V = \frac{V_0}{R_0^2} xy \tag{24}$$

The electric field is obtained by taking the gradient of the potential:

$$E_x = -\frac{\partial V}{\partial x} = -\frac{V_0}{R_0^2} y$$
$$E_y = -\frac{\partial V}{\partial y} = -\frac{V_0}{R_0^2} x \tag{25}$$

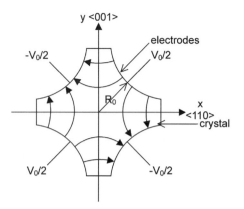

Figure 5 Geometry for generating a linear electric field profile using quadrupole electrodes in KDP-type materials. Crystallographic directions for proper deflector operation are also shown. Optical beam propagation is perpendicular to the page, and the optical electric field polarization is parallel to the x-axis ($\langle 110 \rangle$ direction) (from Ref. 8).

Thus we see that both components of the electric field vary linearly with position. However, if a crystal with symmetry $\bar{4}2m$ is oriented as indicated in Fig. 5, then there is no EO effect for E_x, to the first order. Using the index expressions in Table 1 for KDP, the index gradient for an optical wave polarized along the x-direction ($\langle 110 \rangle$) becomes

$$\frac{dn}{dx} = \frac{n^3 r_{63} V_0}{2R_0^2} \tag{26}$$

The beam displacement and deflection angle at the scanner output can be obtained by substituting Eq. (26) into Eqs. (1) and (3). The results are

$$X(L) = \frac{n^3 r_{63} V_0}{2R_0^2} \frac{L^2}{2} \tag{27}$$

$$\theta_{\text{def}} = \left(\frac{L n^3 r_{63}}{2R_0^2} \right) V_0 \tag{28}$$

Table 1 Eigenvalues and Eigenvectors for KDP With an Electric Field Applied Along the Optic Axis[a]

Eigenvalue	Eigenvector	Index	Δn
$\frac{1}{n_o^2} + r_{63} E_3$	$\langle 110 \rangle$	$n_o - \frac{\Delta n}{2}$	$n_o^3 r_{63} E_3$
$\frac{1}{n_o^2} - r_{63} E_3$	$\langle 1\bar{1}0 \rangle$	$n_o + \frac{\Delta n}{2}$	$n_o^3 r_{63} E_3$
$\frac{1}{n_e^2}$	$\langle 001 \rangle$		0

[a] For consistency with subsequent sections, Δn is the index change between two regions with oppositely oriented field; thus the index change arising from applying a field with a single polarity to a single domain material is $\Delta n/2$.

The deflection sensitivity and pivot point are readily found to be

$$\frac{\theta_{\text{def}}}{V} = \frac{Ln^3 r_{63}}{2R_0^2} \tag{29}$$

$$L_{\text{P,in}} = \frac{L}{2} \tag{30}$$

It is a difficult task to shape a crystal to accommodate hyperbolic electrodes as shown in Fig. 5. Instead, Fig. 6(a) shows a more practical geometry.[10] The field inside the crystal has been computed using the Finite Element Method[11] for KDP. Although the electrodes are not shaped precisely like hyperbolae, the field near the center of the crystal is still approximately linear, as shown in Fig. 6(b). Another approximation of hyperbolic electrodes was developed by Ireland and Ley.[12] In this case, cylindrical electrodes were used.

3.4.2 Graded Index with Constant Spacing

A second option for producing a linearly graded index of refraction depends on depositing a resistive electrode and a conductive electrode on opposite faces of a slab of EO material, as shown in Fig. 7. When the voltage is applied at the leads, current flows across the resistive electrode, which is essentially a resistor, resulting in a graded electrical field within the device. The change in optical index, which is proportional to the electric field in the material, is thus graded according to the voltage.

The difficulties of producing this type of electrode has limited their consideration for use to very thin devices, on the order of tens of microns. The optical beams are not typically round in these scanners, since a relatively wide stripe is needed for the voltage gradient electrodes. Difficulties of coupling optical beams into such wide and thin layers, and their great divergence on the output side add to the difficulties of this approach, although it has been proposed for optical switching for telecommunications use. Electrical heating due to the current flow is another limiting factor.

3.4.3 Graded Index with Constant Spacing and Single Voltage

A third approach to producing a gradient in the index of refraction is illustrated in Fig. 8. The portion of the beam that travels under the root of the conducting top electrode will traverse more material affected by the electrical field than a ray traversing only the tips. This technique suffers from the complex fringing electrical fields around the top electrode. These fields give rise to out-of-plane distortion and other beam quality problems. A wide variety of electrode shapes and spacings are possible,[1] some of which mitigate the effects of fringing fields to a degree.

3.5 Poled Structures

There are two ways of achieving an effective linear gradation in index: grading the electric field, or grading the material properties. The techniques covered above, except the use of multiple discrete inverted prisms, all use device or electrode geometry to effectively grade the electrical field. Grading the material properties is possible using the technique of "poling" or "domain inversion." This process can be performed completely independently of the base material production, making it an effective tool for device fabrication.

Electro-Optical Scanners

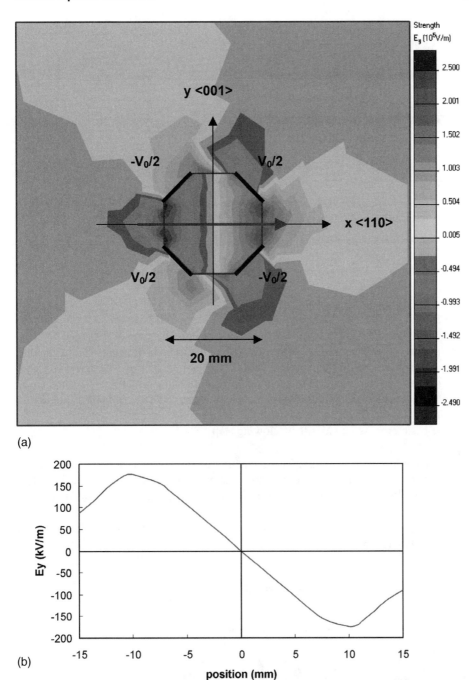

Figure 6 (a) Finite element analysis[11] of the electric field in a KDP crystal shaped like an octagon to approximate a quadrupolar field. The shading indicates the strength of the y component of the electric field. The vertical stripes near the center indicate that E_y is approximately linear with x and independent of y; (b) Calculation of the electric field E_y along the contour shown in part (a). Note the linearity of the field near the origin.

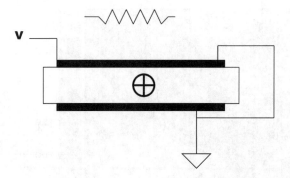

Figure 7 A resistive electrode may be used to produce graded electrical fields. This is a schematic view, seen end on, of such a device.

One salient advantage of poled devices is that they are not typically affected by electrical fringing fields. They are normally constructed with cover electrodes that are large enough to ensure uniform fields over the active region of the device, which may contain 50 or more interfaces. The result is that the beam will have to pass through only two fringing fields, at the entrance and exit, vs. 100 or more in some field-graded devices. The poling fabrication technique is an outgrowth of research directed toward making quasi-phase-matched second harmonic generating gratings.[13–17] It was discovered that by using photolithographic techniques, domain patterns with virtually any shape can be realized in materials such as z-cut lithium niobate, lithium tantalate, and potassium titanyl phosphate and its isomorphs such as rubidium titanyl arsenate.

The basic process includes the following steps:

1. Patterning: photolithographic techniques are used to pattern photoresist on one or both of the crystal surfaces with the shape of the final inverted region. Wafers are often used for greatest compatibility with standard semiconductor processing equipment.
2. Apply poling electrodes: either metal or liquid electrodes are placed on the wafer surface.

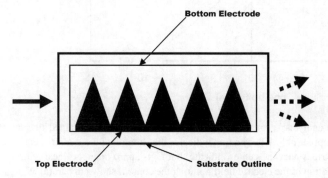

Figure 8 A serrated electrode, all of which is held at a single voltage, has the effect of producing a graded electrical field over the length of the electro-optical device.

Electro-Optical Scanners

3. Poling: high electric fields, greater than the coercive field of the medium or on the order of 20 kV/mm for lithium niobate, are applied. Practitioners use voltage pulses, ramps, and quasi-DC waveforms – there does not appear to be an accepted standard practice at this time.
4. Apply operating electrodes: this is often done using standard photolithography and thin film deposition. The operating electrodes are typically much larger than the poling electrodes to ensure uniform electric field in the poled region, but they may be limited in size due to the desire to minimize the device's electrical capacitance and therefore driver power requirements.
5. Annealing: temperatures ranging from about 200°C to over 1000°C are used, with the appropriate temperature and cycle profile determined by experimentation.

During the "poling" step, atoms are shifted in the crystal lattice, but only in the volumes defined by the photolithographically applied patterns. This effectively "flips" the ferroelectric domains without having to cut, polish, AR coat, and reassemble the crystal.

Using electric field poling techniques, it is possible to drive the inverted domain regions completely through relatively thick substrates – up to 3 mm have been reported for devices fabricated with profile precision of a few microns in RTA (R. Stolzenberger, personal communication, 2003). For more details on the physics of the domain inversion process, see Gopalan *et al.*[18] Using such a process, confinement to a waveguide is possible, but not necessary, and scanners that work for any properly polarized beam passing through the substrate can be made (Fig. 9).[19] For applications not requiring compatibility with waveguide devices, these bulk devices offer easier coupling, lower coupling and propagation losses, and improved beam quality.

3.5.1 Prismatic Poled Structures

The first deflector of this type was a waveguide device fabricated in lithium tantalate, as shown in Fig. 10.[21] Domain inversion was achieved using patterned proton-exchange followed by rapid thermal annealing. This process creates a domain-inverted region extending to a depth of 10–20 μm. The planar waveguide was subsequently formed by

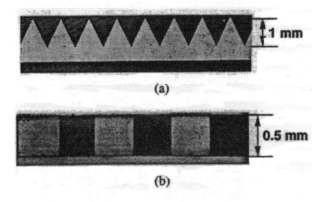

Figure 9 A bulk electro-optic wafer deflector using patterned domain inversion: (a) patterned Ta electrode used to define the geometry of the deflector; (b) etched Y cross-section of the sample showing domain inversion through the thickness of the wafer (from Ref. 20).

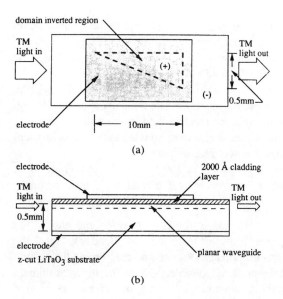

Figure 10 Geometry of the first electro-optic waveguide deflector using patterned domain inversion: (a) top view of the substrate showing the domain-inverted region; (b) cross-section through the prism region (from Ref. 22).

proton exchange in 260° pyrophosphoric acid. A 2000 Å thick layer of SiO_2 was deposited as a cladding layer before deposition and patterning of the final cover electrode, to reduce optical loss in the waveguide. Cylindrical lenses were used to edge-couple light into and out of the scanner.

Improved deflection sensitivity can be achieved by using thinner substrates, or by selectively thinning the substrate below the scanner using pulsed laser ablation.[23] Selective thinning allows the internal field to be increased while maintaining mechanical strength around the border of the substrate.

3.5.2 Rectangular Scanners

The simplest scanner geometry is that for which the prisms are enclosed within a rectangular-shaped region. The general case is illustrated in Fig. 11, where the active region is divided into an arbitrary number of variously shaped prisms. For $\theta_{in} \ll 1$ and $\Delta n \ll n$, the result of applying Snell's law at each interface is a cumulative deflection angle given by[5]

$$\theta_{in} = \sum_{i=1}^{N} \frac{\Delta n_i}{n} \cot \phi_i \tag{31}$$

where Δn_i is the total index change across the ith interface, N is the total number of interfaces in the scanner, and ϕ_i is the angle the ith interface makes with the beam axis. Note that Eq. (31) is valid regardless of the overall shape of the scanner, and each term in the sum is positive since both Δn_i and ϕ_i change signs from one interface to the next. For rectangular scanners with a fixed $|\Delta n_i| = \Delta n$ at each interface, the scanning angle can be

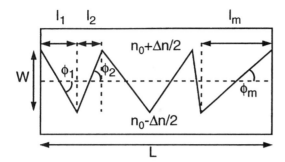

Figure 11 An arbitrarily divided rectangular prism scanner (from Ref. 24).

related to the width W and total length L of the device as follows:

$$\theta_{\text{def}} = n\theta_{\text{in}} = \Delta n \sum_{i=1}^{N} |\cot \phi_i| = \Delta n \sum_{i=1}^{N} \frac{l_i}{W} = \Delta n \frac{L}{W} \qquad (32)$$

Note that this is the same result as would be achieved for a scanner with a constant index gradient, from Eq. (11). We therefore have the somewhat surprising result that the scanning angle does not depend on how many prisms are in the scanner, but only on the ratio L/W! To see why this is, note that as the number of interfaces increases, the angle of incidence becomes closer to normal thereby reducing the refraction at each interface. Consequently, the sum of the effects of all the interfaces is constant. The question of the effects of varying numbers of triangles is considered more closely below from a different point of view.

Optimum Number of Triangles in Rectangular Scanners

Considering the case of one interface, it is apparent that the scanning properties can, in fact, depend on the number of prisms or triangles. For a sufficiently large value of L/W, total internal reflection can occur for one voltage polarity, but not the other. Consequently the scanning properties will be strongly asymmetric with respect to drive voltage polarity. On the other hand, if many interfaces are used, the incident angle at each interface will be sufficiently far from grazing that total internal reflection cannot occur for practical values of drive voltage. Chiu et al.[21] have considered the scanning asymmetry as a function of the number of interfaces using ray-tracing simulations. As shown in Fig. 12, the asymmetry in the scanning properties becomes negligible after about 10–15 interfaces.

Another consideration pertaining to the number of interfaces is the Fresnel transmission through the multiple interfaces. For small numbers of interfaces, the reflection at each interface can be high owing to the near grazing incidence (even total internal reflection is possible, as mentioned above). This reflected light is diverted from the optical path, resulting in low transmission through the device. In the opposite limit of very many interfaces, the beam approaches normal incidence at each interface. Thus the reflection at each interface approaches a finite value, while the number of such reflections continues to increase. Thus the total reflection increases for both large and small numbers of interfaces. An optimum number of interfaces can be found that minimizes the total reflection, or equivalently, maximizes the transmission through the device. For $\Delta n/n \ll 1$

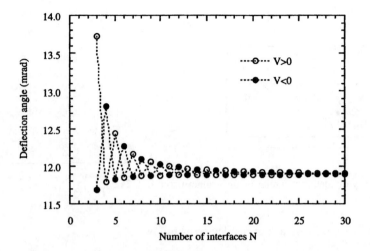

Figure 12 Symmetry of deflection angle vs. the number of interfaces from ray tracing analysis (from Ref. 26).

and $R \ll 1$, the normalized reflected intensity R is approximately given by

$$R = m\left[1 + \left(\frac{L}{mW}\right)^2\right]^2 \left(\frac{\Delta n}{2n}\right)^2 \tag{33}$$

which has a minimum at the optimum number of interfaces:

$$m_{\text{opt}} = \sqrt{3}\frac{L}{W} \tag{34}$$

Interestingly, this condition corresponds to filling the scanner with equilateral triangles. The behavior of the reflected intensity given by Eq. (33) is illustrated in Fig. 13. Examination of this figure shows that although an optimum does exist, the reflected intensity is negligible over the practical range of interfaces large enough to satisfy the symmetry condition discussed above. We conclude that as long as the number of interfaces exceeds 10–15, it is satisfactory to assume that the properties of the scanner are independent of the number of interfaces.

Deflection Sensitivity for Rectangular Scanners

To obtain the deflection sensitivity for a rectangular device, we need only to substitute the expression for Δn into Eq. (32). Using $E_3 = V/h$ in Table 2 for crystals of symmetry $3m$, the result is

$$\frac{\theta_{\text{def}}}{V} = \frac{n_e^3 r_{33}}{h}\frac{L}{W} \tag{35}$$

where h is the thickness of the substrate, and the incident beam is polarized along the z-axis of the crystal (extraordinary wave). Bulk scanners will also work if the incident beam

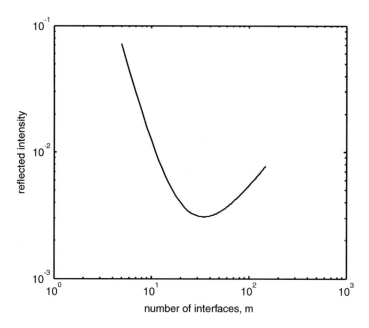

Figure 13 Normalized reflected intensity as a function of number of interfaces for $\Delta n/n = 10^{-4}$, $L/W = 20$, $m_{\text{opt}} = 35$.

is polarized in the plane of a z-cut substrate (ordinary wave), but with reduced deflection. In this case, the deflection sensitivity is given by (Table 2):

$$\frac{\theta_{\text{def}}}{V} = \frac{n_o^3 r_{13}}{h} \frac{L}{W} \tag{36}$$

It should be noted that Eqs. (35) and (36) are applicable to bulk devices; for waveguide devices, the voltage drop across the cladding layer must also be taken into account. This will result in a slight decrease in deflection sensitivity.

Table 2 Eigenvalues and Eigenvectors for $LiNbO_3$ and $LiTaO_3$ with an Electric Field Applied Along the Optic Axis[a]

Eigenvalue	Eigenvector	Index	Δn
$\frac{1}{n_o^2} + r_{13}E_3$	$\langle 100 \rangle$	$n_o - \frac{\Delta n}{2}$	$n_o^3 r_{13} E_3$
$\frac{1}{n_o^2} + r_{23}E_3$	$\langle 010 \rangle$	$n_o - \frac{\Delta n}{2}$	$n_o^3 r_{23} E_3$
$\frac{1}{n_e^2} + r_{33}E_3$	$\langle 001 \rangle$	$n_e - \frac{\Delta n}{2}$	$n_e^3 r_{33} E_3$

[a]Δn is defined as in Table 3. Note that for these crystals $r_{13} = r_{23}$.

Pivot Point Location for Rectangular Scanners

The internal pivot point for rectangular scanners can be obtained directly from Eqs. (13) and (14):

$$L_{P,in} = \frac{L}{2} \tag{37}$$

In practical geometries, there is usually a spacing s between the output plane of the scanner and the edge of the crystal, as shown in Fig. 14. This spacing provides a longer electrical creepage path around the end of the scanner, and allows for cutting and polishing operations to be performed without impinging on the active scanner area. In this case, the location of the pivot point as viewed from outside of the crystal is given by (cf. Eq. 15)

$$L_P = \frac{1}{n}[L_{P,in} + s] \tag{38}$$

3.5.3 Trapezoidal Scanners

One difficulty with the design of rectangular scanners results from the beam displacement within the scanner, as given by Eqs. (12) and (13). Clearly the width of the scanner must be increased to accommodate this displacement, but increasing the width reduces the deflection sensitivity. However, because of the shape of the trajectory, this increase is only needed at the output of the scanner. As discussed earlier, one way to address this is to focus the beam through the scanner so that the reduction in output beam diameter is comparable to the displacement. However, we saw that this technique reduces the number of resolvable spots. Another possibility is to increase the width of the output forming a trapezoidal shape (Fig. 15).

Deflection Sensitivity of Trapezoidal Scanners

The deflection angle for a trapezoidal scanner is given by[5]

$$\theta_{in} = \frac{\Delta n}{n} \frac{L}{W_1 - W_0} \ln\left(\frac{W_1}{W_0}\right) \tag{39}$$

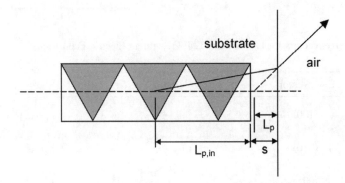

Figure 14 Shift of the apparent location of the pivot point by substrate refraction, including the effect of a spacing s between the output of the scanner and the substrate edge.

Electro-Optical Scanners

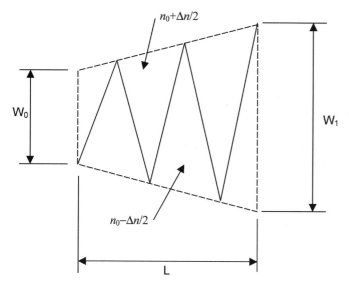

Figure 15 Geometry of a trapezoidal prism scanner.[5]

Re-expressing this in terms of the external deflection angle and substituting for the change in refractive index gives the deflection sensitivity

$$\frac{\theta_{\text{def}}}{V} = \frac{n_e^3 r_{33}}{h} \frac{L}{W_1 - W_0} \ln\left(\frac{W_1}{W_0}\right) \qquad (40)$$

Pivot Point Location for Trapezoidal Scanners

The displacement of the beam at the output of a trapezoidal scanner is given by[5]

$$X(L) = \frac{\Delta n}{n} \frac{L}{W_1 - W_0} \left[\frac{W_1}{W_1 - W_0} \ln\left(\frac{W_1}{W_0}\right) - 1 \right] L \qquad (41)$$

The internal pivot point is obtained using the definition (14) along with Eqs. (39) and (41). The result is

$$L_{P,\text{in}} = \left[\frac{W_1}{W_1 - W_0} - \frac{1}{\ln(W_1/W_0)} \right] L \qquad (42)$$

The pivot point as viewed externally can be computed from Eq. (38), as before.

Comparison of Trapezoidal and Rectangular Scanners

The increase in deflection sensitivity compared to rectangular scanners can be illustrated by considering rectangular and trapezoidal scanners with the same lengths, and the width of the rectangular scanner equal to the average of the input and output widths of the trapezoidal device. Thus, if W_R is the width of the rectangular scanner and W_0, W_1 are the input and output widths of the trapezoidal scanner, we require

$$W_R = \frac{W_0 + W_1}{2} \tag{43}$$

Assuming the same maximum index difference Δn, Fig. 16 shows the improvement gained by the trapezoidal geometry. The improvement is modest (<10%) for $W_0/W_R > 0.5$. Since the usual case is that the beam diameter is much larger than the output displacement, W_0/W_1 and W_0/W_R are normally only slightly smaller than unity, yielding a typical improvement of a few percent.

3.5.4 Horn-Shaped Scanners

Since the scanning sensitivity is diminished as the width increases, the optimal solution is to increase the width gradually so as to track the beam trajectory. If the change in beam diameter through the scanner is neglected, the shape of the scanner can be obtained by simply adding a constant offset to the beam displacement, as shown in Fig. 17(a). The width of the scanner can therefore be written in the form

$$W(z) = W_0 + 2X_{\max}(z) \tag{44}$$

where $X_{\max}(z)$ is the displacement at position z with maximum voltage applied. The factor of 2 accommodates bipolar operation of the scanner. The general shape $W(z)$ has been obtained by Chiu et al.,[5] and is plotted in terms of normalized coordinates in Fig. 17(b). To facilitate use of this curve in designs, values of the normalized width as a function of z are tabulated in Table 3.

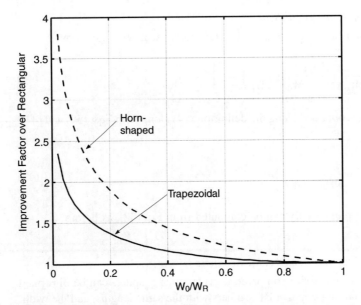

Figure 16 Comparison of scanning sensitivity of rectangular, trapezoidal, and horn-shaped scanners (from Ref. 5).

Electro-Optical Scanners

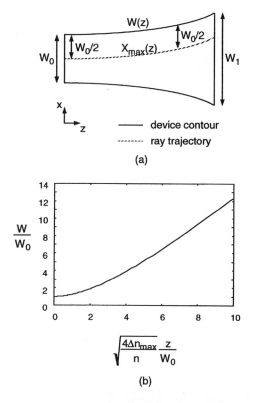

Figure 17 Normalized shape-optimized scanner design curve: (a) reference geometry and (b) normalized shape contour (from Ref. 5). Tabulated values are given in Table 3.

Table 3 Tabulated Values for the Normalized Horn-Shaped Scanner Curve Shown in Fig. 17[a]

Z^*	W^*	Z^*	W^*	Z^*	W^*	Z^*	W^*
0.00	1.00	2.50	2.49	5.00	5.36	7.50	8.84
0.25	1.05	2.75	2.73	5.25	5.68	7.75	9.21
0.50	1.11	3.00	2.99	5.50	6.01	8.00	9.58
0.75	1.21	3.25	3.26	5.75	6.35	8.25	9.96
1.00	1.34	3.50	3.53	6.00	6.69	8.50	10.34
1.25	1.48	3.75	3.82	6.25	7.04	8.75	10.73
1.50	1.65	4.00	4.11	6.50	7.39	9.00	11.11
1.75	1.83	4.25	4.41	6.75	7.75	9.25	11.50
2.00	2.04	4.50	4.72	7.00	8.11	9.50	11.89
2.25	2.26	4.75	5.03	7.25	8.47	9.75	12.29
						10.00	12.68

$Z^* = \dfrac{z}{W_0}\sqrt{\dfrac{4\Delta n_{max}}{n}}$; $W^* = W/W_0$.

[a] These values can be used for designing masks for scanner fabrication.

The shape of the optimal curve allowing for the change in beam diameter through the scanner depends on the gaussian beam parameters. Figure 18 shows a simulation of an optimal scanner with the beam focused on the output plane.[27] The simulation was performed using the beam propagation method (BPM).[28,29]

Deflection Sensitivity of Horn-Shaped Scanners

The deflection angle as a function of Δn for an optimum horn-shaped scanner is

$$\theta_{in} = \frac{\Delta n}{n} \sqrt{\frac{n}{\Delta n_{max}} \ln\left(\frac{W_1}{W_0}\right)} \qquad (45)$$

with a maximum value of

$$\theta_{in,max} = \sqrt{\frac{\Delta n_{max}}{n} \ln\left(\frac{W_1}{W_0}\right)} \qquad (46)$$

where Δn_{max} is the index variation across the scanner with the maximum applied voltage to be used. The deflection sensitivity is obtained by multiplying Eq. (45) by n and substituting for Δn from Table 1. The result for crystals with symmetry $3m$ (e.g., lithium

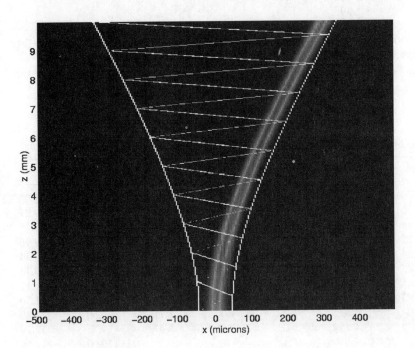

Figure 18 Simulation of horn-shaped scanner operation using the beam propagation method (BPM). Parameters are $W_0 = 92$ μm, $W_1 = 678$ μm, $L = 10$ mm, $\lambda_0 = 0.6328$ mm, $n_e = 2.1807$ (lithium tantalate), and $\Delta n = 2.1 \times 10^{-3}$. The radius of the beam waist is 30 μm and is focused at the output of the scanner (from Ref. 27).

Electro-Optical Scanners

niobate) is

$$\frac{\theta_{\text{def}}}{V} = n_e^2 \sqrt{\frac{r_{33}}{hV_{\max}} \ln\left(\frac{W_1}{W_0}\right)} \tag{47}$$

Pivot Point Location of Horn-Shaped Scanners

The unipolar displacement at the output of the horn scanner is

$$X(L) = \frac{\Delta n}{\Delta n_{\max}} \frac{W_0}{2} \left[\frac{W_1}{W_0} - 1\right] \tag{48}$$

Taking the ratio of Eqs. (48) and (45) gives the pivot point:

$$L_{\text{P,in}} = \frac{\dfrac{W_0}{2}\left[\dfrac{W_1}{W_0} - 1\right]}{\sqrt{\dfrac{\Delta n_{\max}}{n} \ln\left(\dfrac{W_1}{W_0}\right)}} \tag{49}$$

Remarkably, the fact that Δn drops out of this equation means that a well-defined pivot point exists (i.e., $L_{\text{P,in}}$ does not depend on voltage), even in such a complex horn-shaped device.

Comparison of Horn-Shaped Scanners With Trapezoidal and Rectangular Scanners

The deflection sensitivity of a horn-shaped scanner with the same input width, output width, and length as a trapezoidal scanner is shown in Fig. 16, normalized to that of a rectangular scanner with the same average width. The improvement over trapezoidal and rectangular scanners is clear.

A summary of the design equations for rectangular, trapezoidal, and horn-shaped scanners is presented in Table 4.

3.5.5 Domain Inverted Total Internal Reflection Deflectors

Domain inversion in ferroelectric crystals can also be used to create relatively long, straight interfaces within a bulk crystal. Since the domains are antiparallel across the interface, an applied electric field will result in a step change in the index of refraction. If a light beam intersects this interface at very high angles (i.e., near grazing incidence) a state of total internal reflection (TIR) can result (Fig. 19).

Eason and coworkers report an analysis of such a device,[30] with an application to telecommunications switching in mind. Such a device would typically be operated in a digital fashion, ON or OFF, in order to steer a collimated beam into one or another optical fiber (with appropriate collection optics on the end of each fiber). Compared to a scanner composed of many triangles, such a TIR device will exhibit greater deflection for the same voltage and device size. This gives some advantage by shortening the package length since the two beams separate faster providing more room for the output fibers and their collimation optics. A further advantage is that TIR device performance may be engineered to exhibit nearly polarization independent performance, while triangular prisms are strongly polarization dependent.

Table 4 Summary of Design Formulas for Rectangular, Trapezoidal, and Horn-Shaped Scanners

Scanner type	Geometry	Deflection θ_{def}	Output Beam displacement $X(L)$	Pivot point location $L_{P,in}$
Rectangular		$\theta_{def} = \Delta n \dfrac{L}{W}$	$X = \dfrac{1}{2}\dfrac{\Delta n}{n}\dfrac{L^2}{W}$	$L_{P,in} = L/2$
Trapezoidal		$\theta_{def} = \Delta n \dfrac{L}{W_1 - W_0} \ln\left(\dfrac{W_1}{W_0}\right)$	$X(L) = \dfrac{\Delta n}{n}\dfrac{L}{W_1 - W_0}$ $\times \left[\dfrac{W_1}{W_1 - W_0}\ln\left(\dfrac{W_1}{W_0}\right) - 1\right]L$	$L_{P,in} = \left[\dfrac{W_1}{W_1 - W_0} - \dfrac{1}{\ln(W_1/W_0)}\right]L$
Horn-shaped		$\theta_{def} = \Delta n \sqrt{\dfrac{n}{\Delta n_{max}}} \ln\left(\dfrac{W_1}{W_0}\right)$	$X(L) = \dfrac{\Delta n}{\Delta n_{max}}\dfrac{W_0}{2}\left[\dfrac{W_1}{W_0} - 1\right]$	$L_{P,in} = \dfrac{(W_0/2)[(W_1/W_0) - 1]}{\sqrt{(\Delta n_{max}/n)}\ln(W_1/W_0)}$

Electro-Optical Scanners

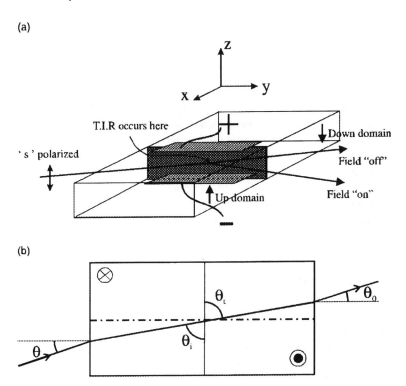

Figure 19 (a) Schematic of domain engineered total internal reflection (TIR) deflector; (b) plan view of scanner showing grazing angle of the input optical beam to the poled interface (from Ref. 30).

3.5.6 Domain Inverted Grating Structures

The component based upon patterned domains that has received the most attention in recent years is the quasi-phase-matched grating for second harmonic generation (SHG).[13–17] The frequency shifting effect occurs in a "periodically poled" region in the crystal, which is produced by forming many precisely spaced parallel domains in materials such as lithium niobate or KTP.[17] For instance, some early efforts were focused on producing blue light from IR laser diodes for use in data storage applications.

In SHG applications, the light propagates perpendicular to the poled stripes, and generally there is no applied electric field. By rotating the incident beam to near grazing incidence and allowing for the application of an electric field, an EO Bragg grating is produced. This is analogous to an AO Bragg deflector where the electric field rather than the acoustic intensity controls the distribution of the light into the various orders.

Gnewuch and coworkers have designed and tested such a device operating at 633 nm.[31] Similar to the TIR device above, the incident light can be switched to two different positions. The advantage of this structure, however, is that it operates at approximately 25 V, vs. 500 V for a prismatic domain poled device operating with a similarly sized optical beam. This offers the potential for enormous electrical power savings during high-frequency operation. Without acoustic velocity issues, the EO version of the Bragg deflector can operate at high speeds and handle higher beam powers than an AO version could.

3.5.7 Other Poled Structures

The power of patterned ferroelectric domain inversion is that a wide variety of optical components can be fabricated by designing masks with appropriate shapes. For example, although patterning the optimal horn-shape photolithographically is straightforward, the fabrication of a horn-shaped scanner by assembling discrete prisms would be prohibitively difficult. Further, with patterned domain inversion multiple components can be easily integrated on a common substrate.

As another example, EO cylindrical lenses can be formed by patterning oval, semicircular, or circular inverted domains.[32,33] Stacks of such lenses are readily combined with a prism scanner.[34] An optical integrated system of this type could be used, for example, to collimate the light from a fiber before entering the scanner. It should be noted, however, that lenses made in this way are cylindrical lenses, and hence only focus in the plane of the substrate. This is of no consequence in planar waveguide devices, but must be considered if bulk operation is desired.

A periodically poled structure can also be integrated with prism scanners to realize a device capable of generating and steering blue light.[35] The SHG conversion efficiency of this bulk integrated device was relatively low, however, since the optical intensity in the grating was limited by the focused beam diameter. To achieve the highest conversion efficiency, the light must be confined to a channel waveguide.

The issue of low conversion efficiency encountered with the first SHG scanner device can be addressed by combining the above elements to form an integrated SHG grating, lens stack, and scanner, as shown in Figs. 20 and 21.[36] The channel waveguide in the grating keeps the optical intensity high to maximize the nonlinear frequency doubling efficiency. The output of the channel waveguide then opens into a planar waveguide, and the beam subsequently diverges. A lens stack is next used to collimate the beam before entering the scanner.

4 ELECTRONIC DRIVERS FOR ELECTRO-OPTIC DEFLECTORS

4.1 Overview

A primary consideration when applying EO deflector devices as part of a system, and an issue often raised by potential users, is the electronic driver. The driver is essentially an amplifier, converting a low-voltage control signal to the higher voltage required to achieve the desired optical index change. It is usually custom-designed for each application, requires specialized circuit design skills and may present safety hazards. Depending on the specific system requirements, the design challenges can vary; however, the high-voltage nature of these devices can be the single most restricting factor, as component availability can restrict the design space. Also, the high-voltage (HV) supplies required to power the driver(s) can be a considerable cost and design challenge.

Other parameters that can influence the design are: switching speeds; repetition rate; switch/component package density requirements; electromagnetic interference (EMI) and radio frequency interference (RFI) considerations; switching power efficiency; and thermal considerations.

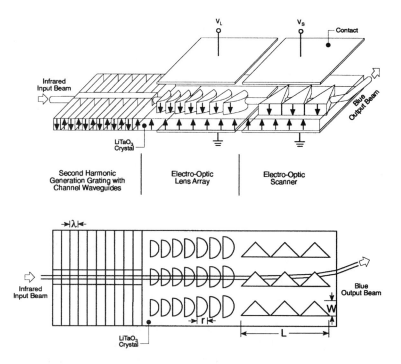

Figure 20 Schematic of an integrated device containing a channel waveguide with an SHG grating, collimating lens stack, and scanner (from Ref. 36).

Figure 21 Photograph of the device described in Fig. 20. Blue light generated by the SHG waveguide can be seen on the left, followed by the lens and scanner electrodes.

Significant strides in recent years in power field effect transistor (FET) and insulated gate dipolar transistor (IGBT) technology have alleviated many practical constraints and enabled power densities and performance that were not possible in earlier years. These devices, developed by Motorola, IXYS, International Rectifier, Toshiba, ST Microelectronics, and others, primarily for the motor drive/control industry, can be well suited for EO scanner drive applications. Also, very fast high-voltage diodes by these same manufacturers and surface mount HV capacitors from Johansson Dielectric and others have driven achievable performance up considerably.

4.2 High-Voltage Power Supplies

Most EO scanner drivers rely on a constant high voltage supply as a subsystem. For instance, a modulator system can be realized by switching one electrode of an EO element between a high voltage and ground while the other electrode is held at ground. Benchtop high-voltage power supplies are readily available from several manufacturers including Spellman, Ultravolt, Trek, and others. The market for application-specific high-voltage supplies is not large and, as such, cost can often be quite high and standard packaging options somewhat limited. In many cases it is desirable to design a custom HV supply that suits the application.

There are several design topologies available for high-voltage power supply design. Whether the input power is AC line voltage or a DC supply, a voltage boost converter of some type needs to be employed. A typical boost converter is discussed and then some higher efficient topologies covered.

4.2.1 Conventional Boost Converters

Switching supply topologies, such as boost converters, can be utilized when a high voltage needs to be generated from a much lower one. Typical boost converters consist of a switching transistor (Q1), usually a FET, with an inductor (L1) connected between the drain and the low voltage supply. Current is transferred discontinuously at the switching frequency, and the stored energy amount and pulse duration are proportional to the output voltage feedback signal (Fig. 22).

The load in a boost converter is usually fed through a rectifying diode. The current in the inductor (L1), $I_{L1}(pk) = (V_{dc} * t_{on})/L1$, ramps up linearly during the ON cycle of the FET. The energy stored is $E = 1/(2 * L1 * I_{L1}^2(pk))$. When the FET is turned off that energy is then transferred to the load via the rectifying diode. A portion of V_{out} is fed back through a pulse width modulation (PWM) converter to control the desired pulse width of the FET drive. Boost converters are typically only used in lower power applications of less than 10 W.

4.2.2 Flyback Converters

In many high-voltage applications with moderate to high power requirements, the size of the inductor in the boost converter needed to store the proper amount of energy becomes unwieldy, and losses become high. In these cases, a transformer can replace the boost converter inductor. This topology is referred to as a flyback converter. A schematic example is shown in Fig. 23.

The basic operation of a flyback converter is that when the FET is turned ON, current ramps up at a rate $di/dt = (V_{dc})/L_{pm}$, where L_{pm} is the magnetizing inductance of the primary winding of transformer T1. When the FET is subsequently turned off the

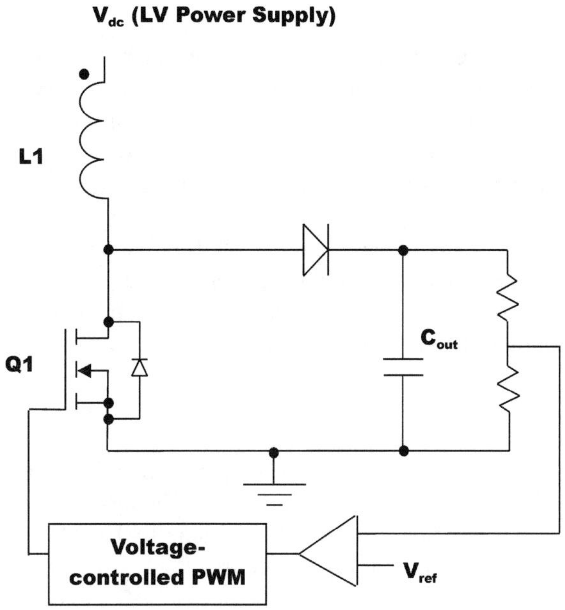

Electro-Optical Scanners

Figure 22 Schematic of conventional boost converter. The driver circuit for the EO element is connected across the output storage capacitor, C_{out}. This type of high voltage supply is typically used in applications requiring less than 10 W of output power.

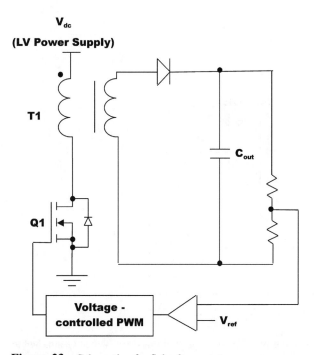

Figure 23 Schematic of a flyback converter.

current has ramped to the $I_{pk} = (V_{dc})*T_{on}/L_{pm}$ thus storing the energy $E = L_{pm}*(I_{pk})^2/2$. With the FET turned OFF, the magnetizing inductance causes an instantaneous reversal in polarities of all windings' voltages and the primary current transfers to the secondary as $I_s = I_{pk}(N_p/N_m)$ where N_p and N_m are the primary and secondary winding count. Using a higher voltage primary V_{dc} can help minimize transformer size and keep the I_{pk} to manageable levels. Also, one can employ multiple winding outputs to increase output voltage as needed or to select one of multiple output levels.

The flyback converter topology can be used for 5–150 W supplies for $V_{out} \leq = 5000$ V. A limiting factor can be that the required current in the primary of the flyback transformer is not excessive while keeping within an acceptable, and realistic, transformer size. Efficiencies of the order of 85% and above are achievable.

4.3 Digital Drivers

Many EO systems such as telecommunications switches and display system optical beam modulators, can be realized using drivers with only two output states. The term "digital driver" is used here, although the voltages being considered can range to 1000 V and above, requiring far different components than digital logic circuits.

4.3.1 Simple Totem Pole Circuits

The simplest and highest speed driver for controlling a capacitive EO device is a "totem pole" or "half bridge" arrangement of FETs where the totem pole is connected across a high-voltage power supply and the device is connected between the FETs to ground (Fig.

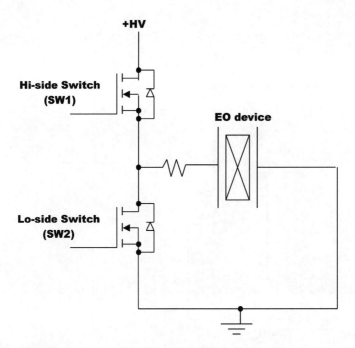

Figure 24 Schematic of a totem pole driver. The +HV lead of this driver would be attached to the high side of C_{out} of a high-voltage power supply.

Electro-Optical Scanners

24). Each FET is turned ON or OFF in turn to set the voltage across the output (load) EO device to either the high or low state. Both FETs are never turned ON at the same time – this would lead to a high current directly from the high voltage supply and would likely destroy the FETs unless proper limiting circuitry is used.

In some cases, the FETs would have resistors in series with them to dissipate some of the charge/discharge energy (without the resistors, nearly all of the energy is dissipated in the FETs during the switching transition, which can cause substantial heating). This will, of course, slow down the charging and discharging edges slightly. This straightforward totem pole results in a very high speed, although very lossy, method to drive the output capacitor, in this case, an EO scanner.

The high-side drive can be easily crafted from a p-channel FET where HV < 200 V. However, many practical applications can require a higher voltage, and such FETs are not available. As such, n-channel FETs are recommended for most applications and can be utilized with a floating high-side gate drive. A simple implementation of this would be a transformer-coupled FET gate drive circuit.

It should be noted that this high-side gate driver must be crafted such that in the "dwell ON state," the gate drive must be able to keep the FET ON for the longest system dwell time, which may be an undetermined duration. If continuous (infinite) dwell is required then a refresh circuit should be added to the transformer-coupled FET gate driver circuit.

Gate drivers for the FETs can be crafted from discrete elements – many of those that were designed for the motor drive industry can be utilized. The main FETs must be selected for low junction capacitances rather than minimal $R_{ds}(on)$ as a primary constraint as these capacitances must be charged and discharged during the switching cycle and can be comparable to the load capacitance of the EO device itself and contribute significantly to circuit losses.

Devices such as the highly integrated totem pole driver ICs from International Rectifier (IR2213), ST Microelectronics (L6285), and others can be utilized for applications up to 1000 V. Beyond that, IGBTs can be employed. These devices have T_{on}/T_{off} propagation delays and these parameters can drift significantly with temperature such that timing control circuitry must take it into account. Switching speeds on the order of 100 V/ns are possible.

The maximum voltage rating of most devices in the circuit can be cut in half, thus greatly increasing the component choices and safety factors, by utilizing a full bridge drive where two totem poles are set up to switch $\frac{1}{2}HV$ each to the load. Efficiency benefits can also be realized as the loss from the two charge cycles can be less than $\frac{1}{2}$ that of switching the entire HV in one step. Further, this topology enables "adiabatic" switching to be utilized such that further efficiency gains can be realized – this concept is covered below.

4.3.2 Adiabatic Drivers

Inherent in traditional switching logic design is the $CV^2/2$ of energy that is dissipated every time a transistor is turned on to charge or discharge a capacitive load. This dissipation is a direct consequence of the fact that, in traditional switching logic configurations, charge for the load is taken from a power or ground rail and that the device to be charged initially sits at a fixed potential very different from that of the rail.

In a simple totem pole application of charging and discharging a load C to a voltage V, the energy dissipated to flip the output is $E = \frac{1}{2}CV^2$. This energy does not depend on the needed time to switch, nor the clock rate, but is strictly related to the energy transfer process. In fact, during the rising transition, the power supply delivers all the charge $Q = CV$ at voltage V, while during the falling transition that charge is returned at zero voltage. So, actually, the energy $E = CV^2$ is drawn from the high-voltage supply, with half of the energy being stored in the load capacitor, and the other half being dissipated in losses.

Put another way, half of this energy is dissipated by the FET in the pull-up network (rising transition) and the other half by the pull-down network (falling transition), independent of how fast the transitions are. To reduce the dissipated energy, only methods that reduce the load capacitances or the supply voltage can be applied, but in any case they are strictly limited by the load C and voltage V.

The term "adiabatic process" is most often applied to the thermodynamic cycle by which a gas, such as air, is heated or cooled by expansion or compression without an external source or sink of heat.

The energy transfer processes in the electronic driver can be done adiabatically if, during the rising (or falling) transition, the power supply delivers (or recovers) the charge to (or from) the load at a potential close to that of the source/supply potential. In other words, in order to implement adiabatic processes in a switching circuit, the switching devices should be turned on only when the source-drain voltage is zero, and source-drain voltage should be changed only while the device is off; and, if possible, given the desired performance of the circuit, any voltage change must be done as gradually as possible.

There are difficulties in implementing this solution, however. First, the logic must be designed so that switching transitions can occur only at suitable times (that is, only when there is no potential drop across the switching devices). Choreographing this timing can add considerable complexity as there are switching delays inherent in the FETs and gate driver elements. FET drivers and control circuitry also have temperature-dependent characteristics. Secondly, zero energy dissipation only occurs with arbitrarily slow switching: with realistic switching rates, the energy savings might not be enough to make up for the additional complexity. Lastly, adiabatic design relies on the assumption that one can efficiently provide the moving supply (in fact a clock) to the circuit that it drives. This last characteristic is not achievable with HV applications and is more suited for logic level designs.

In Fig. 25, note that the two half bridges are configured such that the "lo-side" $-$HV switch is referenced to the $-$HV rail. This is fairly easily implemented as will be discussed later.

The full bridge circuit operates adiabatically as follows. Start with the crystal discharged, that is PH and NL are off (or open) while PL and NH are turned on. To charge the crystal, one of the grounded FETs, PL is turned OFF – this is adiabatic, in that it is done while no current is flowing ($V_{ds} \sim 0$ V). Only a very short period of time is required for it to fully turn off, then immediately turn ON switch PH. Obviously PL must be fully OFF before PH can begin to turn ON to avoid a catastrophic shoot-through condition HV to GND. Note the basic totem pole circuit always switched with V_{ds} at $\sim$ HV, leading to losses in the FET.

This switch transition will result in the following: the $+$HV device, PH, will need to supply energy to charge the Load and the parasitic capacitance of PL, C_{PL}, which includes the capacitance of the protection diode, which is usually a significant component of the total load. Thus the energy required from the $+$HV supply is $(C_{load} + C_{PL})(HV)^2$. Since each HV supply is $\frac{1}{2}$ the voltage of what it was in the simple totem pole arrangement, this translates to $(C_{load} + C_{PL})HV^2/4$ of energy drawn so far.

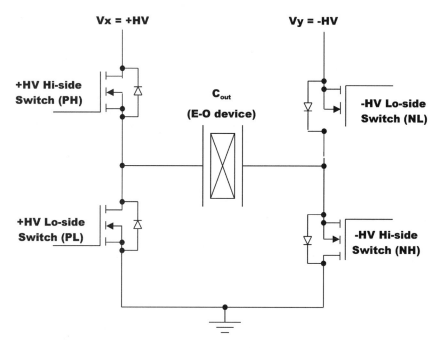

Figure 25 A dual totem pole driver. By using both a positive and a negative high-voltage supply, the absolute voltage level may be halved compared to a simple totem pole circuit. This can simplify some design tasks. This type of circuit can also be controlled to operate "adiabatically," which offers significant power savings.

The next step, to fully charge the crystal, is to turn NH OFF (again while $V_{ds} \sim 0$ V) and turn on NL. As before the $-$HV supply needs to charge up $C_{load} + C_{NH}$. The required energy thus to turn ON the C_{load} is:

$$E = (C_{load})(+HV)^2 + (C_{load})(-HV)^2 \tag{50}$$

Since each HV in this case is one-half the voltage compared to the simple totem pole case, the energy is $(C_{load})(HV/4) + (C_{load})(HV/4)$ or $(C_{load})(HV^2/2)$, a saving of approximately half of the energy (minus that to charge the stray and parasitic capacitances of the additional switching device) for each cycle.

For the adiabatic discharge, NL is turned OFF, then NH is turned ON. This will force the recovery of half the charge in the crystal to go back to $+$HV supply (or $C_{load}*HV^2/4$ of energy is recovered). Also, the parasitic capacitance of Cp(NL) will be dissipated and Cp(NH) will be charged to Vy, requiring Cp(NH)*Vy*2 energy from Vy supply. This total energy is equivalent to Cp(NH)Vy*2/4. Finally, to completely discharge the crystal, we need to turn OFF PH and turn ON switch PL.

Of course, the high-voltage supplies used in an adiabatic driver control scheme must be able to source and sink current – which is not true of most HV supplies. However, additional circuit elements can be added to the flyback converter to accommodate this requirement. Further gains in overall energy efficiency can be realized by making the high-voltage power sink operate as an energy recovery system. One option for this is shown in

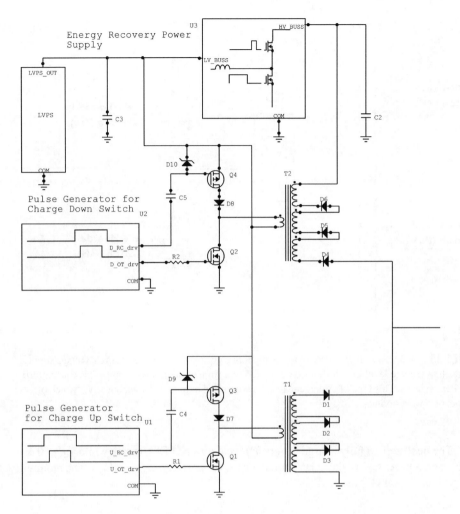

Figure 26 A schematic of flyback high-voltage supply with an energy recovery circuit, allowing for connection to an adiabatic driver circuit.

Fig. 26, where the energy pushed back to the high-voltage rail is fed into a step-down DC–DC converter with the output connected to a system level low voltage bus.[37]

4.4 Analog Drivers

The wide variety of EO applications and elements leads naturally to a wide variety of drivers, especially for "nondigital" applications. Possible designs range from those using densely packaged hybrid circuitry to those using 4000 V, or higher, vacuum tubes for control elements. A variety of IC op-amps are also available up to 600 V, allowing for essentially arbitrary waveforms within the bandwidth of the amplifier.

Analog drivers are generally specified by their operating voltage and bandwidth. The bandwidth of a driver is always specified in relation to the load; the usable bandwidth of the driver will decrease as the load capacitance increases. For this reason, the cabling or other connection between the driver and the EO element must be considered in addition to

Electro-Optical Scanners

the actual EO element load when specifying the driver, since they can sometimes dominate the overall system performance. Efficiency, single-ended or differential input, linearity, and other factors may also be specified.

A sampling of the variety of designs possible is:

- Single-ended: this is the simplest, most straightforward variety of driver. The driver is connected to one electrode on the EO device, with the other usually held at ground. Op-amps are typical single-ended drivers.
- Differential: some amplifiers work best in the middle of their output voltage range. A differential driver is constructed using two amplifiers, each connected to one side of the EO element and each biased to one-half the peak voltage. As the voltage on one is raised, the other is lowered. Differential topologies can be used where deflection in either direction from nominal is required. They also effectively act as a single-ended design but with twice the voltage – which may be useful if FETs at the desired voltage are not available.
- Resonant: since EO elements act as capacitors, they can be coupled with an inductor and driven at the resonant frequency of the pair. Voltage amplification is possible, and standard techniques can be used to synchronize the circuit with other elements of the complete optical system.
- Transformer coupled: voltage amplification is also possible using a transformer. This type of circuit may be operated in a resonant mode, or to produce some other waveform. Transformer coupling is best suited to periodic waveforms since the bandwidth of a transformer can be limited, and special tuning and compensation techniques may be required for complex waveforms. Triangle-wave drive voltages, useful for some display applications, can be done very effectively.

5 PROPERTIES AND SELECTION OF ELECTRO-OPTIC MATERIALS

5.1 General

The optical performance of an EO deflection system depends on the material chosen, the operating electric field, and the characteristics of the beam to be deflected (wavelength, diameter, divergence, M^2 quality factor, and so on). Unfortunately for system designers, there is no one "best" material that meets most application requirements.

Most practical EO materials are crystals with anisotropic properties, and almost all of these are grown by crystal suppliers because they are not abundant in nature. Properties can vary by manufacturer and grade, making it important to work in concert with the material suppliers when selecting a material; they are also good sources of material property information. Extensive lists of material properties can also be found in the literature.[38] Electro-optical properties of some common crystals are shown in Table 5.

Crystal purity, optical quality, internal strain, physical size, doping, domain structure, electrical conductivity, and other quality measures can vary widely from vendor to vendor, from one boule to the next by a single vendor, and within a single boule as well. These and related factors have led the industry to focus on a few materials that are in relatively large-scale production, such as lithium niobate and ADP, which are covered below. Other materials, such as lithium tantalate and KTP, are also covered since they are steadily increasing in quality and availability.

Table 5 Basic Properties of Some Popular EO Materials[a]

Material	EO coefficient r_{ij} (10^{-12} m/V)		Index of refraction	Dielectric constant
PLZT	$r_{13} = 67$		$n_0 = 2.312$	
	$r_{33} = 1340$		$n_e = 2.299$	
LiNbO$_3$	$r_{13} = 9.6$		$n_0 = 2.286$	$\varepsilon_1 = \varepsilon_2 = 78$
	$r_{22} = 6.8$		$n_e = 2.200$	$\varepsilon_3 = 32$
	$r_{33} = 31$			
LiTaO$_3$	$r_{13} = 8.4$		$n_0 = 2.176$	$\varepsilon_1 = \varepsilon_2 = 51$
	$r_{33} = 30.5$		$n_e = 2.180$	$\varepsilon_3 = 45$
KH$_2$PO$_4$ (KDP)	$r_{41} = 8$		$n_0 = 1.507$	$\varepsilon_1 = \varepsilon_2 = 42$
	$r_{63} = 11$		$n_e = 1.467$	$\varepsilon_3 = 21$
KD$_2$PO$_4$ (KD*P)	$r_{63} = 24.1$		$n_0 = 1.502$	$\varepsilon_3 = 50$
			$n_e = 1.462$	
(NH$_4$)H$_2$PO$_4$ (ADP)	$r_{41} = 23.41$		$n_0 = 1.522$	$\varepsilon_1 = \varepsilon_2 = 58$
	$r_{63} = 7.83$		$n_e = 1.477$	$\varepsilon_3 = 14$
Ba$_{0.25}$Sr$_{0.75}$Nb$_2$O$_6$	$r_{13} = 67$		$n_0 = 2.3117$	$\varepsilon_3 = 3400$ (15 MHz)
(SBN, Tc = 395 K)	$r_{43} = 1340$		$n_e = 2.2987$	
	$r_{51} = 42$			
KNbO$_3$	$r_{13} = 28$	$r_{23} = 1.3$	$n_1 = 2.280$	
	$r_{42} = 380$	$r_{33} = 64$	$n_2 = 2.329$	
	$r_{51} = 105$		$n_3 = 2.169$	

[a] All properties are measured at 633 nm (for optical properties), and low frequency (for dielectric constant). All properties are approximate, and should be verified with material vendors during procurement. (from Ref. 38).

Materials with very high EO coefficients, such as bulk form strontium barium niobate (SBN) and PLZT in the form of deposited films, are starting to appear. Other "new" materials are doped forms of lithium niobate or tantalate, and stoichiometric lithium niobate and tantalate. Many of these materials still exhibit performance variations between suppliers – it is best to discuss your needs with several crystal suppliers and perform multiple sample runs to ensure that the proper selection is made.

5.2 ADP, KDP, and Related Isomorphs

Relatively easy commercial growth processes make ADP (NH$_4$H$_2$PO$_4$) and KDP (KH$_2$PO$_4$) popular materials for bulk EO devices. High optical quality crystals can be grown to over 10 cm diameter, and can be cut, polished, and mounted without undue difficulty. In both materials, the hydrogen atoms may be replaced with deuterium. In this case they are referred to as AD*P and KD*P. This replacement results in an increase in the linear electro-optic coefficients by a factor of ~ 2.5.

The resistivity of ADP and related materials is very high, typically $> 10^{10}$ Ωcm when operated near room temperature. As operating temperatures approach the Curie temperature, C_T, the loss tangent and the dielectric constant increase, leading to heating and high power draws on electronic drivers if operated at high speeds. Heating in the crystal also leads to beam distortion via the thermo-optic properties of the material.

One major drawback to using ADP and its isomorphs in EO scanning systems is that the materials are hygroscopic. They are typically housed in a hermetic package that is

5.3 Lithium Niobate and Related Materials

A large class of ferroelectric materials have the form $A^{1+}B^{5+}O_3$ or $A^{2+}B^{4+}O_3$, and are related to the mineral perovskite ($CaTiO_3$). Several of these materials are in mass production for devices based on their piezoelectric properties, such as $LiNbO_3$ for surface acoustic wave filters, which are found in cell phones and a host of other signal processing applications. Czochralski growth is the typical practice, with boule diameters approaching 15 cm for $LiNbO_3$, although 7.5 and 10 cm is more common. During the processing of crystal boules, they are typically poled to form a single domain throughout by heating to a point near the Curie temperature and then applying a DC electric field, which is maintained during the cooldown. This ensures that all crystal domains have uniform orientation – a critical consideration for good optical quality of the ensuing device.

The perovskite materials are not water-soluble, eliminating some of the packaging problems encountered with ADP and similar materials. Another use of perovskites, in particular lithium tantalate, is as a pyroelectric detection element. If precautions are not taken in handling and processing, extremely high voltages can be generated between the faces of wafers. This charging can lead to electrical flashovers, which can damage electrodes or other coatings, or it can damage attached electrical equipment such as drivers or thermocouples. Wherever possible, controlled slow heating and cooling are recommended; the use of air ionizers in the work space also mitigates these effects.

The piezoelectric properties of the perovskites must be considered whenever the deflector will be operated at high speeds. The electrostrictive strain component of the index change can be as large as the EO component if a mechanical resonance is present. The mechanical performance of the entire device assembly – crystal, electrical leads, mounting adhesive and mounting base – must be considered early on, but careful testing is still a requirement.

Lithium niobate ($LiNbO_3$) and lithium tantalate ($LiTaO_3$) are the most common perovskite materials in use. For reasons of producibility and quality, they are typically grown to be slightly lithium-rich – this is referred to as congruently grown niobate or tantalate. These congruent materials exhibit Curie temperatures of 1470 and 890 K, respectively, giving them stable EO properties at room temperature or slightly elevated temperatures. They are also commercially available with fairly consistent properties across several vendors.

Recently, stoichiometric lithium niobate and lithium tantalate have been produced in commercially relevant sizes.[18,39] These materials exhibit lower coercive fields, which leads to easier fabrication of poled devices, and higher EO coefficients, as well as a broader transmission range. Being relatively new, it is best to contact the crystal growers for detailed information on properties and processing practices for stoichiometric materials.

Various dopants can be introduced when growing lithium niobate or tantalate to alter properties for special applications. Magnesium is a common dopant for lithium niobate, added to mitigate photorefractive damage from short wavelengths. This variant of the material is often used when producing second harmonic generation (SHG) or other nonlinear devices, typically in the visible spectrum.

Significant effort has been applied to the problem of domain inversion processing of lithium niobate and, to a lesser extent, lithium tantalate. These efforts were driven primarily by interests in SHGs and related nonlinear devices, but the practices are transferable to the production of scanners, as mentioned previously.

Barium titanate ($BaTiO_3$) and KTN (a solid solution of $KTaO_3$ and $KNbO_3$) belong in the perovskite group of materials. They have good EO properties, but have critical temperatures near or below room temperature. This makes some properties very temperature-dependent, and can result in creating or changing ferroelectric domains simply by handling and processing the crystal. They are not in wide use at this time.

5.4 Potassium Titanyl Phosphate (KTP)

In 1976 the Du Pont Company reported[40] on the growth and properties of the crystalline material $K_xRb_{1-x}TiOPO_4$. The material is ferroelectric and has found use in the production of SHGs and other nonlinear devices. The material is relatively difficult to grow (compared to lithium niobate), although the situation is improving due in part to military interest in the material, with slabs over 40 mm square being produced.

Dopants and special processing are used to produce various grades of KTP, one must check with manufacturers for availability and detailed properties. It is difficult to produce domain-inverted devices in KTP since the coercive field and dielectric strength are very near each other. Problems have also been encountered with relatively high electrical conductivity of the crystals, especially in large flux-grown crystals, further limiting its appeal.

5.5 Other Materials

5.5.1 AB-Type Binary Compounds

The main interest in these materials has been for EO devices in the infrared, particularly at 10.6 μm for use in CO_2 laser systems. GaAs, ZnTe, ZnS, CdS, and CdTe are among the most common materials that are available in large sizes. The EO coefficients of these materials is relatively small, only about 10% of lithium niobate, but their transmission beyond 10 μm may make them useful in some applications, especially for military uses.

5.5.2 Kerr Effect in Liquids

Much attention has been paid in the past to Kerr effect liquids, especially nitrobenzene.[1] The attraction was due to the high purity of materials compared to most crystals, and the basically unlimited size of the resulting device.

Advances in crystal growth techniques have largely mitigated the perceived quality and size advantages a liquid material may offer. In addition, the liquids can exhibit heating, currents, turbulence, and other behavior that affects performance. They are not widely used or considered for application at this time.

5.5.3 Electro-Optic Ceramics in the (Pb, La)(Zr, Ti)O_3 System

Lanthanum-modified lead zirconate titanate (PLZT) ceramic materials have been investigated since 1969 for their EO properties, which can be tailored to a degree by controlling the precise chemical makeup of the material.[1]

Electro-Optical Scanners

These materials are not available in large single crystal form, limiting their application to scanning systems. Scanning devices based on thin films of PLZT have been proposed, but the problems of coupling into and out of such films are daunting.

5.5.4 Other Materials

Significant EO materials development is still ongoing, driven by the reality that almost all current EO systems could be improved by using materials with higher EO coefficients, higher optical damage limits, lower conductivity, or improvements in other technical parameters. Efforts can be roughly characterized as either creating new materials or modifying existing ones.

An example of a relatively new material is strontium barium niobate (SBN). The crystal is grown by the Czochralski method, with boule diameters typically under 50 mm. There are several, slightly different formulations available, and quality and properties can vary even in modestly sized samples. Very high EO coefficients and relatively low Curie temperature make the material attractive for future applications. It is best to contact the material growers for current specifications prior to developing a design with an SBN element.

An example of modifications to a standard material is the development of magnesium doped lithium niobate, available from a variety of suppliers. The doping raises the optical damage threshold in the short wavelength part of the transmission band – a characteristic important for SHG devices, among others.

Careful selection, specification, inspection, and qualification of materials is a key to successful EO system design. Given the continuous improvement in crystal growing practices, inspection techniques, and materials formulation it is imperative to work closely with the materials suppliers during the design process.

5.6 Material Selection

Currently, there are only a handful of materials suitable for use in commercial EO systems. In addition to optical transparency at the desired wavelength, the following factors are common to nearly all applications.

- High electrical resistivity: greater than 10^{10} Ωcm is desired. This requirement stems from the desire for no resistive heating when operating voltages (typically hundreds of volts) are applied to the device. Ion migration can also occur, especially in the presence of DC fields, which can create substantial optical perturbations.
- High optical homogeneity: refractive index variations of less than 1 in 10^6 are desired. This requirement helps to preserve beam quality, and can also be a stand-in for crystal compositional variations.
- Large EO effect: absolute index variations of at least 10^{-4} are desired, with reasonable applied voltage. Excessively high voltages create packaging problems, long-term drift due to ion migration and require high driver power when high-speed operation is required.
- Processability: standard handling and process operations should not impact material quality or device performance. For reasonable cost, the materials must be able to be oriented, cut, polished, AR coated, and mounted with only minor (if any) departure from practices used for other optical materials.

If the device is going to be operated at high frequencies, the dielectric constant and loss tangent become important considerations, with thermal conductivity and temperature dependence of optical properties also needing to be considered.

If the device under consideration will be produced via a domain inversion process, other factors must be added to the list of considerations. Obviously it must be ferroelectric, which implies that piezoelectric and pyroelectric effects must also be taken into consideration. Also, the desire to pole the material will restrict the orientations available. Poled devices should also not be operated near to their coercive field or near the Curie temperature, which can be quite low for some materials, or there is a chance of depoling occurring.

6 ELECTRO-OPTIC DEFLECTION SYSTEM DESIGN PROCESS

Selecting a system design for a particular set of operating parameters is an iterative process, likely requiring thorough analysis of multiple trial designs. The complex interplay of material properties, EO element geometry, electronic power consumption, operating speeds, and the realities of current fabrication methods does not lend itself to a closed form solution, nor are there large catalogs of standard alternatives to choose from.

The recommended process is:

1. Verify that the speed, optical efficiency, or ruggedness of an EO deflector are required. If other technologies such as galvanometers or AO deflectors can be used, they are likely going to triumph in a head-to-head comparison of total system cost and complexity.
2. Consider the wavelength of the laser to be used. Few, if any, materials are optically clear across the entire spectrum of wavelengths available today. Also, many material properties vary with wavelength so it is important to look at the properties of interest at the wavelength of interest.
3. Consider whether the system can be built using a single linear polarization. If not, it is likely that beam splitting and recombining after deflection will be required, adding significant complexity and cost to the system.
4. Review the operating environment in light of safety and reliability of high voltage electrical systems. If moisture and dust are present, the EO scanner will likely need to be placed in a sealed housing, adding length and additional windows to the optical path.
5. Select appropriate design guidelines and safety factors such as electrical creepage distances, dielectric strength, and optical power per unit area on surfaces. These guidelines and safety factors may bound the design options.
6. Generate and analyze trial designs. One aspect that is sometimes overlooked is the mechanical response of the EO material. All attractive EO crystals exhibit piezoelectric responses to some degree. In some cases, high-speed electrical pulses can excite mechanical resonances that create a time-varying strain in the material, which can alter the deflection from that expected or contribute to beam losses.
7. Verify that the selected design can actually be built by consulting with the appropriate material suppliers, electronics designers, optical designers, and engineers familiar with current practice for all manufacturing steps.

7 CONCLUSIONS

Electro-optic scanning systems can be very fast and optically very efficient. This performance often comes at the cost of working near the cutting edge of materials, electronics, and processing technologies. The demands for speed are likely to continue increasing, however, as laser and computing technology continue to advance. To address these demands, a true systems approach should be used when designing an EO scanning system. The overlapping implications of decisions in areas as diverse as high-voltage electronic drivers, mechanical isolation, temperature control, and beam size must be weighed carefully.

Large scale application of EO scanners has not yet occurred due to the difficulties and uncertainties discussed in this chapter. Recent advances in telecommunications switching, computer-to-plate printing, and biomedical imaging applications are driving development of EO devices and electronic drivers at a rapid pace. New EO materials, with new combinations of properties, are also being developed. Such progress in each of the key areas of EO scanning system design and construction may eventually lead to their wider application.

ACKNOWLEDGMENTS

Several people contributed ideas, commentary and references during the preparation of this section. Of particular note is Richard Stolzenberger, PhD, now a consultant. The authors are also grateful for the support (and tolerance) of our spouses. Gerald Marshall, the volume editor, also provided the ongoing encouragement to complete the effort in the face of industry turmoil.

REFERENCES

1. Ireland, C.; Ley, J. Electrooptical scanners. In Marshall, G., Ed.; *Optical Scanning*; Marcel Dekker: New York, 1987; 687–778.
2. Stancil, D.D. Electro-optical scanners. In *Encyclopedia of Optical Engineering*. Marcel Dekker: New York, 2003; 456–474.
3. Lee, S.M.; Hauser, S.M. Kerr constant evaluation of organic liquids and solutions. Rev. Sci. Instruments 1964, *35*, 1679.
4. Kruger, R.; Pepperl, R.; Schmidt, U. Electrooptic materials for digital light beam deflectors. Proc. IEEE 1973, *61*, 992.
5. Chiu, Y.; Zou, J.; Stancil, D.D.; Schlesinger, T.E. Shape-optimized electrooptic beam scanners: analysis, design, and simulation. J. Lightwave Technol. 1999, *17* (1), 108–114.
6. Lotspeich, J.F. Electrooptic light-beam deflection. IEEE Spectrum 1968, *5*, *Feb*, 45–52.
7. Lee, T.C.; Zook, J.D. Light beam deflection with electrooptic prisms. IEEE J. Quantum Electronics 1968, *QE-4* (7), 442–454.
8. Fowler, V.J.; Buhrer, C.F.; Bloom, L.R. Electro-optic light beam deflector. Proc. IEEE 1964, *52* (2), 193–194.
9. Fowler, V.J.; Schlafer, J.A. Survey of laser beam deflection techniques. Appl. Optics 1966, *5* (10), 1675–1682.
10. Kiyatkin, R.P. Analysis of control field in quadrupole optical-radiation deflectors. Opt. Spectrosc. 1975, *38* (2), 209–210.
11. QuickField, for finite element calculations. Retrieved from *www.quickfield.com* March 22, 2004.
12. Ireland, C.; Ley, J. Electrooptical scanners. In Marshall, G., Ed.; *Optical Scanning*, Marcel Dekker: New York, 1987; 752–754.

13. Armstrong, J.A.; Bloembergen, N.; Ducuing, J.; Pershan, P.S. Interactions between light waves in a nonlinear dielectric. Phys. Rev. 1962, *127*, 1918–1939.
14. Fejer, M.M.; Magel, G.A.; Jundt, D.H.; Byer, R.L. 'Quasi-phase-matched second harmonic generation: tuning and tolerances.' IEEE J. Quantum Electronics 1992, *28* (11), 2631–2654.
15. Mizuuchi, K.; Yamamoto, K. Highly efficient quasiphase-matched 2nd harmonic generation using 1st-order periodically domain-inverted $LiTaO_3$ waveguide. Appl. Phys. Lett. 1992, *60* (11), 1283–1285.
16. Wang, Y.; Petrov, V.; Ding, Y.J.; Zheng, Y.; Khurgin, J.B.; Risk, W.P. Ultrafast generation of blue light by efficient second-harmonic generation in periodically-poled bulk and waveguide potassium titanyl phosphate. Appl. Phys. Lett. 1998, *73* (7), 873–875.
17. Ktaoka, Y.; Narumi, K.; Mizuuchi, K. Waveguide-type SHG blue laser for high-density optical disk system. Rev. Laser Eng. 1998, *26* (3), 256–260.
18. Gopalan, V.; Sanford, N.A.; Aust, J.A.; Kitamura, K.; Furukawa, Y. Crystal growth, characterization, and domain studies in lithium niobate and lithium tantalate ferroelectrics. In *Handbook of Advanced Electronic and Photonic Materials and Devices*; Nalwa, H.S., Ed.; Academic Press: New York, 2001; Vol. 4, Ferroelectrics and Dielectrics, 57–114.
19. Li, J.; Cheng, H.C.; Kawas, M.J.; Lambeth, D.N.; Schlesinger, T.E.; Stancil, D.D. Electrooptic wafer beam deflector in $LiTaO_3$. IEEE Photonics Tech. Letts. 1996, *8* (11), 1486–1488.
20. Revelli, J.F. High-resolution electrooptic surface prism waveguide deflector: an analysis. Appl. Optics 1980, *19*, 389–397.
21. Chen, Q.; Chiu, Y.; Lambeth, D.N.; Schlesinger, T.E.; Stancil, D.D. Guided-wave electro-optic beam deflector using domain reversal in $LiTaO_3$. J. Lightwave Technology 1994, *12* (4), 1401–1404.
22. Lee, C.L.; Lee, J.F.; Huang, J.Y. Linear phase shift electrodes for the planar electrooptic prism deflector. Appl. Optics 1980, *19*, 2902–2905.
23. Chen, Q.; Chiu, Y.; Devasahayam, A.J.; Seigler, M.A.; Lambeth, D.N.; Schlesinger, T.E.; Stancil, D.D. Waveguide optical scanner with increased deflection sensitivity for optical data storage. In *SPIE Proc. Series*, Vol. 2338, 1994; Topical Meeting on Optical Data Storage, Dana Point, CA; May 16–18, 1994; 262–267.
24. Sasaki, H.; De La Rue, R.M. Electro-optic multichannel waveguide deflector. Electronics Letts. 1977, *13* (10), 295–296.
25. Chiu, Y.; Burton, R.S.; Stancil, D.D.; Schlesinger, T.E. Design and simulation of waveguide electrooptic beam deflectors. J. Lightwave Technol. 1995, *13* (10), 2049–2052.
26. Takizawa, K. Electrooptic Fresnel lens-scanner with an array of channel waveguides. Appl. Optics 1983, *22* (16), 2468–2473.
27. Fang, J.C.; Kawas, M.J.; Zou, J.; Gopalan, V.; Schlesinger, T.E.; Stancil, D.D. Shape-optimized electrooptic beam scanners: experiment. IEEE Photonics Technol. Lett. 1999, *11* (1), 66–68.
28. Chiu, Y.; Burton, R.S.; Stancil, D.D.; Schlesinger, T.E. Design and simulation of waveguide electrooptic beam deflectors. J. Lightwave Technol. 1995, *13* (10), 2049–2052.
29. Feit, M.D.; Fleck, J.A., Jr. Light propagation in graded-index optical fibers. Appl. Opt. 1978, *17* (24), 3990–3998.
30. Eason, R.; Boyland, A.; Mailis, S.; Smith, P.G.R. Electro-optically controlled beam deflection for grazing incidence geometry on a domain-engineered interface in $LiNbO_3$. Optics Commun. 2001, *197*, 201–207.
31. Gnewuch, H.; Pannell, C.; Ross, G.; Smith, P.G.R.; Geiger, H. Nanosecond response of bragg deflectors in periodically poled $LiNbO_3$. IEEE Photonics Technol. Lett. 1998, *10* (12), 1730–1732.
32. Kawas, M.J. Design and characterization of domain inverted electro-optic lens stacks on $LiTaO_3$. Department of Electrical and Computer Engineering; Carnegie Mellon University, 1996; M.S. Thesis.

33. Kawas, M.J.; Stancil, D.D.; Schlesinger, T.E. Electrooptic lens stacks on LiTaO$_3$ by domain inversion. J. Lightwave Technol. 1997, *15* (9), 1716–1719.
34. Gahagan, K.T.; Gopalan, V.; Robinson, J.M.; Jia, Q.; Mitchell, T.E.; Kawas, M.J.; Schlesinger, T.E.; Stancil, D.D. Integrated electro-optic lens/scanner in a LiTaO$_3$ single crystal. Appl. Optics 1999, *38* (4), 1186–1190.
35. Gopalan, V.; Kawas, M.J.; Gupta, M.C.; Schlesinger, T.E.; Stancil, D.D. Integrated quasi-phase-matched second-harmonic generator and electrooptic scanner on LiTaO$_3$ single crystals. IEEE Photonics Technology Lett. 1996, *8* (12), 1704–1706.
36. Chiu, Y.; Gopalan, V.; Kawas, M.J.; Schlesinger, T.E.; Stancil, D.D.; Risk, W.P. Integrated optical device with second-harmonic generator, electrooptic lens, and electrooptic scanner in LiTaO$_3$. J. Lightwave Technol. 1999, *17* (3), 462–465.
37. Cleland, A.; Gass, H. Energy recirculating driver for capacitive load. Patent Cooperation Treaty application, document #WO 02/14932, August 16, 2001; revised Feb. 21, 2002.
38. Yariv, A. *Optical Electronics in Modern Communications*; Oxford University Press: New York, 1997.
39. Furukawa, Y.; Kitamura, K.; Suzuki, E.; Niwa, K. J. Stoichiometric LiTaO$_3$ single crystl growth by double crucible Czochralski method using automatic powder supply system. Crystal Growth 1999, *197*, 889.
40. Zumsteg, F.; Bierlein, J.; Gier, T. $K_x Rb_{1-x} TiOPO_4$: A new nonlinear optical material. J. Appl. Phys. 1976, *47*, 4980.

14

Multichannel Laser Thermal Printhead Technology

SEUNG HO BAEK, DANIEL D. HAAS, DAVID B. KAY,
DAVID KESSLER, and KURT M. SANGER

Eastman Kodak Company, Rochester, New York, U.S.A.

1 INTRODUCTION

Multiple-channel printheads and printing systems are very important to the graphic arts industry for incorporation into imagesetters, halftone color proofers, and computer-to-plate systems producing large-format halftone images. Laser thermal media, dry and environmentally benign, are widely used in the graphic arts industry. Laser thermal printing is growing because of its many conveniences for digital prepress processes, but recording an image on the media requires high laser energy, about five orders of magnitude more than silver halide (AgX) photographic media. Multiple-channel printheads are used in printing systems to attain higher productivity by compensating for low media sensitivity.

There are several different types of media: laser thermal dye transfer from a donor sheet to a receiver; ablation of dye from its support layer; cross-linking of a coated polymer layer; and changing the phase of an exposed medium. All of these media can provide image resolution as high as 3000 dots per inch (dpi), which is finer than high-quality AgX photographic media.

Three basic configurations of printing systems are: the flying-spot printer; the internal-drum printer;[1] and the external-drum printer. The flying spot, common in desktop printers, uses either a polygonal mirror[2] or galvanometric mirror[3] to deflect the beam from a single-mode laser across a medium. That medium may be held on a flat bed. Multiple-channel printheads are rarely used in flying-spot printers because of the complexity of the optical system.

The internal-drum printer holds the image-recording medium stationary while the beam is deflected through the large field of a rapidly rotating monocentric optical system. A single-channel printhead containing a single-mode laser is typically used in imagesetters exposing highly sensitive AgX media for halftone graphic arts applications.

Rotation of an external drum carrying the image-recording medium provides fast-scan in one direction denoted as the *x*-direction while the printhead moves perpendicularly along the slowscan direction, called the *y*-direction. The laser light propagates along the *z*-direction, perpendicular to both the fastscan and the slowscan directions. Because the optics are used "on axis," the number of resolvable spots is limited only by the extent of printhead travel. The external-drum configuration does not require the use of single-mode lasers as the flying-spot printing system does, so it is much easier and cost-effective to use multiple-channel printhead architecture with multimode, high-power lasers in an external drum printer to achieve high productivity and large image size.

Eastman Kodak Company (Rochester, New York) developed three types of multiple-channel printheads to write 1800–2540 dpi subpixels of 150–200 lpi (lines per inch) halftone dots on laser thermal media. These printheads are: a fiber array on a silicon V-groove[4] pigtailed to multimode diode lasers; independently modulated diode-laser array integrated on a single substrate; and external light valve with a high-power, multimode laser bar. This chapter will present these three different multiple-channel printhead configurations.

2 PHYSICS OF CONTINUOUS-TONE LASER-THERMAL TRANSFER

Low-molecular-weight dyes can be vaporized from a donor and transferred to a receiver by heating the donor hotter than that dye's vaporization temperature. Absorption of focused light induces a large, rapid temperature rise within a small volume of the donor. Controlling the beam irradiance of a laser that is raster-scanned across the donor can produce well-defined images. No further chemical processing is needed, and all material handling can be performed in room light.

This chapter considers laser-thermal donors capable of transferring intermediate amounts of their coating of visible dye to the receiver, enabling the production of continuous-tone images. However, the commercially successful application of these laser-thermal printheads has been reproduction of binary halftone images by darkening sub-pixels of each halftone dot in a pattern specified by a raster-image processor (RIP) as described later in Sec. 4.1, typically darkening all of those subpixels to the same reflective density. A donor with the limited capability of transferring only single density[5,6] would suffice if all customers required the same binary density of the darkened halftone subpixels for every image. The continuous-tone capability of the donors discussed in this chapter is being exploited to enable one laser-thermal halftone proofer and one set of donors to emulate a range of ink densities anticipated from various printing presses and ink sets. In order to explain the sensitivity of these laser-thermal halftone systems to donor properties and optical characteristics of the printhead, the continuous-tone behavior must be analyzed.

Figure 1 depicts a vertical section of a layer of infrared and visible dyes coated in a binder on the bottom surface of a donor sheet. Matte beads hold the donor at a fixed gap from the receiver and minimize sticking of the heat-softened dye layer to the receiver's surface. The sequence of panels illustrates a laser, scanning from left to right, at four successive times. The dye layer is cool when the laser is first activated in the left panel. Accumulation of the laser energy heats the dye in the second panel, with fastest heating in the beam's focus, where laser irradiance is greatest. Dye nearest the laser captures more

Multichannel Laser Thermal Printhead Technology

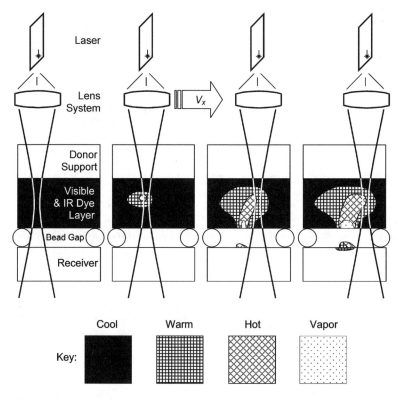

Figure 1 Schematic representation of laser heating and transfer of absorptive donor.

heat than dye farther along the beam's propagation path as a result of the dye's attenuation of the light. Heat builds up at the donor's lower surface in the third panel because insignificant thermal conductivity or convection at the donor–receiver gap causes the donor's lower surface to act as a thermal insulator. If dye at its vaporization temperature acquires its heat of vaporization, this additional energy transforms the hot molten dye into free gas molecules or an aerosol that can condense onto the cooler surfaces of the receiver and surrounding donor. In the fourth panel, dye molecules transferred from the surface uncover greater depths of the dye layer with continued laser irradiance, building up a thicker deposit of dye on the receiver to modulate the image density. Dye condensing on the receiver imparts some of its heat to that surface in the course of image deposition, warming parts of the receiver. Material in the laser's heat trail on the donor cools back to ambient temperature by diffusion of heat to the surrounding unexposed donor.

2.1 Exposure Deposited by a Scanning Laser Beam

An elliptical Gaussian laser beam with standard-deviation radii σ_x and σ_y scanning at speed V_x imposes an irradiance E_{laser}[7,8] at the top of the dye layer $\{z = \ell\}$:

$$\vec{E}_{\text{laser}}[x, y, z, t] = -\hat{z} P_{\text{laser}} \frac{\exp\left[-\frac{1}{2}\left(\frac{x - V_x t}{\sigma_x}\right)^2\right]}{\sqrt{2\pi}\sigma_x} \frac{\exp\left[-\frac{1}{2}\left(\frac{y}{\sigma_y}\right)^2\right]}{\sqrt{2\pi}\sigma_y} \quad \text{for } \ell \leq z$$

(1)

The negative sign preceding the symbol for the unit vector in the z-direction indicates that the irradiance is shining down through the support, into the dye layer, then through the bead gap. The exposure $H[x, y, t]$ is defined as the cumulative energy incident upon a unit area of the donor. Exposure is a property of the beam and its scanning apparatus just outside the material, unencumbered by concern for the fate of the light flux upon entering that material. In the course of writing a single scanline with the irradiance in Eq. (1) activated at $\{t = 0\}$ when centered at the origin $\{x = 0, y = 0\}$, the exposure deposited by time t is

$$H_{\text{single}}[x, y, t] = \int_{t'=0}^{t'=t} |E_{\text{laser}}[x, y, z = \ell, t']| dt' \tag{2}$$

The layer containing visible and infrared dye is typically thinner than 1 μm so that the standard-deviation radii σ_x and σ_y of the beam remain constant throughout the thin dye layer of the donor and retain their same values in air as in a medium of higher refractive index.[9,10] Substituting Eq. (1) for the laser irradiance into Eq. (2) gives

$$H_{\text{single}}[x, y, t] = \frac{P_{\text{laser}}}{2\pi\sigma_x\sigma_y} \exp\left[-\frac{1}{2}\left(\frac{y}{\sigma_y}\right)^2\right] \int_{t'=0}^{t'=t} \exp\left[-\frac{1}{2}\left(\frac{x - V_x t'}{\sigma_x}\right)^2\right] dt' \tag{3}$$

The change of variables:

$$u = \frac{x - V_x t'}{\sigma_x \sqrt{2}} \Rightarrow dt' = \frac{\sigma_x \sqrt{2}}{V_x} du \tag{4}$$

converts the integration over time in Eq. (3) to an integration over the distance traversed by the moving beam while writing the scanline:

$$H_{\text{single}}[x, y, t] = \frac{P_{\text{laser}}}{V_x \sigma_y \pi \sqrt{2}} \exp\left[-\frac{1}{2}\left(\frac{y}{\sigma_y}\right)^2\right] \int_{u=\frac{x-V_x t}{\sigma_x \sqrt{2}}}^{u=\frac{x}{\sigma_x \sqrt{2}}} \exp[-u^2] du \tag{5}$$

The definite integral is recognized as the Error Function:[11]

$$\text{ERF}[v] = \frac{2}{\sqrt{\pi}} \int_{u=0}^{u=v} \exp[-u^2] du \tag{6}$$

The exposure directed upon the donor while writing a single scanline with a Gaussian laser beam is

$$H_{\text{single}}[x, y, t] = \frac{P_{\text{laser}}}{V_x \sigma_y 2\sqrt{2\pi}} \exp\left[-\frac{1}{2}\left(\frac{y}{\sigma_y}\right)^2\right]$$
$$\times \left\{\text{ERF}\left[\frac{x}{\sigma_x \sqrt{2}}\right] - \text{ERF}\left[\frac{x - V_x t}{\sigma_x \sqrt{2}}\right]\right\} \tag{7}$$

Anywhere in the interior of the scanline more than a few beam radii σ_x from its ends, the first error function approaches a value of 1 and the second approaches -1, so that the exposure accumulated throughout most of the scanline is

$$H_{\text{single}}[x, y, t] = \frac{P_{\text{laser}}}{V_x \sigma_y \sqrt{2\pi}} \exp\left[-\frac{1}{2}\left(\frac{y}{\sigma_y}\right)^2\right] \quad \text{for } \sigma_x \ll x \ll (V_x t - \sigma_x) \quad (8)$$

A uniform field exposed with an infinite series of these identical scanlines, spaced a raster distance Y apart, accumulates

$$H_{\text{uniform}}[y] = \frac{P_{\text{laser}}}{V_x \sigma_y \sqrt{2\pi}} \sum_{m=-\infty}^{m=\infty} \exp\left[-\frac{1}{2}\left(\frac{y - mY}{\sigma_y}\right)^2\right] \quad (9)$$

as depicted in Fig. 2.

Because the exposure profile of this uniform field repeats every scanline, the area-averaged exposure $\langle H_{\text{uniform}} \rangle$ can be computed by integrating over a single scanline

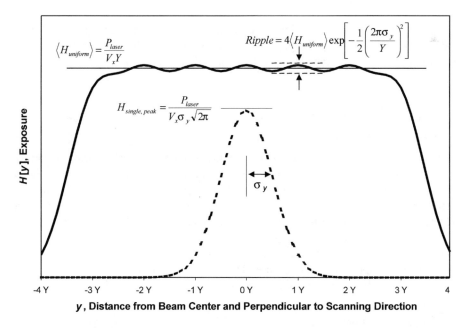

Figure 2 Superposition of Gaussian scanline exposures accumulates a uniform field $\{(\sigma_y\sqrt{[2\pi]})/Y\}$ greater than the peak of the individual scanline, plus sinusoidal ripple.

spacing Y:

$$\langle H_{\text{uniform}} \rangle = \frac{\int_{y=-\infty}^{y=\infty} H_{\text{uniform}}[y]dy}{\int_{y=-\infty}^{y=\infty} dy} \tag{10}$$

$$= \frac{\int_{y=0}^{y=Y} \frac{P_{\text{laser}}}{V_x \sigma_y \sqrt{2\pi}} \sum_{m=-\infty}^{m=\infty} \exp\left[-\frac{1}{2}\left(\frac{y-mY}{\sigma_y}\right)^2\right] dy}{Y} \tag{11}$$

$$= \frac{P_{\text{laser}}}{YV_x \sigma_y \sqrt{2\pi}} \sum_{m=-\infty}^{m=\infty} \left\{ \int_{y=(m-1)Y}^{y=mY} \exp\left[-\frac{1}{2}\left(\frac{y}{\sigma_y}\right)^2\right] dy \right\} \tag{12}$$

$$= \frac{P_{\text{laser}}}{YV_x \sigma_y \sqrt{2\pi}} \int_{y=-\infty}^{y=\infty} \exp\left[-\frac{1}{2}\left(\frac{y}{\sigma_y}\right)^2\right] dy \tag{13}$$

$$= \frac{P_{\text{laser}}}{YV_x} \tag{14}$$

Conversion from Eq. (11) to Eq. (13) is accomplished by recognizing that the portion between $\{y = 0\}$ and $\{y = Y\}$ of the Gaussian centered at $\{y = mY\}$ is identical to the portion between $\{y = (-m)Y\}$ and $\{y = (-[m-1])Y\}$ of the Gaussian centered at $\{y = 0\}$. The tabulated value for the definite integral in Eq. (13) is:

$$\int_{y=-\infty}^{y=\infty} \exp\left[-\frac{1}{2}\left(\frac{y}{\sigma_y}\right)^2\right] dy = \sigma_y \sqrt{2\pi} \tag{15}$$

If the exposure profiles of neighboring scanlines in the raster overlap above 50% of their peak amplitudes $\{Y < 2.4\ \sigma_y\}$, the exposure at any location in the uniform field can be approximated by its area-averaged value $\langle H_{\text{uniform}} \rangle$ with less than 5% sinusoidal ripple.[12]

2.2 Static Approximation for Uniform Heating Throughout the Dye Layer of the Donor

The temperature profile produced by a scanning source generating heat uniformly throughout the dye-layer thickness ℓ, in the absence of thermal diffusion, can be calculated by simply accounting for energy deposition. Multiplying this static temperature by a proportionality factor representing the fraction of heat escaping by thermal diffusion provides an approximation for the temperature in the presence of thermal diffusion.

The absorbed fraction of the light energy elevates the temperature of the donor material above the room temperature T_{ambient} in proportion to the amount and heat capacity $\{\rho\ c_\rho\}$ of material capturing that exposure. The temperature profile at a location in the donor immediately after beam passage is predicted by this static model to be:

$$T[x, y, z, t] = T_{\text{ambient}}$$

$$+ \frac{H[x, y, t](1-R)\eta_{\text{confine}}}{\ell \rho c_\rho} (\textit{Fraction of Light Absorbed}) \tag{16}$$

R is the fraction of incident light reflected by interfaces of the donor before and upon entering the absorptive dye layer. η_{confine} is the fraction of heat remaining within the volume of the donor exposed by the beam, when the dye layer attains its maximum temperature, despite thermal diffusion. The corollary of the Beer–Lambert Law

$$(\textit{Fraction of Light Absorbed}) = A_{\text{IR}} = 1 - 10^{-\varepsilon_{\text{IR}} C_{\text{IR}} \ell} \tag{17}$$

accounts for the portion of light incident on the upper surface of the dye layer $H[x, y, t]$ that is absorbed in the course of transiting the dye thickness ℓ.

Constancy of the beam radii σ_x and σ_y throughout the dye layer thickness $\{\ell < 1\,\mu\text{m}\}$ is confirmed by the irradiance profiles in the leftmost column of Fig. 3, acquired at increments of 12 μm through the plane of best focus for a laser-thermal printer imaging the end of a diode-laser-coupled optical fiber onto the donor. The absolute minimum of beam expansion over a dye layer 1 μm thick would be 1 part per million, according to Kogelnik's derivation for a single-mode laser[13,14] with best-focus beam waist radius $\{\omega_0 = 13\,\mu\text{m}\}$ at wavelength $\{\lambda = 830\,\text{nm}\}$:

$$\omega_{\text{laser}}^2 = \omega_0^2 \left[1 + \left(\frac{\lambda z}{\pi \omega_0^2} \right)^2 \right] \tag{18}$$

in which the waist radius ω_{laser} is defined to be the distance from the center of the beam to the point at e^{-2} of the profile's peak. The waist radius is twice the standard-deviation radius σ_{laser} of the Gaussian, that is, the distance from the beam center to the point at which the irradiance is $e^{-1/2}$ of the profile's peak:[7]

$$\omega_{\text{laser}} = 2\sigma_{\text{laser}} \tag{19}$$

For this static approximation of heat distribution, the temperature is treated as constant across the thickness ℓ of the dye layer, that is, the z-direction, even though the Beer–Lambert Law indicates that more heat is deposited on the side through which the light enters than on the side through which the light departs. This uniform-temperature approximation is relevant for light incident through the support because the side of the light's departure is the outer surface of the dye layer bordering a mild vacuum with extremely small thermal conductivity compared to the donor materials. This outer surface acts as a thermal insulator, causing heat to build up in the nearby material. This build-up of heat at the outer surface occurs for a beam scanning slowly enough that the time for the beam's center to traverse the beam's full width at half-maximum (FWHM) is longer than the characteristic time for heat to diffuse across the dye layer:

$$\left\{ \frac{FWHM_{\text{beam}}}{V_x} = t_{\text{traverse},FWHM} \right\} > \left\{ t_{\text{diffusion}} = \frac{\ell^2}{2\alpha_{\text{dye}}} \right\} \tag{20}$$

implying that the dye layer ℓ is "thermally thin." α_{dye} is the thermal diffusion coefficient in the dye layer including binder, typically about $10^{-7}\,\text{m}^2/\text{s}$ for organic polymers. Heat diffuses into the cooler plastic support and away from its location of most intense

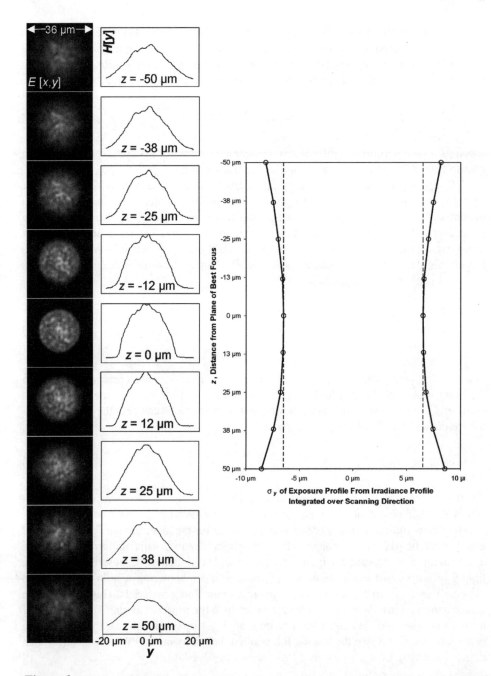

Figure 3 (Left column) Two-dimensional irradiance profiles of one laser beam at planes spaced 12 μm along the light's propagation direction about the plane of best focus ($z = 0$ μm, in the fifth panel from the top). (Middle column) One-dimensional integrals over the fastscan x-direction of the corresponding irradiance profiles; each curve can be considered as the exposure profile deposited by the beam along the slowscan y-direction; note that each curve is not a single slice through the peak of the irradiance profile. (Right column) Gaussian standard-deviation radius of the beam's exposure profile (solid curve) inferred from Gaussian fits to the fastscan integral of the observed irradiance distributions widens with distance from its plane of best focus faster than a single-mode laser (dashed curve) would.

deposition at its entrance into the dye layer, further leveling the temperature profile across the thickness ℓ of the heated dye layer.

The temperature profile produced immediately after the passage of a single scanning beam can be approximated by substituting the exposure profile of Eq. (8) into Eq. (16):

$$T_{\text{single}} = T_{\text{ambient}} + \frac{P_{\text{laser}}(1-R)A_{\text{IR}}\eta_{\text{confine}}}{\sqrt{2\pi}V_x\sigma_y\ell\rho c_\rho} \exp\left[-\frac{1}{2}\left(\frac{y}{\sigma_y}\right)^2\right] \qquad (21)$$

2.3 Additive Density by the Model of Exposure in Excess of Threshold

The model of exposure in excess of threshold assumes that dye is transferred only from regions that retain enough energy from the absorbed laser light to convert the visible dye to its vapor phase.[15] The threshold value $H_{\text{threshold}}$ for energy deposited per unit area:

$$H_{\text{threshold}} = \frac{\ell\rho c_\rho(T_{\text{vap}} - T_{\text{ambient}})}{(1-R)\eta_{\text{confine}}A_{\text{IR}}} \qquad (22)$$

raises the dye's temperature to T_{vap}, potentiating the thickness of the dye for transfer within that region. The exposure in excess of $H_{\text{threshold}}$ constitutes the energy available for transferring the dye molecules from the surface of the donor, as schematized by the diagonally hatched areas in Fig. 4.

The lateral extent $y_{\text{threshold}}$ of donor attaining vaporization temperature, and susceptible to transfer, can be derived by inverting Eq. (21) for T_{vap}:[5,6]

$$y_{\text{threshold,single}} = \sigma_y\sqrt{2\ln\left[\frac{P_{\text{laser}}(1-R)\eta_{\text{confine}}A_{\text{IR}}}{\sqrt{2\pi}V_x\sigma_y\ell\rho c_\rho(T_{\text{vap}} - T_{\text{ambient}})}\right]} \qquad (23)$$

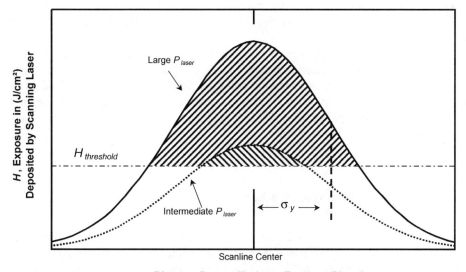

Figure 4 Exposure exceeding the amount to attain vaporization temperature transfers dye from donor to produce density on receiver; shaded area represents exposure in excess of threshold.

The energy exceeding threshold apportions the number of visible dye molecules entering the vapor phase in accordance with that dye's latent heat of vaporization ΔH_{vap}. Note that enthalpy is indicated by a plain capital "H," while exposure is indicated by an italicized capital "*H*." η_{vap} represents the efficiency with which energy in excess of the vaporization threshold is devoted to vaporizing the visible dye as opposed to consumption of this energy by melting or mechanical deformation of the polymeric binder and support, thermal or photolytic decomposition of donor components, or vaporization of the infrared dye and binder that do not contribute to the visible density in the image.

$$\begin{bmatrix} \text{Moles of visible dye} \\ \text{evaporated per unit length} \\ \text{of one raster scan} \end{bmatrix} = \frac{\eta_{vap} \eta_{confine}(1-R)}{\Delta H_{vap}} \int_{y=-y_{threshold}}^{y=y_{threshold}} (H_{single}[y] - H_{threshold}) dy \quad (24)$$

The average number of moles of visible dye deposited onto a unit area of the receiver by a succession of identical scanlines at raster spacing Y apart in the y-direction is the ratio between (the moles vaporized per unit length of one scanline) and (this raster spacing):

$$\begin{bmatrix} \text{Average number of moles} \\ \text{of visible dye deposited per} \\ \text{unit area of receiver surface} \end{bmatrix} = \frac{\begin{bmatrix} \text{Moles of visible dye} \\ \text{evaporated per unit length} \\ \text{of one raster scan} \end{bmatrix}}{Y} \quad (25)$$

The number of dye molecules originating within the raster width on the donor but landing outside that raster width on the receiver is compensated by the dye originating from other rasters on the donor but landing within this raster width on the receiver.

Equation (25) is a molecularly based justification for declaring that laser-thermal dye transfer obeys "additive density"[4,15] along the slowscan direction. In other words, the density profiles transferred by successive scanlines add if the donor is permitted to cool substantially to ambient temperature between fastscans. This basic property of the laser-thermal imaging is a significant departure from the "additive exposure" mechanism of silver halide imaging,[16,17] in which the exposure uniquely determines the image density, regardless of the time sequence for depositing that exposure on photosensitive materials, obeying "reciprocity."

2.4 Optical Density Transfer Predicted by Model of Exposure in Excess of Threshold

If the density profiles of visible dye on the receiver, produced by single scans, overlap their neighbors sufficiently, the density of a large area on the receiver is well approximated by its average value. The density profiles might overlap sufficiently if the laser-beam width σ_y were greater than the raster spacing Y or if the visible dye were to spread laterally by a significant fraction of that raster spacing during the dye's transfer from donor to receiver. The transmissive optical density is directly proportional to the extinction coefficient ε_{vis}

for a uniform thickness of visible dye deposited upon a unit area:

$$Transmissive\ Optical\ Density = \varepsilon_{vis} \begin{bmatrix} \text{Average number} \\ \text{of moles of visible} \\ \text{dye deposited} \\ \text{per unit area of} \\ \text{receiver surface} \end{bmatrix} \quad (26)$$

Equation (26) for the image density on the receiver can be evaluated by substitution of Eqns. (8), (22), (25), and (26) into Eq. (24) integrated between the limits explicitly stated in Eq. (23), and using the definition of the error function in Eq. (6):

$$Transmissive\ Optical\ Density_{single}$$

$$= \frac{P_{laser}(1-R)\eta_{confine}\eta_{vap}\varepsilon_{vis}}{V_x Y \Delta H_{vap}} \text{ERF}\left\{\sqrt{\ln\left[\frac{P_{laser}(1-R)\eta_{confine}A_{IR}}{\sqrt{2\pi}V_x\sigma_y\ell\rho c_\rho(T_{vap}-T_{ambient})}\right]}\right\}$$

$$- 2\frac{\eta_{vap}\varepsilon_{vis}\sigma_y\lambda\rho c_\rho(T_{vap}-T_{ambient})}{\Delta H_{vap} Y A_{IR}}\sqrt{2\ln\left[\frac{P_{laser}(1-R)\eta_{confine}A_{IR}}{\sqrt{2\pi}V_x\sigma_y\ell\rho c_\rho(T_{vap}-T_{ambient})}\right]}$$

$$(27)$$

Table 1 contains examples of the optical and thermal properties appearing in Eq. (27) for two experimental donors.

The latent heat of vaporization ΔH_{vap} in Table 1 is adopted from the value extrapolated from measurements of the vapor pressure of a laser-thermal transfer dye subliming over the temperature range 180–240°C. Extrapolation of these vapor pressures to atmospheric pressure produces 600°C as an estimate of the laser-thermal-transfer visible dye's vaporization temperature. The visible dyes used in the donors begin to lose weight in the range 280–310°C during thermogravimetric analysis. A value of $\{T_{vap} = 360°C\}$ produces good agreement between the onset of experimentally measured reflectance density of laser-thermal-transfer-imaged receivers and that predicted by Eq. (27). The mass density and specific heat are the values for the polyethylene terephthalate[18,19] polymeric support, averaged over the temperature range 20–300°C in order to represent the thermal environment of primarily polymeric binder in the dye layer and polymeric support with a single constant value for each of these parameters.

Replacement of the ratio $\{P_{laser}/V_x\}$ by the product $\{\langle H_{uniform}\rangle Y\}$ justified by Eq. (14), and replacement of the definition of $H_{threshold}$ in Eq. (22), permit re-expression of Eq. (27):

$$Transmissive\ Optical\ Density_{single}$$

$$= \langle H_{uniform}\rangle\left\{\frac{(1-R)\eta_{vap}\eta_{confine}\varepsilon_{vis}}{\Delta H_{vap}}\right\}\text{ERF}\left\{\sqrt{\ln\left[\frac{\langle H_{uniform}\rangle}{\sqrt{2\pi}}\left\{\frac{Y}{\sigma_y}\frac{1}{H_{threshold}}\right\}\right]}\right\}$$

$$- 2\left\{\frac{\sigma_y}{Y}H_{threshold}\right\}\left\{\frac{(1-R)\eta_{vap}\eta_{confine}\varepsilon_{vis}}{\Delta H_{vap}}\right\}\sqrt{2\ln\left[\frac{\langle H_{uniform}\rangle}{\sqrt{2\pi}}\left\{\frac{Y}{\sigma_y}\frac{1}{H_{threshold}}\right\}\right]}$$

$$(28)$$

Table 1 Values of Physical Parameters for Single-Beam Scans Modeled by Exposure in Excess of Threshold

T_{vap}, vaporization temperature of visible dye		360°C		
$T_{ambient}$, temperature of environment surrounding donor		20°C		
R, reflectance from donor interfaces preceding and upon light's entry into absorptive dye layer		6%		
ρ, mass density of dye layer and support		1380 kg/m³ equivalent to 1.38 g/cm³		
c_ρ, specific heat of dye layer and support		1885 J/kg°C equivalent to 0.45 cal/g°C		
ε_{vis}, transmissive extinction coefficient of visible dye		39,000 D_T/[cm (mol/L)]		
ΔH_{vap}, latent heat of vaporization of visible dye		115,000 J/mol equivalent to 27.5 kcal/mol		
η_{vap}, fraction of absorbed light energy devoted to vaporizing visible dye		3%		
σ_y, Gaussian standard-deviation radius of laser exposure profile perpendicular to scanning direction		6.5 µm		
Y, raster spacing between neighboring scanlines perpendicular to scanning direction in completed image		10 µm		
ℓ, dye layer thickness	0.3 µm "Thin donor"			0.8 µm "Thick donor"
A_{IR}, fraction of incident infrared laser light absorbed by dye layer	70%			70%
V_x, scanning velocity of laser beam with respect to the donor surface	11.8 m/s	16.9 m/s	23.7 m/s	10.5 m/s
$\eta_{confine}$, fraction of heat remaining in plane of dye layer normal to scanning direction containing the hottest temperature	55%	60%	64%	75%

containing two recurring clusters of terms:

$$\left\{\frac{(1-R)\eta_{vap}\eta_{confine}\varepsilon_{vis}}{\Delta H_{vap}}\right\} \quad \text{and} \quad \left\{\frac{\sigma_y}{Y}H_{threshold}\right\}$$

Only $\eta_{confine}$ occurs in both clusters, explicitly in the first cluster but implicitly in $H_{threshold}$ according to Eq. (22). Except for $\eta_{confine}$, inaccuracy in one of the parameters of a cluster could be offset by compensating changes limited to other parameters of the same cluster without affecting the curve shape. For instance, if the value chosen for ΔH_{vap} is lower than the visible dye exhibits during laser-thermal transfer, an equally good fit to data by Eqs. (27) or (28) could be obtained with compensatingly lower values for either η_{vap} or ε_{vis} than actually apply under these conditions. Individual terms within each cluster are highly correlated to the other terms in that cluster; varying two highly correlated parameters

Multichannel Laser Thermal Printhead Technology

simultaneously can cause failure of data fitting by "least squares," that is, minimization of the sum of the squares of the deviation of the fitting function from data points.

2.5 Model of Exposure in Excess of Threshold Compared with Experimental Data

The Williams–Clapper transform[20,21] is used to convert the transmissive density vs. exposure, that is, "D_T vs. H" curve predicted by Eq. (28), into reflective density anticipated to be observed by a 0–45° reflectometer measuring the clear receiver containing visible dye and laminated to a paper backing, plotted in Fig. 5. The Williams–Clapper transform accounts for two significant effects of reflectance:

1. The 2.13× longer optical path of light incident through the receiver thickness perpendicular to the paper, reflected back at 45° to the reflectometer's photodetector, and passing a second time through that receiver layer containing image dye; and
2. The multiple reflections of light striking the air interface at 28° from inside the receiver layer and returning toward the paper.

The contribution of multiple reflections to the light reaching the reflectometer's photodetector is significant only for low optical densities because a small amount of transferred dye in the receiver attenuates the light traversing the long effective pathlength of multiple reflections.

Figure 5 demonstrates agreement between the individual symbols representing experimental data and the D_R vs. H curves predicted by the Williams–Clapper transformed model of exposure in excess of threshold for two different thicknesses of dye layer,

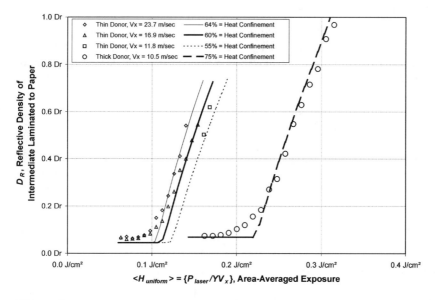

Figure 5 Characteristic curve for laser-thermal-dye transfer exposed by a single beam $\{\sigma = 6.5\ \mu m\}$ on thin and thick donors with intermediate laminated to paper observed in [0–45°] reflectance (individual symbols) compared to curves computed for model of exposure in excess of threshold for $\{T_{vap} = 360°C\}$.

a "thin donor" with dye layer only $\{\ell = 0.3\,\mu m\}$ thick, and a "thick donor" with $\{\ell = 0.8\,\mu m\}$ thick dye layer, both on 100 μm (4 mil) thick polyethylene terephthalate support and both dye layers absorbing the same fraction $\{A_{IR} = 70\%\}$ of incident infrared laser light. Intermediate amounts of image density can be produced by controlling the laser exposure deposited on the donor. Less exposure is required by the thinner dye layer than by the thick dye layer to attain the same image density because the thin donor contains less material to be heated in its dye layer.

Only a small fraction $\{\eta_{vap} = 3\%\}$ of the energy beyond that needed to heat the visible dye to its vaporization point seems to be devoted to transferring the visible dye from the donor to the receiver. The rest of that energy presumably vaporizes volatile molecules, further heats nonvolatiles, decomposes some constituents, and mechanically deforms the donor. The accuracy of η_{vap} estimated from the fit of experimental data with the model of exposure in excess of threshold is influenced by compensating inaccuracies in the values used for the visible dye's latent heat of vaporization ΔH_{vap} and extinction coefficient ε_{vis}, the donor reflectance R preceding light's establishment in the dye layer, and the fraction $\eta_{confine}$ of heat retained in the dye layer.

Odai et al.[22] have inferred the density profile for a single scanline by fitting a uniform field's D vs. H curve with a polynomial, then numerically converting the exposures deposited by a single beam's Gaussian profile into their corresponding densities, with provision for broadening the beam's standard-deviation radius by a proportionality constant to account for thermal diffusion. This numerical inversion relies on the unstated assumption of additive density of successive scanlines without providing a relation between the physical properties of the dyes and the consequent image density.

2.6 Dynamic Temperature Profile Produced by Absorption of a Scanning Gaussian Beam

The heat source $g[x, y, z, t]$ produced by absorption of the scanning laser and by subsequent thermal diffusive cooling generates a temperature distribution $T[x, y, z, t]$ at time t and position $[x, y, z]$ in the donor layer[23] that is the solution of the partial differential equation for heat flow[24–26]

$$\nabla^2 T[x, y, z, t] + \frac{1}{k} g[x, y, z, t] = \frac{\rho c_\rho}{k} \frac{\delta}{\delta t} T[x, y, z, t] \qquad (29)$$

in a uniform, homogeneous, isotropic semi-infinite donor material with constant, temperature-independent specific heat c_ρ, thermal conductivity k, and mass density ρ, for the boundary condition of negligible heat flow across the external surface of the donor:

$$\left.\frac{\delta}{\delta z} T[x, y, z, t]\right|_{z=0} = 0 \qquad (30)$$

and the initial condition of the donor material at room temperature:

$$T[x, y, z, t = 0] = T_{ambient} \qquad (31)$$

The general solution in three spatial dimensions and time for this partial differential equation in an infinite homogeneous medium with constant thermal conductivity, specific heat, and mass density is:[25,27]

$$T[x, y, z, t] = T_{\text{ambient}} + \frac{1}{\sqrt{[4\pi\alpha t]^3}} \int_{x'=-\infty}^{x'=\infty} \int_{y'=-\infty}^{y'=\infty} \int_{z'=-\infty}^{z'=\infty} \\ \times \left\{ \begin{array}{l} (T[x', y', z', t=0] - T_{\text{ambient}}) \\ \times \exp\left[-\frac{(x-x')^2 + (y-y')^2 + (z-z')^2}{4\alpha t}\right] \end{array} \right\} dx'dy'dz' \\ + \frac{1}{\rho c_p} \int_{t'=-\infty}^{t'=t} \frac{1}{\sqrt{[4\pi\alpha(t-t')]^3}} \int_{x'=-\infty}^{x'=\infty} \int_{y'=-\infty}^{y'=\infty} \int_{z'=-\infty}^{z'=\infty} \\ \times \left\{ \begin{array}{l} (g[x', y', z', t']) \\ \times \exp\left[-\frac{(x-x')^2 + (y-y')^2 + (z-z')^2}{4\alpha(t-t')}\right] \end{array} \right\} dx'dy'dz'dt' \tag{32}$$

in which the variable α is the thermal diffusivity of the dye-layer material:

$$\alpha = \frac{k}{\rho c_p} \tag{33}$$

This solution can be extended to a semi-infinite medium by adding an image source located so that the thermally insulating boundary is the plane of mirror symmetry.[28] This symmetric source distribution can be attained by adding to the right side of Eq. (32) a term identical to the right side of Eq. (32) but with every appearance of z replaced by $\{-z\}$. Symmetry of heating by the real source and its identical image source ensures that no net heat flows across the plane at the location of the insulating boundary to satisfy Eq. (30).

Uniform infrared dye concentration C_{IR} exponentially attenuates the laser irradiance of Eq. (1) penetrating the dye layer along the z-direction of beam propagation by the Beer–Lambert Law with the extinction coefficient ε_{IR}:

$$\vec{E}_{\text{laser}}[x, y, z, t] = \vec{E}_{\text{laser}}[x, y, z=\ell, t](1-R)10^{-\varepsilon_{\text{IR}} C_{\text{IR}}(\ell-z)} \quad \text{for } 0 < z < \ell \tag{34}$$

The beam's energy loss is the dye layer's heat gain. The material's absorption converts the irradiance E_{laser} lost per unit distance traversed through the dye layer into a heat source $g[x, y, z, t]$ in units of (power per unit volume):

$$g[x, y, z, t] = \frac{\delta}{\delta z} E_{\text{laser}}[x, y, z, t] \\ = \left\{ \frac{\varepsilon_{\text{IR}} C_{\text{IR}} \ln[10] P_{\text{laser}}(1-R)}{2\pi\sigma_x \sigma_y} 10^{-\varepsilon_{\text{IR}} C_{\text{IR}}(\ell-z)} \\ \times \exp\left[-\frac{1}{2}\left(\frac{x-V_x t}{\sigma_x}\right)^2\right] \exp\left[-\frac{1}{2}\left(\frac{y}{\sigma_y}\right)^2\right] \right\} \\ \text{for } 0 < z < \ell; \quad t_{\text{on}} < t < t_{\text{off}} \tag{35}$$

A Gaussian beam is a felicitous choice of light distribution for supplying heat to a thermally diffusive medium because the natural solution to the heat-flow equation is a Gaussian temperature profile. The Gaussians corresponding to the heat source and to the diffusion can be combined and integrated directly by completing the square in the exponential argument of Eq. (32) to obtain the dynamic temperature profile in the absorptive donor:[29–31]

$$T_{\text{scanned donor}}[x, y, z, t] = T_{\text{ambient}} + \frac{P_{\text{laser}}(1-R)\varepsilon_{\text{IR}} C_{\text{IR}} \ln[10]}{4\pi\rho c_p}$$

$$\times \int_{t'=t_{\text{on}}}^{t'=t_{\text{off}}} \left\{ \frac{\exp\left[-\frac{1}{2}\frac{(x-V_x t')^2}{2\alpha(t-t')+\sigma_x^2}\right]}{\sqrt{2\alpha(t-t')+\sigma_x^2}} \frac{\exp\left[-\frac{1}{2}\frac{y^2}{2\alpha(t-t')+\sigma_y^2}\right]}{\sqrt{2\alpha(t-t')+\sigma_y^2}} \right.$$

$$\times \left[10^{-\varepsilon_{\text{IR}} C_{\text{IR}}\{\ell-z-\ln[10]\varepsilon_{\text{IR}} C_{\text{IR}} \alpha(t-t')\}} \right.$$

$$\times \left(\text{ERF}\left[\frac{z+2\ln[10]\varepsilon_{\text{IR}} C_{\text{IR}} \alpha(t-t')}{\sqrt{4\alpha(t-t')}}\right] \right.$$

$$\left. -\text{ERF}\left[\frac{z-\ell+2\ln[10]\varepsilon_{\text{IR}} C_{\text{IR}} \alpha(t-t')}{\sqrt{4\alpha(t-t')}}\right] \right)$$

$$+ 10^{-\varepsilon_{\text{IR}} C_{\text{IR}}\{\ell+z-\ln[10]\varepsilon_{\text{IR}} C_{\text{IR}} \alpha(t-t')\}}$$

$$\times \left(\text{ERF}\left[\frac{z+\ell-2\ln[10]\varepsilon_{\text{IR}} C_{\text{IR}} \alpha(t-t')}{\sqrt{4\alpha(t-t')}}\right] \right.$$

$$\left. \left. -\text{ERF}\left[\frac{z-2\ln[10]\varepsilon_{\text{IR}} C_{\text{IR}} \alpha(t-t')}{\sqrt{4\alpha(t-t')}}\right] \right) \right] \right\} dt' \quad (36)$$

2.7 Hottest Location in Donor Lags Instantaneous Beam Center

An example of the temperature profile on the outer surface of the laser-scanned donor, computed by numerical integration of Eq. (36) over the laser-exposure time, is plotted in the upper panel of Fig. 6 for a Gaussian beam scanning from left to right. The hottest location in the donor inevitably trails the instantaneous beam center because the location in the donor at the beam center has only received light from the leading half of the Gaussian beam. The trailing half of the beam subsequently attempts to make that location twice as hot; but some of the heat deposited by the leading half of the beam has diffused away by then, reducing the maximum temperature attained. For the scanning conditions and materials of the D_R vs. H curves plotted in Fig. 5, the lag is about $1.5\sigma_x$ for the $\{\ell = 0.3\ \mu\text{m}\}$ thin dye layer and about $2.3\sigma_x$ for the $\{\ell = 0.8\ \mu\text{m}\}$ donor. Faster scanning speed and lower thermal diffusion coefficient increase the lag of the hottest location in the donor behind the instantaneous beam center.

2.8 Temperature Uniformity Throughout Dye-Layer Thickness and η_{confine}

At slow to moderate scanning speeds, the hottest temperature occurs at the exposed outer interface of the dye layer as a result of the insulating character of that interface and in spite of the initial deposition of more heat at the internal interface of the dye layer where it

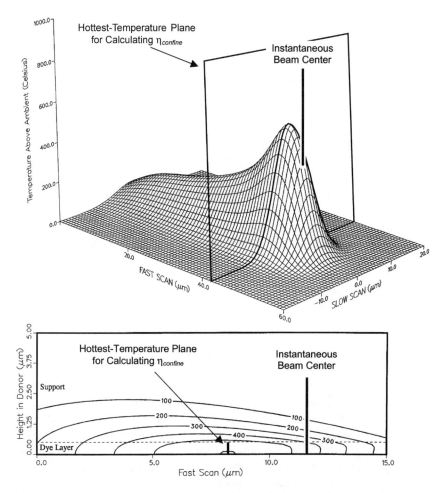

Figure 6 Moving hot spot and thermal tail produced in the donor by a scanning laser beam.

contacts the cooler support. The model of exposure in excess of threshold assumes uniform temperature profile throughout the thickness ℓ of the dye layer in the plane normal to the scanning direction, that is, in the $x-z$ plane of the lower panel of Fig. 6, containing the hottest location anywhere in the donor. Calculations with Eq. (36) disclose that the dynamic temperature distribution varies by less than 18% throughout the thickness ℓ of the dye layer for any of the scanning conditions and material properties of the D_R vs. H curves in Fig. 5, with most of the descent to room temperature occurring within the adjacent 5 μm of the polymeric support.

The values appearing in Table 1 for η_{confine} are calculated as the ratio of {the temperature increment above ambient computed with thermal diffusion by Eq. (36)} to {the adiabatic temperature rise, that is, the temperature anticipated in the absence of thermal diffusion}, averaged over the plane Y wide normal to the scanning direction from the top to the bottom of the ℓ thick dye layer. This plane of hottest temperatures, along each strip parallel to the scanning direction in Fig. 6 at a single instant, represents the greatest concentration of energy ever available for transferring dye in the presence of thermal diffusion. Visible dye will either transfer at this instant of attainment of the highest temperature throughout the dye-layer thickness ℓ, or will have transferred earlier, but an insignificant amount of visible dye will transfer later as this location cools.

2.9 Reciprocity Failure

The leftward 9% progression in exposure of the three D_R vs. H curves for thin donor in Fig. 5 for each $\sqrt{2}$ increase of the scanning velocity is an illustration of "reciprocity failure:" less exposure is required at faster scanning speed to transfer the same image density. The term "reciprocity" is adopted from silver halide imaging,[32] meaning that the image density remains constant if the power of a stationary light pulse deposited on the photographic film is reduced as the reciprocal of lengthening that pulse's duration. If reciprocity were obeyed, the image density would depend only upon exposure, the product of power and time. The temperature dependence of laser-thermal imaging renders its reciprocity failure easy to understand: a specific quantity of heat can elevate a specific volume of material to the hottest temperature by depositing that heat as quickly as possible, minimizing the opportunity for diffusion to carry heat away. Two ways to increase the rate of heat deposition by scanning a spot are to increase the scanning velocity V_x or to decrease the beam size σ_x along the scanning direction while concomitantly increasing the laser power P_{laser} proportionately in order to maintain constant exposure H. Note that increasing the scanning velocity V_x in Eq. (27) for the model of exposure in excess of threshold does not directly change the predicted image density because this model implicitly assumes reciprocity, subsuming V_x into $\langle H_{\text{uniform}} \rangle$ in Eq. (28). Reciprocity enters into the model of exposure in excess of threshold only through the term η_{confine}. The values of η_{confine} for these three scanning speeds with the $\{\ell = 0.3\,\mu\text{m}\}$ thin donor in Table 1 exhibit the identical 9% progression as the D_R vs. H curves, confirming the prediction by Eq. (36) of reciprocity failure in laser-thermal imaging. Thick donor exhibits a smaller reciprocity failure of about 3% for each $\sqrt{2}$ increase of the scanning velocity, similarly agreeing with its progression of η_{confine} computed for scanning speeds near $\{V_x = 10\,\text{m/s}\}$ across an $\{\ell = 0.8\,\mu\text{m}\}$ thick dye layer.

Odai et al. [33] detected increasing transferred density for shorter pulse widths while increasing the laser power to maintain constant exposure. Dlott and colleagues[34,35] demonstrated a factor of 20 reduction in the threshold exposure necessary to transfer the full thickness of a colorant layer by varying the duration of a stationary pulse over seven orders of magnitude, apparently approaching the lower adiabatic[36,37] limit of threshold exposure with pulses shorter than 1 picosecond.

2.10 Warm-Up Transient of the Donor

For the same reason that the hottest location in the donor lags the beam center, the hottest temperature ever attained in the donor occurs sometime after the activation of the beam at t_{on}. The location of the beam center at initiation can never be the hottest location in the donor for a constant scanning velocity because that point will only receive half of the possible exposure. Points farther along the center of the scanline will receive greater fractions of the potential exposure and reach hotter temperatures until a balance called the "quasi-stationary state"[24,38] is attained between deposition of energy in newly encountered material and loss of heat by diffusion. Subsequently maintaining constant laser power and scanning speed causes the temperature profile to propagate across the material while retaining its local shape and peak height. For the intermediate scanning velocity with the thin donor in Fig. 5, the beam traverses a distance slightly greater than the waist diameter of its irradiance profile after activation in Fig. 7 to reach that quasi-stationary state as calculated by Eq. (36). The quasi-stationary state is reached in a shorter distance with slower scanning or in a material with a larger thermal diffusion coefficient, but that distance is not affected by the laser power within the constraint of constant parameters for the heat-flow differential equation, Eq. (29). Less image density is anticipated to be produced in the first pixel after initiating the laser than in subsequent pixels. Alternatively stated, the edge of density production is offset by some fraction of the waist diameter of the beam beyond the location of beam center at the instant of its activation. This warm-up transient of the donor is a separate phenomenon from any laser start-up transients.

2.11 Depth of Focus

The necessity of heating the visible dye to high temperature in order to induce transfer requires that the laser beam be tightly focused. Broadening of the beam caused by movement of the dye layer away from the plane of best focus reduces the transferred image density. The series of irradiance profiles in increments of 12 μm, relative to the plane of best focus, illustrates the beam spread in the leftmost column in Fig. 3. The center column of Fig. 3 portrays the corresponding exposure profiles expected to be deposited by this beam at these distances from the plane of best focus. Each exposure profile is computed by summing the irradiance profile along strips parallel to the scanning direction. Note that these exposure profiles in the center column are not graphs of the leftmost column's irradiance profile along one x-distance. The beam spread plotted as the solid curves in the rightmost column is much greater than the 0.5% that a single-mode laser

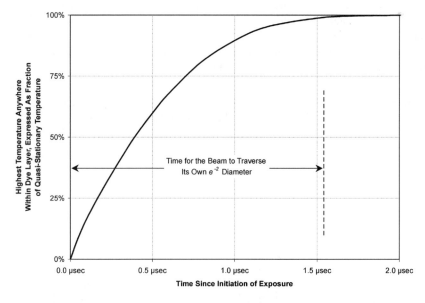

Figure 7 Warm-up transient for $\{\sigma_x = 6.5\ \mu m\}$ beam scanning at $\{V_x = 16.87\ m/s\}$ across $\{\ell = 0.3\ \mu m\}$ thin donor absorbing 70% of the infrared laser light requires approximately the amount of time for the beam to scan across its own waist diameter.

would exhibit over the 50 μm range from best focus, according to Eq. (18) and plotted as the dashed curves.

Each of these exposure profiles in the center column of Fig. 3 is fitted by a Gaussian standard-deviation radius $\sigma_y[z]$ reported in Table 2, with Eq. (8) as the fitting function. Each value of $\sigma_y[z]$ is submitted to the model of exposure in excess of threshold including η_{confine} calculated by Eq. (36) on $\{\ell = 0.3\,\mu\text{m}\}$ thin donor and Williams–Clapper[20] transformed to predict the reflective image density plotted as one of the junctures between solid line segments in Fig. 8. These predicted densities, based upon beam size, are compared with image density measured for uniform fields written with a single diode-laser beam scanned at $\{V_x = 23.7\,\text{m/s}\}$ across thin donor displaced at a sequence of 10 μm increments from the plane of best focus. The 129 mJ/cm^2 exposure used in the model is slightly lower than the 135 mJ/cm^2 exposure used to obtain the experimental densities in order to match the peak density, facilitating visual comparison of the distances for fall-off of the measured image density and for the model's predictions.

Instead of using the size of the aerial irradiance distribution to define the depth of focus,[39] the distance from best focus causing the image density to drop a specific amount is used.[40] This choice of criterion reflects the practice that the plane of best focus is the location producing highest uniform density, not necessarily the plane exhibiting the narrowest waist. Fitting the peak region of the {density vs. distance from best focus} curve to a parabola indicates that limiting density variation to less than $0.1 D_R$ requires the dye layer of the donor to be maintained within a 50 μm range about the plane of best focus throughout the image. This demand justifies the use of mechanically precise apparatus for translating donors of consistent thickness in laser-thermal printers producing images with continuously adjustable laser-thermal-transferred density.

The experimental conditions used to obtain the data in Fig. 8 and the model of exposure in excess of threshold pertain to transfer of partial thickness of the dye layer. Koulikov and Dlott[41,42] have published data and analysis for complete removal of the dye layer within a radius by a stationary pulse. Some recommendations for increasing apparent depth of focus deduced for complete removal of a colorant layer by a stationary pulse may not be suited to printers discussed in this chapter transferring only part of the colorant layer by scanning a continually activated beam to produce images.

Table 2 Beam Size and Efficiency of Heat Confinement for Predicting Density Dependence Upon Distance From Plane of Best Focus by the Model of Exposure in Excess of Threshold

Distance from plane of best focus z (μm)	Standard-deviation radius of exposure profile computed from irradiance profile $\sigma_y[z]$ (μm)	Calculated for $\sigma_y[z]$ with $\{\ell = 0.3\,\mu\text{m}\}$ and $\{V_x = 23.7\,\text{m/s}\}$ η_{confine} (%)
−50	8.2	61
−37	7.5	62
−25	7.0	63
−12	6.6	64
0, best focus	6.5	64
12	6.6	64
25	6.8	64
37	7.4	63
50	8.5	61

Multichannel Laser Thermal Printhead Technology

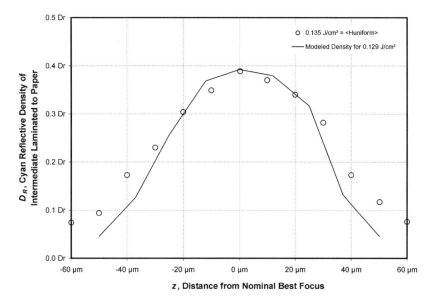

Figure 8 "Singles" depth of focus on $\{\ell = 0.3\,\mu\text{m}\}$ experimental thin donor compared with model of exposure in excess of threshold for $\{T_{\text{vap}} = 360°\text{C},\ V_x = 23.7\ \text{m/s}\}$ exposure profiles collected for one channel through focus.

3 EXPOSURE IN EXCESS OF THRESHOLD WITH MULTIPLE SOURCES

Thermal interactions can occur between the temperature profiles produced in the donor by simultaneous exposure with multiple sources. These interactions are separate from thermally mediated influence of one source upon another source's emission, as might occur if multiple sources are mounted upon the same substrate or heat exchanger.

3.1 Exposure Profile of a Swath by a Multichannel Printhead

Simultaneous printing by equally spaced lines from multiple sources deposits a nearly uniform exposure profile as depicted in Fig. 2. This exposure profile, including ripple at its lowest spatial frequency $\{1/Y\}$, is[12]:

$$H_{\text{multiple}}[y] \approx \frac{P_{\text{laser}}}{V_x Y}\left(1 + 2\exp\left[-\frac{1}{2}\left(\frac{2\pi\sigma_y}{Y}\right)^2\right]\cos\left[2\pi\frac{y}{Y}\right]\right) \tag{37}$$

For exposures near $H_{\text{threshold}}$, the part of the ripple exceeding the threshold determines $y_{\text{threshold,multiple}}$:

$$y_{\text{threshold,multiple}} = \frac{Y}{2\pi}\arccos\left[\frac{1}{2\exp[-(1/2)(2\pi\sigma_y/Y)^2]}\right.$$
$$\left.\times\left\{\frac{(V_x Y \ell \rho c_p (T_{\text{vap}} - T_{\text{ambient}}))}{(P_{\text{laser}}(1-R)\eta_{\text{confine}}A_{\text{IR}})} - 1\right\}\right] \tag{38}$$

Exposures well above threshold transfer dye throughout the raster width Y; and their ripple has no effect as a result of the averaging over the raster. When printing with multiple sources, Eq. (24) for the amount of visible dye transferred from one raster width of the donor becomes:

$$\begin{bmatrix} \text{Moles of visible} \\ \text{dye evaporated} \\ \text{per unit length} \\ \text{of one raster} \\ \text{scan by} \\ \text{multiple sources} \end{bmatrix}$$

$$= \begin{cases} \dfrac{\eta_{\text{vap}} \eta_{\text{confine}} (1-R)}{\Delta H_{\text{vap}}} \\ \quad \times \displaystyle\int_{y=-y_{\text{threshold}}}^{y=y_{\text{threshold}}} (H_{\text{multiple}}[y] - H_{\text{threshold}}) \, dy & \text{for threshold within ripple} \\[2ex] \dfrac{\eta_{\text{vap}} \eta_{\text{confine}} (1-R)}{\Delta H_{\text{vap}}} \\ \quad \times \displaystyle\int_{y=-(1/2)Y}^{y=(1/2)Y} \left[\dfrac{P_{\text{laser}}}{V_x Y} - H_{\text{threshold}} \right] dy & \text{for ripple entirely above threshold} \end{cases}$$

(39)

The analog of Eq. (28) for the transmissive density transferred by simultaneous printing with overlapping multiple sources is:

Transmissive Optical Density$_{\text{multiple}}$

$$= \begin{cases} \dfrac{\eta_{\text{confine}}(1-R)}{\pi} \dfrac{\varepsilon_{\text{vis}} \eta_{\text{vap}}}{\Delta H_{\text{vap}}} \\ \quad \times \left(\begin{array}{l} \sqrt{ \left\{ 2\langle H_{\text{uniform}}\rangle \exp\left[-\dfrac{1}{2}\left(\dfrac{2\pi\sigma_y}{Y}\right)^2\right] \right\}^2 - \{\langle H_{\text{uniform}}\rangle - H_{\text{threshold}}\}^2 } \\ + \{\langle H_{\text{uniform}}\rangle - H_{\text{threshold}}\} \\ \times \left\{ \pi - \arccos\left[\dfrac{\{\langle H_{\text{uniform}}\rangle - H_{\text{threshold}}\}}{2\langle H_{\text{uniform}}\rangle \exp\left[-\dfrac{1}{2}\left(\dfrac{2\pi\sigma_y}{Y}\right)^2\right]} \right] \right\} \end{array} \right) \\ \hfill \text{for threshold within ripple} \\[2ex] \eta_{\text{confine}}(1-R) \dfrac{\varepsilon_{\text{vis}} \eta_{\text{vap}}}{\Delta H_{\text{vap}}} (\langle H_{\text{uniform}}\rangle - H_{\text{threshold}}) \\ \hfill \text{for ripple entirely above threshold} \end{cases}$$

(40)

Multichannel Laser Thermal Printhead Technology

The lengthy first expression "for threshold within ripple" only applies to the small range of exposures near the threshold of dye transfer.

The brief second expression "for ripple entirely above threshold" applies to the preponderance of the density vs. exposure curve. This brief second expression embodies the intuitive linear dependence of image density upon the exposure in excess of threshold, upon the efficiency of heat confinement, and upon the extinction coefficient of the visible dye, that has guided these derivations of transferred density.

3.2 Tilted Printhead

The projected beams from the optical fibers in a multiple-source printer,[43] obeying the conditions of Table 1, are spaced $\{d = 60 \ \mu m\}$ between centers on the donor. One way to space the scanlines $\{Y = 10 \ \mu m\}$ apart, while maintaining the $\{\sigma_y = 6.5 \ \mu m\}$ standard-deviation radius of each writing spot, is to tilt the printhead[44–50] as sketched in Fig. 9.

Writing with multiple sources simultaneously enables some sources to exploit the skirts of their neighboring sources' exposure distributions, energy that would be squandered by writing with a single source. But the tilt of the printhead in Fig. 9 allows some of that energy to escape before the successive source exposes its location. Near the middle of a swath of a multiple-source printhead, $\{\sigma_y = 6.5 \ \mu m\}$ beams on a $\{Y = 10 \ \mu m\}$ raster deposit a nearly uniform exposure that is $\{(\sigma_y \sqrt{[2\pi]})/Y = 1.8 \times\}$ as high as the

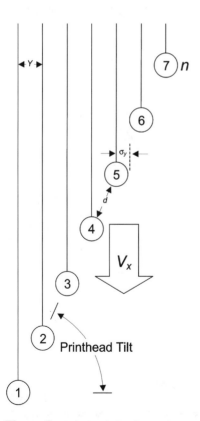

Figure 9 Printhead tilted to reduce Y, spacing between scanlines, while maintaining constant Gaussian-beam standard-deviation radius σ_y.

Gaussian peak deposited by a single source illustrated in Fig. 2. Computation of the dynamic temperature profile for a multiple-source printhead tilted at 80.4° and scanning at $\{V_x = 5.5 \text{ m/s}\}$ across an $\{\ell = 0.8 \, \mu\text{m}\}$ thick donor predicts $\{\eta_{\text{confine,multiple}} = 51\%\}$ for sources in the interior of the swath by averaging the temperature increment over the Y-wide plane across the scanline, spanning the dye-layer thickness and normal to the scanning direction at the hottest spot, then taking the ratio of this averaged dynamic temperature increment to the averaged adiabatic temperature increment for multiple sources. Note that this Y-wide plane of hottest temperatures is tugged about $\frac{1}{4}Y$ toward the preceding beam's track by the asymmetry of the preceding skirt across the track of the beam of interest, and that this hottest-temperature plane lags the center of the beam of interest by $1.6\sigma_x$ for these printing conditions.

A single source would retain $\{\eta_{\text{confine,single}} = 70\%\}$ of the heat within the $\{Y = 10 \, \mu\text{m}\}$ scanline centered on its beam at $\{V_x = 5.5 \text{ m/s}\}$ across the $\{\ell = 0.8 \, \mu\text{m}\}$ thick donor. $\eta_{\text{confine,multiple}}$ is lower than $\eta_{\text{confine,single}}$ because some of the leading neighbor's heat escapes before arrival of the beam of interest. However, the fraction of area-averaged exposure devoted by the multiple-source printhead to heating along this beam's scanline more than offsets the relative efficiency of heat confinement. The net advantage predicted by the model of exposure in excess of threshold is that the multiple-source printhead transfers the same density while depositing only 80% as much exposure as the single source requires, as shown by comparison of the solid and dotted curves of Fig. 10. The experimental data plotted as individual symbols in Fig. 10 seem to indicate an even larger advantage, requiring only about 63% as much exposure from the multiple sources to attain the same image density as for the single source. The closed symbols in Fig. 10 represent measurements of reflective density of uniform fields written with 18

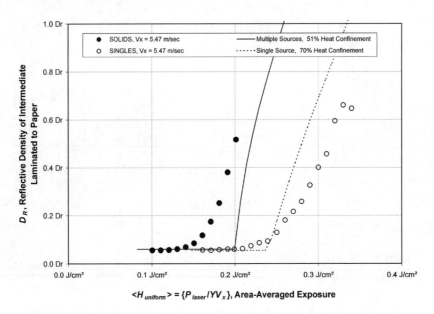

Figure 10 "Solids" require only 63% of "singles" exposure to attain the same density, as compared to 80% predicted by the model of exposure in excess of threshold for the $\{\ell = 0.8 \, \mu\text{m}\}$ thick donor.

beams simultaneously onto receiver laminated to paper, and the solid curve is the prediction of Eq. (40) for the 18-beam printhead tilted at 80.4°, using the parameter values for $\{\ell = 0.8\,\mu\text{m}\}$ thick donor in Table 1, except $\{V_x = 5.5\,\text{m/sec}\}$. The open symbols are measurements with a single beam, and the dashed curve is the prediction of Eq. (27) for a single beam.

3.3 Nearest-Neighbor Interaction

Table 3 helps to rationalize the advantage enjoyed by simultaneous writing with multiple sources in the rightmost column compared to a single source in the center column. About 70% of the heat deposited within the strip from $\{-\frac{1}{2}Y \text{ to } \frac{1}{2}Y\}$ centered on the scanline is available when the hottest part of the profile arrives and about 56% of the beam's energy falls within that strip for both writing schemes, therefore the multiple sources must be at least as effective as the single source for transferring dye. Heating of this strip by the preceding neighbor is the predominant advantage exploited by the multiple sources: 44% of that preceding neighbor's heat from a 21% portion of its exposure skirt remains in this strip when the hottest part of the beam arrives. This nearest-neighbor interaction is predicted to require only 80% of the single-beam exposure to generate the same rise in

Table 3 Exposure and Efficiency of Heat Confinement for a Beam in a Multichannel Printhead Compared to that Beam Used in a Single-Source Printhead

V_x		5.5 m/s
σ_y		6.5 μm
Y		10 μm
ℓ, dye layer thickness		0.8 μm, "thick donor"
A_{IR}		70%
	Single beam	Multiple beam printhead tilted 80.4°
Fraction of total energy deposited by a single scanline contained within strip Y wide centered on beam of interest	56% of energy deposited by one beam	56% of energy deposited by one beam
Efficiency of heat confinement for beam of interest	$\eta_{\text{confine,single}} = 70\%$	$\eta_{\text{confine,self,80°tilt}} = 70\%$
Fraction of total energy deposited by 1st neighboring scanline over strip Y wide centered on beam of interest		21% of energy deposited by one beam
Efficiency of heat confinement for heat from 1st neighboring scanline over strip Y wide centered on beam of interest		$\eta_{\text{confine,1st neighbor,80°tilt}} = 44\%$
Fraction of total energy deposited by 2nd neighboring scanline over strip Y wide centered on beam of interest		1% of energy deposited by one beam
Efficiency of heat confinement for heat from 2nd neighboring scanline over strip Y wide centered on beam of interest		$\eta_{\text{confine,2nd neighbor,80°tilt}} = 42\%$

dye-layer temperature. The second-neighbor interaction is predicted to be an insignificant 1% further reduction of requisite exposure because only 1% of that second neighbor's beam energy falls within this strip. The trailing beams are assumed to not contribute to dye transferred by the beam of interest for these experimental conditions of beam size, spacing, and thermal diffusion because the dynamically calculated temperature in the hottest plane, lagging $1.6\sigma_x$ behind the center of the beam of interest, is unchanged by activating or deactivating the trailing beams.

The experimental data's advantage of requiring only 63% area-averaged exposure for multiple-source printing to attain the same density as single-source printing cannot be matched by the model of exposure in excess of threshold, even in the case of perfect efficiency of heat confinement $\{\eta_{\text{confine}} = 1\}$. The 21% of the first nearest neighbor's beam energy and 1% of the second nearest neighbor's beam energy superimposed on 56% of the beam of interest would still require 72% of single-beam-printing area-averaged exposure to attain the same image density, more than the 63% found by the experimental measurements. Some reasons for disagreement of the model of exposure in excess of threshold for multiple-source printing with experimental data might be:

1. Warming of the dye layer by the leading neighbor's beam for about six times the traversal time of a beam across its own irradiance's waist diameter may soften the binder and facilitate the escape of visible dye from the interior of that layer, increasing η_{vap} or reducing T_{vap} for multiple-source, compared to single-source printing.
2. Temperature dependence of the volume heat capacity $\{\rho\, c_\rho\}$ of the donor may distort the thermal distribution from Eq. (36)'s analytical temperature profile based upon the assumption of constant heat capacity throughout the infinite half-space.
3. More light falls outside the writing spot and onto the neighboring scanline than accounted for by the Gaussian fitted to the exposure profile.
4. The trailing neighboring beam might contribute significantly to dye transferred from the beam of interest's scanline as a result of physical deformation of the dye layer or to temperature-dependence of its thermal properties.

3.4 Dummy Laser

The leading laser in the swath of the tilted printhead of Fig. 9 does not have the advantage of a thermal tail from any other beam. This leading laser would transfer less dye if operated at the same power as the other lasers in the printhead and would cause the artifact of a light line in the image. This swath edge artifact can be avoided by operating this leading laser as a "dummy laser"[48,50] at a power just below the threshold for transferring dye on its own. The second laser in the printhead enjoys nearly the same advantage of preheating by this preceding "dummy" beam as each subsequent beam does from its first leading neighbor.

3.5 Interleaving

Interleaving or interlacing[44,47,51–53] enables use of a printhead with writing spots spaced farther apart than the desired scanline spacing while avoiding the need to tilt the printhead. Interleaving can reduce interactions of nearest neighbors[54] while maintaining constant printhead scanning speed or step size in the slowscan direction. Figure 11 illustrates interleaving of the $\{N = 5\}$ writing spots for a printhead with writing spot spacing, d, three

Multichannel Laser Thermal Printhead Technology

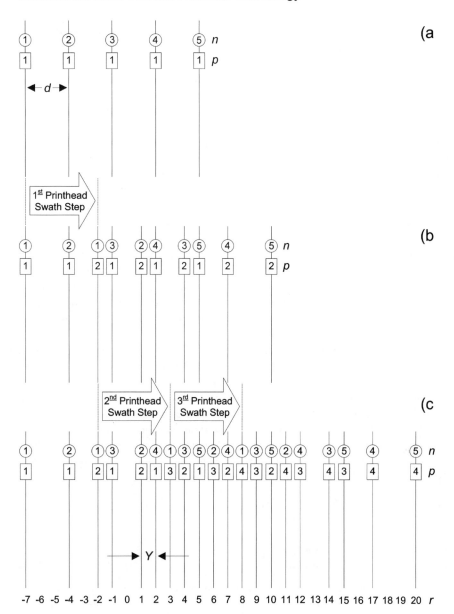

Figure 11 $\{p = 4\}$ swaths of $\{N = 5\}$ sources interleaved by $\{b = 3\}$ with printhead swath steps of $\{S = 5Y\}$.

times as large as the ultimate scanline spacing Y. On successive passes, scanlines are written in the gaps between scanlines from previous swaths. The necessity of producing three scanlines in the slowscan distance spanned by one source spacing d means that an interleaving factor b of three is to be used.

$$b = \frac{d}{Y} \qquad (41)$$

In order to avoid multiple exposure of the same scanline with different sources on different passes, the number of independent writing sources in the printhead, N, and the interleaving factor, b, must be relatively prime, that is, integers N and b must not have any common divisors other than the number 1. For any integer value of b, the printhead containing N sources spaced $\{d = bY\}$ apart moves the same distance S between writing successive swaths as an N-source printhead with the desired source spacing Y would, namely $\{S = NY\}$.

Each numeral in a circle in Fig. 11 represents the unique index n of the source in the N-source printhead responsible for exposing that scanline in the image. The numeral in the square indicates the printhead pass, p, during which that scanline is exposed. The row of numerals across the bottom of Fig. 11 indicates the raster line, r, in the completed image represented by that scanline.

The top panel labeled "a" in Fig. 11 depicts the first swath for an $\{N = 5\}$ printhead interleaved $\{b = 3\}$. The printhead simply writes scanlines spaced $\{d = 3Y\}$ apart, three times farther apart than desired in the final image. The middle panel labeled "b" shows that the printhead has stepped $\{S = NY = 5Y\}$ along the slowscan direction and exposed a second swath of scanlines $\{d = 3Y\}$ apart. But the scanline written by the first-indexed source $\{n = 1\}$ in the printhead on this second swath $\{p = 2\}$ is located two-thirds of the way from the scanline written by the second-indexed source $\{n = 2, p = 1\}$ and that by the third-indexed source $\{n = 3, p = 1\}$ on the first pass.

The bottom panel labeled "c" depicts the raster pattern after the fourth swath consequent to the second and third printhead steps. The second printhead step of $\{S = 5Y\}$ places the scanline written by the first-indexed source on this third swath $\{n = 1, p = 3\}$ midway between the scanline written by the fourth-indexed source on the first swath $\{n = 4, p = 1\}$ and that by the third-indexed source on the second swath $\{n = 3, p = 2\}$. This third swath completes a stretch of seven contiguous scanlines at the desired raster spacing, Y, extending from scanline $\{n = 2, p = 2\}$ constituting the first useful raster line $\{r = 1\}$ to scanline $\{n = 4, p = 2\}$ as raster line $\{r = 7\}$. The "filling order" is established by the $\{p = b = 3\}$ swath:

i. source $\{n = 2\}$ on pass $\{p + 1\}$
ii. source $\{n = 4\}$ on pass $\{p\}$
iii. source $\{n = 1\}$ on pass $\{p + 2\}$
iv. source $\{n = 3\}$ on pass $\{p + 1\}$
v. source $\{n = 5\}$ on pass $\{p\}$

This filling order is unique to the combination of the number of sources in the printhead, N, and the interleaving factor, b. Electronic circuits have been devised for routing the data to the correct laser-driver current supply.[55] The rightmost source $\{n = N\}$ in any N-source printhead always writes the $\{r = N\}$ scanline in the completed raster on the first printhead pass $\{p = 1\}$, so that $\{n = N\}$ is always last in the filling order. Swath b is the first opportunity to produce N or more contiguous lines in the completed raster.

Changing the values of either N or b requires determining the filling order anew, possibly by the same technique used to construct Fig. 11.

The principal disadvantages of interleaving are: scanlines are not written in their ordinal sequence; part of the leading edge and trailing edge of the scanned area are never filled in, requiring at least $(b - 1)$ extra scans to complete an image; and production of a desired density requires as much area-averaged exposure as single-source printing because the nearest-neighbor interaction is insignificant as a result of the greater spacing of scanlines exposed during a single swath. This last point of insignificant nearest neighbor interaction can be an advantage for interleaving because nearest-neighbor artifacts can be avoided,[54] such as difficulty balancing a printhead or accentuation of temperature profile shifts by varying scanline spacing. Because interleaving avoids the nearest-neighbor interaction, no lasers need to be underutilized as "dummy" lasers.

3.6 Limitation Imposed by Reciprocity Failure on Applicability of MTF

Modulation transfer function (MTF) analysis is defined in the exposure space of the subsystems producing any image, not in the final density space viewed by an observer. Because MTF is defined in the exposure space, the MTF of each subsystem that linearly affects the final exposure profile can be multiplied by the MTFs of all other linear subsystems in a laser printer.[56] This multiplicative property of linear subsystem MTFs is called "cascading." Additive exposure of silver halide imaging obeying reciprocity allows a unique exposure profile to be inferred from the density map of the image. This ability to transform a density profile into its causative exposure profile permits full generality of cascading MTFs to be applied to an imaging system obeying reciprocity, such as silver halide imaging.

Laser thermal imaging violates reciprocity partially during each swath because some of the heat escapes during exposure of each writing spot and during the delay between successive superimposed writing spots. Reciprocity is violated completely between swaths because the cooldown of the donor to room temperature means that repeated exposures just below threshold would not produce any density. The MTF may be applicable to image modulation in the fastscan direction because the deposited-light-energy effects are cumulative. Applicability of MTF along the slowscan direction for laser-thermal-dye transfer with significant reciprocity failure might be limited to exposure profiles that are only linear perturbations from the profile and scanning conditions for which the MTF was measured. The MTF measured from a test pattern of small-amplitude sinewaves written by a single source might not correctly predict the spatial frequency characteristics in an image of large-amplitude squarewaves written by a multiple-source, laser-thermal printer using the same sized writing spot. The MTF of an additive-density mechanism might cascade with other MTFs as a result of fortuitous compensation of higher-order harmonics engendered by nonlinear response to the original exposure, as demonstrated for a second-generation silver halide image printed from original silver halide transparency by Lamberts.[16]

4 FIBER ARRAY PRINTHEAD

Kodak Approval Digital Color Proofing Systems have been successful applications of digital printing since their introduction in 1991. These systems use laser thermal technology to image cyan, magenta, yellow, and black dyes onto an intermediate sheet that the customer then laminates to his paper stock. The high writing resolution of laser thermal technology does an excellent job simulating the halftone printing process. The resultant print is a standard in the printing industry.

4.1 Tonescale by Halftone Dots

Unlike a continuous tone image such as a silver halide print, a halftone image is written at a constant density level. Tonescale is obtained by grouping many pixels together using a center-weighted halftone. The first step in the process uses a raster image processor to convert each color in the artwork into a high-resolution binary bitmap. Figure 12 shows a typical printed black and white image composed of halftone dots. Changing the size of the halftone dots produces the tonescale. This process may be replicated on press by imprinting more or less ink to create the tonescale. Color images are printed on press using cyan, magenta, yellow, and black inks. A separate printing plate of halftone dots is exposed for each ink to create the tonescale for its corresponding color.

To simulate the printing process, a high-resolution bitmap of each of the color planes is generated. Figure 13 shows the raster-image processing step of sampling a cyan plane into a high-resolution bitmap. Next, each color bitmap is imaged with the corresponding color donor onto the intermediate at a constant exposure. Finally, the intermediate image is laminated to the customer's paper stock, resulting in the color halftone proof.

The contrast sensitivity of the human eye, published by Van Nes and Bouman,[57] peaks at about 3 to 6 cycles per degree. At a normal viewing distance of 300 mm, the transverse distance subtended by 1 degree of view is 5.2 mm, so that 3 to 6 cycles/degree corresponds to an optimum spatial frequency of 0.58–1.15 cycles/mm. If the viewing distance is variable, this frequency range needs to be expanded. If the viewer is likely to use magnification to view the image, the critical frequencies should be increased. In practice, a trained viewer may be able to detect up to 4 cycles/mm, while few untrained viewers are able to detect 2 cycles/mm. Other factors, including the smoothness of the paper and the incident illumination, significantly impact the visibility of artifacts at these spatial frequencies.

Figure 12 Example of a halftone image.

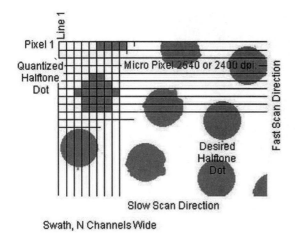

Figure 13 Rasterizing the cyan halftone plane into a high-resolution bitmap.

At the peak spatial frequency, the contrast sensitivity is 500:1 or 0.2% luminance variation of the image. This is an extremely small tolerance to hold and is difficult to meet in most digital imaging systems. For printing systems on rough paper, a more practical limit is just below 2%. For transparencies with bright illumination, the 0.2% limit is required. Firth et al. [58] describe a continuous-tone laser printer designed to meet the 0.2% limit writing onto silver halide photographic media.

4.2 Optical Fiber Array for Conveying Light from Lasers

Figure 14 shows a sketch of the optical path of a fiber V-groove printing system. In this system, 50 μm core step index fibers of 0.22 numerical aperture (NA) are glued into V-grooves[59,60] etched in silicon[61,62] at 130 μm spacing between centers. A printing lens

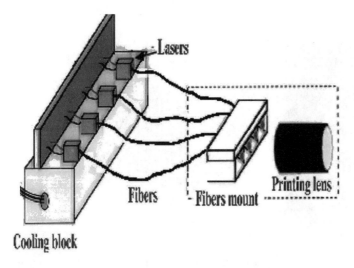

Figure 14 Fiber V-groove optical path.

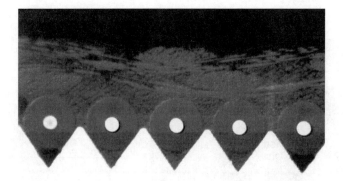

Figure 15 Portion of a V-groove fiber array.

with a demagnification of 2.2 : 1 and an input numerical aperture of 0.25 NA images the array of fibers onto a drum holding the donor over the intermediate. The polished surface of the fiber-optic V-groove array is shown in Fig. 15. Each optical fiber is coupled to a multimode diode laser. Each diode laser emits up to 400 mW at 0.12 NA from the fiber. The printing lens transmits 80% of this energy into a writing spot with {2.2 × 0.12 = 0.264 NA}. This printing lens passes energy up to and including an output numerical aperture of 0.55 NA. Scattering of energy in the fiber-by-fiber irregularities, crimps, or sharp bends in the path converts the light into higher-order modes having higher numerical aperture. The lower numerical aperture modes have a larger depth of focus, so energy coupled into higher-order modes increases the sensitivity to focus position. Energy outside the lens input numerical aperture is clipped by the lens. Fiber movement may cause the energy output from the lens to vary, resulting in an increase in optical power noise.

The left and right halves of Fig. 16 show two- and three-dimensional views, respectively, of a multimode laser beam generated from the 50 μm fiber core imaged through the 2.2 : 1 printing lens, captured using a beam analyzer model LBA-100A (Spiricon, Inc., Logan, UT).

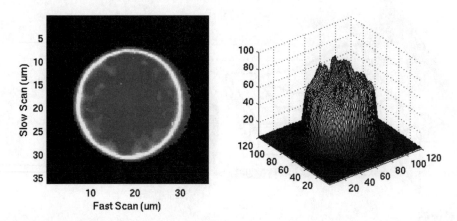

Figure 16 Two- and three-dimensional Spiricon images of an individual-spot irradiance after the 2.2 : 1 lens.

Multichannel Laser Thermal Printhead Technology

Care is required coupling the optical fiber to the diode laser. Optical reflections returning to the diode laser cause mode hopping, changing the output power and wavelength.[15,63] Multiple diode-laser channels reduce the sensitivity to individual laser noise as each channel may be considered an uncorrelated random noise source. Multiple diode-laser elements, constituting each laser, further reduce the individual channel noise. Images created using a single laser channel exhibit artifacts as a result of laser start-up transients caused by changes in diode laser temperature, optical power, wavelength, or mechanical position. Additional channels break up the appearance of these artifacts. Modulation of the laser with image data is another method of breaking up the artifacts in the prints, making them less noticeable.

4.3 Control of the Diode Laser

The diode-laser output power and wavelength is a function of operating temperature. The operating temperature increases as a function of "on" time. The cooling circuit attempts to maintain a constant average temperature. However, temperature rises of 5–9°C (9–16°F) between the "off" and "on" state are normal because of the time constants involved between the diode laser and the detection and cooling subsystem. This is most noticeable as streaks or tears in images with irregularly shaped solid areas.

Diligence is required when designing a stable current source to drive and switch the diode laser. Diode-laser optical power is directly proportional to current exceeding the threshold current. We have used two methods to drive diode-laser current. The first method uses a 12-bit digital-to-analog converter (DAC) driving a high-speed current amplifier to deliver current to each channel. The second method uses a constant-current source that is switched between the diode laser and a dummy load using a high-speed, field-effect transistor (FET). The latter design[64] is able to maintain current within 0.1%, while achieving an optical rise time of 45 ns.

4.4 Tracks in Donor Produced by Fiber-Coupled Laser Channels

Figure 17 shows a thick donor, similar to Kodak Approval Digital Color Proofing Film DC02/Cyan Donor, exposed with individual diode-laser channels consisting of 50 μm core fiber spaced 130 μm apart, 2.2 : 1 lens demagnification, with the V-groove array tilted to achieve a 2540 dpi writing resolution. Each channel sweeps across the donor from top to bottom of the picture. Every fourth channel is activated for only the time to scan 50 μm, then deactivated for the time to scan 250 μm. After scanning 25 μm with all channels deactivated, every fourth neighboring channel is activated. This process is repeated through all four sets of every fourth laser, resulting in an exposure track showing the structure within the spot appearing as an oval with its long axis running down the page along the scanning direction. Notice that at 2540 dpi writing resolution, corresponding to 10 μm between channel centers projected on the donor, the nominally 25 μm round laser spot has cleaned out a trace approximately 20 μm wide. This donor was overexposed to clearly show the outline of each laser beam. Normally, the exposure is set so that the line width is approximately 14 μm. For a thermal medium, the line width is exposure-dependent because the width of the energy deposited in excess of the thermal threshold increases with the exposure as predicted by Eq. (23). Fiber movement noise rotates and changes the structure within the spot, resulting in a different line trace within the image.

Figure 18 shows a thick donor exposed by 18 writing channels and two dummy outrigger channels, constituted by a V-groove array of 50 μm core fiber with 130 μm

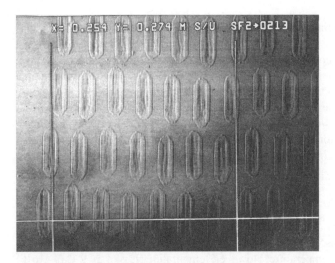

Figure 17 Thick donor overexposed with a 28 channel printhead using 50 μm fiber core, 130 μm fiber-to-fiber spacing, 2.2:1 lens demagnification, and tilting of the V-groove array to achieve 2540 dpi line-to-line spacing along the slowscan y-direction, creating records of individual channels.

fiber-to-fiber spacing, a 2.2:1 reduction lens, and the V-groove array tilted to achieve 1800 dpi writing resolution in the slowscan y-direction. There is one dummy channel on each side of the writing swath. The channels are counted from the right-hand side of the image. The laser power is set slightly above the level adequate for a single activated laser to transfer dye at the central peak of its irradiance profile, so that the nearest-neighbor interaction enables two neighboring activated lasers to transfer significantly more dye than the sum of each laser alone. The faint light trace in the dark upper-right area of Fig. 18 is

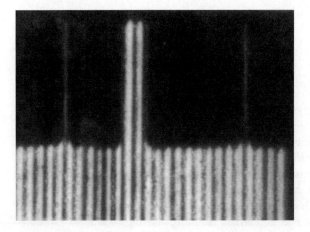

Figure 18 Thick donor exposed with 18 writing channels and two dummy outrigger channels, consisting of a V-groove array with 50 μm core fiber and 130 μm fiber-to-fiber spacing, a 2.2:1 lens, and the V-groove array tilted to achieve 1800 dpi writing resolution in the slowscan y-direction. Channels 5, 15, 16, and 17 are shown leading the group of 18 writing channels. Channels are counted starting at the right-hand side of the image.

Multichannel Laser Thermal Printhead Technology

channel 5 turned on ahead of the group as the printhead scans down the page. Channel 5 also produces the faint light trace in the dark upper-left area of Fig. 18 during the subsequent swath of the printhead scanning down the donor. When channel 5 merges into the other channels being energized in the lower half of Fig. 18, the peak exposure in the donor appears to move between channels. This nearest-neighbor interaction is caused by the optical beam being larger than the line-to-line spacing, resulting in the peak energy deposited being greatest between channels. The widest leading feature in the dark upper half of Fig. 18 is composed of channels 15, 16, and 17. The faint lines within this widest leading feature are the individual channels with the two lighter areas composed of the overlap between the two groups: (channels 15 and 16) and (channels 16 and 17).

Dummy channels as described by Baek and Mackin[50] are used to increase colorant transfer by the ends of the swath to the same level as by the swath interior when knitting swaths together. In Fig. 18, these dummy channels are used next to channels 18 and 1. The dummy channels aid production of the 18th bright line starting from the right of Fig. 18 in the area where all 18 channels are energized. Without the dummy channels, insufficient colorant would be transferred at the swath boundaries, resulting in a white-line defect on the print, and a dark-line defect on the written donor.

4.5 Scanline Spacing

The channel-to-channel spatial frequency at writing resolutions of 1800 dpi and 2540 dpi is higher than the range of visual sensitivity. Low-frequency spot placement error falls within the range of visible sensitivity. Channel-to-channel spacing requirements are driven by the need to avoid creating a white-space or dark-space error between successive channels. In a reflection print, the swath-to-swath spacing requirement is determined by eliminating the white-space error between swaths. Dark-space errors are harder to see, resulting in a higher tolerance for too much overlap. Many practical systems are colorant-limited; exposure beyond saturation does not result in any image-luminance error because image density is clipped at the maximum amount of colorant available.

We can measure the fiber-to-fiber placement error, as shown in Fig. 19, in both the along-array and cross-array directions. The spot-to-spot spacing is the distance between the channel centers calculated from along-array and cross-array measurements. The array is tilted to achieve the desired spot-to-spot spacing. Adjusting the angle of the array compensates for variations in lens magnification. The data for each channel are digitally delayed to align pixels to a line normal to the fastscan direction on the proof.

4.6 Area-Averaged Exposure Calculated from Printer Properties

The scanline spacing, Y, can be replaced by the reciprocal of the writing resolution expressed in dpi. The writing velocity, V_x, can be replaced by the product of the drum circumference and drum rotation rate, called "drum speed," expressed in revolutions per minute (RPM), to calculate the exposure from Eq. (14) as shown in Eq. (42).

$$Area\text{-}Averaged\ Exposure = \frac{Average\ Laser\ Power \times Writing\ Resolution}{Drum\ Circumference \times Drum\ Speed} \quad (42)$$

Area-averaged exposure is adjusted for the entire proof by selecting the laser power and drum speed. Three methods are used to adjust exposure. The first method sets the

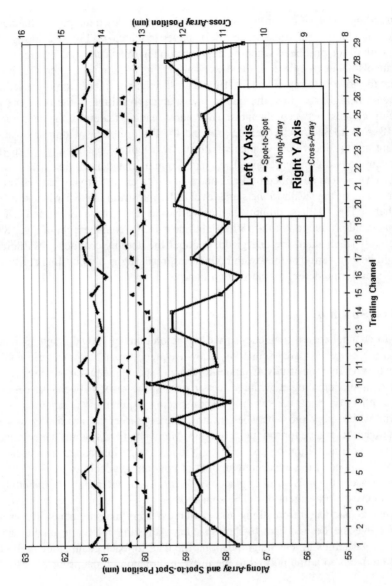

Figure 19 Distance between successive aerial writing spots for a 30-channel V-groove-array fiber-optic printhead with 50 μm core fibers spaced 130 μm apart, measured after the 2.2 : 1 reduction objective lens using a Spiricon Beam Analyzer.

Multichannel Laser Thermal Printhead Technology

lasers to a fixed power level and adjusts the drum speed to modulate the exposure. The second method fixes the drum speed to a known level then adjusts laser power to achieve the desired exposure. The third method first sets the drum speed at the fastest possible quantized value within the limitation of available laser power, then refines the exposure by trimming laser power. A digital servo loop utilizing a drum encoder is used to maintain drum speed within 0.05%.

4.7 Media Responses

The sensitometric responses of various thick donors, imaged with a 28 channel fiber printhead containing 50 μm core fibers and a 2.2 : 1 lens demagnification at 2540 dpi writing resolution and a maximum laser power of 345 mW, are shown in Fig. 20. The thick donor responses shown are similar to the response obtained using Kodak Approval Digital Color Proofing Films. Each color has a linearly increasing response section that is used to adjust the solid density on the print. Variations in exposure result in changes in reflective density. The steeper the sensitometric slope, the more sensitive the medium is to variations in exposure, demanding tighter tolerances on the factors that control exposure in order to maintain constant density throughout an image. The sensitometric curves differ among the various donor colors due to differences in dye-layer thickness, concentrations of the absorptive infrared dyes, vaporization temperatures and extinction coefficients of the visible dyes, and efficiencies of devoting absorbed laser energy to thermal transfer of those visible dyes. A change in mechanism above 0.2 J/cm^2 indicated by bubbles in the donor transfers more visible dye than vaporization does, producing steeper slopes in the density vs. exposure curves. The curves flatten at high exposure upon transferring most of the dye available on the donor. Imaging thin donor on the same printing system produces the sensitometric responses shown in Fig. 21. The thin donor responses shown are similar to the response obtained using Kodak Approval Digital Proofing Media, Type 2. Thin donor results in lower density range with reduced sensitometric slope. This relaxes the machine requirements to produce artifact-free images while sacrificing range of density adjustability.

4.8 Spot Overlap, Optical Crosstalk, and Their Effects

While larger spots hide spot placement errors and increase throughput in thermal media, they also lower the system MTF and make it more difficult to balance or adjust the swath response to hide the intensity and placement errors because interactions are extended across several neighboring writing spots instead of only the nearest neighbor.

Figure 22 shows the density vs. exposure response while writing with one, two, three, or more laser channels as disclosed by Sanger et al.[65,66] To expose a "Singles" pattern, the printhead moves one scanline in the slowscan y-direction per revolution of the drum while writing with a single channel. To expose "Pairs," the printhead moves two scanlines per revolution of the drum while writing with two adjacent channels. Similarly, "Triples" is exposed with three adjacent channels while moving three lines per revolution of the drum. The increase in output density between "Singles," "Pairs," "Triples," and more than three laser channels is a result of the optical writing spot being larger than the scanline spacing, eliciting nearest-neighbor interaction. The energy from a single laser is used less efficiently than the energy from two or more lasers at a time when the optical writing spot is more than one raster line wide. The increase in efficiency is significant in determining the overall throughput of the system. The price for this efficiency is the

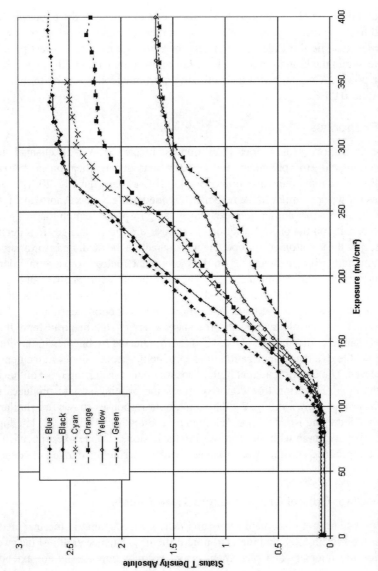

Figure 20 Typical thick donor sensitometric curves exposed using 28 writing channels, two dummy channels, 50 μm core fiber spaced 130 μm apart in a V-groove array, 2.2:1 lens demagnification, 2540 dpi writing resolution, and 345 mW per channel maximum at the film plane.

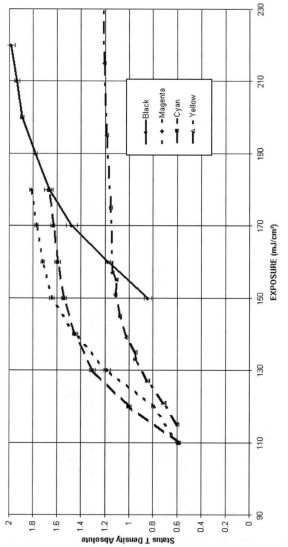

Figure 21 Kodak Approval Digital Proofing Media Type 2 typical thin-donor sensitometric curves exposed using 28 writing channels, two dummy channels, 50 μm core fiber spaced 130 μm apart in a V-groove array, 2.2 : 1 lens demagnification, 2540 dpi writing resolution, and 345 mW per channel maximum at the film plane.

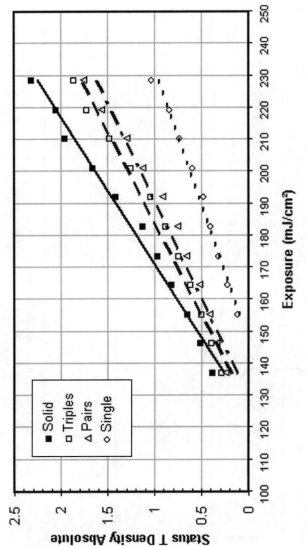

Figure 22 "Single," "Pair," "Triple," and 18-channel "Solid" patches exposed on thick donor using 28 writing channels, two dummy channels, 50 μm core fiber spaced 130 μm apart in a V-groove array, 2.2:1 lens demagnification, 2540 dpi writing resolution, and 345 mW per channel maximum at the film plane.

Multichannel Laser Thermal Printhead Technology

increase in the nearest-neighbor interaction among the channels and a decrease in contrast modulation around the rim of the halftone dot on the print.

4.9 Analysis of Balancing in the Presence of Channel-to-Channel Crosstalk

One model of the crosstalk between channels shown in Eq. (43) might be inverted for an exact solution to the balance problem.

$$\overline{Pair\ Density} = h \cdot \overline{\overline{A}} \cdot \overline{Single\ Exposure} \qquad (43)$$

where *Pair Density* is a 1×28 element vector, h is a constant, A is a 28×28 element matrix, and *Single Exposure* is a 1×28 element vector, and a single bar over a variable indicates that it is a vector while a double bar indicates a two-dimensional matrix.

The pair density is equal to a constant times the exposure of the two adjacent channels used within the pair. For a 28-channel printer, there would be 28 adjacent pair combinations, including the first and last channel. Each row in the A matrix of Eq. (44) contains two values corresponding to the two channels used to make the corresponding pair. The values in each row are expected to be the same for equal effects channel-to-channel:

$$\overline{\overline{A}} = \begin{bmatrix} 1 & 0 & 0 & \cdots & 1 \\ 1 & 1 & 0 & & 0 \\ 0 & 1 & 1 & & 0 \\ \vdots & & & \ddots & \vdots \\ 0 & 0 & 0 & \cdots & 1 \end{bmatrix} \qquad (44)$$

There may be a difference between the leading and trailing channel if the leading channel consumes infrared dye so that the trailing channel is less efficient. In this case, the coefficient of the leading channel may be set to level a and the coefficient of the trailing channel to level b resulting in the A matrix of Eq. (45) for computing pair densities with leading and trailing channel effects:

$$\overline{\overline{A}} = \begin{bmatrix} b & 0 & 0 & \cdots & a \\ a & b & 0 & & 0 \\ 0 & a & b & & 0 \\ \vdots & & & \ddots & \vdots \\ 0 & 0 & 0 & \cdots & b \end{bmatrix} \qquad (45)$$

We may further modify the A matrix if the leading and trailing channel effects are expected to change for each group of channels, resulting in the A matrix shown in Eq. (46). Equation (46) might be expected if the channel-to-channel spacing is significantly different for each pair of adjacent channels.

$$\overline{\overline{A}} = \begin{bmatrix} b & 0 & 0 & \cdots & a \\ c & d & 0 & & 0 \\ 0 & e & f & & 0 \\ \vdots & & & \ddots & \vdots \\ 0 & 0 & 0 & \cdots & z \end{bmatrix} \qquad (46)$$

The A matrices of Eqs. (44) and (45) are singular and do not have inverses. When attempting to solve Eq. (47) for the changes in exposure required to balance the pair densities, the inverse A matrix has a determinant of zero.

$$\Delta \overline{Single\ Exposure} = \overline{\overline{A}}^{-1} \cdot \overline{Pair\ Error} \qquad (47)$$

where $\Delta Single Exposure$ is a 1×28 vector, A^{-1} is a 28×28 element matrix, and *Pair Error* is a 1×28 vector equal to $\{\langle Pair \rangle - Pair)$. If the relationship between lines in Eq. (46) is the same so that the ratio of a to b equals the ratio of c to d, then Eq. (46) does not have an inverse either.

Iterating through Eq. (47) a few times, we observe that the solution oscillates channel-to-channel. Our conclusion is that the exact solution is unstable and a stable filter must be used to balance the printhead. All systems with crosstalk may have a similar problem.

Equation (43) may be expanded to solve for combinations of three adjacent channels in terms of "Pairs" to account for crosstalk that is more than two channels wide. The matrices for "Triples" are also singular and symmetric. Higher-order matrices will exhibit the same properties. We have obtained good-looking solids using multiple-channel arrays by measuring "Triples" and "Quads" responses, smoothing out the error signal, and applying the smoothed error signal to the exposure setting of each channel. The disadvantage of this technique is the production of streaks in light tints exposed with few channels turned on at a time. There is a compromise between the amount of correction applied while imaging solid areas and while writing with a single channel.

4.10 Balance Requirements

To test how accurate the balance needs to be, we perturbed the laser power levels dependent upon the position of the laser in the array. The first pass has no exposure change. The second pass changes the exposure across the array using a cosine wave starting on channel 1 with a given magnitude. The third pass changes the exposure across the array using a cosine wave that peaks on channel 3. This continues until the 14th pass, where the cosine wave peaks on channel 27. The cosine series shows that, in a solid image, we can see a $2\ mJ/cm^2$ peak cosine wave superimposed across a balanced array. We could not see a $1\ mJ/cm^2$ peak cosine wave series. The average exposure used was approximately $180\ mJ/cm^2$ on thick donor imaged with 28 writing channels and two dummy channels, a maximum of 345 mW per channel, with 50 μm core fiber spaced 130 μm apart, a 2.2 : 1 lens demagnification, and the V-groove array tilted to achieve 2540 dpi writing resolution, resulting in a solid cyan Status T density of 1.35 D_R on McCoy 80 lb. gloss paper.

An exposure series was designed as a rectangular series with a pulse one channel wide, changing phase on each pass. Another exposure series was designed with different pulse series using two channels, three channels, and up to and including seven channels. Each series was normalized to unit area so that the first series using two channels had a magnitude of $0.5\ mJ/cm^2$. It is difficult to assign to a specific channel the cause of change in a uniform area with the one-channel pulse width, even at $20\ mJ/cm^2$ peak exposure increment. However, an entire swath becomes more visible with as little as $5\ mJ/cm^2$ area exposure increment throughout that swath, compared to the surrounding uniform field.

We designed a cosine exposure series with 7.1 cycles/mm spatial frequency, twice that of the swath (3.6 cycles/mm). The series with 1 mJ/cm^2 cosine at 7.1 cycles/mm, superimposed on a uniform field created visible banding. The swath in solid areas containing this 1 mJ/cm^2 cosine at 7.1 cycles/mm showed high frequency noise, although it is impossible to discern 7.1 cycles/mm by eye. Positive- and negative-ramp exposure series show that a ramp with a peak magnitude of 1 mJ/cm^2 creates a visible swath in a uniform field.

5 INDEPENDENTLY MODULATED LASER ARRAY ON A SINGLE SUBSTRATE

Even though the fiber-pigtailed multichannel printhead described in the previous section is simple and straightforward technically, the total cost of the fiber-pigtailed laser printhead becomes excessively expensive as the number of channels increases. The electronics cost to support the printhead also grows rapidly as the number of channels increases. About 30 fiber-pigtailed channels is a reasonable compromise because the ratio of the printhead cost to total system cost is less than 20%.

To reduce the printhead cost further, we developed a new printhead technology by grouping over 100 single-mode diode lasers on the same substrate into 10 channels. Sophisticated arrays of microlenslets overlap the beams from each group of these single-mode diode-lasers into one spot to gain optical power and redundancy of emitters for reliability.

5.1 Optical Configuration of Monolithic Multichannel Printhead

The optical layout[43,67–69] is shown in Fig. 23 for the along-array direction and in Fig. 24 for the cross-array direction.

The 160 single-mode diode lasers are gathered into 10 groups, separated by 250 μm space between groups. Each channel, composed of 16 single-mode diode lasers, is 750 μm

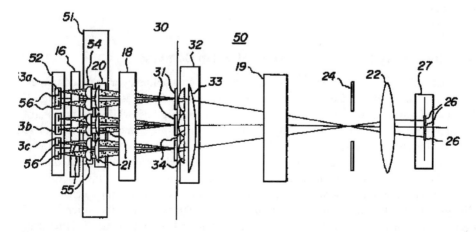

Figure 23 Optical scheme in the along-array direction of the monolithic multichannel printhead, reprinted from U.S. Patent 5,619,245.

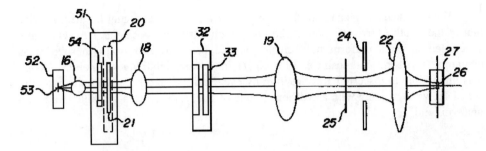

Figure 24 Cross-array optical scheme of the monolithic multichannel printhead, modified from U.S. Patent 5,619,245.

wide. Each of the 10 channels is driven as an ensemble by one modulatable current driver. The total length of all of the single-mode diode lasers in the array is 10 mm.

The fill factor of the array is very low, about 6%.[23,43] The challenge of optical system design is to increase the fill factor to nearly 100% on the medium without losing much optical power and without much optical-system complexity. Another requirement is that all 16 single-mode lasers overlap at the printing spot of their channel. Because the 16 single-mode lasers within a channel are mutually incoherent, each channel behaves optically like a multimode laser.

In addition to optical power gain for each channel, reliability is increased by overlapping as many as 16 single-mode diode lasers. If one single-mode diode laser degrades in power or ceases emitting, the others can compensate the power loss by slightly increasing their emitted power. The multichannel printhead-system reliability and robustness are improved significantly as a result.

The light from 160 single-mode diode lasers, fabricated monolithically on a single substrate, are first collected by a fiber lens stretched along the array, shown as the round object item 16 nearest the diode-laser array on the leftmost side in Fig. 24.[69] The fiber lens reduces the numerical aperture of each beam from the diode lasers in the cross-array direction. The fiber lens is aligned with the diode lasers throughout the entire length of the monolithic array.

5.2 Along-Array Numerical Aperture of Beam Combiner

Each successive group of 16 emitters is formed into a channel using a beam combiner's microlenslet array oriented in the along-array direction, as shown as item 51 in Fig. 23. The along-array direction runs from top to bottom of Fig. 23 while the cross-array direction is in and out of the paper. Only 2 of the 16 emitters in each channel are depicted in Fig. 23 for clarity. The left half of the beam combiner's microlenslet array is composed of a collimating submicrolenslet for every emitter. The neighboring 16 collimated beams are directed into the literal input aperture of one beam combiner's focusing microlenslet for each channel on the right half of the beam combiner's collimating microlenslet array. Filling the literal aperture at the focusing microlenslet's input causes that microlenslet's output numerical aperture to be filled with light at its subsequent focus. This focusing microlenslet overlaps the 16 beams onto one spot for each channel on an intermediate plane located 50 mm to the right of the light source.

Multichannel Laser Thermal Printhead Technology

A field lens near this intermediate plane diverts the beams so that they pass through a field stop[1] at the entrance pupil of the printing lens. If only a field lens were used, a line of separated images of the beam combiner's focusing microlenslets would be projected onto that stop; slight misplacement of that stop would severely vignette at least one of the outermost laser channels but not affect the central channels. Attaching a field lenslet array to the left side of the field lens in Fig. 23 causes each of the beam-combiner focusing microlenslet's images to be superimposed onto all of the others at the stop of the printing lens, equalizing the vignetting for all of the channels. The printing lens images the field lenslet array onto the medium.

This optical system can form either a round spot or an elliptical spot, according to the relative magnifications in the along-array and cross-array directions, as shown in Fig. 25. A two-cylinder system in the cross-array direction produces an elliptical spot with a 2:1 aspect ratio on the media, as in the upper strip of Fig. 25, whereas a three-cylinder system produces a round spot as in the lower strip.

5.3 "Smile" in the Relative Positions of the Writing Spots

Because of a diode-laser fabrication error, not all 160 single-mode laser diodes on the same substrate emit beams from the same plane. This inaccuracy of the diode-laser light sources projects onto the medium an arch in the sequence of writing spots known as "smile." Smile creates banding when the printhead is set at an angle to write at a desired resolution higher than the printhead pitch as schematized by Fig. 26.[54,55,68] Smile is simply a lack of straightness in the array over its length, as shown in Fig. 27.

Ideally, the array would have no smile, but our design can accommodate smile of up 4 μm peak-to-peak without a significant image defect such as banding. The smile of the light source can create image defects such as banding or curved line, reducing the yield of laser diode arrays, the image quality, and the depth of focus of the printhead.

5.4 Smile Corrector

We have developed a simple, yet useful, optical component known as a "smile corrector,"[70] shown in Fig. 28, which can correct most, if not all, of the smile in an array. This smile corrector consists of 16 glass plates, each approximately 2 mm thick and 1 mm wide, inserted into the path before the last cylinder where the beams are approximately collimated. By properly tilting each of the plates, the position of that beam at the pupil is shifted to correct the smile at the medium's plane.

5.5 Cooling the Monolithic Diode-Laser Array

Many diode lasers are operating in a small area, therefore thermal cooling can be a serious problem. We developed a compact thermal control system to maintain the operating temperature of the submount 1.2 cm (0.5 in.) from the laser junctions of a continuously emitting printhead within $\pm 0.1°C$, using a closed-loop temperature-control system. A thermoelectric cooler (TEC) actively cools the diode lasers and exchanges heat with the air circulated by a cooling fan among 4 cm (1.6 in.) long cooling fins pictured on the right side of Fig. 29.

Figure 25 Elliptical beam (top) and round beam (bottom) writing spots projected onto the image-recording medium.

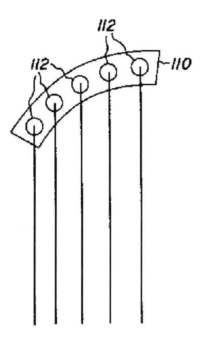

Figure 26 "Smile" in writing-spot location with tilted array causes scanlines to be too close together on one side of swath and too far apart on the other; reprinted from U.S. Patents 5,724,086 and 5,808,605.

5.6 Flexures Between the Monolithic Diode-Laser Array and the Optics Tube

The diode lasers and heat exchanger are connected through a set of flexures to the optical lens assembly in a tube. The flexures are flexible pieces of sheet metal that, when tightly bolted in place, become rigid supports coupling the laser assembly with six degrees of adjustment, three translating and three rotating, to the optics tube. Figure 29 shows all the components just before the final assembly.

After the fiber lens and combining lenslet array are aligned with the monolithic diode-laser array inside the diode-laser-array enclosure, optical power vs. current is measured for each channel of an array both at the combining plane, 50 mm from the diode lasers, and at the focal plane intended to be coincident with the surface of the written medium. At the combining plane, operating current (I_{op}) is set to obtain 850 mW from each of the 10 channels. Typical efficiency of the optical system from diode laser to image-recording medium is approximately 70% for the printhead. Theoretical calculation

Figure 27 Smile exhibited in aerial irradiance distribution of writing spots in plane of thermal donor.

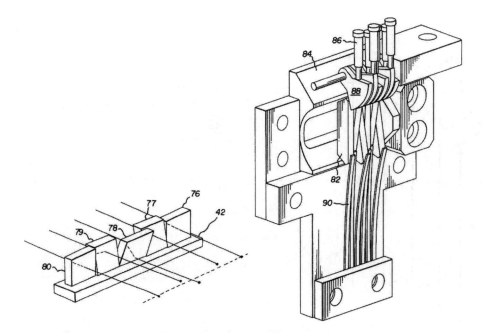

Figure 28 "Smile corrector" for improving colinearity of writing spots by displacing laser beams in a collimated segment of the optical path; schematic of tilting on left, apparatus on right, reprinted from U.S. Patent 5,854,651.

Figure 29 Monolithic multiple-diode-laser printhead assembly.

suggests that the system optical efficiency should be about 80–85%, implying that about 10% additional loss is observed. There are several unaccounted losses, such as excessive loss in the combining lenslet array, optical misalignment, and contamination of optical surfaces by dirt.

5.7 Causes of Banding with the Monolithic Diode-Laser Array

Multi-channel printing systems possess several potential causes of image banding that do not exist in single-channel printing systems, such as unequal scanline spacing within a swath, swath-to-swath misregistration, cyclical spot-placement error, and spot size difference. If too many adjacent lasers in one channel emit too little light, their writing-spot centroid shifts as a function of focus position, resulting in apparent dot-pitch change as well as banding in the written images. The number of defective emitters is limited to fewer than four in each channel and to fewer than 10 in an entire array during acceptance inspection prior to printhead assembly. Severe optical aberration can cause banding by expanding the writing spots on one side of the array, while intensifying the writing spots on the other side, creating a density wedge at the visually detectable spatial frequency of the swath, 1 cycle/mm. Other printer subsystems may contribute to banding, such as cyclical lead screw errors, vibration, flutter in motion-control mechanisms, and media-coating nonuniformity. Isolation and correction of each contribution to banding is a challenging process.

5.8 Lifetime of the Monolitihic Multiple-Diode-Laser Printhead

Each channel of the printhead comprises 16 emitters. When some emitters of each channel fail, the fewer remaining emitters must be driven harder to compensate for the lost power. This redundancy lengthens the printhead lifetime. Each emitter has a mean time between failures (MTBF) of 400,000 hours caused by abrupt random termination of emission, called "catastropic failure," and mean time to failure (MTTF) of 400,000 hours caused by gradual deterministic degradation of its emission efficiency. End of the life of a channel is declared when 30% more than initial current is required to maintain the same optical power on the medium. The end of life of any single channel signals the end of life of the entire 10-channel printhead.

Redundancy's extension of the multiple-channel-printhead life can be illustrated by considering a printhead needing all 160 emitters to operate in order for the entire printhead to be serviceable. The probability of all 160 emitters surviving, and therefore $S_{\text{printhead}}$ of the printhead surviving, is related to the probability S_{emitter} of any one emitter surviving:

$$S_{\text{printhead}} = [S_{\text{emitter}}]^{160} \qquad (48)$$

The surviving fraction of emitters is dominated by random catastrophic failure early in its life, following a first-order exponential decay:

$$S_{\text{emitter}}[t] = \exp\left[-\frac{t}{MTBF_{\text{emitter}}}\right] \qquad (49)$$

A 95% survival of the printheads would require 99.968% survival of emitters in this single-failed-emitter scenario, which would occur after only 128 hours of printhead activation with 400,000 hour $\text{MTBF}_{\text{emitter}}$. Redundancy of three extra emitters in each channel permits more than 95% of the printheads to survive 5000 hours, in spite of the limitations of catastrophic failure, slow degradation, and dissipation of consequently increasing ohmic heat, as long as air cooler than 28°C is circulated by the fan through the heat exchanger.

6 GEN III MULTICHANNEL LASER THERMAL PRINTHEAD

The term "Gen III" denotes a multichannel printhead with many more channels than the optical-fiber-coupled printheads or monolithic diode-laser-array printheads described in the previous sections of this chapter.

6.1 Gen III Technology: Laser Thermal Printheads with Hundreds of Channels

Laser thermal printheads with a large number of channels have been developed over the past decade. The printheads were developed for laser thermal (LT) computer-to-plate (CTP) platesetters. The more channels in a printhead, the less laser power needed from each channel to maintain the same printer productivity, so that the speed of the printhead relative to the medium can be slower. In the case of media mounted on a rotating drum, this means a lower RPM of the drum while exposing the media, and less time to accelerate the drum to that lower RPM. In a flatbed CTP, more channels allow lower printhead velocity while exposing the media and less time to reverse the printhead direction to that lower printhead velocity.

Gen III printheads deliver a significant amount of laser power to the media, typically 12–20 W, compared to a total of about 10 W from the fiber-optically coupled lasers and 6 W from the monolithic diode-laser array. The printheads typically have 200–256 channels, with each channel delivering about 45–90 mW. The channels of light stitch seamlessly because they are derived from a continuous line of laser beam illumination, and this is a very valuable characteristic of the printhead. Channel density at the media is typically 1000 dpi, 1270 dpi, or 2400 dpi, depending on the application. Gen III printheads are defined in this chapter as using a semiconductor laser-bar source, followed by illumination optics, a light valve or spatial light modulator, and post-optics including printing lens, all housed in a single enclosure. Figure 30, modified from a Kodak patent,[71] shows the principle of combining these components into a multichannel printhead. Printheads can be assembled and aligned as a unit, then shipped or stored for later installation into the platesetter.

6.2 Laser Bar Sources, and Collimating and Combining Optics

Gen III printheads use semiconductor diode-laser bars. One example of a diode laser bar has a monolithic array of 19 broad area emitters, each 150 μm wide at a pitch of 500 μm (or 650 μm). Vendors of these laser sources include Coherent, Inc. (Santa Clara, CA), and Spectra-Physics Semiconductor Lasers (Tucson, AZ). The laser emissions are multimode in the along-array direction; an irradiance distribution is shown in Fig. 31 as the dashed curve with divergence typically 10° (FWHM). In the cross-array direction, each beam is emitted from the narrow active epitaxial layer thinner than 1 μm. The irradiance in the cross-array direction has a nearly Gaussian profile, and the divergence angle is much larger, typically 35–40° FWHM as a result of diffraction.

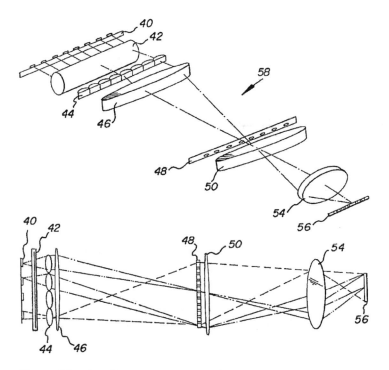

Figure 30 Gen III type printhead, depicting the diode-laser-bar source (#40), light-collecting cross-array lens (#42), cylindrical combining collimating microlenslet array (#44), cylindrical combining focusing lens (#46), spatial light modulator (#48), and spherical printing lens (#54) in perspective view in the upper panel and in the along-array direction in the lower panel, reprinted from U.S. Patent 5,521,748.

The cross-array beams from a laser bar can be collimated with a single, short-focal-length (fl ∼ 1 mm) aspheric cylinder lens designated as item #42 in Fig. 30. The beams in the along-array direction can be roughly collimated by a beam combiner's cylindrical collimating lenslet array with focal lengths (fl ∼ 2 mm) shown in Fig. 30 as item #44 similar to a single channel of the optical configuration described previously in Sec. 5.2. When these two anti-reflection (AR) coated optical elements are aligned to a diode-laser bar, typically about 75% of the power can be collected and provided to the next optical element. A third lens, the beam combiner's cylindrical focusing lens shown as item #46 in Fig. 30, can be added in the along-array direction to combine the images of the 19 emitters into a single image, typically at the beam-combiner focusing-lens focal plane. The combined irradiance distributions of the 19 emitters shown as the solid curve in Fig. 31 are more uniform than the dashed curve portraying the irradiance from a single emitter. Another way to combine the beams without the need of a third lens is discussed in the Gelbart[72] patent. Additional uniformizing optics are usually required after the combined beams to provide a reasonably uniform beam irradiance, with about 10% fluctuation, for the line of illumination presented to the spatial light modulator. Examples of uniformizing optics are shown by Kurtz[73] and in patents by Kurtz and Kessler[74] and Moulin.[75]

The diode-laser bar beam's Lagrange value in the along-array direction (L_a) differs from the Lagrange value in the cross-array direction. The along-array Lagrange value is

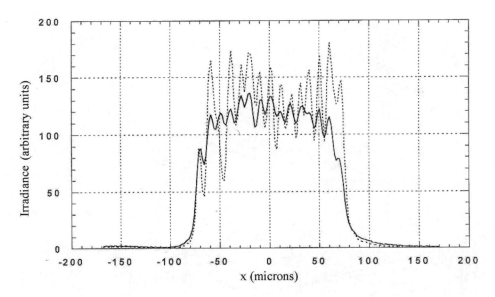

Figure 31 Normalized irradiance profile from a single emitter of a diode-laser bar in the along-array direction is the dashed curve. Normalized irradiance profile from combining all 19 emitters from a diode-laser bar is the solid curve.

useful for tracking the beam numerical aperture (NA) through the rest of the optical system:

$$L_a = (Number\ of\ Emitters) \frac{(Width\ of\ Emitter)}{2} (Beam\ NA) \qquad (50)$$

Because the Lagrange is conserved throughout the rest of the beam path, the NA can be calculated at any plane where the beam half-width is known. The Lagrange of a small to moderate numerical aperture Gaussian beam is approximated by $\{\lambda/\pi\}$, but this cross-array beam is best tracked using Gaussian beam propagation formulas.

6.3 Spatial Light Modulator or Light Valve Array

The spatial light modulator has 200 or more independently addressable channels and must be able to sustain a laser beam power density of about $1\ kW/cm^2$. This laser beam power density is rather high, and depends on modulator-pixel size, fill factor, and optical efficiency. The modulator must also provide a minimum contrast ratio of 10 : 1 (the ratio of channel "on" irradiance to the channel "off" irradiance), and work well at diode-laser-bar wavelengths, typically between 800 and 840 nm. In addition, modulated beam rise- and fall-times should be less than 2 μs. To meet these requirements, patents propose various candidates including microelectromechanical (MEM),[76] and electro-optical[77] technologies.

6.4 Gen III Laser Thermal Printheads in CTP Platesetters

The main area of application of Gen III laser thermal printheads is in the graphic arts field as CTP platesetters. These machines print halftone images of a document (text and photos) onto thermally sensitive polymer media that is coated onto a thin aluminum substrate. After being imaged, the plate is sent through a solution to, for example, remove the unexposed regions in a negative working plate, followed by washing and sometimes baking for very long press runs. The plates are then used on printing presses that ink them and transfer the image to paper.

7 CONCLUSIONS

Since the late 1980s, environmental concerns have demanded new printing technology that is environmentally conscious and dry, free from any wet chemical processing. Laser-thermal printing meets this customer desire while providing the same image quality as silver halide in terms of resolution and density for reflective and transparent images as well as printing plates. Laser thermal's requirement for much more laser energy than silver halide's can be satisfied by several multichannel printhead architectures.

The transfer of visible dye by the donor's absorption of the scanned beam energy can be predicted by the static model of exposure in excess of threshold in conjunction with the calculation of the dynamic temperature distribution in the dye layer. Influences of thermal properties of the donor materials, extinction coefficients of both visible and infrared dyes, and laser-scanning conditions upon the amount of dye transferred by laser exposure are enunciated by the model. Shift of the single-beam-printing D vs. H curve to lower exposures with reduction of dye layer thickness agrees with experimental data. Single-beam-printing depth of focus of about 50 µm can be explained by the change in the D_R vs. H curve with beam enlargement as the donor is displaced from the plane of best focus. Only a small fraction of the exposure in excess of threshold seems to be devoted to vaporization and transfer of the visible dye, with the remainder of that energy consumed by heating, vaporizing, decomposing, or deforming other donor constituents.

The spacing between beams from a multiple-source printhead can be adjusted by tilting the printhead or by interleaving scanlines. A different expression for the model of exposure in excess of threshold applies to writing with a multiple-source printhead than with a single-source printhead. Multiple-beam printing's ability to produce the same density as single-beam printing with less area-averaged exposure predicted by the combination of the calculated dynamic temperature profile and the model of exposure in excess of threshold apparently underestimates the multiple-beam printing's advantage as exhibited by experimental data.

Potential causes of artifacts engendered by multichannel printheads can be mitigated by techniques of dummy lasers, interleaving, lenslet arrays, smile correctors, balancing the powers of the individual channels, and other adjustments discussed in this chapter.

ACKNOWLEDGMENTS

The authors thank Charles D. DeBoer (Integrated Nano-Technologies LLC, Henrietta, NY), an early advocate of the potential of laser-thermal-dye transfer, and James C. Owens (Torrey Pines Research, Carlsbad, CA), an innovator in the use of optical fibers in laser imaging, for reviewing this manuscript. Contibutors to laser-thermal imaging are gratefully acknowledged: Kevin Hsu and Joe Kaukeinen for mounting optical fibers in V-grooves; Jack Harshbarger, Ronald Firth, Richard Anderson, Sanwal Sarraf, Chris Goldsmith, and Tim Tredwell as champions of projects bringing that potential to fruition; Richard Henzel, Derek Chapman, Stephen Neumann, Mitchell Burberry, Lee Tutt, Linda

Kaszczuk, and Glenn Pearce for creating donors; Ian Hodge and Gary W. Byers for guidance on thermal properties of the donors; Roger Kerr and Glenn Hawn for the precision mechanical subsystems; and Thomas A. Mackin, Michael Schultz, John Gentzke, and William R. Markis for devising electronics and programs that control the printers.

REFERENCES

1. Hopkins, R.E.; Stephenson, D. Optical systems for laser scanners. In *Optical Scanning*; Marshall, G.F. Ed.; Marcel Dekker: New York, 1991; 29 pp.
2. Sherman, R.J. Polygonal scanners: applications, performance, and design. In *Laser Beam Scanning: Opto-Mechanical Devices, Systems, and Data Storage Optics*; Marshall, G.F., Ed.; Marcel Dekker: New York, 1985.
3. Montagu, J. Galvanometric and resonant low inertia scanners. In *Laser Beam Scanning: Opto-Mechanical Devices, Systems, and Data Storage Optics*; Marshall, G.F., Ed.; Marcel Dekker: New York, 1985.
4. Sarraf, S.P.; DeBoer, C.D.; Haas, D.D.; Jadrich, B.S.; Connelly, R.; Kresock, J. Laser thermal printing. 9th International Conference of Advances in Non-Impact Printing Technologies, Yokohama, Japan. The Society for Imaging Science and Technology: Springfield, VA, 1993; Paper Oct 6-11-p1, 358–361.
5. Maydan, D. Micromachining and image recording on thin films by laser beams. AT&T Tech. J. 1971, *50*, 1761–1789 (Equation 14).
6. Lee, I-Y.S.; Tolbert, W.A.; Dlott, D.D.; Doxtader, M.M.; Foley, D.M.; Arnold, D.R.; Ellis, E.W. Dynamics of laser ablation transfer imaging investigated by ultrafast microscopy. J. Imaging Sci. Technol. 1992, *36*, 180–187 (Equation 3b).
7. Marshall, G.F. Gaussian laser beam diameters. In *Laser Beam Scanning: Opto-Mechanical Devices, Systems, and Data Storage Optics*; Marshall, G.F., Ed.; Marcel Dekker: New York, 1985.
8. Jeunhomme, L.B. *Single-Mode Fiber Optics*; Marcel Dekker: New York, 1990; 17–23.
9. Yariv, A. *Quantum Electronics*; Wiley: New York, 1967; 129 pp (Derivation appears in Yariv, A. *Solutions Manual for Optical Electronics*, 4th Ed.; Saunders College Publishing Division of Holt, Rinehart, and Winston: Philadelphia, 1991; 27–29).
10. Kessler, D.; Shack, R.V. yy' Diagram, a powerful optical design method for laser systems. Appl. Opt. 1992, *31*, 2692–2707.
11. Abramowitz, M.; Stegun, I.A. *Handbook of Mathematical Functions*; Dover: New York, 1965; 297 pp (Equation 7.1.1).
12. Haas, D.D. Contrast modulation in halftone images produced by variations of scanline spacing. J. Imaging Technol. 1989, *15*, 46–55.
13. Kogelnik, H.; Li, T. Laser beams and resonators. Appl. Opt. 1966, *5*, 1550–1567 (Equation 20).
14. Kogelnik, H. Propagation of laser beams. In *Applied Optics and Optical Engineering*; Shannon, R.P., Wyant, J.C., Eds.; Academic Press: New York, 1979; Vol VII, 155–190 (Equations 85 and 86).
15. DeBoer, C.D.; Sarraf, S.P.; Weber, S.; Jadrich, B.S.; Haas, D.D.; Kresok, J.; Burberry, M. Instant transparencies by laser dye transfer. IS&T 46th Annual Conference. The Society for Imaging Science and Technology: Springfield, VA, 1993; 201–203.
16. Lamberts, R.L. Sine-wave response techniques in photographic printing. J. Opt. Soc. Amer. 1961, *51*, 982–987.
17. Thomas, W., Jr., Ed. *SPSE Handbook of Photographic Science and Engineering*; Wiley: New York, 1973; 424 pp.

18. Schramm, R.E.; Clark, A.F.; Reed, R.P. *A Compilation and Evaluation of Mechanical, Thermal, and Electrical Properties of Selected Polymers*; AEC Order No. SAN-70-113; National Bureau of Standards, US Department of Commerce: Boulder, CO, 1973.
19. Ringwald, E.L.; Lawton, E.L. Physical constants of poly(oxyethyleneoxyterephthaloyl) (poly(ethylene terephthalate)). In *Polymer Handbook*, 2nd Ed.; Brandrup, J., Immergut, E.H., Eds.; Wiley: New York, 1975; III-216, III-217, V-71–V-78.
20. Williams, F.C.; Clapper, F.R. Multiple internal reflections in photographic color prints. J. Opt. Soc. Amer. 1953, *43*, 595–599.
21. Shore, J.D.; Spoonhower, J.P. Reflection density in photographic color prints: generalizations of the Williams-Clapper Transform. J. Imaging Sci. Technol. 2001, *45*, 484–488.
22. Odai, Y.; Kitamura, T.; Kokado, H.; Katoh, M. Resolution performance of images printed by laser dye thermal transfer. J. Imaging Sci. Technol. 1996, *40*, 271–275.
23. DeBoer, C.D. Laser thermal media; the new graphic arts paradigm. J. Imaging Sci. Technol. 1998, *42*, 63–69.
24. Schneider, P.J. *Conduction Heat Transfer*; Addison-Wesley: Reading, MA, 1955; 282–290.
25. Özisik, M.N. *Boundary Value Problems of Heat Conduction*; Dover: New York, 1968; 6, 80–82.
26. Körner, T.W. *Fourier Analysis*; Cambridge University Press: Cambridge, U.K., 1988; 274–281.
27. Paterson, S. The conduction of heat in a medium generating heat. Phil. Mag. 1941, *32*, 384–392.
28. Carslaw, H.S.; Jaeger, J.C. *Conduction of Heat in Solids*, 2nd Ed.; Clarendon Press: Oxford, 1959; 277 pp.
29. Inoue, F.; Itoh, A.; Kawanishi, K. Thermomagnetic writing in magnetic garnet films. Japan J. Appl. Phys. 1980, *19*, 2105–2114 (Equation 3.6).
30. Bartholomeusz, B.J. Thermal response of a laser-irradiated metal slab. J. Appl. Phys. 1988, *64*, 3815–3819 (Equation 22).
31. McDaniel, T.W.; Bartholomeusz, B.J. Modeling the magneto-optical recording processes. In *Handbook of Magneto-Optical Data Recording*; McDaniel, T.W., Victora, R.H., Eds.; Noyes Publications: Westwood, NJ, 1997; 641 pp.
32. Hamilton, J.F. Reciprocity failure and the intermittency effect. In *The Theory of the Photographic Process*, 4th Ed.; Mees, C.E., James, T.H., Eds.; Macmillan: New York, 1977; 133 pp.
33. Odai, Y.; Kitamura, T.; Kokado, H.; Katoh, M.; Irie, M. Printing system and printing characteristics of laser dye thermal transfer using a high-power laser. J. Imaging Sci. Technol. 1996, *40*, 412–416.
34. Hare, D.E.; Rhea, S.T.; Dlott, D.D. New method for exposure threshold measurement of laser thermal imaging materials. J. Imaging Sci. Technol. 1997, *41*, 588–593.
35. Koulikov, S.G.; Dlott, D.D. Time-resolved microscopy of laser photothermal imaging. Opt. Photonics News 2000, *June*, 26–32.
36. DeBoer, C.D. Digital imaging by laser induced transfer of volatile dyes. 7th International Conference of Advances in Non-Impact Printing Technologies. The Society for Imaging Science and Technology: Springfield, VA, 1991; Vol. 2, 449–452.
37. Habbal, F. Imaging with high-power lasers. Opt. Photonics News 2002, *September*, 54–57.
38. Jakob, M. *Heat Transfer*; Wiley: New York, 1949; Vol. 1, 343 pp.
39. Sirohi, R.S.; Kothiyal, M.P. *Optical Components, Systems, and Measurement Techniques*; Marcel Dekker: New York, 1991; 95 pp.
40. Levene, M.L.; Scott, R.D.; Siryj, B.W. Material transfer recording. Appl. Opt. 1970, *9*, 2260–2265.
41. Koulikov, S.G.; Dlott, D.D. Focus fluctuations in laser photothermal imaging. J. Imaging Sci. Technol. 2000, *44*, 1–12.

42. Dlott, D.D. Focus fluctuations in laser-materials interactions. Opt. Photonics News 2002, *September*, 35–37.
43. Kessler, D. Optical design issues of multi-spots laser thermal printing. IS&T/OSA Annual Conference; The Society for Imaging Science and Technology: Springfield, VA, 1996; 215–219.
44. Minoura, K. Optical Mechanical Scanning Using Several Light Beams. U.K. Patent Application GB 2,069,176A, August 19, 1981.
45. Kitamura, T. Beam Recording Apparatus Effecting the Recording by a Plurality of Beams. U.S. Patent 4,393,387, July 12, 1983.
46. Tsukada, T.; Morinaga, K. Optical Exposure Unit for Electrophotographic Printing Device. U.S. Patent 4,435,064, March 6, 1984.
47. Haas, D.D.; Owens, J.C. Single-mode fiber printheads and scanline interleaving for high-resolution laser printing. Proc. SPIE. Society of Photo-Optical Instrumentation Engineers: Bellingham, WA, 1989; Vol. 1079, 420–426.
48. Baek, S.H.; DeBoer, C.D. Scan Laser Thermal Printer. U.S. Patent 5,168,288, December 1, 1992.
49. Guy, W.F.; Mackin, T.A. High Resolution Thermal Printers Including a Print Head with Heat Producing Elements Disposed at an Acute Angle. U.S. Patent 5,258,776, November 2, 1993.
50. Baek, S.H.; Mackin, T.A. Thermal Printer Capable of Using Dummy Lines to Prevent Banding. U.S. Patent 5,278,578, January 11, 1994.
51. Tsao, S. Apparatus for Arranging Scanning Heads for Interlacing. U.S. Patent 4,232,324, November 4, 1980.
52. Haas, D.D. Method of Scanning. U.S. Patent 4,900,130, February 13, 1990.
53. Starkweather, G.K.; Dalton, J.C. Error Reducing Raster Scan Method. U.S. Patent 5,079,563, January 7, 1992.
54. Haas, D.D.; Mackin, T.A.; Sanger, K.M.; Sarraf, S.P. Interleaving Thermal Printing with Discontiguous Dye-transfer Tracks on an Individual Multiple-source Printhead Pass. U.S. Patent 5,808,655, September 15, 1998.
55. Mackin, T.A.; Haas, D.D.; Sanger, K.M. Printhead Having Data Channels with Revisable Addresses for Interleaving Scan Lines. U.S. Patent 5,724,086, March 3, 1998.
56. Yip, K-L.; Muka, E. MTF analysis and spot size selection for continuous-tone laser printers. J. Imaging Technol. 1989, *15*, 202–212.
57. Van Nes, F.L.; Bouman, M.A. Spatial modulation transfer in the human eye. J. Opt. Soc. Amer. 1967, *57*, 401–406.
58. Firth, R.R.; Kessler, D.; Muka, E.; Naor, M.; Owens, J.C. A continuous-tone laser color printer. J. Imaging Technol. 1988, *14*, 78–89.
59. Hsu, K.; Owens, J.C.; Sarraf, S.P. Fiber Optic Array. U.S. Patent 4,911,526, March 27, 1990.
60. Kaukeinen, J.Y.; Tyo, E.M. Method of Making a Fiber Optic Array. U.S. Patent 4,880,494, November 14, 1989.
61. Miller, C. *Optical Fiber Splices and Connectors*; Marcel Dekker: New York, 1986; 266–267.
62. Mentzer, M.A. *Principles of Optical Circuit Engineering*; Marcel Dekker: New York, 1990; 61, 301–307.
63. Sarraf, S.P. Apparatus and Method for Eliminating Feedback Noise in Laser Thermal Printing. U.S. Patent 5,420,611, May 30, 1995.
64. Spurr, R.W.; Baek, S.H.; Mackin, T.A.; Markis, W.R.; Sanger, K.M. Laser Diode Controller. U.S. Patent 5,966,394, October 12, 1999.
65. Sanger, K.M.; Mackin, T.A.; Schultz, M.E. Method and Apparatus for the Calibration of a Multichannel Printer. U.S. Patent 5,291,221, March 1, 1994.
66. Sanger, K.M.; Mackin, T.A.; Schultz, M.E. Method of calibrating a multichannel printer. U.S. Patent 5,323,179, June 21, 1994.
67. Kessler, D.; Simpson, J. Multi-beam Optical System Using Lenslet Arrays in Laser Multi-beam Printers and Recorders. U.S. Patent 5,619,245, April 8, 1997.

68. Wyatt, S.K. Integration and characterization of multiple spot laser array printheads. In *Laser Diode and LED Applications III*; Proc. SPIE 3000; Linden, K.J., Ed.; 1997; 169–177.
69. Kessler, D.; Endriz, J.G. Optical Means for Using Diode Laser Arrays in Laser Multibeam Printers and Recorders. U.S. Patent 5,745,153, April 28, 1998.
70. Kessler, D.; Blanding, D.L. Optically Correcting Deviations from Straightness of Laser Emitter Arrays. U.S. Patent 5,854,651, December 29, 1998.
71. Sarraf, S.P. Light Modulator with a Laser or Laser Array for Exposing Image Data. U.S. Patent 5,521,748, May 28, 1996.
72. Gelbart, D. Apparatus for Imaging Light from a Laser Diode onto a Multi-channel Linear Light Valve. U.S. Patent 5,517,359, April 14, 1996.
73. Kurtz, A.F. Design of a laser printer using a laser array and beam homogenizer. Proc. SPIE, Society of Photo-Optical Instrumentation Engineers: Bellingham, WA, 2000; Vol. 4095, 147–153.
74. Kurtz, A.F.; Kessler, D. Laser Printer Using a Fly's Eye Integrator. U.S. Patent 5,923,475, July 13, 1999.
75. Moulin, M. Illumination System and Method for Spatial Modulators. U.S. Patent 6,137,631, October 24, 2000.
76. Reznichenko, Y.; Kelly, H.A. Optical Imaging Head Having a Multiple Writing Beam Source. U.S. Patent 6,229,650 B1, May 8, 2001.
77. Nutt, A.C.G.; Ramanujan, S.; Revelli, J.F., Jr. Modulator for Optical Printing. U.S. Patent 6,211,997 B1, April 3, 2001.

Glossary

ALAN LUDWISZEWSKI

Ion Optics, Inc., Waltham, Massachusetts, U.S.A.

There are many terms and concepts that are unique to optical scanning, or have specific meaning within the context of scanning systems. This chapter provides definitions for the most commonly used terms. It is not meant to provide an exhaustive discussion of the topics, some of which could comprise a chapter in themselves, but rather to provide a concise explanation of an unfamiliar term. Many of the terms listed here are described in significantly more detail elsewhere in this text.

Accuracy

The maximum difference between the commanded and actual beam position. This includes any nonlinearities, hysteresis, noise, encountered drifts, resolution, and other factors.

Acousto-optic deflector

A device that utilizes an acoustic wave propagating through an optical medium to generate a refractive index wave. This periodic refractive index acts as a sinusoidal grating, defracting a laser beam into several orders. The angular position of the first-order defracted beam is varied by changes to the acoustic drive frequency.

Back EMF

The voltage produced by an inductor in opposition to a change in current or magnetic field. It is interesting to note that the back EMF constant of a galvanometric scanner is equivalent to the torque constant:

$$e/\omega = \tau/I$$

where e is the back EMF, ω is the angular velocity, τ is the torque, and I is the drive current.

Bandwidth

The maximum frequency at which a system can track a sinusoidal input with an output attenuated to no less than 0.707 (-3 dB point) of the command. For open loop frequency responses with a phase margin of $90°$, the open loop crossover frequency equals the closed loop bandwidth. For other phase margins, the relationship is not as straightforward. A complete treatment of these relationships can be found in most control theory texts.

Beam diameter

The width of a laser beam to a specified level in intensity, which for a Gaussian beam also corresponds to the same magnitude of excluded power. The most commonly specified beam diameter is the $1/e^2$ diameter, which is the diameter to the points where the beam intensity has dropped to 13.5% of peak intensity. For a Gaussian beam, 13.5% of the beam power is outside of the $1/e^2$ diameter. Other commonly used beam diameters include the $1/e$ diameter, to a 37% intensity, which is commonly used in safety standards; the full width at half maximum (FWHM) beam diameter, to the 50% intensity points; and the full power beam diameter, which is specified to the $1/e^8$ point and excludes only 0.034% of the total beam power.

Beam expander

An optical system used to increase beam diameter and thereby reduce beam divergence. Frequently used to achieve a smaller focused spot for a given focal distance.

Beam quality

A measure of the closeness to ideal of a laser beam. Typically expressed as the M^2 value, which is the ratio of an achieved focused spot diameter to that which would be created by a theoretically ideal Gaussian beam. This is also the ratio of divergence angles of the actual and ideal beams.

Critical angle

The angle at which light striking a boundary between a more refractive medium a less refractive medium experiences total internal reflection.

Damping, Mechanical

Loss mechanisms that result in converting mechanical motion into heat. Typical losses include those due to bearing friction, aerodynamic losses, mechanical hysteresis (due to material stress/strain behavior), and magnetic hysteresis. At the microradian and submillisecond level, typical galvanometric scanner losses do not fit a linear model and different mechanisms are dominant in different frequency ranges. Thus, for galvan-

Glossary

ometers, the "damping coefficient" is actually a complex function of frequency and amplitude, and typically not specified other than as an approximate value.

Demagnetization current

The peak drive current a galvanometric scanner can withstand without demagnetizing the scanner's permanent magnets.

Divergence

The angle at which a laser beam spreads in the far-field, typically expressed in milliradians. The beam divergence may be listed as either the full angle or as a half angle.

Drift, Position detector

The change in relationship between the output of the position detector and the position of the output shaft in a galvanometric scanner. It consists of the sum of the gain drift, null drift, and uncorrelated drifts such as those caused by mechanical shifts internal to the device.

Drift, Position detector gain

The change in scale factor of the position detector in a galvanometric scanner. The angular magnitude of this drift is dependent on the shaft position. This parameter is typically specified in terms of the ratio of change in detector output per unit temperature or time (ex. ppm/°C or %/1000 hrs).

Drift, Position detector zero

The drift of the electrical zero of a galvanometric scanner position detector with temperature or time. Drift is specified in units of angle per unit of environmental influence (ex. μrad/°C or μrad/1000 hrs).

$F-\theta$ lens

An objective lens used in preobjective scan systems that has a focal length that varies as a function of incident angle such that a flat focus field is created and the spot displacement is a linear function of focal length and incident angle: $X = f \cdot \theta$. In scanning systems, $F-\theta$ lens input pupil is generally located at the scan mirror, or between the two mirrors in a two-axis system.

Facet errors

Errors in the placement of reflective facets in a polygon scanner. Errors can be in tilt, angle, radius, and flatness. Tilt typically refers to an error in the facet angle relative to the axis of rotation; that is, a deviation from the facet being parallel to the axis of rotation in a regular polygon mirror or a deviation from the design tilt angle in a pyramidal scanner. Tilt error is also frequently referred to as pyramidal error. The facet angle is the angle a facet subtends at the axis of rotation. Neglecting any rounding at the interface of adjacent facets, the sum of all facet angles will add up to 360°. Any individual facet's error from the nominal angle ($n/360°$) is a facet angle error, often referred to as the dividing error. A related but different measure of mirror accuracy is the facet-to-facet angular error measured from the midpoint of adjacent facets. Another critical specification is the tolerance in the center-to-facet radial distance measured at the closest point of each facet. This is sometimes referred to as the "height" of the facet and should not be mistaken for the width and length dimensions of a facet.

Hysteresis

Hysteresis describes the nonlinear dependence of the state of a physical system on the history of the system. For example, a scanner will frequently settle to a slightly different position when approached from different directions.

Jitter

Velocity noise disturbances during a raster scan create irregular positioning errors known as jitter. This is most easily pictured by thinking of a raster scan system laying down a series of evenly spaced pixels. Scan-to-scan deviations of individual pixels from their expected location is jitter. Jitter is described in units of scan angle and is often expressed as the standard deviation of the maximum jitter error observed in each scan line of a large number of consecutive scans.

L/R time constant

The ratio of inductance to resistance, expressed in seconds. In response to a step change in voltage, the time required for the current through an inductor-resistor network to achieve 63% of its final value. This can be a performance-limiting factor when driving galvanometers with voltage sources. When a current source output stage is used, this will not affect performance until the limits of the voltage compliance are reached. When designing output stages, consideration must also be given to the back EMF when establishing the voltage compliance requirements.

Lag time

The lag time of a galvanometric scanning system is defined as the tracking error divided by the slew rate. Owing to characteristics of the controller, stage saturation, aerodynamics, and other nonlinearities, lag time is not a constant, but it is approximately so for slew rates below the maximum slew rate. Lag time is an indication of the relative speed of a scanner and the ultimate performance obtained with a given load and tuning.

Magnetic hysteresis

The memory property of magnetic material exhibited by the lack of change in induction resulting from an initial increase or decrease in magnetomotive force. The result is often seen as a difference in settling times and positions when a given command angle is approached from different directions.

Mechanical null position

The steady-state position of an unpowered galvanometric or resonant scanner. In resonant scanners or galvanometric scanners that incorporate flexure bearings or a torsion bar restraint, the mechanical null position is readily apparent, but not highly repeatable at a fine scale due to magnetic imbalances and hysteresis effects. Scanners without mechanical rotor constraints, such as many moving magnet and moving coil scanners, do not have any defined mechanical null position.

Nonlinearity, Best fit straight line

This method of quantifying nonlinearity involves finding a first-order linear function that is the closest approximation to the measured data. The nonlinearity is then calculated as the maximum observed deviation from this line. This results in the smallest measurement of nonlinearity.

Nonlinearity, Pinned center

Pinned center nonlinearity uses a straight line that intersects a given datum point, such as the mechanical or electrical zero position of a scanner, and has a slope that best approximates that of the measured data. Nonlinearity is then calculated from this reference.

Nonlinearity, Pinned endpoint

Pinned endpoint nonlinearity uses a line drawn through the minimum and maximum measured values as the reference transfer function. The maximum nonlinearity is then measured from this reference.

Null, Electrical

The position of a scanner corresponding to zero output on the position detector.

Null, Mechanical

The steady-state position of an unpowered scanner. This position is determined by the torsion spring or flexure, if any, and the magnetic spring of the scanner. In many scanners without a torsion spring or flexure, the magnetic spring is not of sufficient magnitude to overcome frictional forces and establish a clear mechanical null position.

Optical resolution

The optical resolution of a scanning system can be described as the number of separately resolvable spots that can be produced or discerned. For diffraction limited optics, this is dependent upon the aperture width in the scan direction, the aperture shape factor, the wavelength of the source, and the total scan angle.

Overshoot

The amount of overscan that occurs prior to a scanner settling at a new location. Overshoot is frequently dependent on command factors such as step size and drive waveform as well as servo tuning.

Pixel clock

An electronic circuit used to linearly place pixels in a target field in spite of variations in the scanned beam velocity. Velocity variations can originate from sources including inconsistencies in scanner velocity, either intrinsic, such as in resonant scanners, or jitter related such as drive noise or torque disturbances, or due to optical nonlinearities. Pixel clocks typically phase lock to the scan frequency and can be triggered off single point in-field detectors, a grating clock, or scanner position or velocity signals.

Preobjective scanning

Systems where the scanner is located before the final focusing lens. Frequently preobjective systems incorporate $F-\theta$, telecentric, or other field-flattening objective lens.

Postobjective scanning

Systems where the scanner is located after the final focusing lens. Postobjective systems frequently are used when the scan field is a curved surface, such as in internal drum systems. Alternatively, a moving element can be incorporated to maintain focus over a flat working field. This is a popular approach for systems with very large target fields or large scan apertures.

Power density

The amount of radiant energy incident on a surface. Laser beams are typically not uniform in power density, so care must be taken to consider the peak power density and not the average power density when specifying optical component damage thresholds.

Repeatability

The inaccuracies in final position achieved in response to identical commanded positions. Repeatability can be specified as unidirectional or bidirectional depending upon whether the commanded position is always approached from the same direction. Bidirectional repeatability will include any system hysteresis in the measurement.

Resolution

Resolution is the ability to discern individual spots in the target field. This term is frequently confused with accuracy, which includes gain and offset drifts, noise, resolution, and other factors. Depending on system design, the limit to resolution may be due to optical considerations (optical resolution), digital addressability, position sensor signal-to-noise ratio, or other factors.

Resonance, Cross axis

Structural resonances that cause motion perpendicular to the scan axis are referred to as cross-axis resonances. These resonances may be accentuated by poor mirror design, can cause periodic wobble or system instabilities, and may limit achievable system bandwidth.

Resonance, Torsional

An on-axis resonance that appears in scanners due to the distributed masses on a compliant rotor shaft or the flexible coil of a d'Arsonval system. These resonances can appear as periodic jitter and may cause servo stability issues due to the resonant peak created in the

Glossary

scanner's transfer function. Mirror design and mounting frequently have a significant effect on the torsional resonance frequency and amplitude.

Resonant scanner

A scanner that moves a mirror in a sinusoidal motion at a single, resonant frequency. Resonant scanners typically use the mass of the mirror in conjunction with a mechanical spring to establish the resonant frequency and can achieve a greater amplitude · frequency·aperture product than other limited-rotation scanners.

Scan angle, Maximum

The maximum angle a scanner can achieve, usually specified in optical angle.

Scan angle, Mechanical

The angular excursion of a scanner's shaft. One half of the optical scan angle.

Scan angle, Optical

The angular excursion of a scanned optical beam. Twice the mechanical angle.

Scan angle, Peak

The maximum angle a scanner, or scanned beam, travels with a given raster or vector command. Usually specified in optical angle.

Scan angle, Useful

The portion of the scan over which useful work is done. Usually ties to another criteria (eg., velocity linearity) that must be maintained during this period. Usually specified in optical angle.

Scan aperture

The size and shape of the optical beam that will be passed by a scanner without vignetting. Larger scan apertures generally allow a system to maintain better beam quality at the expense of speed and cost of the scanner.

Scan efficiency, Angular

The ratio of useful scan angle to peak scan angle in raster scanned systems. Typically expressed as a percentage.

Scan efficiency, Interval

The ratio of time spent transversing the useful scan angle to the total scan period in raster scanned systems. Typically expressed as a percentage.

Settling time

Settling time quantifies the ability of a scanning system to move to a given location and stabilize to within a certain error band. This error band may be expressed as a percentage of the full field or of the step, or alternatively, as a specific angular error band. The angle that must be traversed before settling at the new location can have a major effect on settling time and should be indicated when settling time is specified. When settling time is specified with the error band referenced as a fraction of the executed step, care must be taken since for small steps the settling error band may be below the resolution of the system.

Signal-to-noise ratio

The signal-to-noise ratio (SNR) indicates the relative magnitude of full-scale sensor output to sensor noise. This can be a RMS, peak, or peak-to-peak measurement. RMS measurements have the most statistically complete data and are often considered the most convenient for calculations based on SNR. This parameter is also generally specified in decibels. The sensor noise is a function of frequency and the frequency band that it is measured over must be included in any specification of this parameter.

Slew rate

The angular velocity of a scanner. Generally specified in optical angle per unit time.

Spring constant

The torque per unit angle created by a mechanical or magnetic spring. Typically used in referring to the centering force in the rotor of a galvanometric or resonant scanner.

Telecentric lens

An objective lens used in preobjective scan systems that has a focal length that varies as a function of incident angle such that a flat focus field is created and the spot displacement is a linear function of incident angle. Most significantly, a telecentric lens has the characteristic that the beam chief ray is perpendicular to the image plane across the entire image plane. Telecentric lenses are generally the most expensive flat-field lens and are only used in applications where the variations in spot shape, field distortion, and power density created by changes in incident angle or focus are unacceptable.

Torque constant

The measurement of torque output of a galvanometer, per unit current. Torque constant is most often specified in newton · meters per amp or dyne · centimeters per amp. The torque constant is important in calculating the relationships between current, load inertia, and scanner acceleration. Care must be taken in using this parameter since it is will tend to vary slightly over angle and with current. Careful controller design and analysis may be required in order to achieve desired performance over the range of gains encountered due to these nonlinearities. Most often this will have the greatest effect on settling performance, bandwidth, and stability.

Torque constant nonlinearity

The inconsistency in torque constant as a function of angle or current. Ideally the torque constant would be constant with angle and a linear function of current. Often this parameter is specified as the maximum percentage error that may be seen over angle and up to a defined maximum current.

Torque-to-inertia ratio

This key figure of merit defines the maximum achievable angular acceleration for a given scanner and load combination. The relationship between angular acceleration and torque is given by the equation: $\alpha = \tau/I$.

Tracking error

The difference, in angle, between the actual and commanded positions. Tracking error must be specified in conjunction with a slew rate and is typically expressed in terms of optical angle. Control system tuning has a considerable effect on the value of this parameter.

Velocity nonlinearity

The inconsistency of velocity over the maximum scan angle or other defined portion of the scan, typically defined as a percentage of the nominal velocity.

Vignetting

The loss of light due to an optical element or aperture's finite size. Typically vignetting refers to an aperture that obstructs a cone of light in a noncircularly symmetric way.

Wobble

Cross-axis motion of a scanner during scan. Generally specified in terms of an optical angle representing the standard deviation of the maximum repeatable or nonrepeatable wobble error measured in each scan line of a large number of consecutive scans.

Wobble, Nonrepeatable

Random or non-scan-synchronous cross-axis motion of a scanner. This will appear as perturbations perpendicular to the scan motion in a line scan or as jitter in the perpendicular axis of an *X-Y* system.

Wobble, Repeatable

Consistent cross-axis motions of a scanner during a line scan. Often caused by bearing runout, this motion can appear as consistent irregularities in a line scan system or as nonlinearity in an *X-Y* system. Wobble-correcting optics can be employed in line scan systems to reduce scan-related wobble artifacts. Repeatable bow or S-shaped cross-axis distortion of a line can frequently be the result of system misalignment.

ACKNOWLEDGMENTS

I would like to express my gratitude to Shepard Johnson of GSI Lumonics and Dr. Irina Puscasu of Ion Optics for their suggestions and review of this material. The foundation for much of this chapter originates from Standards for Oscillatory Scanners published in SPIE

Proceedings Vol. 1454. That work was expanded greatly though use of the publications and references listed; however, all errors are my responsibility. I would also like to thank Gerald Marshall for his patience and perseverance in bringing together the large number of authors required to produce this comprehensive work.

REFERENCES

1. Beiser, L. Short Course Notes from the SPIE/IS&T. *Laser Scanning Technology in Scanners and Printers*, Symposium on Electronic Imaging Science and Technology, San Jose, CA, 1991.
2. Gottlieb, M.; Ireland, C.L.M.; Ley, J.M. *Electro-Optic and Acousto-Optic Scanning and Deflection*; Marcel Dekker: New York, 1983.
3. Hopkins, R.E.; Stephenson, D. Optical systems for laser scanners. In *Optical Scanning*; Marshall, G.F., Ed.; Marcel Dekker: New York, 1991.
4. Ludwiszewski, A. *Standards for Oscillatory Scanners*. SPIE Proceedings, Bellingham, WA, 1995; Vol. 1454, 174–185.
5. Marshall, G.F. Gaussian laser beam diameters and divergence. In *Optical Scanning*; Marshall, G.F., Ed.; Marcel Dekker: New York, 1991.
6. Saleh, B.E.A.; Teich, M.C. *Fundamentals of Photonics*; John Wiley & Sons: New York, 1991.
7. Schlessinger, M. *Infrared Technology Fundamentals*, 2nd Ed.; Marcel Dekker: New York, 1995.
8. Sherman, R.J. Polygon scanners: applications, performance, and design. In *Optical Scanning*; Marshall, G.F., Ed.; Marcel Dekker: New York, 1991.
9. Smith, W. *Modern Optical Engineering: The Design of Optical Systems*, 2nd Ed.; McGraw Hill: New York, 1990.

Index

Aberration, 16, 17, 73, 80–94, 103–108, 132, 135, 276, 531–533, 537–540, 562, 595
Acoustic, 508, 600–605, 607–641, 643–653, 655, 658–661
  array, 644
  beam, 603–605, 620–622, 633, 643–644, 646
  cell, 619
  power flow, 607
  propagation, 604
  transducer, 600, 627, 637, 650, 655, 661
  traveling wave lens, 625
  wave, 600–605, 607, 609–610, 613, 616, 618–619, 622–627, 637, 643–644, 648, 650–651, 658, 660, 703
Acousto-optic, 599–601, 607–609, 611–613, 617–620, 622–625, 627–633, 637, 643, 648, 650, 652–653, 655, 657–660, 662, 666
  modulator, 295, 617–618, 628–629, 633, 655
  scanner, 599, 617–618, 630, 632, 637, 655, 658–659, 666
Across-scan. *See* Cross-scan
Actuator, 421, 498–501, 507, 508, 569
Additive
  density. *See* Density, additive. *See* Exposure, additive

Adiabatic driver, 697–700
Aerodynamic
  bearing. *See* Air bearing
  bearing comparison of, 348, 352, 371
Aerostatic
  bearing comparison of, 348, 352, 371
  bearing. *See* Air bearing
Air bearing, 272, 280, 282, 284, 287, 294, 299, 306, 309, 311, 317, 338–339, 345–371, 382–384
  comparison of, 346–352
Air lubricated bearing, 345–371, 382–384
Airy disk, 76–81, 250,
  distribution, 250, 562
  pattern. *See* Airy distribution
Aliased image, 151, 173, 174
Allowable wave aberration, 567
Along-array numerical aperture. *See* Numerical aperture, along-array
Along-scan. *See* In-scan
Amorphous, 557, 632–633
Amplifier, 328, 332, 600, 661, 692, 700, 743
Amplitude distribution, 553, 562
Analog drivers, 334, 447, 700
Anamorphic, 75, 95–98, 560
Anisotrophy, 609
  diffraction, 609–610, 612
Anti-aliasing, 158, 231
Anti-reflection, (AR coatings), 295, 538, 650

Aperture, 9, 19, 22, 24, 32, 34, 65, 249, 250, 265, 288, 292, 295, 374–375, 391, 415, 496, 499, 506, 523, 526–527, 529, 531, 548, 605, 618–620, 623, 625–629, 632, 650, 655, 657, 659, 661
  stop, 75, 86–88
Area-average exposure. *See* Exposure, area-average
Array transducers, 643, 645–646, 648
Array, 502, 629, 643, 645, 647–649, 657, 712, 739, 741–745, 752–762
Aspherical
  lens, 548, 761
    objective lens, 566
  surfaces, lenses, 74, 85, 132
Aspherical pressed glass (APG), 565
Astigmatic focusing, 46–48, 58, 60–61, 63, 580
Astigmatism, 46–48, 58, 60–61, 63, 90–93, 97–98, 133–135, 383, 546, 548, 565, 657
  general, 3, 16, 52–53, 62–63
  simple, 13, 46–48, 59–62, 63
Asymmetric
  divergence, 15, 47–48, 52, 59–62, 64
  points, 490, 496
  waists, 13, 47–48, 52, 59–62, 64
Auto-focus scanner, 525–527
Automatic gain control (AGC), 531
Auxiliary beam, 41, 62
Axial color, 88–89, 103–108
Azimuthal propagation plots, 48–52, 62

Balance, dynamic, 285, 306, 369, 444, 541
Balanced armature, 437, 453, 444–447
Ball bearing comparison of, 347–348, 431
Ball bearing, 280, 282, 284, 287, 294, 311, 333, 345–348, 375, 377–379, 382, 384, 431, 479–481
Banding, 179–182, 275, 284, 292–294, 383
  artifact in image, 310, 752–753, 755, 757, 759–762
Bandwidth, 102, 131, 150–153, 305–306, 444, 479, 612, 614–615, 617–625, 630, 632, 637, 640, 643, 647–648, 651, 654, 657–659, 660–662, 665, 700
Bar code   39, 72, 123–124, 513–514
  characteristics, 510–514, 516, 524–525, 527, 535
  stitching, 512, 546
  truncated codes, 511, 512

Beam
  caustic surface, 50–51, 52, 64
  combiner, 754–755, 760–761
  deflection, 102, 515, 523, 671, 711–712
  diameter (beam width), 1, 8, 12, 13, 20–23, 26–32, 291, 292, 385, 387–388, 391, 393, 395, 397, 404, 406, 415, 524, 526, 620, 624, 627, 657, 661, 713, 717, 720, 728–729, 736
  expander, 71, 98
  footprint, 291, 292, 388–389, 391, 395, 406, 713, 729–730, 733–736, 743–745, 753–754
  gaussian, 10, 13, 64, 524, 620, 625–626, 655, 671, 713–717, 724–727, 733, 736, 760–763
  idealized, 15, 64
  propagation analyzer, 45–46, 55, 64
  propagation constant, 15, 41, 51–53, 56, 60–62, 64, 65, 67
  quality, 15–16, 25, 26, 31, 33–36, 54, 62, 64, 65
  real, 15, 64, 725
  splitter, 554
  standard-deviation radius, 713–720, 722, 724–726, 728, 730–734, 736
  waist, 1, 11–12, 14, 62, 67, 102, 291, 292, 524, 620, 622, 624, 671, 672, 717, 724, 728–730, 736
  width. *See* Beam diameter
Beam-lens transform, 12, 16–20, 44, 51, 59–62
Beam-waist diameter, radius. *See* Beam, waist
Bearing, 345–372, 375–384, 431–435, 437
  aerodynamic, 282, 346, 348, 349–350, 352, 361–370, 372
  aerostatic, 282, 346, 348, 352–361, 368–371
  air, 272, 280, 282, 284, 287, 294, 299, 306, 309, 311, 317, 338–339, 345–371, 382–384
  ball, 280, 282, 284, 287, 294, 311, 333, 345–348, 375, 377–379, 382, 384, 479–481
  comparison of, 347–348, 478–481
  friction, 304–305, 317, 332, 351–351, 355–356, 361, 478, 480
  gas, 346, 348–352, 356, 361–363, 364, 382
  preload, 377–378, 437, 480
  stiffness, 282, 306, 316, 348, 355–356, 360, 481, 492, 494

Index

Bendix pivot, 432, 493–494
Bessel function, 245, 603, 608
Bimetal flexure, 479
Bimorph
   flexure, 499, 500, 508
   pivot, 99–500
Binary
   imaging, 198–214, 201, 230–231
   scanned imaging, 712, 739
Bi-optic scanners, 518–519
Birefringent, 542, 609–612, 669
Blue noise, 164
Blur, 157–158, 173, 185–187
Bonding film, 637
Bow, 274, 286, 288, 296. *See also* Smile corrector
Bragg, 599, 604–606, 608, 610, 612–616, 618, 620–625, 627, 643–645, 648, 658, 691, 692
   angle, 124–126, 536–537, 606, 610, 612–615, 618, 622, 643, 648, 660
   cell, 599, 604, 618, 623, 627, 643, 645, 658
   diffraction, 606–607, 614–616, 620, 658
   equation, 124, 613–614
   error, 541
   planes, 540–541
Bulk refractive index, 540–541

Calibrated focal length, 84, 117–118
Camera
   digital, 21, 26, 27, 28, 32, 173–174, 182
   resolution, 26, 30, 32, 173–174
Capacitance, 679, 697–700
Carrier-to-noise ratio (CNR), 577
Cascaded pivot, 497–498
CD-R, 556
Center-of-scan, 395–396, 411
Channels, 629, 655–657, 711–712, 731, 743–745, 747, 751–755, 757, 759–763
Characteristic curve, 158–260, 723, 726, 728
Chirp lens, 627–628, 661
Chroma of color, 172–173
Chromatic
   aberrations, 88–90, 132
   dispersion, 89, 561
Chromaticity diagrams, 169, 170, 244–245
CIELAB and CIELUV, 172–173
Cindrich-type holographic, 125

Circle
   circumscribed, 386–387, 393, 397, 400–401, 404, 406, 408
   inscribed, 400–401
Clip-points, 22, 26, 27–28
Coating, 277–280, 286, 287, 294, 295, 366, 641, 650
Codabar, 513, 514
Coil windings, 425–427
Collective and dispersive surfaces, 92
Collimated incident beam, 71–72, 78, 86, 95–96
Collimating
   lens, 127, 130, 133–134, 561. *See also* Collimating optic
   optic, 754–755, 760–761
Color
   appearance models, 172
   gamut, 169
   imaging, 166–173, 496, 711, 739, 747
   management, 233–234
   matching function, 168–169, 244,
   purity, 170, 244,
   value, 172
   visual response, 167–169,
Colorimetry 169–173
Coma, 90–93, 108
Combining optics, 754–755, 757, 759–761
Commercial profilers, 2, 21, 25–26, 27
Commision Internationale de l'Eclairage (CIE), 168–169
   standard observer, 168, 244–245
Compact disk (CD), 551, 555
Comparator, phase, 331–332, 334, 339, 341
Compression, 600, 603–605, 651–653
Computer-to-plate (CTP), 707, 711, 740, 760, 762–763
Confidence level in psychometrics, 235,
Confocal microscopes, 468–471
Conjugates, 76
Constant acceleration, 475
Continuous tone (contone), 293–295, 712, 730, 739–741
Contrast, 159–160, 607, 618, 751, 762
   sensitivity functions (CSF), 194–195, 251, 740–741, 745, 759
Controller
   gain, 305, 331–332, 341
   phase-lock, 304, 310, 318, 334, 336, 339, 341
Convolution error, 23–25, 27, 64, 66, 67
Cooling of printhead, 743, 755

Coordinates, 385–386, 390–391, 393, 395, 397, 399–400, 404, 415
Copper density, 426
Coquin, 644, 647
Cosine fourth law, 440
Counter rotating resonant systems, 454
Critical
    angle focusing, 582
    frequency, image quality, 222
Cross-array numerical aperture. *See* Numerical aperture, cross-array
Cross-axis motions, 439, 481, 492
Cross-flexure bearings, 431–435, 444–475, 493–496
Cross-scan, 95–99, 121–123, 661
Cross-scan error, 176–181, 284, 288, 293, 352, 439, 745
Cross-scan error correction, 126–128
Cross-scan error, active correction, 284, 293
Cross-scan error, passive correction, 284, 293, 751–753
Crossed-axis pivot, 431–435, 444–475, 493–496
Crosstalk
    between lasers, 731, 747, 751–752,
    due to scanline overlap, 731, 735–736, 738–739, 744–745, 747, 751–752,
Crystallization, 557
Curie point, 421, 427, 506, 558, 702, 703, 705
Curve-fits of beam propagation data, 15, 38, 42, 44–45, 717, 729
Customer research methods, 235–236
Cylinder surfaces, 96–98, 755, 761

Damping, 316, 332, 352, 357, 369
Debye-Sears, 604–605, 607
Deflection speed, 665
Deflector, 515, 518, 531, 609, 615, 617–618, 620, 623, 625, 627, 643, 648, 655, 661–662, 665
Defocus, 567
Degenerate modes, 6, 31, 65
Density,
    additive, 719–721, 724, 730, 739
    mass, 721, 724–725
    optical, 159, 520, 712–713, 719–724, 728–730, 732–736, 738–739, 745, 747, 751–752, 759, 763
    wedge, 184, 199–200, 759
Depoling, 706

Depth-of-field, 512, 515, 517, 519, 520, 523–530, 534–536, 543, 546
Depth-of-focus, 2, 20, 19–20, 57–59, 82–84, 93, 97, 102–103, 717, 729, 730, 742, 755, 763
Design aperture, 78–80, 82–83, 103, 496, 499, 506
Detail rendition, 160–164, 201–208
Detectability metrics, binary imaging, 201–202, 740, 759
Detective quantum efficiency, 217–218
Detector
    facet, 285, 342
    phase, 305, 334, 340–341
Device independence, 234
Diameter conversions, 3, 21, 32–38, 44, 54–56, 62, 64
Diamond turning, machining, 271–273, 276–277, 286, 295, 440
Dichromated gelatin (DCG), 534, 536–541
Dielectric, 277, 278, 287, 601, 642–643, 667, 706
    constant, 601, 642–643, 702
Differential
    driver, 701
    phase detection, 585
Diffraction, 15, 24, 34, 36, 62, 290, 601–613, 615–618, 620–625, 627, 630–631, 643, 648, 657, 659, 661, 690
    angle, 548
    efficiency, 534–538, 540, 548, 608, 617, 620, 622–625, 628, 630–631, 648, 657, 659
    limit, 13, 15, 76–78, 250, 620
    order, 124–125, 605, 613, 615
Diffractive overlay, 24, 64
Digital
    driver, 696–700
    images, structure, 145–150
    imaging, fundamental principles, 144–157, 739–741
    photography imaging performance, 173–174, 181–182, 185
    quality factor, 228–229
Digital audio disk (DAD), 551
Diode laser, 48, 130–133, 712, 717, 730, 742–743, 753–755, 757, 759–760
    failure of, 759–760
    bar, 712, 760–762
Displacement in pivots, 480–481, 501
Display and image quality, 141, 144–145, 182, 220–222

Index

Distinguishable grey levels, 153–157
Distortion, 74, 84–94, 105, 177, 270, 276, 283, 295, 372, 374, 383, 440–466, 625, 643, 650, 655, 657
Dithered systems, 164–198
Divergence, 1, 12, 14, 18–19, 38, 57, 62, 277, 624–625, 673, 701
Domain inversion, 678–680, 704
Dominant wavelength, 170, 244
Donor, 711–714, 716–717, 719–721, 724, 726–736, 739–740, 742–745, 747, 752, 763–764
Donut mode, 7, 8, 21–22, 24, 25, 26, 34, 65
Drift, 630
Driver
  brushless, 334–335, 338–339, 341
  motor, 316, 334–335, 338–339, 341, 421
Drum, 711–712, 742, 745, 747, 760
Dummy laser, 736, 739, 743–745, 752, 763
Duty cycle, 280, 290–292, 447–452, 481
DVD-RAM-RW, 552
DVD-ROM, 552
Dye, 711–714, 716–717, 719–730, 732–736, 739, 744, 747, 751, 763
Dynamic
  deformations, 270, 271, 437, 498
  performances, 432, 444–448, 506, 508
  ranges, 160, 529
  viscosity air, 351

Edge noise, 149, 210–213, 231–233
Elastic strain, 601
Electronic drivers, 467, 692–701
Electro-optic
  crystals, 701–706
  effect, 667–669
  grating scanner, 691
  lenses, 692
Embedded gaussian, 13–15, 32, 37, 54, 62, 65
Encoder
  accuracy, 305–307, 327
  data, 305–307, 327–329
  disk, 307, 376, 377
  optical, 274, 281, 282, 287, 294, 304–307, 318, 326, 328–329, 334–335, 338, 341, 361
Energy
  gap, 571
  storage in flexures, 479, 498
Entrance pupil, 73, 80, 86–88, 94–95

Equivalent
  cylindrical beam, 4, 48–52, 54, 56–57, 59, 61, 62, 64
  radial mode, 50
Erasable optical disk, 556
Error
  diffusion, 164, 198
  function, 11, 714–715, 721
  signal detection, 578
Eschler, 648
Exposure, 629, 713–716, 719–724, 726–736, 738–740, 744–745, 747, 751–753, 763
  additive, 720, 739
  area-average, 715–716, 721, 728, 734–738, 745, 763
Externally pressurized air bearings. *See* Aerostatic bearings
Extinction ratio, 295, 624–625, 650

Facet
  angle A, 274, 386–387, 405
  chord, 389, 391
  radius, 274–276, 285, 294
  variation, 276, 294
  width, 387, 389, 395
    tangential, 288, 388
Facet-to-axis variance, 274, 275
Facet-to-facet
  angle variance, 274
  error, 274, 383
  tangential angle, 405, 407
Far-field, 12, 14, 38–39, 46–48, 52, 65, 76–77, 585, 657, 671, 672
  pattern, 587
Fast random access, 591
Feed beam, 78, 95–99, 290, 292
Feedback,
  position, 316–317, 328, 505, 506
  velocity, 304, 318, 328, 341
Ferroelectric crystals, 508, 679, 680, 692, 703–705
Fiducial marks, 429
Field curvature, 88, 90–94, 105
Field lens. *See* Lens
Fifth order distortion, 90
Fizeau interferometer, 565
Flare light, 185
Flaw detection, 133, 490
Flexure
  cost, 479, 485, 492, 507
  fatigue limit, 481–485

grain, 486–488
  rotational spring rate, 482
  stress risers, 482, 485–488
Flexures, 431–435, 444–475, 478–481, 757
Flying objective scanning, 457, 469–472
F-number, 2, 20, 76–79, 515, 626, 628. *See also* Numerical aperture
Focal depth. *See* Depth-of-focus
Focus, 617, 619–620, 625–628, 650, 659, 661, 712, 717, 729–730, 742, 754–755, 757, 759, 761, 763
Focusing, 620, 626, 628
  drive mechanism, 569
  error, 578
  servo, 578
Foster, 626
Foucault focusing, 582
Four-cuts method, 3, 37–45, 65
Fourier, 2, 621
  transform, 2, 193–194, 621
Fourier-Bessel transform, 562
Fractional change in beam diameter, 40
Frequency-selective visual model, 228
Fretting corrosion, 435
F-tan θ, 84–88
F-theta (F-θ), 84–88, 288–289
Full-width-at-half-maximum (FWHM), 79, 81, 553, 656, 717, 760
Fully reversed stress, 481
Fundamental mode, 2, 4, 7–13, 21–22, 65

Gain
  control loop, 305, 328, 332, 341
  drift, 428, 429, 473
Galvanic couples, 486–480
Gas bearing, 345–346, 348–353, 356, 361–363, 369, 382, 384
  compressibility number, 363, 364–365
  lubricated bearings, 346, 348–349, 352
Gaussian, 10, 65, 249, 291, 292, 595, 620, 624–626, 629, 655, 724, 730, 736, 760
  beam, 10, 76–80, 524, 620, 625–626, 655, 713–714, 716–717, 726, 762
Gel-swell, 540–541
Gen III printhead, 760–762
Generator, reference, 307, 339–341
Germanium, 632, 658
Ghost images, 288, 294, 295, 404–405, 407, 409–413, 415
  inside the image format, 412, 415

  outside the image format, 386, 404, 413
  stationary, 386, 404–405, 407, 409–411, 415
Glass types, 74, 102–108
Golf club scanner, 463–466
Gouy phase shift, 12, 14, 52
Graded index, 674–676
Granularity, graininess, microuniformity, 154–157, 165, 212–214, 222–224
Grating, 272, 273, 522, 532, 534, 541, 548, 601, 605, 691
Grey wedges, 199
  noise, 200, 209–210

Half-mirror, 561
Halftone, 160–165, 175, 177, 179–181, 293, 711–712, 739–740, 751, 762
  printers, 293, 711
  system response, 160–165, 200–201
  tonal non-uniformity, 177, 179–181
Hall sensor, 311, 326, 375–376
Heat
  confinement, 717, 722, 724, 726–728, 730, 733–736, 739
  diffusion, 713, 716–717, 724–729, 736
  dissipation, 426, 506–507, 716–717, 739, 755
Helium neon (HeNe) laser, 13, 509, 513–514
HeNe laser. *See* Helium neon laser
Hermite-gaussian function, 1, 4, 31, 65, 66
High voltage drivers, 696–701
Higher order mode, 7, 20, 34
Histogram analysis, 182
Hologram
  construction, 130–131, 520–522, 538–542
  fringe spacing (grating spacing), 124–125, 532, 534
Holographic
  bar code scanners/scanning, 124, 130–135, 520, 523, 526, 531, 533–534, 537, 541, 543
  deflector (Holographic disc), 126–133, 371, 375, 520, 522–523, 526–527, 529–531, 533–539, 541, 546, 548
Holographic optical element (HOE), 124, 520–522, 531, 533, 539
Horn-shaped scanner, 686–689
Hue of colors, 169–172
Human visual system (HVS)
  spatial frequency response, 194, 195, 251–252, 254, 740–741, 745, 759
Hunting, speed, 281, 285, 316–318, 328

Index

Hybrid bearing, 368–369, 378–379
Hydrodynamic oil bearings comparison of, 346–348
Hydrogen embrittlement, 487–488
Hyperbolic propagation plot, 2, 12, 14, 15, 31, 38, 42, 46, 53, 67

ICC profile, 234
Image
    distortions, 176–179, 440–444
    format, 386, 404–405, 407–413, 415
    format field angle, 407, 413
    format scan duty cycle. *See* Scan duty cycle
    irradiance, 80
    plane, 266, 287, 290, 291, 293, 307, 404–405, 413, 415, 656, 712–713, 717–718, 729–731
    processing, 194, 229–235, 740–741
    quality, 80, 139–263, 310
    quality circle, 143–144, 235
    quality literature, 140
    quality models, 143–144, 223
    quality, summary measures of, 214–229
Imager,
    military, 333–334
    thermal, 333–336,
Impedance matching, 655
Incident beam, 553, 559
    collimated, 385–386, 532–533, 617
    location, 408
    offset angle, 385–386, 395–397, 400–413
Index modulation (Δn), 536–537, 540
Induced moving coil scanner, 453
Inductor, 640
Information content and capacity, 154–157, 225–228
Infrared, 265, 276, 514, 632–633, 657, 659, 711–712, 714, 720, 724–725, 747
    laser scanning, 657, 711–712, 739, 741–742, 753–754, 760
In-scan, 97–98
Instantaneous center-of-scan (ICS), 386, 395, 404, 411
    coordinates, 397–399
    loci, 395–397, 400–401, 404

Integer M^2 values, 7, 31, 33, 66
Integrating cavity effect, 185
Integrator, 328, 332
Intensity distribution, 562, 563
Interleaved 2 of 5, 513–514
Interleaving, 736–739, 763

Internal drum scanner, 346, 350
International Organization for Standards (ISO), 3, 20, 31–32, 37, 45, 62, 253–254
Invariant, beam, 13, 15, 62, 65, 761–762
Irradiance profile, 2, 4, 9, 13, 21, 62, 717–718, 729–730, 744, 760
Irradiance, 712–714, 717, 725, 729–730, 736, 744, 760–762
Isotropic, 486, 501, 601, 609–610, 612, 629, 658
    AO interaction, 601

Jaggies, 231–233
JBIG compression, 230
Jitter, 165, 177, 180, 281–284, 288, 294, 377, 382, 473
    speed, 311, 328, 335, 339
    transport, 309
JPEG, 230–231
Just noticeable difference (JND), 236–237

Kerr
    effect, 666, 704
    rotation, 559
Knife-edge profile, 11, 21–28

L*a*b*, 172
LaGrange invariant, 761–762. *See also* Optical invariant
Laguerre-gaussian function, 2, 4, 31, 65, 66
Laser, 1, 42, 48, 53, 54, 56, 265, 268–269, 280, 288, 294, 307, 477, 490, 560, 571, 599–600, 650, 655, 657–658, 661, 711–725, 736, 737, 739, 741–747, 752
    array, 712, 753–755, 757, 759–760, 763
    diodes. *See* Diode laser
    noise, 30, 44–45, 576, 743
    printers, 265, 266, 292, 346, 379, 655, 711–712, 739, 741–742, 753–754, 760
    radar, 288, 657
    structure, 571
Laser-detector-unit (LDU), 569
Laser-thermal-dye transfer, 711–713, 719–724, 727–730, 732–736, 739, 744–745, 747, 763
Lateral color, 88–89
Lens, 186, 250, 252, 755
    designs, 117–123, 288, 741–745, 747, 752, 745–755, 759–763

field 755
F-theta. *See* F-theta (F-θ)
Lenslet array, 753–755, 757, 759, 761, 763
Light
    sources, standard, 170–171, 244–246
    wave, 559, 602–603, 613
Line bow, 124–127, 274, 296, 496. *See also* Smile corrector
Line edge noise range metric, binary imaging, 210–213
Line
    fidelity metrics, binary imaging, 202–203
    jitter, 310–311, 326, 474
    scan, 307–308, 310, 714–716, 720, 724, 728, 730–731, 733–740, 745–747, 759
    spread function, 77, 81, 134, 146, 186–187
Lithium, 632–633, 641, 643, 651–652
    niobate, 632–633, 641–642, 645, 651–652, 673, 678, 701–704
    niobate transducer, 642, 645, 652
    tantalate, 641, 701–704
Lossy compression, 229–231
Lowest order mode, 7, 13, 65, 66
$L^*u^*v^*$, 171
LZ and LZW compression, 230

M^2, 2, 3, 4, 8, 13–16, 20, 31–46, 62, 64, 65, 67, 77
Machine-readable diameter definitions, 26, 31
Magnetic
    bearing, 346–348, 379–382
    material, 419, 426
Magneto-optical (MO), 558
Magnification effects, 174–175, 745, 755
Maréchal's
    criterion, 564
    equation, 596
Mark, 502, 552
Mason equivalent circuit, 638–639
Mass density. *See* Density, mass
Material
    figure of merit, 438–439, 484
    processing, 467
    properties, 438–439, 483, 630, 717, 721–722, 724, 734
Meitzler, 639, 641
Melting temperature, 557, 720
MEMs
    flexure scanner, 498–507
    pivot, 505
Microlenslet array, 754–755

Microscopy, 468–471
Mid-position, 385–389, 393–394, 396–397, 400, 404, 406, 408
Mirror, 265–283, 372–374, 382–383, 435–440, 494, 499, 650, 656, 711
    beryllium, 270, 309, 372, 374
    conventionally polished, 276, 277, 286, 711
    cross axis misalignment, 436
    diamond turned, 273, 277, 286
    irregular polygonal, 266, 268
    material, 270–273, 372–374, 438–439
    polygon, 265–283, 300–306, 711
    prismatic polygonal, 266, 267
    pyramidal polygonal, 266, 268
Mirror-facet, 265–283, 386–387, 405, 407
Mirror-facet plane, 386, 394
Mixed mode, 8, 13–16, 31, 33–35, 37, 62, 66, 712, 742, 754, 760
MOCVD, 574
Mode
    fractions, 7, 8, 14, 33–37, 62
    hops, 130–131, 743
    order number, 6, 31, 66
    or spot pattern, 4–6, 11, 52, 60, 66
    pure, 4–6, 31, 33, 66, 711–712, 717–718, 729, 753–755
Modulation transfer function (MTF) 83, 157, 185, 187–194, 245–251, 739, 747
    approaches for engineering analysis, 187–194
    curve, area under, 220
    diffraction limited lens, 250
    double exponential, 250, 252
    equations and graphs, 245–250
    gaussian spread function, 249
    photographic films, 251, 254
    uniform disk, 245, 249
    uniform, sharply bounded spread functions, 245, 247, 249
    uniform slit, 247, 249
Modulator, 294, 295, 599, 617–618, 623–624, 628–629, 633, 655–657, 662, 712, 760–762
    phase, 328–329, 712, 760–762
Moles, 720
Monogon, 265, 266, 268, 286, 290, 302, 305, 372–375, 382–383
Monolithic diode-laser array, 712, 753–760, 763
Monomorph flexure, 500, 502, 504
Morphological image processing, 231
Mosaic transducer, 643–644

Motion non-uniformity, 176–179, 180–181, 759
Motor, 265, 280–282, 285–286, 294, 345, 361, 366, 375, 377, 381–382
  brushless DC, 311, 318–322, 324–325, 375–376
  commutation, 318, 323–324, 334, 375
  hysteresis-synchronous, 281, 311–317, 328, 375–376
  model, 319–327
  rotor, 311–312, 375–376
  speed, 313, 315, 317–319, 745, 747
  torque, 313–316, 318–320, 376, 421
Multi-function plate (MFP), 546, 548

Near-field, 12, 14, 38–39, 66, 76–77
Noise, 165–166, 197–198, 209–214
  effects in information content, 225, 228
  equivalent quanta, 218–219
  in imaging systems, 47, 742–743, 753
  paper, 165, 528
  quantization, 197
Noise-clip option, 30, 66
Non-linear enhancement, 231–233,
Nonlinearity, 182–183, 191, 739
Nonorthogonal systems, 16, 52–53
Non-uniformity in imaging, 177–181, 244, 759
Normalizing gaussian, 15, 54, 56, 66
Null drift, 427–429
Numerical aperature (NA), 63, 67, 76, 553, 564, 742, 754–755, 760–762
  aperture, along-array, 742, 754–755, 760–762
  aperture, cross-array, 742, 754–755, 760–762
Nyquist frequency, 150–152, 191

Objective lens, 288–290, 294, 385, 391–392, 397, 404, 407–411, 415, 554, 561, 565, 745
  optical axis, 386, 391–392, 397, 404, 406
Objective scanning, 72
Offset
  angle, 392, 396–397, 400, 402, 404–407, 409
  distance, 392, 406–408
Oil lubricated bearings comparison of, 346, 348
$1/e^2$ beam diameter (width), 1, 2, 9–10, 20–23, 25–28, 32, 62, 64, 76–79, 81–83, 292, 524, 624, 657

$1/e^2$-diameter. *See* $1/e^2$ beam diameter (width)
"1-space" and "2-space", 17, 19, 57–62
Optical
  axis, 385, 391, 397
  beam, 609, 617, 620–625, 713–716, 720, 745
  density. *See* Density, optical
  disk, 551, 556
  drive, 592
  fiber, 63, 712, 717, 733, 739, 741–744, 747, 752–754, 763
  invariant, 75–80
  pick-up, 568
  pulse, 624
  Read/Write, 552
  scanner, 54, 55, 416–476, 477–47, 659, 711
  scanning, 8, 382, 416–476, 481, 496, 502, 504, 506, 508, 711
Optical-fiber printhead, 712, 733, 736, 739, 741–745, 747, 751–753, 760, 763
Optical transfer function (OTF), 187–188, 191
Optics distortion, 382–383
Oscillating scanners, 416–476
Oscillator,
  PWM, 339–341
  reference, 318, 327, 339
Oscillatory pivot, 497
Overwriting, 557

Paddle scanner, 461–463
Paired comparison scaling techniques, 238–242
Pairs, 747, 750–752
Paper noise. *See* Noise, paper
Parasitic capacitance, 698, 699
Paraxial rays, 4, 13, 16, 52, 66
Partial dotting in halftones, 163–164
PCR disk, 557
Perception, 143, 169, 223, 235–236
Periodic poling, 691, 692
Permittivity, 667
Perovskite, 703–704
Petzval radius, 86–88, 93–94
Phase
  change, 601–603
  detector, 304–305, 331–334, 339, 341
  effects in image quality, 176–177, 191–192

lock, 304–305, 310, 317, 319, 330–334, 339, 341
and space quadrature, 6
transfer function, 188, 331
Phase change rewritable (PCR), 556
Phonon, 602, 605–606
Photoelastic, 600–601, 608–609, 630, 632–633
   constant, 600–601, 608, 631–632
   effect, 600–601, 631–632
   interaction, 601
   tensor components, 632
Photographic film and images, 153–156, 165–166, 187, 251
Photomask, 655–657, 659, 661
Photon, 13, 53, 601, 605–607
Photopolymer, 536–537
Photorefractive damage, 703
Photoresist, 534–535, 541–542, 641, 762–763
Pick-up, 500, 560
   optics, 560
Piezoelectric, 637–639, 640–642, 651
   pivot, 505
Pinhole profile, 21–27, 62, 66
Pit, 552
Pitfalls in M^2 measurement, 38, 40, 41, 44, 45, 62
Pits pattern, 553
Pivot
   degrees of freedom, 478, 481
   force linearity, 478–481
   hysteresis, 478
   life, 478, 481, 485
   load capability, 478
   point, 670–671
   restoring force, 478
   temperature limit, 478
   transparency, 479
Pixel,
   clock, 308–309, 311, 326–327
   density (size), 26, 173, 712, 729
   registration, 284, 288, 294, 296, 308–309, 310–311, 326–327, 745
   spacing. *See* Scan frequency effects
   spacing non-uniformity, 736–739, 745, 759
Planckian radiator, 245
Plane linear diffraction grating (PLDG), 124, 522
Platesetter, 711–712, 760, 762–763
Plating, nickel, 271
Pockel's effect, 668

Point spread function, 81, 186. *See also* Line spread function
Polarization, 280, 535, 538, 558, 571, 609–611, 613, 633, 661
   angle, 559
Polarizing
   beam splitter, 561
   hologram, 569
Poling, 676–679
Polygon, 265–268, 279–280, 282–295, 361–362, 366, 368–369, 372, 379–383
   diameter, 99–101, 292, 372
   facets, 99–101, 276, 289, 290, 292–295, 372, 383
   inertia, 304–305, 339, 341
   speed, 282, 287, 304–306, 309–311, 323, 329, 372
Polygonal
   mirror, 270–274, 280, 283, 285, 288, 289, 656
   scanner, 265, 266, 281–287, 296, 385–389, 393–397, 398–401, 404–409, 415, 655
   scanning, 268, 385–386, 405, 415
Position transducer, 421–427
Post-objective scanning, 72, 288–290, 294, 458–466
Power-in-the-bucket, 22, 66
P-polarized beam, 535, 561, 569
Preferences in imaging performance, 143–144, 222–225
Pre-objective scanning, 73, 288, 290, 385–416, 457
Principal
   diameters, 21, 66
   planes, lens, 17, 18, 42
   propagation planes, 13, 14, 46–48, 59, 62, 66
Print contrast signal (PCS), 514
Printer,
   continuous tone, 293–295, 712–739
   halftone, 160–165, 293, 711–712, 739–741, 760–763
   laser, 292, 346, 379, 712
Printhead, 711–712, 731, 733–736, 738–739, 745, 747, 752–755, 757, 759–763
   balance, 747, 750–753
Prism, 50, 265, 372–375, 383
Prismatic poled structures, 679–690
Probabilities in image quality evaluation, 225–227, 236

Index

Profile connection space in color management, 234
Profiler, 21, 25–26, 45–46, 66, 742
Propagation constants, 13, 15, 41, 51–53, 56, 60–62, 64, 65, 67
Propagation plot, 12, 28, 49–51, 64, 67
Psychometrics, 235–243
Pulse-width modulation, 232, 339–341, 728
Pupil shift, 293
Push-Pull, 585
Pyramidal error, 274, 275, 284, 383
Pyroelectric, 703, 706
PZT flexure, 499–501

Quadratic
   electro-optic effect, 668–669
   ratio of astigmatic diameters, 58, 60–61
Quality factor, 222–223, 228
Quantization, 145–147, 153–157, 191–193
Quartz, 613, 619, 630, 632, 641, 652

Radial access, 591
Radio wave, 603
   modulation, 603
Radiometric information from pictures, 154
Radius of curvature, 11–16, 46–48, 50–51
Ragged or structured edges, 177–179
Raman-Nath, 603–604, 607, 612, 614, 625
Random access, 448, 552, 599
   scanning, 448
Rapid crystallization, 557
Rapid quenching, 557
Rare earth magnets, 376, 419–427
Raster
   distortion, 177, 747, 755, 757
   scanning, 448–452, 480, 711–712
   spacing, 176–179, 715–716, 720–721, 727, 731–740, 745, 747, 759
Raster input scanner (RIS), 145
Raster output scanner (ROS), 145, 711–712
Rayleigh range, 1, 11–13, 37, 40–41, 43, 48, 52, 57, 62, 67
Read-only optical disk, 554
Receiver, 711–713, 720–721, 723–724, 735
Reciprocity failure, 720, 728, 739, 745
Reference
   frequency, 308, 310, 328, 339
   generator, 304–305, 310, 319, 331, 334, 340
Reflectance, 277–280, 285, 509, 514, 722–723
   uniformity, 179–182, 279

Reflected incident beam axis, 392
Refraction of air, 441
Refractive index, 51, 520, 536, 540, 600–603, 608–609, 626–627, 630–632, 714
Regression models of image quality, 143–144, 223–224, 228
Relative aperture, 76–88. *See also* F-number
Relative intensity of noise (RIN), 576
Relaxed design, 74
Removability, 552
Replication, 552
Resolution, 76, 80–82, 85, 100, 174, 222, 224, 234, 268, 270, 275, 290, 599–600, 617–619, 623, 630, 659, 661, 711, 739–740, 743–745, 747, 755
   criterion, 204–205, 524–525
Resolution enhancement technology (RET), 232
Resolvable spots, 288, 617, 619, 623–625, 661, 712
Resolving power, 193, 204–208
Resonance, 357, 369–371, 419, 444, 447, 475–476, 506, 508, 648
   cross axis, 444–447
Resonant scanner, 453–455, 508
Resonator, 6, 8, 15, 16, 48, 62–63, 67
Restoration of digital images, 231
RF driver, 643
Ribbon coil, 425–426
Rise time, 623–624, 650, 743
Rolled materials for flexures, 488–490
Rotary
   flying objective microscope, 471–472
   scanner, 345–384
Rotation axis, 386–387, 392–397, 405–406, 415
   offset distance. *See* Offset
Rotational actuator, 569
Rules of thumb, 23, 92

Sampled tracking, 587
Sampling, 145–153, 173–174, 176, 192
   of images, 740–741
   phases, 148–150, 186–187, 191–193, 199, 201–208
   theorem, 150–153
Saturation of colors, 171, 173
Saw-tooth drive signal, 450
Scales for psychometrics, 236–239

Scan
  duty cycle, 99–101, 288, 387–389, 393, 395, 404, 406–409, 413, 415, 447–452
  efficiency. *See* Scan duty cycle
  frequency effects, 173–176
  heads, 457
  jitter, 126–127, 165, 177, 180, 311, 318, 328, 335, 339, 377, 382, 473
  line/s, 512, 514–519, 529, 531–534, 536, 543–544, 547, 548
  linearity, 84–85, 112, 310, 661
  tracking, 527, 530–531
Scan-angle, 82, 84, 95–101, 113–116, 125
  multiplication, 531–533
Scan-axis, 265, 290, 385–387, 391–393, 395–397, 400–408, 415
Scanline, 274, 276, 288, 290, 293, 296, 712–717, 720, 724, 726–729, 733–739, 745, 747, 759, 763
  spacing. *See* Scan frequency effects, and also Raster spacing
Scanned field image
  format, 386
  plane, 404–405
Scanner, 54, 55, 140–142, 144–145, 177, 181–188, 192, 194–197, 198, 200–201, 211, 265–270, 275, 280–290, 293–296, 345–384, 417–476, 509–548, 599, 612–615, 617, 627–628, 630, 632, 637, 655, 657–659
  film, 302, 306–310, 326, 338
  jitter, 281–284, 288, 294, 311, 318, 328, 335, 339, 377, 382, 473
  moving magnet, 421–427
  resolution, 173–176, 739
  specification, 275, 282–286, 309–310, 423–427
  speed, 303, 305–306, 317, 347–348, 362, 368, 370, 726, 728
  tolerances, 180–181, 309–311
Scanning
  architectures, 144, 457, 733–739
  postobjective, 289, 290, 294, 457
  preobjective, 288, 457
  raster, 265, 268, 307, 450
Scatter, 273, 276, 277, 286, 288, 294, 295, 538, 541, 650
Scattering of light in paper, 165
Scratch and dig, 276, 277, 286, 287
Screens, halftone, 161–165

Second-moment diameter, 2, 3, 20, 22, 28–31, 36–37, 53, 56, 62, 67
Self-acting air bearings. *See* Aerodynamic bearings
Selwyn law for granularity, 154–155
Semiconductor laser, 48, 571, 760–762
Sensitometry, 158–160, 711, 747
Sensor, commutation, 311, 323–324
Servo compensation, 328, 331, 341, 747
Shaft
  frequency, 371
  synchronous whirls, 369–370
  wobble, 374, 382–383, 431
Shaped electrodes, 669
Siedel aberration, 595
Signal-to-noise ratio (SNR), 45, 215–217, 219
Silver halide, 534, 536–537, 711, 720, 728, 739–741
Single-mirror TABS, 457
Singles, 747, 751–752
Sinusoidal test target for MTF, OTF analysis, 188–191
Sittig, 639–640
Slit
  detection, 586
  profile, 21–28, 62, 67, 249
Smile corrector, 755, 763
Sound wave, 599, 605, 613, 648
Spatial phase detection, 582
Spectral locus, 170, 244
Spectrophotometer, 280
Speed
  regulation, 304, 309–311, 316–317, 319, 329, 334–336, 747
  stability, 283, 284, 304, 309–311, 316–317, 319, 329, 334–336, 747
Spherical aberration, 90–91
Spiral groove bearing, 364–367
S-polarized beam, 535, 538, 561, 569
Spot
  ellipticity, 21, 46–48, 52, 59–61, 519, 713–714, 755
  size detection (SSD), 580
Spot-invariant. *See* Waist-invariant
Spread function, 157, 173, 186–188. *See also* Line spread function
  and information capacity, 225–226
Spurious response, 158
Square root integral (SQRI) for imaging performance, 220–221
Square wave analysis for MTF, OTF, 739

Index

Standard errors for proportions, 242–243
Standard-deviation radius. *See* Beam standard-deviation radius
Starred mode, 6–7, 65, 66
Start of scan (SOS), 308
Statistical
  regression in imaging performace, 223
  significance, 241–243
Stereolithography, 4, 54–55, 57–58, 62, 63
Stiffness constant, 484, 607
Stigmatic beam, 9, 13, 67
Stray light, 185, 295
Streaks, 179–180, 743, 752
Strehl
  ratio, 81
Strehl definition (SD), 562
Stressed design, 74
Structured background, 1, 77–178
Subjective quality factor (SQF), 222–223
Substrate noise. *See* Noise, paper
Subtractive color, 166-167
Surface
  figure, 274, 276, 277, 286
  relief phase media, 534–535
  roughness, 273, 276, 277, 285, 286, 528
Swath, 731, 733–734, 736–739, 744–745, 747, 752–753, 759
  balance, 736, 739, 745, 747, 751–753, 759
  spacing, 737–738, 745, 753, 759
System bandwidth, 150, 305–306, 444–462, 472
System modulation transfer acutance (SMT acutance), 222

TABS, 457–466
Tachometer,
  feedback, 308, 328, 334
  optical, 304, 308, 329, 334, 336
Telecentric scan, 121
Temperature, 348–349, 351–352, 366, 426–429, 630, 652–653, 655, 657, 712–713, 716–717, 719–721, 724–729, 731–734, 736, 739, 743, 755, 763
  coefficient, 427, 630–631, 721
  dependence, 130–131, 427, 721, 736
Tensor equation, 601
Tensor, 601, 632
Test patterns, standard, 206–207, 245, 247–248, 739, 747, 752
Thermal
  diffusion. *See* Heat diffusion
  drift, 427–429, 473

impedance, 419, 45–427, 506–507
Third-order aberrations, 90–93
3-beam, 584
Threshold, 146–150, 198–201, 650, 719–724, 727–728, 730–734, 736, 739, 743, 763
  detectability curve, 219, 740
  scales in psychometrics, 236
Thresholding an image, 146–150, 198–201
Tilt, 564, 597, 603, 605, 644, 648, 733–736, 743–745, 752, 755, 763
Tilted surfaces, 108
Time-bandwidth product, 618, 623, 661
Times diffraction limit number (TDL), 5, 62, 65, 67
Tone reproduction, 158–160, 182–185, 199–201, 739–740
Tonescale, 160, 182–185, 712, 739–740
Torque
  motor, 313–316, 318–320, 421
  ripple, 318–320, 326, 329, 334, 336
  waveform, 318–320, 323
Torsional
  pivot, 454–455, 493
  resonance, 445, 447, 506
Totem pole driver, 696–697
Track, dynamic, 275, 283, 284, 286–287, 293, 383, 734, 743
Tracking, 643, 762
  drive mechanism, 569
  error, 583
Transducer array, 629, 644–645, 648–649
Transducer, 427–431, 508, 600, 603, 608–609, 618–620, 622–625, 627, 629–630, 637–638, 640–644, 646–655, 657–659, 661
  bonding, 650–651
Transformation constant, 18, 20, 44, 57, 59
Transformer
  coupling, 454–455
Transformer-coupled driver, 454–455, 697, 701
Transverse
  mode, 4–6, 8, 66, 446–447
  stiffness, 446–447, 481
Trapezoidal scanners, 448–452, 684–686
Traversal time, 714, 717, 729, 736
Triangular tooth scanning, 451
Triples, 747, 752
Tristimulus value, 171–172
Truncation ratio, 77–80
Tunable resonant scanners, 454

Tunnels, scan or scanning, 520, 531
Twisted beams, 3, 4, 16, 52–53, 62
Two-axis beam steering, 54, 457–466

Ultrasonics, 508
Uniform Grocery Product Code Council, 510
Uniqueness of M^2, 16, 33–36
Universal product code (UPC), 509–513, 515–520, 544
Unlubricated pivot, 478
$u'v'$ chromaticity diagram, 170, 172–173, 245

Vacuum, 277–280, 651–653, 655, 717
Vaporization, 712–713, 719–722, 724, 747, 763
Variable
  light-collection aperture, 523, 526–527, 529, 531
Variable-aperture diameter, 22, 25, 27–28, 62, 66, 67
Vector, 605–607, 609–612, 615–616, 620, 714, 751–752
  scanning, 448
V-groove, 712, 741–744, 752, 763
Videodisk, 551
Vignetting aperture, 75, 99–100
  stop, 755
Virtual pivot, 478,
Visible laser diode (VLD) 48, 509, 525, 537, 546, 548
Visual system, 141–144, 153–154, 167–169, 174, 180–181, 194–195, 219–221, 227–229, 234–237, 251–252, 254, 740, 745, 759
Volume phase media, 536–538

Waist
  diameter. *See* Beam waist
  location, 1, 11, 13, 18, 62, 68
Waist-divergence product.
  See Waist-invariant
Waist-invariant, 13, 15, 76–77
Warm-up transient, 728–729, 736
Warner, 613, 615
Wave
  aberrations, 562
  equation, 4, 8, 11, 12, 13, 68
Wavefront curvature, 10–14, 46–48, 53
Wavelength, 270, 273, 276, 278, 280, 286, 288, 290, 509, 513–514, 524–526, 536–539, 541, 600–605, 608, 613, 617, 623, 626–627, 631, 637, 643, 651, 654–655, 657–658, 661, 717, 743, 762
Wedge prism, 561
Westwind Air Bearings Ltd., 349, 365, 384
Williams-Clapper transform, 723, 730
Winding,
  Delta, 318, 321–323, 326
  WYE, 321–323, 326
Wire-suspended actuator, 569
Wobble, 282, 284, 374, 382–383, 431–435, 480, 533
  correction, 126–128, 292–294, 374, 382–383
Wobbling, 582
Write-once disk system, 555

Zero translation flexure pivot, 423–435, 481